Offshore Wind Energy

Contents

About the Editors

Emre Uzunoglu

Emre Uzunoglu (Ph.D.) is a Senior Consultant at KONGSTEIN in Hamburg, Germany, and a collaborator with the Hydrodynamics Group at Instituto Superior Técnico in Lisbon, Portugal. In recent years, he has primarily focused on strategy projects in the offshore wind industry while maintaining academic connections through research initiatives. Before moving into consultancy, he obtained his PhD in Naval Architecture and Ocean Engineering from the University of Lisbon, where he specialised in offshore wind platform design methodologies.

Antonio Souto-Iglesias

Antonio Souto Iglesias (Prof) is a Naval Architect, PhD, a Full Professor of Hydrodynamics at Universidad Politécnica de Madrid (UPM), and the Director of the Model Basin Research Group (CEHINAV) at the Naval Architecture School (ETSIN). He is also the Vice Dean of Academic Strategy at ETSIN and a member of the ITTC (International Towing Tank Conference) Seakeeping Committee. His research interests include experimental and numerical methods in hydrodynamics. He has been the Principal Investigator (PI) of 70 industry projects, mostly on ship hydrodynamics and offshore renewables dynamics. He has published articles in *Ocean Engineering, Renewable Energy, Applied Ocean Research, Marine Structures*, and similar journals. His scholar h-index is 37. Together with R. Guanche (Institute of Environmental Hydraulics of the University of Cantabria, Spain) and F. Huera-Huarte (University of Rovira i Virgili, Spain), he chaired the 37th International Conference on Ocean, Offshore & Arctic Engineering, OMAE 2018, in Madrid.

Carlos Guedes Soares

Carlos Guedes Soares (Prof) is a Distinguished Professor of the Engineering School (Instituto Superior Técnico) of the University of Lisbon and Scientific Coordinator of the Centre for Marine Technology and Ocean Engineering (CENTEC). He concluded his postgraduate studies at the Massachusetts Institute of Technology, USA, in 1976, and at the Norwegian Institute of Technology of the University of Trondheim in 1984. He has supervised and co-supervised more than 75 PhD students and has approximately 40,000 citations in Web of Science. He is Co-Editor-in-Chief of the *Journal of Marine Science and Application* and is a member of the Editorial Board of more than 15 journals. He is a Fellow of SNAME (Society of Naval Architects and Marine Engineers), RINA (Royal Institution of Naval Architects), IMarEST (Institute of Marine Engineering, Science and Technology), ASME (American Society of Mechanical Engineers), and the Portuguese Engineering Association (Ordem dos Engenheiros). Additionally, he is a Fellow of the Portuguese Academy of Engineering and is the Chairman of WEGEMT, the European Association of Universities in Marine Technology (comprising 40 Universities).

Journal of
Marine Science
and Engineering

MDPI

Editorial

Special Issue on Offshore Wind Energy

E. Uzunoglu [1], A. Souto-Iglesias [2] and C. Guedes Soares [1,*]

1 Centre for Marine Technology and Ocean Engineering (CENTEC), Instituto Superior Técnico, Universidade de Lisboa, 1049-001 Lisbon, Portugal; emre.uzunoglu@centec.tecnico.ulisboa.pt
2 CEHINAV, DACSON, ETSIN, Universidad Politécnica de Madrid, 28040 Madrid, Spain; antonio.souto@upm.es
* Correspondence: c.guedes.soares@centec.tecnico.ulisboa.pt

As the impact of fossil fuels on the planet becomes clear, the world is increasingly focusing on renewable energy sources. For instance, by 2030, the European Union aims to supply 42.5 per cent of its energy demand through renewables, followed by becoming a climate-neutral continent by 2050. These targets require mature technologies scaled to meet the demand. In this regard, wind and solar have advantages over alternatives such as wave and tidal energy.

After decades of onshore use for electricity generation, offshore wind started with the Vindeby Farm in Denmark. Eleven turbines of 450 kW with rotor diameters of 35 m were installed on monopiles in 1991. The farm stayed in service for 26 years, being decommissioned in 2017. Within those years, turbine capacities reached 15 MW, increasing by a factor of 33. While this capacity increase is also a function of efficiency, it primarily relates to larger turbine sizes due to the Betz limit.

Moving these structures into deeper waters is beneficial from both energy production and social standpoints. From an energy production perspective, a significant part of the world's wind resource is in deeper waters. The unhindered wind allows a consistent flow regime for the turbine to operate more efficiently. Regarding the social impact, turbines located farther offshore address concerns such as the visual impact. However, fixed bottom structures become economically unfeasible when operating at depths beyond 60 m. Hence, floating support structures are required, introducing their own economic and technological challenges.

The life cycle of floating wind units, consisting of design, fabrication, installation, maintenance, and decommissioning, is similar to their oil and gas counterparts. However, the turbine's presence couples the hydrodynamic and aerodynamic challenges. Thus, the design of these structures becomes a truly interdisciplinary field. Another significant difference is that the turbines must be installed in higher numbers. For example, a 500 MW wind farm would result in more than 30 turbines at a capacity of 15 MW each. Hence, the technological issues and costs multiply compared to installing one oil and gas extraction unit. Installation, operation, and maintenance planning must consider these factors, including the interaction of structures in a farm. For these reasons, while floating wind technology can borrow some of the cross-industry experience, innovation is critical in making this new technology feasible.

This Special Issue addresses the abovementioned topics through 21 papers. It starts by discussing the economics of floating wind and how the technology integrates into the current energy mix. These works are followed by evaluating the design factors to turn floating wind into a viable alternative for the future. Continuing through the order of the farm's life cycle, installation and logistics are presented, followed by topics concerning operation and maintenance.

This Special Issue opens by providing an overview of the opportunities and market needs of the floating wind industry [contribution 1]. Adopting this technology as a part of the renewable energy strategy requires political support and the right market environment.

Citation: Uzunoglu, E.; Souto-Iglesias, A.; Guedes Soares, C. Special Issue on Offshore Wind Energy. *J. Mar. Sci. Eng.* **2024**, *12*, 816. https://doi.org/10.3390/jmse12050816

Received: 25 January 2024
Accepted: 7 April 2024
Published: 14 May 2024

The assessment in the mentioned work examines the state of the art and highlights the areas for improvement.

The current grid infrastructure's fit to renewable energy sources is one of the components requiring an adaptation. Unlike consistent energy sources where the output can be set, the grid must deal with inconsistent power output from the wind resource. Evaluation of Europe [contribution 2] shows that several countries' current power systems have higher readiness levels to incorporate the incoming technology. Wind farms can easily contribute to the energy mix in these cases.

Once the contribution and integration options are clarified, economic and technical perspectives can be evaluated with the method outlined in [contribution 3]. The work provides 18 case studies for the CENTEC-TLP platform, covering three electric tariffs and the costs of capital. Beyond providing answers for a particular platform, it offers a robust approach to estimating the economics of a wind farm.

Integrating floating wind energy at a feasible cost requires suitable designs considering motion dynamics, structural safety, and reliability. The following sections cover these topics, starting with the hydrodynamic design. A SPAR platform is investigated in [contribution 4] as one of the options, providing a detailed breakdown of the design and dynamics for a 10 MW rated turbine. The numerical evaluation of motions for these platforms is sensitive to the modeling approaches [contribution 5] used in the analysis software. This case is significant for all platform types, such as barges, and the complexity increases for multi-use platforms [contribution 6].

Floating wind turbines involve handling the interaction between aerodynamics and hydrodynamics. The importance of such interaction is demonstrated in [contribution 7], highlighting how the responses of the platform can be improved through novel control strategies. Once the dynamics are covered, structural design must be performed to conform with the safety and reliability requirements, as outlined in [contribution 8]. The study evaluates a TLP-Barge hybrid platform, identifying that specific locations may need detailed attention.

Verifying a system's design, especially for newer technologies, necessitates experimental work. For floating wind turbines, the testing approaches bring topic experts together to deliver a solution for the structure's coupled dynamics, considering the points discussed in [contribution 9]. Once a suitable approach is found, testing processes are undertaken, and the results are reevaluated numerically [contribution 10]. The revision provides an opportunity to improve the numerical methods [contribution 11]. In this regard, isolating and examining each component may be necessary. For instance, mooring lines may need to be adapted to renewable energy needs [contribution 12].

Fabrication and installation follow the design stage. Accordingly, the next sections of this Special Issue focus on an overview of and solutions to installing floating wind farms. The farm layout and the complex logistics involved are covered in [contribution 13]. When installing individual units, one option is to develop novel installation vessels to ensure safety during transport and installation [contribution 14]. In most cases, the mooring lines must be pre-installed (pre-lay) for these methods to work. The pre-lay process depends on careful planning and optimization, accounting for weather windows [contribution 15]. Operation and maintenance become the primary concerns after the installation phase. At those stages, understanding and estimating the failure rates of the components [contribution 16] and prioritizing failures necessitate suitable methods [contribution 17]. Thus, preventative measures, as exemplified in [contribution 18], can be applied.

Regardless of prevention, the wind farm is subject to maintenance, where the size of the installation area, combined with the logistic setup, leads to planning challenges. Using Petri-Nets and Monte-Carlo simulations for this purpose is a solution described in [contribution 19]. Given the difficulty of working offshore within weather windows, allocating service vessels with high-efficiency methods [contribution 20] becomes another area of interest. In the case that damage occurs despite these precautions, its correct quantification [contribution 21] allows having the best approach to reducing the overall impact on the environment and the wind farm.

With these works, this Special Issue covers a significant part of a floating wind farm's life cycle. It starts with the economics and proceeds to the design and analysis processes, followed by installation logistics. The operation of the installed farm is discussed from a maintenance planning perspective. These contributions aim to provide an understanding of the developments within this novel field and contribute to its standardization and large-scale implementation. It is noted, of the 21 papers included, 15 are the result of the European project ARCWIND, which helps to provide consistency in this Special Issue.

Funding: This research received no external funding.

Conflicts of Interest: The authors declare no conflict of interest.

List of Contributions:

1. Díaz, H.; Serna, J.; Nieto, J.; Soares, C.G. Market Needs, Opportunities and Barriers for the Floating Wind Industry. *J. Mar. Sci. Eng.* **2022**, *10*, 934. https://doi.org/10.3390/jmse10070934.
2. NAmaro, N.; Egorov, A.; Glória, G. Fostering Offshore Wind Integration in Europe through Grid Connection Impact Assessment. *J. Mar. Sci. Eng.* **2022**, *10*, 463. https://doi.org/10.3390/jmse10040463.
3. Filgueira-Vizoso, A.; Castro-Santos, L.; Iglesias, D.C.; Puime-Guillén, F.; Lamas-Galdo, I.; García-Diez, A.I.; Uzunoglu, E.; Díaz, H.; Soares, C.G. The Technical and Economic Feasibility of the CENTEC Floating Offshore Wind Platform. *J. Mar. Sci. Eng.* **2022**, *10*, 1344. https://doi.org/10.3390/jmse10101344.
4. Guo, X.; Zhang, Y.; Yan, J.; Zhou, Y.; Yan, S.; Shi, W.; Li, X. Integrated Dynamics Response Analysis for IEA 10-MW Spar Floating Offshore Wind Turbine. *J. Mar. Sci. Eng.* **2022**, *10*, 542. https://doi.org/10.3390/jmse10040542.
5. Srinivas, A.; Robertson, B.; Gadasi, J.B.; Simpson, B.G.; Lomónaco, P.; Ilzarbe, J.M.B. Impact of Limited Degree of Freedom Drag Coefficients on a Floating Offshore Wind Turbine Simulation. *J. Mar. Sci. Eng.* **2023**, *11*, 139. https://doi.org/10.3390/jmse11010139.
6. Zhai, Y.; Zhao, H.; Li, X.; Shi, W. Hydrodynamic Responses of a Barge-Type Floating Offshore Wind Turbine Integrated with an Aquaculture Cage. *J. Mar. Sci. Eng.* **2022**, *10*, 854. https://doi.org/10.3390/jmse10070854.
7. Shah, S.A.A.; Gao, B.; Ahmad, I.; Ullah, H.; Ahmed, N.; Saeed, A. Adaptive Backstepping Integral Sliding Mode Control for 5DOF Barge-Type OFWT under Output Constraint. *J. Mar. Sci. Eng.* **2023**, *11*, 492. https://doi.org/10.3390/jmse11030492.
8. Zavvar, E.; Abdelwahab, H.S.; Uzunoglu, E.; Chen, B.-Q.; Soares, C.G. Stress Distribution on the Preliminary Structural Design of the CENTEC-TLP under Still Water and Wave-Induced Loads. *J. Mar. Sci. Eng.* **2023**, *11*, 951. https://doi.org/10.3390/jmse11050951.
9. Hmedi, M.; Uzunoglu, E.; Zeng, C.; Gaspar, J.F.; Soares, C.G. Experimental Challenges and Modelling Approaches of Floating Wind Turbines. *J. Mar. Sci. Eng.* **2023**, *11*, 2048. https://doi.org/10.3390/jmse11112048.
10. Hmedi, M.; Uzunoglu, E.; Medina-Manuel, A.; Mas-Soler, J.; Vittori, F.; Pires, O.; Azcona, J.; Souto-Iglesias, A.; Soares, C.G. Experimental Analysis of CENTEC-TLP Self-Stable Platform with a 10 MW Turbine. *J. Mar. Sci. Eng.* **2022**, *10*, 1910. https://doi.org/10.3390/jmse10121910.
11. Hallak, T.S.; Soares, C.G.; Sainz, O.; Hernández, S.; Arévalo, A. Hydrodynamic Analysis of the WIND-Bos Spar Floating Offshore Wind Turbine. *J. Mar. Sci. Eng.* **2022**, *10*, 1824. https://doi.org/10.3390/jmse10121824.
12. Lopez-Olocco, T.; González-Gutiérrez, L.M.; Calderon-Sanchez, J.; Loureiro, A.M.; Ynocente, L.S.; Barrio, A.B.; Valdés, N.V. Experimental and Numerical Study of the Influence of Clumped Weights on a Scaled Mooring Line. *J. Mar. Sci. Eng.* **2022**, *10*, 676. https://doi.org/10.3390/jmse10050676.
13. Díaz, H.; Soares, C.G. Approach for Installation and Logistics of a Floating Offshore Wind Farm. *J. Mar. Sci. Eng.* **2022**, *11*, 53. https://doi.org/10.3390/jmse11010053.
14. Hassan, M.; Soares, C.G. Dynamic Analysis of a Novel Installation Method of Floating Spar Wind Turbines. *J. Mar. Sci. Eng.* **2023**, *11*, 1373. https://doi.org/10.3390/jmse11071373.

15. Altuzarra, J.; Herrera, A.; Matías, O.; Urbano, J.; Romero, C.; Wang, S.; Soares, C.G. Mooring System Transport and Installation Logistics for a Floating Offshore Wind Farm in Lannion, France. *J. Mar. Sci. Eng.* **2022**, *10*, 1354. https://doi.org/10.3390/jmse10101354.
16. Li, H.; Peng, W.; Huang, C.-G.; Soares, C.G. Failure Rate Assessment for Onshore and Floating Offshore Wind Turbines. *J. Mar. Sci. Eng.* **2022**, *10*, 1965. https://doi.org/10.3390/jmse10121965.
17. Li, H.; Teixeira, A.P.; Soares, C.G. An Improved Failure Mode and Effect Analysis of Floating Offshore Wind Turbines. *J. Mar. Sci. Eng.* **2022**, *10*, 1616. https://doi.org/10.3390/jmse10111616.
18. Fu, X.; Sheng, M. Research on Structural Failure Analysis and Strengthening Design of Offshore Wind Turbine Blades. *J. Mar. Sci. Eng.* **2022**, *10*, 1661. https://doi.org/10.3390/jmse10111661.
19. Lotovskyi, E.; Teixeira, A.P.; Soares, C.G. Availability Analysis of an Offshore Wind Turbine Subjected to Age-Based Preventive Maintenance by Petri Nets. *J. Mar. Sci. Eng.* **2022**, *10*, 1000. https://doi.org/10.3390/jmse10071000.
20. Silva, L.M.R.; Li, H.; Soares, C.G. Service Operation Vessels Fleet Size and Mix Location Routing for the Maintenance of an Offshore Floating Wind Farm. *J. Mar. Sci. Eng.* **2023**, *11*, 664. https://doi.org/10.3390/jmse11030664.
21. Bashir, M.; Xu, Z.; Wang, J.; Soares, C.G. Data-Driven Damage Quantification of Floating Offshore Wind Turbine Platforms Based on Multi-Scale Encoder–Decoder with Self-Attention Mechanism. *J. Mar. Sci. Eng.* **2022**, *10*, 1830. https://doi.org/10.3390/jmse10121830.

Journal of
Marine Science and Engineering

MDPI

Review

Market Needs, Opportunities and Barriers for the Floating Wind Industry

Hugo Díaz [1], José Serna [2], Javier Nieto [2] and C. Guedes Soares [1,*]

1 Centre for Marine Technology and Ocean Engineering (CENTEC), Instituto Superior Tecnico, Universidade de Lisboa, 1049-001 Lisbon, Portugal; hugo.martinez@centec.tecnico.ulisboa.pt

2 Esteyco, S.A., Avenida de Burgos, 12b-bajo, 28036 Madrid, Spain; jserna@esteyco.com (J.S.); javier.nieto@esteyco.com (J.N.)

* Correspondence: c.guedes.soares@centec.tecnico.ulisboa.pt

Abstract: This paper reviews the status of floating wind energy expansion, market needs, opportunities, and barriers. Even more expensive than many other generation technologies currently, the floating wind will contribute to the decarbonization of Europe. This document assesses the market strategies available to develop floating wind farms in Europe. The study includes four main phases in addition to the overview of the current state-of-the-art: a technology review, market outlook, opportunities, and commercialization barriers. During its development, the offshore wind has moved from experimentation to a final design (Semisubmersible/barge, Tension Leg Platform, and Spar).

Keywords: barriers; floating wind; market needs; opportunities

Citation: Díaz, H.; Serna, J.; Nieto, J.; Guedes Soares, C. Market Needs, Opportunities and Barriers for the Floating Wind Industry. *J. Mar. Sci. Eng.* **2022**, *10*, 934. https://doi.org/10.3390/jmse10070934

Academic Editor: Decheng Wan

Received: 20 June 2022
Accepted: 4 July 2022
Published: 7 July 2022

1. Introduction

The rapid economic and technological development of society causes an ever-increasing energy dependence. The Energy Information Administration estimates a world energy consumption increase of 50% by 2050 [1]. However, energy production comes at a cost. The consequences of hydrocarbon use are now becoming more present due to the environmental state deterioration. As the energy crisis becomes more pressing, society searches for ways to save the planet.

The European Union (EU) is one of the main consumers of energy. The total electricity consumption in 2019 was over 2428 MWh [2]. The electricity sources are petroleum products (36%), natural gas (22%), renewable energy (15%), and nuclear energy and solid fossil fuels (13%) [3]. Wind energy includes 35% of renewables, representing almost 5% of the total electricity generation [4]. The major wind power producers are Germany and Spain, while total capacity is still increasing [5]. However, offshore wind provides some advantages over onshore wind.

The total available offshore wind resource is more significant than the onshore ones. Moreover, the topography of some coastal regions is not appropriate to host land-based wind power [6]. Additionally, onshore wind farm locations are positioned in regions with scarce populations. In contrast, the optimal offshore sites are near the North and West coasts, areas with a high population density [7]. Offshore turbines reduce visual and noise impact and the "Not In My Backyard" (NIMBY) phenomenon. However, only Northwestern Europe has a large-scale offshore wind energy industry [8]. This is because the current fixed-bottom technology is only economically feasible in shallow waters.

The floating technology enables the extraction of wind resources from previously expensive or complex areas to fixed-bottom turbine installation. These floating concepts have revived many countries' interests in offshore wind energy [9].

Europe has set ambitious renewable energy goals to reduce its environmental impact [10]. The vast untapped offshore wind resources will play an important role in fulfilling the energy demands to achieve the renewable energy targets of the EU.

This paper analyzes the readiness of the European Atlantic coast for a floating wind turbine market from the social, technological, legal, and environmental points of view. The general industry and technology trends are described. Recommendations are given to the government to prepare the industry and change policies. Moreover, the suitability of a floating wind market from an economic and technological perspective is analyzed for the EU. This generates the following question: What are the needs, opportunities, and barriers for the floating market in the EU?

The paper follows the following scheme. Section 2 presents a brief description of the operational floating farms. Section 3 reviews the literature related to floating technologies. Section 4 explains the roadmaps of floating wind energy, and Section 5 focuses on the main aspects of the floating wind market. Section 6 summarizes the findings and discusses their implications for the policy and the literature.

2. Floating Wind Farms

Concept development and testing at sea have been a continuum since the installation of the 2.3 MW Hywind and 2 MW WindFloat1 prototypes. Nowadays, several concepts are on sea tests from Europe to Asia. However, these developments have allowed Scotland and Portugal to count on operational floating wind farms.

Scotland

- Hywind wind farm.

Commissioned in 2017, it was the first floating wind farm worldwide and is composed of Norwegian technology (Hywind). The farm comprises five turbines, Siemens SWT-6.0-154, with a total capacity of 30 MW. The installation achieved an average capacity factor of 57.1% in 2021. Located in an average water depth of 110 m, the distance to shore is around 25 km.

- Kincardine wind farm.

The Kincardine floating farm was the third floating wind farm. Currently, this farm is dubbed the world's largest with a total capacity of 50 MW and consists of five Vestas V164-9.5 MW and one V80-2 MW turbine, each installed on WindFloat semi-submersible platforms. Located 15 km offshore Aberdeenshire, the site achieves water depths from 60 to 80 m. Kincardine will generate over 200 GWh of green electricity annually, enough renewable electricity to power more than 50,000 Scottish households.

Portugal

- Windfloat wind farm.

In July 2020, three floating turbines were installed outside Viana do Castelo, Portugal. With three MHI Vestas V164-8.4 MW turbines, the WindFloat Atlantic project has 25 MW of installed capacity. In the first year of operation, WindFloat Atlantic reached 75 GWh, which is enough to supply 60,000 inhabitants. Located 20 km offshore, WindFloat platforms are anchored to the seabed at a water depth of 100 m.

The three wind farms are the first examples of floating wind in a fast-emerging market (see Appendices A and B) that show the advantages of the floating cleantech.

3. Technology Review

There has been an outbreak of floating foundations in the last decade, revealing several concepts, some of them have reached the demonstration phase or commercial pre-series. This is motivated by the good prospects that the floating wind has to open the possibility to harness the deep-water areas out of the reach of the traditional bottom fixed technologies. Furthermore, floating foundations can be installed in deep and remote waters with stronger wind resources [11–13].

Floating structures are also less intrusive to the seabed than bottom-fixed systems. These technologies provide the potential for a simple standardization and mass production in the long term. The floating technologies are divided in mainly four stability philosophies (see Figure 1): spars, semi-submersibles, barges, and TLPs [14,15].

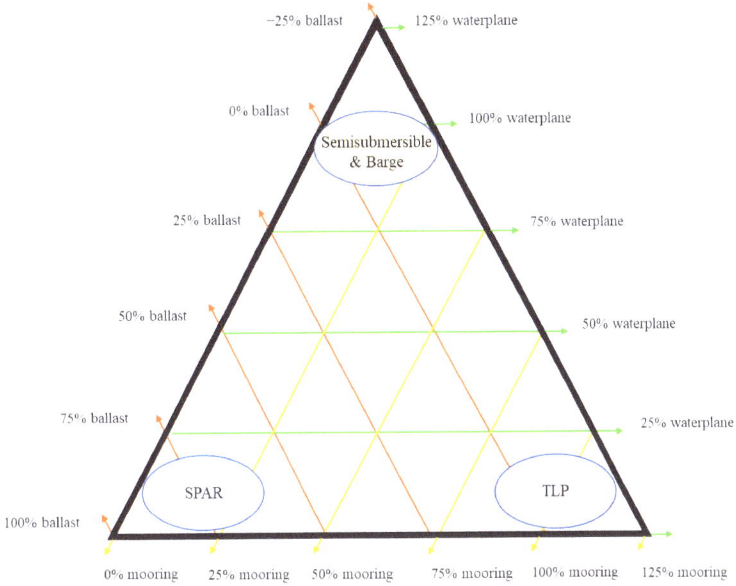

Figure 1. Floating foundations stability triangle.

The spar platforms are ballast stabilized, meaning that the gravity center is below the buoyancy center. The semi-submersibles and barge floaters are water plane stabilized, meaning that they provide sufficient inertia at the water plane. The semi-submersible concepts comprise several large columns interconnected by submerged pontoons, providing hydrostatic stability and additional buoyancy. Finally, tension leg platforms (TLPs) are stabilized through their station-keeping system that is in tension. The TLP is anchored to the ocean floor with taut leg mooring lines and drive/suction piles as a buoyant structure. This mooring system needs to support the enormous vertical loads created on the mooring lines.

The station-keeping system, in general, holds the platforms in place. The system can be segmented into three main categories, catenary mooring, taut mooring, and tendons (TLPs).

The bottom-fixed structures are not economically feasible for depths above 50 m. In this case, floating concepts are an ideal solution [16]. Based also on geographical characteristics, the mooring systems represent an important parameter in the conceptualization of a floating project.

In a catenary mooring system, the line has a significant length lying on the seabed until reaching the anchor. The catenary mooring system restoring force is provided by the weight of the hanging mooring line; therefore, the lines are typically made of chains or steel rope or others to which clump weights are added.

The mooring line joins at the anchor at an angle in a taut mooring system without a segment lying on the seabed. The elasticity of the mooring lines provides the restoring force. These systems are often made of synthetic ropes. The tendons might be considered a particular case of the taut mooring system; however, tendons installed in TLPs are usually made of steel tubes that provide little axial flexibility. Therefore, the vertical motions are controlled.

Finally, it is worth mentioning the anchor system that is the end of the station-keeping system. Based on soil conditions and the type of mooring system, there are several anchor possibilities, gravity anchors, pile anchors, free-fall anchors, suction anchors, fluke anchors, plate anchors, and grouted rock anchors. Some of the previous floating structures have been supported by Hywind Scotland, WindFloat Atlantic, and Kincardine Scotland pre-commercial floating wind arrays [17–19].

Despite the high costs associated with floating offshore wind energy [20], adequate industry and government support could facilitate the economic feasibility [21]. Additionally, the floating technologies may benefit from an expansion that generates cost reduction. Moreover, FOWT addresses a sustainable business course target of 7.3 to "double the global rate of improvement in energy efficiency by 2030" [22]. The inherent variability of renewable energies can be overpassed with a more compatible and broader system.

Several experts confirm the technical viability of FOWT [23–27]. However, the global feasibility of floating wind was not previously assessed. This study analyzes the drivers for and barriers to taking up FOWT (see Table 1). The case studies in the Atlantic coast and North Sea areas were chosen for three main reasons. First is the scope of the Arcwind project [28]. Second, commissioning the major projects (Hywind Scotland, WindFloat Atlantic, and Kincardine Scotland) could help new countries interested in exploring floating wind. Third, Spain, France, and Portugal have the largest number of floating concepts in development in countries on the European Atlantic coast. The Arcwind project is based on the development of three concepts in different locations. The concepts are provided by Saitec (Barge solution) [29], Centec (TLP solution) [16], and Esteyco (Spar solution) [30] (see Figure 2).

Table 1. Main opportunities and barriers of floating foundations.

Foundation	Opportunities	Barriers
Semisubmersible	High technology readiness levels (TRLs) Global market position Several turbines integration in a foundation Easy geometries Full system easy transportation Depth independence Simple mooring system	Internal forces dependent on geometries Expensive ballast system High competition between developers Heavy structure Large dimensions for future 15 MW+ turbines
Barge	Low-cost production Easy geometries Depth independence Simple mooring system Motion reduction Flat without interspaces Full system easy transportation	No global market position Launching technique Heavy structure Large dimensions for future 15 MW+ turbines
Conventional Spar	High TRL Easy mass production with tower synergies Material usage Simple mooring system	Final assembly in the position No global market position Construction complexity
TLP	Low-cost production Light structure Not soil dependent for anchor system Suitable for extreme sea states Excellent stability Full system easy transportation	Expensive mooring system Redundant moorings No global market positions Low TRL Suitable intermediate depths

Figure 2. Floating wind platforms involved in the Arcwind project. (**a**) SATH® platform concept (Semisub), (**b**) TELWIND® platform concept (Spar), and (**c**) CENTEC-TLP platform concept (TLP).

4. Floating Wind Power in Numbers

This section is focused on presenting the offshore wind market perspective up to 2050. Firstly, a short review of the process of the development of a wind farm provides the time scale. After, the market outlook is briefly described based on two main sources, the Global Wind Energy Council [31] and the WindEurope [32], who among other tasks issued periodically reports about the wind market.

4.1. Offshore Wind Farm Development Stages

WindEurope incorporated in their last report a narrative that explains the different steps and the timeline for developing a wind farm which is considered interesting to better understand the magnitudes that will later be exposed [33].

The different phases (see Figure 3) that need to be completed until producing energy from an offshore wind farm are:

- **Leasing:** This comprises selecting wind farm locations, managing the environment, maritime spatial planning (MSP), and others. The leasing task is complete with adjudicating the right to explore the wind resource in a given space to a developer as part of a competitive process. Typically, it takes 2 years.
- **Consenting:** This comprises the obtaining of construction and operation permits. The early site surveys and the front-end engineering design parallel the permitting process. The process takes 2 + 2 years to obtain the data and consent. Some projects can take significantly longer due to economic or environmental reasons. Obtaining the licenses to install and operate the project means the end of the consenting step.
- **Financial close:** A detailed design and favorable financial investment are included in this step. Moreover, the power purchase agreement is mandatory in this phase. A time slot of 2 years from consent is required due to the competitive auctions. The possibility of failing at this stage is high; nevertheless, the impact on development rates is not considered.
- **Installation:** This step includes offshore and onshore construction, grid connection, manufacture, and pre-assembly of components before installation. Almost 3 years are required to complete this phase. The wind farm is considered commissioned after starting with generating and transmitting power back to shore.

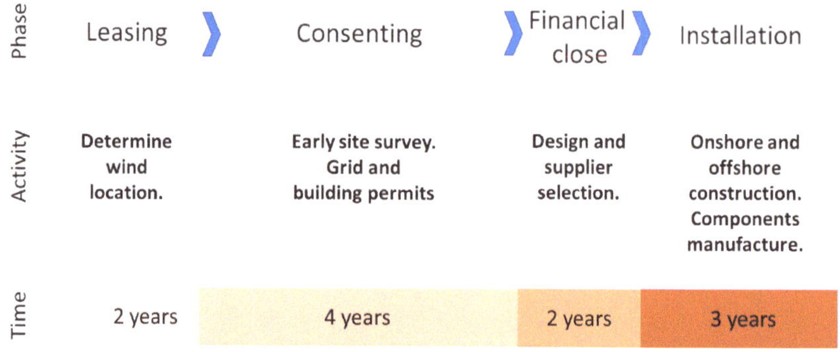

Figure 3. Floating farm development schedule. Adapted from [32].

4.2. Global Wind Energy Council

The Global Wind Energy Council (GWEC) is the international association for the wind sector. GWEC represent 1500 institutions in more than 80 countries. The list of representatives includes developers, manufacturers, suppliers, research institutes, associations, electricity providers, and finance and insurance institutions.

The members are also all of the national wind industry trade associations from emerging and developed markets.

The first report of GWEC was launched in 2019 under the title Global Offshore Wind Report [34]. The document provides a comprehensive analysis of the market, and this issue indicates the relevance of offshore wind within the whole wind industry.

The study shows a total installation of 23 GW, representing a growth of 21% each year since 2013. During 2017 and 2018, more than 4 GW of new capacity was installed, growing up to 8% of the total new installations during both years. China was the largest offshore market in 2018 based on new constructions, followed by the UK and Germany.

The report also provides a market outlook. The experts expect double-digit growth based on current policies and expected tenders and auctions. The scenario could be realistic based on growth in China and other Asian markets and considering annual installations of 15–20 GW after 2025 to reach 165 GW of new installed capacity globally in 2030. This prognostic brings us a total installed capacity of 200 GW.

Regionally, around 80 GW are attributed to Europe by 2030, around 100 GW to Asia (China, Taiwan, South Korea, Vietnam, Japan, and India), and around 10 GW to the US (this is increased to 22 GW by the US Energy Information Administration).

Currently, the offshore wind market is dominated by fixed structures. Nevertheless, based on upcoming projects, the floating structure type will experiment with a significant growth rate during the incoming years [35].

Particularly, as seen in Figure 4, the prospect is to progress from virtually nothing to date to around 500 MW in 2023 and 12 GW+ by 2030. The total annual offshore wind capacity needs to increase from 28 GW/year in 2030 to 45 GW/year in 2050, distant from the 4.5 GW of power available in 2018 (see Figure 5).

Hopefully, the floating technology may contribute to keeping Europe's prevalence in the offshore market for a longer period (and forever in terms of technology ownership), in opposition to the apparent trend of Asian markets, bound to lead the market worldwide in terms of installed power in about a decade or so. The previsions for the Asian markets indicate a total capacity of 100 GW by 2030 and 600 GW by 2050.

China will dominate the Asian market, with an installed capacity of 56 GW by 2030 and 382 GW by 2050. In the US, an installed capacity of 23 GW by 2030 and 164 GW by 2050 could be achieved.

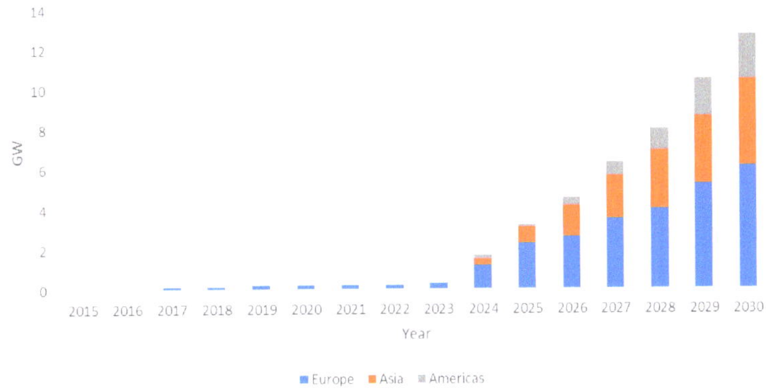

Figure 4. Floating offshore wind market outlook. Adapted from [34].

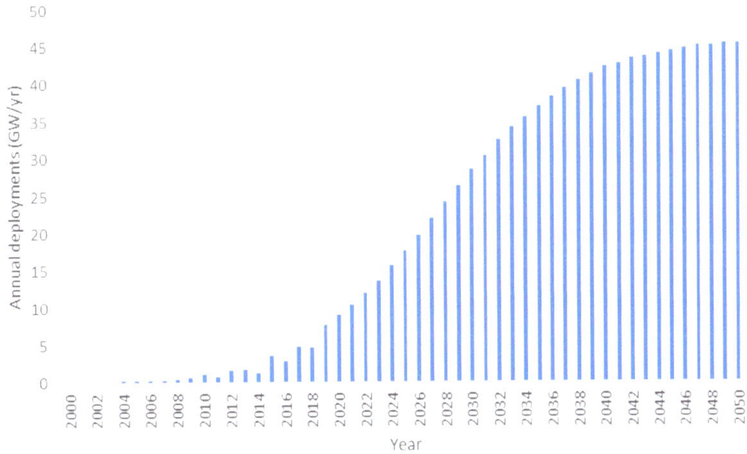

Figure 5. Annual offshore wind capacity, historical data, and projections. Adapted from [36].

4.3. Wind Europe

With a presence in more than 35 countries, WindEurope is an organization that actively promotes wind power in Europe. Of the over 400 members, there are wind turbine manufacturers, component suppliers, contractors, research institutes, renewables associations, electricity providers, developers, consultants, and finance and insurance companies.

WindEurope actively supports members' requirements and needs to further their development, offering the sector the best networking and learning opportunities.

In November 2019, WindEurope issued the report called "Our energy, our future" that includes an outlook of the offshore wind market that is briefly described in this section [37].

To achieve carbon-neutral by 2050, the European Commission needs between 230 and 450 GW of offshore wind energy by 2050. The 2030 policy framework can facilitate 89 GW by 2030. However, the bases to enable higher installation levels need the national governments' support.

The report has investigated most of the areas in northern Europe. The following map shows the area division considered by WindEurope for elaborating the study.

Areas considered in the Atlantic Arc Region are ES01, ES02, ES03, PT01, PT02, PT03, FR01, FR02, UK01, UK05, IE01, and IE02 (see Figure 6). It is very relevant that most of these areas are to be destined for floating wind given the depths available. Table 2 indicates the percentage of the areas with depths over 50 m, which is a good threshold for the reach of the bottom fixed wind. The areas are divided as well depending on the estimated LCOE. The report uses the following LCOE bands:

- Below EUR 50/MWh in 2030—Very low LCOE.
- Between EUR 50/MWh and EUR 65/MWh in 2030—Low LCOE.
- Between EUR 65/MWh and EUR 80/MWh in 2030—Medium LCOE.
- Over EUR 80/MWh in 2030—High LCOE.

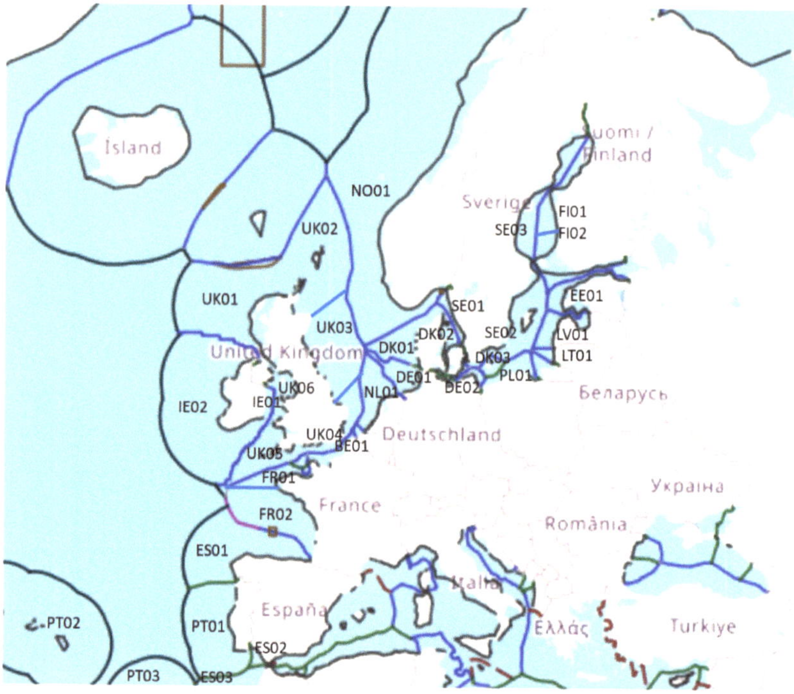

Figure 6. European water subregions. Adapted from [32].

As it can be seen, there is a great portion in these areas that is going to be indicated for FOWT. To be precise, the sum is 77% of the offshore area.

Table 2. Floating share in each sector analyzed.

Sub Region	Capacity Allocated (GW)	Offshore Area (km^2)	Portion of Area with d > 50 m			
			Very Low LCOE	Low LCOE	Mid LCOE	High LCOE
IE01	6.7	1340	0%	60%	0%	0%
IE02	15.5	3100	0%	11%	37%	48%
FR01	20.9	4180	4%	45%	14%	0%
FR02	19.6	3920	0%	23%	52%	15%
UK01	2.6	520	2%	49%	21%	4%
UK05	8.5	1700	0%	35%	25%	2%

In the Atlantic Region, the WindEurope report estimates that the increase in consenting power, area, and installation rates need to be indicated in the following table. These rates will allow generating the 85 GW required to obtain the overall 450 GW in Europe for fulfilling the aim of being carbon-neutral for 2050.

Just under 2 GW (370 km^2) per year is needed (see Table 3), and this rises to 3.9 GW (or 770 km^2) annually through the 2030s. Therefore, to achieve the needed increase, the consenting rate requires double by 2025 and quadruple in the 2030s.

Table 3. Required installation rates in the Atlantic region.

	Average Rate						
	2019–2025	2026–2030	2031–2035	2036–2040	2041–2045	2046–2050	Total 2050
Installed GW	1.1	1.8	2.7	3.9	3.9	3.4	85
Consented GW	1.9	3.4	3.9	3.8	3.2	3.1	
Consented km^2	370	680	770	750	630	620	

One of the most important conclusions of the WindEurope report is that floating wind will soon be as simple as other foundations available. WindEurope predicts full commercialization of the floating technologies. This organization estimates that the share of offshore floating wind will be between 100 and 150 GW of the 450 GW required by 2050.

4.4. *European Commission: An EU Strategy to Harness the Potential of Offshore Renewable Energy for a Climate-Neutral Future*

The EU has set ambitious targets for reducing greenhouse gas emissions in the power sector [38]. A reduction of 55% by 2030 and complete decarbonization by 2050 is planned. The offshore wind sector (as low carbon technology) will be a key player in the European Union's decarbonization goals. Since the first turbine was installed in the marine environment 25 years ago, offshore wind has provided energy for millions worldwide.

Today, offshore wind competes with fossil fuels. The power of seas, such as tidal currents, waves, floating photovoltaic, and algae biofuels, are technologies in development.

The European Union's seas have a vast potential. The technological and physical potential is crucial to achieving the 2030 carbon emission targets and climate neutrality by 2050. The offshore wind industry could achieve this target only with the 3% of the European maritime space also satisfying the EU Biodiversity Strategy goals.

The EU target of 300 GW of offshore wind energy by 2050 needs a sector re-scaling in the next 30 years. This expansion requires multiplying the capacity by nearly 30 times and an investment estimated at up to EUR 800 billion. Nonetheless, some obstacles should be outgrown to ensure that all players can support this increase in deployment rate.

The EU strategy needs to address barriers and challenges and specific policy solutions adapted to the various stages of technology development and regional contexts. The Europe sea basins have the potential and conditions to develop offshore renewable energy.

5. Floating Wind Energy Market

One of the crucial aspects of floating wind energy development is the associated technology. The floating wind energy market comprises many stakeholders and investors, such as floating technology developers, utility companies, equipment manufacturers, contractors, and energy producers.

The construction of floating farms requires several industrial segments and companies in each sector. Noteworthy are the companies that produce floating platforms, offshore wind turbines, wind towers, cables, mooring systems, and installation companies.

Over the past years, the competitiveness of the technological market for floating wind components has increased [9]. Consequently, the increased competition has reduced manufacturers' profit margins and sale values. The floating wind market expansion increases the technology sales (see Table 4).

Currently, seven major wind turbine manufacturers (Siemens, Senvion, Bard, Vestas, Bonus, Bordtank, and Wind World) stand out, which in 2018 operated 77% of the worldwide market [39]. The leading developers involved in floating foundations are presented in Table 4 and Appendices A and B.

As indicated in previous sections, the share of the Atlantic Region is almost 80% of the wind offshore areas. The water depths in the Atlantic region are ideal for floating wind and studies estimate that 85 GW could be deployed in the next decades. In line with the previous prognosis, some recent studies based on sustainable use of the maritime space considered that the Atlantic coast of Portugal, Spain, and France could host 32 GW of floating wind [6].

Table 4. Europe floating wind projects.

Project	Country	Pipeline Status	COD	Capacity (MW)	Water Depth (m)	Developer	Turbine Rating (MW)	Substructure
Eolink 2/10 scale prototype	France	Installed	2018	0.2	10	Eolink	0.2	Semisubmersible
Floatgen project	France	Installed	2018	2	33	Ideol	2	Barge
Groix Belle Ille	France	Approved	2021	24	62	Eolfi	6	Semisubmersible
Provence Grand Large	France	Approved	2021	24	30	EDF	8	Tension Leg Platform
Eolmed	France	Approved	2021	24	62	Ideol	6.2	Barge
Les Eoliennes Flotant du Golfe du Lion	France	Approved	2021	24	71	Engie, EDPR, Caisse de Depots	6	Semisubmersible
GICON Schwimmendes Offshore Fundament SOF Pilot	Germany	Financial close	2022	2.3	37	GICON	2.3	Tension Leg Platform
Hywind Demo	Norway	Installed	2009	2.3	220	Unitech offshore	2.3	Spar
TetraSpar Demostrator	Norway	Financial close	2019	3.6	200	Innogy, Shell, Stiesdal	3.6	Semisubmersible
Hywind Tampen	Norway	Permitting	2022	88	110	Equinor	8	Spar
NOAKA	Norway	Planning	2023	-	130	Equinor. Aker, BP	-	-
Windfloat Atlantic	Portugal	Financial close	2019	25	50	Windplus	8	Semisubmersible
DemoSath—Bimep	Spain	Approved	2020	2	68	Saitec offshore technologies	-	Semisubmersible
X1 Wind prototype—Plocan	Spain	Approved	2021	-	62	X1 wind	-	Tension Leg Platform
Floating Power Plant—Plocan	Spain	Approved	2021	-	62	FPP	8	Hybrid wave power semisubmersible
Hwind Scotland Pilot Park	UK	Installed	2017	30	100	Equinor	6	Spar
Dounreay Tri	UK	Approved	2021	10	76	Hexicon	5	Semisubmersible
Kinkardine Offshore wind farm—Phase 1	UK	Installed	2018	2	62	Cobra	2	Semisubmersible
Kinkardine Offshore wind farm—Phase 2	UK	Under construction	2020	50	62	Cobra	9.5	Semisubmersible

In the scope of the Arcwind project, more than 80 floating wind farm locations were determined and characterized (see Table 5). Of all of them, 6 locations were characterized to host floating platforms due to their suitability to host the Esteyco (Spar), Saitec (Semisubmersible), and Centec (TLP).

Table 5. Main floating wind farms proposed in the scope of the Arcwind project.

Country	Location	Coordinates (WGS84)		Wind Farm Capacity (MW)	Turbines	Area (km²)	Platform
France	Lannion	49.2	−3.6	470	47	62	Telwind [1]
Spain	Gran Canaria	27.8	−15.3	120	12	25.37	
Portugal	Figueira da Foz	40.2	−9.4	700	70	110	Sath [2]
Scotland	A15	58.75	−6	180	18	22	
Spain	Ribadeo	44	−7.3	880	88	405	C-TLP [3]
Ireland	F15	52.7	−10.5	250	25	30.3	

[1] Esteyco Spar; [2] Saitec Semisubmersible; [3] Centec TLP.

5.1. Research and Development Trends and the Future of the Floating Wind Industry

The floating wind platforms are designed for extreme sea conditions, allowing installation and maintenance time reduction. Establishing coastal facilities could solve the logistical constraints of the sector. The floating turbine construction and maintenance may be realized on land and transported to the farm. New designs, larger rotors, advanced materials, and taller towers suggest a new field for innovation. The new technological improvements focus on six areas: component weight and costs, transport and assembly, monitoring and control, turbine reliability, grid integration, and optimizing performance. The most ambitious research and development (R&D) plan is trying to develop a rotor blade longer than 200 m for a 50 MW offshore wind turbine, 2.5 times longer and 6 times more potent than turbines currently in operation [40,41]. The integration of floating concepts with other energy technologies (wave energy converters or solar panels) is currently in the R&D process.

5.2. The Port Activities in the Floating Wind Industry

Seaport services include handling goods at port terminals, towing services, piloting services, and industrial services. Moreover, the port can offer manufacturers test facilities, training centers, rental warehouses, offices, and operation centers [42].

The ports are currently adapting their business to support floating wind and other marine renewable energies. These entities adapt their infrastructure to stock the components, assemble the units, and stock the foundations providing cost reduction and efficiency [43].

Several studies analyzed the types of ports and the associated production process [44]. The type of port used to establish floating wind activities will depend on the installation strategy, costs, the distance from the manufacturing centers, and the distance to the wind farm [45].

The seaport can play a crucial role in the offshore wind production process. Managing all stages at the local and temporal levels is essential for the activity's profitability. The primary function associated with the port is the transportation of various equipment from land to the sea. However, some ports could assume the role of manufacturing and assembly, which raises this activity's potential in these entities.

5.3. The Environmental Considerations

A stable wind resource is essential for offshore wind projects. Wind variability is a significant factor. A steadier wind can provide a more constant energy output. These conditions are present on the European Atlantic coast.

Suitable offshore sites near populated centers mean new opportunities. The offshore farms near the final consumer reduce transmission costs and energy losses. At the same time, the farm's visibility from the shore needs to be considered. Moreover, the weather patterns are essential to avoid possible coastal natural disasters. The frequency of these events can significantly impact the project cost and electricity cost—these aspects need to be considered when searching for suitable locations. One of the primary purposes of the Arcwind project was to identify appropriate areas from a sustainable point of view. The Not in My Backyard (NIMBY) phenomenon is at a maximum in the site selection of floating wind farms.

5.4. Social Aspects

The social aspects of floating energy projects can be the deciding factor in the success of a project. A negative public attitude towards floating wind farms can delay or cancel a project. Some projects in the EU illustrate the importance of public acceptance. The NIMBY opposition against wind turbine farms is related to the turbines' aesthetics and any possible pollution [46]. Factors such as the culture, economy, political stance, location, and coast use determine the public attitude [47]. The opposition against wind farms comes from people linked to the ocean/coast jobs. Local community members fear the project will damage the environment, disrupt commercial activity, and lower property prices. An excellent opportunity to achieve a floating project implementation is guaranteeing the involvement of the local public and social entities in the process of wind farm acceptance.

5.5. Legal Aspects

Legal aspects are one of the biggest concerns when it comes to the development of the floating wind industry. Europe has had around 5 years of commercial experience in floating wind projects. So far, several commercial leases at the Atlantic Coast have been submitted by energy suppliers to the corresponding organizations (national and regional). Moreover, stakeholders can pressure the project by filing lawsuits. The outcome of these lawsuits is often uncertain and can heavily delay projects and run up costs. Another main concern during the permitting is the public resistance because of visual impact. The visual effect of an offshore installation is taken into account during the permitting process.

The permitting process for an offshore wind farm in European waters is complicated due to the duplicity of documents at the national and regional levels.

Furthermore, the rules are not specific to the floating sector, which means that laws that are not specific need to be fulfilled. This type of project requires a clear regulatory framework between the regulatory bodies to facilitate the fast implementation for the energy developers.

5.6. Technological Aspects

The technological environment is crucial to benefit from the opportunities of the floating wind market. Several obstacles related to floating turbine technology can emerge.

Some technical challenges are inherent to renewable energy solutions. The existing energy infrastructure decides how easily energy from offshore turbines can be integrated into the grid, and the lack of infrastructure dramatically increases the cost. The supply chain and the manufacturing environment on the European coast are other technical factors to consider.

5.7. Floating Offshore Wind Barriers for Commercialization

Most barriers are to be solved by the policymakers who need to realize the huge opportunities that the FOWT represent, in terms of job creation, environmental commitments, and Europe's future. All available reports concur in the following barriers for FOWT:

1. **Maritime Spatial Planning**: There is a need to increase the rate of site allocation and development for achieving, at least in Europe, the environmental commitments of the Paris Climate Agreement. This is in hands of the policymakers that need to move quickly to put in motion the long process required for a wind farm development.

2. **Environmental Impact**: Although the comparison with the oil and gas business is clear, it is required to ensure that focusing the energy transition on the offshore wind industry is the right decision.
3. **Multiple Use**: The future of offshore wind goes towards the sharing of the sea between users.
4. **Expand the grid offshore and onshore**: The electricity grid infrastructure in Europe needs to be able to absorb the power foreseen to be installed. This will require, most likely, the collaboration between countries, which requires a legal framework that still needs to be constructed.
5. **Stable Rates**: The policymakers need to stabilize the energy rates for allowing the project to capture investments. This will increase the supply chain that will feel more comfortable investing in the necessary upgrading of vessels, ports, and onshore facilities that could be amortized in longer periods.
6. **Mobilizing investments**: WindEurope, in the frame of Europe's objective for 2050, estimates that CAPEX needs to be threefold for offshore wind farms and transmission grids. Their estimate requires to go from EUR 6 bn per year in 2020, to more than EUR 21 bn per year in 2025. In 2030, the investments will need to be around EUR 45 bn. In total, this means spending around 10% of the total infrastructure budget across Europe, on offshore wind.
7. **Technology Readiness**: The TRL achieved by different concepts such as Equinor's spar, Ideol's barge, or PPI's semisubmersible, encourage the experts to believe that this will not be a drawback for the FOW future. However, for other concepts, there is still a long way to walk, but the de-risking process shall be faster every day.
8. **Industrialization**: The designers need to intensify their efforts for thinking about the ways to allow an industrial process that allows the cost competitiveness of offshore wind energy. The step from demonstrator or pre-series to full commercialization needs to be taken. Thinking about these fabrication, transport, and installation procedures will allow the governments to know how to upgrade the harbor facilities.
9. **Scalability**: The turbine sizes are in continuous growth, and it is usually overtaking the foundation designs. The recent GE 12 MW turbine is a tremendous challenge for the foundation designers. In light of the speed in which these new turbines have reached the market, it is expected that the exigencies of the floaters will be very demanding in the future.
10. **High Levelized Cost of Energy**: In recent years, significant cost reductions in the onshore and bottom fixed offshore wind sectors were witnessed. FOW is anticipated to follow a similar downward trend with a cost decrease of 38–50% leading to 2050, following the suggestions of IEA experts. There are several other factors which may also lead to further cost reductions.
    - **(a)** One of the main advantages of FOW is the positioning in areas with higher average wind speeds, allowing to harness the best possible wind resources without depth constraints. The capacity factor can thus be improved and lead to increased electricity generation. With higher capacity factors, the levelized cost of energy (LCOE) will be reduced.
    - **(b)** Technology that allows a cost-effective exchange of large turbine components offshore when floating foundation structures are moving due to wave motion. The maintenance of turbines in FOWT is a non-solved issue, the players in the industry are demanding cost-effective solutions for the large correctives of the turbines in the FOW.

      For spars, whose final configuration is too deep to be taken into the harbor, this would imply such prohibitive costs that it simply cannot be considered a valid option for future large commercial floating farms. Proof of this is the contest of ideas to solve this problem proposed in 2017 by Equinor. This initiative has, however, failed to provide any concept promising and ready enough to be field-tested and demonstrated, and as of today Equinor openly admits that

LCM interventions for their SPAR floating wind turbines are an issue yet to be solved.

With semisubmersibles, whose draught is much smaller, the full de-installation and reinstallation of complete units might be considered a viable alternative for a commercial wind farm. They would be towed to the harbor to use onshore cranes for Large Corrective Operations. Even so, the cost of the mooring and mooring large structures, disconnecting and reconnecting the dynamic riser power cables, decommissioning and recommissioning the turbine and electrical systems are very high, and required the mobilization of multiple specialized vessels and need very long operative weather windows leading to prolonged unproductive periods.

(c) In the same framework as the previous, the industry is looking for cost-effective solutions for maintaining floating offshore foundations due to the capability of towing the structure to the port.

(d) Cost-effective manufacturing, installation, and maintenance of the large volume of mooring lines and anchors in floating wind farms. Mooring systems and their installations are important cost contributors, particularly given the large volume of mooring lines and anchors that must be installed.

(e) Cost-effective monitoring and inspection of a large number of mooring lines, cables, and foundation structures. Current solutions are based on Remotely Operated Vehicles (ROVs) or divers, feasible solutions inherited from the oil and gas industry due to the small number of lines.

Despite all the challenges mentioned above, the good performance of the FOWT must be remembered. FOW could cement Europe's leadership in renewables globally [33].

6. Discussion and Conclusions

The energy system plays an essential role in reducing greenhouse gas emissions. Europe is one of the leaders in renewable energy globally and has set the very ambitious goal of 40% renewables by 2030. The vast offshore wind resources off the Atlantic coast are being eyed as the potential next renewable energy source for the European Union. This opens the question of the barriers and opportunities for the floating wind market. Therefore, the social, technological, environmental, and legal aspects relevant to floating offshore wind energy were investigated. The document demonstrates that the offshore environment offers excellent opportunities for floating concepts.

Floating wind energy heavily depends on political support, especially in the beginning. The support of the national governments is highly unclear under some financial uncertainty. However, the European Union is actively engaged in realizing offshore wind projects through initiatives such as the Interreg and H2020 funds. Furthermore, the EU has a vast economy to support such projects and is attractive to developers because of its high national electricity prices. The technological environment in the EU can facilitate development. The European industry can support the growth of this new technology, and the jobs related to manufacturing are available locally. A renewable energy mix with floating wind energy seems logical. The biggest challenge is the highly complex regulatory environment that can significantly increase costs and slow down projects. Finally, local opposition against such projects is not uncommon in the EU and can increase financial uncertainty.

In the future, it will be interesting to learn more from the pre-commercial floating projects currently commissioned in Europe and the UK. Overall, the opportunities for a floating wind market in Europe seem bright, and the present threats are deemed surmountable.

Author Contributions: Conceptualization, J.S. and J.N.; formal analysis, J.S. and J.N.; writing—original draft preparation, J.S., J.N. and H.D.; writing—review and editing, C.G.S.; project administration, C.G.S. All authors have read and agreed to the published version of the manuscript.

Funding: This work was conducted within the ARCWIND project—Adaptation and implementation of floating wind energy conversion technology for the Atlantic region (http://www.arcwind.eu/, accessed on 1 June 2022), which is co-financed by the European Regional Development Fund through the Interreg Atlantic Area Programme under contract EAPA 344/2016. This work contributes to the Strategic Research Plan of the Centre for Marine Technology and Ocean Engineering (CENTEC), which is financed by the Portuguese Foundation for Science and Technology (Fundação para a Ciência e Tecnologia—FCT) under contract UIDB/UIDP/00134/2020.

Institutional Review Board Statement: Not applicable.

Informed Consent Statement: Not applicable.

Data Availability Statement: Not applicable.

Conflicts of Interest: The authors declare no conflict of interest.

Appendix A. Asian Floating Wind Projects

Project	Country	Pipeline Status	COD	Capacity (MW)	Water Depth (m)	Developer	Turbine Rating (MW)	Substructure
Fukushima floating offshore wind farm phase 1	Japan	Installed	2013	2	120	Marubeni Corporation	2	Semisubmersible
Fukushima floating offshore wind farm phase 2	Japan	Installed	2015	5	120	Marubeni Corporation	5	Semisubmersible
Sakiyama	Japan	Installed	2016	3	100	TODA Corporation	2	Spar
Kitakyusu (NEDO)	Japan	Under construction	2019	3	70	NEDO—Ideol	3	Semisubmersible
Hitachi Zosen	Japan	Permitting	2024	400	-	Equino—Hitachi	-	Semisubmersible
Macquaire	Japan	Planning	2025	500	100	Macquaire	-	-
Ulsan	South Korea	Financial close	2019	0.75	15	Consortium	0.75	Semisubmersible
Donghae KNOC	South Korea	Planning	2027	-	-	Equinor—KNOC	-	-
Ulsan shell	South Korea	Planning	2027	200	-	Shell—Hexicon	-	Semisubmersible
Ulsan Macquaire	South Korea	Planning	2027	200	-	Macquaire	-	-
Ulsan SK	South Korea	Planning	2027	200		SK—E&S	-	-
Ulsan KF	South Korea	Planning	2027	200	-	KF Wind—PPI	-	Semisubmersible
Floatinf W1N	Taiwan	Planning	2025	500	-	Eolfi—Cobra	-	-

Appendix B. American Floating Wind Projects

Project	Country	Pipeline Status	COD	Capacity (MW)	Water Depth (m)	Developer	Turbine Rating (MW)	Substructure
Castle Wind	USA	Planning	2027	1000	900	EnBW	8	Semisubmersible
Redwood coast	USA	Planning	2025	150	550	EDPR—PPI	8	Semisubmersible
Aqua Ventus I	USA	Planning	2022	12	100	University of Maine	6	Semisubmersible
Oahu North	USA	Planning	2027	400	850	AW Wind	6	Semisubmersible
Oahu South	USA	Planning	2027	400	600	AW Wind	6	Semisubmersible
Progression Wind	USA	Planning	2027	400	650	Progression Wind	6	Semisubmersible

References

1. U.S. Energy Information Administration. *International Energy Outlook 2021*; U.S. Energy Information Administration: Washington, DC, USA, 2021.
2. Eurostat. *Energy Balance Sheets 2017 Data*, 2019th ed.; Eurostat: Luxembourg, 2019; ISBN 978-92-76-08714-4.
3. Eurostat Eurostat Data Browser. Available online: https://ec.europa.eu/eurostat/en/ (accessed on 5 November 2021).
4. Eurostat. *Energy Data 2020 Edition*; European Union: Luxembourg, 2020; ISBN 978-92-76-20629-3.
5. Wind Europe. *Wind Energy in Europe 2020 Statistics and the Outlook for 2021–2025*; Wind Europe: Brussels, Belgium, 2021.
6. Díaz, H.; Guedes Soares, C. An integrated GIS approach for site selection of floating offshore wind farms in the Atlantic continental European coastline. *Renew. Sustain. Energy Rev.* **2020**, *134*, 110328. [CrossRef]
7. Díaz, H.; Fonseca, R.B.; Guedes Soares, C. Site selection process for floating offshore wind farms in Madeira Islands. In *Advances in Renewable Energies Offshore*; Guedes Soares, C., Ed.; Taylor & Francis: London, UK, 2019; pp. 729–737.
8. Akhtar, N.; Geyer, B.; Rockel, B.; Sommer, P.S.; Schrum, C. Accelerating deployment of offshore wind energy alter wind climate and reduce future power generation potentials. *Sci. Rep.* **2021**, *11*, 11826. [CrossRef] [PubMed]
9. Watson, S.; Moro, A.; Reis, V.; Baniotopoulos, C.; Barth, S.; Bartoli, G.; Bauer, F.; Boelman, E.; Bosse, D.; Cherubini, A.; et al. Future emerging technologies in the wind power sector: A European perspective. *Renew. Sustain. Energy Rev.* **2019**, *113*, 109270. [CrossRef]
10. European Commission the European Green Deal. *Eur. Comm.* **2019**, *53*, 24. [CrossRef]
11. Salvação, N.; Bernardino, M.; Guedes Soares, C. Assessing the offshore wind energy potential along the coasts of Portugal and Galicia. In *Developments in Maritime Transportation and Exploitation of Sea Resources*; Guedes Soares, C., Lopez Pena, F., Eds.; Taylor & Francis Group: London, UK, 2014; pp. 995–1002. [CrossRef]
12. Castro-Santos, L.; Bento, A.R.; Silva, D.; Salvação, N.; Guedes Soares, C. Economic Feasibility of Floating Offshore Wind Farms in the North of Spain. *J. Mar. Sci. Eng.* **2020**, *8*, 58. [CrossRef]
13. Salvacao, N.; Guedes Soares, C. Resource Assessment Methods in the Offshore Wind Energy Sector. In *Floating Offshore Wind Farms*; Castro-Santos, L., Diaz-Casas, V., Eds.; Springer International Publishing: Berlin/Heidelberg, Germany, 2016; pp. 121–141.
14. Uzunoglu, E.; Karmakar, D.; Guedes Soares, C. Floating offshore wind platforms. In *Floating Offshore Wind Farms*; Castro-Santos, L., Diaz-Casas, V., Eds.; Springer International Publishing: Berlin/Heidelberg, Germany, 2016; pp. 53–76.
15. Wang, C.M.; Utsunomiya, T.; Wee, S.C.; Choo, Y.S. Research on floating wind turbines: A literature survey. *IES J. Part A Civ. Struct. Eng.* **2010**, *3*, 267–277. [CrossRef]
16. Uzunoglu, E.; Guedes Soares, C. Hydrodynamic design of a free-float capable tension leg platform for a 10 MW wind turbine. *Ocean Eng.* **2020**, *197*, 106888. [CrossRef]
17. Equinor Hywind Scotland. Available online: https://www.equinor.com/ (accessed on 5 November 2021).
18. Principle Power Windfloat. Available online: www.principlepowerinc.com (accessed on 5 November 2021).
19. Cobra Group Kincardine Floating Offshore Wind Farm. Available online: https://www.grupocobra.com/en/proyecto/kincardine-offshore-floating-wind-farm/ (accessed on 5 November 2021).
20. Díaz, H.; Rodrigues, J.M.; Guedes Soares, C. Evaluation of an offshore floating wind power project on the Galician coast. In Proceedings of the International Conference on Offshore Mechanics and Arctic Engineering—OMAE, Trondheim, Norway, 25–30 June 2017; Volume 10. [CrossRef]
21. Díaz, H.; Rodrigues, J.M.; Guedes Soares, C. Preliminary cost assessment of an offshore floating wind farm installation on the Galician coast. In *Progress in Renewable Energies Offshore*; Guedes Soares, C., Ed.; Taylor & Francis Group: London, UK, 2016; pp. 843–850.
22. International Bank for Reconstruction and Development/The World Bank. *The Energy Progress Report 2018*; International Bank for Reconstruction and Development/The World Bank: Washington, DC, USA, 2018.
23. Bashetty, S.; Ozcelik, S. Review on dynamics of offshore floating wind turbine platforms. *Energies* **2021**, *14*, 6026. [CrossRef]
24. Huijs, F.; De Bruijn, R.; Savenije, F. Concept design verification of a semi-submersible floating wind turbine using coupled simulations. *Energy Procedia* **2014**, *53*, 2–12. [CrossRef]
25. Jiang, Z.; Yttervik, R.; Gao, Z.; Sandvik, P.C. Design, modelling, and analysis of a large floating dock for spar floating wind turbine installation. *Mar. Struct.* **2020**, *72*, 102781. [CrossRef]
26. Matha, D.; Sandner, F.; Molins, C.; Campos, A.; Cheng, P.W. Efficient preliminary floating offshore wind turbine design and testing methodologies and application to a concrete spar design. *Philos. Trans. R. Soc. A Math. Phys. Eng. Sci.* **2015**, *373*, 20140350. [CrossRef] [PubMed]
27. Uzunoglu, E.; Guedes Soares, C. A system for the hydrodynamic design of tension leg platforms of floating wind turbines. *Ocean Eng.* **2019**, *171*, 78–92. [CrossRef]
28. ARCWIND. Project Arcwind. Available online: http://www.arcwind.eu/ (accessed on 5 November 2021).
29. Baita-Saavedra, E.; Cordal-Iglesias, D.; Filgueira-Vizoso, A.; Morató, À.; Lamas-Galdo, I.; Álvarez-Feal, C.; Carral, L.; Castro-Santos, L. An economic analysis of an innovative floating offshore wind platform built with concrete: The SATH® platform. *Appl. Sci.* **2020**, *10*, 3678. [CrossRef]
30. Cartelle-Barros, J.J.; Cordal-Iglesias, D.; Baita-Saavedra, E.; Filgueira-Vizoso, A.; Couñago-Lorenzo, B.; Vigara, F.; Cortés, C.; Cerdán, L.; Nieto, J.; Serna, J.; et al. Pathways to bring the costs down of floating offshore wind farms in the Atlantic Area. *Wind Energy Sci. Discuss.* **2019**, *2019*, 1–12.

31. Global Wind Energy Council Green Recovery Data and Analysis. Available online: https://gwec.net/ (accessed on 5 November 2021).
32. WindEurope Policy. Available online: https://windeurope.org/ (accessed on 5 November 2021).
33. Wind Europe. *Floating Offshore Wind Vision Statement*; Wind Europe: Brussels, Belgium, 2017.
34. Global Wind Energy Council. *GWEC Global Wind Report 2019*; Global Wind Energy Council: Brussels, Belgium, 2019.
35. Global Wind Energy Council. *Global Offshore Wind: Annual Market Report 2020*; Global Wind Energy Council: Brussels, Belgium, 2020.
36. IRENA. *Future of Wind: Deployment, Investment, Technology, Grid Integration and Socio-Economic Aspects*; IRENA: Abu Dhabi, United Arab Emirates, 2019.
37. Wind Europe. *Floating Offshore Wind Energy—A Policy Blueprint for Europe*; Wind Europe: Brussels, Belgium, 2019.
38. European Union. *An EU Strategy to Harness the Potential of Offshore Renewable Energy for a Climate Neutral Future*; European Union: Luxembourg, 2020.
39. Díaz, H.; Guedes Soares, C. Review of the current status, technology and future trends of offshore wind farms. *Ocean Eng.* **2020**, *209*, 107381. [CrossRef]
40. Yao, S.; Chetan, M.; Griffith, D.T.; Escalera Mendoza, A.S.; Selig, M.S.; Martin, D.; Kianbakht, S.; Johnson, K.; Loth, E. Aero-structural design and optimization of 50 MW wind turbine with over 250-m blades. *Wind Eng.* **2021**, *46*, 273–295. [CrossRef]
41. Schütt, M.; Anstock, F.; Schorbach, V. Progressive structural scaling of a 20 MW two-bladed offshore wind turbine rotor blade examined by finite element analyses. *J. Phys. Conf. Ser.* **2020**, *1618*, 052017. [CrossRef]
42. Host, A.; Skender, H.P.; Mirković, P.A. The perspectives of port integration into the global supply Chains—The case of North Adriatic ports. *Pomorstvo* **2018**, *32*, 42–49. [CrossRef]
43. Junqueira, H.; Robaina, M.; Garrido, S.; Godina, R.; Matias, J.C.O. Viability of creating an offshore wind energy cluster: A case study. *Appl. Sci.* **2021**, *11*, 308. [CrossRef]
44. Cooper, B.D.; Marrone, J.F. Port requirements to support offshore wind development in North America. In *Ports 2013 Success Through Diversification, Proceedings of the 13th Triennial International Conference, Seattle, WA, USA, 25–28 August 2013*; ASCE: Reston, VA, USA, 2013; pp. 1473–1482. [CrossRef]
45. Díaz, H.; Guedes Soares, C. A Multi-Criteria Approach to Evaluate Floating Offshore Wind Farms Siting in the Canary Islands (Spain). *Energies* **2021**, *14*, 865. [CrossRef]
46. Firestone, J.; Kempton, W.; Lilley, M.B.; Samoteskul, K. Public acceptance of offshore wind power: Does perceived fairness of process matter? *J. Environ. Plan. Manag.* **2012**, *55*, 1387–1402. [CrossRef]
47. Nordman, E.; VanderMolen, J.; Gajewski, B.; Isely, P.; Fan, Y.; Koches, J.; Damm, S.; Ferguson, A.; Schoolmaster, C. An integrated assessment for wind energy in Lake Michigan coastal counties. *Integr. Environ. Assess. Manag.* **2015**, *11*, 287–297. [CrossRef] [PubMed]

Journal of
*Marine Science
and Engineering*

MDPI

Article

Fostering Offshore Wind Integration in Europe through Grid Connection Impact Assessment

Nuno Amaro *, Aleksandr Egorov and Gonçalo Glória

R&D NESTER—Centro de Investigação em Energia REN State Grid, 2835-659 Sacavém, Portugal;
aleksandr.egorov@rdnester.com (A.E.); goncalo.gloria@rdnester.com (G.G.)
* Correspondence: nuno.amaro@rdnester.com

Abstract: Floating offshore wind energy is one of the solutions that can foster the ongoing climate transition in Europe. The ARCWIND project aims to contribute to this topic by considering multiple research activities designed to contribute to the development of multiple floating technologies, identifying high-potential deployment areas while considering their economic viability and the impact that these would have in existing power systems. Regarding the latter activity, a two-step methodology was implemented to first calculate the nodal capacity that existing electricity networks have to absorb energy from these potential new wind farms and secondly to assess the impact at the point of connection. This assessment is performed by identifying grid reinforcement needs, verifying the impact on short circuit current levels and measuring the impact on the existing energy mix at countrywide level. This article includes the description of this methodology as well as its application to six different use cases covering five European countries: Portugal, Spain, France, the United Kingdom, and Ireland. Results obtained seem to indicate that, in most cases, the current power systems have enough capacity for the possible connection of new floating offshore wind farms without major reinforcement needs and that these wind farms can have a major contribution to the countries' energy mix and to the achievement of established climate targets.

Keywords: nodal capacity; offshore wind; RES integration; ARCWIND

Citation: Amaro, N.; Egorov, A.; Glória, G. Fostering Offshore Wind Integration in Europe through Grid Connection Impact Assessment. *J. Mar. Sci. Eng.* **2022**, *10*, 463. https://doi.org/10.3390/jmse10040463

Academic Editors: Emre Uzunoglu, Antonio Souto-Iglesias and Carlos Guedes Soares

Received: 21 February 2022
Accepted: 23 March 2022
Published: 24 March 2022

Publisher's Note: MDPI stays neutral with regard to jurisdictional claims in published maps and institutional affiliations.

1. Introduction

Renewable energy sources are one of main solutions for decreasing carbon emissions in power systems; thus, contributing to the ongoing energy transition in Europe and other regions of the world [1]. A growing penetration of renewable energy sources and the technological developments leading to the advent of smart grids is directly correlated with an increasing number of grid connection requests, from the side of energy producers. This represents the distributed nature of renewable energy sources, when compared to fossil fuel energy generation units. One example of this positive correlation is Portugal, where the number of renewable energy generation units, also including Distributed Energy Resources (DER), have been increasing gradually. Renewable energy has already been used to supply around 50% of electricity consumption consecutively over recent years and the number of new connection requests and new connection projects from generators have been also increasing accordingly [2].

This increase in grid connection requests from generation units represents different technological challenges that need to be considered, particularly in order to safely integrate DER at a large scale [3]. From a grid planning perspective, one of these challenges relates to the requirement that System Operators (SOs) must be able to determine the capacity that each grid node has to absorb new energy sources, defined as nodal capacity, both in normal and under contingency conditions. This process allows SOs to verify which are the grid areas more prone to congestion issues and helps with the decision-making process related to asset management, investment, and grid expansion needs. In some countries, regulatory authorities impose that SO's need to calculate and publish the nodal capacity of the different nodes in the power system, in the current and considered expansion scenarios in their grid development plans. In Portugal, REN, the Portuguese Transmission System Operator (TSO) calculates these capacities using a zonal approach (joint capacity of different nodes in the same zone), and publishes them in the Portuguese Transmission System network development plan [4].

However, the definition and calculation of nodal capacities does not have a univocal solution. SOs normally use in-house developed tools whose methodologies are not published, and different scientific groups deal with this process using different approaches as well. Such methods include the utilization of Optimal Power Flow (OPF) tools [5,6] and different meta-heuristics such as Genetic Algorithms and Particle Swarm Optimization techniques [7–10]. However, most of these works are related to the calculation of optimal locations for deployment of DER and not as a systemic analysis for the calculation of nodal capacities of grid nodes.

In previous research [11,12] we developed a Nodal Capacity Allocation tool, allowing a systemic vision on this topic. This methodology calculates the nodal capacity values of multiple user-defined grid nodes. Depending on the objective of the study and grid intrinsic characteristics, nodal capacities of the candidate nodes can be calculated using different approaches. The tool uses genetic algorithms to maximize the total value of power absorbed at grid level. Obtained results always represents valid technical solutions in terms of AC power flow and voltage profile (user-defined voltage limits), allowing it to comply with existing planning and operational criteria.

Wind is currently one of the largest energy resources used in Europe. In 2020, wind accounted for 16% of the electricity consumed in the EU27+UK area, with a total installed capacity of 220 GW [13]. From these, 194 GW are onshore solutions but offshore wind already represents an important share of deployed installations and during 2020 it represented a total of 20% of new installations commissioned [13]. In the offshore wind field, fixed installations still represent the vast majority of deployed solutions, due to the technical advancements when compared to floating wind solutions. Nonetheless, due to the nature of their coastal areas (mainly depth), different regions and countries need to focus on floating wind solutions, as is the case of most Atlantic areas in Europe. Multiple floating wind solutions are already installed or currently being tested in a commercial stage, as is the case of WindFloat Atlantic, with a total installed capacity of 25 MW over the Portuguese Coast [14]. In this context, the project ARCWIND aims to contribute to the development of floating wind solutions in multiple European Atlantic countries.

One of the research activities included in ARCWIND relates to the need to perform grid integration studies for potential floating wind installations in selected areas of interest. In this research activity, a two-step methodology was put into place. This methodology considers the calculation of the nodal capacity that power systems have to receive new connection requests, particularly from offshore wind solutions, and uses different metrics to perform a grid impact assessment study by evaluating possible grid expansion needs and the impact in terms of contribution to the existing energetic mix in each area under study.

In this paper, we present the outcomes of this activity, reporting the nodal capacities and grid impact assessment performed in six different areas included in five European countries: Portugal, Spain, France, Ireland, and Scotland. A previous study including the calculation of the nodal capacity of selected candidates nodes in the Portuguese Transmission Network was already published in [15] and results herein reported for Portugal only include the grid assessment study outcomes. For the other countries, the results of both are included. This work aims provide a contribution to decision makers to foster the development of the considered power systems and floating wind solutions, assessing how these will impact the power system developments and also its contribution to the achievement of current climate targets set at the European level.

2. ARCWIND Project

ARCWIND is an R&D project funded by the European Regional Programme INTERREG Atlantic Area. It aims to contribute to the development of floating offshore wind technologies, focused on its application in European Atlantic areas in Portugal, Spain, France, the United Kingdom, and Ireland [16].

The project is composed of a set of multi-disciplinary activities, aimed at creating sound contributions to the development of these technologies. These activities include: (i) a multi-criteria methodology for wind assessment and site planning decision-making processes; (ii) laboratory size prototype development and testing of three different technologies of floating wind platforms; (iii) array design, installation logistics, and maintenance planning studies; (iv) electric grid integration related studies; (v) cost estimation and economic viability studies.

Considering the current development of floating offshore wind solutions and the potential impact that these might have in power systems, in the medium and long term, it is necessary to perform different studies related to the integration of these new sources in existing power systems. Under the scope of ARCWIND, a two-step methodology was considered, aiming at reaching two complementary goals: the calculation of the nodal capacities of candidate nodes in existing power systems and the grid analysis of future scenarios considering floating offshore wind farms in the project's selected areas.

The application of a multi-criteria methodology for wind assessment and site planning allowed to identify different areas of interest in the five countries considered in ARCWIND [17,18]. Considering 10 MW wind turbines, the maximum theoretical installed capacity was calculated for each one for these areas. These values, presented in Table 1, and together with the location of the areas are the only required input data for the grid impact assessment studies performed within the scope of the project.

Table 1. Identified areas of high potential and their maximum installed capacity.

Country	Area Name	Maximum Installed Capacity (MW)
Portugal	Figueira da Foz	700
Spain	Ribadeo	880
Spain	Gran Canaria	120
France	Lannion	470
Ireland	Galway Bay	98
Scotland	Orkney	154

3. Grid Connection Impact Assessment Methodology

In order to evaluate the impact that the connection of the potential wind farms would have on the existing power systems, a two-step methodology was implemented. These two steps are complementary and aim to evaluate the impact at local and national levels.

The nodal capacity calculation process aims at identifying the best possible grid node to perform the connection of the offshore wind farm, starting from a set of pre-established candidates.

After the best candidate grid node is identified and selected, the second step assumes that a wind farm with the previously calculated installed capacity is connected to this node and further impact assessment studies are performed. This methodology is applied to all studied countries and high potential areas.

3.1. Nodal Capacity Calculation

Nodal capacity can be defined as the capacity that each grid node has to receive new power sources in safe operational conditions. The nodal capacities were calculated by applying the methodology described in [11,12]. The methodology has two main requisites:

- the existence of a model of the power system under analysis, which allows the execution of steady state simulations (power flow);

- Manual selection of candidate nodes, which are the network buses (or substations) where the capacity will be calculated.

Nodal capacities are calculated by connecting new generation units to the candidate nodes and increasing generation levels until overloads are obtained in the network. This represents an optimization problem that is solved using genetic algorithms with a target function of increasing both the local and system level nodal capacities, through the evaluation of the fitness function of individual solutions in the population.

The selection of candidate nodes is highly correlated with the goal of the study to be performed. In this particular case, because the goal is to study the possible connection of new offshore wind farms (based on floating technologies), the following criteria was applied to all areas under study, in order to select candidate nodes:

- Geographic location of the substation;
- Adequacy of voltage level of the substation to the wind farm projected installed capacity;
- Exclusion of client substations (e.g., industrial facilities or generation unit substations);
- Exclusion of substations that can generate high congestion (e.g., if there is a single line connecting the substation to the remaining power system).

3.2. *Connection Impact Assessment*

The second step of the implemented methodology consists of a verification of the impact on the point of connection. This analysis is performed as follows:

- Selection of the best candidate node. This selection is performed taking into consideration the calculated nodal capacities for the different candidate nodes, together with the criteria for selection of the candidate nodes themselves. Therefore, for example, if two candidate nodes provide enough capacity to receive energy generated in the offshore wind farm, the one closer to the wind farm projected location is selected as the best candidate.
- Identification of grid expansion needs. In cases where the best candidate (the one with a higher nodal capacity) does not have enough capacity to absorb the wind farm energy in safe operational conditions (without voltage issues or line congestion), the necessary grid expansion needs will be indicated.
- Short circuit calculation and impact on protection systems. Three-phase short circuit currents are calculated for the selected grid node and neighboring ones in order to assess the impact of the connection of the wind farm on the increase in short circuit current levels. These are of upmost importance as protection systems (including circuit breakers) are designed based on short circuit current levels and its increase might result in the need to refurbish part of these systems. If deemed necessary, these refurbishment needs are highlighted.
- Impact on the energy mix of the country. This evaluation is performed in terms of the contribution that the designed wind farms would have to the total installed capacity in the country. This impact is calculated through the analysis of current installed capacities. Additionally, projected offshore wind farms are assumed to have a capacity factor similar to the average wind capacity factor for each country under analysis in order to verify how the energy generated from these could contribute to their total year demand.

4. ARCWIND Use Cases

The study of the impact on the grid connection was performed for all areas considered in ARCWIND. The project includes five target countries, covering all Atlantic areas of Europe in six different regions. High potential areas were identified in all targeted countries as presented in Table 1. These are translated directly into six use cases, which were analyzed by applying the methodology described in the last section.

4.1. Power System Models

The methodology implemented to calculate the grid impact on the point of connection of the designed wind farms required the existence of power system models, which is a requisite to perform the nodal capacity calculation process. These models need to allow the execution of steady state simulations, namely power flows, in order to check the existence of possible congestion in the system, which determine the nodal capacities.

As the targeted areas are from different countries and geographic areas, individual power system models had to be created or adapted for each use case individually. Table 2 contains a summary of power system models, highlighting their sources and means used to validate them.

Table 2. Power system models considered in ARCWIND.

Country/Area	Source	Validation
Portugal	REN—internal TSO dataset	Verification of congestion
Spain—Ribadeo	ENTSO-E dataset [19]	Format conversion, verification of congestion
Spain—G. Canaria	In-house development	Model development and power flow validation using REE operational data [20,21]
France	ENTSO-E dataset [19]	Format conversion, verification of congestion
Ireland	In-house development	Model development and power flow validation using EirGrid operational data [22,23]
Scotland	In-house development	Model development and power flow validation using NG-ESO operational data [24,25]

As can be seen in Table 2, used power system models have different sources. This also represents a different validation effort in order to ensure that each model has the consistency required to perform this study.

The power system model used for Portugal was directly provided by REN, the Portuguese TSO. This model is used by REN in their internal studies and therefore was already validated. The only required activity was the execution of preliminary simulations in order to verify that the system did not include any congestion, which is a requisite of the nodal capacity calculation tool.

The power systems for Spain (continental) and France were obtained from ENTSO-E, the European Network of Transmission System Operators for Electricity, which uses models provided by the different TSOs that compose this association. This model is available through the signature of an NDA which allows its utilization for R&D purposes without disclosing the model itself [19]. These models are also officially used by REE and RTE, the Spanish and French TSO, respectively. Because the model is provided by ENTSO-E in CGMES format, a conversion was performed in order to obtain a similar model in PSSE format, which is the required one for the nodal capacity tool. In addition to this conversion, and similarly to the Portuguese case, a few changes were made in the existing power flows in order to eliminate existing congestion. These included the increase in rated power of transformers far away from the area under study and do not have any measurable impact on the outcomes of the performed simulations. These three models are realistic models used by the TSOs and include detailed representations of the corresponding power systems, with realistic parametrization of its components (e.g., lines, transformers, and load levels).

The models for Gran Canaria, Scotland, and Ireland were completely developed in the scope of this activity. This need derived from the fact that there are no public models for these power systems. The models were created taking into account public information provided by the TSOs of these countries, namely REE, NG-ESO, and EirGrid. The creation of these models included two main activities:

- Modeling realistic grid topologies;
- Modeling operational (load and generation) scenarios.

Grid topologies were modeled taking as input the public network maps for transmission systems of these areas. As one example, Figure 1 depicts the map of the Gran Canaria transmission network which was replicated in the created model. As there is no public information available, line lengths were calculated using the physical distance between the substations and average parameters were used for their characterization (e.g., line resistance and reactance). These average parameters were obtained from realistic characteristics of overhead lines for the considered voltage levels. Additionally, these maps were also used to locate main generation units, as can also be seen in Figure 1.

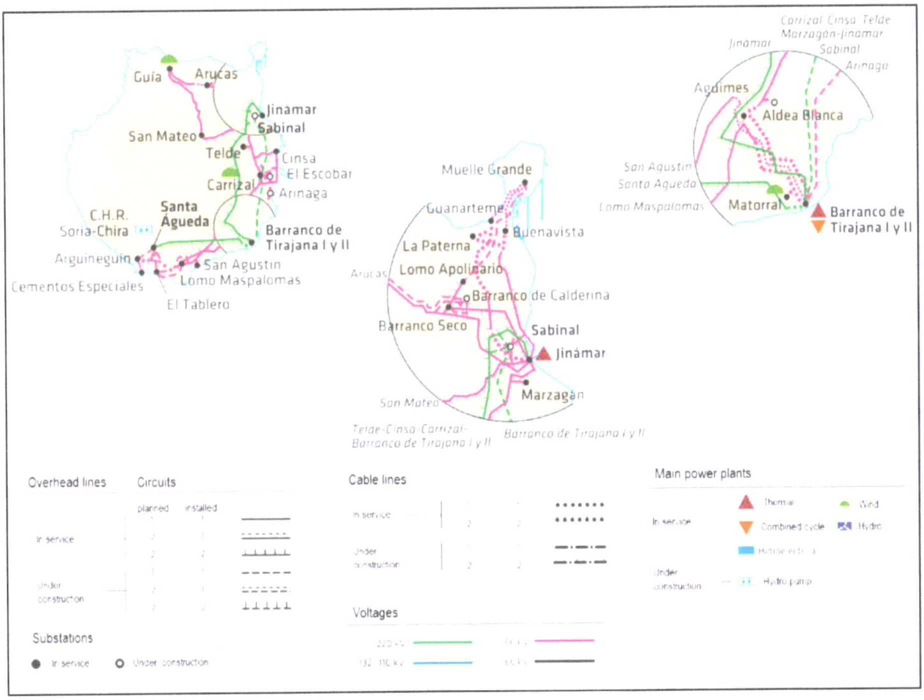

Figure 1. Gran Canaria transmission network map. Source: [26].

After the creation of the grid topologies, other technical parameters were obtained from the operational data of the TSOs in order to create realistic operational scenarios to be considered by the tool. These consider the load and generation levels. Since the location of most important generators is already included in the grid maps (see example of Figure 1), the generation values were included according to each generation unit main characteristics. Complementarily, load levels were split among the different substations either using the data provided by the TSOs (when there are available load levels per area) or using a direct correlation with urban density, increasing load in urban areas. A final validation of these models included the simulation of power flows in different operational conditions (e.g., using contingency analysis) in order to assess their robustness and accuracy when compared to operational data available from the TSOs. Figure 2 depicts an example of a power flow performed in Gran Canaria, where one can see the power flow and voltage profile of a small portion of the network.

The considered power systems are heterogeneous, being different in their size, energetic mix, and load levels, which is a natural consequence of the countries' differences. In order to allow the analysis performed in this study, Table 3 includes a high-level description of the characteristics of each one. Values included herein reflect the situation as of 31 December 2020 and were collected from TSO public databases.

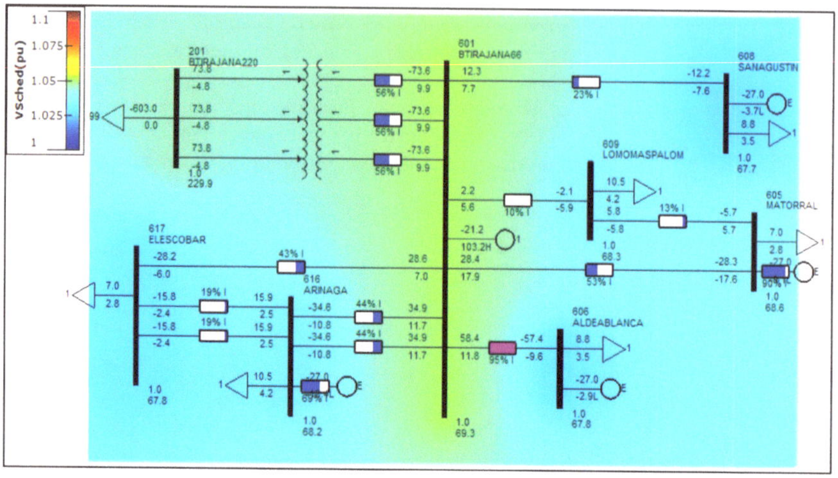

Figure 2. Power flow analysis in the Gran Canaria network.

Table 3. Power system models considered in ARCWIND.

Country/Area	Total Installed Capacity (GW)	Wind Installed Capacity (GW)	Total Load Supplied in 2020 (TWh)
Portugal	20.4	5.2 (0 offshore)	48.8
Spain—Continental	105.6	27 (0 offshore)	249.99
Spain—G. Canaria	1.26	0.194 (0 offshore)	3.34
France	132.2	16.6 (14 MW offshore)	460
Ireland	12.4	4.25 (0 offshore)	30
Scotland	12.2	9.35 (0.98 GW offshore)	30.3

4.2. Candidate Grid Nodes

The selection of a set of candidate nodes for each location followed the principles described in Section 3.1. As an example of this selection process, the methodology is explained herein for the selection of candidate nodes in the case of Spain (Ribadeo). Other use cases followed a similar methodology and candidate nodes for all cases are included in Table 4.

Table 4. Candidate nodes including substation name and voltage levels for all use cases.

Country/Area	Substations	Voltage Level
Portugal	Estarreja (220 kV), Lavos, Paraimo	400 kV
Spain—Continental	Boimente, El Palo, Pesoz, Xove	400 kV
Spain—G. Canaria	Barranco de Tirajana	220 kV
France	La Martyre, Pleyber Christ, Plaine-Haute, Rospez, Tregueux	220 kV
Ireland	Tullabrack, Moneypoint, Slievecallan	110 kV
Scotland	Cassley, Dounreay	132 kV

Figure 3 depicts the electricity transmission network near the high potential area for floating offshore wind farm deployment in Ribadeo, Spain, with this area highlighted as well. A high level model of the network topologies for this and other areas is available in [27].

The process of selection of candidate nodes for this location followed the next steps. Considering the targeted installed capacity of 880 MW, only 400 kV substations were considered as candidates. This is also in line with the available network assets in this area (where red lines correspond to 400 kV voltage level and green to 220 kV voltage level).

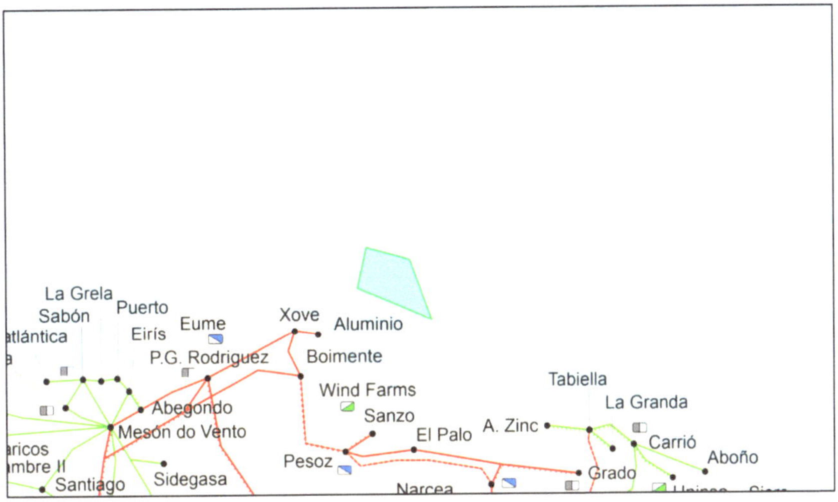

Figure 3. Spanish electricity transmission network near Ribadeo.

Based on this analysis, seven potential candidate nodes were identified: the four included in Table 4 and P.G. Rodriguez, Aluminio, and Sanzo. However, P.G. Rodriguez and Aluminio were excluded because there is a major generation facility and an industrial load facility connected to these substations, respectively. Sanzo was excluded as this is the main connection point of several onshore wind farms that already exist in the area. Thus, the application of these criteria, which was described in Section 3.1, resulted in the candidate nodes included in Table 4.

5. Results

The calculation of the nodal capacity in selected grid nodes as well as a connection impact assessment was performed for the six use cases considered in ARCWIND.

The nodal capacity calculation process, described in Section 3.1, was applied to the different sets of candidate nodes in order to identify the best point of connection (the best candidate) for projected wind farms in each one of the identified high potential areas. Table 5 includes the results of this analysis for all candidate nodes in the six use cases.

The best candidates were selected based on the available nodal capacity and the geographic location to the high potential area. These are indicated in bold in Table 5.

In the cases of Portugal, France, Ireland, and Scotland, the nodes identified as best candidates are the ones with a higher nodal capacity. In this context, the nodes considered as best candidates are proposed as the selected grid nodes for the connection of the offshore wind farms, in the case of a future connection and are the ones selected for the connection impact assessment.

In Spain (Ribadeo), the substation of Boimente was selected because it had a large capacity (almost two times the targeted installed one in the offshore area) and was closer to this area than the Pesoz substation which would be the other potential best candidate. In the case of Gran Canaria, a single node was considered as a candidate due to its location being close to the identified high potential area and including an adequate voltage level. Obtained values indicate that the Barranco de Tirajana substation has sufficient nodal capacity for the connection of the possible offshore wind farm.

In order to demonstrate the accuracy and applicability of the nodal calculation process, Figure 4 depicts the results obtained for the simulations performed in the Spanish transmission system, near the Ribadeo area. Nodal injections are highlighted in green and it can be seen that there are multiple lines with power flows near their maximum rated capacity, which acts as the limiting factor for the nodal capacity calculation process.

Table 5. Nodal capacity calculation results.

Country/Area	Candidate Node	Nodal Capacity (MW)
Portugal	Estarreja (220 kV)	129
	Lavos (400 kV)	**139**
	Paraimo (400 kV)	112
Spain—Continental	**Boimente (400 kV)**	**1505**
	El Palo (400 kV)	844
	Pesoz (400 kV)	1846
	Xove (400 kV)	781
Spain—G. Canaria	**Barranco de Tirajana (220 kV)**	**603**
France	La Martyre (220 kV)	587
	Pleyber Christ (220 kV)	571
	Plaine-Haute (220 kV)	243
	Rospez (220 kV)	**776**
	Tregueux (220 kV)	462
Ireland	Tullabrack (110 kV)	205
	Moneypoint (110 kV)	**380**
	Slievecallan (110 kV)	219
Scotland	Cassley (132 kV)	235
	Dounreay (132 kV)	**316**

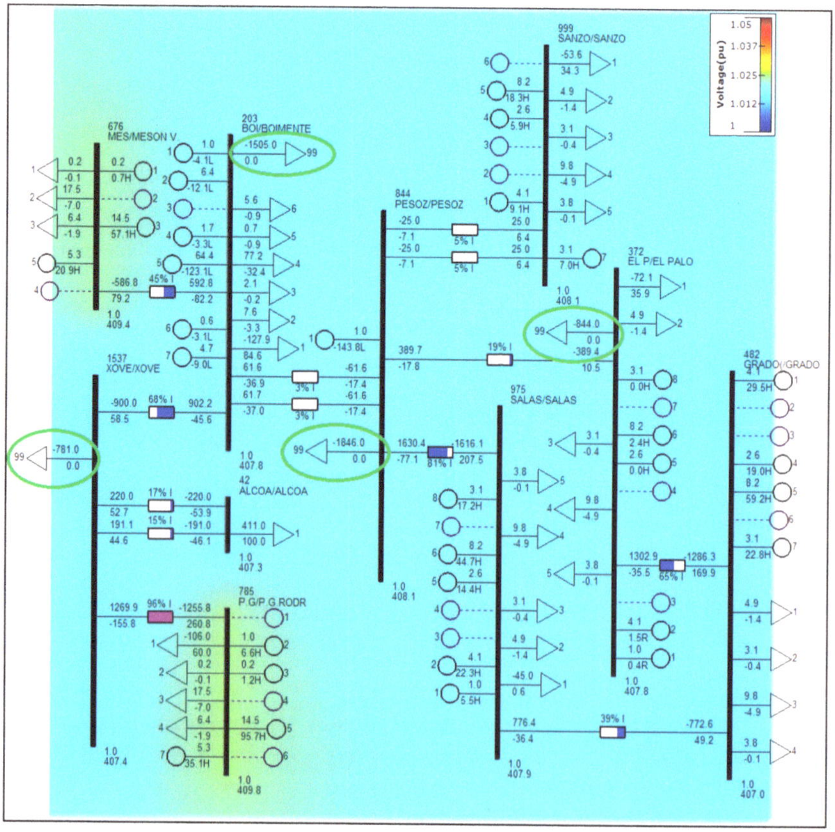

Figure 4. Nodal capacity calculation results for Spain (Ribadeo area).

After the identification of the best candidate nodes, the analysis of the impact of the connection in these nodes was performed. This set of simulations, particularly focused in simulating short circuits, considers that an equivalent generator is connected to the selected grid node, with a power generation corresponding to the wind farm maximum capacity, as presented in Table 1. In order to measure the impact of the connection, three-phase short circuits were simulated in the point of connection and neighbor substations in two different conditions: without the connection of the wind farm (base case) and with the wind farm connected. This allows to measure the contribution of the wind farm to short circuit current levels, which can dictate the need to refurbish protection system components as the circuit breakers. The neighbor substations do not always match the previously selected candidate nodes exactly. This is related to the grid topologies.

The outcome of the calculation of short circuit current (SCC) levels is presented in Table 6. This analysis does not include the cases of Scotland and Ireland as the created grid models (which are based on average parameters and open-source data) do not allow for the necessary parametrization and short circuit values obtained would not have any physical meaning.

Table 6. Short circuit current analysis.

Country/Area	Substation	Base Case SCC (kA)	Wind Farm Case SCC (kA)	Delta
Portugal	Lavos	16.44	18.25	9%
	Batalha	15.27	16.78	9.1%
	Paraimo	17.98	19.81	9.1%
Spain—Continental	Boimente	22.54	23.27	3.2%
	El Palo	18.61	18.76	0.8%
	Pesoz	21.51	21.75	1.1%
	Salas	21.80	21.92	0.6%
	Sanzo	21.16	21.40	1.1%
	Xove	20.15	20.58	2.1%
Spain—G. Canaria	Barranco de Tirajana	4.42	5.23	15.8%
France	Rospez	9.61	9.65	0.4%
	Plaine-Haute	23.21	23.36	0.6%
	Tregueux	8.15	8.20	0.9%
	Brennilis	12.08	12.42	0.8%

Results obtained for Spain (Ribadeo) and France indicate that there are marginal increases in short circuit current levels, which should not have any impact on the existing protection systems. In the case of Gran Canaria, the increase of 15.8% in the short circuit current levels could result in the need to refurbish the circuit breakers. In the case of Portugal, the connection of the wind farm would result in a maximum increase of 9.1% which will be further discussed in Section 6.

The last activity performed within the scope of this work aimed to verify the impact on the energy mix of each country as well as possible contributions to renewable energy generation to fulfill yearly demand values. This analysis assumes that the floating offshore wind farms would have similar capacity factors to the current ones at national level, which can be considered as a conservative approach (as most installed wind power is onshore). Table 7 contains the results of this analysis, for the five different countries. As can be seen, the projects wind farms would have significant impacts on the different countries, supplying electric load with non-negligible values, having a significant impact on the increase in renewable energy generation in the five countries and consequent reduction in energy generated from fossil fuels.

Table 7. Impact on energy mix.

Country/Area	Impact on Current Wind Installed Capacity (%)	Annual Energy Generation (TWh)	% of Load Supplied (TWh)
Portugal	13.5%	1.7	3.5%
Spain—Continental	3.3%	2.2	0.9%
Spain—G. Canaria	61.9%	0.3	9.0%
France	2.8%	1.0	0.2%
Ireland	2.3%	0.3	0.8%
Scotland	1.6%	0.4	1.2%

6. Discussion

The obtained results seem to indicate that the offshore wind farms designed in ARCWIND could contribute actively to the generation of wind energy; thus, contributing to the ongoing decarbonization process in the five studied countries.

In general, performed studies indicate that it seems to be possible to integrate the designed wind farms into existing power systems, with no grid reinforcement needs in the case of Spain (Ribadeo), France, Scotland, and Ireland. In the case of Portugal, the integration would only be possible with significant grid expansion measures. Since Lavos was considered as the best candidate node and selected for connection impact assessment, preliminary results obtained using the set of operational scenarios considered in this study indicate the need to reinforce the lines between Lavos and Batalha and Paraimo substations. In the case of Gran Canaria, although the nodal capacity calculation process indicates that the wind farm could be connected without the need for grid reinforcements, there is some operational risk in this connection. The installed capacity (120 MW) corresponds to 54% of the minimum load and 22% of peak load verified in 2020. This represents possible curtailment needs for the wind farm to retain the operational security of the system.

In order to further demonstrate the validity of the implemented methodology and obtained results, a comparison with official data from REN, the Portuguese TSO, is presented. This was performed for the nodal capacity and short circuit current calculation methods.

REN biannually publishes the current capacity of the network. These are included in the network development plan, whose last version, from 2021 and reflecting the period of 2022–2031, is available in [4]. The methodology used from REN is slightly different from the one presented in this document. REN publishes the nodal capacities using a zonal approach, i.e., for a set of geographically close nodes and not for individual ones. Since no public information is available for individual nodes, it can be assumed that the individual capacities for each node are the average value for that zone, being added to obtain the total zonal value.

The nodes considered as candidate nodes in the study performed in ARCWIND belong to two different zones. The zonal capacity of these two zones is presented in Table 8. This table also includes the individual nodal capacities calculated in this study.

Table 8. Nodal capacity calculation in Portugal: comparison with REN data.

Zone	Node	REN Zonal Capacity (MW)	ARCWIND Capacity (MW)
Zone A	Paraimo	340 (113.3 average per node)	112
	Estarreja		119
	Mourisca		N/A
Zone B	Lavos	696 (174 average per node)	139
	Pombal		N/A
	Batalha		N/A
	Rio Maior		N/A

Comparing the values calculated in ARCWIND with those given by REN, it can be concluded that the methodology used in this work provides valid technical results. The differences obtained for the different nodes (1.1% in Paraimo, 13.8% in Estarreja, and 20.1% in Lavos) result from the fact that this study only considers a small subset of four operational scenarios (generation and load conditions) while REN uses dozens of different scenarios in their studies.

In addition to the zonal capacities, REN also includes in their network expansion plans the maximum and minimum short circuit currents calculated for the considered operational scenarios. These are given in Table 9, together with the previously calculated values in this study. As can be seen, both SCC levels for the base case and considering the connection of the wind farm are within the interval given by REN, which demonstrates the applicability of the methodology. Since these are within the interval defined by REN, even if there is an increase of 9.1% in short circuit currents with the connection of the wind farm, there is no need for refurbishment of protection systems.

Table 9. Short circuit current calculation in Portugal: comparison with REN data.

Substation	Base Case SCC (kA)	Wind Farm Case SCC (kA)	REN SCC Interval (kA)
Lavos	16.44	18.25	(15.7–22.8)
Batalha	15.27	16.78	(14.8–19.8)
Paraimo	17.98	19.81	(16.8–20.9)

7. Conclusions

This work presented a study related to the impact assessment at the grid connection point for different sized floating offshore wind farms considered in the scope of the ARCWIND project. This study included two main activities related to the calculation of the capacity that the networks have to absorb the power generated by those wind farms and the impact that they would have on the system, both at local and national level.

Results obtained seem to indicate that, with the exception of the Portuguese Case, all other considered networks have enough capacity to perform the connection of the wind farms without the need for major grid reinforcements. Looking into possible operation constraints, the only case that could result in curtailment needs is the connection of the wind farm in Gran Canaria Island, due to the considered wind farm size when compared to the operational conditions of the power system (load levels). Additionally, the calculation of short circuit currents also shows a possible need to refurbish part of the existing protection devices in Gran Canaria. Other cases would not need any change in these systems.

The installation of these floating wind farms, given that the necessary environmental and economic viability studies were positive, would have an important technical contribution to the ongoing energy transition process, by substantially increasing the installed wind capacity and generation levels in the five countries.

Author Contributions: N.A. contributed with the development of the methodology, writing of the paper, and analysis of obtained results. A.E. contributed with the creation of some of the power system models and the execution of the simulations for the nodal capacity calculation process. G.G. performed the short circuit simulation and analysis. All authors have read and agreed to the published version of the manuscript.

Funding: This research was partially funded by INTERREG Atlantic Area through the project ARCWIND (EAPA 344/2016).

Institutional Review Board Statement: Not applicable.

Informed Consent Statement: Not applicable.

Conflicts of Interest: The authors declare no conflict of interest.

References

1. IEA. *Global Energy & CO_2 Status Report 2020*; IEA: Paris, France, 2021.
2. REN, Redes Energéticas Nacionais. *Technical Data 2020*; REN: Lisbon, Portugal, 2021.
3. Lopes, J.A.P.; Hatziargyriou, N.; Mutale, J.; Djapic, P.; Jenkins, N. Integrating distributed generation into electric power systems: A review of drivers, challenges and opportunities. *Electr. Power Syst. Res.* **2007**, *77*, 1189–1203. [CrossRef]
4. REN. *Plano de Desenvolvimento e Investimento da Rede Nacional de Transporte (PDIRT) 2022-31*; REN: Lisbon, Portugal, 2021.
5. Harrison, G.P.; Wallace, A.R. Optimal power flow evaluation of distribution network capacity for the connection of distributed generation. *IEE Proc. Gener. Transm. Distrib.* **2005**, *152*, 115. [CrossRef]
6. Vovos, P.; Harrison, G.P.; Wallace, R.; Bialek, J. Optimal power flow as a tool for fault level constrained network capacity analysis. *IEEE Trans. Power Syst.* **2004**, *20*, 734–741. [CrossRef]
7. Keane, A.; Ochoa, L.F.; Member, S.; Borges, C.L.T.; Dent, C.; Harrison, G.P. State-of-the-Art techniques and challenges ahead for distributed generation planning and optimization. *IEEE Trans. Power Syst.* **2015**, *28*, 1493–1520. [CrossRef]
8. Kanwar, N.; Gupta, N.; Niazi, K.R.; Swarnkar, A. Optimal distributed generation allocation in radial distribution systems considering customer-wise dedicated feeders and load patterns. *J. Mod. Power Syst. Clean Energy* **2015**, *3*, 475–484. [CrossRef]
9. Gupta, P.; Pandit, M.; Kothari, D.P. A review on optimal sizing and siting of distributed generation system: Integrating distributed generation into the grid. In Proceedings of the 6th IEEE Power India International Conference (PIICON), Delhi, India, 5–7 December 2014; pp. 1–6. [CrossRef]
10. Meena, N.K.; Parashar, S.; Swarnkar, A.; Gupta, N.; Niazi, K.R. Improved elephant herding optimization for multiobjective DER accommodation in distribution systems. *IEEE Trans. Ind. Inform.* **2018**, *14*, 1029–1039. [CrossRef]
11. Amaro, N.; Reis, F. Evolutionary capacity allocation tool to safe grid integration of renewable energy sources. In Proceedings of the 2018 International Conference on Smart Energy Systems and Technologies (SEST), Seville, Spain, 10–12 September 2018; pp. 1–6. [CrossRef]
12. Amaro, N.; Carrola, F.; Reis, F. Optimizing nodal capacity allocation using risk assessment of element failure rate. In Proceedings of the 2020 IEEE 14th International Conference on Compatibility, Power Electronics and Power Engineering (CPE-POWERENG), Setubal, Portugal, 8–10 July 2020; pp. 44–49. [CrossRef]
13. WindEurope. Wind Energy in Europe 2020 Statistics and the outlook for 2021–2025. 2021. Available online: https://windeurope.org/intelligence-platform/product/wind-energy-in-europe-in-2020-trends-and-statistics/ (accessed on 16 February 2022).
14. EDP. WindFloat Project 2018. Available online: https://www.edp.com/en/windfloat (accessed on 16 February 2022).
15. Amaro, N.; Egorov, A.; Reis, F. Grid Capacity for Floating Offshore Wind Integration—The Portuguese Case. In *Developments in Renewable Energies Offshore*, 1st ed.; CRC Press: London, UK, 2020; p. 6. [CrossRef]
16. Arcwind Consortium. *Arcwind Project Website, 2017*. Available online: http://www.arcwind.eu/ (accessed on 17 February 2022).
17. Salvação, N.; Bentamy, A.; Soares, C.G. Estimating the offshore wind energy along the Portuguese coast using wrf and satellite data. In *Advances in Renewable Energies Offshore, Proceedings of the 3rd International Conference Renew Energies Offshore, RENEW 2018, Lisbon, Portugal, 8–10 October 2018*; CRC Press: Boca Raton, FL, USA, 2018; pp. 703–710.
18. Remmers, T.; Cawkwell, F.; Desmond, C.; Murphy, J.; Politi, E. The Potential of Advanced Scatterometer (ASCAT) 12.5 km Coastal Observations for Offshore Wind Farm Site Selection in Irish Waters. *Energies* **2019**, *12*, 206. [CrossRef]
19. ENTSO-E. ENTSO-E Grid Model for TYNDP 2018. 2018. Available online: https://www.entsoe.eu/publications/statistics-and-data/#entso-e-on-line-application-portal-for-network-datasets (accessed on 17 February 2022).
20. REE. REData-datos del Sistema Elétrico. 2021. Available online: https://www.ree.es/es/datos/aldia (accessed on 17 February 2022).
21. REE. *El Sistema Eléctrico Canario*; REE: Lisbon, Portugal, 2016.
22. EirGrid. EirGrid Online Library. 2020. Available online: https://www.eirgridgroup.com/library/ (accessed on 18 February 2022).
23. EirGrid. Smart Grid Dashboard. 2021. Available online: http://smartgriddashboard.eirgrid.com/ (accessed on 17 February 2022).
24. National Grid ESO. Electricity Ten Year Statement. 2021. Available online: https://www.nationalgrideso.com/research-publications/etys (accessed on 20 February 2022).
25. National Grid ESO. Live System Information. 2021. Available online: https://data.nationalgrideso.com/system/live-system-information (accessed on 20 February 2022).
26. REE. *Plan de Desarrollo de la Red de Transporte de Energía Eléctrica 2015–2020*; REE: Lisbon, Portugal, 2015.
27. ENTSO-E. Transparency Platform. Available online: https://transparency.entsoe.eu/ (accessed on 20 February 2022).

Journal of
Marine Science and Engineering

MDPI

Article

The Technical and Economic Feasibility of the CENTEC Floating Offshore Wind Platform

Almudena Filgueira-Vizoso [1], Laura Castro-Santos [2,*], David Cordal Iglesias [3], Félix Puime-Guillén [4], Isabel Lamas-Galdo [5], Ana Isabel García-Diez [2], Emre Uzunoglu [6], Hugo Díaz [6] and C. Guedes Soares [6,*]

1 Escola Politécnica de Enxeñaría de Ferrol, Campus Industrial de Ferrol, Departamento de Química, Universidade da Coruña, Esteiro, 15471 Ferrol, Spain
2 Escola Politécnica de Enxeñaría de Ferrol, Campus Industrial de Ferrol, Departamento de Enxeñaría Naval e Industrial, Universidade da Coruña, Esteiro, 15471 Ferrol, Spain
3 Escola Politécnica de Enxeñaría de Ferrol, Campus Industrial de Ferrol, Universidade da Coruña, Esteiro, 15471 Ferrol, Spain
4 Campus Industrial de Ferrol, Departamento de Empresa, Facultade de Economía e Empresa, Elviña, Universidade da Coruña, 15071 A Coruña, Spain
5 Escola Politécnica de Enxeñaría de Ferrol, Campus Industrial de Ferrol, Departamento de Ciencias da Navegación e Enxeñaría Mariña, Universidade da Coruña, Esteiro, 15471 Ferrol, Spain
6 Centre for Marine Technology and Ocean Engineering (CENTEC), Instituto Superior Tecnico, Universidade de Lisboa, 1049-001 Lisbon, Portugal
* Correspondence: laura.castro.santos@udc.es (L.C.-S.); c.guedes.soares@centec.tecnico.ulisboa.pt (C.G.S.)

Citation: Filgueira-Vizoso, A.; Castro-Santos, L.; Iglesias, D.C.; Puime-Guillén, F.; Lamas-Galdo, I.; García-Diez, A.I.; Uzunoglu, E.; Díaz, H.; Guedes Soares, C. The Technical and Economic Feasibility of the CENTEC Floating Offshore Wind Platform. *J. Mar. Sci. Eng.* **2022**, *10*, 1344. https://doi.org/10.3390/jmse10101344

Academic Editor: Raúl Guanche García

Received: 1 September 2022
Accepted: 16 September 2022
Published: 21 September 2022

Abstract: This paper defines a methodology for the economic feasibility analysis of a floating offshore wind farm composed of tensioned leg platforms, which are part of the EU ARCWIND research project. In this context, the phases and subphases of its life-cycle process are considered to deal with aspects such as bathymetry, characteristics of the platforms, distance from the farm to shore, distance from the farm to port and offshore wind speed. All the costs and other external parameters such as capital cost, electric tariff, interest rate, percentage of financing and corporate tax have been analysed to calculate the internal rate of return, net present value, discounted pay-back period and levelized cost of energy of the farm. This work studies a farm composed of TLP offshore wind platforms designed by CENTEC and located at Ribadeo in Spain. Results indicate the costs and the economic feasibility of this platform for deep waters. They indicate that the platform is economically feasible for the location selected.

Keywords: CENTEC-TLP; Tensioned Leg Platform; offshore wind; economic feasibility; *IRR*; *NPV*; LCOE

1. Introduction

Wind energy is produced onshore and offshore. Taking advantage of the fact that 70% of the planet's surface is covered by water and that a small percentage of wind energy is offshore, for example, in Europe in 2020, 16.4% of electricity demand was covered with wind energy, and within this 13.4% was onshore [1]; only the remaining 3% was offshore.

The production of offshore wind energy is made by fixed and floating platforms. Within the fixed ones, mention can be made of monopiles [2–4], tripod [5–7], tripilote, gravity [8,9] and jacket [10], and within the floating ones the main ones are: Semi-submersible [11], TLP (tensioned leg platform) [12] and Spar [13].

This paper describes a method of determining the economic feasibility analysis of a floating offshore wind farm [14–17] composed of CENTEC-TLP (tension leg platform) devices, which are part of the ARCWIND research project. For this purpose, the phases and subphases of its life-cycle process were considered, taking into account several inputs for this implementation: bathymetry, characteristics of the platforms, distance from farm to shore, distance from farm to port and offshore wind speed. Therefore, the method proposed allows the user vary some important inputs in order to know their influence in the main economic parameters. This is important because the majority of the studies consider general costs but not specific ones. All the costs and other external parameters such as capital cost, electric tariff, interest rate, percentage of financing and corporate tax have been taken into account to calculate the cash flows of the farm. This was the basis for calculating the main economic feasibility parameters: internal rate of return (*IRR*), net present value (*NPV*), discounted pay-back period (DPBP) and levelized cost of energy (LCOE) of the considered offshore wind farm [18–24]. Different researchers have developed methodologies for calculating the LCOE, and this calculation can be applied to different sectors, for example, Ueckerdt et al. [21] analyse as system LCOE allow the economic comparison of generating technologies and deriving optimal quantities in particular for VRE (variable renewable sources), Johnston et al. [22] review the challenges of accurately estimating levelized cost of energy (LCOE) for offshore wind outlining differing approaches to calculating LCOE, the factors influencing this, and the impact of variation in LCOE calculation. This paper studies a farm composed of the TLP offshore wind platform designed by CENTEC and located at Ribadeo in the northwest of Spain. Results will indicate the economic feasibility of the TLP platform.

2. Materials and Methods

2.1. Technical Feasibility

The CENTEC-TLP is a steel hull with a DTU-10MW turbine. The platform is detailed in [25]. It may be regarded as a free-float capable TLP. It attains this behaviour through a towing mode that resembles the behaviour of a barge platform. Once towed to the location, it is then installed to perform as a TLP [26].

The tension leg platform is overall lighter than other hulls as it replaces the reliance on hull form to attain stability using mooring lines. When it comes to the offshore wind sector, most works [12,27] try to use the SeaStar hull form [28] to keep the low mass advantage. This idea originates from the older MIT/NREL design [29]. However, this hull form is limited in application. Even in the oil and gas sector, only 5 of over 30 TLPs prefer this geometry [30], with the latest installation being in 2007. Despite a large number of designs, the impracticality of the hull form is apparent: there are no demo projects with TLPs despite both SPAR and Semisubmersible hull forms [31] being installed and operational as wind farms already.

Despite the large number of works using the SeaStar, the conventional TLP form is Conoco Phillips' Hutton in 1984, similar to a floating production system (e.g., [32]) regarding its use of ring pontoons underwater. This hull form is more suitable for addressing the technical challenges imposed by the wind industry. TLP's disadvantages are primarily associated with the stages up to when it is installed and operational. Its instability without its mooring lines forces alternative solutions such as additional barges.

At this point, it should be considered that multiple offshore wind platforms need to be installed on farms. For this reason, the oil-and-gas-sector experience does not translate well into the offshore wind case. It may be economically feasible to carry one platform to the installation location at a high price for oil and gas. On the other hand, when it comes to offshore wind, this operation may need to be repeated up to 100 times, depending on the size of the wind farm. Hence, the hiring costs of specialized vessels accumulate. For similar reasons, recent studies emphasize improving production speeds using modular production methods (e.g., tension leg platforms as explained in [33,34]).

Another problem lies in the towing locations. The assumption that the structure needs to be towed into the installation area using standard vessels after being assembled at the shore, signifies that it needs to have small motions in waves. The movements should be limited so that the high centre of gravity of the turbine and the tower do not impose stability issues. It needs to also do that at a low draft so that the hull can be towed from most ports. For instance, a draft of 20 m would significantly limit the locations where the platform can be towed from considering that the average draft of ports is below 10 m. Another consideration is the commercial availability of mooring lines used and the maximum stress allowance imposed by [35]. The goal is to adhere to regulations without requiring specially produced mooring lines.

The CENTEC-TLP addresses the problems listed above through its hull form. It acts as a stable barge when towed, and once installed, it performs as a TLP. Its dynamics in both forms have been experimentally and numerically verified [26,36,37].

2.2. Economic Feasibility

2.2.1. Economic Feasibility Parameters Analysed

The objective of this section is to carry out the economic feasibility analysis of the TLP platform. The economic viability of the project can be calculated by different highly effective methods. To determine the economic viability of this project, the Cash Flow is calculated throughout its useful life, as well as various economic parameters:

- Net Present Value (*NPV*).
- Internal Rate of Return (*IRR*).
- Discounted Pay-Back Period (DPBP).
- Levelized Cost Of Energy (LCOE)

2.2.2. Net Present Value (*NPV*)

The Net Present Value (*NPV*), in €, is the net value of the cash flows of the floating offshore wind farm, taking into account its discount from the beginning of the investment [38–40]. It is dependent on the cash flow in year t (CF_t), the discount rate (r) and the initial investment (G_0), as Equation (1) is shown.

$$NPV = -G_0 + \sum_{t=1}^{n} \frac{CF_t}{(1+r)^t} \tag{1}$$

The discount rate (r) considered for the financed project is the *WACC* (weighted average cost of capital). It was calculated using the equation.

- $NPV > 0$. The investment will generate earnings above the required return (r). This will imply that the acceptance of the project is recommended
- $NPV < 0$. The investment produces returns below the required minimum return (r). It is not recommended to accept the project
- $NPV = 0$. The project does not add monetary value above the required profitability (r). The decision must be based on other criteria such as obtaining a better position in the market.

2.2.3. Internal Rate of Return (*IRR*)

The internal rate of return is the interest generated by the project throughout its useful life. Mathematically, it is defined as the discount rate that cancels the *NPV*, that is, the interest rate that makes the future flow of funds financially equivalent to the initial outlay.

$$-G_0 + \sum_{t=1}^{n} \frac{CF_t}{(1+IRR)^t} = 0 \tag{2}$$

The economic feasibility of the project will depend on the *IRR*:

- $IRR <$ k. The profitability obtained from the project is less than the minimum required, so the investment is not recommended.
- $IRR >$ k. The profitability of the project is above the minimum required, therefore, it is recommended to decide to invest.
- $IRR =$ k. The profitability is the same as that required, the same happens as in the case where the $NPV = 0$, the decision is conditioned by other factors

2.2.4. Discounted Pay-Back Period (DPBP)

Discounted pay-back period (DPBP), in years, considers the cash flow of each year with the respective discount rate and adds it to all the previous cash flows with their respective discount rate, accumulating its *NPV*. When this sum is equal to or greater than the initial investment, this is the year of the DPBP, as Equation (3) is shown. The best DPBP is as low as possible.

$$\sum_{t=1}^{n} \frac{CF_t}{(1+r)^t} \geq G_0 \tag{3}$$

The conditions regarding the feasibility of the project are:

- DPBP <<< *t*. The initial outlay takes less time to recover than the life of the project (t). Accept project.
- DPBP = *t*. The initial outlay takes to recover the same as the life of the project (t). Indifferent.
- DPBP > *t*. The initial outlay takes longer to recover than the life of the project (t). Reject the project.

2.2.5. Levelized Cost of Energy (LCOE)

The levelized cost of energy (*LCOE*), in €/MWh, has been calculated considering the total costs of the farm (LCC_t) in €, the energy produced by the floating offshore wind farm (E_t) in MWh/year and the capital cost of the project (r) [41].

$$LCOE = \frac{\sum_{t=0}^{t=N_{farm}} \frac{LCC_t}{(1+r)^t}}{\sum_{t=0}^{t=N_{farm}} \frac{E_t}{(1+r)^t}} \tag{4}$$

The total costs of the farm for the year t LCC_t are calculated in the next sub-section of the paper (LCC_t is CAPEX + OPEX). On the other hand, the energy production of the floating offshore wind farm (E_n) in kWh/year has been estimated by multiplying the energy produced by each offshore wind turbine by the number of wind turbines (NWT) [41]. The energy produced by one offshore wind turbine is dependent on the total hours per year (factor 8.76), the reduction due to losses (η), the power curve of the offshore wind turbine ($P_{PC}(v)$) and the Weibull density probability function ($p_{Weibull}$), whose value will depend on the wind speed (v), the wind scale parameter (c_w) and the wind shape parameter (k_w) of the location of the European Atlantic Arc selected.

$$E_n = NWT \cdot 8.76 \cdot \eta \cdot \int_0^{v_{cut-out}} P_{PC}(v) \cdot p_{Weibull}(v; c_w; k_w) dv \tag{5}$$

where:

$$p_{Weibull}(v; c_w; k_w) = \frac{k_w}{c_w} \cdot \left(\frac{v}{c_w}\right)^{k_w - 1} \cdot e^{-\left(\frac{v}{c_w}\right)^{k_w}} \tag{6}$$

The power curve (P_{PC}) will be integrated into three zones which depend on the power of the wind turbine in MW and on the characteristic wind speed (in m/s) [41]: cut-in wind speed, rated wind speed and cut-out wind speed ($v_{cut-out}$).

2.2.6. Costs

The total cost of the farm (see Equation (5)) is necessary for calculating the LCOE and some aspects of the cash flow [14,42,43]. The costs are divided into the phases in which the wind farm passes during its life cycle. Each phase is formed by several subphases, which are determined by the components or factors that must be considered to obtain the costs in each case. In this way, the total cost of of an offshore wind farm is obtained from Equation (1) (Castro-Santos & Diaz-Casas, 2014).

$$LCC = C_1 + C_2 + C_3 + C_4 + C_5 + C_6 \tag{7}$$

where:

- C_1: Conception and definition cost.
- C_2: Design and development cost.
- C_3: Manufacturing cost.
- C_4: Installation cost.
- C_5: Exploitation cost.
- C_6: Dismantling cost.

It is important to notice that there are costs that exist during all the years of the project such as the exploitation cost and other costs that are considered only in the initial years of the project, when it is been built (conception and definition cost, design and development cost, manufacturing cost and installation cost). Finally, there are costs, such as the dismantling cost, that are expended in the last year of the project (N_{farm}).

The main cost items are shown in Table 1.

Table 1. Main variables.

Variable	Concept	Units
EPCIC	EPCIC stands for Engineering, Procurement, Construction, Installation & Commissioning (contract)	%
OffshoreSiteR	Offshore site renting	€
Nplatforms	Number of floating platforms	platforms
C_{emet}	Meteorological structure cost	€
C_{smet}	Meteorological sensors cost	€
C_{samet}	Auxiliary meteorological systems cost	€
EvaluationReportCost	Evaluation report cost	€/MW
TurbineP	Power of the wind turbine	MW
GeophysicalDailyCost	Geophysical campaign cost (bathymetry, sub bottom profiler, garbage, wreck detections, etc.)	€/day
GeophysicalDays	Duration of the geophysical campaign	days
GeotechnicalDailyCost	Geotechnical campaign cost (CPT: cone penetration tests)	€/day
GeotechnicalDays	Duration of the geotechnical campaign	days
$C_{insurance}$	Insurances coefficient	€/MW
TravellingCosts	Travelling costs of the preparation of the project	-
Staff	Number of staff of the project enterprise	staff
AverageCostY	Staff average cost per year	€/year
Years0	Number of years until year 0	years
OfficeRentingM	Office renting per month	€/month
RatioWTGPriceCosts	Ratio between wind turbine generator price and its costs	-
Nplatforms	Number of floating platforms	platforms
FEED	Drafting Costs	€
DetailDesignCosts	Detailed design costs	€
TankTestingInplace	Tank testing-inplace cost (based on other similar projects)	€
TankTestingTransport	Tank testing-transport cost (based on other similar projects)	€
PreliminaryWorksYardCosts	Preliminary works at yard Costs	€
OneDrydockCost	One dry dock costs	€
$N_{drydocks}$	Number of drydocks	docks
C_{3241}	Renting on harbour area cost for foundations	€
C_{3242}	Renting on harbour area cost for generator	€
C_{3243}	Renting on harbour area cost for office	€
C_{3244}	Renting on wet harbour area cost	€
PreparingSeabedCost	Cost of preparing the seabed for one unit	€
TotalMaterialCostPlatform	Total cost of the materials of the platforms considering scale economy	€
CraneDrydockConstructionCostYT	Crane dry dock construction cost	€
CraneDrydockSupportCostYT	Crane dry dock support cost	€
HarbourInternalTowingYears	Number of years neccesary for the harbour internal towings	years
TugMobDemobCost	Cost of mob or demob the tug	€
BollarPullTugDR	Daily rate associated with a bollard pull tug (65 tonnes)	€/day

Table 1. *Cont.*

Variable	Concept	Units
$N_{DiversPlat}$	Number of divers per platform	divers
$DiverDailyCost$	Daily cost associated with one diver	€/day
$DiverDaysIntervention$	Days that the diver need for each intervention	days
$N_{AccessesPlat}$	Number of accesses per platform	-
$PlatformAccessCost$	Platform access cost (1 unit)	€/platform
$N_{InternalsTower}$	Number of internals per tower	-
$TowerInternalsCost$	Tower internals cost (1 unit)	€/platform
$N_{JTubesPlat}$	Number of J-tubes per platform	j-tubes
$JTubeCost$	J-tube cost (per unit)	€/j-tube
$ReductionFactorBallast$	Reduction factor (economy of scale) for the ballast system	-
$TotalBallastCostPlat$	Total Ballast Cost platform	€
$N_{SetsCommunicationPlat}$	Number of sets (communication system) per platform	-
$ReductionFactorCommunication$	Cost of 1 communication system	€
$CommunicationSystemCost1$	Reduction factor (economy of scale) for the communication system	-
$N_{SetsLightingPlat}$	Number of sets (lighting system) per platform	-
$LightingSystemCost1$	Cost of 1 lighting system	€
$ReductionFactorLighting$	Reduction factor (economy of scale) for the lighting system	-
$PlatformMonitoringCost$	Platform monitoring cost (inclinometer and others) (1 platform)	€/platform
$StructuralMonitoringCost$	Structural monitoring cost	€
$C_{33}+C_{34}$	Mooring and anchoring manufacturing cost	€
C_{351}	Electric cable manufacturing cost	€
C_{352}	Substation manufacturing cost	€
C_{411}	Cost of installation of turbines at port	€
C_{412}	Cost of the transport of the turbines	€
C_{413}	Cost of installation of turbines offshore	€
C_{42}	Cost of installing the floating TLP platforms	€
C_{abarge}	Cost of installing mooring and anchoring of a barge per day	€/day
C_{atug}	Cost of installing mooring and anchoring of a tug per day	€/day
C_{aaMOD}	Cost of labour per day for installing mooring and anchoring	€/day
$C_{aapumpsanddivers}$	Cost of divers	€/day
$N_{anchoring}$	Number of anchors	anchors
$T_{instbarge}$	Time of installing mooring and anchoring	Anchors/day
$BurialCableCost1Plat$	Cost associated with the burial cable intertidal area for 1 platform	€/platform
$HDDCost1Plat$	Cost associated with the HDD horizontal drilling for 1 platform	€/platform
$DynamicCableInstallationCost1Plat$	Cost associated with the dynamic cable installation for 1 platform	€/platform
$ExportCableInstallationCost$	Export cable installation cost	€
$CableMechanicalProtectionCost$	Cable mechanical protection cost	€
$CableBurialIntertidalAreaCost$	Cable burial (intertidal area) cost	€
$InterarrayCableCommissioningCost$	Interarray cable commissioning cost	€
$ExportCableCommissioningCost$	Export cable commisioning cost	€
C_{51}	Cost of assurance	€
C_{521}	Data acquisition (SCADA) cost	€
C_{522}	SAP & Maritime coordination costs	€
C_{523}	Meteorological prediction cost	€
C_{524}	Administration cost	€
C_{531}	Turbine maintenance cost	€
C_{532}	Export cable and grid connection maintenance	€
C_{533}	Interarray cable survey and repairs cost	€
C_{534}	Substructure maintenance	€
N_{farm}	Number of years of life-cycle of the farm	Years
C_{54year}	Onshore logistics costs per year	€/year
$C_{551year}$	Workboats costs per year	€/year

Table 1. Cont.

Variable	Concept	Units
$C_{552year}$	Helicopter costs per year	€/year
$C_{553year}$	Crane barge service costs per year	€/year
$C_{554year}$	Offshore accommodation (if any) cost	€/year
$PD_{turbine}$	Percentage of dismantling turbine and platform	-
$PD_{mooring}$	Percentage of dismantling mooring	-
PD_{cable}	Percentage of dismantling cable	-
$PD_{substation}$	Percentage of dismantling substation	-

Conception and definition costs (C_1) depend on legal aspects (C_{11}) and offshore wind farm certification costs (C_{12}). Particularities and differences between legal aspects exist for the countries where the offshore wind farm could be located. Therefore, it would be advisable to go deeper into each case so that the costs would vary.

$$C_1 = C_{11} + C_{12} \tag{8}$$

The legal aspects' cost (C_{11}) depends on the offshore site renting (*OffshoreSiteR*) and the number of wind devices (*Nplatforms*).

$$C_{11} = OffshoreSiteR \cdot Nplatforms \tag{9}$$

The offshore wind farm certification cost (C_{12}) depends on the cost of the offshore wind farm certification (C_{121}), the meteorological research cost (C_{122}), the environmental impact assessment (C_{123}), the geophysical campaign cost (C_{124}), the geotechnical campaign cost (C_{125}) and the initial insurances project cost (C_{126}).

$$C_{12} = C_{121} + C_{122} + C_{123} + C_{124} + C_{125} + C_{126} \tag{10}$$

where:

$$C_{121} = (1 + EPCIC) \cdot OffshoreSiteR \cdot Nplatforms \tag{11}$$

$$C_{122} = (1 + EPCIC) \cdot (C_{emet} + C_{smet} + C_{samet}) \tag{12}$$

$$C_{123} = (1 + EPCIC) \cdot EvaluationReportCost \cdot N_{platforms} \cdot TurbineP \tag{13}$$

$$C_{124} = (1 + EPCIC) \cdot GeophysicalDailyCost \cdot GeophysicalDays \cdot N_{platforms} \tag{14}$$

$$C_{125} = (1 + EPCIC) \cdot GeotechnicalDailyCost \cdot GeotechnicalDays \cdot N_{platforms} \tag{15}$$

$$C_{126} = C_{insurance} \cdot N_{platforms} \cdot TurbineP \tag{16}$$

The design and development cost (C_{21}) depends on the travelling costs for the preparation of the project (*TravellingCosts*), the number of staff for the project enterprise (*Staff*), the staff average cost per year (*AverageCostY*), the number of years until year 0 of the project (*Years0*) and the office renting per month (*OfficeRentingM*).

$$C_2 = C_{21} \tag{17}$$

where:

$$C_{21} = (1 + TravellingCosts) \cdot (12 \cdot Staff \cdot AverageCostY \cdot Years0 \cdot OfficeRentingM) \tag{18}$$

On the other hand, manufacturing costs (C_3) will be explained from the sub-phases that constitute an offshore wind farm. The manufacturing costs depend on generator manufacturing costs (C_{31}), floating platform manufacturing cost (C_{32}), mooring manufacturing cost (C_{33}), anchoring manufacturing cost (C_{34}) and electric system manufacturing cost (C_{35}).

$$C_3 = C_{31} + C_{32} + C_{33} + C_{34} + C_{35} \tag{19}$$

The generator manufacturing cost (C_{31}) depends on the costs of turbine components (*ComponentsCost10MWturbine*), the ratio of power and size of the turbine considered (*RatioWTGPriceCosts*) and the number of platforms (*Nplatforms*).

$$C_{31} = ComponentsCost10MWturbine \cdot RatioWTGPriceCosts \cdot Nplatforms \tag{20}$$

The floating platform manufacturing cost (C_{32}) depends on the platform engineering cost (C_{321}), the tank testing cost (C_{322}), the manufacturing facilities—civil work costs (C_{323}), the manufacturing facilities—renting on harbour area cost (C_{324}), the manufacturing facilities—other costs (C_{325}), the materials and manufacturing costs of externals (C_{326}), the substructure assembly cost of external works (C_{327}), the harbour's internal towing cost (C_{328}), the divers (harbour operations) cost (C_{329}), the external access cost (C_{3210}), the tower internals costs (C_{3211}), the J tubes-cable access cost (C_{3212}), the ballast (if any) system cost (C_{3213}), the communication system cost (C_{3214}), the emergency system (if any) cost (C_{3215}), the lighting and other signs cost (C_{3216}), the monitoring system cost (C_{3217}), the safety equipment cost (C_{3218}) and the crew transfer vessels during construction cost (C_{3219}).

$$\begin{aligned} C_{32} = C_{321} + C_{322} + C_{323} + C_{324} + C_{325} + C_{326} + C_{327} + C_{328} + C_{329} + C_{3210} + C_{3211} + C_{3212} + C_{3213} \\ + C_{3214} + C_{3215} + C_{3216} + C_{3217} + C_{3218} + C_{3219} \end{aligned} \tag{21}$$

where:

$$C_{321} = FEED + DetailDesignCosts \tag{22}$$

$$C_{322} = TankTestingInplace + TankTestingTransport \tag{23}$$

$$C_{323} = PreliminaryWorksYardCosts + OneDrydockCost \cdot N_{drydocks} \tag{24}$$

$$C_{324} = C_{3241} + C_{3242} + C_{3243} + C_{3244} \tag{25}$$

$$C_{325} = N_{platforms} \cdot PreparingSeabedCost \tag{26}$$

$$C_{326} = TotalMaterialCostPlatform \tag{27}$$

$$C_{327} = CraneDrydockConstructionCostYT + CraneDrydockSupportCostYT \tag{28}$$

$$C_{328} = HarbourInternalTowingYears \cdot \left(TugMobDemobCost + \frac{N_{platforms}}{HarbourInternalTowingYears} \cdot 2.5 + TugRentingDaysY \right) \tag{29}$$

$$C_{329} = N_{platforms} \cdot N_{DiversPlat} \cdot DiverDailyCost \cdot DiverDaysIntervention \tag{30}$$

$$C_{3210} = N_{platforms} \cdot N_{AccessesPlat} \cdot PlatformAccessCost \tag{31}$$

$$C_{3211} = N_{InternalsTower} \cdot TowerInternalsCost \cdot N_{platforms} \tag{32}$$

$$C_{3212} = 0 \tag{33}$$

$$C_{3213} = N_{platforms} \cdot N_{JTubesPlat} \cdot JTubeCost \tag{34}$$

$$C_{3214} = 0 \tag{35}$$

$$C_{3215} = ReductionFactorBallast \cdot N_{platforms} \cdot TotalBallastCostPlat \tag{36}$$

$$C_{3216} = N_{platforms} \cdot N_{SetsCommunicationPlat} \cdot ReductionFactorCommunication \cdot CommunicationSystemCost1 \tag{37}$$

$$C_{3217} = 0 \tag{38}$$

$$C_{3218} = N_{platforms} \cdot N_{SetsLightingPlat} \cdot LightingSystemCost1 \cdot ReductionFactorLighting \tag{39}$$

$$C_{3219} = N_{platforms} \cdot PlatformMonitoringCost + StructuralMonitoringCost \tag{40}$$

The mooring manufacturing cost (C_{33}) depends on the mooring lines engineering and project management (PM) cost (C_{331}), the tendon engineering and PM cost (if any) (C_{332}), the mooring lines manufacturing cost (C_{333}), the connector manufacturing cost (C_{334}), the chain stopper, fairlead manufacturing cost (C_{335}), the bending shoe fairlead, rotatory chain fairlead or similar cost (C_{336}), the mooring lines monitoring cost (C_{337}), the temporary buoys and/or other auxiliary components cost (C_{338}) and the tendon manufacturing and shipping cost (if any) (C_{339}):

$$C_{33} = C_{331} + C_{332} + C_{333} + C_{334} + C_{335} + C_{336} + C_{337} + C_{338} + C_{339} \tag{41}$$

The anchoring manufacturing cost (C_{34}) depends on the anchoring engineering and PM cost (C_{341}) and the anchor construction cost (C_{342}):

$$C_{34} = C_{341} + C_{342} \tag{42}$$

The electric system manufacturing cost (C_{35}) depends on the electric cable manufacturing cost (C_{351}) and the substation manufacturing cost (C_{352}):

$$C_{35} = C_{351} + C_{352} \tag{43}$$

Given the characteristics of offshore wind farms, related to the remote location from the coast, the installation of the components is an important part of its life cycle. In addition, several aspects related to the floating condition of the TLP platform must be considered to define an effective installation strategy from an economic point of view.

The costs of this phase are divided according to the elements of which an offshore wind farm is composed. In this sense, the installation costs (C_4) depend on generator installation costs (C_{41}), platforms installation costs (C_{42}), mooring and anchoring installation costs (C_{43}) and electric cable installation costs (C_{44}):

$$C_4 = C_{41} + C_{42} + C_{43} + C_{44} \tag{44}$$

The cost of installing the offshore wind turbines (C_{41}) is divided into the installation at port (C_{411}), the transport (C_{412}) and the installation offshore (C_{413}):

$$C_{41} = C_{411} + C_{412} + C_{413} \tag{45}$$

The cost of installing the floating TLP platforms (C_{42}) depends on the onshore commission cost (C_{421}), the offshore platform commissioning cost (C_{422}) and other costs-lifting equipment and miscellaneous costs (C_{423}):

$$C_{42} = C_{421} + C_{422} + C_{423} \tag{46}$$

The cost of installing the mooring and anchoring of the TLP platforms (C_{43}) depends on the anchor installation cost (C_{431}), the mooring line installation cost (C_{432}), the tendon installation cost (if any) (C_{433}) and other costs-miscellaneous (C_{434}):

$$C_{43} = \left(C_{abarge} + C_{atug} + C_{aaMOD} + C_{aapumpsanddivers}\right) \cdot \frac{N_{anchoring}}{T_{instbarge}} \tag{47}$$

The cost of installing the electric system (C_{44}) depends on the cost associated with the burial cable intertidal area for 1 platform (*BurialCableCost1Plat*), the cost associated with the HDD horizontal drilling for 1 platform (*HDDCost1Plat*), the cost associated with the dynamic cable installation for 1 platform (*DynamicCableInstallationCost1Plat*), the number of platforms (*Nplatforms*), the export cable installation cost (*ExportCableInstallationCost*), the cable mechanical protection cost (*CableMechanicalProtectionCost*), the cable burial (intertidal area) cost (*CableBurialIntertidalAreaCost*), the inter array cable commissioning cost (*InterarrayCableCommissioningCost*) and the export cable commissioning cost (*ExportCableCommissioningCost*):

$$\begin{aligned} C_{44} = \Big(N_{platforms}\cdot\ & (BurialCableCost1Plat + HDDCost1Plat \\ & + CableOnshoreCivilWorksCost1Plat)) \\ & + \Big(DynamicCableInstallationCost1Plat \cdot N_{platforms}\Big) \\ & + ExportCableInstallationCost + CableMechanicalProtectionCost \\ & + CableBurialIntertidalAreaCost + InterarrayCableCommissioningCost \\ & + ExportCableCommissioningCost \end{aligned} \tag{48}$$

The operation and maintenance costs are an important part of the life cycle of an offshore wind farm since it will be the most durable phase in time and in which many factors must be considered: the logistic capacity, distance to the coast, the reliability of the turbine components, etc.

Operation and maintenance costs (C_5) can be disaggregated into assurance costs (C_{51}), administration and operations cost (C_{52}), maintenance cost (C_{53}), onshore logistics (C_{54}) and offshore logistics (C_{55}). Materials, replacements, consumables and others are maintenance costs that must be considered. The costs of offshore logistics operations depend on the use of barges, marine cranes and helicopters:

$$C_5 = C_{51} + C_{52} + C_{53} + C_{54} + C_{55} \tag{49}$$

where:

$$C_{52} = C_{521} + C_{522} + C_{523} + C_{524} \tag{50}$$

$$C_{53} = C_{531} + C_{532} + C_{533} + C_{534} \tag{51}$$

$$C_{54} = N_{farm} \cdot C_{54year} \tag{52}$$

$$C_{55} = N_{farm} \cdot C_{55year} \tag{53}$$

where:

$$C_{55year} = C_{551year} + C_{552year} + C_{553year} + C_{554year} \tag{54}$$

Once the life of the components that make up the offshore wind farm has come to an end, the last phase of the life cycle begins the dismantling [44]. For this task, it is required to disconnect the mooring system. In addition, heavy decoupling operations of floating turbines must be carried out, as well as the transport of structures to the coast. Therefore, dismantling costs (C_6) depend on generator dismantling costs (C_{61}), the floating platform dismantling costs (C_{62}), mooring and anchoring dismantling costs (C_{63}) and electric systems dismantling costs (C_{64}):

$$C_6 = C_{61} + C_{62} + C_{63} + C_{64} \tag{55}$$

where:

$$C_{61} = PD_{turbine} \cdot (C_{41} + C_{42}) \tag{56}$$

$$C_{62} = 0 \tag{57}$$

$$C_{63} = PD_{mooring} \cdot C_{43} \tag{58}$$

$$C_{64} = PD_{cable} \cdot C_{44cable} + PD_{substation} \cdot C_{44sube} \tag{59}$$

3. Case Study

The CENTEC-TLP is composed of lower pontoons that help to stabilize the transport phase and act as the primary buoyancy bodies in the installed form (see Figure 1). The overall width of the platform is 49.5 m, with a draft of 20 m when installed. During the towing phase, the draft lowers to 3.85 m, making it possible to place the platform into the water in most ports without being concerned with port-side draft limitations. The structural integrity is provided by the support braces. It uses 3 mooring lines per corner. Detailed sizing information of the platform is available in [25].

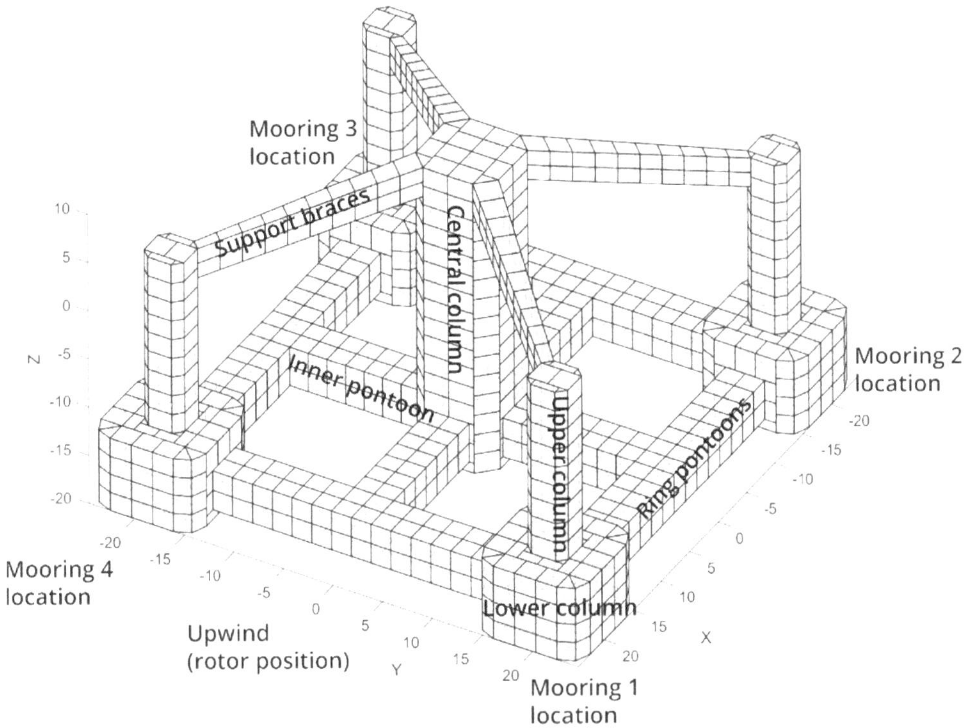

Figure 1. Hull form of CENTEC—TLP.

Regarding the offshore wind turbine, the DTU 10 MW Reference Wind Turbine has been selected. Its main characteristics are: 10 MW of rated power, 4 m/s cut in wind speed, 25 m/s cut out wind speed, 11.4 m/s rated wind speed, 3 blades, 178.3 m rotor diameter, 5.6 m hub diameter and 119 m hub height.

An objective of the Arcwind project is to identify optimal locations for floating wind projects in the European Atlantic region promising offshore wind development area, as identified in previous studies [45–48] Nine locations have been defined in the Spanish coastal area and further known constraints have been identified, the locations selection methodology is presented and discussed by Díaz and Guedes Soares [49].

These nine locations are subsequently ranked using an in-house evaluation modelling tool against several criteria to establish the optimum location within Spain. The main characteristics of the Ribadeo location are shown in Table 2.

Table 2. Main characteristics of the Ribadeo floating wind farm.

Item	Value	Units
Water depth	150	m/s
Wave conditions	2.15	m
Marine currents	0.51	m/s
Distance to local electrical grid	18.60	km
Distance from coastal facilities	89.50	km
Distance from shore	17.61	km
Distance from maritime routes	1.50	km
Distance from protected areas	10.92	km
Area of the territory	405	km^2
Wind farm capacity	880	MW
Number of 10 MW turbines	88	-

The nine zones identified during the site selection study [49] as most suitable for the development of commercial-scale floating wind farms, were pragmatically characterized for their compliance with a set of defined technical and environmental parameters. The key constraints considered immovable for offshore wind farm development were as follows; offshore wind resource, the Exclusive Economic Zone (EEZ), water depths, proximity to other maritime activities and distance to environmental protected areas. Further constraints were also considered within the analysis, such as the proximity to underwater lines, proximity to maritime routes, visual impact, distance to the electrical grid and wave conditions. Due to the significant level of constraints considered, exclusion zones were established. This approach considers the main factors that prevent the sustainable deployment of a floating wind farm within the European waters guaranteeing the viability of the achieved locations.

The resultant areas after applying the exclusion criteria cover 1275 km^2 of the total area evaluated necessitating a secondary assessment of these areas based on the proposed evaluation criteria. Therefore, the entire space proposed would suppose a total installation of around 500 floating wind turbines. Figure 2 identifies some of the restrictions involved in the process of location identification. It shows the identified nine locations based on the site selection assessment.

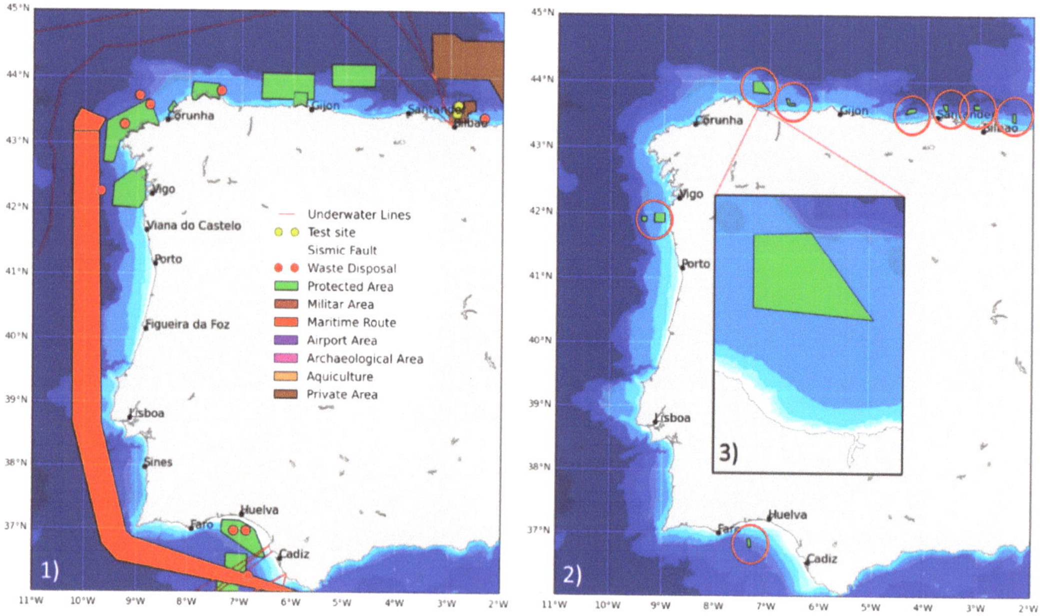

Figure 2. Floating wind map in Spain. (1) Some restrictions present in Spanish waters. (2) Floating wind farm proposed. (3) Ribadeo floating wind farm.

The best location within Spain's EEZ was selected using multicriteria decision-making techniques that ranked the floating wind farm areas proposed based on criteria and experts' opinions [50–52]. Applying these assumptions to the nine locations, the Ribadeo location was identified as the most suitable for the installation of floating turbines. Figures 1–3 visualize this area.

This study is focused on the technical and economic feasibility of the CENTEC-TLP concept for the Ribadeo location (see Figure 2). Ribadeo is located on the Galician Coast, Northwest of Spain (latitude, longitude: 43.837, −7.326). The wind farm location is characterized by an average wind velocity of 9.3 m/s and wind potential of 4923 h/yr. Moreover, other details that show the huge potential of the area are presented in Table 1. Figure 2 shows three maps: (1) the map of restrictions to install offshore wind farms in Spanish waters;

(2) the floating offshore wind farms proposed (in green); and (3) the Ribadeo floating offshore wind farm, which is the final location selected for this study.

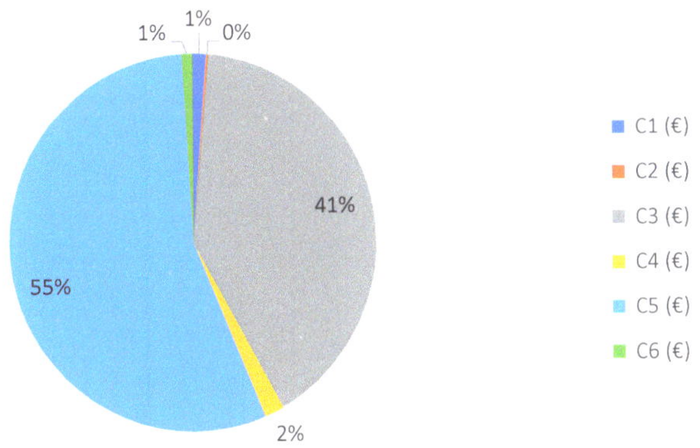

Figure 3. Total cost results. Being C1 (conception and definition cost), C2 (design and development cost), C3 (manufacturing cost), C4 (installation cost), C5 (exploitation cost) and C6 (dismantling cost).

Several scenarios were considered in the present case study for the different locations of the farms. These scenarios were defined considering three electric tariffs (50 €/MWh, 100 €/MWh and 150 €/MWh), due to the permanent change in the electric rates in Europe during the last months [53], and three capital costs (6%, 8% and 10%) [54].

The initial variables are shown in Table 3.

Table 3. Main variables.

Variable	Value	Units
EPCIC	0	%
OffshoreSiteR	100,000	€
Nplatforms	88	platforms
C_{emet}	2,338,480	€
C_{smet}	467,696	€
C_{samet}	116,924	€
EvaluationReportCost	2500	€/MW
TurbineP	10	MW
GeophysicalDailyCost	15,000	€/day
GeophysicalDays	2	days
GeotechnicalDailyCost	35,000	€/day
GeotechnicalDays	3	days
$C_{insurance}$	33,877.2	€/MW
TravellingCosts	0.15	-
Staff	15	staff
AverageCostY	65,000	€/year
Years0	7	years
OfficeRentingM	4000	€/month
RatioWTGPriceCosts	-	-
Nplatforms	88	platforms
FEED	3,877,520	€
DetailDesignCosts	5,028,000	€
TankTestingInplace	125,000	€
TankTestingTransport	75,000	€
PreliminaryWorksYardCosts	5,000,000	€
OneDrydockCost	3,550,000	€

Table 3. *Cont.*

Variable	Value	Units
$N_{drydocks}$	1	docks
C_{3241}	17,104,140.9	€
C_{3242}	82,125	€
C_{3243}	54,750	€
C_{3244}	862,312.5	€
$PreparingSeabedCost$	200,000	€
$TotalMaterialCostPlatform$	312,052,35	€
$CraneDrydockConstructionCostYT$	2,940,000	€
$CraneDrydockSupportCostYT$	1,900,000	€
$HarbourInternalTowingYears$	2	years
$TugMobDemobCost$	3000	€
$BollarPullTugDR$	18,000	€/day
$N_{DiversPlat}$	2	divers
$DiverDailyCost$	1500	€/day
$DiverDaysIntervention$	2	days
$N_{AccessesPlat}$	1	-
$PlatformAccessCost$	175,000	€/platform
$N_{InternalsTower}$	1	-
$TowerInternalsCost$	125,000	€/platform
$N_{JTubesPlat}$	1	j-tubes
$JTubeCost$	40,000	€/j-tube
$ReductionFactorBallast$	0.8	-
$TotalBallastCostPlat$	0	€
$N\_SetsCommunicationPlat$	1	-
$ReductionFactorCommunication$	75,000	€
$CommunicationSystemCost1$	0.8	-
$N_{SetsLightingPlat}$	1	-
$LightingSystemCost1$	20,000	€
$ReductionFactorLighting$	0.8	-
$PlatformMonitoringCost$	20.000	€/platform
$StructuralMonitoringCost$	90.000	€
$C_{33} + C_{34}$	268,712,67	€
C_{351}	60,901,417	€
C_{352}	15,000.000	€
C_{411}	132,285	€
C_{412}	7,002,858.21	€
C_{413}	14,744,889	€
C_{42}	12,414,616	€
C_{abarge}	7500	€/day
C_{atug}	22,502	€/day
C_{aaMOD}	5656	€/day
$C_{aapumpsanddivers}$	0	€/day
$N_{anchoring}$	1056	anchors
$T_{instbarge}$	3	Anchors/day
$BurialCableCost1Plat$	25,000	€/platform
$HDDCost1Plat$	0	€/platform
$DynamicCableInstallationCost1Plat$	100,000	€/platform
$ExportCableInstallationCost$	8,000,000	€
$CableMechanicalProtectionCost$	80,000	€
$CableBurialIntertidalAreaCost$	1,000,000	€
$InterarrayCableCommissioningCost$	100,000	€
$ExportCableCommissioningCost$	100,000	€
C_{51}	22,618,239	€
C_{521}	6,750,000	€
C_{522}	33,000,000	€
C_{523}	3,575,000	€
C_{524}	19,250,000	€
C_{531}	630,300	€

Table 3. *Cont.*

Variable	Value	Units
C_{532}	16,654,000	€
C_{533}	19,250,000	€
C_{534}	115,500,000	€
N_{farm}	25	Years
C_{54year}	1,210,000	€/year
$C_{551year}$	5,500,000	€/year
$C_{552year}$	4,950,000	€/year
$C_{553year}$	17,600,000	€/year
$C_{554year}$	33,000,000	€/year
$PD_{turbine}$	70%	-
$PD_{mooring}$	90%	-
PD_{cable}	10%	-
$PD_{substation}$	90%	-

4. Results

4.1. Technical Overview

The technical validation of CENTEC-TLP was carried out both numerically and experimentally. This section summarises the previous publications discussing the performance of the TLP before proceeding with the economics. The towing dynamics are explained in [26]. It was tested experimentally in waves up to 3 m of significant wave height with speeds up to 5 knots. The motions of the platform under these scenarios show that the pitching angle is limited to 0.7 degrees. Given that the platform has stability up to 10 degrees, its motion dynamics allow a large weather window to tow it to the location. Without the platform motions being a limiting factor, other concerns such as the installation process and the owing performance of the tugboats gain prevalence. The platform also has low wave resistance, allowing it to be towed using lower-powered tugboats.

After the platform is towed and installed, the primary factor for tension leg platforms is the mooring line tensions. The dynamics explained in [55] show that the platform can withstand the 50-year extreme weather in the selected installation area as well as the below and above rated operational conditions. As the structure is a TLP, its pitching angles are limited to values under a unit degree. Hence, it is not affected by the platform pitch related power production issues as explained in [56]. The verification of both the towing and motion performance leaves the platform's economic feasibility as the deciding factor, which is detailed in the following sections.

4.2. Economic Results

Regarding the total costs of the farm (see Figure 3), the farm has a C1 of EUR 54.16 M, a C2 of EUR 8.23 M, a C3 of EUR 1769.83 M, a C4 of EUR 74.85 M a C5 of EUR 2419.85 M and a C6 of EUR 39.51 M. They represent the 1.2% of C1, the 0.2% of C2, the 40.5% of C3, the 1.7% of C4, the 55.4% of C5 and the 0.9% of C6.

Regarding LCOE (see Figure 4), this case study gives values of EUR 76.67/MWh, EUR 71.71/MWh and EUR 67.03/MWh for an electric tariff of EUR 150/MWh and 10, 8 and 6% of capital cost respectively. In addition, it gives values of EUR 62.77/MWh, EUR 58.42/MWh and EUR 54.33/MWh for an electric tariff of EUR 100/MWh and 10, 8 and 6% of capital cost respectively. finally, it gives values of EUR 80.08/MWh, EUR 73.89/MWh and EUR 68.11/MWh for an electric tariff of EUR 50/MWh and 10, 8 and 6% respectively.

The LCOE usually does not depend on the electric tariff considered. However, in the present model, it depends indirectly because the LCOE has been calculated considering the WACC (weighted average cost of capital), which depends directly on the own resources of the enterprise. These own resources have been calculated considering the total initial investment (C1, C2, C3, C4 and C6) and the investment in working capital. Particularly, the investment in working capital has been calculated considering an investment of two months of the incomes of the project (tariff*energy produced), which makes the LCOE depend indirectly on the electricity tariff.

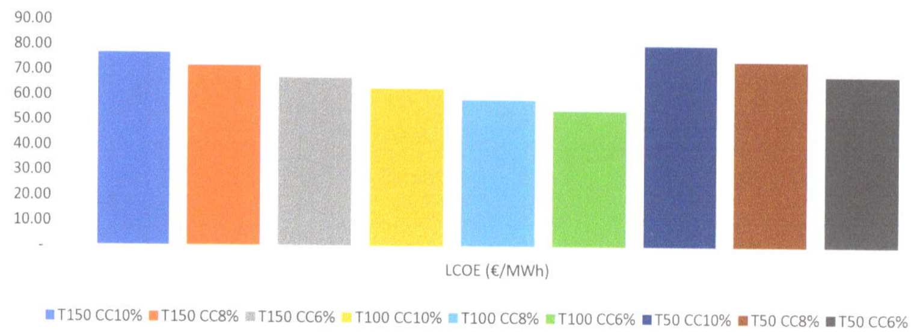

Figure 4. LCOE results.

Regarding *NPV* (see Figure 5), this case study gives values of EUR 1800.44 M, EUR 2180.77 M and EUR 2636.45 M for an electric tariff of EUR 150/MWh and 10, 8 and 6% of capital cost respectively. In addition, it gives values of EUR 258.33 M, EUR 330.86 M and EUR 418.11 M for an electric tariff of EUR 100/MWh and 10, 8 and 6% of capital cost respectively. Finally, it gives values of EUR −898.35 M, EUR −849.96 M and EUR −782.90 M for an electric tariff of EUR 50/MWh and 10, 8 and 6% respectively. As it is observed, a higher capital cost gives values of *NPV* lower. Therefore, the farm will be economically feasible in terms of *NPV* for a tariff of EUR 150/MWh and a tariff of EUR 100/MWh, because its value is higher than 0.

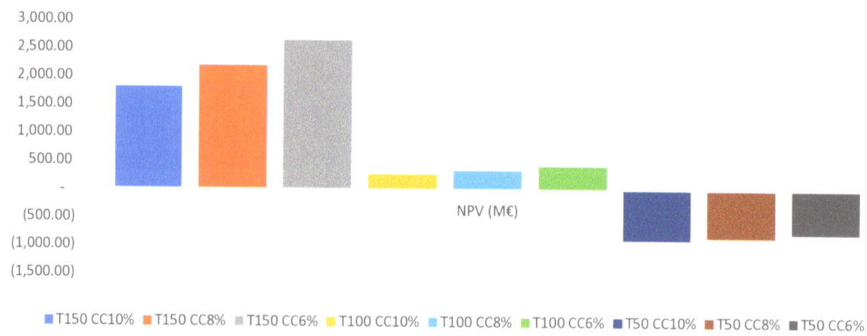

Figure 5. *NPV* results.

Regarding *IRR* (see Figure 6), this case study gives values of 14.74% for an electric tariff of EUR 150/MWh, 11.25% for an electric tariff of EUR 100/MWh and 0.44% for an electric tariff of EUR 50/MWh, being its value independent on the capital cost considered. Therefore, the farm will be economically feasible in terms of *IRR* for a tariff of EUR 150/MWh and a tariff of EUR 100/MWh, because its value is higher than the WACC considered.

Regarding DPBP (see Figure 7), this case study gives values of 9 years, 9 years and 8 years for an electric tariff of EUR 150/MWh and 10, 8 and 6% of capital cost respectively. In addition, it gives values of 13 years, 13 years and 12 years for an electric tariff of EUR 100/MWh and 10, 8 and 6% of capital cost respectively. Finally, it gives values of 28 years for all the capital costs considered (10, 8 and 6%). As is observed, a higher electric tariff gives values of DPBP lower. Therefore, the farm will be economically feasible in terms of DPBP for a tariff of EUR 150/MWh and for a tariff of EUR 100/MWh, because its values (from 8 to 13 years) are lower than the life cycle of the project (25 years).

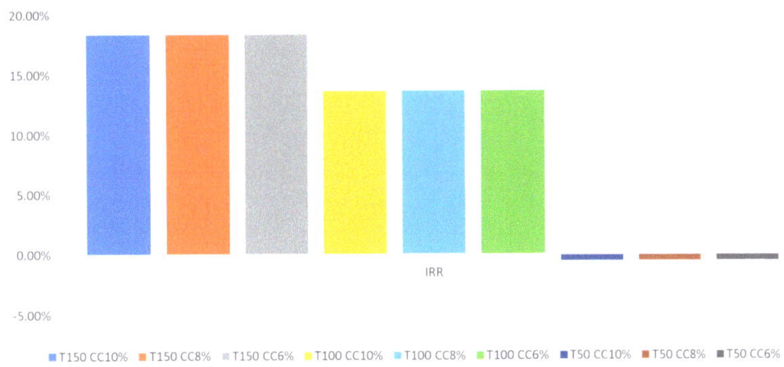

Figure 6. *IRR* results.

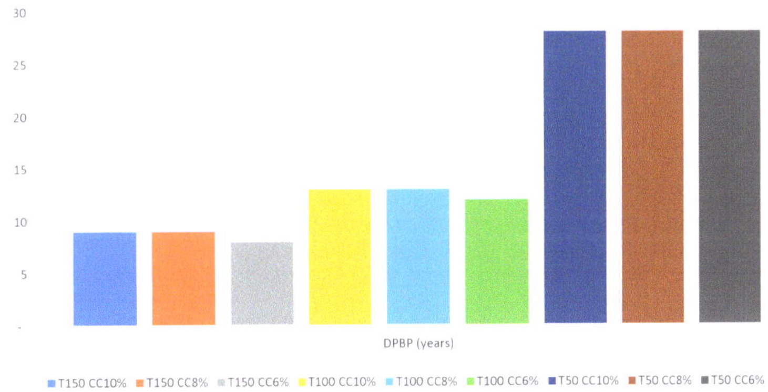

Figure 7. DPBP results.

5. Discussion

This work has described the methodology for calculating the economic feasibility of a floating offshore wind farm composed of TLP platforms. For this purpose, firstly the life-cycle costs have been analysed. In this context, the costs have been divided into the costs of each life-cycle stage of the process: conception and definition cost, design and development cost, manufacturing cost, installation cost, exploitation cost and dismantling cost.

Secondly, the main economic feasibility indicators have been calculated taking into account the cash flows of the project: internal rate of return (*IRR*), net present value (*NPV*), discounted pay-back period (PBP) and levelized cost of energy (LCOE).

The case study analysed is based on the TLP platform designed by the CENTEC. One farm of 880 MW has been considered for this platform located along the Atlantic Coast of the European Union. It has been located in Ribadeo (the North Area of the North-West region of Galicia, in Spain). Eighteen case studies have been analysed considering different scenarios in terms of three different electric tariffs and three different costs of capital.

Results indicate how important is the electric tariff to determine the values of *IRR*, *NPV* and DPBP. The best result is obtained for an electric tariff of EUR 150/MWh and a cost of capital of 6%. It has values of 18.34% of *IRR*, EUR 2636.45 M of *NPV* and 8 years of DPBP. Regarding their values of LCOE, the farm has a minimum value of EUR 54.33/MWh. It makes the platform economically feasible for the location selected because the internal rate of return is higher than the capital cost, the net present value is higher than zero and the discounted pay-back period is lower than the life cycle of the project.

The values of LCOE are similar to other authors, which go from GBP 49.28/MWh to GBP 72.74/MWh for the minimum values of Anastasia Ioannou et al. [57] or which go from USD 60/MWh to USD 110/MWh for Maira Bruck et al. [58], for instance.

The method proposed is important to help all the stakeholders (investors, facilities, enterprises, etc.) to do deep analysis in economic terms of the TLP platform selected.

Author Contributions: Conceptualization, A.F.-V., D.C.I., F.P.-G., I.L.-G., A.I.G.-D. and L.C.-S.; Investigation, A.F.-V., D.C.I., L.C.-S., E.U., H.D.; Methodology, A.F.-V., D.C.I., L.C.-S., E.U., H.D.; Resources, A.F.-V., D.C.I., L.C.-S., E.U., H.D.; Supervision, C.G.S.; Writing—original draft, A.F.-V., L.C.-S., D.C.I., F.P.-G., I.L.-G., A.I.G.-D., E.U., H.D., C.G.S.; Writing—review & editing, C.G.S. All authors have read and agreed to the published version of the manuscript.

Funding: This research was funded by EAPA_344/2016 ARCWIND project, co-financed by the European Regional Development Fund through the Interreg Atlantic Area Programme. This work contributes to the Strategic Research Plan of the Centre for Marine Technology and Ocean Engineering (CENTEC), which is financed by the Portuguese Foundation for Science and Technology (Fundação para a Ciência e Tecnologia—FCT) under contract UIDB/UIDP/00134/2020.

Institutional Review Board Statement: Not applicable.

Informed Consent Statement: Not applicable.

Data Availability Statement: Not applicable.

Conflicts of Interest: The authors declare no conflict of interest.

References

1. WindEurope. *Wind Europe Wind Energy in Europe 2020. Statistics and the Outlook for 2021–2025*; WindEurope: Brussels, Belgium, 2020.
2. Negro, V.; López-Gutiérrez, J.S.; Esteban, M.D.; Alberdi, P.; Imaz, M.; Serraclara, J.M. Monopiles in Offshore Wind: Preliminary Estimate of Main Dimensions. *Ocean Eng.* **2017**, *133*, 253–261. [CrossRef]
3. Gideon, R.A.; Bou-Zeid, E. Collocating Offshore Wind and Wave Generators to Reduce Power Output Variability: A Multi-Site Analysis. *Renew. Energy* **2021**, *163*, 1548–1559. [CrossRef]
4. Yeter, B.; Garbatov, Y.; Guedes Soares, C. Uncertainty Analysis of Soil-Pile Interactions of Monopile Offshore Wind Turbine Support Structures. *Appl. Ocean Res.* **2019**, *82*, 74–88. [CrossRef]
5. Plodpradit, P.; Dinh, V.N.; Kim, K. Du Coupled Analysis of Offshore Wind Turbine Jacket Structures with Pile-Soil-Structure Interaction Using FAST v8 and X-SEA. *Appl. Sci.* **2019**, *9*, 1633. [CrossRef]
6. AREVA Technical Data the Evolution of the Proven Offshore M5000-135.5000. Available online: http://windturbine.co.kr/homepage/data/file/submenu4_1/622063319_JICPZj5z_AREVAwind_M5000_uk.pdf (accessed on 20 September 2022).
7. Yeter, B.; Garbatov, Y.; Guedes Soares, C. Fatigue Damage Assessment of Fixed Offshore Wind Turbine Tripod Support Structures. *Eng. Struct.* **2015**, *101*, 518–528. [CrossRef]
8. Esteban, M.D.; Couñago, B.; López-Gutiérrez, J.S.; Negro, V.; Vellisco, F. Gravity Based Support Structures for Offshore Wind Turbine Generators: Review of the Installation Process. *Ocean Eng.* **2015**, *110*, 281–291. [CrossRef]
9. Wu, X.; Hu, Y.; Li, Y.; Yang, J.; Duan, L.; Wang, T.; Adcock, T.; Jiang, Z.; Gao, Z.; Lin, Z.; et al. Foundations of Offshore Wind Turbines: A Review. *Renew. Sustain. Energy Rev.* **2019**, *104*, 379–393. [CrossRef]
10. Lee, Y.S.; González, J.A.; Lee, J.H.; Kim, Y.I.; Park, K.C.; Han, S. Structural Topology Optimization of the Transition Piece for an Offshore Wind Turbine with Jacket Foundation. *Renew. Energy* **2016**, *85*, 1214–1225. [CrossRef]
11. Cermelli, C.; Roddier, D.; Aubault, A. WindFloat: A Floating Foundation for Offshore Wind Turbines—Part II: Hydrodynamics Analysis. In Proceedings of the Ocean Engineering; Ocean Renewable Energy, Ocean Space Utilization, Parts A and B., ASME, Honolulu, HI, USA, 31 May–5 June 2009; Volume 4, pp. 135–143.
12. Oguz, E.; Clelland, D.; Day, A.H.; Incecik, A.; López, J.A.; Sánchez, G.; Almeria, G.G. Experimental and Numerical Analysis of a TLP Floating Offshore Wind Turbine. *Ocean Eng.* **2018**, *147*, 591–605. [CrossRef]
13. Rahman, I.M.M.; Begum, Z.A.; Yahya, S.; Lisar, S.; Motafakkerazad, R.; Cell, A.S.-P. *Advances in Renewable Energy and Sustainable Systems*; Nova Science Publishers: Hauppauge, NY, USA, 2016; ISBN 6312317269.
14. Castro-Santos, L.; Filgueira-Vizoso, A.; Carral-Couce, L.; Formoso, J.Á.F. Economic Feasibility of Floating Offshore Wind Farms. *Energy* **2016**, *112*, 868–882. [CrossRef]
15. Castro-Santos, L.; Filgueira-Vizoso, A.; Lamas-Galdo, I.; Carral-Couce, L. Methodology to Calculate the Installation Costs of Offshore Wind Farms Located in Deep Waters. *J. Clean. Prod.* **2018**, *170*, 1124–1135. [CrossRef]
16. Castro-Santos, L.; deCastro, M.; Costoya, X.; Filgueira-Vizoso, A.; Lamas-Galdo, I.; Ribeiro, A.; Dias, J.M.; Gómez-Gesteira, M. Economic Feasibility of Floating Offshore Wind Farms Considering Near Future Wind Resources: Case Study of Iberian Coast and Bay of Biscay. *Int. J. Environ. Res. Public Health* **2021**, *18*, 2553. [CrossRef]

17. Kikuchi, Y.; Ishihara, T. Upscaling and Levelized Cost of Energy for Offshore Wind Turbines Supported by Semi-Submersible Floating Platforms. *J. Phys. Conf. Ser.* **2019**, *1356*, 012033. [CrossRef]
18. Castro-Santos, L.; Filgueira-Vizoso, A.; Piegari, L. Calculation of the Levelized Cost of Energy and the Internal Rate of Return Using GIS: The Case Study of a Floating Wave Energy Farm. In Proceedings of the ICCEP 2019—7th International Conference on Clean Electrical Power: Renewable Energy Resources Impact, Otranto, Italy, 2–4 July 2019.
19. Castro-Santos, L.; Filgueira-Vizoso, A.; Lamas-Galdo, I.; Álvarez-Feal, C.; Carral-Couce, L. Influence of the Discount Rate in the Economic Analysis of a Floating Offshore Wind Farm in the Galician Region of the European Atlantic Area. In Proceedings of the Volume 10: Ocean Renewable Energy, American Society of Mechanical Engineers, Madrid, Spain, 17–22 June 2018.
20. Cordal-Iglesias, D.; Filgueira-Vizoso, A.; Baita-Saavedra, E.; Graña-López, M.Á.; Castro-Santos, L. Framework for Development of an Economic Analysis Tool for Floating Concrete Offshore Wind Platforms. *J. Mar. Sci. Eng.* **2020**, *8*, 958. [CrossRef]
21. Ueckerdt, F.; Hirth, L.; Luderer, G.; Edenhofer, O. System LCOE: What Are the Costs of Variable Renewables? *Energy* **2013**, *63*, 61–75. [CrossRef]
22. Johnston, B.; Foley, A.; Doran, J.; Littler, T. Levelised Cost of Energy, A Challenge for Offshore Wind. *Renew. Energy* **2020**, *160*, 876–885. [CrossRef]
23. Pawel, I. The Cost of Storage—How to Calculate the Levelized Cost of Stored Energy (LCOE) and Applications to Renewable Energy Generation. *Energy Procedia* **2014**, *46*, 68–77. [CrossRef]
24. Review: Costo Nivelado de La Energía (LCOE): Nuevo Enfoque Para El Análisis de Proyectos de Energía. Available online: https://repositorio.utdt.edu/handle/20.500.13098/11261 (accessed on 1 September 2022).
25. Uzunoglu, E.; Guedes Soares, C. Hydrodynamic Design of a Free-Float Capable Tension Leg Platform for a 10 MW Wind Turbine. *Ocean Eng.* **2020**, *197*, 106888. [CrossRef]
26. Mas-Soler, J.; Uzunoglu, E.; Bulian, G.; Guedes Soares, C.; Souto-Iglesias, A. An Experimental Study on Transporting a Free-Float Capable Tension Leg Platform for a 10 MW Wind Turbine in Waves. *Renew. Energy* **2021**, *179*, 2158–2173. [CrossRef]
27. Robertson, A.N.; Jonkman, J.M.; Goupee, A.J.; Coulling, A.J.; Prowell, I.; Browning, J.; Masciola, M.D.; Molta, P. Summary of Conclusions and Recommendations Drawn from the DeepCWind Scaled Floating Offshore Wind System Test Campaign. In Proceedings of the ASME 2013 32nd International Conference on Ocean, Offshore and Arctic Engineering, American Society of Mechanical Engineers, Nantes, France, 9–14 June 2013; p. V008T09A053.
28. Kibbee, S.; Chianis, J.; Davies, K.; Sarwono, B. A Mini-Platform for Deepwater—the SeaStar TLP. In Proceedings of the 4th Offshore Symposium on Tension Leg Platform Technology, Houston, TX, USA, 24–25 February 1995; The Society of Naval Architects and Marine Engineers: Houston, TX, USA, 1995.
29. Sclavounos, P.D.; Tracy, C.; Lee, S. Floating Offshore Wind Turbines: Responses in a Seastate, Pareto Optimal Designs and Economic Assessment. In Proceedings of the 27th International Conference on Offshore Mechanics and Arctic Engineering, Estoril, Portugal, 15–20 June 2007; pp. 1–20. [CrossRef]
30. Souza, R.D.; Aggarwal, R. The Tension Leg Platform Technology—Historical and Recent Developments. In Proceedings of the Offshore Technology Conference, Rio de Janeiro, Brasil, 29–31 October 2013; pp. 1–18.
31. Roddier, D.; Cermelli, C.; Aubault, A.; Weinstein, A. WindFloat: A Floating Foundation for Offshore Wind Turbines. *J. Renew. Sustain. Energy* **2010**, *2*, 033104. [CrossRef]
32. Lee, J.Y.; Koo, B.J.; Samsung Heavy Industries Co. Ltd; Clauss, G.; Berlin, T.U. Automated Design of a Tension Leg Platform with Minimized Tendon Fatigue Damage and Its Verification by a Fully Coupled Analysis. *Ship Technol. Res.* **2007**, *54*, 11–27. [CrossRef]
33. Yatish, T.S. Modular Wind Energy Systems. In *Modular Systems for Energy and Fuel Recovery and Conversion*; CRC Press Taylor & Francis Group: Boca Raton, FL, USA, 2019; pp. 371–409. ISBN 978-0-367-23512-3.
34. Zavvar, E.; Chen, B.Q.; Uzunoglu, E.; Guedes Soares, C. Stress Distribution on the CENTEC-TLP in Still Water and Rated Wind Speed. In *Trends in Maritime Technology and Engineering*; CRC Press: Boca Raton, FL, USA, 2022; pp. 519–527.
35. *DNVGL-ST-0119*; Floating Wind Turbine Structures. DNV-GL: Bærum, Norway, 2018.
36. Vittori, F.; Pires, O.; Azcona, J.; Uzunoglu, E.; Guedes Soares, C.; Zamora Rodríguez, R.; Souto-Iglesias, A. Hybrid Scaled Testing of a 10MW TLP Floating Wind Turbine Using the SiL Method to Integrate the Rotor Thrust and Moments. In *Developments in Renewable Energies Offshore*; Guedes Soares, C., Ed.; Taylor & Francis Group: London, UK, 2021; pp. 417–423. ISBN 9780367681319.
37. Uzunoglu, E.; Soares, C.G. An Integrated Design Approach for a Self-Float Capable Tension Leg Platform for Wind Energy. In *Developments in Maritime Technology and Engineering*; Guedes Soares, C., Santos, T.A., Eds.; Taylor & Francis Group: London, UK, 2021; pp. 673–682. ISBN 9780367773779.
38. Short, W.; Packey, D.; Holt, T. *A Manual for the Economic Evaluation of Energy Efficiency and Renewable Energy Technologies*; National Renewable Energy Laboratory: Golden, CO, USA, 1995.
39. Castro-Santos, L.; Martins, E.; Guedes Soares, C. Cost Assessment Methodology for Hybrid Floating Offshore Renewable Energy Platforms. *Renew. Energy* **2016**, *97*, 866–880. [CrossRef]
40. Castro-Santos, L.; Silva, D.; Bento, A.R.; Salvação, N.; Guedes Soares, C. Economic Feasibility of Floating Offshore Wind Farms in Portugal. *Ocean Eng.* **2020**, *8*, 58. [CrossRef]
41. Castro-Santos, L.; Diaz-Casas, V. Economic Influence of Location in Floating Offshore Wind Farms. *Ocean Eng.* **2015**, *107*, 13–22. [CrossRef]

42. Castro-Santos, L.; Bento, A.R.; Silva, D.; Salvação, N.; Guedes Soares, C. Economic Feasibility of Floating Offshore Wind Farms in the North of Spain. *J. Mar. Sci. Eng.* **2020**, *8*, 18. [CrossRef]
43. Green, R.; Vasilakos, N. The Economics of Offshore Wind. *Energy Policy* **2011**, *39*, 496–502. [CrossRef]
44. Colmenar-Santos, A.; Campíñez-Romero, S.; Pérez-Molina, C.; Mur-Pérez, F. Repowering: An Actual Possibility for Wind Energy in Spain in a New Scenario without Feed-in-Tariffs. *Renew. Sustain. Energy Rev.* **2015**, *41*, 319–337. [CrossRef]
45. Salvação, N.; Guedes Soares, C. Wind Resource Assessment Offshore the Atlantic Iberian Coast with the WRF Model. *Energy* **2018**, *145*, 276–287. [CrossRef]
46. Salvação, N.; Bernardino, M.; Guedes Soares, C. Assessing Mesoscale Wind Simulations in Different Environments. *Comput. Geosci.* **2014**, *71*, 28–36. [CrossRef]
47. Campos, R.M.; Guedes Soares, C. Spatial Distribution of Offshore Wind Statistics on the Coast of Portugal Using Regional Frequency Analysis. *Renew. Energy* **2018**, *123*, 806–816. [CrossRef]
48. Campos, R.M.; Guedes Soares, C. Assessment of Three Wind Reanalyses in the North Atlantic Ocean. *J. Oper. Oceanogr.* **2017**, *10*, 30–44. [CrossRef]
49. Díaz, H.; Guedes Soares, C. An Integrated GIS Approach for Site Selection of Floating Offshore Wind Farms in the Atlantic Continental European Coastline. *Renew. Sustain. Energy Rev.* **2020**, *134*, 110328. [CrossRef]
50. Díaz, H.; Guedes Soares, C. A Multi-Criteria Approach to Evaluate Floating Offshore Wind Farms Siting in the Canary Islands (Spain). *Energies* **2021**, *14*, 865. [CrossRef]
51. Diaz, H.; Guedes Soares, C. A Novel Multi-Criteria Decision-Making Model to Evaluate Floating Wind Farm Locations. *Renew. Energy* **2021**, *185*, 431–454. [CrossRef]
52. Diaz, H.; Palos Teixeira, A.; Guedes Soares, C. Application of Fuzzy and Monte Carlo Analytic Hierarchy Processes for Ranking Floating Wind Farm Locations. *Ocean Eng.* **2022**, *245*, 110453. [CrossRef]
53. Rokicki, T.; Bórawski, P.; Gradziuk, B.; Gradziuk, P.; Mrówczyńska-Kamińska, A.; Kozak, J.; Guzal-Dec, D.J.; Wojtczuk, K. Differentiation and Changes of Household Electricity Prices in EU Countries. *Energies* **2021**, *14*, 6894. [CrossRef]
54. Baita-Saavedra, E.; Cordal-Iglesias, D.; Filgueira-Vizoso, A.; Morató, À.; Lamas-Galdo, I.; Álvarez-Feal, C.; Carral, L.; Castro-Santos, L. An Economic Analysis of An Innovative Floating Offshore Wind Platform Built with Concrete: The SATH® Platform. *Appl. Sci.* **2020**, *10*, 3678. [CrossRef]
55. Uzunoglu, E.; Guedes Soares, C. Response Dynamics of a Free-Float Capable Tension Leg Platform for a 10 MW Wind Turbine at the Northern Iberian Peninsula. In *Developments in Renewable Energies Offshore*; Guedes Soares, C., Ed.; Taylor & Francis Group: London, UK, 2021; pp. 408–416. ISBN 9780367681319.
56. Tran, T.T.; Kim, D.H. The Platform Pitching Motion of Floating Offshore Wind Turbine: A Preliminary Unsteady Aerodynamic Analysis. *J. Wind Eng. Ind. Aerodyn.* **2015**, *142*, 65–81. [CrossRef]
57. Ioannou, A.; Angus, A.; Brennan, F. Stochastic Prediction of Offshore Wind Farm LCOE through an Integrated Cost Model. *Energy Procedia* **2017**, *107*, 383–389. [CrossRef]
58. Bruck, M.; Sandborn, P.; Goudarzi, N. A Levelized Cost of Energy (LCOE) Model for Wind Farms That Include Power Purchase Agreements (PPAs). *Renew. Energy* **2018**, *122*, 131–139. [CrossRef]

Journal of
Marine Science and Engineering

MDPI

Article

Integrated Dynamics Response Analysis for IEA 10-MW Spar Floating Offshore Wind Turbine

Xiaojiang Guo [1,†], Yu Zhang [2,†], Jiatao Yan [3], Yiming Zhou [1], Shu Yan [1], Wei Shi [4,5] and Xin Li [6,*]

1 Huaneng Clean Energy Research Institute, Beijing 102209, China; xj_guo@qny.chng.com.cn (X.G.); ym_zhou@qny.chng.com.cn (Y.Z.); s_yan@qny.chng.com.cn (S.Y.)
2 Department of Civil Engineering, Faculty of Infrastructure Engineering, Dalian University of Technology, Dalian 116024, China; zyu0608@mail.dlut.edu.cn
3 South Branch of China Huaneng Group Co., Ltd., Guangzhou 510620, China; b_zhang@qny.chng.com.cn
4 Deepwater Engineering Research Centre, State Key Laboratory of Coastal and Offshore Engineering, Dalian University of Technology, Dalian 116024, China; weishi@dlut.edu.cn
5 Research Institute, Dalian University of Technology, Shenzhen 518057, China
6 Institute for Earthquake Engineering, Faculty of Infrastructure Engineering, Dalian University of Technology, Dalian 116024, China
* Correspondence: lixin@dlut.edu.cn; Tel.: +86-41184708709
† These authors contributed equally to this work.

Abstract: Wind energy in the deep-sea area is more abundant and the capacity of wind turbines can be made larger. Therefore, the research on deep-sea floating offshore wind turbines will be the primary strategy for wind energy exploitation in the future. The spar-type platform depends on the characteristics of a small water plane, deep draft, and good stability, which has been applied to the commercial development of deep-sea wind energy. In the next ten years, the 10-MW wind turbine will become the mainstream class installed in the floating offshore wind turbine farm. Thus, it is very necessary to conduct a comprehensive and in-depth study on the 10-MW spar type floating offshore wind turbine. The direct-drive 10-MW offshore wind turbine was proposed by the International Energy Agency (IEA) in Wind Task 37 in 2019. In this paper, a spar-type platform is designed to support the IEA 10-MW reference wind turbines, and a nonlinear aero-hydro-servo-elastic numerical model is established using the FAST tool (which is developed by the National Renewable Energy Laboratory, NREL). Then, the accuracy of the wind turbine and the sensitivity of the controller are verified, and the natural periods of the floating offshore wind turbine are obtained by free-decay tests. The natural periods of the platform in six degrees-of-freedom are found to be within the range recommended by the design standard. The measured wind and wave data of the target site close to Fujian Province of China are used to evaluate the performance of the floating offshore wind turbine under the 100-, 50-, 5-, and 2-year-return stochastic weather conditions. The results indicate that the design of the spar platform is reasonable and has excellent hydrodynamic performance.

Keywords: floating offshore wind turbine; IEA 10-MW; spar type platform; FAST; dynamic response; integrated analysis

Citation: Guo, X.; Zhang, Y.; Yan, J.; Zhou, Y.; Yan, S.; Shi, W.; Li, X. Integrated Dynamics Response Analysis for IEA 10-MW Spar Floating Offshore Wind Turbine. *J. Mar. Sci. Eng.* **2022**, *10*, 542. https://doi.org/10.3390/jmse10040542

Academic Editors: Emre Uzunoglu, Antonio Souto-Iglesias and Carlos Guedes Soares

Received: 11 February 2022
Accepted: 11 April 2022
Published: 14 April 2022

1. Introduction

In recent decades, marine renewable energy has been paid more and more attention by governments worldwide, among which wind energy is the most promising clean energy. The development and utilization of wind energy will significantly slow down global warming and reduce CO_2 emissions [1,2]. The medium to exploit wind energy is the wind turbine and the primary offshore wind turbines in service have bottom-fixed foundations at present. To reduce and eliminate noise and visual pollution, the exploitation of offshore wind energy will move from shallow waters (water depths smaller than 50 m) to deeper waters in future research. Meanwhile, the costs of wind turbine construction and installation will increase with the water depth for the bottom-fixed foundations. Thus, the floating foundations will become an inevitable choice [3–5]. The commercial advantages of floating offshore wind turbines (FOWTs) in the deep-sea area are obvious, such as larger-scale wind turbines, superior wind resources in the deep sea, less noise, and visual pollution [6–8]. At present, the main types of offshore wind turbines are spar platform, tension-leg platform (TLP), semi-submersible platform, and barge platform. The spar platform has many advantages compared to the other three platforms. It has a better hydrodynamic performance because of the characteristics of a small water plane, deep draft, lower center of gravity; during the installation phase, many challenges of tension-leg installation will arise, while it is relatively convenient for the spar. Moreover, the spar platform is a simple structure and it has a lower steel consumption and construction cost. Therefore, the spar-type platform has been applied to the commercial exploitation of deep-water wind energy [9–12].

With the consideration of spar-type FOWTs developed earliest, it has become the most mature technology. At present, the spar-type FOWTs have been used for commercial power generation in England. The world's first spar-type FOWT prototype test at sea is called the Hywind Demo (Figure 1), and this test project began in 2009 and lasted eight years. The power capacity of the wind turbine is 2.3-MW, the draft depth is 100 m, the hub height is 65 m, the rotor diameter is 85 m, and the water depth is 220 m. The Hywind Demo produced 7.3 million kWh electricity in its first test year and successfully withstood wind speed of 40 m/s and wave height of 19 m over the test course of eight years [13]. In 2015, six years after the Hywind Demo prototype test, Equinor (called Statoil before), considering that the concept of Hywind was mature and cost-effective, could be used for commercial development. In 2017, the world's first commercial floating offshore wind farm, the Hywind Scotland (Figure 2), underwent construction. The power of the turbine is 6-MW, the draft depth is 78 m, the hub height is 98 m, the diameter of the rotor is 154 m, and the water depth is 95–120 m. The average wind speed is 10 m/s, and the average wave height is 1.8 m in this sea area. We can find that the draft depth and water depth of Hywind Scotland are all less than Hywind Demo, but the other performances of Hywind Scotland are all significantly superior to Hywind Demo [14].

Figure 1. Hywind Demo.

Figure 2. Hywind Scotland.

If the scale of Hywind Scotland is relatively small and still in an experimental phase, then the upcoming project Hywind Tampen in Norway will further demonstrate the reliability of the Hywind scheme. The Hywind Tampen project is different from the wind farm constructed previously. The Norwegian residents are not the consumer of the electricity generated by the Hywind Tampen farm. The five Oil & Gas platforms, Gullfaks and Snorre, that are nearby will be the main electricity user (Figure 3). The construction of the Hywind Tampen wind farm began on 1 October 2020 and is expected to be completed in the second half of 2022. The power of the turbine is 8-MW, the diameter of the rotor is 167 m, and the water depth is 260–300 m. In addition, the foundation (spar platform) of the Hywind Tampen project will be constructed with concrete instead of steel, which will lead to a decrease of 40% in cost compared with Hywind Scotland. Hywind is the first one commercial floating wind turbine technology, but the costs are still significantly higher than the bottom-fixed foundations located in shallow water. Thus, there are still many efforts that need to be made before the construction investment is acceptable and the FOWTs technology is suitable for the deep-water [15]. In addition to the spar-type floating platform, the semisubmersible floating platform has also been used in offshore wind farms as pilot park projects, such as the 25 MW offshore wind farm WindFloat Atlantic project [16] and 50 MW Kincardine project [17].

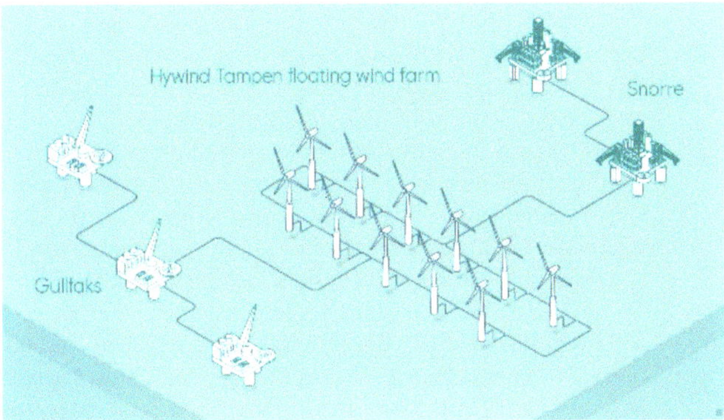

Figure 3. Hywind Tampen.

From the introductions above, it can be concluded that the 10-MW class wind turbines will be the trend of FOWTs in the next ten years. This is because large-scale wind turbines are believed to be advantageous from an investment and installation perspective [18]. Therefore, the research on the 10-MW FOWTs has great significance and will be a guide for future practical projects. In 2013, Technical University of Denmark (DTU) published the DTU 10-MW Reference Wind Turbine [19], later, in 2019, IEA also proposed an offshore direct-drive 10-MW wind turbine. However, due to the DTU 10-MW published previously, there is a lot of research on it.

Many efforts have been performed by scholars of Norwegian University of Science and Technology (NTNU) on 10-MW wind turbines, they designed many foundations to support the DTU 10-MW reference wind turbines (RWT), including bottom-fixed foundations, semi-submersible type floater [20,21], TLP type floater [22,23], spar type floater [24–26]. Son et al. [27] designed a new WindFloat type semi-submersible platform to support the DTU 10-MW wind turbine. The OpenFAST and OrcaFAST are used to predict the performance of the WindFloat, and the numerical results show that the platform has an excellent hydrodynamic performance. Shin et al. [28,29] also designed a semi-submersible platform to support the DTU 10-MW RWT and a scale ratio of 1:90 model test was performed at the same time, the characteristics such as the effective amplitude response operator (RAO) and six degree of freedoms (DOFs) motion of platform under regular waves and irregular waves were examined. Especially the response of the platform under the action of the 2nd order wave effect was focused on in their research. Leble and Barakos [30] investigated the possibility of a 10-MW wind turbine entering a vortex ring state during pitching and yawing motion. The results show that the vortex ring state is existent during oscillations and has a strong effect on the frequency and amplitude of wind turbines. Jeong and Kim [31] based on the computational fluid dynamics method to explore the flatback airfoil effect on the 10-MW wind turbine blade.

Considering the low material costs and expected high durability, concrete is treated as an alternative to steel for the construction material for floating platforms of wind turbines, and Equinor also plans to use concrete as substructures for the Hywind Tampen project. Oh et al. [32] designed a spar-type platform using prestressed concrete to support the DTU 10-MW; the diameter of the spar is 16 m and the wall thickness is 0.45 m. Serafeim et al. [33] used the passive load control methods to reduce the blade weight of DTU 10-MW. Through the optimization design of the structure, the blade mass and the bending moment at the root of the blade are reduced by 10% and 5%, respectively. Bredmose et al. [34] designed a Triple Spar platform, which is a hybrid between spar and semi-submersible with the advantages of both of them. A 1:60 scale model test was performed under the acting of wind and wave, and the active blade pitch control first was interesting. From the above literature review, we can know that the research on DTU 10-MW is relatively abundant, while the research on IEA 10-MW is scarce, this is because the DTU 10-MW was released early (in 2013) than IEA 10-MW (in 2019). Recently, the IEA 10-MW has been widely investigated and has often been used as a benchmark; hence, it has also been used in this research. The main aim of this research is to verify the feasibility of the spar platform to support the IEA 10MW based on Chinese local environmental conditions.

The current design of the floating platform supporting a 10-MW wind turbine is of significant interest to the industry as it is a necessary milestone to develop deeper water sites commercially. However, academic research on 10-MW wind turbines is not very abundant. In this paper, the wind turbine version we used is the IEA 10-MW RWT and the spar platform and mooring system are redesigned based on Bachynski's research, while the OC3-Hywind [35] is used as a reference at the same time. The combination of IEA 10-MW RWT and this newly designed spar platform is then studied. The environmental conditions of the target site close to Fujian Province of China are used to evaluate the performance of the FOWT. Thus, to implement this research, the organization structure of this paper is organized as follows:

In Section 2, the parameters of the IEA 10-MW wind turbine and the spar platform and mooring line are introduced. In Section 3, the numerical model of the IEA 10-MW spar type FOWT is established in FAST v8 [36] and the accuracy of the model and the sensitivity of the controller are verified. In Section 4, the platform motion, nacelle acceleration, and the mooring tension of the numerical model under some typical cases are performed. The main conclusions and future works are presented in Section 5.

2. Specification of IEA 10-MW and Spar Platform

2.1. Design Parameters of the IEA 10-MW

From the IEA report [37], it still has several areas for potential improvement for DTU 10-MW. The controller, for example, still needs to be sufficiently developed as the drivetrain design is not detailed, and although it is a better direct-drive version, the structural design of the blade is not detailed. Thus, the main objective of the IEA Wind Task 37 is to solve these common questions. In 2019, the IEA 10-MW RWT was released, and the overall characteristics of the offshore turbine designed for the IEA Wind Task 37 are based on the DTU 10-MW RWT. The turbine is developed to have a rated electrical power of 10-MW, and this turbine is a direct-drive version generator. The key parameters of the IEA 10-MW compared to the DTU 10-MW RWT are listed in Table 1. The steady-state operational data for the 10-MW turbine are listed in Table 2.

Table 1. Key parameters of the IEA 10-MW compared to the DTU 10-MW RWT.

Parameter	Value	Comment
Wind Regime	IEC class 1A	Same as DTU 10-MW RWT
Rotor Orientation	Clockwise rotation—Upwind	Same as DTU 10-MW RWT
Control	Variable Speed Collective Pitch	Same as DTU 10-MW RWT
Cut-in wind speed	4 m/s	Same as DTU 10-MW RWT
Cut-out wind speed	25 m/s	Same as DTU 10-MW RWT
Rated wind speed	11 m/s	Optimized
Rated electrical power	10 MW	Same as DTU 10-MW RWT
Number of blades	3	Same as DTU 10-MW RWT
Rotor Diameter	198.0	Optimized
Airfoil series	FFA-W3	Same as DTU 10-MW RWT
Hub Diameter	4.6 m	Reduced from 5.4 m
Hub Height	119.0 m	Same as DTU 10-MW RWT
Drivetrain	Direct-drive	Changed from Medium Speed, Multiple-Stage Gearbox
Minimum Rotor Speed	6.0 rpm	Same as DTU 10-MW RWT
Maximum Rotor Speed	8.68 rpm	Constrained by max tip speed
Gearbox Ratio	N/A	Direct-drive
Maximum Tip Speed	90.0 m/s	Same as DTU 10-MW RWT
Hub Overhang	7.1 m	Same as DTU 10-MW RWT
Shaft Tilt Angle	6.0 deg.	Increased from 5 deg
Rotor Precone Angle	−4 deg.	Increased from −2.5 deg
Blade Prebend	6.2 m	Increased from 3.2 m
Blade Mass	47,700 kg	12% increase from DTU 10-MW RWT
Nacelle Mass	542,600 kg	-
Tower Mass	628,442 kg	Provisional, same as DTU 10-MW RWT

Table 2. Operational summary of the IEA 10-MW rotor.

Parameter	Value	Parameter	Value
Maximum C_P (–)	0.49	Min Rotor Speed (rpm)	6.0
Design tip speed ratio (TSR) (–)	10.58	Design Pitch (deg)	0
Rated Mechanical Power (MW)	10.6383	Omega Rated (rpm)	8.68
Rated Torque (MNm)	11.704	Max Tip Speed (m/s)	90.0
Cut-in Wind Speed (m/s)	4.0	Cut-out Wind Speed (m/s)	25.0
Wind Speed Region II1/2 (m/s)	8.5	Rated Wind Speed (m/s)	10.75

2.2. Parameters of the Spar Platform and Mooring System

Here we are using a spar platform designed based on the OC3 Hywind and NTNU [38], and the draft is the same as the original OC3 spar. But the diameters of two cylinders are increased to provide enough buoyancy. The bottom of the steel cylinder is filled with concrete ballast to achieve the expected draft. The density of concrete ballast is 2600 kg/m^3, and the density of steel is 7850 kg/m^3. The wall thickness of the steel cylinder is 6 cm. The project of the mooring system is also the same as the OC3 Hywind, which consists of three catenary lines. An added yaw stiffness from the mooring system is represented by a rotational spring. Based on previous experience, if the fairleads are located at a depth of the total center of gravity (CoG) of the system, it will reduce the coupling between surge and pitch motion; thus, the fairlead depth is designed as 77.0 m below still water level (SWL). Based on the reference [38], the platform mass was modified from 1.21×10^7 kg to 1.381×10^7 kg, and the mooring line mass density was modified from 233.1198 kg/m to 245.4 kg/m. Other parameters can be found in Tables 3 and 4 for detailed information.

Table 3. Parameters of the spar platform.

Parameter	Value
Water depth (m)	320
Draft (m)	120
Elevation to tower base above SWL (m)	10
Depth to top of taper below SWL (m)	4
Depth to bottom of taper below SWL (m)	12
Diameter above taper (m)	8.5
Diameter below taper (m)	13
Mass including ballast (kg)	1.381×10^7
Displacement (m^3)	1.53×10^4
Roll moment of inertia about CoG (kg·m^2)	9.6899×10^9
Pitch moment of inertia about CoG (kg·m^2)	9.6899×10^9
Yaw moment of inertia about CoG (kg·m^2)	3.3589×10^8
Vertical CoG below SWL (m)	95.5
Vertical CoB below SWL (m)	62.3

Table 4. The parameters of the mooring line.

Parameter	Value
Radius to anchors (m)	855.2
Unstretched mooring line length (m)	902.2
Radius to fairleads from spar centerline (m)	6.5
Mooring line diameter (m)	0.09
Equivalent mooring line mass density (kg/m)	245.4
Equivalent mooring line axial stiffness (N)	3.84×10^8
Fairlead depth below SWL (m)	70.0
Yaw spring stiffness (Nm/rad)	1.48×10^8

3. Model Validation

3.1. Modeling in FAST

The numerical model of the spar FOWT was built using the fully coupled aero-hydro-servo-elastic-mooring simulation tool FAST. The external force acting on the FOWT system mainly comes from the aerodynamic loads acting on the wind turbine and the hydrodynamic loads acting on the platform, as well as the mooring tension from the mooring system. The periodic aerodynamic forces and randomly fluctuating aerodynamic forces, which are calculated in the AeroDyn module within the FAST code. The influence of the wake and the blade airfoil aerodynamics are calculated based on Blade Element Momentum (BEM) theory. The effect of the tower on the blade is accounted for in the AeroDyn module based on a combination of potential flow and tower shadow models.

The hydrodynamic and hydrostatic induced load and the mooring tension are calculated by HydroDyn and MoorDyn, respectively. The additional linear damping model in OC3 Hywind is used here to calculate the additional damping force. The viscous drag coefficient, C_D, is taken to be 0.7, according to DNVGL-ST-0437 [39]. The typical wave spectrum, JONSWAP, was used to determine the irregular wave condition in the HydroDyn input file. The hydrodynamic load is calculated by the potential flow theory. The hydrodynamic-added-mass and -damping matrices and hydro-static matrix were obtained from the hydrodynamic analysis tool, ANSYS AQWA [40]. The hydrodynamic coefficients output files from AQWA are transformed to WAMIT for-mat and fed into the FAST main program.

3.2. The Operational Indicators of Turbine

In this section, the 6 DOFs of the floating platform are limited to zero, then the generator power, blade pitch angle, rotor speed, and rotor thrust of the IEA 10-MW wind turbine are calculated under the action of different steady-state wind. The direction of wind along with the positive of the x-axis. Figure 4 is the schematic diagram of the coordinate system in FAST.

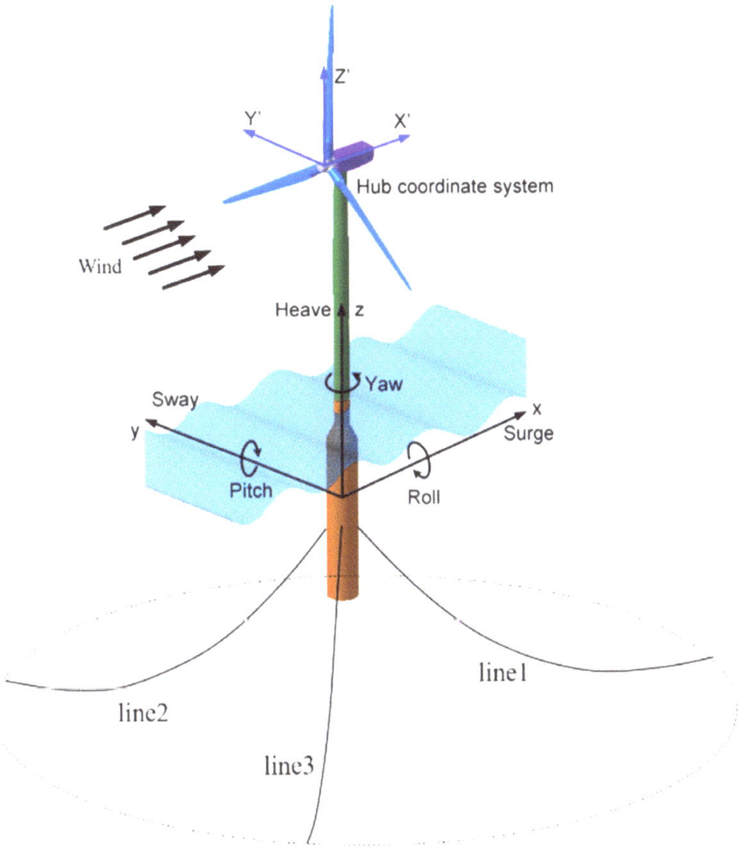

Figure 4. A sketch of the coordinate system in FAST.

It can be seen from Figure 5 that when the wind speed is less than the rated wind speed, the control strategy is to keep the blade pitch angle always at 0° and to increase the rotor speed to achieve the optimum generator power. When the wind speed reaches the rated value, the blade pitch angle controller starts working, the generator power, rotor thrust, and rotor speed reach the maximum value; then, as the wind speed continues to increase, the generator power and rotor speed remain unchanged, the blade pitch angle increases with the increase in wind speed, but the rotor thrust decreases with the increase in wind speed. As IEA 10MW is designed as a direct drive, thus, the rotor speed is lower than DTU 10MW. The rotor thrust of IEA 10MW reaches the maximum when the wind speed is 10.5 m/s, while DTU 10MW is 11.5 m/s.

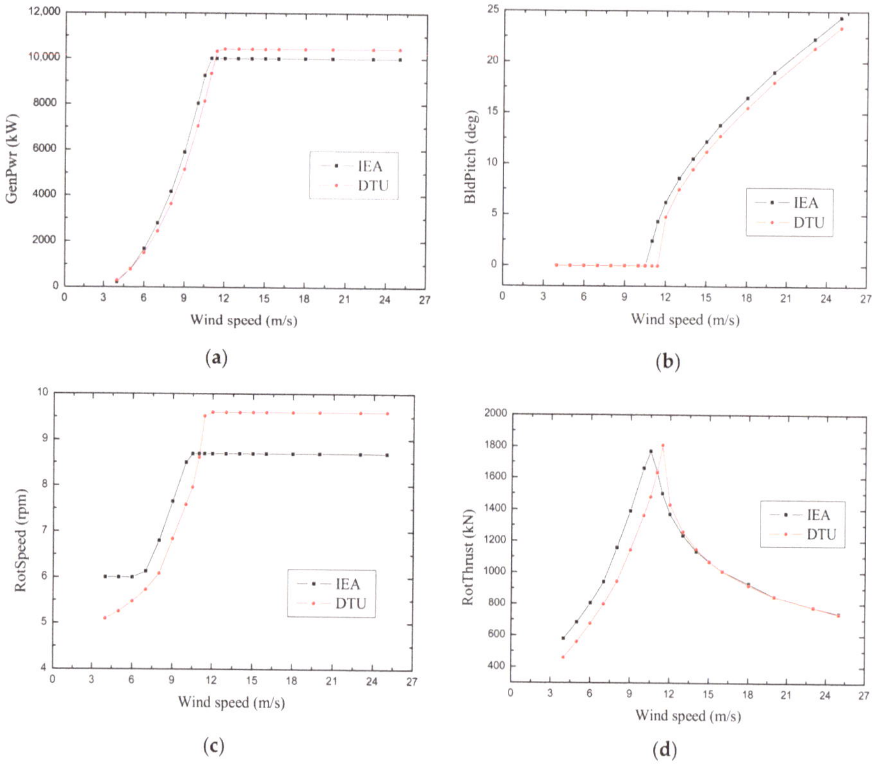

Figure 5. Steady-state performance and operation of the IEA 10-MW offshore wind turbine: (**a**) Generator power vary with the wind speed; (**b**) Bladed pitch angle varies with the wind speed; (**c**) Rotor rotation speed varies with the wind speed; (**d**) Rotor thrust varies with the wind speed.

3.3. Controller Sensitivity

The ROSCO controller [41] is used for the IEA 10-MW wind turbine; here, its sensitivity is tested. The first case is a 9 m/s steady-state wind with a pulse of 5 m/s; the second case is a step steady-state wind from a 10 m/s jump to 12 m/s. The six DOFs of platform motion are free. The responses of the controller are shown in Figure 6. From the results, we can see that the controller reacts quickly to the wind change; thus, the controller is sensitive.

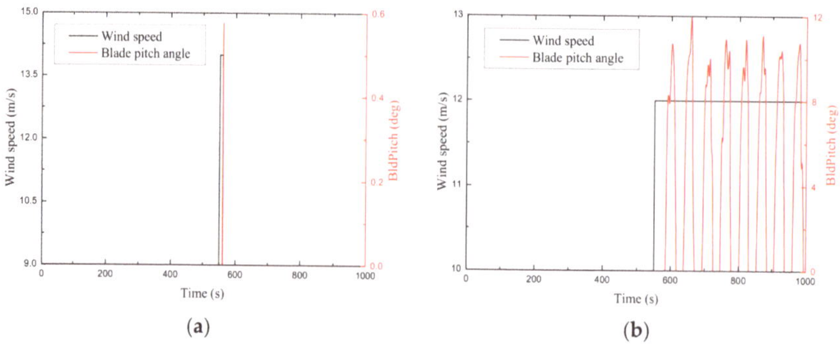

Figure 6. Blade pitch angle control with the wind speed varying: (**a**) Steady-state wind with a pulse; (**b**) Steady-state wind with a jump.

3.4. Natural Periods

The major energy contained in ocean waves usually ranges from 5 to 25 s of the wave spectral period; thus, in the design process, the natural periods of the FOWT should avoid this particular wave spectral range. This is because keeping the natural periods of FOWT out of the major wave spectral period can avoid the resonance problem. Usually, the natural period of the heave motion of the spar platform is around 20 to 25 s and the pitch and roll motion are around 30 s. The natural periods of the spar platform can be obtained by decay analysis. Due to the symmetry of the spar structure, only surge, heave, pitch, and yaw need to be considered for the decay tests. The decay curves are shown in Figure 7 and the natural periods are listed in Table 5. According to the performance of the decay test, the natural periods of the IEA 10-MW spar type FOWT established in this paper are acceptable.

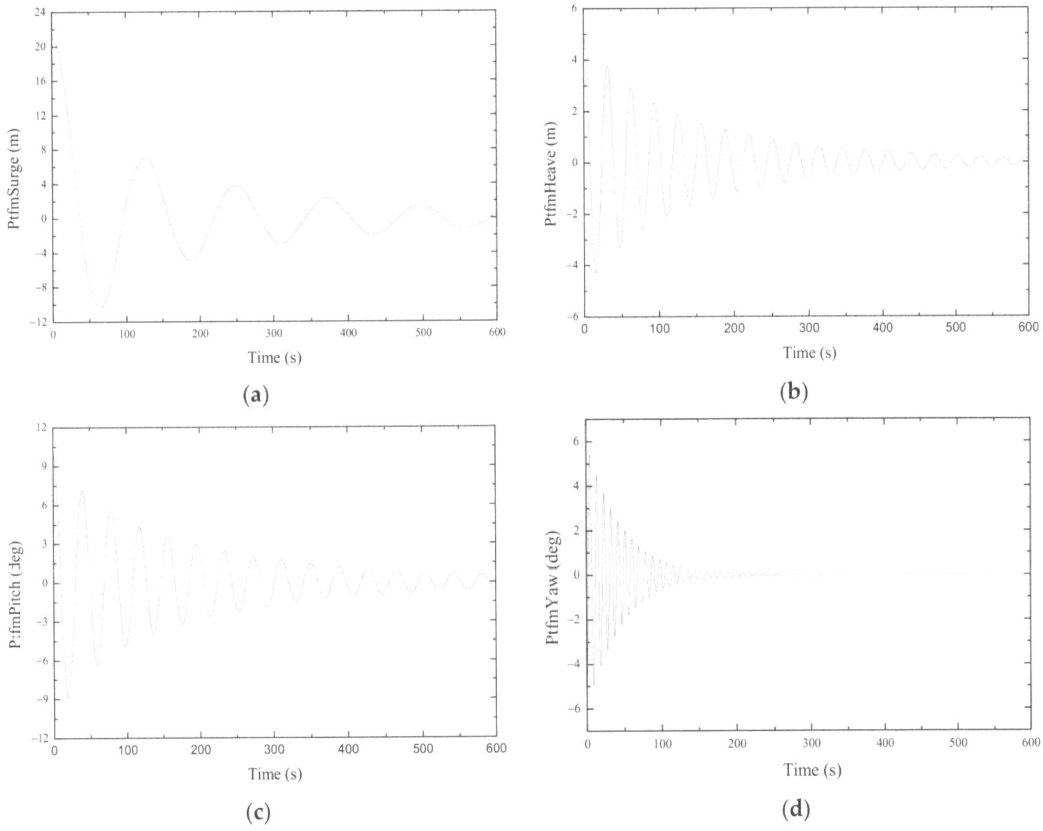

Figure 7. Time series of free decay tests: (**a**) Platform surge motion; (**b**) Platform heave motion; (**c**) Platform pitch motion; (**d**) Platform yaw motion.

Table 5. The natural periods of the IEA 10-MW spar type FOWT from decay tests.

DOFs		Surge	Heave	Pitch	Yaw
Natural period (s)	IEA 10-MW	103.8	33.4	26.3	8.1
	DTU 10-MW	103.3	31.3	35.5	7.6
	DNVGL [42]	~100 (catenary)	25–40	25–40	5–20

4. Results and Discussions

4.1. Environmental Parameters

Under the operation condition, the FOWT is subjected to the combined action of wind and wave loads. Thus, the performance of the platform motion response should deserve more attention under the coupled action of wind and waves. To comprehensively investigate the survivability of the FOWT, the operational and extreme environmental conditions for 100-, 50-, 5-, 2-years return periods are considered. The JONSWAP spectrum defines the random waves; there are three current models in FAST; here, we use the Near-Surface model. The waves and current models are plotted in Figure 8. The turbulence wind is generated by TurbSim [43], the Normal Turbulence Model (NTM), and the class C defined in the IEC standard for turbulence wind is used here for offshore condition turbulent [44]. The wind speed in hub height and the speed distributed over the whole wind farm are shown in Figure 9. The significant wave height, the wave peak periods, the current speed, and the mean wind speed are listed in Table 6. A total of 24 typical dynamic load cases (LC) are defined; these LCs are determined based on the wind and wave data measured at the target site close to Fujian Province, China, provided by Huaneng Clean Energy Research Institute. The current velocity is defined based on Zhang's research [45]. Only one direction of the wind, wave, and current is considered because the spar platform is a cylinder, the direction of the wind, wave, and current is traveling along the *x*-axis positive. For LC13 to LC24, the turbine is set to be parked and the pitch angles of the blade are feathered to 90° as the mean wind speed exceeds the cut-out speed. Each case is run for 4800 s, while the first 1200 s are removed to exclude the transient effect of the startup.

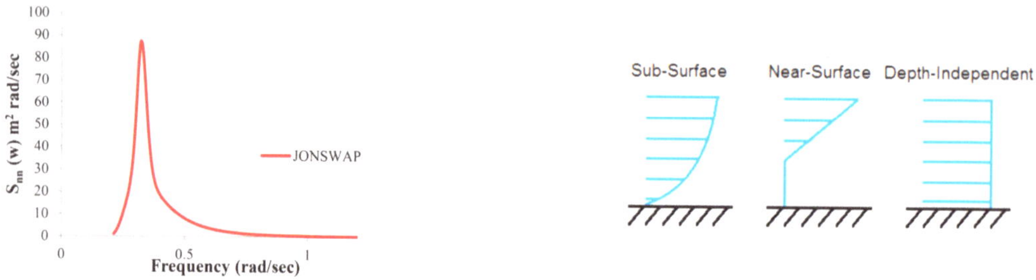

(a) (b)

Figure 8. The wave and current information. (**a**) JONSWAP spectrum of waves; (**b**) standard current sub-models.

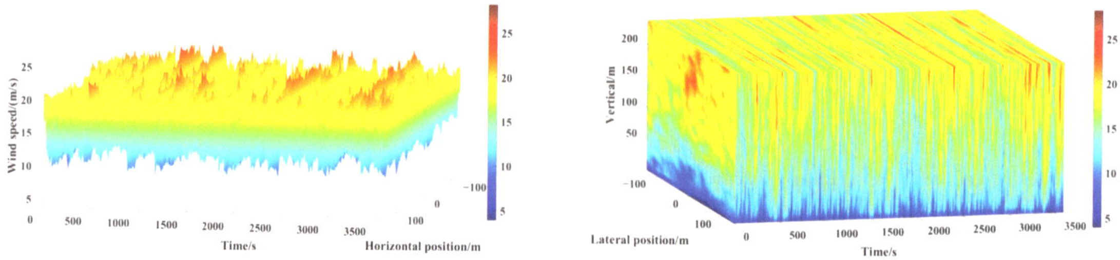

(a) (b)

Figure 9. The turbulence wind generated in TurbSim: (**a**) Wind speed in hub height; (**b**) wind speed distributed over the whole wind farm.

Table 6. The parameters of wind-wave combined dynamic load cases.

Return Period (Year)	Load Case No.	Significant Wave Height (m)	Wave Peak Period (s)	Current Speed (m/s)	Mean Wind Speed (m/s)	Turbine Status
2	LC1	4.22	8.70	1.38	19.17	operating
	LC2	3.68	7.10		17.40	
	LC3	3.43	9.83		13.60	
	LC4	3.81	10.4		15.64	
	LC5	3.16	8.77		11.00	
	LC6	4.34	12.2		18.76	
5	LC7	5.10	11.1	1.51	20.00	operating
	LC8	6.21	9.80		24.92	
	LC9	5.94	8.20		21.67	
	LC10	5.47	13.6		23.37	
	LC11	4.99	12.2		25.15	
	LC12	6.42	10.0		22.45	
50	LC13	8.96	13.5	2.00	34.16	parked
	LC14	8.45	10.4		36.78	
	LC15	8.13	9.30		40.05	
	LC16	8.69	16.4		35.55	
	LC17	10.16	13.8		49.01	
	LC18	9.07	11.5		37.79	
100	LC19	8.61	10.3	2.20	51.80	parked
	LC20	8.45	9.30		44.18	
	LC21	8.83	9.90		59.29	
	LC22	11.38	10.6		60.02	
	LC23	9.38	11.4		41.45	
	LC24	10.76	16.3		70.00	

Firstly, the effect of the 2nd order wave loads on the spar platform is studied. The results are shown in Figure 10 and from the figures, we can see that both with or without considered 2nd order wave loads have little influence on the spar surge and pitch motion responses, while only the maximum and minimum values have slight differences, the mean-values, and standard deviation are equal. Thus, the 2nd order wave loads will be neglected in all dynamic load cases.

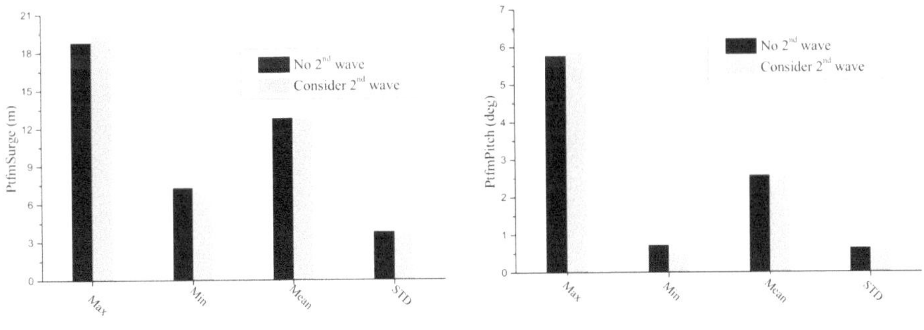

(a) (b)

Figure 10. The statistics values of platform responses with or without considered 2nd order wave load: (**a**) Platform surge motion; (**b**) platform pitch motion.

4.2. Platform Motions Response

The maximum (Max), minimum (Min), mean value (Mean), and standard deviation (STD) of the platform motions in 6 DOFs for the 24 load cases are shown in Figures 11–16. The results demonstrate that this FOWT can survive in the 100-year return extreme weather conditions. Moreover, some conclusions can also be drawn. Considering the wave and wind are all propagating along the direction of x-axis positive; thus, the amplitude of the platform surge and pitch are very significant and the amplitude of the platform motion in the other 4 DoFs are relatively small. For surge and pitch motion, when the wind turbine is under operating, the platform motion responses are larger than that when the wind turbine is under parked. The maximum surge motion nearly reaches 25 m, the pitch angle nearly reaches 9°, and the surge amplitude is less than 1/10 of the water depth. For sway and heave and roll motion, when the wind turbine is under operating, the platform motion responses are smaller than that when the wind turbine is under parked. This is because the thrust and overturning moment generated by the wind leads to a significant platform motion. It also illustrates that the operation condition is the most dangerous condition for FOWTs. For yaw motion, because of the rotor gyroscopic effect, the platform motion responses of the wind turbine under operating are larger than when parked. In addition, the differences in platform motion responses with or without current are not significant. For all 24 load cases, the platform motion responses in LC3, LC4, LC5, and LC24 are the largest.

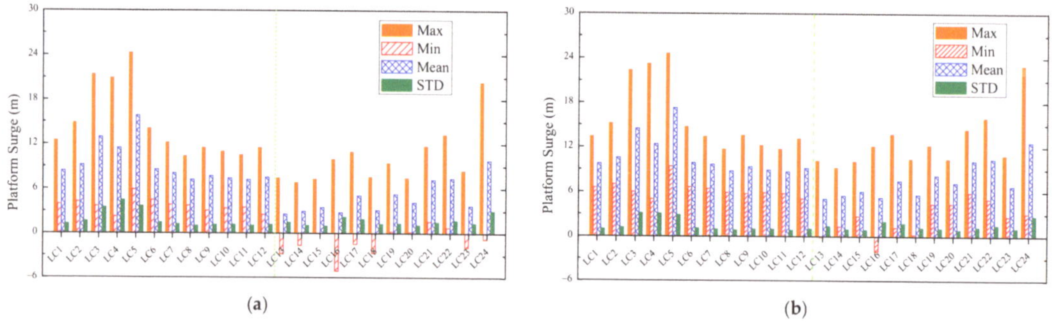

Figure 11. Statistics of platform surge motion response for (**a**) without current; (**b**) with current.

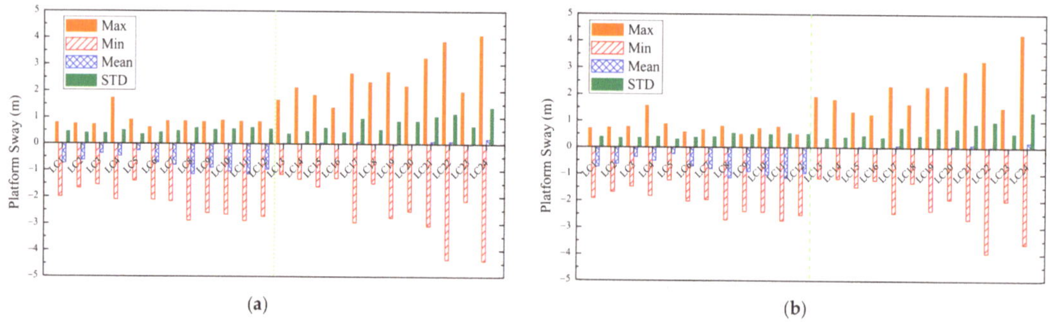

Figure 12. Statistics of platform sway motion response for (**a**) without current; (**b**) with current.

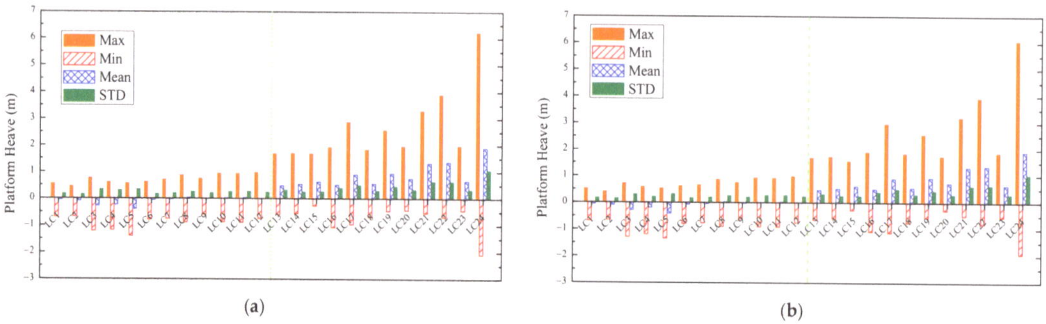

Figure 13. Statistics of platform heave motion response for (**a**) without current; (**b**) with current.

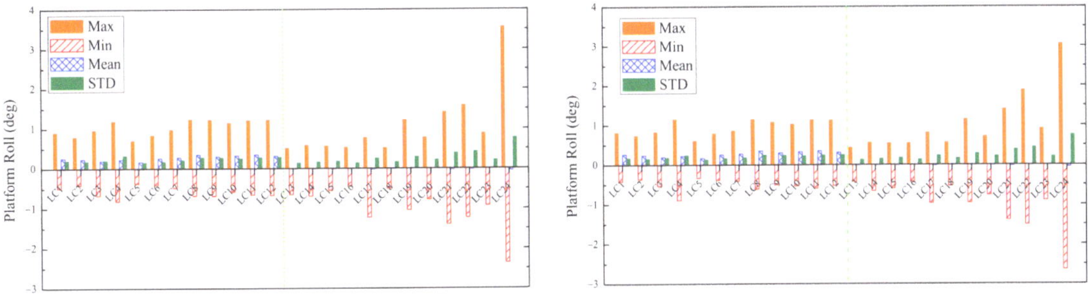

(a) (b)

Figure 14. Statistics of platform roll motion response for (**a**) without current; (**b**) with current.

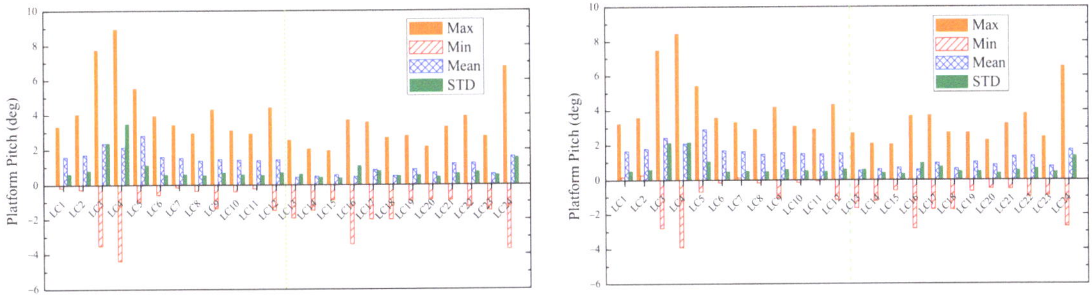

(a) (b)

Figure 15. Statistics of platform pitch motion response for (**a**) without current; (**b**) with current.

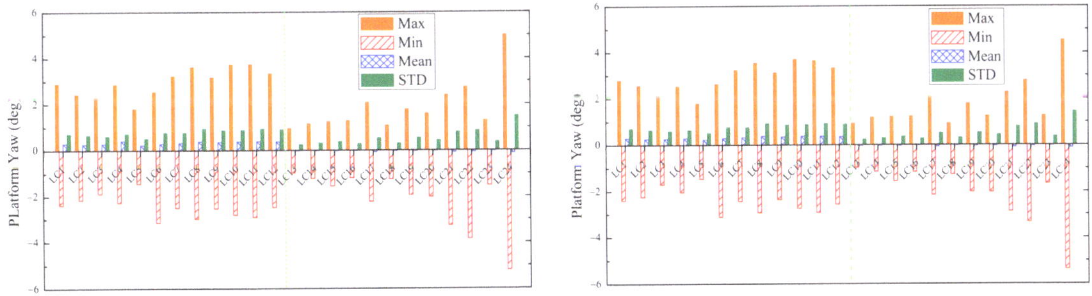

(a) (b)

Figure 16. Statistics of platform yaw motion response for (**a**) without current; (**b**) with current.

4.3. Nacelle Acceleration and Mooring Tension

The maximum, minimum, mean value, and standard deviation of the nacelle acceleration in x, y, and z-direction for the 24 load cases are shown in Figures 17–19. It can be observed that the amplitude of nacelle acceleration in the x-direction is larger than in the y-direction; the nacelle acceleration in the z-direction is the lowest. The nacelle accelerations in 100- and 50-year return stochastic weather conditions are generally larger than in 2- and 5-year return stochastic weather conditions. The largest value of nacelle acceleration is nearly 4.5 m/s appears in LC22 in the x-direction. The nacelle acceleration of LC24 is the largest in the y-direction. The mean values are all nearly zero. The variation pattern of nacelle acceleration is a little different from the platform motion for 24 load cases. The differences in nacelle acceleration responses with or without current are not significant.

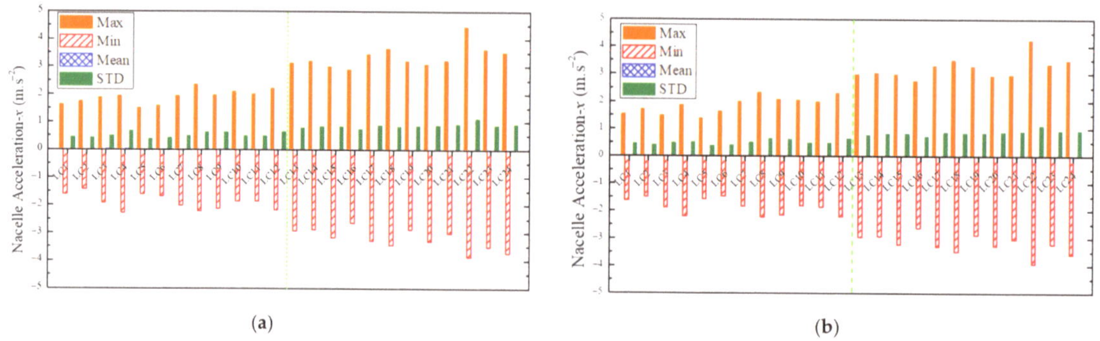

Figure 17. Statistics of nacelle acceleration response in x-direction for (**a**) without current; (**b**) with current.

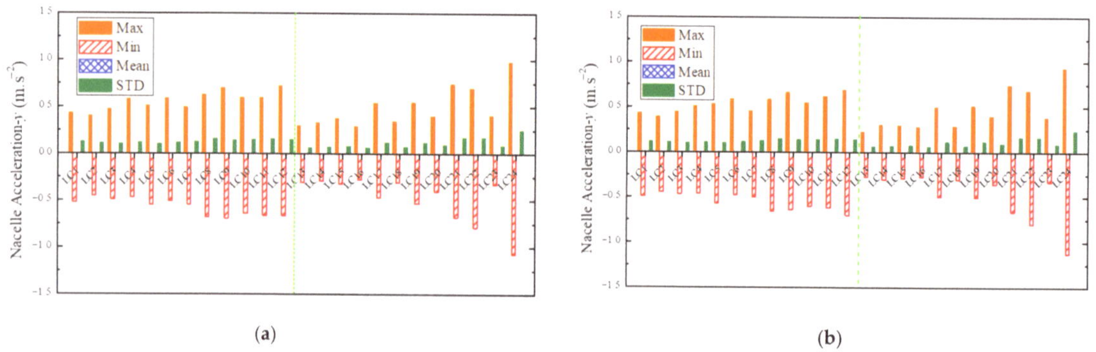

Figure 18. Statistics of nacelle acceleration response in y-direction for (**a**) without current; (**b**) with current.

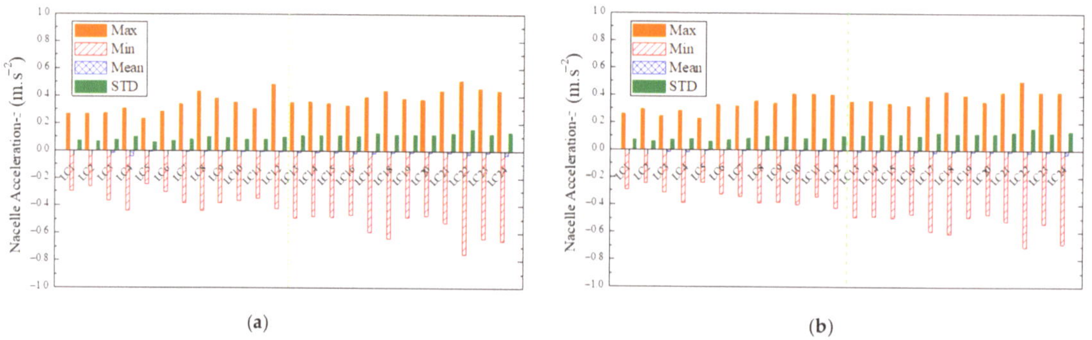

Figure 19. Statistics of nacelle acceleration response in z-direction for (**a**) without current; (**b**) with current.

The maximum, minimum, mean value, and standard deviation of the mooring tension of lines 1, 2, and 3 for the 24 load cases are shown in Figures 20–22. From Figure 4, we can see that line 1 is in the direction of wave propagation, the lines 2 and 3 are in the reverse direction of wave propagation. Thus, the mooring tension of lines 2 and 3 are greater than line 1. For line 1, the mooring tension of 100- and 50-year return stochastic weather conditions are generally larger than in 2- and 5-year return stochastic weather conditions. The largest mooring tensions appear in LC5 and LC24 for lines 2 and 3. Though lines 2 and 3 are arranged symmetrically on both sides of the x-axis, the mooring tension between them is a little different and this is because the platform has displacement in the y-direction, which can be seen in Figures 12 and 14. Moreover, we can also conclude that the fluctuation of the mooring tension is not significant, the standard deviation is close to zero and the maximum, minimum, and mean values are very close to each other. The fairlead position is near the CoG of the whole system; hence, the change of the mooring tension can reflect that the displacement of the CoG of the whole system is not very large. The differences in nacelle acceleration responses with or without current are not significant.

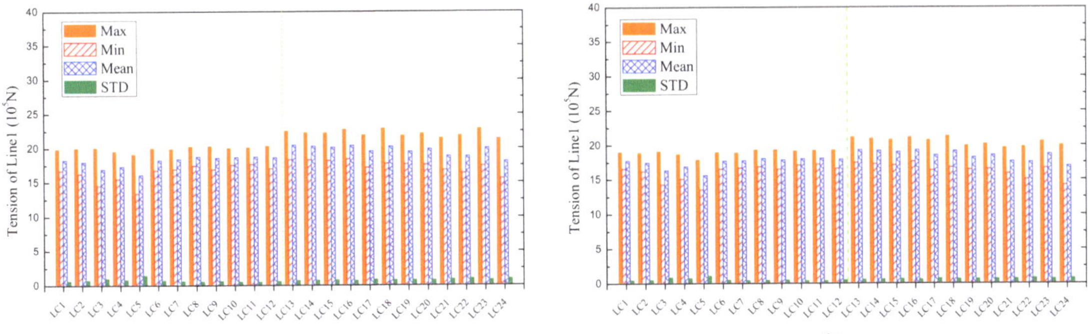

(a) (b)

Figure 20. Statistics of mooring tension of line 1 for (**a**) without current; (**b**) with current.

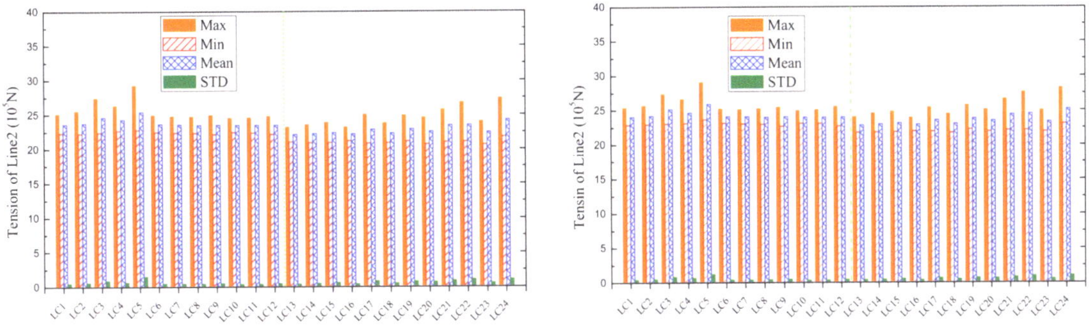

(a) (b)

Figure 21. Statistics of mooring tension of line 2 for (**a**) without current; (**b**) with current.

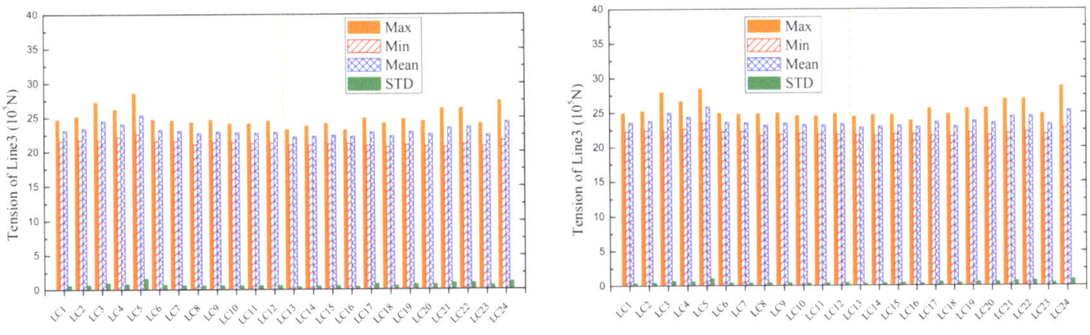

(a) (b)

Figure 22. Statistics of mooring tension of line 3 for (**a**) without current; (**b**) with current.

5. Conclusions

In recent years, offshore wind energy has experienced a huge development. Floating wind turbines may become competitive for water depths larger than 50 m. In this paper, a spar-type platform is used to support the IEA 10-MW RWT, and the numerical model is established in FAST. First, the correctness of the wind turbine and the sensitivity of the controller are verified. Then, the natural periods are obtained by the free decay test; the results have shown that the natural periods of this model avoid the major wave spectral period range. Finally, the fully coupled integrated load analysis is performed and 24 typical dynamic load cases are selected to test the stability of the spar type IEA 10-MW FOWT. The maximum motion response amplitude of the platform occurs at the 2-year-return stochastic weather conditions, and the motion response is almost within the acceptable scope. The nacelle acceleration and the mooring tension significantly respond at 100- and 50-year-return stochastic weather conditions. The comparison study of with and without current is implemented and the results show that the influence of the current is not significant. Furthermore, for the spar type platform, the effect is slight of platform motion response with or without considering the 2nd order wave loads. Throughout the development trend of FOWTs, 10-MW capacity wind turbines will become the mainstream of the next stage, so we should pay more attention and research to it.

Author Contributions: Conceptualization, X.G. and W.S.; methodology, J.Y. and W.S.; software, Y.Z. (Yu Zhang); formal analysis, S.Y.; investigation, W.S. and X.L.; resources, Y.Z. (Yiming Zhou); writing—original draft preparation, X.G., Y.Z. (Yu Zhang), and W.S.; writing—review and editing, X.G., Y.Z., X.L. and W.S. All authors have read and agreed to the published version of the manuscript.

Funding: This research was funded by the China Huaneng Group Science and Technology Fund (HNKJ20-H53) and the National Natural Science Foundation of China (Grant No. 51939002, 52071058). This work is also partially supported by special funds for promoting high-quality development from the department of natural resources of the Guangdong province (GDNRC [2020]016) and the Central Guidance on Local Science and Technology Development Fund of Shenzhen (2021Szvup018).

Institutional Review Board Statement: Not applicable.

Informed Consent Statement: Not applicable.

Data Availability Statement: Not applicable.

Conflicts of Interest: The authors declare no conflict of interest.

References

1. Barthelmie, R.J.; Pryor, S.C. Climate Change Mitigation Potential of Wind Energy. *Climate* **2021**, *9*, 136. [CrossRef]
2. Zhang, X.; Ma, C.; Song, X.; Zhou, Y.; Chen, W. The impacts of wind technology advancement on future global energy. *Appl. Energy* **2016**, *184*, 1033–1037. [CrossRef]
3. Pham, T.D.; Dinh, M.C.; Kim, H.M.; Nguyen, T.T. Simplified floating wind turbine for real-time simulation of large-scale floating offshore wind farms. *Energies* **2021**, *14*, 4571. [CrossRef]
4. Wang, Y.; Shi, W.; Michailides, C.; Wan, L.; Kim, H.; Li, X. WEC shape effect on the motion response and power performance of a combined wind-wave energy converter. *Ocean Eng.* **2022**, *250*, 111038. [CrossRef]
5. Ren, Z.; Verma, A.S.; Li, Y.; Teuwen, J.J.; Jiang, Z. Offshore wind turbine operations and maintenance: A state-of-the-art review. *Renew. Sustain. Energy Rev.* **2021**, *144*, 110886. [CrossRef]
6. Possner, A.; Caldeira, K. Geophysical potential for wind energy over the open oceans. *Proc. Natl. Acad. Sci. USA* **2017**, *114*, 11338–11343. [CrossRef]
7. Chen, C.; Ma, Y.; Fan, T. Review of model experimental methods focusing on aerodynamic simulation of floating offshore wind turbines. *Renew. Sustain. Energy Rev.* **2022**, *157*, 112036. [CrossRef]
8. Le, C.; Zhang, J.; Ding, H.; Zhang, P.; Wang, G. Preliminary design of a submerged support structure for floating wind turbines. *J. Ocean Univ. China* **2020**, *19*, 49–66. [CrossRef]
9. Peng, L.; Zhao, H.; Li, X. New ductile fracture model for fracture prediction ranging from negative to high stress triaxiality. *Int. J. Plast.* **2021**, *145*, 103057. [CrossRef]
10. Borgen, E. Floating Wind Power in Deep Water-Competitive with shallow-water Wind Farms. *Mod. Energy Rev.* **2010**, *2*, 49–53.
11. Ren, Y.; Vengatesan, V.; Shi, W. Dynamic Analysis of a Multi-column TLP Floating Offshore Wind Turbine with Tendon Failure Scenarios. *Ocean Eng.* **2022**, *245*, 110472. [CrossRef]
12. Jonkman, J. Dynamics of offshore floating wind turbines-model development and verification. *Wind Energy* **2019**, *12*, 459–492. [CrossRef]
13. Hywind Demo. Available online: https://www.equinor.com/en/what-we-do/floating-wind/hywind-demo.html (accessed on 1 December 2021).

14. Equinor and ORE Catapult Collaborating to Share Hywind Scotland Operational Data. Available online: https://www.equinor.com/en/news/2019-11-28-hywind-scotland-data.html (accessed on 1 December 2021).
15. Roddier, D.; Cermelli, C.; Aubault, A.; Peiffer, A. Summary and con-clusions of the full life-cycle of the WindFloat FOWT prototype project. In Proceedings of the ASME 2017 36th International Conference on Ocean, Offshore and Arctic Engineering, Hamburg, Germany, 5–10 June 2017.
16. Ramirez, L.; Fraile, D.; Brindley, G. *Offshore Wind in Europe: Key Trends and Statistics 2020*; WindEurope Intelligence Platform: Brussels, Belgium, 2021.
17. Hywind Tampen: The World's First Renewable Power for Offshore Oil and Gas. Available online: https://www.equinor.com/en/what-we-do/hywind-tampen.html (accessed on 1 December 2021).
18. Hofmann, M.; Sperstad, I.B. Will 10 mw wind turbines bring down the operation and maintenance cost of offshore wind farms. *Energy Procedia* **2014**, *53*, 231–238. [CrossRef]
19. Bak, C.; Zahle, F.; Bitsche, R.; Kim, T.; Yde, A.; Henriksen, L.C.; Natarajan, A.; Hansen, M. *DTU 10-MW Reference Wind Turbine*; DTU Wind Energy Report-I-0092; DTU: Kongens Lyngby, Denmark, 2013.
20. Wang, Q. Design and Dynamic Analysis of a Steel Pontoon-Type Semi-Submersible Floater Supporting the DTU 10-MW Reference Turbine. Master's Thesis, Delft University of Technology, Delft, The Netherlands, 2014.
21. Castro, I.R. Design of a 10-MW Wind Turbine Rotor Blade for Testing of a Scaled-Down Floating Offshore Support Structure. Master's Thesis, Delft University of Technology, Delft, The Netherlands, 2017.
22. Tian, X. Design, Numerical Modelling and Analysis of TLP Floater Supporting the DTU 10MW Wind Turbine. Master's Thesis, Norwegian University of Science and Technology, Trondheim, Norway, 2016.
23. Crozier, A. Design and Dynamic Modeling of the Support Structure for a 10-MW Offshore Wind Turbine. Master's Thesis, Norwegian University of Science and Technology, Trondheim, Norway, 2011.
24. Jmh, A.; Eeb, A.; Jrram, B. Integrated design optimization of spar floating wind turbines. *Mar. Struct.* **2020**, *72*, 102771.
25. Wang, S.; Nejad, A.R.; Bachynski, E.E.; Moan, T. Effects of bedplate flexibility on drivetrain dynamics: Case study of a 10-MW spar type floating wind turbine. *Renew. Energy* **2020**, *161*, 808–824. [CrossRef]
26. Hegseth, J.M.; Bachynski, E. A semi-analytical frequency domain model for efficient design evaluation of spar floating wind turbines. *Mar. Struct.* **2019**, *64*, 186–210. [CrossRef]
27. Son, D.; Pinguet, R.; Roddier, D. Global Sizing of the WindFloat for a 10 MW Generic Wind Turbine. In Proceedings of the ASME 2018 1st International Offshore Wind Technical Conference, San Francisco, CA, USA, 4–7 November 2018.
28. Ahn, H.; Shin, H. Experimental and numerical analysis of a 10-MW floating offshore wind turbine in regular waves. *Energies* **2020**, *13*, 2608. [CrossRef]
29. Yu, Y.; Pham, T.D.; Shin, H.; Ha, K. Study on the motion characteristics of 10-MW superconducting floating offshore wind turbine considering 2nd order wave effect. *Energies* **2021**, *14*, 6070. [CrossRef]
30. Leble, V.; Barakos, G. 10-MW wind turbine performance under pitching and yawing motion. *J. Sol. Energy Eng.* **2017**, *139*, 041003. [CrossRef]
31. Jeong, J.H.; Kim, S.H. CFD investigation on the flatback airfoil effect of 10 MW wind turbine blade. *J. Mech. Sci. Technol.* **2018**, *32*, 2089–2097. [CrossRef]
32. Oh, S.; Inoue, H.; Takahashi, Y. Structural design of a prestressed-concrete spar-type floater for 10mw wind turbines. *J. Phys. Conf. Ser.* **2020**, *1669*, 012012. [CrossRef]
33. Serafeim, G.P.; Manolas, D.I.; Riziotis, V.A.; Chaviaropoulos, P.K. Lightweight optimal rotor design of a 10-MW scale wind turbine using passive load control methods. *J. Phys. Conf. Ser.* **2020**, *1618*, 022061. [CrossRef]
34. Bredmose, H.; Lemmer, F.; Borg, M.; Pegalajar-Jurado, A.; Mikkelsen, R.F.; Larsen, T.S. The triple spar campaign: Model tests of a 10mw floating wind turbine with waves, wind and pitch control. *Energy Procedia* **2017**, *137*, 58–76. [CrossRef]
35. Jonkman, J. *Definition of the Floating System for Phase IV of OC3*; Technical Report; National Renewable Energy Laboratory (NREL): Golden, CO, USA, 2010.
36. Jonkman, J.M.; Buhl, M.L., Jr. *FAST User's Guide*; Technical Report No. NREL/EL-500-38230; National Renewable Energy Laboratory (NREL): Golden, CO, USA, 2005.
37. Bortolotti, P.; Tarres, H.C.; Dykes, K.L.; Merz, K.; Sethuraman, L.; Varelst, D.; Zahle, F. *IEA Wind Task 37 on Systems Engineering in Wind Energy WP2.1 Reference Wind Turbines*; U.S. Department of Energy Office of Scientific and Technical Information: Oak Ridge, TN, USA, 2019.
38. Xue, W. Design, Numerical Modelling and Analysis of a Spar Floater Supporting the DTU 10MW Wind Turbine. Master's Thesis, Norwegian University of Science and Technology, Trondheim, Norway, 2016.
39. DNVGL. *DNVGL-ST-0437: Loads and Site Conditions for Wind Turbines*; DNV GL AS: Bærum, Norway, 2016.
40. ANSYS Inc. *ANSYS AQWA, Librium. NAUT and Tether Manuals*; ANSYS Inc.: Canonsburg, PA, USA, 2017.
41. ROSCO 2.5.0 Documentation. Available online: https://rosco.readthedocs.io/en/develop/source/install.html (accessed on 10 February 2022).
42. DNVGL. *DNVGL-RP-0286: Coupled Analysis of Floating Wind Turbines*; DNVGL-RP-0286; DNV GL AS: Bærum, Norway, 2019.
43. Jonkman, B.J. *TurbSim User's Guide: Version 2.00.00 [R]*; National Renewable Energy Laboratory (NREL): Golden, CO, USA, 2016.

44. IEC. *61400-1. Wind Turbines-Part 1: Design Requirements for Offshore Wind Turbines*; International Electrotechnical Commission: Geneva, Switzerland, 2005.
45. Zhang, W.; Zhou, Z.; Pradhan, D.; Wang, P.; Jin, H. Design considerations of drag anchors in cohesive soil for floating facilities in the South China sea. *Mar. Struct.* **2022**, *81*, 103101. [CrossRef]

Journal of
Marine Science and Engineering

MDPI

Article

Impact of Limited Degree of Freedom Drag Coefficients on a Floating Offshore Wind Turbine Simulation

Arjun Srinivas [1,2], Bryson Robertson [1,*], Jonah Benjamin Gadasi [1], Barbara Gwynne Simpson [3], Pedro Lomónaco [1] and Jesús María Blanco Ilzarbe [2]

[1] School of Civil and Construction Engineering, Oregon State University, Corvallis, OR 97331, USA
[2] Faculty of Engineering, Universidad del País Vasco/Euskal Herriko Unibertsitatea, 48013 Bilbao, Spain
[3] School of Civil and Environmental Engineering, Stanford University, Stanford, CA 94305, USA
* Correspondence: bryson.robertson@oregonstate.edu

Abstract: The worldwide effort to design and commission floating offshore wind turbines (FOWT) is motivating the need for reliable numerical models that adequately represent their physical behavior under realistic sea states. However, properly representing the hydrodynamic quadratic damping for FOWT remains uncertain, because of its dependency on the choice of drag coefficients (dimensionless or not). It is hypothesized that the *limited* degree of freedom (DoF) drag coefficient formulation that uses only translational drag coefficients causes mischaracterization of the rotational DoF drag, leading to underestimation of FOWT global loads, such as tower base fore-aft shear. To address these hydrodynamic modeling uncertainties, different quadratic drag models implemented in the open-source mid-fidelity simulation tool, OpenFAST, were investigated and compared with the experimental data from the Offshore Code Comparison Collaboration, Continued, with Correlation (OC5) project. The tower base fore-aft shear and up-wave mooring line tension were compared under an irregular wave loading condition to demonstrate the effects of the different damping models. Two types of hydrodynamic quadratic drag formulations were considered: (1) member-based dimensionless drag coefficients applied only at the translational DoF (namely *limited-DoF* drag model) and (2) quadratic drag matrix model (in dimensional form). Based on the results, the former consistently underestimated the 95th percentile peak loads and spectral responses when compared to the OC5 experimental data. In contrast, the drag matrix models reduced errors in estimates of the tower base shear peak load by 7–10% compared to the *limited-DoF* drag model. The underestimation in the tower base fore-aft shear was thus inferred be related to mischaracterization of the rotational pitch drag and the heave motion/drag by the *limited-DoF* model.

Keywords: floating offshore wind; semi-submersible; OpenFAST; quadratic drag; OC5

Citation: Srinivas, A.; Robertson, B.; Gadasi, J.B.; Simpson, B.G.; Lomónaco, P.; Ilzarbe, J.M.B. Impact of Limited Degree of Freedom Drag Coefficients on a Floating Offshore Wind Turbine Simulation. *J. Mar. Sci. Eng.* **2023**, *11*, 139. https://doi.org/10.3390/jmse11010139

Academic Editor: Constantine Michailides

Received: 5 December 2022
Revised: 24 December 2022
Accepted: 4 January 2023
Published: 7 January 2023

1. Introduction

Climate change is driving a need for alternative energy technologies [1]. As a result of the long-term temperature goals from the Paris Agreement, governmental policies worldwide are striving towards decarbonization [2]. To meet decarbonization objectives, many policies will rely on the utilization of offshore wind resources as a zero-carbon renewable energy source. Offshore wind is estimated to reach about 40% of the total wind energy production in 2050, with global installed offshore wind capacity expected to improve from 29 GW in 2019 to 1748 GW in 2050 [3]. Floating offshore wind (FOWT), in particular, is anticipated to generate 15% of all offshore wind energy by 2050, accounting for 264 GW [3].

According to the Global Wind Energy Council (GWEC), offshore wind is a low-cost, reliable technology that delivers significant economic benefits from manufacturing through operation, and can be deployed rapidly at scale [4,5]. Currently, the largest existing floating wind farm is Hywind Tampen located in the Norwegian North Sea, with a spar-type platform having a total capacity of 88 MW. The latest offshore wind projects have a capacity factor (the ratio of the electrical energy actually produced to the electrical energy that would have been produced at continuous rated power operation for the same period) of 40–50%, matching the values of an efficient thermal power plant (gas fired or coal fired) in some regions [6,7]. Floating offshore wind has gained momentum, because deep-water installations are subject to abundant and consistent wind, with at least four times as much ocean surface space available compared to bottom-fixed wind types [3].

To support their design and assessment, reliable numerical models need to be able to adequately represent the hydrodynamic loading of FOWT under realistic sea states. The International Energy Agency (IEA) Wind Technology Collaboration Program (TCP) was founded in 1977 to focus on the planning, execution, research, and development of large-scale wind system projects [8]. Under the IEC Wind TCP Task 23 and Task 30, the Offshore Code Comparison Collaborative (OC3) and its subsequent extensions aim to improve understanding of the global behavior of floating offshore wind, including the assessment of existing numerical modeling tools needed for the analysis, design, and assessment of wind turbine technologies [9]. The OC3 and subsequent OC4 projects verified numerical modeling tools through code-to-code comparisons of different offshore wind systems using wind and/or wave load cases. This comparative analysis has contributed to enhanced comprehensibility of FOWT dynamics and modeling techniques. Additionally, it allowed for better understanding of the impact of various model approximations, leading to an improved standard for FOWT modeling [10].

The Offshore Code Comparison, Collaboration, Continued with Correlation (OC5) project was an extension of this work, wherein mid-fidelity numerical tools were validated by comparing numerical responses against experimental data [11]. A 1:50 FOWT model used in the OC5 experiments was subjected to wind and wave loads simultaneously and individually at the Maritime Research Institute Netherlands (MARIN) offshore wave basin (the full-scale FOWT setup is shown in Figure 1). For the wave-only loading cases, the numerical tools compared with the OC5 experimental data consistently underestimated global forces for tower base fore-aft shear and up-wave mooring line tension (refer to Figure 1c). In the frequency domain, the tower base fore-aft shear and up-wave mooring line tension responses were significantly underestimated at low frequencies outside the linear wave excitation regime, associated with the pitch (0.03 Hz) and surge (0.01 Hz) natural frequencies of the OC5 FOWT.

Robustly characterizing the hydrodynamic response is critical to the reliable design and assessment of FOWT. Underestimates from the numerical models found in the OC5 project could be attributed to experimental model uncertainties and/or numerical model inaccuracies [12,13]; e.g., experimental model uncertainties originate from uncertainties in system excitation, system properties, scale effects, and accuracy and precision of the measurement equipment [14]. Numerical model inaccuracies generally relate to the model approximations of the real system physics, which are dependent upon certain aspects, such as the type of the examined load case, calculation of system non-linearities, the representation of system viscous drag, and the like. These inaccuracies can be diminished through building well-performing computational grids, apart from other considerations. Numerically representing the hydrodynamic response for FOWT is closely linked to the selection of realistic drag force models [15]—both linear and quadratic models. For example, many mid-fidelity numerical models implement quadratic drag in two ways: (1) application at the member level, entailing strip-wise application of the Morison drag term that depends on the dimensionless drag coefficients [16–19], and (2) application of global level drag as a matrix [14,20–24]. The strip-wise drag force calculation accounts for relative changes in velocity between the body and fluid, while the latter matrix approach is based only on the body motion (calculated by multiplying the drag coefficients, in dimensional form, from the drag matrix with the square of the body velocity).

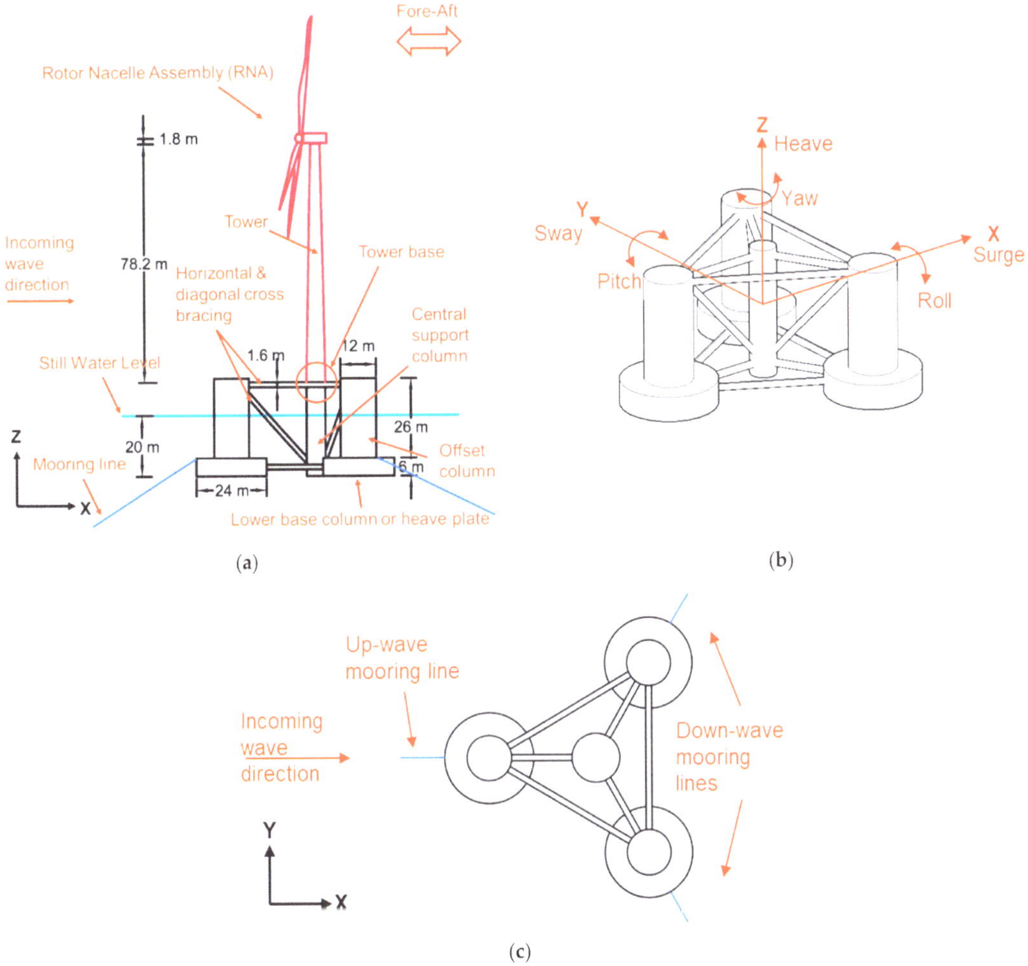

Figure 1. OC5 project FOWT model details: (**a**) FOWT full-scale model side view; (**b**) semi-submersible foundation degrees of freedom; (**c**) semisubmersible foundation top view showing the mooring line setup.

The use of a tuned linear and quadratic drag matrix, along with potential flow theory, has shown improved estimates against experimental data, particularly for low-frequency pitch motion and response in past investigations [14,20]. The tuning is performed using the system free decay tests in the desired degree of freedom (surge, heave or pitch). However, a tuned linear (P) and quadratic (Q) drag matrix, or the so-called PQ approach, was not studied in the previous OC5 validation campaign [14]. It is hypothesized that underestimates of the global hydrodynamic loads, particularly the tower base fore-aft shear, is due to mischaracterization of the rotational degree of freedom (DoF) hydrodynamic quadratic drag, because tower base moment-induced shear on a FOWT is strongly influenced by the rotational DoF pitch caused by the wave excitation [13], refer to Figure 1a,b. Similarly, up-wave mooring line tension is strongly influenced by the translational DoF surge, refer to Figure 1b,c. Both platform pitch and surge are excited by differences in hydrodynamic pressure along the wave, which were strong for this FOWT platform [25].

Herein, a numerical model of the OC5 FOWT was studied under wave-only excitation to understand and characterize the impact of hydrodynamic quadratic drag modeling on system response. Two types of hydrodynamic quadratic drag modeling were employed [11,22]: (1) member-based dimensionless drag coefficient (*limited-DoF*) drag model, and (2) quadratic drag matrix model (in dimensional form). The tower base fore-aft shear and up-wave mooring line tension were analyzed in both time and frequency domains to study the effect of *limited-DoF* drag. To inform the selection of an appropriate hydrodynamic damping model in numerical analyses of FOWT, results were used to: [i] assess the impact of enhanced rotational DoF hydrodynamic quadratic drag on FOWT response, and [ii] characterize how FOWT response can be affected by the source and method of tuning the quadratic drag matrix using the PQ and Faltinsen's methods [26].

2. Background on the OC5 FOWT Numerical-Experimental Campaign

For mid-fidelity modeling, the OC5 FOWT can be considered a large volume floating structure, wherein the structure is large relative to the length of the water particle excursion. However, there is a lot of open volume in-between the constituent elements of the platform, and each of them are small in comparison to the wave. It is thus important to consider both the large volume and slender body loading on the semisubmersible platform. This means that there is both a large perturbation of the incident flow field and flow separation at the slender elements of the platform, with the drag forces also important to the FOWT response near resonance. When limiting the effects of hydrodynamic drag forces up to second-order terms, a significant share of linear and quadratic damping originates from radiation damping and viscous drag (from flow separation and skin friction), respectively [27]. Thus, many mid-fidelity models, e.g., OpenFAST, represent the platform hydrodynamics using potential flow-theory-based panel methods augmented with viscous effects using quadratic drag.

2.1. Experimental Campaign

Mid-fidelity models currently lack enough accuracy to replace physical model testing [28], necessitating experimental testing and associated numerical validation. Table 1 summarizes the structural and hydrodynamic properties of the OC5 FOWT platform [29]. The OC5 FOWT substructure consisted of a semisubmersible-type platform (refer to Figure 1), which is currently the most common type of FOWT full-scale prototype installation due to its low-cost transportation and installation and wide range of depth feasibility [30–33]. The OC5 semisubmersible platform included three offset columns with a large diameter lower base (acting as heave plates), a central support column below the superstructure (wind turbine and tower), and a series of horizontal and diagonal cross bracing [29]. The platform used a catenary mooring configuration consisting of a chain connecting the floating body to the seabed. For the mooring characteristics, interested readers can refer to the OC5 semisubmersible floating system definition paper [29].

Table 1. Structural and hydrodynamic properties of the OC5 semisubmersible platform (full-scale).

OC5 Phase II Semisubmersible Platform Structural and Hydrodynamic Properties		
Particulars	**Units**	**Value**
Complete system mass	kg	1.396×10^7
Draft	m	20.00
Displacement	m^3	13,917.00
CM [1] location below SWL [2]	m	8.07
System roll inertia about CM	kg-m^2	1.395×10^{10}
System pitch inertia about CM	kg-m^2	1.555×10^{10}
System yaw inertia about CM	kg-m^2	1.369×10^{10}
System surge natural frequency	Hz	0.01
System pitch natural frequency	Hz	0.03

Table 1. *Cont.*

OC5 Phase II Semisubmersible Platform Structural and Hydrodynamic Properties		
Particulars	**Units**	**Value**
Water density, ρ	kg/m^3	1025.00
Water depth, d	m	200.00
Displaced water volume	m^3	13,917.00
Center of buoyancy below SWL	m	1.32×10
Static buoyancy force	N	1.399×10^8
Hydrostatic restoring in heave	N/m	3.836×10^6
Hydrostatic restoring in roll	N-m/rad	-3.776×10^8
Hydrostatic restoring in pitch	N-m/rad	-3.776×10^8

[1] CM is the center of mass of the full system. [2] SWL is the still water line.

2.2. Numerical Campaign

OpenFAST is a computationally efficient and well-established floating offshore wind numerical simulation package, that is frequently used for numerical verifications and experimental validation [23,34–38]. Developed by the National Renewable Energy Laboratory (NREL), OpenFAST (formerly FAST or Fatigue, Aerodynamics, Structures, and Turbulence) is one of the most commonly used offshore wind turbine simulation software in academia and industry. Herein, OpenFAST was used to perform the hydrodynamic quadratic drag studies of the OC5 FOWT.

OpenFAST allows for nonlinear coupling between the different dynamics of the FOWT, including the aerodynamics (AeroDyn), hydrodynamics (HydroDyn), control system (ServoDyn), structural dynamics (ElastoDyn), and mooring dynamics (MoorDyn), among others. The coupling is performed through modularization of the software components, wherein each module controls input/output parameters and constraints at the interface between them, representing the different aspects of the FOWT dynamics. Entire modules can be turned on/off depending on the modeling purposes. Herein, the platform hydrodynamics is under study, and thus the wave-only excitation loading using the irregular wave case LC3.3 (refer to Table 2) was used, such that the modules HydroDyn, ElastoDyn, and MoorDyn were only turned on in the numerical simulation.

Table 2. FOWT irregular wave-only load case.

OC5 FOWT Irregular Wave Loading					
Load Case	**Description**	**Rotor [rpm]**	**Blade Pitch [deg]**	**Wave Condition**	**Simulation Time [min]**
LC 3.3	Operational wave	0.00	90.00	$H_{m0} = 7.1$ m, $T_p = 12.1$ s, $\gamma = 3.0$	176.00

Real sea state has waves that are irregular and random in shape, height, speed of propagation, and direction, and such an irregular sea state is usually defined using a wave-frequency spectrum (power spectral density function of the vertical sea surface displacement) [39]. The proposed LC3.3 load case represents the real sea state, defined using the JONSWAP spectrum parameters: significant wave height (H_{m0}), peak period (T_p), and peak enhancement factor (γ). Thus, the results presented here can be considered similar to other realistic sea state loads defined by a wave frequency spectrum.

3. Hydrodynamic Modelling of FOWT and Drag Forces

Within fluid mechanics, hydrodynamics deals with the study of isovolume fluids, i.e., incompressible and non-dilatable, and forces acting on immersed solid bodies. In general, hydrodynamic forces can be classified by viscous or inviscid effects, both of which contain first- and second-order wave forcing (neglecting higher-order forces). For the OC5 FOWT with catenary mooring, the wave excitation and radiation forces encompass the inviscid effects, whereas the drag and mooring forces include the viscous effects of the FOWT hydrodynamics.

It should be noted that the semisubmersible platform is considered a rigid body with six DoFs (refer to Figure 1b) and the tower was modeled as elastic in the numerical simulation. The hydrodynamics of the platform were modeled using the potential flow theory-based panel method, along with the addition of quadratic drag through either (1) dimensionless drag coefficients, i.e., Morison drag terms applied using distributed strip theory, or (2) a drag matrix [40]. An external potential flow solver called WAMIT (Wave Analysis at Massachusetts Institute of Technology) was used to model the platform as a 3D diffracting body [41]. Linear potential flow theory solves the radiation-diffraction problem by generating the hydrodynamic coefficients (added mass and radiation damping) in the frequency domain solver WAMIT, which in turn are used to solve the FOWT system dynamics in the time domain solver OpenFAST. WAMIT allows for first- and second-order potential theory solutions from the FOWT submerged body analysis.

The experimental apparatus effects were also considered in the numerical simulation. From earlier validation of the OC5/OC6 numerical models against experimental data, it was found that there is a linear damping effect from the experimental apparatus that needs to be considered [15,27,42]. Additionally, the cable bundle in the experimental setup was found to change the system dynamics from past investigations; hence, a pre-load and linear stiffness in the surge and sway directions needs to be added to the hydrodynamic model [42,43].

Applying Newton's second law of motion to the FOWT subjected to wave-only loading, the governing equations of motion can be formulated according to Equation (1) [17,44]:

$$[M]\ddot{\mathbf{\Phi}} = F_{hydostatic} + F_{excitation} + F_{radiation} + F_{mooring} + F^{(2)} + F_{drag} \tag{1}$$

where $\ddot{\mathbf{\Phi}}$ is the second time derivative of the generalized floating body motion in each DoF (three translations and three rotations); $\boldsymbol{M}$ is the mass matrix of the FOWT system; $\boldsymbol{F}_{hydrostatic}$ are the hydrostatic forces; $\boldsymbol{F}_{excitation}$ are the first-order wave excitation forces; $\boldsymbol{F}_{radiation}$ are the radiation forces; $\boldsymbol{F}_{mooring}$ are the mooring forces; $\boldsymbol{F}^{(2)}$ are the second-order wave excitation forces; and $\boldsymbol{F}_{drag}$ are the quadratic drag forces.

Note that the wave excitation forces ($\boldsymbol{F}_{excitation}$) consist of a Froude–Krylov force ($\boldsymbol{F}_{FK}$) created by the undisturbed incident wave field, and a diffraction force ($\boldsymbol{F}_{D}$) created by the interaction of the body with the incident wave field. The pressure fields associated with the undisturbed incident wave and diffracted wave are considered as $\boldsymbol{p}_{FK}$ and $\boldsymbol{p}_{D}$, respectively. Then, the Froude–Krylov and diffraction forces are obtained by integrating the pressure field over the submerged portion of the body as seen in the equations below.

$$F_{FK} = \iint_{S_B} p_{FK}\boldsymbol{n}dS \tag{2}$$

$$F_{D} = \iint_{S_B} p_{S}\boldsymbol{n}dS \tag{3}$$

where S_B is the body surface with $\boldsymbol{n}$ as its normal vector.

The mooring forces ($F_{mooring}$) were calculated in the MoorDyn module within OpenFAST. MoorDyn uses a dynamic mooring line model that accounts for internal axial stiffness and damping forces, weight and buoyancy forces, hydrodynamic forces from Morison's equation, and vertical spring-damper forces from contact with the seabed. The mooring line is discretized into 'L' evenly sized line segments connecting L + 1 node points. Line segments are assumed to be rigid and massless with identical properties of unstretched length, diameter, density, and Young's modulus, while the mass is concentrated at the node points (lumped mass). Hydrodynamic loads are calculated at the node points [45].

Of particular relevance are the second-order wave excitation force and drag force. The OpenFAST hydrodynamic model characteristics used in this paper are detailed in Appendix A.

$F^{(2)}$ represents the second-order wave excitation forces acting on the platform. These forces are proportional to the square of the wave amplitude and are a result of the combination of neighboring frequencies in random waves [46]. It consists of a frequency-dependent, but time-independent mean drift force, difference frequency wave drift force, and sum frequency wave force [16], also known as triads. The last two forces result from the interaction between pairs of wave harmonics, and the forces harmonically change in time with difference or sum frequency, commonly referred to as the wave sub- and super-harmonics [16,47,48]. Drift force can be visualized at the particle level, as the fluid orbital motions do not follow a closed trajectory (first order effect), meaning there is mass transport. The mean and slowly varying (difference frequency loads) drift forces are very relevant for large moored floating structures like the OC5 FOWT. This is because these structures have long natural periods (low natural frequency) in its surge (≈0.01 Hz) and pitch (≈0.03 Hz) motions and the low-frequency second-order forces excite them (particularly the difference frequency component) [46]. It should be noted that the sum-frequency wave forces are generally expected to be more important for tension leg platforms (TLPs) than a catenary moored FOWT [39]. The second-order wave excitation forces are computed in the panel method solver—WAMIT—in the form of quadratic transfer functions (QTFs). These QTFs are imported to HydroDyn solver of OpenFAST and are used to apply the mean and slowly varying drift forces. Equation 4 represents the general form of modeling the second-order wave excitation forces from the QTFs. Herein, full summation and difference QTF are considered in the FOWT numerical simulation.

$$F^{(2)}(t) = \Re\left\{\sum_{j}^{N}\sum_{k}^{N} a_j a_k Q^{(2)}(\omega_j, \omega_k) e^{i(\omega_j \pm \omega_k)t}\right\} \tag{4}$$

where a_j and a_k are the amplitudes of the wave component j and k, respectively, of an irregular sea state with N wave components, $e^{i(\omega_j \pm \omega_k)t}$ is the complex number representation of a wave, $\Re\{*\}$ is the real part of the complex number $*$, ω_j and ω_k are the wave frequencies for wave component j and k, respectively, $Q^{(2)}(\omega_j, \omega_k)$ is the QTFs, which are six DoF vectors of complex coefficients that are a function of the frequency and the relative heading directions of each wave pair [49].

3.1. Drag Force Modeling Methodologies and Representations

The drag force on the floating body, F_{drag}, results from the flow field. The steady/unsteady motion of a floating body with sharp corners results in flow separation, creating vortices. There is a transfer of energy from the body motion to the generated vortices. The vortices could be shed even for floating bodies with rounded corners, provided the body is small compared to water particle excursion. The associated force has a damping effect on the body.

The drag force can be applied to the floating body either using Morison's drag term (as shown in Equation (5)) at the member level or by using the global quadratic drag matrix, as mentioned previously.

$$F_{drag} = \frac{1}{2}\rho_w C_D A \left|v - \dot{\mathbf{\Phi}}\right| \cdot \left(v - \dot{\mathbf{\Phi}}\right) \tag{5}$$

where ρ_w is the density of water, C_D is the member-based dimensionless drag coefficient, A is the projected area (typically the projection of the body on a plane perpendicular to the direction of motion), v is the fluid velocity, and $\dot{\mathbf{\Phi}}$ is the body velocity vector.

Generally, the dimensionless drag coefficients (in the Morisons drag term—detailed below) and the quadratic drag matrix are tuned using experimental data. Herein, the drag matrix is tuned using free decay data from both the experimental and OpenFAST numerical models. Two methods of drag coefficient calculation are investigated, namely the Faltinsen's and the PQ method [26,29]. Note that the experimental free decay test data was unavailable. However, drag coefficients calculated using the PQ method on the experimental free decay were available from OC5 published literature [11]. Different drag force models used in this study are discussed in detail in the subsequent sections.

3.1.1. Morison's Drag Force

Application of the member-based drag model calculates the force using the dimensionless drag coefficient, C_D, which is determined empirically and is a function of the Reynolds number. The calculation uses the Strip theory, where the body is discretized into a number of strips along the depth with the Morison's drag term applied to each strip, as shown in Figure 2. Under the Strip theory, the cylindrical members are discretized into 1 m long line segments (strips or elements) along the length with the ends of segments representing the nodes. The distributed transverse drag forces are computed at each node and integrated along the depth to obtain the total transverse drag force. Linear wave theory is used to derive the water particle orbit and kinematics at the various depths to aid in the calculation of the drag force [50].

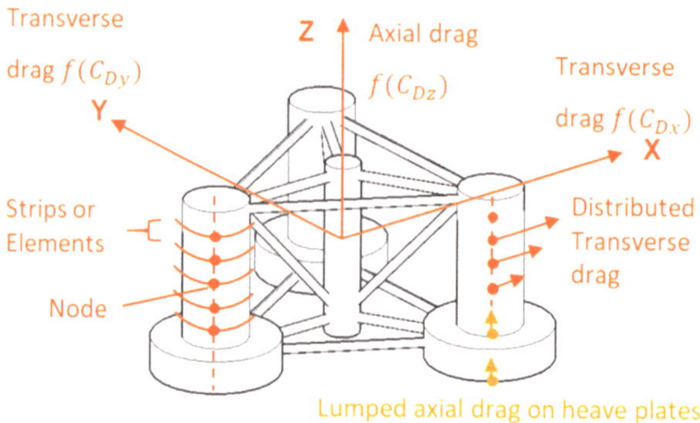

Figure 2. Strip-wise discretization of a typical submerged cylindrical body for Morison drag term application.

The dimensionless drag coefficients were applied in the transverse (x,y) directions along each strip of a member. The mathematical formulation for the application of Morison drag term along the transverse directions (x,y) in the HydroDyn module of OpenFAST is given by Equation (6). Note that HydroDyn does not distinguish the drag between the transverse directions (x,y).

$$F_{drag[transverse]} = \frac{1}{2} C_{Dx,y} \rho_w D \left| \boldsymbol{v_{rx,y}} \right| \boldsymbol{v_{rx,y}} \tag{6}$$

where $\boldsymbol{F_{Drag\ [transverse]}}$ is the transverse drag force per unit length along the line segment $[\frac{N}{m}]$, D is the cylinder diameter, 'C_{Dx}' and 'C_{Dy}' are the member-based dimensionless drag coefficients along the translational directions x and y, respectively, ρ_w is the density of water $[\frac{kg}{m^3}]$, $\boldsymbol{v_r} = v - \dot{\boldsymbol{\Phi}}$ is the relative velocity between body and fluid $[\frac{m}{s}]$.

In the axial (z) direction, the dimensionless drag coefficients can be applied either to the top and bottom end nodes of the heave plates (double-sided) or to the bottom end node of the heave plate (single-sided). They are calculated as lumped loads at the member end nodes (axial drag is only calculated at user-defined member end nodes in HydroDyn) [40]. The mathematical formulation for the application of Morison drag term along the axial direction (z) on one side of the heave plate is given by Equation (7).

$$F_{drag[axial]} = \frac{1}{2} C_{Dz} \rho_w \boldsymbol{A_{heave}} |v_{rz}| v_{rz} \tag{7}$$

where $\boldsymbol{A_{heave}}$ is the heave plate face cross-sectional area, 'C_{Dz}' is the member-based dimensionless drag coefficient along the translational direction z.

Two commonly used methods of drag force application using Morison are discussed herein, referred to as Model Typical and Model A, and shown in Table 3. An earlier study on axial drag coefficient tuning (free decay based) for the Model Typical type of drag force application resulted in only a single-sided drag coefficient applied at the bottom of the heave plate [42]. Model Typical uses a single-valued dimensionless quadratic drag coefficient 'C_D' applied to all submerged FOWT members. The values used in the former OC5 test campaign are shown in Table 3 [42]. However, several studies have shown that this type of dimensionless quadratic drag tuning results in the mismatch of system motion response at irregular sea states (realistic sea states) [15,27,51,52]. Hence, it was not further studied herein, but is described to show potential modeling choices.

Table 3. Drag force application in OpenFAST using the Morison drag term.

Model Name	Description
Model Typical	Single-valued transverse C_D (dimensionless) applied to all members in the transverse and axial direction. $C_{Dx} = C_{Dy} = 1.2$; $C_{Dz} = 3.9$ (Single-sided: Applied at the bottom of heave plate) [42].
Model A	Different C_D (dimensionless) applied to different sized (diameter 'D') members in the transverse and single-valued C_D in axial direction. $C_{Dx,y}$ $(D = 1.6\ m) = 0.63$, $C_{Dx,y}$ $(D = 6.5\ m) = 0.56$, $C_{Dx,y}$ $(D = 12\ m) = 0.61$, $C_{Dx,y}$ $(D = 24\ m) = 0.68$, and $C_{Dz} = 4.8$ (Double-sided: Applied at the top and bottom of heave plate)

Model A, referred to as the *limited-DoF* model, uses member-based dimensionless 'C_D' value application. As shown in Table 3, Model A only allows entries for the limited translational DoF drag coefficients (transverse and axial member directions). OpenFAST approximates the drag effects in rotation from the values of translational drag coefficients, which assumes a finely discretized slender rotating body, as in the proprietary software ProteusDS and ORCAFlex [53]. However, the FOWT platform consists of non-slender geometries (such as heave plates), where this approximation is invalid.

Model A transverse drag coefficients ($C_{Dx,y}$) use published values from the OC4 phase II definition paper [21], which are flow-regime-averaged $C_{Dx,y}$ values calculated for each member (including heave plates) by interpolating the experimental data found in [54]. Initially, the wave particle velocity is calculated along the depth of the submerged platform column using the linear wave theory. The Reynolds number range for each column member was then formulated to be $Re = 10^5 - 10^7$ for an arbitrary set of periodic sea states (from mild to extreme) defined in the OC4 phase II definition paper. Interpolating the experimental data found in [54], the $C_{Dx,y}$ values were formulated to be in a range between 0.3 and 1.2, for the defined flow regime. The single drag coefficient value for a particular column (as shown in Table 3) can be obtained in two steps. First, the flow-regime-averaged $C_{Dx,y}$ is defined along the depth of the submerged platform column and then this depth-varying $C_{Dx,y}$ is averaged again.

The axial drag coefficient (C_{Dz}) for Model A is applied at the top and bottom end nodes of the heave plates, wherein the heave plates are assumed to be flat with flow normal to its face. The C_{Dz} value shown in Table 3 was also published in the OC4 phase II definition paper. It was found by matching a FAST and OrcaFlex coupled simulation with DeepCwind experimental data. Note that the DeepCwind consortium, OC4, and OC5 projects use the same semisubmersible platform.

3.1.2. Global Drag Force Matrix

The second model for the drag force representation uses the dimensional hydrodynamic quadratic drag coefficients applied globally using a 6×6 drag matrix. This application allows for explicit definition of the quadratic drag in all six degrees of freedom (refer to Equations (8) and (9) and can consider both the coupled and uncoupled effects from different DoFs. Herein, any coupled effects in the quadratic drag matrix were neglected, i.e., only the diagonal terms of the drag matrix are populated.

The global quadratic drag-matrix-based models are referred to as the Model B variants. The drag force application using these models are based on the following equation:

$$\boldsymbol{F_{Add}} = \boldsymbol{F_0} - [\boldsymbol{C}]\boldsymbol{\Phi} - [\boldsymbol{B_{linear}}]\dot{\boldsymbol{\Phi}} - [\boldsymbol{B_{quad}}]\left|\dot{\boldsymbol{\Phi}}\right|\dot{\boldsymbol{\Phi}} \quad (8)$$

where $\boldsymbol{F_{Add}}$ denotes the generic addition of preload, stiffness, and damping (linear and quadratic) to the FOWT platform; $\boldsymbol{F_0}$ is the 6×1 static load vector for preload; $[\boldsymbol{C}]$ is the 6×6 linear restoring matrix (linear stiffness matrix); $[\boldsymbol{B_{linear}}]$ is the linear drag matrix; $[\boldsymbol{B_{quad}}]$ is the quadratic drag matrix; and $\boldsymbol{\Phi}$ is the generalized motion vector representing the six DoFs. When the FOWT numerical model does not require any force tuning in the form of additional preload, stiffness, and linear drag matrices, the generic additional force $\boldsymbol{F_{Add}}$ transforms to drag force $\boldsymbol{F_{drag}}$ as shown in Equation (9).

$$\boldsymbol{F_{Add}} = \boldsymbol{F_{drag}} = \begin{bmatrix} B_{quad\ xx} & 0 & 0 & 0 & 0 & 0 \\ 0 & B_{quad\ yy} & 0 & 0 & 0 & 0 \\ 0 & 0 & B_{quad\ zz} & 0 & 0 & 0 \\ 0 & 0 & 0 & B_{quad\ Mx} & 0 & 0 \\ 0 & 0 & 0 & 0 & B_{quad\ My} & 0 \\ 0 & 0 & 0 & 0 & 0 & B_{quad\ Mz} \end{bmatrix} \left|\dot{\boldsymbol{\Phi}}\right|\dot{\boldsymbol{\Phi}} \quad (9)$$

where $B_{quad\ xx}$, $B_{quad\ yy}$, $B_{quad\ zz}$ are the quadratic hydrodynamic drag along the translational DoFs and $B_{quad\ Mx}$, $B_{quad\ My}$, $B_{quad\ Mz}$ are the quadratic hydrodynamic drag along the rotation DoFs.

The drag matrix assembly involves two steps: [i] numerical- or experimental-based free decay test is performed for a particular DoF to define individual wave crest amplitudes, and [ii] constituent linear and quadratic coefficients are extracted from the free decay data and used with the system properties and infinite frequency added mass (the limiting value of added mass at zero period; at this limit there are no diffraction effects).

The Model B variants, namely B1, B2, and B3, differ in the source and method of tuning the drag matrix as shown in Table 4. Model B1 represents more a conventional tuning of the drag matrix based on the experimental OC5 data using the PQ method. Models B2 and B3 implement a novel tuning of the drag matrix based on the Model A decay results using the PQ and Faltinsen's methods, respectively.

Table 4. Overview of the numerical models with different drag force application.

Model Name	Description
Model A	Member level drag force application with dimensionless drag coefficients applied to the different sized members in the transverse and axial direction
Model B1	Global level drag force application using a drag matrix assembled from the OC5 experimental free decay data using the PQ method
Model B2	Global level drag force application using a drag matrix assembled from the OC5 numerical Model A free decay data using the PQ method
Model B3	Global level drag force application using a drag matrix assembled from the OC5 numerical Model A free decay data using the Faltinsen's method

The drag matrix tuning methodologies (PQ and Faltinsen's) differ in the equation of motion representation, constituent coefficient relation, and drag coefficient (with dimension) calculation. Interested readers can refer to the detailed description of the PQ method in [29], and the Faltinsen's method in [26].

In physical model testing, free decay tests are commonly used to obtain drag coefficients and natural frequencies of the system. Hydrodynamic systems inherently have non-linearities due to the effect of viscous flows [46]. The single degree of freedom equation of motion for such a system with non-linear damping can be represented by Equation (10):

$$(M + A_\infty)\ddot{\Phi}(t) + (B_{linear})\dot{\Phi}(t) + (B_{quad})\left|\dot{\Phi}(t)\right|\dot{\Phi}(t) + (k)\Phi(t) = 0 \tag{10}$$

where $(M + A_\infty)$ is the total mass (or mass moment of inertia) accounting for the system and the infinite frequency added mass component—[kg], [kg.m^2]; B_{linear} is the linear drag coefficient—[Ns/m], [Nms/rad]; B_{quad} is the quadratic drag coefficient—[Ns^2/m^2], [Nms^2/rad^2]; k is the hydrostatic stiffness coefficient—[N/m], [Nm/rad]. All the coefficients of Equation (10) are in the generalized single degree of freedom motion Φ.

The consecutive crest amplitudes (ϕ_1, ϕ_2, and so on) from the free decay test are extracted using the individual wave definition, where two successive down (or up) crossings define the individual wave. Figure 3a show the individual definition using the zero-down crossing method.

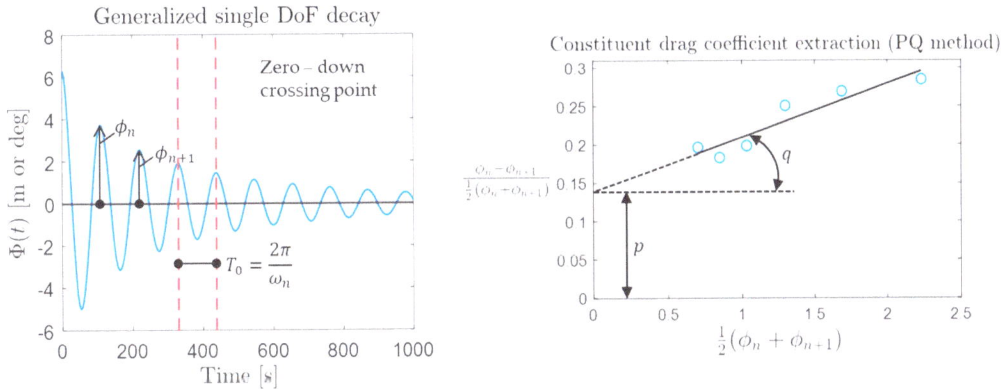

(a) (b)

Figure 3. PQ method constituent coefficient extraction using least square curve fit of the crest amplitudes: (**a**) free decay test showing the wave crest amplitudes; (**b**) straight line curve by least square method to obtain the constituent coefficients 'p' and 'q'.

The constituent linear and quadratic drag coefficients with dimension were then extracted using the consecutive crest amplitude values (ϕ_1, ϕ_2, and so on) from the free decay tests using both the PQ and Faltinsen's methods [26,29].

PQ Method

For the PQ method, the constituent drag coefficients, 'p' and 'q,' correspond to the linear and quadratic parts of the system hydrodynamic drag, respectively. The latter coefficient is used to assemble the quadratic drag matrix. The PQ uses the same equation of motion denoted by Equation (10).

In the PQ method, the linear and quadratic constituent coefficients are determined by the relation shown in Equation (11).

$$\frac{\phi_n - \phi_{n+1}}{\frac{1}{2}(\phi_n + \phi_{n+1})} = p + \frac{1}{2}(\phi_n + \phi_{n+1}) \cdot q \tag{11}$$

As shown in Figure 3, the left side of Equation (11) is plotted against the mean motion amplitude term $\frac{1}{2}(\phi_n + \phi_{n+1})$. A straight-line curve fit by least squares is used and the constituent coefficients 'p' and 'q' are calculated from the y-intercept and slope of the fitted straight line.

Faltinsen's Method

The single degree of freedom equation of motion under the Faltinsen's method is expressed in Equation (12):

$$\ddot{\Phi} + (\widetilde{p})\dot{\Phi} + (\widetilde{q})\left|\dot{\Phi}\right|\dot{\Phi} + \omega_n^2\Phi = 0 \tag{12}$$

The coefficients $\widetilde{p}$ and $\widetilde{q}$ denote the linear and quadratic constituent coefficients, respectively. Again, the latter coefficient is used to form the quadratic drag matrix.

Assuming that the damping is constant with respect to the amplitude of oscillation, the linear and quadratic constituent coefficients can be determined from the relation given in Equation (13).

$$\frac{2}{T_0}\log\left(\frac{\phi_{n-1}}{\phi_{n+1}}\right) = \widetilde{p} + \frac{16}{3}\frac{\phi_n}{T_0}\widetilde{q} \tag{13}$$

where $\frac{T_0}{2}$ is the half period between ϕ_n and ϕ_{n+1} for any n^{th} oscillation. A straight-line curve fit similar to the PQ method is used. The constituent coefficients $\widetilde{p}$ and $\widetilde{q}$ can then be read out from Figure 4 in the form of the y-intercept and slope of the curve fitted line.

$$\widetilde{B_{linear}} = \widetilde{p}\cdot(M + A_\infty);\ \widetilde{B_{quad}} = \widetilde{q}\cdot(M + A_\infty) \tag{14}$$

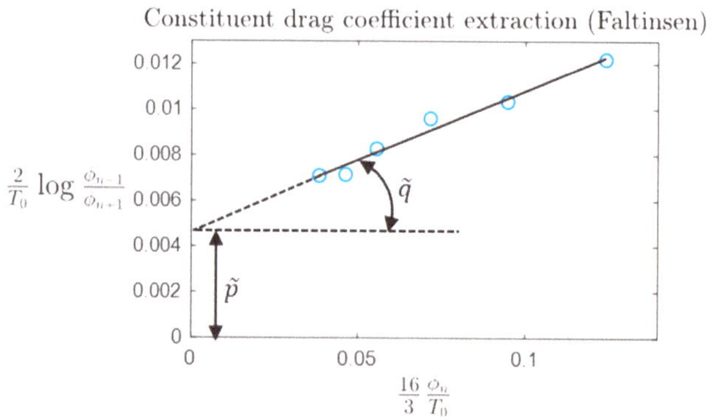

Figure 4. Faltinsen's constituent coefficient extraction using least-square curve fit of the crest amplitudes.

After obtaining the constituent coefficients from the Faltinsen's methodology, multiplying the total mass (consisting of the system mass and the infinite frequency added mass) by the quadratic constituent coefficient gives the quadratic drag coefficient B_{quad} with units of [Ns^2/m^2] or [Nms^2/rad^2]. The process is repeated for each DoF motion. The drag matrix with diagonal terms can then be populated.

4. Results

The results from the OpenFAST numerical simulations using the four different hydrodynamic quadratic drag models (Models A, B1-B3) are presented below for the OC5 FOWT; subjected to the LC3.3 irregular wave-only loading (refer to Table 2) defined using the JONSWAP spectrum parameters: significant wave height (H_{m0} = 7.1 m), peak period (T_p = 12.1 s), and peak enhancement factor (γ = 3.0). Results were analyzed in both the time and frequency domains and compared in terms of global tower base fore-aft shear and up-wave mooring line tension. The time domain history plots for tower base fore-aft shear and up-wave mooring line tension are shown in Appendix B.

4.1. Time-Domain Results

The time-domain signal of the tower base fore-aft shear and up-wave mooring line tension were analyzed in terms of mean, coefficient of variation (CoV), and 95-percentile peak load. Mean loads were obtained by dividing the sum of the loads at each time step by the number of time steps. The CoV was calculated by normalizing the standard deviation using the mean, which can be useful for characterizing the dispersion of data around the mean and comparing the variation between different signals whose mean are significantly different from each other [55]. The 95-percentile peak load was obtained by first finding out the local peaks of the time series data and then removing the top 5% of the local peak values data. The top 5% of the local peaks was arbitrarily chosen to remove the data outliers and to be consistent with the original OC5 test campaign [11].

4.1.1. Mean loads and Coefficient of Variation (CoV)

Tables 5 and 6 tabulate the mean loads of the tower base fore-aft shear and up-wave mooring line tension, respectively, for both the experiment and numerical models.

Table 5. Mean and Coefficient of Variation (CoV) values for tower base fore-aft shear.

	Experiment	Model A	Model B1	Model B2	Model B3
Mean loads [kN]	−4.54	2.51	3.24	3.22	3.25
Coefficient of variation	75.70	101.36	91.83	93.84	90.71

Table 6. Mean and coefficient of variation (CoV) values for up-wave mooring line tension.

	Experiment	Model A	Model B1	Model B2	Model B3
Mean loads [kN]	1233.85	1255.07	1254.41	1254.87	1254.59
Coefficient of variation	0.15	0.12	0.11	0.11	0.10

Mean and CoV of Tower Base Fore-Aft Shear

In the case of tower base fore-aft shear, the mean loads for the numerical models were relatively different from the experiment, regardless of the type of damping model. The difference in mean loads between the numerical models and the experiment may be attributed to the experimental setup and its uncertainties as described before in Section 2 (Numerical-Experimental Campaign). The absolute value of the mean is considered in the calculation of the tower base fore-aft shear CoV.

Within the numerical models (Model A, B1-B3), the drag matrix models (Model B1-B3) tended to have similar mean values for tower base fore-aft shear, with differences of ≈0.01 kN to 0.03 kN between each other. The member-based drag coefficients applied in Model A tended to have a slightly smaller mean compared to the drag matrix models. Models using a drag matrix resulted in similar levels of CoV, with Model B2 having the highest degree of dispersion from its mean between the drag matrix models (≈2 to 3).

Mean and CoV of Up-Wave Mooring Line Tension

In the case of up-wave mooring line tension, the mean loads for all the models were large, on the order of 120 MT, with difference in means between the experiment and numerical models of ≈20 kN to 21 kN. Differences are due to the experimental setup, where the effects of bundled cables in the experiment were represented by adjustments in the numerical model with additional preload and stiffness in the surge and sway directions. The CoV values of the experiment and numerical models were similar, on the order ≈ 0.03 to 0.05 difference.

Between numerical models (Model A, Model B1-B3), the mean loads were similar for the up-wave mooring line tension, with differences of ≈0.66 kN (negligible compared to mean). All the numerical models had similar levels of data dispersion. The drag matrix models, in particular, resulted in CoV very close to each other.

In aggregating these observations, the application of the drag coefficients (dimensionless or not) in the numerical models affects the mean loads and coefficient of variation, and is expected. Similar mean loads and CoV were observed across the drag matrix models for both tower base fore-aft shear and up-wave mooring line tension, because, regardless of the source and method of drag matrix tuning, the global quadratic drag application to the system remains the same for all the drag matrix models.

4.1.2. Percent Error Results

The percent error between the numerical and experimental models in terms of the 95-percentile peak load for the tower base fore-aft shear and up-wave mooring line tension are plotted in Figure 5.

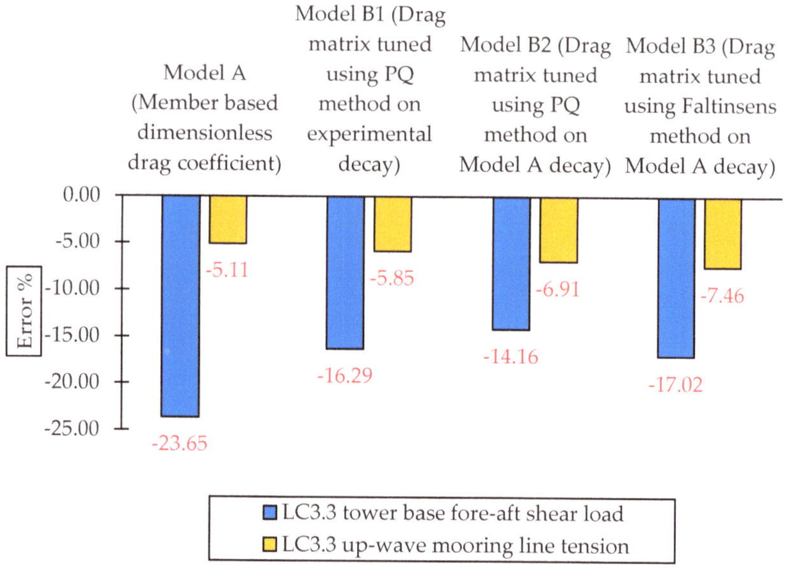

Figure 5. LC3.3, 95-percentile peak load error % of the numerical models compared to the experiment.

The negative values on the bar plot indicate underestimation of the loads by the numerical models (Model A, Model B1-B3). Estimates of the tower base shear loads significantly improved when using the drag matrix models (Models B1-B3) compared to the member-based drag coefficient model (Model A)—an improvement (error percent reduction between numerical models) of 7.36%, 9.49%, and 6.63% for Models B1, B2, and B3, respectively, when compared to Model A.

Improved estimates of tower base fore-aft shear in the pitch direction result from improved representation of drag in the rotational DoFs using the drag matrix models. The Model B variants can explicitly define 6 DoF drag to better characterize rotational DoF drag, particularly pitch. In the case of up-wave mooring line tension, the percent error was similar between numerical models, with negligible decrements of 0.74%, 1.8%, and 2.35% for Models B1, B2, and B3, respectively, when compared to Model A. Unlike the tower base for-aft shear, rotational DoF pitch drag does not have a significant impact on the up-wave mooring line tension.

The error percent values brought about by the drag matrix models (B1-B3) for the tower base shear and up-wave mooring line tension are acceptable, given the mid-fidelity level of numerical analysis.

4.2. Frequency-Domain-Based Results

In the frequency domain, results are shown using Power Spectral Density (PSD) plots, which are commonly used for random time-domain signals, such as global OC5 FOWT loads, such as tower base shear and up-wave mooring line tension [56]. The PSD, $S_f(\omega)$, is the Fourier transform of the autocorrelation function $R_f(\tau)$ as seen in Equation (16). The PSD plots the signal's power content against frequency, ω [57]:

$$S_f(\omega) = \int_{-\infty}^{\infty} R_f(\tau) e^{i\omega\tau} d\tau \tag{15}$$

where $R_f(\tau) = \lim_{T\to\infty} \frac{1}{T} f(t) f(t+\tau) dt$; $f(t)$ is the time-domain signal. At $\tau = 0$, $R_f(0) = \lim_{T\to\infty} \frac{1}{T} f^2(t) dt = \frac{1}{2\pi} \int_{-\infty}^{\infty} S_f(\omega) d\omega$. The PSD was calculated by multiplying the signal fast Fourier transform (FFT) by its complex conjugate, resulting in a spectrum where the real part is the amplitude of each harmonic with the signal's units squared, $(*)^2$. The amplitude was then normalized by the frequency step (inverse of total measured duration), $df = \frac{1}{t(N)-t(1)}$; where N is the total signal length. The resulting PSD has units of $(*)^2/Hz$.

4.2.1. Power Spectral Density (PSD) of Platform Motion

The platform motion PSDs in surge, heave, and pitch are plotted in Figure 6a–c, respectively.

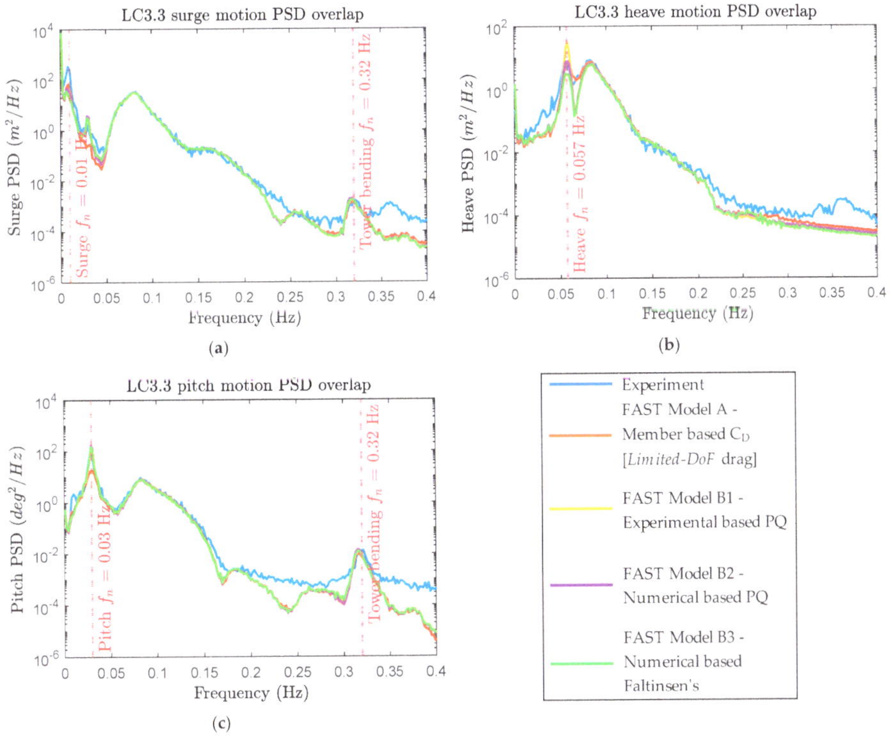

Figure 6. OC5 FOWT platform motion power spectral density (PSD) for LC3.3: (**a**) Surge motion PSD; (**b**) heave motion PSD; (**c**) pitch motion PSD.

For the surge motion PSD, the frequency response amplitudes for the system natural frequencies at surge (0.01 Hz), pitch (0.03 Hz), and tower bending (0.32 Hz) are shown for both experimental and numerical model (Models A, B1-B3). All the models underestimated the response amplitude for surge natural frequency, with the *limited-DoF* Model A better representing the low-frequency surge response compared to the drag matrix models (B1-B3). In contrast, the amplitude associated with surge-induced pitch was not well represented by Model A and was overestimated by the drag matrix models. All the models estimated similar amplitudes for tower bending and were close to the experimental results.

For the heave motion PSD, frequency response amplitudes at the heave natural frequency (0.057 Hz) are discussed across the numerical models. The *limited-DoF* Model A and drag matrix model B2 (drag matrix tuned based on Model A decay using the PQ method) had similar amplitudes for heave. Both models were closer to the experimental results compared to other models because Model A uses drag coefficients tuned to the heave motion based on the experimental data, and Model B2 tunes the quadratic drag matrix based on Model A decay. Models B1 and B3 over- and underestimate the response amplitude at the heave natural frequency, respectively. There is no response amplitude recorded at the tower bending natural frequency by the experimental and numerical models.

For the pitch motion PSD, frequency response amplitudes at the pitch (0.03 Hz) and tower bending (0.32 Hz) natural frequencies are discussed across the numerical models. Model A most underestimated the amplitude associated with pitch, whereas the drag matrix models (B1-B3) consistently remained very close to the experimental values for the pitch response. Thus, the drag matrix models were more capable of characterizing the pitch drag. All the numerical models closely matched the response amplitude at the pitch-induced tower bending natural frequency with the experimental results.

4.2.2. PSD of Tower Base Fore-Aft Shear

For the LC3.3 irregular wave loading, Figure 7a shows the whole spectrum plot for tower base fore-aft shear. To show details, sub figures Figure 7b,c zoom into the two main regions of concern, namely near the pitch and tower bending frequencies.

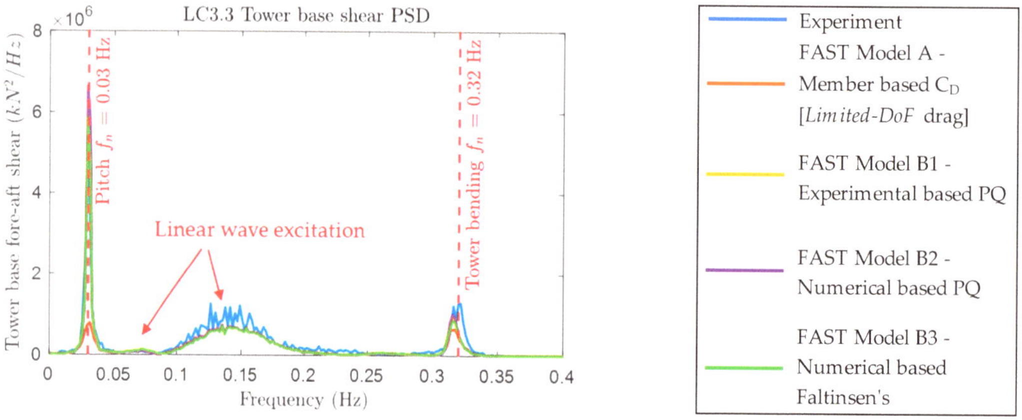

(**a**)

Figure 7. *Cont.*

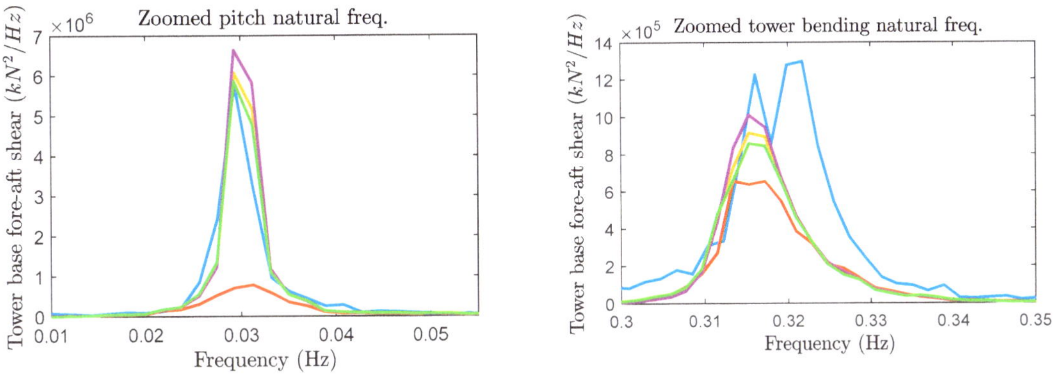

(**b**) (**c**)

Figure 7. LC3.3 tower base fore-aft shear force power spectral density (PSD): (**a**) full spectrum view; (**b**) spectrum zoomed near pitch natural frequency; (**c**) spectrum zoomed near tower bending natural frequency.

At the frequency associated with pitch, Model A significantly underestimated the amplitude for low-frequency pitch when compared to the experiment, due to the use of member-based dimensionless drag coefficient. In contrast, the drag matrix models (B1-B3) only marginally overestimated the amplitude for low-frequency pitch. Model B2 produced the highest estimate, in which the drag matrix was tuned from Model A decay using the PQ method.

At the frequency associated with tower bending, all the models underestimated the response. Model B2 has the highest amplitude, followed by models B1, B3, and A.

4.2.3. PSD of Up-Wave Mooring Line Tension

For the LC3.3 irregular wave loading, Figure 8a plots the up-wave mooring line tension PSD. Similar to tower base fore-aft shear, sub-plots in Figure 8b,c zoom in on the two main regions of concern, namely near the surge and linear wave excitation frequencies.

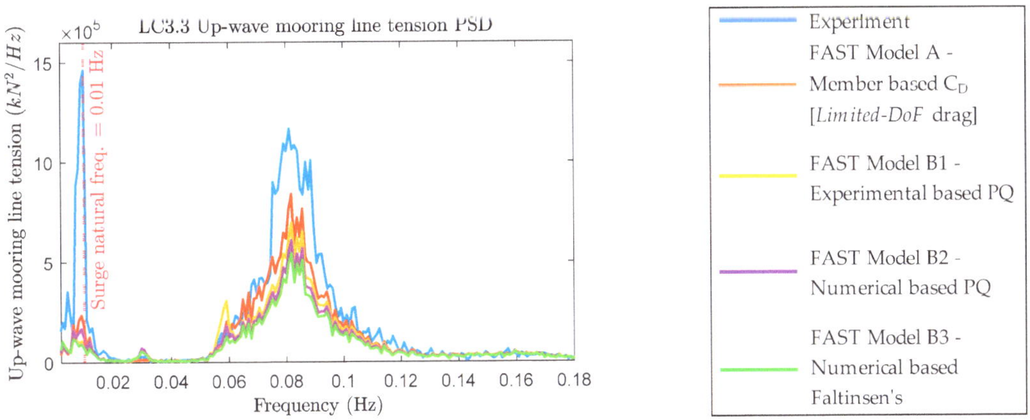

(**a**)

Figure 8. *Cont.*

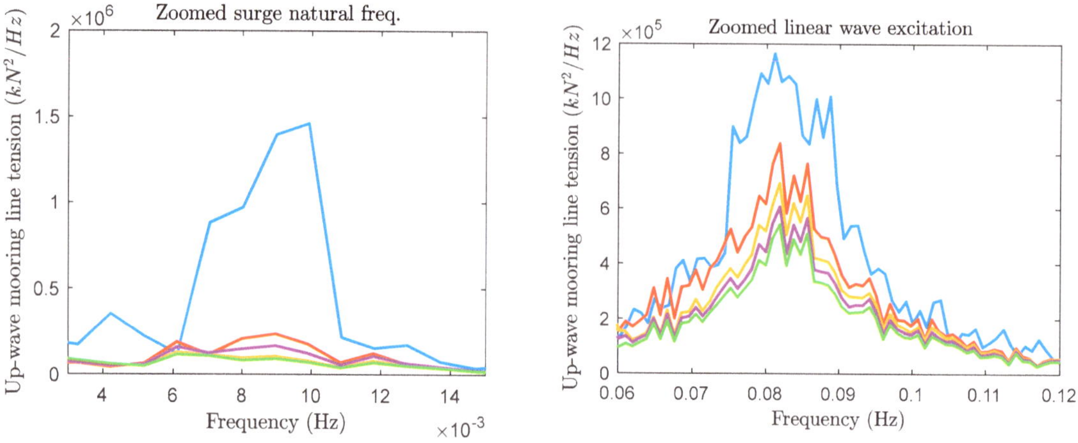

(b) (c)

Figure 8. LC3.3 up-wave mooring line tension power spectral density (PSD): (**a**) full spectrum view; (**b**) spectrum zoomed near surge natural frequency; (**c**) spectrum zoomed near linear wave excitation.

All the numerical models clearly underestimated the amplitude at the frequencies associated with surge and linear wave excitation. Additionally, response was close together for all the numerical models. Model A had the largest amplitude among the numerical models, followed by models B2, B1, and B3. Trends observed here were similar to those observed for the surge-induced response, because the platform surge motion is a primary driver of up-wave mooring line tension.

A similar trend can be observed for the zoomed-in region for the linear wave excitation. All the numerical models underestimated the response amplitude and were close together. Model A has the largest response amplitude among the numerical models, followed by Models B1, B2, and B3. This trend is similar to the error percent results, wherein, the model error percent followed—Model A < B1 < B2 < B3. Note that large response amplitude corresponds to lower error percent. It can be thus inferred that the time-domain-based percent error results and the frequency-domain-based up-wave mooring line tension PSD plots yielded the same results, giving consistency and confidence to the approach.

5. Discussion

From the results, when the pitch drag is more properly represented, the platform pitch motion improves, as observed in the power spectral density plots. Irrespective of the type of excitation force, the platform pitch motion results in bending moment on the tower base due to the presence of a heavy rotor nacelle assembly (RNA) at the tower top of the FOWT, with induced shear in the tower base as well. Thus, models improving rotational DoF drag in pitch lead to significantly improved estimates of the tower base fore-aft shear, as shown by the 95-percentile of the peak load and spectral response. In particular, using the drag matrix models (B1, B2, and B3) resulted in near 7% to 10% improvements in estimates of the 95-percentile peak loads when compared to Model A for the tower base fore-aft shear.

Of all the numerical models, Model B2 (PQ method based) resulted in the smallest percent error (14%) when compared to the experimental results in terms of the 95-percentile peak loads of tower base fore-aft shear. Model B2 also provided the best estimates of the heave motion, closely followed by Model A, because the heave drag in Model B2 was tuned to the Model A decay, which was based on the experimental data. Thus, two of the main drivers for underestimating the 95-percentile peak loads of tower base fore-aft shear arise from mischaracterizing the rotational DoF quadratic pitch drag and translational DoF heave drag of the platform. With improved characterization of the platform pitch and heave drag, estimates of the tower base fore-aft shear loads were best between the models. Note, all Model B variants (B1-B3) still possessed similar levels of 95-percentile peak loads and heave motion spectral response, with Model A and B2 resulting in the best estimate of response amplitude at the heave natural frequency. Thus, it can also be inferred that underestimates of the 95-percentile peak loads depend more on mischaracterizing rotational DoF pitch drag than mischaracterizing translational heave drag.

No significant differences were observed between the different drag matrix models, irrespective of the source and method of tuning. Between the drag matrix Model B variants, the numerical-based tuning proposed in Model B2 was at par or at times improved compared to the experiment-based tuning in Model B1 for the 95-percentile peak loads of tower base fore-aft shear. Note, this improvement was small. When compared to the 95-percentile peak load of the experiment, Model B2 resulted $\approx 2\%$ reduction in percent error (or improved load estimate) than Model A. Thus, tuning of Model B2 using the PQ method based on the free decay of a *limited-DoF* drag Model A, best characterized both rotational DoF pitch and translational DoF heave drag, leading to the best estimates of tower base fore-aft shear.

All the numerical models underestimated the 95-percentile peak load and spectral response of the up-wave mooring line tension. However, estimates were close together, indicating other modeling assumptions besides the drag model may also be at play. With respect to the experimental results, Model A resulted in a percent error of 5.11% and the drag matrix model B variants had a percent error in the range of $\approx$5.9–7.5%. The similarity suggests that surge damping and surge-affected second-order wave excitation forces were similar between Model A and the Model B variants.

Underestimates of the up-wave mooring line tensile forces could be alleviated through tuning the surge (transverse) quadratic drag [15,58] and/or tuning the second-order wave excitation forces [59] or by improving the numerical representation of the mooring system. The latter would address mischaracterizing the quadratic transfer functions (QTFs) generated in the potential flow solver WAMIT, leading to mischaracterizing wave mean and slow drift forces. To improve estimates of the second-order wave excitation forces, the QTFs from the potential flow solver could be better tuned using experiments or CFD to improve estimates of second-order wave excitation forces, assuming the effects of higher-order forces are negligible. A tuned QTF potential flow model with viscous drag accounted for by the quadratic drag matrix (tuned based on free decay) would likely result in better estimates of tower base fore-aft shear and up-wave mooring line tension peak loads along with their spectral response. The formulation of a numerical model with a tuned QTF and drag matrix is also independent of geometry, and thus versatile. These effects on surge-affected up-wave mooring line tension need to be studied in further detail.

6. Conclusions

This study aimed to numerically understand the impact of *limited* degree of freedom (DoF) drag coefficients on a floating offshore wind turbine (FOWT) simulation. Currently, the adequate representation of the hydrodynamic quadratic damping conditions for FOWT remains uncertain, because of its dependency on the type of application of the drag coefficients (dimensionless or not). To address these uncertainties associated with the FOWT hydrodynamic modeling, the open-source mid-fidelity numerical simulation tool, OpenFAST, was used to investigate and compare different hydrodynamic quadratic drag models against the experimental data from the Offshore Code Comparison, Collaboration, Continued, with Correlation (OC5) project. Two types of quadratic drag formulation were considered: (1) member-based dimensionless drag coefficient (*limited-DoF*) model and (2) quadratic drag matrix model (in dimensional form). The former is referred to as the Model A configuration, and the latter are the Model B variants, namely B1, B2, and B3, which differ in the source and method of tuning the drag matrix. Model B1 represents more conventional tuning of the drag matrix based on the experimental OC5 free decay data using the PQ method. Models B2 and B3 implement a novel tuning of the drag matrix based on the Model A decay data using the PQ and Faltinsen's methods, respectively.

Model A, which employed *limited-DoF* drag coefficients, consistently underestimated the 95-percentile peak loads and spectral responses for both tower base fore-aft shear and up-wave mooring line tension. In Model A, the Morison drag term is applied using the distributed strip theory and drag is calculated normal to the member orientation (in one direction). However, drag does affect multiple DoFs because rotational drag effects are approximated from translational drag coefficients. Explicitly defining rotational DoF drag, particularly pitch, using the drag matrix models enabled significantly improved estimates of the 95-percentile and spectral response for tower base fore-aft shear.

When the pitch drag was more properly represented, the platform pitch motion improved, leading to significantly improved estimates of the tower base fore-aft shear.

In comparing the Model A and Model B variants, mischaracterizing the rotational DoF quadratic pitch drag and translational DoF heave drag of the platform resulted in underestimates of the 95-percentile peak loads of tower base fore-aft shear. These underestimates depended more on mischaracterizing rotational DoF pitch drag than mischaracterizing translational heave drag.

Herein, valuable insight into the quantification of the effects of better characterization of pitch drag on a FOWT numerical simulation is presented. Novelty lies in the use of numerical models (Model B2 and B3) with drag coefficients tuned based on the *limited-DoF* drag model (Model A) free decay in addition to the experiment-based drag-coefficient-tuned model (Model B1). The proposed novel method of tuning (Model B2 and B3) produced global load approximations that are on a par or even improved (tower base fore-aft shear) compared to the conventional experimental free decay-based tuning (Model B1). Tuning of Model B2 using the PQ method based on the free decay of a *limited-DoF* drag (Model A) led to the best estimates of tower base fore-aft shear, because it best represented both pitch and heave drag.

All the numerical models underestimated the 95-percentile peak load and spectral response of the up-wave mooring line tension. However, estimates were close together, indicating that other modeling assumptions besides the drag model may also be at play.

Although this study highlighted differences between damping models, there is a limit to the capabilities of mid-fidelity numerical analyses. As they neglect some aspects of physical response to maintain computational efficiency, mid-fidelity numerical models, such as OpenFAST, must be calibrated with experimental data or higher fidelity computational fluid dynamics (CFD) models to best represent physical behavior. Within the hybrid hydrodynamic numerical model (potential flow theory with quadratic drag), there is a mischaracterization of both the drag force application and/or second-order wave excitation forces by the potential flow solver WAMIT. Herein, the drag force mischaracterization was studied in terms of the impact of *limited-DoF* drag. It is hypothesized that the drag models used herein alongside a better representation of second-order wave excitation forces would also result in improved numerical estimates, because surge, heave, and pitch motions are complementary to each other. Thus, future studies could improve the models presented herein even further with additional improvements to the potential flow model.

Author Contributions: Conceptualization, A.S. and B.R.; methodology, A.S. and J.B.G.; formal analysis, A.S., writing—original draft, A.S.; writing—review and editing, A.S., B.R., B.G.S., P.L. and J.M.B.I.; validation, P.L. and J.B.G.; visualization, A.S.; supervision, B.R., B.G.S. and P.L.; funding acquisition, J.B.G. All authors have read and agreed to the published version of the manuscript.

Funding: This work was partially supported by the Research Group of the Basque Government (IT1514-22) and the U.S. Department of Energy (DOE) under award number DE-EE0008960, titled "Coupled Aerodynamic and Hydrodynamic Hybrid Simulation of Floating Offshore Wind Turbines".

Institutional Review Board Statement: Not applicable.

Informed Consent Statement: Not applicable.

Data Availability Statement: Some or all data, models, or code that support the findings of this study are available from the corresponding author upon request. The findings, opinions, recommendations, and conclusions in this paper are those of the authors alone and do not necessarily reflect the views of others, including the sponsors.

Conflicts of Interest: The authors declare no conflict of interest.

Abbreviations

The following abbreviations are used throughout this manuscript:

A	Projected area
B	Drag coefficients
C	Additional linear stiffness
CFD	Computational Fluid Dynamics
CM	Center of Mass
CoV	Coefficient of variation
C_D	Drag coefficient
D	Cylinder diameter
DoF	Degree of Freedom
F_{Add}	Generic Force Addition term
F_{Drag}	Drag Force term
F_0	Preload
FFT	Fast Fourier Transform
FOWT	Floating Offshore Wind Turbine
GWEC	Global Wind Energy Council
H_{m0}	Significant Wave Height
IEA	International Energy Agency
L	Mooring line segments
M	Mass matrix
MARIN	Maritime Research Institute Netherlands
N	Total Signal Length
NREL	National Renewable Energy Laboratory
OC3	Offshore Code Comparison Collaborative progr.
OC4	Offshore Code Comparison Collaboration, Continued progr.
OC5	Offshore Code Comparison Collaboration, Continued, with Correlation progr.
p	Linear constituent coefficient
PQ	Linear (P) and Quadratic (Q)
PSD	Power Spectral Density
q	Quadratic constituent coefficient
QTF	Quadratic Transfer Functions
$\Re$	Real part of the complex number
Rf	Autocorrelation function
RNA	Rotor Nacelle Assembly
S_B	Body surface
SWL	Still Water Level
T	Period
t	Time
TCP	Technology Collaboration Program
TLP	Tension Leg Platform
v	Fluid velocity
WAMIT	Wave Analysis at Massachusetts Institute of Technology
Greek Symbols	
k	Hydrostatic stiffness coefficient
ω	Wave frequency
Φ	Generalized floating body motion in each DoF
ψ	Crest amplitude value
%	Percent error

Appendix A. OpenFAST Hydrodynamic Model Characteristics

The mid-fidelity FOWT numerical simulation tool, OpenFAST, was used to study the impact of *limited-DoF* drag coefficients. The characteristics of the hydrodynamic model used in the HydroDyn module of OpenFAST are summarized below.

- The hydrodynamic model uses the potential flow model (WAMIT solver) augmented with quadratic Morison drag terms or a 6 × 6 quadratic drag matrix, which allows for the inclusion of the radiation-diffraction solution and viscous drag (quadratic).
- In case of second-order wave kinematics, full sum and difference second-order WAMIT QTF is used. Wave stretching is not included. It is essential to include full difference QTF for moored FOWT analysis [28,39,44]. Even though the hydrodynamic model using full difference QTF underestimates the loads, those models are the closest to the experiment [28].
- Additional preloads ($\boldsymbol{F_0}$ in N) in surge and sway were included using the generic force addition term $\boldsymbol{F_{Add}}$, as shown below, to account for the influence of measurement cable bundle [42,44].

$$\boldsymbol{F_{Add}} = \boldsymbol{F_0} = \begin{bmatrix} 1.40 \times 10^5 \\ 1.00 \times 10^4 \\ 0 \\ 0 \\ 0 \\ 0 \end{bmatrix} \tag{A1}$$

- Additional linear stiffness ($\boldsymbol{C}$ in N/m) in surge and sway were included to account for the influence of measurement cable bundle using the generic force addition term $\boldsymbol{F_{Add}}$ [42,44].

$$\boldsymbol{F_{Add}} = [\boldsymbol{C}]\boldsymbol{\Phi} = \begin{bmatrix} 5000 & 0 & 0 & 0 & 0 & 0 \\ 0 & 7000 & 0 & 0 & 0 & 0 \\ 0 & 0 & 0 & 0 & 0 & 0 \\ 0 & 0 & 0 & 0 & 0 & 0 \\ 0 & 0 & 0 & 0 & 0 & 0 \\ 0 & 0 & 0 & 0 & 0 & 0 \end{bmatrix} \boldsymbol{\Phi} \tag{A2}$$

- All the numerical models (Model A, B1-B3) have the inclusion of the same linear drag matrix to account for external damping from the experimental apparatus [42].

$$\boldsymbol{F_{Add}} = [\boldsymbol{B_{linear}}]\dot{\boldsymbol{\Phi}} = \begin{bmatrix} 3.67 \times 10^4 & 0 & 0 & 0 & 0 & 0 \\ 0 & 1.60 \times 10^4 & 0 & 0 & 0 & 0 \\ 0 & 0 & 1.18 \times 10^4 & 0 & 0 & 0 \\ 0 & 0 & 0 & 7.68 \times 10^7 & 0 & 0 \\ 0 & 0 & 0 & 0 & 7.14 \times 10^7 & 0 \\ 0 & 0 & 0 & 0 & 0 & 6.34 \times 10^7 \end{bmatrix} \dot{\boldsymbol{\Phi}} \tag{A3}$$

Appendix B. Time Domain History Plots for Tower Base Fore-Aft Shear and Up-Wave Mooring Line Tension

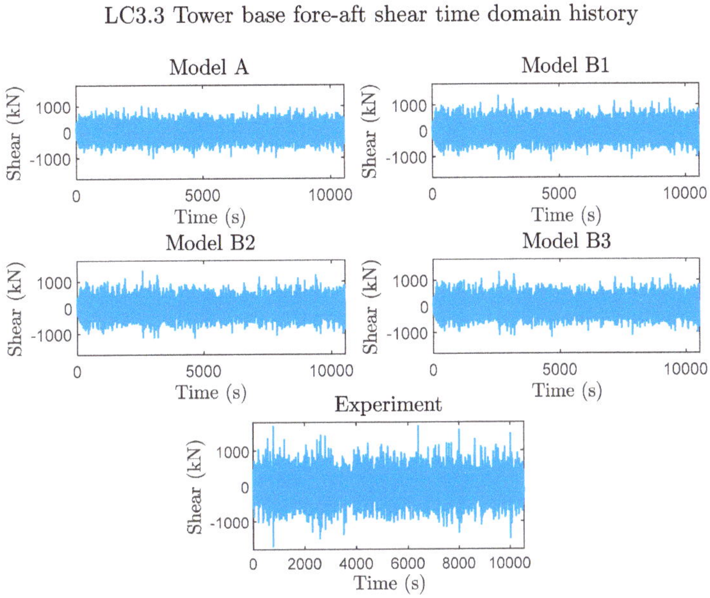

(a)

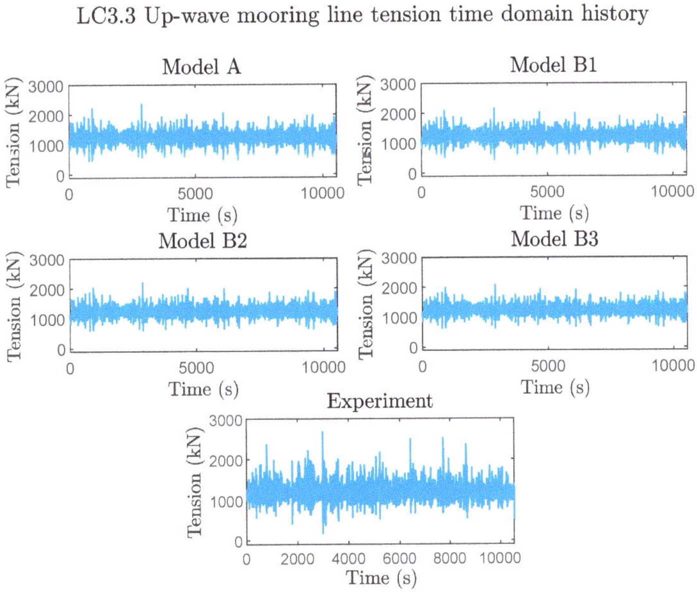

(b)

Figure A1. LC3.3 tower base fore-aft shear and up-wave mooring line tension time domain history: (**a**) tower base fore-aft shear time domain history; (**b**) up-wave mooring line tension time domain history.

References

1. United Nations. Renewable Energy—Powering a Safer Future. Available online: https://www.un.org/en/climatechange/raising-ambition/renewable-energy (accessed on 7 October 2022).
2. United Nations. The Paris Agreement. Available online: https://unfccc.int/process-and-meetings/the-paris-agreement/the-paris-agreement (accessed on 4 December 2023).
3. DNV. *Floating Offshore Wind: The Next Five Years*; DNV: Oslo, Norway, 2022.
4. Global Wind Energy Council (GWEC). *Floating Offshore Wind—A Global Opportunity*; Global Wind Energy Council (GWEC): Brussels, Belgium, 2022.
5. Office of Energy Efficiency and Renewable Energy. The Future of Offshore Wind Is Big—Literally. 2021. Available online: https://www.energy.gov/eere/wind/articles/future-offshore-wind-big-literally (accessed on 14 July 2022).
6. International Energy Agency (IEA). *Offshore Wind Outlook 2019*; International Energy Agency (IEA): Paris, France, 2019; Available online: https://www.iea.org/reports/offshore-wind-outlook-2019 (accessed on 14 July 2022).
7. Cottura, L.; Caradonna, R.; Ghigo, A.; Novo, R.; Bracco, G.; Mattiazzo, G. Dynamic modeling of an offshore floating wind turbine for application in the mediterranean sea. *Energies* **2021**, *14*, 248. [CrossRef]
8. International Energy Agency (IEA). *Global Wind Energy Research Collaboration*; International Energy Agency (IEA): Paris, France; Available online: https://iea-wind.org/ (accessed on 23 September 2022).
9. International Energy Agency (IEA). *Task 30 | IEA Wind TCP*; International Energy Agency (IEA): Paris, France; Available online: https://iea-wind.org/task30/ (accessed on 23 September 2022).
10. Robertson, A.; Jonkman, J.; Vorpahl, F.; Popko, W.; Qvist, J.; Frøyd, L.; Chen, X.; Azcona, J.; Uzunoglu, E.; Guedes Soares, C.; et al. Offshore Code Comparison Collaboration Continuation within IEA Wind Task 30: Phase II Results Regarding a Floating Semisubmersible Wind System. In Proceedings of the 33rd International Conference on Ocean, Offshore and Arctic Engineering 2014, San Francisco, CA, USA, 8–13 June 2014. [CrossRef]
11. Robertson, A.N.; Wendt, F.; Jonkman, J.M.; Popko, W.; Dagher, H.; Gueydon, S.; Qvist, J.; Vittori, F.; Azcona, J.; Uzunoglu, E.; et al. OC5 Project Phase II: Validation of Global Loads of the DeepCwind Floating Semisubmersible Wind Turbine. *Energy Procedia* **2017**, *137*, 38–57. [CrossRef]
12. Uzunoglu, E.; Soares, C.G. On the model uncertainty of wave induced platform motions and mooring loads of a semisubmersible based wind turbine. *Ocean Eng.* **2018**, *148*, 277–285. [CrossRef]
13. Beardsell, A.; Alexandre, A.; Child, B.; Harries, R.; Beyond, D.M. OC5—Further advances in floating wind turbine modelling using Bladed. *J. Phys. Conf. Ser.* **2018**, *1102*, 012023. [CrossRef]
14. Robertson, A.; Bachynski, E.E.; Gueydon, S.; Wendt, F.; Schünemann, P. Total experimental uncertainty in hydrodynamic testing of a semisubmersible wind turbine, considering numerical propagation of systematic uncertainty. *Ocean. Eng.* **2020**, *195*, 106605. [CrossRef]
15. Wang, L.; Robertson, A.; Jonkman, J.; Yu, Y.-H. OC6 phase I: Improvements to the OpenFAST predictions of nonlinear, low-frequency responses of a floating offshore wind turbine platform. *Renew Energy* **2022**, *187*, 282–301. [CrossRef]
16. Zhang, L.; Shi, W.; Karimirad, M.; Michailides, C.; Jiang, Z. Second-order hydrodynamic effects on the response of three semisubmersible floating offshore wind turbines. *Ocean Eng.* **2020**, *207*, 107371. [CrossRef]
17. Kvittem, M.I.; Bachynski, E.E.; Moan, T. Effects of Hydrodynamic Modelling in Fully Coupled Simulations of a Semi-submersible Wind Turbine. *Energy Procedia* **2012**, *24*, 351–362. [CrossRef]
18. Bayati, I.; Jonkman, J.; Robertson, A.; Platt, A. The effects of second-order hydrodynamics on a semisubmersible floating offshore wind turbine. *J. Phys. Conf. Ser.* **2014**, *524*, 012094. [CrossRef]
19. DNV DNVGL-RP-C205: Environmental conditions and environmental loads. 2010. Available online: https://rules.dnv.com/docs/pdf/DNVPM/codes/docs/2010-10/rp-c205.pdf (accessed on 4 December 2023).
20. Gueydon, S. Aerodynamic Damping on a Semisubmersible Floating Foundation for Wind Turbines. *Energy Procedia* **2016**, *94*, 367–378. [CrossRef]
21. Robertson, A.; Jonkman, J.; Masciola, M.; Song, H.; Goupee, A.; Coulling, A.; Luan, C. *Definition of the Semisubmersible Floating System for Phase II of OC4*; National Renewable Energy Lab: Golden, CO, USA, 2014. [CrossRef]
22. Burmester, S.; Vaz, G.; Gueydon, S.; el Moctar, O. Investigation of a semi-submersible floating wind turbine in surge decay using CFD. *Ship Technol. Res.* **2020**, *67*, 2–14. [CrossRef]
23. Coulling, A.J.; Goupee, A.J.; Robertson, A.N.; Jonkman, J.M.; Dagher, H.J. Validation of a FAST semi-submersible floating wind turbine numerical model with DeepCwind test data. *J. Renew. Sustain. Energy* **2013**, *5*, 023116. [CrossRef]
24. Gueydon, S.; Duarte, T.; Jonkman, J. Comparison of Second-Order Loads on a Semisubmersible Floating Wind Turbine. In Proceedings of the ASME 2014 33rd International Conference on Ocean, Offshore and Arctic Engineering. Volume 9A: Ocean Renewable Energy, San Francisco, CA, USA, 8–13 June 2014. [CrossRef]
25. Todalshaug, J.H. Hydrodynamics of WECs. In *Handbook of Ocean Wave Energy. Ocean Engineering & Oceanography*; Springer: Cham, Switzerland, 2017; pp. 139–158. [CrossRef]
26. Faltinsen, O. *Sea Loads on Ships and Offshore Structures*; Cambridge University Press: New York, NY, USA, 1990. Available online: https://www.osti.gov/biblio/5464335 (accessed on 4 December 2023).
27. Böhm, M.; Robertson, A.; Hübler, C.; Rolfes, R.; Schaumann, P. Optimization-based calibration of hydrodynamic drag coefficients for a semisubmersible platform using experimental data of an irregular sea state. *J. Phys. Conf. Ser.* **2020**, *1669*, 012023. [CrossRef]

28. Otter, A.; Murphy, J.; Pakrashi, V.; Robertson, A.; Desmond, C. A review of modelling techniques for floating offshore wind turbines. *Wind Energy* **2022**, *25*, 831–857. [CrossRef]
29. Robertson, A.; Jonkman, J.; Wendt, F.; Goupee, A.; Dagher, H. Definition of the OC5 DeepCwind Semisubmersible Floating System. 2016. Available online: https://a2e.energy.gov/api/datasets/oc5/oc5.phase2/files/oc5.phase2.model.definition-semisubmersible-floating-system-phase2-oc5-ver15.pdf (accessed on 4 December 2023).
30. Karimirad, M.; Michailides, C. V-shaped semisubmersible offshore wind turbine: An alternative concept for offshore wind technology. *Renew Energy* **2015**, *83*, 126–143. [CrossRef]
31. Gao, S.; Zhang, L.; Shi, W.; Wang, B.; Li, X. Dynamic Responses for WindFloat Floating Offshore Wind Turbine at Intermediate Water Depth Based on Local Conditions in China. *J. Mar. Sci. Eng.* **2021**, *9*, 1093. [CrossRef]
32. Zhao, Z.; Shi, W.; Wang, W.; Qi, S.; Li, X. Dynamic analysis of a novel semi-submersible platform for a 10 MW wind turbine in intermediate water depth. *Ocean Eng.* **2021**, *237*, 109688. [CrossRef]
33. Li, C.; Zhuang, T.; Zhou, S.; Xiao, Y.; Hu, G. Passive Vibration Control of a Semi-Submersible Floating Offshore Wind Turbine. *Appl. Sci.* **2017**, *7*, 509. [CrossRef]
34. Jonkman, J.M. Dynamics of offshore floating wind turbines-model development and verification. *Wind Energy* **2009**, *12*, 459–492. [CrossRef]
35. Jonkman, J.; Buhl, M. Development and Verification of a Fully Coupled Simulator for Offshore Wind Turbines. In Proceedings of the 45th AIAA Aerospace Sciences Meeting and Exhibit, Reno, NV, USA, 8–11 January 2007. [CrossRef]
36. Kim, J.; Shin, H. Validation of a 750 kW semi-submersible floating offshore wind turbine numerical model with model test data, part II: Model-II. *Int. J. Nav. Archit. Ocean. Eng.* **2020**, *12*, 213–225. [CrossRef]
37. Browning, J.R.; Jonkman, J.; Robertson, A.; Goupee, A.J. Calibration and validation of a spar-type floating offshore wind turbine model using the FAST dynamic simulation tool. *J. Phys. Conf. Ser.* **2014**, *555*, 012015. [CrossRef]
38. Driscoll, F.; Jonkman, J.; Robertson, A.; Sirnivas, S.; Skaare, B.; Nielsen, F.G. Validation of a FAST Model of the Statoil-hywind Demo Floating Wind Turbine. *Energy Procedia* **2016**, *94*, 3–19. [CrossRef]
39. DNV DNV-RP-0286 Coupled Analysis of Floating Wind Turbines. 2019. Available online: https://www.dnv.com/energy/standards-guidelines/dnv-rp-0286-coupled-analysis-of-floating-wind-turbines.html (accessed on 26 September 2022).
40. Jonkman, J.M.; Robertson, A.N.; Hayman, G.J. HydroDyn User's Guide and Theory Manual. National Renewable Energy Laboratory. 2015. Available online: https://www.nrel.gov/wind/nwtc/assets/downloads/HydroDyn/HydroDyn_Manual.pdf (accessed on 4 December 2023).
41. WAMIT Wamit User Manual, Version 6. 2017. Available online: https://www.wamit.com/manualupdate/v74_manual.pdf (accessed on 30 May 2022).
42. Wendt, F.F.; Andersen, M.T.; Robertson, A.N.; Jonkman, J.M. Verification and Validation of the New Dynamic Mooring Modules Available in FAST v8. 2016. Available online: https://www.osti.gov/biblio/1328087 (accessed on 4 December 2023).
43. Wendt, F.; Robertson, A.; Jonkman, J. FAST Model Calibration and Validation of the OC5-DeepCwind Floating Offshore Wind System Against Wave Tank Test Data. *Int. J. Offshore Polar Eng.* **2019**, *29*, 15–23. [CrossRef]
44. Duarte, T.; Sarmento, A.; Alves, M.; Jonkman, J.; Sarmento, A. State-Space Realization of the Wave-Radiation Force within FAST: Preprint. 2013. Available online: http://www.osti.gov/bridgeonlineordering:http://www.ntis.gov/help/ordermethods.aspx (accessed on 28 September 2022).
45. Hall, M. MoorDyn User's Guide. 2017. Available online: http://www.matt-hall.ca/files/MoorDyn%20Users%20Guide%202017-08-16.pdf (accessed on 4 December 2023).
46. Chakrabarti, S.K. *The Theory and Practice of Hydrodynamics and Vibration*; World Scientific: River Edge, NJ, USA, 2002.
47. Pessoa, J.; Fonseca, N.; Soares, C.G. Experimental and Numerical Study of the Depth Effect on the First Order and Slowly Varying Motions of a Floating Body in Bichromatic Waves. In Proceedings of the International Conference on Offshore Mechanics and Arctic Engineering—OMAE, Shanghai, China, 6–11 June 2010; pp. 723–731. [CrossRef]
48. Fonseca, N.; Pessoa, J.; Soares, C.G. Calculation of Second Order Drift Forces on a FLNG Accounting for Difference Frequency Components. In Proceedings of the International Conference on Offshore Mechanics and Arctic Engineering—OMAE, Estoril, Portugal, 15–20 June 2009; Volume 6, pp. 369–377. [CrossRef]
49. DSA Wave Drift Load Modelling, Documentation—Theory—Validation. 2017. Available online: https://dsaocean.com/wp-content/uploads/2017/12/WaveDriftDocumentation.pdf (accessed on 24 May 2022).
50. Zhang, D.; Paterson, E.G. A study of wave forces on an offshore platform by direct CFD and Morison equation. *E3S Web Conf.* **2015**, *5*, 04002. [CrossRef]
51. Lemmer, F.; Yu, W.; Cheng, P.W.; Pegalajar-Jurado, A.; Borg, M.; Mikkelsen, R.F.; Bredmose, H. The TripleSpar Campaign: Validation of a Reduced-Order Simulation Model for Floating Wind Turbines. In Proceedings of the ASME 2018 37th International Conference on Ocean, Offshore and Arctic Engineering. Volume 10: Ocean Renewable Energy, Madrid, Spain, 17–22 June 2018. [CrossRef]
52. Robertson, A.N.; Gueydon, S.; Bachynski, E.; Wang, L.; Jonkman, J.; Alarcón, D.; Amet, E.; Beardsell, A.; Bonnet, P.; Boudet, B.; et al. OC6 Phase I: Investigating the underprediction of low-frequency hydrodynamic loads and responses of a floating wind turbine. *J. Phys. Conf. Ser.* **2020**, *1618*, 032033. [CrossRef]
53. Dynamic Systems Analysis ProteusDS Manual v2.29.2. Victoria, BC, Canada. 2016. Available online: https://dsaocean.com/downloads/documentation/ProteusDS%202015%20Manual.pdf (accessed on 4 December 2023).
54. Catalano, P.; Wang, M.; Iaccarino, G.; Moin, P. Numerical simulation of the flow around a circular cylinder at high Reynolds numbers. *Int. J. Heat Fluid Flow* **2003**, *24*, 463–469. [CrossRef]

55. Coefficient of Variation Meaning and How to Use It. Available online: https://www.investopedia.com/terms/c/coefficientofvariation.asp (accessed on 12 October 2022).
56. Siemens What is a Power Spectral Density (PSD)? 10 July 2020. Available online: https://community.sw.siemens.com/s/article/what-is-a-power-spectral-density-psd (accessed on 27 October 2022).
57. Meirovitch, L. *Fundamentals of Vibrations*; McGraw-Hill: Boston, MA, USA, 2001.
58. Kvittem, M.I.; Berthelsen, P.A.; Eliassen, L.; Thys, M. Calibration of Hydrodynamic Coefficients for a Semi-Submersible 10 MW Wind Turbine. In Proceedings of the ASME 2018 37th International Conference on Ocean, Offshore and Arctic Engineering. Volume 10: Ocean Renewable Energy, Madrid, Spain, 17–22 June 2018. [CrossRef]
59. Lopez-Pavon, C.; Watai, R.A.; Ruggeri, F.; Simos, A.N.; Souto-Iglesias, A. Influence of Wave Induced Second-Order Forces in Semisubmersible FOWT Mooring Design. *J. Offshore Mech. Arct. Eng.* **2015**, *137*, 031602. [CrossRef]

Journal of
Marine Science and Engineering

MDPI

Article

Hydrodynamic Responses of a Barge-Type Floating Offshore Wind Turbine Integrated with an Aquaculture Cage

Yuting Zhai [1,2], Haisheng Zhao [1,2,*], Xin Li [1,2,*] and Wei Shi [1]

1 State Key Laboratory of Coastal and Offshore Engineering, Dalian University of Technology, Dalian 116024, China; d12006089@mail.dlut.edu.cn (Y.Z.); weishi@dlut.edu.cn (W.S.)

2 School of Hydraulic Engineering, Faculty of Infrastructure Engineering, Dalian University of Technology, Dalian 116024, China

* Correspondence: hzhao@dlut.edu.cn (H.Z.); lixin@dlut.edu.cn (X.L.)

Abstract: The dynamic responses of a new structure combining a barge-type floating offshore wind turbine and an aquaculture cage is investigated numerically. First, a 5 MW barge-type floating offshore wind turbine with an aquaculture cage (FOWT-AC) is designed and the numerical model is established in ANSYS-AQWA. The numerical model of the barge-type FOWT-AC is then checked, and the natural periods of the six degrees of freedom motion satisfy the recommendations of the DNV specification. Based on the reasonable model, the comparison study of dynamic responses between the barge-type FOWT-AC and FOWT under the environmental conditions of the South China Sea is carried out, and it is observed that the FOWT-AC produces a basically lower standard deviation of the motion responses. To investigate the new structure of the barge-type FOWT-AC deeply, the analyses of second-order hydrodynamic response, typical environmental conditions and the mooring line breaking scenario are carried out. The simulation results show that the second-order wave loads increase the dynamic response of the barge-type FOWT-AC slightly unless it causes resonance for the structure. In addition, the motion responses of the floating structures increase significantly when the currents are applied, especially when the aquaculture cage is integrated into the barge-type FOWT. When one of the mooring lines connected to the offshore or onshore side of the platform breaks, the presence of the aquaculture cage results in a smaller standard deviation in the motion responses of the coupled structure, which means that the barge-type FOWT-AC structure is more stable.

Keywords: offshore wind turbine; aquaculture cage; barge type; motion response; mooring system

Citation: Zhai, Y.; Zhao, H.; Li, X.; Shi, W. Hydrodynamic Responses of a Barge-Type Floating Offshore Wind Turbine Integrated with an Aquaculture Cage. *J. Mar. Sci. Eng.* **2022**, *10*, 854. https://doi.org/10.3390/jmse10070854

Academic Editors: Eugen Rusu, Emre Uzunoglu, Antonio Souto-Iglesias and Carlos Guedes Soares

Received: 28 April 2022
Accepted: 16 June 2022
Published: 22 June 2022

1. Introduction

The development and utilization of wind power started to gain attention since the problems of environmental pollution and the energy crisis became serious. The offshore wind turbine is certainly an excellent choice because of the higher speed of the wind and the lower turbulence of offshore wind farms compared to land-based wind farms. The cost of building traditional fixed wind turbines increases rapidly as the wind power development gradually moves to the deep sea [1], thus floating wind turbines are beginning to be studied by scholars in various countries. Compared with the onshore wind turbine, the floating offshore wind turbine (FOWT) is a dynamic system with a higher degree of freedom because it is simultaneously subjected to the wind, currents and wave loads, and is constrained by a mooring system [2].

At present, the main forms of floating offshore wind turbines are the spar-type, semi-submersible type, tension leg type [3], barge-type [4], innovative type platforms [5], etc. The spar-type platform has a large draft and high-water depth requirement, and its advantages are simple structural form and it is unconditionally stable [6]. The semi-submersible type platform has the advantages of simple installation and good towing, while its structure is complicated and expensive to build [7]. The TLP-type platform has been limited in its development and use due to the high cost of installation [8]. The barge-type platform has the advantages of simple structure, low cost, and long life, particularly suitable for a water depth of 50~100 m due to its low draft feature [9].

As described in the references [10,11], dynamic wave-structure interaction may result in dynamic amplification/reamplification of the motion response of the floating structure; thus, it is important to study the hydrodynamic performance or the dynamic response of these floating platforms under various environmental loads. Utsunomiya et al. [12] developed a dynamic analysis tool to study spar-type offshore wind turbines and compared the analysis results of the software with those of the experiments. The results showed that the developed program has captured the main behavior of the FOWT. The study of Chen et al. [13] found that the semi-submersible FOWT is more sensitive to wave loading compared to Spar-buoy FOWT. However, when vortex excitation motion (VIM) occurs, the simulation underestimates the current load and lateral flow response because the simulation code does not take into account the effects of VIM. Cao et al. [14] found that the pitch resonant responses differed more significantly between the Newman model and the full QTF model compared to the surge resonance response by studying a semi-submersible 10 MW wind turbine. Therefore, the use of the Newman approximation is inaccurate for second-order calculations, and the high-frequency response appears in the tower top shear spectrum only when the full QTF is used. In the past, the hydrodynamic performance of the barge platform had some disadvantages such as poorer stability compared with other platforms; then IDEOL company [15] developed the concept of "damping pool" (called as "moon pool" as well) to improve the hydrodynamic performance, reduce motion response, and lower construction costs of barge-type platforms. Currently, various studies of barge-type wind turbines have begun to grow, and all currently designed barge-type wind turbine platforms are starting to come with a moon pool. For example, a barge-type FOWT with a moon pool had been designed and their numerical results using AQWA agreed well with the experimental results; meanwhile, wave-frequency motion was found to induce motion primarily in the vertical plane [14]. Whether the breakage [16–19] of mooring lines affects the normal operation of the 5 MW barge-type FOWT was explored by analysis tool F2A, indicating that shutdown is a beneficial measure to protect the platform safety from mooring line breakage [20].

The aquaculture farming below the water level is also an effective use of marine resources in addition to the utilization of wind energy above the water. Currently, high-yield deep-sea farming has become a popular industry, and more than 20 countries and regions around the world are dedicated to the field of aquaculture [21]. Norway, as one of the most developed fishing countries in the world, has already been a world leader in the field of aquaculture equipment design with the aquaculture production increasing to 10 times as much as that of 1990. Particularly in recent years, several famous deep-sea farms have accelerated the development of deep-sea aquaculture production. A 79 m diameter rigid semi-submersible fish farm, Arctic Offshore Farming, was proposed in 2017, consisting of a support structure with pontoons and braces, a mooring system and net cages [22]. One of the most famous aquaculture farm is the OceanFarm1 [23] developed by SalMar in Norway.

The fish farms are subject to currents and waves causing motion responses or fatigue damage to the aquaculture cage; thus, a hydrodynamic analysis of the aquaculture cage is essential. By comparing the numerical results and experimental results in tanks, a method for the structural analysis of aquaculture nets was developed and it was concluded that the resistance load and cage volume depended on the size and weight of the net system [24]. Based on the lumped-mass method and rigid-body kinematics theory, Xu et al. [25] established a mathematical model of a gravity cage system attacked by irregular waves and simulated the hydrodynamic response of the cage system. It was found that the cage system had no significant tumbling motion, and the cage system tended to self-profile against longer waves at high frequencies.

Currently, the upfront investment in wind power is high, and the annual income from power generation alone is still far less than the total investment in offshore wind farms. In order to compensate it, some scholars have come up with the idea of combining offshore wind turbines with aquaculture cage structures. For example, He [26] proposed the multi-use platform concept of combining 10 MW wind turbines with aquaculture in 2015. The earlier research on floating offshore wind turbines with a fish farming cage in China was conducted at Tsinghua University [27]. In order to explore the motion performance, a floating offshore wind turbine with a steel fish-farming cage (FOWT-SFFC) was designed, and the random response and nonlinear dynamic performance of the structure were studied simultaneously; it is noted that the large motion of the pitch has little effect on the nonlinearity of the viscous drag and the mooring interaction [27,28]. Chu and Wang [28] were concerned with the hydrodynamic response of a novel offshore fish farm in the combined floating spar wind turbine and fish cage. An offshore floating multi-purpose platform supporting a 10 MW wind turbine on one side of this barge platform with a fish cage was designed for Blue Growth Farm. Numerical simulation results showed that the dynamic responses of the multi-purpose platform were excellent [29,30]. Moreover, there are other multi-purpose platform concepts that combine wind turbines, aquaculture systems and wave energy converters [31,32] or solar energy [33] into these multi-purpose structures to synthesize various resources.

The coupled structure of a wind turbine with an aquaculture cage can not only use the sea space efficiently, but can also achieve greater benefits with a smaller investment. However, this new type of coupling structure is limited in its development because of its novel orientation. Therefore, it is necessary to propose a reasonable new structure of floating offshore wind turbine integrated with an aquaculture cage (FOWT-AC). Currently, most of the studies on this coupled structure are based on semi-submersible or spar type platforms, whereas the barge-type platform is convenient for transportation, simple in structure, and large in platform area, which is also conducive to the installation of aquaculture control unit supplying constant power for remote and autonomous fish farming operations. There is little research on the barge-type floating offshore wind turbines integrated with an aquaculture cage. Therefore, a novel coupled structure combining a barge-type FOWT and an aquaculture cage is designed in this study.

Numerical simulation is an effective means in analyzing the hydrodynamic characteristics of floating structures, and various numerical methods have been developed for the wave-structure interaction of flexible, floating or moving marine structures. For example, the traditional Boundary Element Method (BEM) [34] can be used to calculate the linear steady-state response of large floating bodies in regular waves; the Smoothed Particle Hydrodynamics (SPH) method [35] can solve the distortion and stretching of the medium by solving partial differential equations, and even combine it with the Finite Element Method (FEM) or BEM to evolve into the coupled SPH-FEM/DEM [36,37]; furthermore, the Finite Volume Method (FVM) [38], Finite Difference Method or even arbitrary Lagrangian-Eulerian Method [39] are widely used for the flow field analysis or dynamic analysis of marine structures. To investigate the hydrodynamic responses of the integrated structure and the effect of the presence of the aquaculture cage on the structure, the FEM-based numerical model is developed by ANSYS-AQWA and the numerically simulated results for the barge-type FOWT-AC and barge-type FOWT are compared in detail.

The remainders of this paper are organized as follows: Section 2 introduces the theoretical methods used in this research; Section 3 presents descriptions of the FOWT-AC, simulation environment and model validation. In Section 4, the results and discussion of this study are reported. Section 5 presents the conclusions of this work. Finally, Section 6 presents the future work.

2. Theoretical Methods

The FOWT-AC is a complex multi-body coupled structure with six freedom degrees, which is mainly subject to the environment loads of wind, waves and currents. The interaction of the coupled structure and environmental loads results in a complicated dynamic response, including elastic deformation of the blades and tower, the rigid body motion of the platform, the dynamic response of the mooring system, and so on. Therefore, the analysis of the coupled structure involves hydrodynamics, structural dynamics, the control of the wind turbine and many other aspects of the theoretical methods.

2.1. Equation of Motion in the Time Domain

To calculate the responses of such a complex system, which is regarded as a rigid body in this paper, the time-domain equation of the FOWT-AC can be obtained in the following equation when the three-dimensional potential flow theory is used,

$$(\boldsymbol{M}+\boldsymbol{A}_{\infty})\ddot{\boldsymbol{x}}(t)+\int_{0}^{t}\boldsymbol{K}(t-\tau)\dot{\boldsymbol{x}}(\tau)d\tau+\boldsymbol{C}\boldsymbol{x}(t)=\boldsymbol{F}_{wind}(t)+\boldsymbol{F}_{wave}(t)+\boldsymbol{F}_{curr}(t)+\boldsymbol{F}_{moor}(t) \quad (1)$$

where $\boldsymbol{x}$, $\dot{\boldsymbol{x}}$ and $\ddot{\boldsymbol{x}}$ represent the six degrees of freedom (DOFs) displacement, velocity and acceleration of the FOWT-AC platform; $\boldsymbol{M}$ is mass matrix; $\boldsymbol{A}_{\infty}$ is the infinite-frequency added mass matrix; $\boldsymbol{K}$ represents the wave-radiation-retardation kernel matrix; $\boldsymbol{C}$ is defined as hydrostatic stiffness matrix; $\boldsymbol{F}_{wind}$ is the wind loads on the FOWT-AC obtained by wind speed-thrust curve; $\boldsymbol{F}_{wave}$ and $\boldsymbol{F}_{cuur}$ are the wave excitation load and current force, respectively, which are calculated by potential flow theory and Morison equation; $\boldsymbol{F}_{moor}$ represents the restoring force of the mooring system.

$$\boldsymbol{K}(t)=\frac{2}{\pi}\int_{0}^{\infty}\boldsymbol{b}(\omega)\cos(\omega t)d\omega \quad (2)$$

where $\boldsymbol{b}$ is the linear radiation damping matrix.

2.2. Potential Flow Theory

The total excitation loads which come from incident waves act on the FOWT-AC structure including first- and second-order wave excitation forces. If the wave steepness of the incident wave is small relative to the size of the structure such as the platform of FOWT-AC, the velocity potential can be written in the form of a series expansion:

$$\boldsymbol{\Phi}(x,y,z,t)=\boldsymbol{\Phi}^{(1)}(x,y,z,t)+\boldsymbol{\Phi}^{(2)}(x,y,z,t)+\cdots+\boldsymbol{\Phi}^{(n)}(x,y,z,t)+\cdots, \quad (3)$$

where $\boldsymbol{\Phi}^{(1)}$, $\boldsymbol{\Phi}^{(2)}$, $\boldsymbol{\Phi}^{(n)}$ are the first-order, second-order and nth-order velocity potentials, respectively.

For any sea state, it is assumed that the waves are superimposed by Airy linear waves. The first-order total velocity potential can also be seen as a superposition of the first-order velocity potentials of different frequency wave components, the specific formula of which is expressed as follows:

$$\boldsymbol{\Phi}^{(1)}(x,y,z,t)=\mathrm{Re}\sum_{j}\phi_{j}^{(1)}(x,y,z)e^{i\omega_{wave,j}t}, \quad (4)$$

where ω_{wave} is the wave frequency; i and j are the ith and jth incident wave.

Similarly, the expression for the second-order total potential function containing sum and difference frequencies is as follows:

$$\boldsymbol{\Phi}^{(2)}(x,y,z,t)=\mathrm{Re}\sum_{j}\sum_{i}[\phi_{ji}^{+}(x)e^{i(\omega_{wave,j}+\omega_{wave,i})t}+\phi_{ji}^{-}(x)e^{i(\omega_{wave,j}-\omega_{wave,i})t}], \quad (5)$$

Once the potential function is obtained, the hydro-dynamic pressure (p) at any point of the flow field can be obtained according to Bernoulli's Equation (6), and the wave force acting on the structure can be obtained by integration:

$$\frac{\partial \boldsymbol{\Phi}}{\partial t} + \frac{1}{2}(\nabla \boldsymbol{\Phi}) \cdot (\nabla \boldsymbol{\Phi}) + gz + \frac{p}{\rho} = 0, \quad (6)$$

$$\boldsymbol{F}_{wave} = \int_{S_B} p(x, y, z, t)\boldsymbol{n} dS, \quad (7)$$

where S_B is the wet surface of the structure at any moment; p is the hydro-dynamic pressure; $\boldsymbol{n}$ is the normal vector of the structure surface.

2.3. Mesh Grouping Method

Due to the large size of the aquaculture cage, the large number of netting lines increase the modeling difficulty and also grow the computational volume significantly when the nets are modeled according to the prototype. Therefore, most of the netting models use the mesh grouping method to reduce the number of netting lines. In order to ensure the accuracy of the calculations after mesh grouping, the following principles need to be observed:

(1) The nets should be kept equal in quantity before and after grouping.
(2) Similar geometric scales should be maintained before and after grouping.
(3) The nets before and after grouping should be kept equal to the hydrodynamic force.

The above three conditions can be expressed as three equations as follows:

$$M_{grouping} = M, \quad (8)$$

$$A_{grouping} = \sum A, \quad (9)$$

$$(EA_{section})_{grouping} = \sum EA_{section}, \quad (10)$$

where M for the quality of the netting; A and $A_{section}$ represent the projected area and cross-sectional area of the netting; and E represents the Young's modulus of the netting.

The paramount characteristic for netting without knots is the solidity ratio (S_n). The S_n is defined as the area of nets to the area of the plane. From Figure 1, the solidity ratio can be calculated as follows:

$$S_n = \frac{2d}{\lambda} - \left(\frac{d}{\lambda}\right)^2, \quad (11)$$

where λ represents the mesh size, and d is rope diameter.

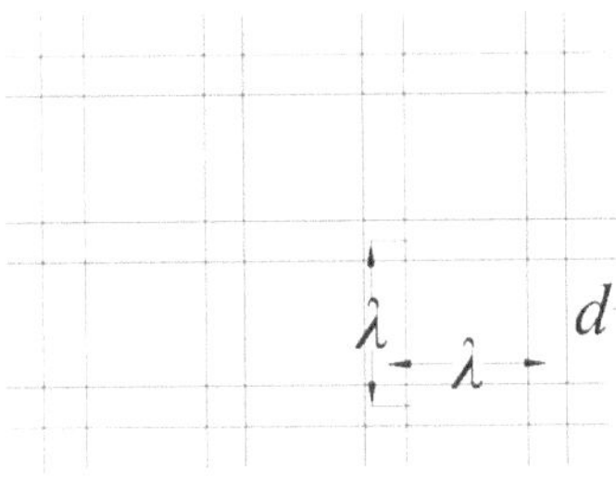

Figure 1. Equivalent netting.

In order to ensure that the hydrodynamic forces before and after grouping are similar, it is necessary to ensure that the real nets and the equivalent nets have the same C_d and densities S_n.

2.4. Hydrodynamic Force of Netting

The two main simplified models for calculating the hydrodynamic force of netting are the Morison model and the Screen model. The Screen model is originally developed for steady flow, while the Morison model is for a wider range of situations. As opposed to potential flow theory, the Morison equation is the method that applies when the diameter of the structure is small relative to the wavelength. Therefore, the Morison equation is adopted herein for simulating the hydrodynamic force of the netting and columns of the cage due to their diameter being small relative to the wavelength. The Morison equation can be written as the following expression,

$$F(t) = \frac{1}{2}\rho C_d D u(t)|u(t)| + (1 + C_m)\rho \frac{\pi D^2}{4} \dot{u}(t), \quad (12)$$

where $u(t)$ is the horizontal velocity at the wave water particles; $\dot{u}(t)$ is the horizontal acceleration at the wave water particles; C_m is the additional mass coefficient, and C_d is the drag coefficient.

Because the quality of the netting is quite small, the inertial forces on the netting are small, and only drag forces are considered [40]. A Morison model force diagram for two orthogonal equivalent net lines is shown in Figure 2, where A is the projected area of the net line; U_{rel} is the horizontal relative velocity of water particles and structure. By simply superimposing the drag force vectors on all the net lines, the drag force applied to the entire netting can be obtained.

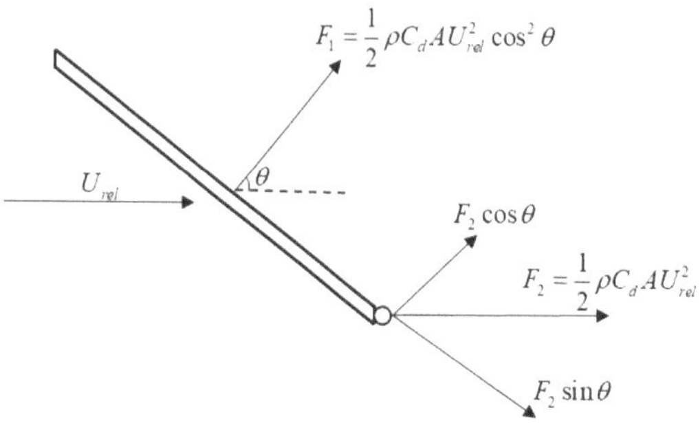

Figure 2. Force diagram for the Morrison model of orthogonal lines.

2.5. Mooring Lines Tension

Because the quasi-static method considers the steady and slow drifting motion of the floating structure by ignoring the dynamic effects, the quasi-static calculation can shorten calculation period and guarantee a certain accuracy when inertia forces have less influence. In the numerical simulation of floating wind turbines, the quasi-static method is already a mature method. For example, Zhao et al. [41,42] used this method in a semi-submersible wind turbine, and Yu et al. [43] employed it in the spar wind turbine, and reasonable numerical results were obtained. Hence, the mooring lines tension is calculated based on the quasi-static equilibrium method in this study, and the mass of the line, elastic tension and subsea friction are considered while the bending stiffness, the inertia and the damping of the mooring system are ignored [44]. It is beneficial for the safe design of the integrated structure since relatively conservative results are predicted.

A single mooring line with undercover length in a local coordinate system are shown in Figure 3. The anchor is the origin and the positive direction of the z-axis is vertical. The horizontal and vertical components of mooring lines effective tension (H_F, V_F) at the fairlead are calculated by the following equation [43]:

$$X_F(H_F, V_F) =$$
$$L - \frac{V_F}{\omega} + \frac{H_F}{\omega}\ln\left[\frac{V_F}{H_F} + \sqrt{1 + (\frac{V_F}{H_F})^2}\right] + \frac{H_F L}{EA_m} + \frac{C_B\omega}{2EA_m}\left[-(L - \frac{V_F}{\omega})^2 + (L - \frac{V_F}{\omega} - \frac{H_F}{C_B\omega})\cdot\max(L - \frac{V_F}{\omega} - \frac{H_F}{C_B\omega}, 0)\right] \quad (13)$$

$$Z_F(H_F, V_F) = \frac{H_F}{\omega}\left[\sqrt{1 + (\frac{V_F}{H_F})^2} - \sqrt{1 + (\frac{V_F - \omega L}{H_F})^2}\right] + \frac{1}{EA_m}(V_F L - \frac{\omega L^2}{2}), \quad (14)$$

$$F_{moor} = \sum\sqrt{H_F^2 + V_F^2}, \quad (15)$$

where L is the length of the mooring line; ω is the density of the mooring line in the water; A_m is the cross-sectional area of the mooring line; EA_m is the extensional stiffness; C_B is the friction coefficient of the seabed connection section of the mooring line.

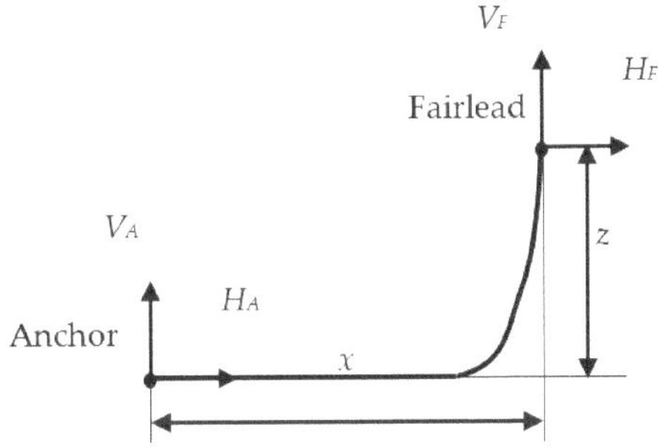

Figure 3. Mooring system model.

3. Numerical Model and Model Testing

3.1. Design of the FOWT-AC

The overall conceptual diagram of the barge-type FOWT-AC for this study is shown in Figure 4. The barge-type platform adopted refers to a similar concept in reference [16], which supports the National Renewable Energy Laboratory (NREL) 5 MW baseline wind turbine. Their specific parameters are shown in Tables 1 and 2. The aquaculture cage is directly connected to the bottom of the barge platform, which comprises eight side columns, sixteen diagonal columns and nets, forming a square shape cage with internal space of 60 m length and 30 m height, as shown in Figure 5. The wind turbine and the aquaculture cage share the platform to build workplaces, so the aquaculture cage can use the buoyancy of the platform as well as the electricity generated by the wind turbine directly for operation and maintenance. The selection of nets refers to the parameters of Econet of OceanFarm1 [23]. The solidity ratio S_n, mesh size λ and rope diameter d of the aquaculture cage are set to 0.16, 5 m and 0.4 m. The range of the damping coefficient in the model of the aquaculture cage is [0.4–1.2] [23], which is mainly determined from the Reynolds value. The value used in this study is 0.8 [23] according to the common sea state. The main parameters of the aquaculture cage can be found in Table 3. The designed water depth of the numerical simulation is selected as 50 m. In this case the mooring system belongs to a shallow water mooring. Due to the complexity of the shallow water mooring, eight mooring lines are applied to keep the coupled structure in position, which features two sets of upwind lines and two sets of downwind lines. The mooring line adopts a gearless catenary type, and it is Grade 4. Table 4 lists the detailed parameters of the mooring system.

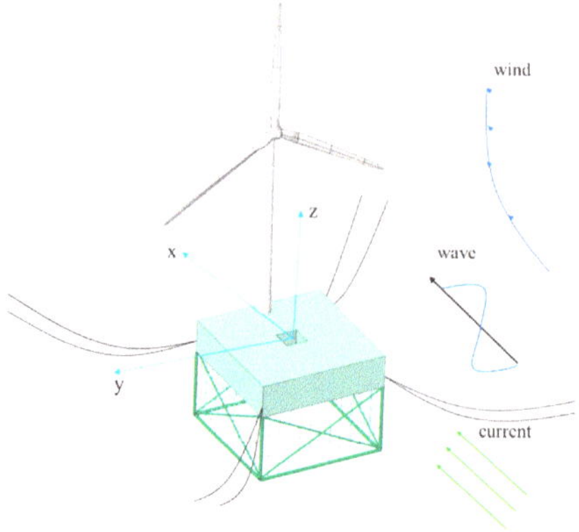

Figure 4. Conceptual diagram of the barge-type FOWT with aquaculture cage.

Table 1. Main parameters of the barge-type platform.

Parameters	Value
Barge size (W × L × H) (m)	60 × 60 × 15
Moonpool (W × L × H) (m)	10 × 10 × 15
Draft (m)	10
Mass (kg)	32,200,282
COG (m)	(−0.37, 0, −4.86)
Roll inertia (kg/m^3)	9,660,000,000
Pitch inertia (kg/m^3)	9,660,000,000
Yaw inertia (kg/m^3)	19,300,000,000

Table 2. Specifications of the NREL 5 MW wind turbine.

Parameters	Value
Rotor, nacelle, tower mass (t)	110, 240, 347.46
Hub height (m)	90
COG (m)	(20.0, 0.0, 70.6)
Rotor, hub diameter (m)	126,3
Roll inertia (kg/m^3)	240,000,000
Pitch inertia (kg/m^3)	240,000,000
Yaw inertia (kg/m^3)	480,000,000
Rotor, nacelle, tower mass (t)	110, 240, 347.46

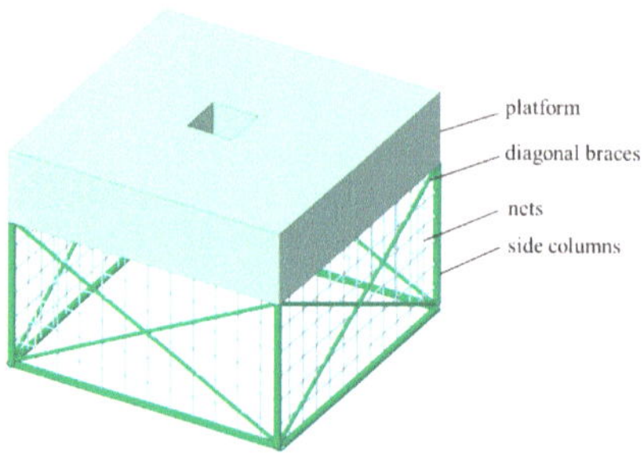

Figure 5. Details of the aquaculture cage.

Table 3. Main parameters of the aquaculture cage.

Parameters	Value
Diameter of side column, diagonal column and equivalent net (m)	1.5, 0.8, 0.4
The thickness of the side column and diagonal column (m)	0.02
Diameter of the equivalent net (m)	0.4
Distance of adjacent equivalent net (m)	5
Weight of equivalent net (g/m2)	590

Table 4. Properties of the mooring system.

Parameters	Value
Number of mooring lines	4 × 2
Anchor between adjacent lines	5°/85°
Water depth (m)	50
Radius to anchor form the platform centerline (m)	408
Unstretched mooring line length (m)	375
Mooring line diameter (m)	0.162
Catalog breaking load (kN)	9319
Equivalent mooring line mass density (kg/m)	522.25

3.2. Numerical Model

These coupled time-domain analyses for the barge-type FOWT-AC can be conducted by ANSYS-AQWA. ANSYS-AQWA [45] is mainly based on the Morison equation and 3D potential flow theory for simulation analysis, which allows the user to apply histories via dynamic link libraries and calculate the motion position, velocity and force of the structure accordingly. However, the 3D effects generated by the diagonal columns in the aquaculture cage are not captured by the simulation. Besides, the effect of wave energy loss on the onshore columns of the aquaculture cage is ignored.

Since the focus of this study is on the hydrodynamic performance and motion response of the platform, the wind turbine only plays the roles of gravity and the transfer of wind thrust and moment from the blades to the platform, so the upper parts of the structure can be simplified. In this work, the blades are simplified to a disk with the same wind sweeping area, and the dynamic link library is used to apply the thrust force and wind tilting moment that vary in real-time with the relative wind speed. The speed-thrust curve is shown in Figure 6. And the mesh model of the numerical simulation simplified is presented in Figure 7.

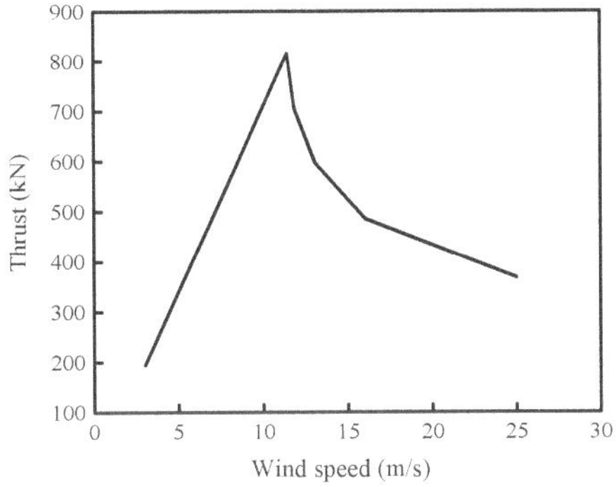

Figure 6. Diagram of the wind speed-thrust curve.

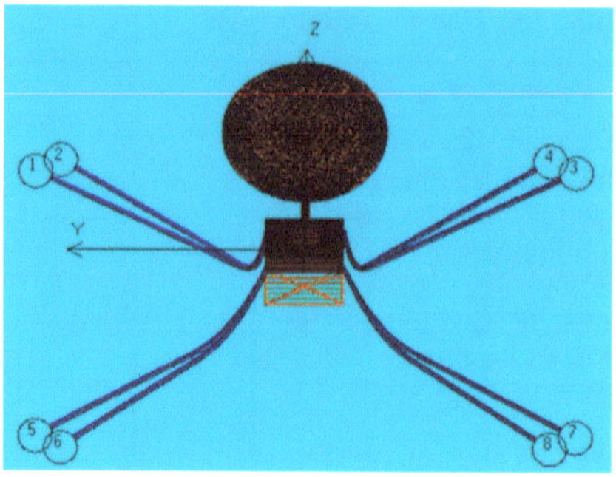

Figure 7. The mesh model of the numerical simulation simplified, where mooring line 1 to 8 are marked.

The flow chart of numerical simulation for barge-type FOWT-AC is illustrated in Figure 8. The FOWT-AC structure is composed of four parts from top to bottom, namely, wind turbine, barge platform, aquaculture cage and mooring system, and is coupled by three external loads of wind, wave and currents with its own structural motions. The wind turbine and above-water parts of the platform are subjected to the aerodynamics, while the underwater parts of the platform, aquaculture cage and mooring system suffer hydrodynamics. As indicated previously, the wind loads are set by the dynamic link library using user-force; the hydrodynamics of the platform and aquaculture cage are simulated by the potential flow theory and Morison equation; the tension of the mooring system is calculated by quasi-static theory [45]. Moreover, the operating state will be determined by the control system according to the dynamic response state of the wind turbine. The coupling effects mentioned in the above determine the final dynamic characteristics of the integrated FOWT-AC system.

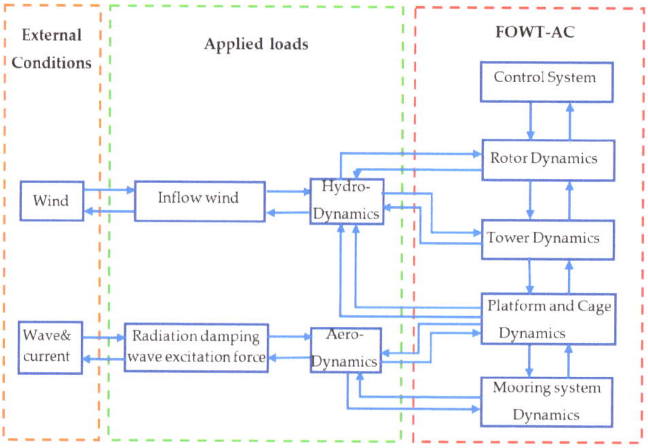

Figure 8. Flow chart of numerical model.

3.3. Environmental Conditions

To investigate the effect of hydrodynamic performance of the barge-type FOWT-AC, the environmental conditions (ECs) in this study are selected mainly by using the irregular wave, steady wind and current. And they are divided into eight ECs, as detailed in Table 5. The JONSWAP spectrum is used to generate the irregular wave time series by indicating the significant wave height (H_s) and spectral peak period (T_p). Three values of wave height are selected including the normal and extreme sea states in the South China Sea. And three values of wind speed are selected: below rated wind speed, rated wind speed and greater than rated wind speed. Two current conditions with and without currents are set up to investigate the effect of currents on the structure. Each simulation lasts for 4200 s, and the results of the first 600 s are eliminated to mitigate the startup transient effects. The wind, current and wave directions are input along the positive x-axis in the numerical calculations just as Figure 4. The schematic layout of mooring lines is illustrated in Figure 9. The ratio of wavelength-to-structural length (or width) has a critical effect on the hydrodynamic loads applied on a marine platform [39]; for example, the Morison equation is generally used when the ratio of the characteristic size of the structure to wavelength is less than 0.2, while the potential flow theory should be used for large size structures when the effect of radiation damping cannot be neglected. In this shallow water case, according to the dispersion relationship, it is known that the wavelength range in the above wave period cases is between 115 m and 210 m. The ratios of characteristic length of the columns of cage to the wavelength are between 0.004 and 0.013; thus, the Morison equation method is employed for the columns of the cage, while the potential flow theory is adopted for the hydrodynamic analysis of the barge-type platform due to the ratios of structural length to wavelength being located between 0.3 and 0.52.

Table 5. Environmental conditions.

	Wind Speed (m/s)	H_s (m)	T_p (m)	Current Speed (m/s)	Number of the Broken Line	Turbine Status
EC1	11.4	3.2	5.7	1	no	Operating
EC2	8	3.2	5.7	1	no	Operating
EC3	25	3.2	5.7	1	no	Shutdown
EC4	11.4	1.67	5.17	1	no	Operating
EC5	11.4	5.52	9.4	1	no	Operating
EC6	11.4	3.2	5.7	0	no	Operating
EC7	11.4	3.2	5.7	1	#3	Operating
EC8	11.4	3.2	5.7	1	#8	Operating

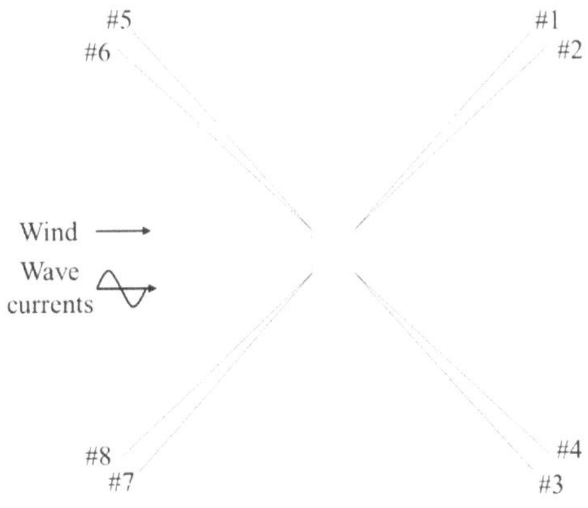

Figure 9. The schematic layout of mooring line 1 to 8.

3.4. Model Validation

Because of the powerful performance of ANSYS-AQWA, some researchers have chosen it to study FOWT. These models were built with a panel model of the large structure, Morison elements of the slender structure, and a quasi-static theory of the mooring system. Nallayarasu and Saravanapriya [46] built a Spar-type wind turbine model by ANSYS-AQWA and compared the numerical simulation results of the model with the experimental results and concluded that the experimentally measured RAOs agrees very well with the numerically simulated results with a maximum difference of 15%. By comparing the results of the 1:30 experimental model with the numerical model built using ANSYS-AQWA, Ruzzo et al. [47] noted that the experimental results in the peak frequency of the RAO were similar to the numerical prediction. Rajeswari and Nallayarasu [48] compared hydrodynamic responses of the experimental and the numerical results of the semi-submerged floaters with three columns and four columns, and a difference of less than 10% except the surge motion was observed. Based on the studies of these scholars it can be determined that the numerical modeling of floating wind turbines can get excellent results through the modeling theory of ANSYS-AQWA. For the new structures of floating wind turbine with a fish cage, Chu and Wang [28] modeled the Spar-type platform with a fish cage by means of ANSYS-AQWA and obtained reasonable results. The above references verify the validity of the theoretical methods implemented in ANSYS-AQWA in performing the hydrodynamic analyses of the floating wind turbine and the integrated structure.

To guarantee the mesh quality of the numerical model for the barge-type FOWT-AC, the comparison of the second-order drift force between the near-field and far-field is carried out. This is because the near-field method employs the potential flow theory to resolve the second-order drift force by integrating over the surface of a wet element, whereas the momentum theorem is adopted by the far-field method. Therefore, the near-field results are dependent on the mesh quality. When the results of the near-field and far-field methods follow the same trend with little difference, the mesh for hydrodynamic calculations of the floating structure is considered to meet the requirements. Figure 10 shows the comparison of second-order drift force results for the barge-type FOWT-AC and barge-type FOWT in the near and far fields. It can be seen that the results for the near and far fields of the same structure are approximately identical, representing the reliability of the mesh quality.

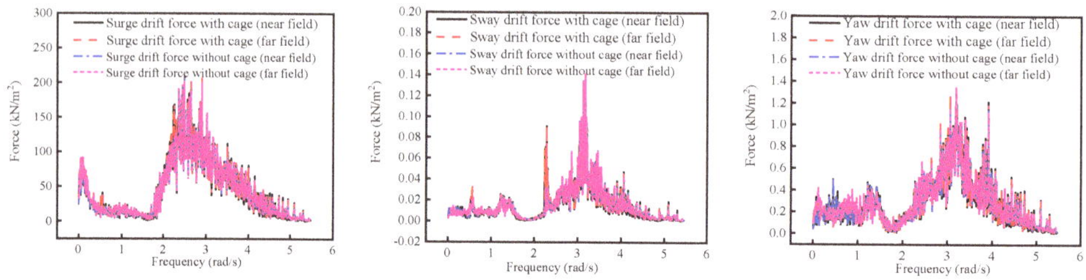

(a) (b) (c)

Figure 10. Comparison of second-order drift force results for the barge-type FOWT-AC and FOWT in the near and far fields: (**a**) Comparison of second-order surge drift force; (**b**) Comparison of second-order sway drift force; (**c**) Comparison of second-order yaw drift force.

After the numerical model is affirmed, AQWA-LINE is used to calculate the natural periods of heave and pitch, and the free decays of the surge and yaw are conducted by AQWA-DRIFT to determine their natural periods for the floating structures with and without the aquaculture cage, respectively. As can be seen from the results presented in Table 6, the natural periods of several degrees of freedom have increased due to the presence of the aquaculture cage, except for the yaw. The change of these periods is mainly due to the increase in the overall weight of the structure and the downward shift of the center of gravity. In addition, the natural periods of these two integrated structures are in accordance with the range of natural periods of barge-type FOWT platforms proposed by the DNV-RP−0286 [49] specification, indicating that the design of the floating structures is reasonable.

Table 6. The natural periods of the barge-type FOWT-AC and barge-type FOWT.

	With Cage	Without Cage
Surge	45.59 s	44.28 s
Heave	9.98 s	9.96 s
Pitch	10.82 s	10.09 s
Yaw	51.66 s	51.66 s
Surge	45.59 s	44.28 s

4. Results and Discussion

4.1. Effect of Second-Order Wave Loads on Motion Responses of Coupled Structures

Second-order wave loads are irregular forces composed of three components, such as second-order mean drift force, sum- and difference-frequency wave loads. Although second-order wave loads are smaller in magnitude than first-order wave loads, they can cause large drifts in marine structures if applied over a long period of time. And the sum- and difference-frequency wave loads can excite the floating structures' eigen-frequencies, resulting in significant oscillation, which may cause damage to the floating structures. Therefore, the study of second-order wave-excitation forces transfer functions (QTFs) is quite necessary.

Second-order difference-frequency and sum-frequency QTFs of the barge-type FOWT with and without an aquaculture cage for a zero-degree wave heading are presented in Figures 11 and 12. 100 frequencies within 0~2 rad/s are considered in the calculation of wave-excitation forces transfer functions, thus the dimension of the QTF is 100×100. It is noted that the QTFs of the barge-type FOWT are more evenly distributed than those of the barge-type FOWT-AC with smaller maximum magnitude values except for difference-frequency QTFs of heave. Although the maximum value of the sum-frequency is greater than the difference-frequency, the overall QTFs of the sum-frequency is distributed in a relatively small amplitude range. Therefore, the sum-frequency has less influence on these coupled structures than the difference-frequency. Besides, the full-field difference-frequency QTFs graph exhibits an apparent symmetry about the anti-diagonals (solid black line), which is related to the mean-drift load at $\omega_i = \omega_j$, thereby resulting in a static load.

Figure 13 shows a comparison of motions for the barge-type FOWT-AC and barge-type FOWT at first order and first order with QTF conditions to investigate the effect of second-order wave forces on floating structures. It is found in Figure 13a that the presence of the second-order wave loads makes the mean value and standard deviations of the surge smaller. Figure 13b shows that the presence of the second-order wave loads makes the mean value of the pitch larger and the standard deviations of the pitch smaller. From Table 7, it can be seen that the changes caused by second-order wave forces are less than 3% both in mean and standard deviation of pitch, while the changes of standard deviation of the surge are up to about 20%. The variation of the motions is due to the Morison drag force coming from the aquaculture cage. But the effect of second-order wave loading on the responses of the coupled structures is not on the same order of magnitude as the first-order wave loading. Therefore, the accuracy of the results of first-order wave forces is sufficient in the subsequent analysis.

Table 7. Percentage comparison results of motion response under 1st and 1st +QTF wave loads.

	Mean of Surge	Std of Surge	Mean of Pitch	Std of Pitch
With cage	2.88%	19.88%	2.73%	0.77%
Without cage	3.51%	21.13%	1.95%	2.00%

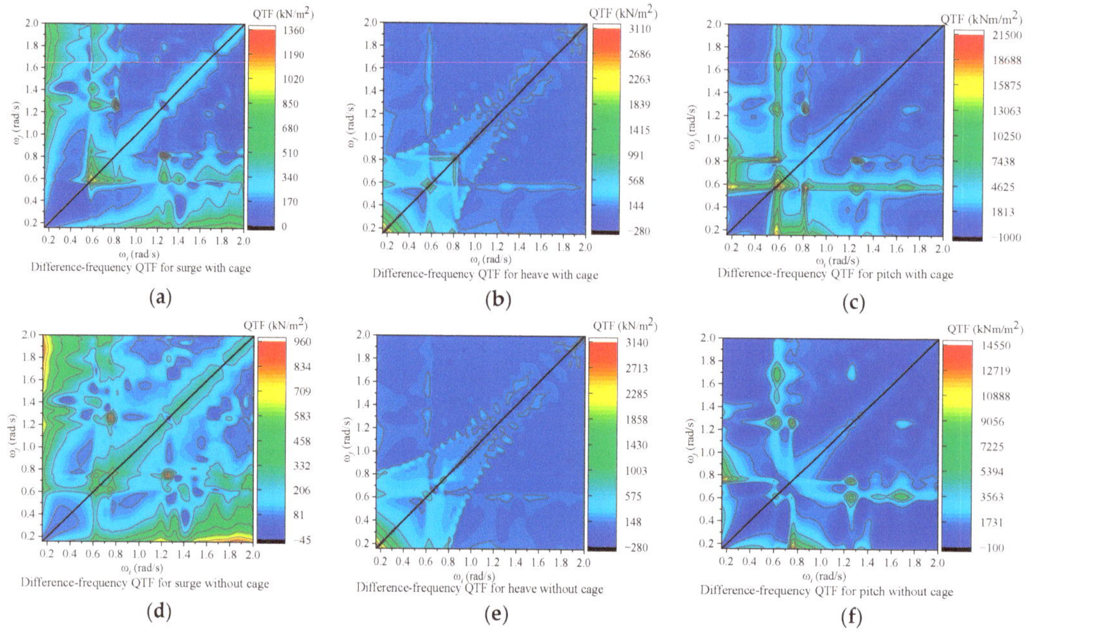

Figure 11. Second-order difference-frequency QTFs of the barge-type FOWT-AC and barge-type FOWT for a zero-degree wave heading: (**a**) Surge of FOWT-AC; (**b**) Heave of FOWT-AC; (**c**) Pitch of FOWT-AC; (**d**) Surge of FOWT; (**e**) Heave of FOWT; (**f**) Pitch of FOWT.

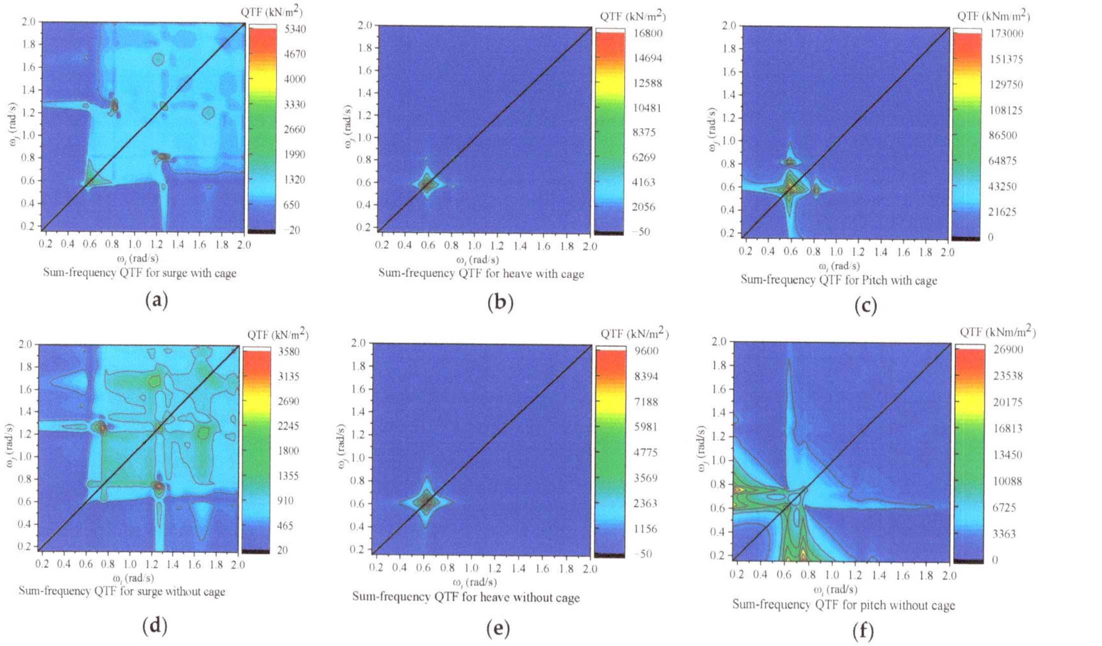

Figure 12. Second-order sum-frequency QTFs of the barge-type FOWT-AC and barge-type FOWT for a zero-degree wave heading: (**a**) Surge of FOWT-AC; (**b**) Heave of FOWT-AC; (**c**) Pitch of FOWT-AC; (**d**) Surge of FOWT; (**e**) Heave of FOWT; (**f**) Pitch of FOWT.

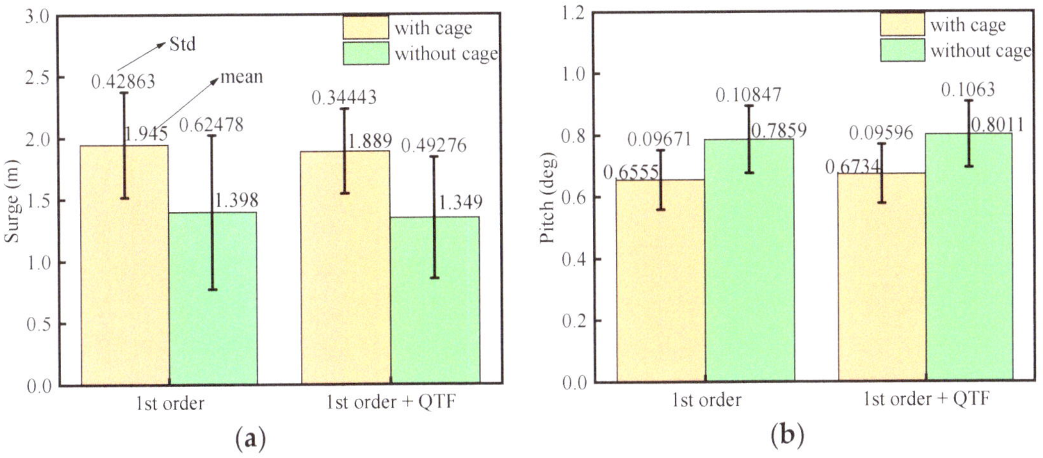

Figure 13. Comparison of motions for the barge-type FOWT-AC and barge-type FOWT at first order and first order with QTF conditions: (**a**) Mean and standard deviation of surge; (**b**) Mean and standard deviation of pitch.

4.2. Dynamic Analysis

For floating offshore structures, the various external loads and configuration difference are directly reflected in the motion response of the structure, thus the study of motion responses in six degrees of freedom is quite important. Herein, the environmental conditions with wind, wave and current in the same direction are chosen as the ideal environmental condition, which is the maximum external load case subjected by the structure. To explore the motion responses of the barge-type FOWT-AC and FOWT in the ideal environmental state, the time series results of 100 s after stable movement are outputted for EC1 as shown in Figure 14.

From Figure 14, it can be found that the presence of the aquaculture cage does have an effect on the motion responses of the structure. From Figure 14a, it is clear that the presence of the aquaculture cage makes the fluctuation amplitude of the surge motion smaller, but the mean value increases due to the Morison drag force on the nets. For roll, sway and yaw motions, Figure 14b,c,f show that the presence of the aquaculture cage reduces the fluctuation amplitude of the motions, but the mean value is still zero without change and the range of motion is small because the environmental loads are all inputted along the x-axis. As shown in Figure 14d, the fluctuation amplitude of the pitch motion is almost unchanged, but the overall motion angle becomes smaller due to the presence of the aquaculture cage. Thanks to the structure's eight suspension lines, there is sufficient recovery to allow for a small range of variation in pitch motion. Therefore, this results in the angle of pitch balance being less than 1° under the wind tilting moment. It can be seen in Figure 14e that the motion amplitude of the heave motion is almost unchanged when the aquaculture cage is present. It is also observed that the range of fluctuation in heave is smaller due to the large area of the platform baseboard compared to semi-submersible, spar, etc. In summary, the maximum value of the motion response of the barge-type FOWT is always larger than that of the barge-type FOWT AC except for the surge motion. Therefore, more attention should be paid to the surge motion in the design of the barge-type FOWT-AC to ensure that their leveling meets the specifications.

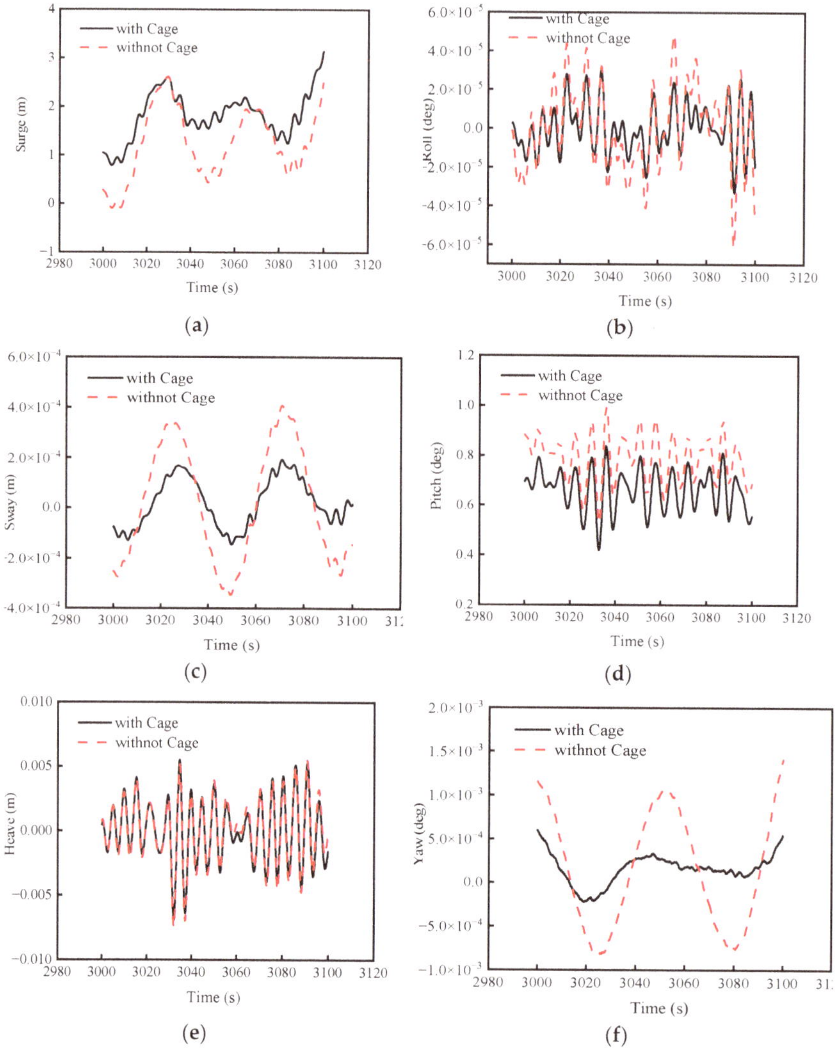

Figure 14. Comparison diagram of the motion response of the barge-type FOWT-AC and barge-type FOWT under EC1: (**a**) Time histories of surge motion; (**b**) Time histories of roll motion; (**c**) Time histories of sway motion; (**d**) Time histories of pitch motion; (**e**) Time histories of heave motion; (**f**) Time histories of yaw motion.

In order to better study the impact of the presence of aquaculture cages on the coupled structures, comparative analyses of surge and pitch motions of the floating structure with and without an aquaculture under varying wind, wave and current are carried out.

Comparisons of surge and pitch motions for the barge-type FOWT-AC and barge-type FOWT at different wind speeds are given in Figure 15. It can be seen from Figure 15a that the average value of the surge for barge-type FOWT-AC is larger than that of barge-type FOWT, but the standard deviation is opposite to the average value at different wind speeds. For Figure 15b, it can be observed that the average value and standard deviation of the pitch for barge-type FOWT is larger than that of barge-type FOWT-AC. This results trend is also consistent with Figure 14. The corresponding trend of surge and pitch motions for the same structure at different wind speeds is consistent with the variation of the wind speed-thrust curve. Combining Figures 14 and 15, it can be found that the standard deviation of the six-degree-of-freedom motion responses of barge-type FOWT-AC at different wind speeds is smaller than that of the barge-type FOWT. As a result, the barge-type FOWT-AC is more stable in the operating operation. This is mainly because the presence of the aquaculture cage increases the hydrodynamic force on the coupled structure below the floating center.

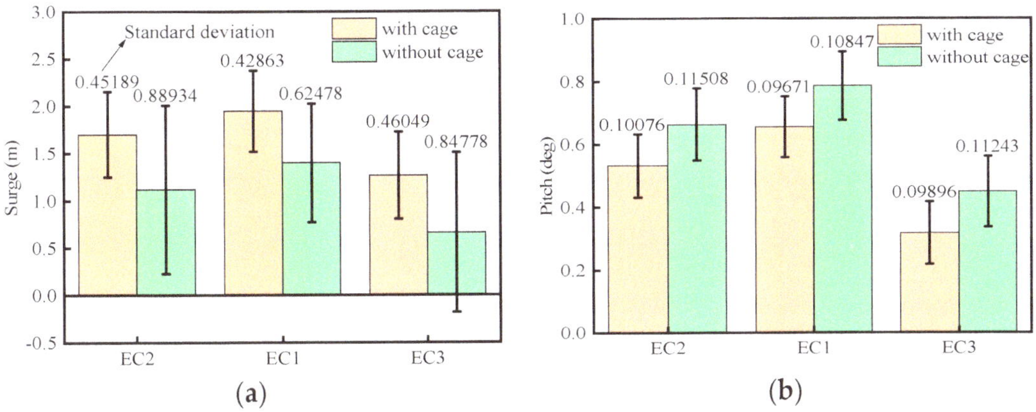

Figure 15. Comparison of motions for the barge-type FOWT-AC and barge-type FOWT at different wind speeds: (**a**) Mean and standard deviation of surge; (**b**) Mean and standard deviation of pitch.

Figure 16 shows a comparison of surge and pitch motions for the barge-type FOWT-AC and barge-type FOWT at different wave heights. From Figure 16a, it can be found that the mean and standard deviation values of the surge increase with the raising of wave height. From Figure 16b, it can be seen that the mean value of pitch decreases with the raising of the wave height, while the standard deviation values of pitch increase with the raising of wave height. The wave height of EC5 reaches 5.52 m, which already belongs to the extreme conditions. In this case, the mean value of surge for the barge-type FOWT-AC is 2.915 m, the standard deviation is 1.707 m, so the surge is less than 10% of the water depth, which is in line with the specifications the vast majority of the time. The mean value of pitch for the barge-type FOWT is 0.660°, the standard deviation is 4.812°, so the pitch is less than 10°, which is in line with the specification. Moreover, the comparison results in the means and standard deviations of surge and pitch for the structures with and without an aquaculture cage are the same as the conclusion obtained in the above for different wind speeds and wave conditions, indicating the stability of the designed floating structures.

Figure 17 illustrates that the presence of currents has a strong influence on the dynamic response of coupled structures, especially for the structure with an aquaculture cage. Figure 17a shows the offset of the equilibrium position of surge for the barge-type FOWT-AC with the current being 1 m/s is 3.54 times larger than that of no current, while it is 2.5 times larger for the case of the barge-type FOWT. From Figure 17b, it can be found that the mean value for the pitch of the barge-type FOWT-AC with the current being 1 m/s is 1.51 times larger than that of no current, while it is 1.73 times larger for the case of the barge-type FOWT. Therefore, the presence of currents significantly increases the mean value of the surge and pitch motions of the coupled structures. The presence of currents and the aquaculture cage decreases the standard deviation of the surge and pitch motion responses because the loads under the floating center of the structure in favor of the structural equilibrium are increased. The presence of currents has an effect on the amplitude of the motion and the equilibrium position of the floating structure [50], and these may be broadly associated with the potential flow and viscous effects [51].

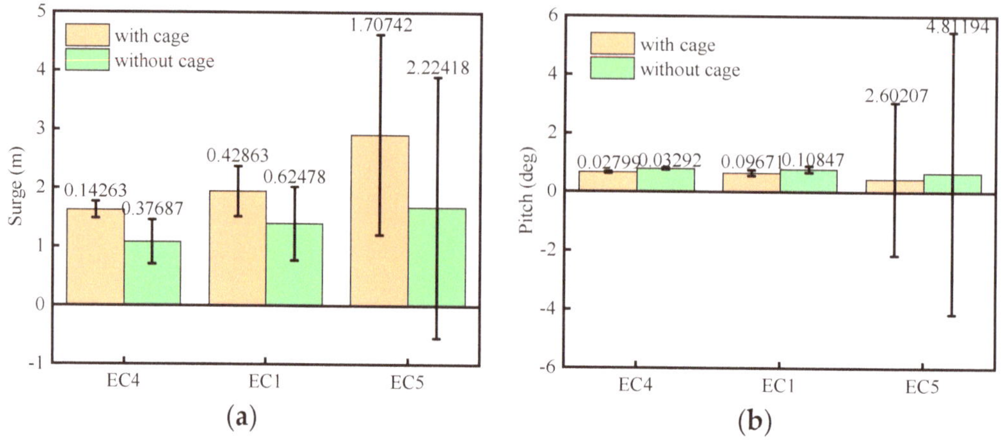

Figure 16. Comparison of motions for the barge-type FOWT-AC and barge-type FOWT at different wave height: (**a**) Mean and standard deviation of surge; (**b**) Mean and standard deviation of pitch.

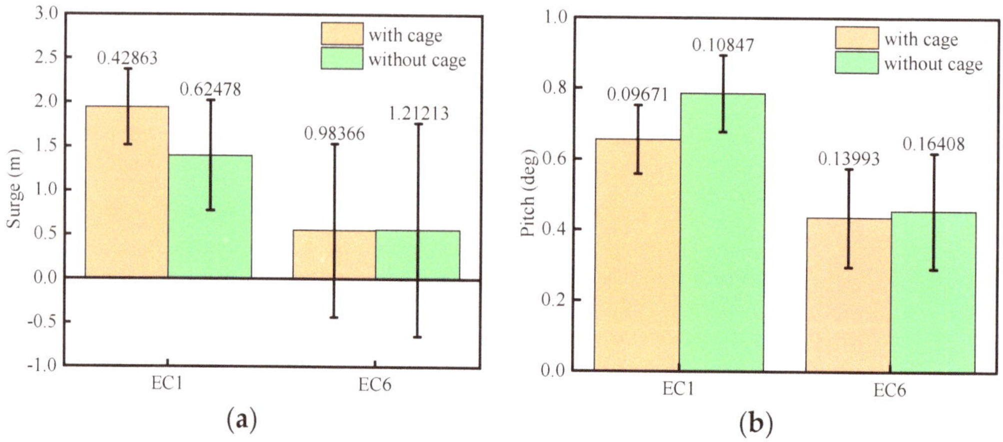

Figure 17. Comparison of motions for the barge-type FOWT-AC and barge-type FOWT in the presence or absence of currents: (**a**) Mean and standard deviation of surge; (**b**) Mean and standard deviation of pitch.

4.3. Dynamic Analysis of Coupled Structures with Mooring Line Failure

For the coupled structures proposed, the mooring lines can be divided into two groups, upstream and downstream, according to the environment conditions given in this study for the wind, wave and current with the same direction. The downstream suspension lines #1–#4 are in a looser state and therefore have less tension, while the upstream suspension lines #5–#8 are in a tighter state and therefore have more tension. Mooring line breakage is one of the most frequent failure conditions of FOWTs. To investigate the effect of mooring line breakage on the tension of other mooring lines and the motion response of the platform, the mooring lines located upstream (#8) and downstream (#3) are selected for the simulation of failure conditions in this paper.

As shown in Figure 18, breakage in one mooring line causes less tension on its diagonal mooring lines, more tension on the other mooring line at the same fairlead, and little change in tension on the remaining mooring lines. When mooring line #3 breaks, the mean values of tension of lines #1, #5 and #6 are smaller than that in the intact state, and the mean values of tension of lines #2, #4, #7 and #8 are larger than that in the intact condition, where the change of line #4 is the largest. Under the condition of mooring line #3 breakage, the change of mean value of line #4 tension is within 20% compared with the normal operating condition, but the change of standard deviation is up to 127%. At this time, the mooring line #8 is used as the maximum tension of the remaining mooring lines, and the maximum amplitude of tension is 2297 kN for the barge-type FOWT-AC, which is smaller than the catalog breaking load of the mooring line 9319 kN. Hence, the rest of the mooring lines can still bear the force of the system normally with line #3 broken. When mooring line #8 breaks, the mean values of tension of lines #1–#4 are smaller than those in the intact state, and the mean values of tension are larger for lines #5–#7 than those in the intact state, where the change in tension for mooring line #7 is significantly larger by observing the change of mean value being within 60% and the standard deviation reaching up to 152%. At this time, mooring line #7 withstands the maximum tension of the whole mooring system, and the maximum amplitude of tension is 4316 kN, which is still smaller than the catalog breaking load of the mooring line 9319 kN. These variations increased a lot relative to line #3 breakage, because line #8 is subjected to more tension upstream before breaking.

Figure 19 shows the translations and rotational motions of the barge-type FOWT-AC under the mooring line #3 and line #8 breakage conditions. Mooring line breakage has a significant effect on the motion response of the coupled structure, especially when mooring line #8 breaks. The change of heave motion is not obvious because the result of heave motion is a change relative to the waterline, but the actual center of gravity has shifted upward after the mooring line breaking. When line #3 breaks, the offset of the equilibrium position of surge and pitch have decreased to a certain extent compared to the normal operating conditions. The motions of sway, roll and yaw have gone from almost zero to a certain magnitude due to the symmetry breaking under the condition of line #3 breakage. The offset of the equilibrium position of surge and pitch have increased to a certain extent compared to the intact state when the line #8 breaks. More notable sway, roll and yaw motions are caused by the mooring #8 breakage compared with the breakage of mooring line #3. Especially yaw, when mooring line #8 breaks, the motion magnitude value is enhanced more than 24 times compared with the breakage of mooring line #3.

Figure 20 presents the comparison of motion responses of the barge-type FOWT-AC and barge-type FOWT for different mooring line breakage. The standard deviation is smaller at the intact state for the two structures than for the failure state, and the standard deviation for the case of barge-type FOWT-AC is smaller than that of barge-type FOWT, implying that the barge-type FOWT-AC is more stable after mooring line breakage. The mean values of surge and pitch motions in the case of mooring line breakage are larger than those in the intact state for the two floating structures.

The presence of the aquaculture cage increases the relative variation of the surge and decreases the relative variation of the pitch when the mooring line #3 breaks, while the opposite change occurs when mooring line #8 breaks. In general, the most affected motion response of the coupled structure is the surge motion under the case of the mooring line #8 breakage, which has a mean value of 3.386 m and a standard deviation of 0.449 m.

The redundancy mooring system plays an important role in ensuring the safety of the entire coupled structure after a single mooring line breaks. When mooring line #3 breaks, its fairlead is restrained by mooring line #4 alone, so the tension of mooring line #4 will be increased. As shown in Figure 21, when the mooring line breaks, the tension of other mooring lines do not become larger or smaller immediately, but gradually changes according to the variable trend of the previous mooring line tension. This also shows that as long as the tension is less than the breaking strength, even if one mooring line breaks, the rest of the mooring lines will not be brittle fracture due to the sudden increase of tension [52,53].

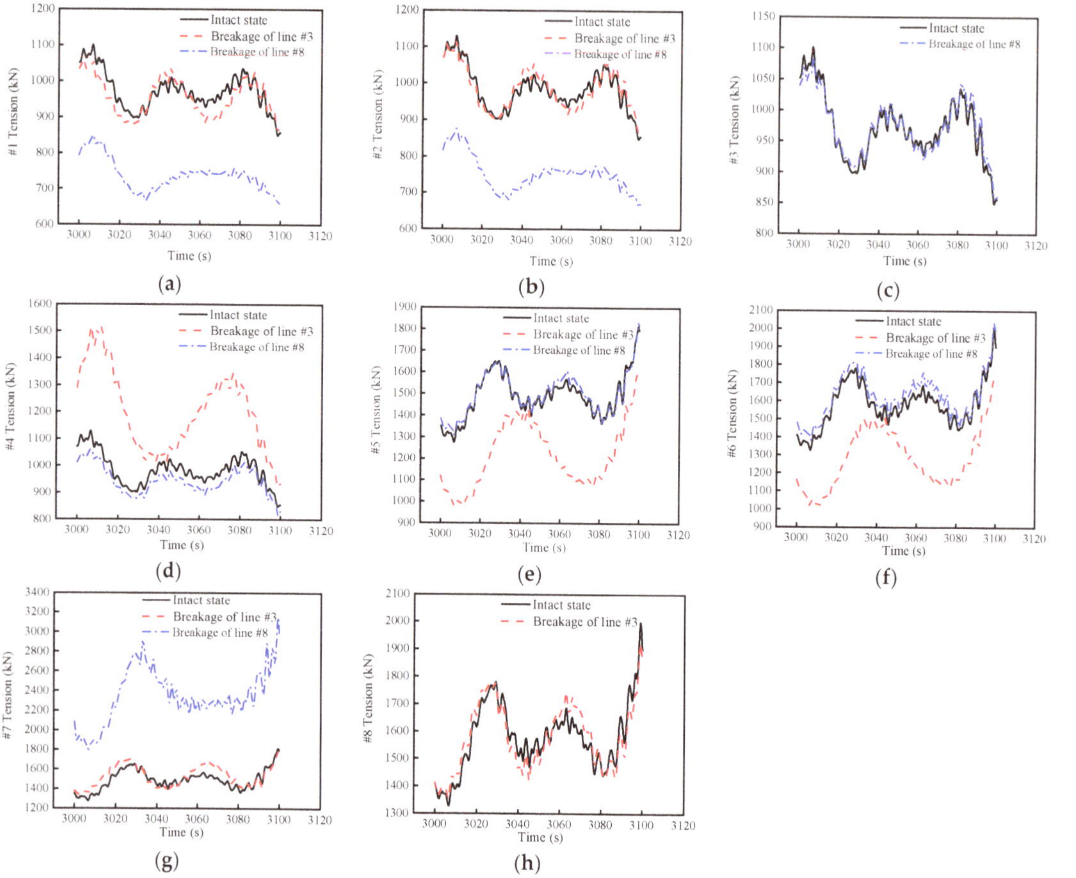

Figure 18. Mooring tension under different ECs for barge-type FOWT-AC: (**a**) Time histories of #1 tension; (**b**) Time histories of #2 tension; (**c**) Time histories of #3 tension; (**d**) Time histories of #4 tension; (**e**) Time histories of #5 tension, (**f**) Time histories of #6 tension; (**g**) Time histories of #7 tension; (**h**) Time histories of #8 tension.

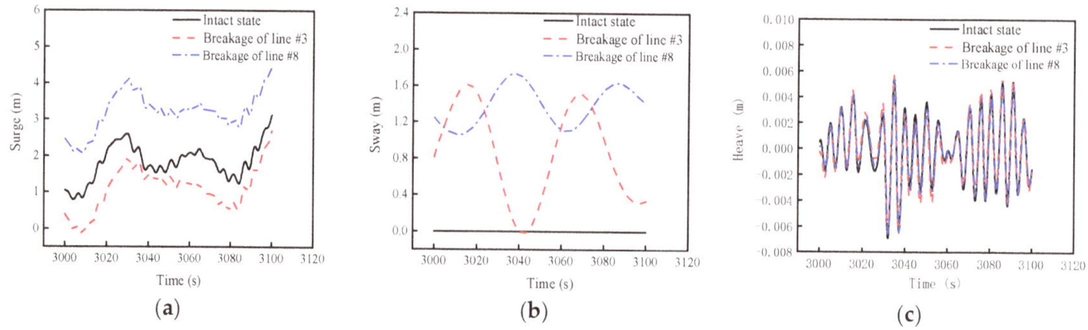

Figure 19. *Cont.*

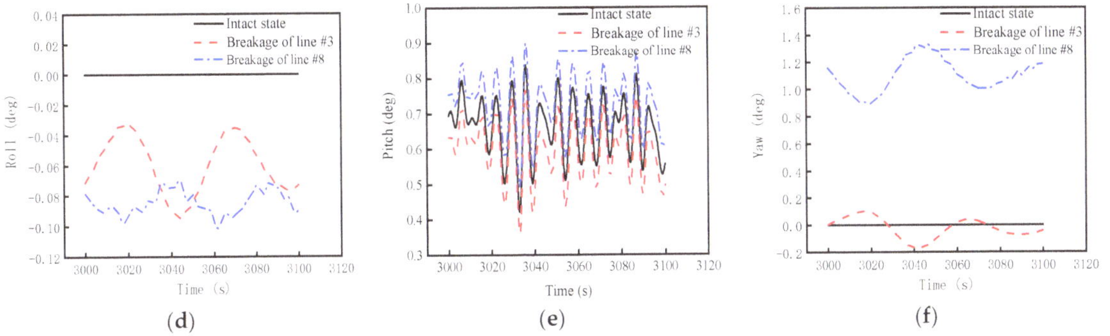

Figure 19. Motion response under different ECs for barge-type FOWT-AC: (**a**) Time histories of surge; (**b**) Time histories of sway; (**c**) Time histories of heave; (**d**) Time histories of roll; (**e**) Time histories of pitch; (**f**) Time histories of yaw.

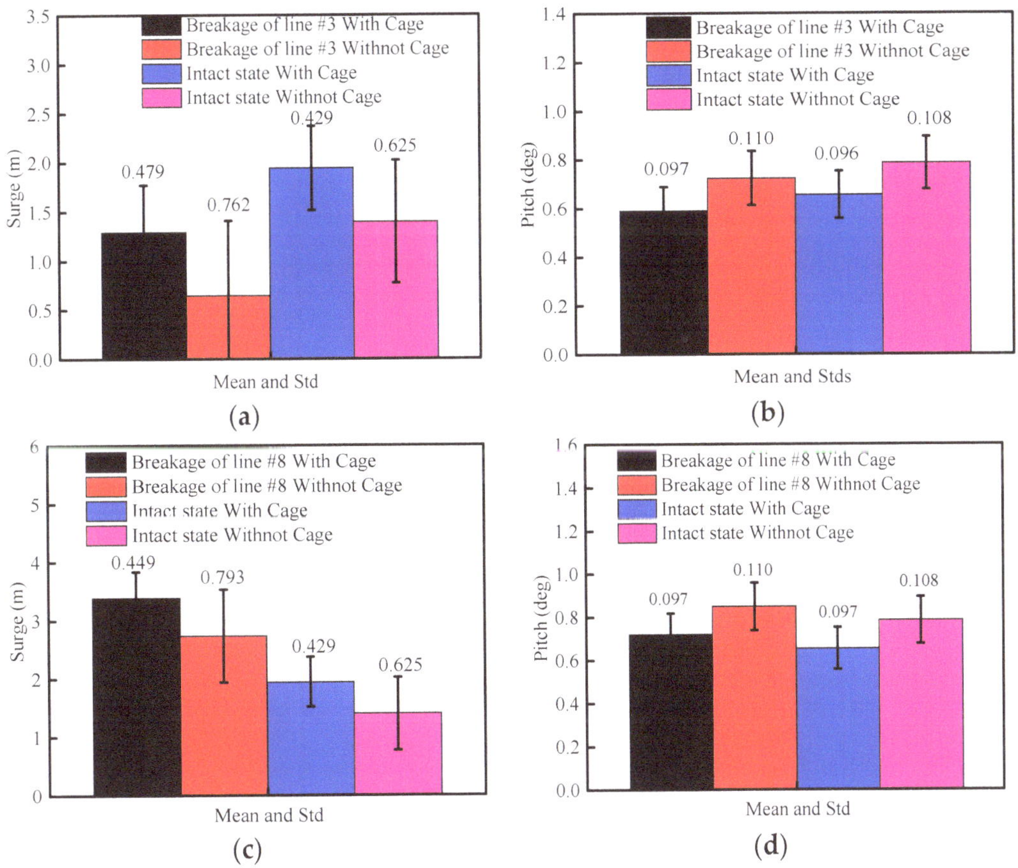

Figure 20. Comparison of motions for the barge-type FOWT-AC and barge-type FOWT under different ECs: (**a**) Surge at breakage of #3; (**b**) Pitch at breakage of #3; (**c**) Surge at breakage of #8; (**d**) Pitch at breakage of #8.

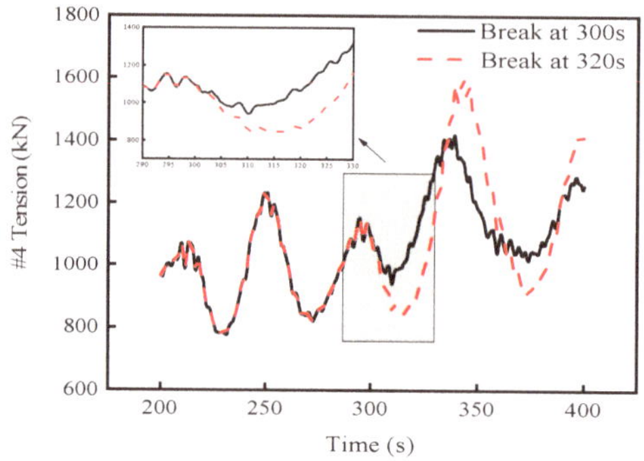

Figure 21. Comparison of #4 tension when #3 breaks at different moments.

5. Conclusions

This paper focuses on proposing a new barge-type FOWT-AC structure and comparing the numerical simulation results with the barge-type FOWT considering first- and second-order wave loads, wind loads and current loads of the environmental conditions in the South China Sea at normal operating conditions and fault conditions. The effects of the aquaculture cage on the dynamic responses of the barge-type FOWT are explored. From this study, the following conclusions are drawn:

(1) The presence of the aquaculture cage has improved the natural period of the heave, pitch, and roll. Except for surge, the maximum values of the motions of the barge-type FOWT-AC are smaller than barge-type FOWT. There is a reduction in the standard deviation of the motion response due to the presence of the aquaculture cage. Therefore, the barge-type FOWT-AC is generally more stable during normal operation.

(2) The effect of second-order wave excitation on the dynamic responses of floating structures is considered in detail, and it is noted that the differential-frequency relative to the sum-frequency has a larger range of effects on the coupled structure. Because the second-order wave loads are not of the same order of magnitude as the first-order wave loads, they have little effect on the motion response of the coupled structures when the structural resonance is not induced.

(3) By comparing the influence of the presence or absence of currents, it can be concluded that the presence of currents has a strong influence on the dynamic response of coupled structures, making the motion response increase significantly because of the increasing of the external loads on the barge-type FOWT-AC. The surge motion of the barge-type FOWT-AC can even reach up to 3.54 times greater when the current is 1 m/s than when there is no current, so the effect of currents cannot be neglected in the hydrodynamic analysis.

(4) The breakage of one mooring line causes less tension on its diagonal mooring lines, more tension on the other mooring line at the same fairlead, and little change in tension on the remaining mooring lines. When the upstream mooring line breaks, it has a greater effect on the motion response of the coupled structure and the mooring tension than the accident of downstream mooring line breakage. More importantly, compared to barge-type FOWT, the standard deviation of the barge-type FOWT-AC is smaller and relatively more stable after mooring line breakage.

6. Future Work

In this paper, some hydrodynamic characteristics and motion responses of the FOWT-AC structure are initially investigated using numerical simulations, but there are still many inadequacies to be investigated in the future.

First of all, the occurrence of extreme environmental conditions has increased due to climate change. The probability of breaking waves occurring that are induced by extreme conditions will certainly increase. The external loads on marine structures under the action of breaking waves and unbroken waves are different by considering the air entrapment effect [54] and the significant impulsive/slamming loads [55] exerted by breaking waves. Therefore, it is important to investigate the changes in the motion response of marine structures subjected to breaking waves. However, predicting the air entrapment effect [54] and high-frequency impulsive/slamming loads [55] accurately in dealing with the breaking wave scenario is a challenge, and whether the currently used simulation methods are still available will be considered in future works.

Secondly, the wave-in-platform loads are a very complex research field due to the large horizontal and uplift forces observed in past studies [56–61], the nonlinear interaction of the wave with the air below the deck [59,60], and the application of a large overturning moment [55,61] that could lead to a concentration of forces in the offshore supports. None of these complex loads were addressed in this research study, but they will be considered in future research works.

Finally, in the current study, the wave direction is normal for the offshore surface of the barge platform, which is an idealized case where the FOWT-AC structure can withstand the maximum external load. However, the actual wave and wind loads may propagate from any angle; thus, load input conditions for different angles will also be investigated in future works.

Author Contributions: Conceptualization, Y.Z., H.Z., X.L. and W.S.; methodology, Y.Z.; software, Y.Z.; validation, Y.Z. and H.Z.; formal analysis, Y.Z.; investigation, Y.Z. and H.Z.; data curation, Y.Z.; writing—original draft preparation, Y.Z.; writing—review and editing, H.Z. and X.L.; supervision, H.Z., X.L. and W.S.; project administration, W.S.; funding acquisition, X.L. All authors have read and agreed to the published version of the manuscript.

Funding: This research work was financially supported by the National Natural Science Foundation of China, Grant No. is 51939002.

Institutional Review Board Statement: Not applicable.

Informed Consent Statement: Not applicable.

Data Availability Statement: Not applicable.

Conflicts of Interest: The authors declare no conflict of interest.

References

1. Cao, Q.; Xiao, L.; Cheng, Z.; Liu, M. An Experimental Study on Dynamic Behavior of a New Concept of 10 MW Semi-Submersible Wind Turbine. *Ocean. Eng.* **2020**, *232*, 109138. [CrossRef]
2. Leble, V.; Barakos, G. Demonstration of a Coupled Floating Offshore Wind Turbine Analysis with High-Fidelity Methods. *J. Fluids Struct.* **2016**, *62*, 272–293. [CrossRef]
3. Oguz, E.; Clelland, D.; Day, A.H.; Incecik, A.; López, J.A.; Sánchez, G.; Almeria, G.G. Experimental and Numerical Analysis of a TLP Floating Offshore Wind Turbine. *Ocean Eng.* **2018**, *147*, 591–605. [CrossRef]
4. Ding, Q.; Chun, L.I.; Cheng, X.; Zhou, Y.E. Performance of Barge Platform for Floating Wind Turbines. *J. Univ. Shanghai Sci. Technol.* **2015**, *5*, 425–432.
5. Li, J.; Tang, Y.; Yeung, R.W. Effects of Second-Order Difference-Frequency Wave Forces on a New Floating Platform for an Offshore Wind Turbine. *J. Renew. Sustain. Energy* **2014**, *6*, 033102.
6. Jonkman, J. *Definition of the Floating System for Phase IV of OC3*; No. NREL/TP-500-47535; National Renewable Energy Lab. (NREL): Golden, CO, USA, 2010.
7. Robertson, A.; Jonkman, J.; Masciola, M.; Song, H.; Goupee, A.; Coulling, A.; Luan, C. *Definition of the Semisubmersible Floating System for Phase II of OC4*; No. NREL/TP-5000-60601; National Renewable Energy Lab. (NREL): Golden, CO, USA, 2014.
8. Ren, Y.; Vengatesan, V.; Shi, W. Dynamic Analysis of a Multi-column TLP Floating Offshore Wind Turbine with Tendon Failure Scenarios. *Ocean Eng.* **2022**, *245*, 110472. [CrossRef]
9. Ikoma, T.; Tan, L.; Moritsu, S.; Aida, Y.; Masuda, K. Motion Characteristics of a Barge-Type Floating Vertical-Axis Wind Turbine with Moonpools. *Ocean Eng.* **2021**, *230*, 109006. [CrossRef]
10. Anagnostopoulos, S.A. Dynamic response of offshore platforms to extreme waves including fluid-structure interaction. *Eng. Struct.* **1982**, *4*, 179–185. [CrossRef]
11. Istrati, D.; Buckle, I. G: Effect of fluid-structure interaction on connection forces in bridges due to tsunami loads. In Proceedings of the 30th US-Japan Bridge Engineering Workshop, Washington, DC, USA, 21–23 October 2014.

12. Utsunomiya, T.; Yoshida, S.; Ookubo, H.; Sato, I.; Ishida, S. Dynamic Analysis of a Floating Offshore Wind Turbine Under Extreme Environmental Conditions. *J. Offshore Mech. Arct. Eng.* **2014**, *136*, 21901. [CrossRef]
13. Chen, J.; Hu, Z.; Fei, D. Comparisons of Dynamical Characteristics of a 5 MW Floating Wind Turbine Supported by a Spar-buoy and a Semi-submersible using Model Testing Methods. *J. Renew. Sustain. Energy* **2018**, *10*, 053311. [CrossRef]
14. Cao, Q.; Xiao, L.; Guo, X.; Liu, M. Second-Order Responses of a Conceptual Semi-Submersible 10 MW Wind Turbine Using Full Quadratic Transfer Functions. *Renew. Energy* **2020**, *153*, 653–668. [CrossRef]
15. Choisnet, T.; Favre, M.; Lyubimova, M.; Rogier, E. A Robust Concrete Floating Wind Turbine Foundation for Worldwide Applications. In Proceedings of the Grand Renewable Energy Conference, Tokyo, Japan, 27 July–1 August 2014; p. O-WdOc-2-7.
16. Chuang, T.; Yang, W.; Yang, R. Experimental and Numerical Study of a Barge-Type FOWT Platform Under Wind and Wave Load. *Ocean Eng.* **2021**, *230*, 109015. [CrossRef]
17. Peng, Z.; Zhao, H.; Li, X. New ductile fracture model for fracture prediction ranging from negative to high stress triaxiality. *Int. J. Plast.* **2021**, *145*, 103057. [CrossRef]
18. Zhao, H.; Lie, S.; Zhang, Y. Fracture assessment of mismatched girth welds in oval-shaped clad pipes subjected to bending moment. *Int. J. Press. Vessel. Pip.* **2018**, *160*, 1–13. [CrossRef]
19. Zhao, H.; Lie, S.; Zhang, Y. Fatigue assessment of cracked pipes with weld misalignment by using stress intensity factors. *Int. J. Fatigue* **2018**, *116*, 192–209. [CrossRef]
20. Yang, Y.; Bashir, M.; Li, C.; Wang, J. Investigation on Mooring Breakage Effects of a 5 MW Barge-Type Floating Offshore Wind Turbine Using F2A. *Ocean Eng.* **2021**, *233*, 108887. [CrossRef]
21. Zhao, Y.; Guan, C.; Bi, C.; Liu, H.; Cui, Y. Experimental Investigations on Hydrodynamic Responses of a Semi-Submersible Offshore Fish Farm in Waves. *J. Mar. Sci. Eng.* **2019**, *7*, 238. [CrossRef]
22. Li, L.; Ong, M.C. A Preliminary Study of a Rigid Semi-Submersible Fish Farm for Open Seas. In Proceedings of the ASME 2017 36th International Conference on Ocean, Offshore and Arctic Engineering, Trondheim, Norway, 25–30 June 2017; American Society of Mechanical Engineers: New York, NY, USA, 2017; p. V009T12A044.
23. Jin, J.; Su, B.; Dou, R.; Luan, C.; Gao, Z. Numerical Modelling of Hydrodynamic Responses of Ocean Farm 1 in Waves and Current and Validation Against Model Test Measurements. *Mar. Struct.* **2021**, *78*, 103017. [CrossRef]
24. Moe, H.; Fredheim, A.; Hopperstad, O.S. Structural Analysis of Aquaculture Net Cages in Current. *J. Fluids Struct.* **2010**, *26*, 503–516. [CrossRef]
25. Xu, T.J.; Dong, G.H.; Zhao, Y.P.; Li, Y.C.; Gui, F.K. Analysis of Hydrodynamic Behaviors of Gravity Net Cage in Irregular Waves. *Ocean Eng.* **2011**, *38*, 1545–1554. [CrossRef]
26. He, W.; Yttervik, R.; Olsen, G.P.; Ostvik, I.; Jimenez, C.; Impelluso, T.; Schouten, J.; Bellotti, G. A Case Study of Multi-Use Platform: Aquaculture in Offshore Wind Farms. *Poster no 54, EWEA Offshore 2015 Copenhagen*, 10–12 March 2015. Available online: https://www.vliz.be/projects/mermaidproject/images/PDFs/EWEA_PO054_FinalSubmission15022015.pdf (accessed on 27 April 2022).
27. Lei, Y.; Zhao, S.X.; Zheng, X.Y.; Li, W. Effects of Fish Nets on the Nonlinear Dynamic Performance of a Floating Offshore Wind Turbine Integrated with a Steel Fish Farming Cage. *Int. J. Struct. Stab. Dyn.* **2020**, *3*, 2050042. [CrossRef]
28. Chu, Y.I.; Wang, C.M. Hydrodynamic Response Analysis of Combined Spar Wind Turbine and Fish Cage for Offshore Fish Farms. *Int. J. Struct. Stab. Dyn.* **2020**, *9*, 2050104. [CrossRef]
29. Liang, L.; Ruzzo, C.; Collu, M.; Gao, Y.; Arena, F. Analysis of the Coupled Dynamic Response of an Offshore Floating Multi-Purpose Platform for the Blue Economy. *Ocean Eng.* **2020**, *217*, 107943.
30. Wang, Y.; Shi, W.; Michailides, C.; Wan, L.; Kim, H.; Li, X. WEC shape effect on the motion response and power performance of a combined wind-wave energy converter. *Ocean Eng.* **2022**, *250*, 111038. [CrossRef]
31. Li, L.; Collu, M.; Ruzzo, C.; Failla, G.; Abhinav, K.A.; Arena, F. Analysis of the Coupled Dynamics of an Offshore Floating Multi-Purpose Platform: Part A—Rigid Body Analysis. In Proceedings of the International Conference on Offshore Mechanics and Arctic Engineering, Scotland, UK, 9–14 June 2019; American Society of Mechanical Engineers: New York, NY, USA, 2019.
32. Abhinav, K.; Collu, M.; Ke, S.; Binzhen, Z. Frequency domain analysis of a hybrid aquaculture-wind turbine offshore floating system. In Proceedings of the ASME 2019 38th International Conference on Ocean, Offshore and Arctic Engineering, Scotland, UK, 9–14 June 2019; American Society of Mechanical Engineers: New York, NY, USA, 2019.
33. Zheng, X.; Zheng, H.; Lei, Y.; Li, Y.; Wei, L. An offshore floating wind–solar–aquaculture system: Concept design and extreme response in survival conditions. *Energies* **2020**, *13*, 604. [CrossRef]
34. Belibassakis, K.A. A boundary element method for the hydrodynamic analysis of floating bodies in variable bathymetry regions. *Eng. Anal. Bound. Elem.* **2008**, *32*, 796–810. [CrossRef]
35. Westphalen, J.; Greaves, D.M.; Raby, A.; Hu, Z.Z.; Causon, D.M.; Mingham, C.G.; Omidvar, P.; Stansby, P.K.; Rogers, B.D. Investigation of wave-structure interaction using state of the art CFD techniques. *Open J. Fluid Dyn.* **2014**, *4*, 18. [CrossRef]
36. Ren, B.; Jin, Z.; Gao, R.; Wang, Y.X.; Xu, Z.L. SPH-DEM modeling of the hydraulic stability of 2D blocks on a slope. *J. Waterw. Port Coast. Ocean. Eng.* **2014**, *140*, 04014022. [CrossRef]
37. Hasanpour, A.; Istrati, D.; Buckle, I. Coupled SPH–FEM Modeling of Tsunami-Borne Large Debris Flow and Impact on Coastal Structures. *J. Mar. Sci. Eng.* **2021**, *9*, 1068. [CrossRef]
38. Tsui, Y.Y.; Huang, Y.C.; Huang, C.L.; Lin, S.W. A finite-volume-based approach for dynamic fluid-structure interaction. *Numer. Heat Transf. Part B Fundam.* **2013**, *64*, 326–349. [CrossRef]

39. Xiang, T.; Istrati, D. Assessment of Extreme Wave Impact on Coastal Decks with Different Geometries via the Arbitrary Lagrangian-Eulerian Method. *J. Mar. Sci. Eng.* **2021**, *9*, 1342. [CrossRef]
40. Zhao, Y.; Li, Y.; Dong, G.; Gui, F.; Wu, H. An Experimental and Numerical Study of Hydrodynamic Characteristics of Submerged Flexible Plane Nets in Waves. *Aquac. Eng.* **2008**, *38*, 16–25. [CrossRef]
41. Zhao, Z.; Wang, W.; Shi, W.; Li, X. Effects of Second-Order Hydrodynamics on an Ultra-Large Semi-Submersible Floating Offshore Wind Turbine. *Structures* **2020**, *28*, 2260–2275. [CrossRef]
42. Zhao, Z.; Shi, W.; Wang, W.; Qi, S.; Li, X. Dynamic analysis of a novel semi-submersible platform for a 10 MW wind turbine in intermediate water depth. *Ocean Eng.* **2021**, *237*, 109688. [CrossRef]
43. Yu, M.; Hu, Z.; Xiao, L. Wind-Wave Induced Dynamic Response Analysis for Motions and Mooring Loads of a Spar-Type Offshore Floating Wind Turbine. *J. Hydrodyn.* **2015**, *26*, 865–874.
44. Jonkman, J.M. *Dynamics Modeling and Loads Analysis of an Offshore Floating Wind Turbine*; University of Colorado at Boulder: Boulder, CO, USA, 2007.
45. ANSYS. *AWQA User Manual*; ANSYS: Canonsburg, PA, USA, 2011.
46. Nallayarasu, S.; Saravanapriya, S. Experimental and Numerical Investigation on Hydrodynamic Response of Spar with Wind Turbine Under Regular Waves. *Int. J. Ocean. Clim. Syst.* **2013**, *4*, 239–260. [CrossRef]
47. Ruzzo, C.; Fiamma, V.; Nava, V.; Collu, M.; Failla, G.; Arena, F. Progress on the Experimental Set-Up for the Testing of a Floating Offshore Wind Turbine Scaled Model in a Field Site. *Wind Eng.* **2016**, *40*, 455–467. [CrossRef]
48. Rajeswari, K.; Nallayarasu, S. Hydrodynamic Response of Three- and Four-Column Semi-Submersibles Supporting a Wind Turbine in Regular and Random Waves. *Ships Offshore Struct.* **2021**, *16*, 1050–1060. [CrossRef]
49. *DNV DNV-RP-0286*; Coupled Analysis of Floating Wind Turbine. DNV: Oslo, Norway, 2019.
50. Song, G.; Bin, T. Wave Elevation for an Array of Columns Under Wave–Current Interactions. *J. Offshore Mech. Arct. Eng.* **2022**, *144*, 021201. [CrossRef]
51. Isaacson, M.; Baldwin, J. Wave–current effects on large offshore structures. *Can. J. Civ. Eng.* **1989**, *16*, 543–551. [CrossRef]
52. Zhao, H.; Lie, S.; Zhang, Y. Elastic-plastic fracture analyses for misaligned clad pipeline containing a canoe shape surface crack subjected to large plastic deformation. *Ocean Eng.* **2017**, *146*, 87–100. [CrossRef]
53. Zhao, H.; Lie, S.; Zhang, Y. Strain-based J-estimation scheme for fracture assessment of misaligned clad pipelines with an interface crack. *Mar. Struct.* **2018**, *61*, 238–255. [CrossRef]
54. Peregrine, D.H.; Bredmose, H.; Bullock, G.; Obrhai, C.; Müller, G.; Wolters, G. Water wave impact on walls and the role of air. *Coast. Eng.* **2004**, *2005*, 4005–4017.
55. Istrati, D.; Buckle, I.; Lomonaco, P.; Yim, S. Deciphering the tsunami wave impact and associated connection forces in open-girder coastal bridges. *J. Mar. Sci. Eng.* **2018**, *6*, 148. [CrossRef]
56. Douglass, S.; Chen, Q.; Olsen, J.; Edge, B.; Brown, D. *Wave Forces on Bridge Decks*; U.S. Department of Transportation, Federal Highway Administration, Office of Bridge Technology: Washington, DC, USA, 2006.
57. Allsop, W.; Cuomo, G.; Tirindelli, M. New prediction method for wave-in-deck loads on exposed piers/jetties/bridges. *Coast. Eng.* **2006**, *2007*, 4482–4493.
58. Denson, K.H. *Wave Forces on Causeway Type Coastal Bridges*; Water Resources Research Institute, Mississippi State University: Stackville, MS, USA, 1978.
59. Azadbakht, M.; Yim, S.C. Effect of trapped air on wave forces on coastal bridge superstructures. *J. Ocean Eng. Mar. Energy* **2016**, *2*, 139–158. [CrossRef]
60. Istrati, D.; Buckle, I. Role of trapped air on the tsunami-induced transient loads and response of coastal bridges. *Geosciences* **2019**, *9*, 191. [CrossRef]
61. Xiang, T.; Istrati, D.; Yim, S.C.; Buckle, I.G.; Lomonaco, P. Tsunami loads on a representative coastal bridge deck: Experimental study and validation of design equations. *J. Waterw. Port Coast. Ocean Eng.* **2020**, *146*, 04020022. [CrossRef]

Journal of
Marine Science and Engineering

MDPI

Article

Adaptive Backstepping Integral Sliding Mode Control for 5DOF Barge-Type OFWT under Output Constraint

Syed Awais Ali Shah [1], Bingtuan Gao [1,*], Irfan Ahmad [2], Hameed Ullah [3], Nigar Ahmed [4] and Anjum Saeed [5]

1 School of Electrical Engineering, Southeast University, Sipailou 2, Nanjing 210096, China
2 Department of Automatic Control and Systems Engineering, University of the Basque Country–UPV/EHU, Rafael Moreno 3, 48013 Bilbao, Spain
3 Department of Electrical Engineering and Information Technology, University of Naples Federico II–UNINA, 80125 Naples, Italy
4 Faculty of Electrical Engineering and Computing, University of Zagreb, Unska 3, 10000 Zagreb, Croatia
5 Department of Electrical Engineering, CEME, National University of Sciences and Technology, Islamabad 46000, Pakistan
* Correspondence: gaobingtuan@seu.edu.cn

Abstract: This article presents a new control solution for a dynamical model of a translational oscillator with a rotational actuator (TORA) based on multi-body dynamics for a barge-type offshore floating wind turbine (OFWT). TORA has been employed as an active structural control strategy. The solution of bounding the output movements of platform pitch and tower bending angle to a certain limit, along with mitigating the OFWT vibrations due to environmental disturbances and uncertainties, is presented in this novel control framework. This new control algorithm consists of a high-gain observer (HGO)-based adaptive backstepping integral sliding mode control (ISMC) and a barrier Lyapunov function (BLF). This guarantees satisfying the constraints on the states and effectively resolves the problem of the unavailability of the system states. The proposed control law based on the BLF has been compared with an adaptive backstepping ISMC to show the efficiency of the output-constraint control scheme. Through MATLAB/SIMULINK numerical simulations and their numeric error table, the effectiveness of the proposed control scheme has been examined. The results confirm the validity and efficiency of the proposed control approaches.

Keywords: adaptive backstepping ISMC; barge-type OFWT; barrier Lyapunov function; output constraint control; TORA; UMS

Citation: Shah, S.A.A.; Gao, B.; Ahmad, I.; Ullah, H.; Ahmed, N.; Saeed, A. Adaptive Backstepping Integral Sliding Mode Control for 5DOF Barge-Type OFWT under Output Constraint. *J. Mar. Sci. Eng.* **2023**, *11*, 492. https://doi.org/10.3390/jmse11030492

Academic Editors: Emre Uzunoglu, Antonio Souto-Iglesias and Carlos Guedes Soares

Received: 31 January 2023
Revised: 20 February 2023
Accepted: 22 February 2023
Published: 24 February 2023

1. Introduction

Wind power is considered to be among the most promising renewable energy sources and has attracted widespread world interest in recent years, due to the sustainability, cost-effectiveness, and low environmental impact of wind generators [1]. Production of offshore wind energy is becoming more appealing as the need for renewable energy grows, with benefits such as higher wind speeds, fewer noise constraints, and fewer visual impacts in marine environments [2]. The two most common forms of offshore wind turbines are bottom-fixed and floating wind turbines. Monopile, gravity-based, and suction bucket structures are the most popular nearshore bottom-fixed wind-based turbines, whereas the best-known floating offshore wind turbine platforms are the spar, tension leg, and barge types [3,4]. Offshore floating wind turbine systems are still in the early stages of development, and they are being tested in the laboratory or real-world scenarios utilizing proof-of-concept tests.

Environmental stress leads to severe vibration effects on offshore platforms, causing deck platform failure, structural failure, operational failure, and even injury to workers. Currently, two methodologies are generally used to reduce structural loads and extend the life of offshore wind turbines [5–8]. The first method uses blade pitch control to minimize the thrust force of the rotor, while the second technique involves vibration suppression devices which have three modes: passive, semi-active, and active. Passive control methods are characterized by using constant parameters and the absence of the need for external energy. A tuned mass damper (TMD) is an example of a passive technique, which is used to absorb vibrational energy by setting one of the natural frequencies of a vibrated system. The passive TMD control technique is the main foundation of the research on OFWT stabilization [9,10]. In comparison with passive strategies, the semi-active control method offers higher performance since it can be adapted over time owing to the additional sensors and control algorithms. Prior authors [11,12] have proved that the semi-active varying-damper-and-spring system is determined to be the most successful overall, managing vibrations just as efficiently with half the mass of a passive TMD. The combination of mass and an actuator is called an active mass damping (AMD) system. A few studies have applied the AMD control strategy to OFWTs [13,14].

The underactuated mechanical system (UMS), with its improved system adaptability, efficiency improvements, and cost-saving, has long been a fascinating topic for researchers [15–19]. The underactuated TORA is an AMD control approach, in which the oscillating cart as a transnational un-actuated part is connected with the actuator known as the rotational eccentric proof mass [20]. For TORA stabilization, numerous control techniques have been devised [21,22]. Furthermore, when dealing with any practicable stabilizing issues, such as OFWTs, there is often a discrepancy between the dynamical model and the real plant used for controller design. Parasitic/unmodeled dynamics, plant uncertainties, and external unknown disturbances result from these inconsistencies. Designing a closed-loop control law is a difficult task, especially in the presence of disturbances and perturbations [23,24]. To deal with these problems, robust control algorithms are considered the best solutions [25]. Sliding mode control (SMC) is a renowned robust nonlinear control strategy in academia and industry for dealing with large and abrupt changes in system dynamics [26–28]. Chattering that appears on the input of control law is one of SMC's drawbacks. The integral sliding mode control (ISMC) ensures a free-reaching phase surface so that the total system behavior accomplishes the desired invariance. It has high robustness properties against external disturbances, and it can overcome system uncertainty, particularly in nonlinear systems [29]. Backstepping is another nonlinear control strategy that is used to split a high-order nonlinear system into subsystems lower than the original system order. It provides virtual control input and "steps back" all the system until the entire control law design is accomplished [30]. For perturbed nonlinear systems, integrating the backstepping design and ISMC is an alternate robust control strategy. Other authors [31,32] have provided adequate knowledge of adaptive control laws in their research works. The inaccessibility of the actual OFWT system states is also an issue. Researchers [33–35] have proposed control strategies by using HGO for estimating the system's states and stable zero dynamics minimum-phase nonlinear systems respectively. Therefore, inspired by all the above-mentioned robust control techniques, this article aims to design a hybrid control algorithm that can provide all system state estimations and can mitigate disturbances/uncertainties at the same time.

The barge-type TORA-based multiple dynamics of the OFWT model, which we recently published [14], is used as an active structure control strategy in this article, where control input is provided to the proof-mass actuator for controlling the rotor's position. This successfully stabilized the OFWT's vibration movement. Limiting the platform pitch and the tower bending angle of the outputs of the OFWT is also a significant problem, which will be solved in this article. A barrier Lyapunov function (BLF) is a possible solution, which can handle the output constraints of platform pitch and tower bending angle. Therefore, this article presents a novel control framework for OFWT stabilization under output constraint by applying a high-gain observer (HGO)-based adaptive backstepping integral sliding mode control based on the barrier Lyapunov function. Furthermore, the proposed control law based on the BLF has been compared with adaptive backstepping ISMC to show the efficiency of the new output constraint control scheme. Through MATLAB/SIMULINK simulations, the effectiveness of the proposed control scheme has been examined. The results confirm the validity and efficiency of the proposed control approaches.

The following sections are included in the rest of the paper: Section 2 presents the dynamical model of a barge-type TORA-based OFWT. The methodology of the new adaptive robust control law is presented in Section 3. The high-gain observer with stability analysis is formulated in Section 4 for the TORA-based proposed OFWT. The MATLAB/SIMULINK results, as well as the research work conclusions, are presented sequentially in Sections 5 and 6.

2. Dynamical Model of Barge-Type TORA-Based OFWT

The barge-type TORA-based five degrees of freedom (5DOF) OFWT model has been taken from our recent research article [14], as shown in Figure 1. It is capable of generating five megawatts. Towards the direction of the negative z-axis, the rotating proof mass (18 m × 6 m × 6 m) has been attached outside of the nacelle. The hub height of the tower is 90 m. The dimension of the barge-type platform attached with mooring lines is (40 m × 40 m × 10 m). The overall OFWT vibratory motion has been considered a linear movement component of the TORA, whereas the proof mass circular motion represents the TORA controllable actuated component. In this way, the five degrees of freedom OFWT model has become an underactuated mechanical system. The designed dynamic system consists of proof-mass rotational movement, tower fore–aft bending, platform pitch motion, and two linear surge and heave displacements. In this research work, the generator's dynamic behavior, degrees of freedom, rotor, and gearbox will not be used according to the designed model [13]. For the fore–aft and pitch motion of the tower and the platform, respectively, the approximated small-angle is used despite the most severe wave and wind cases, so that the equilibrium position not does exceed 10 degrees [5]. It is considered that the proof-mass actuator in the proposed OFWT is unable to cross 90 degrees out of its state of equilibrium due to the small-angle approximation.

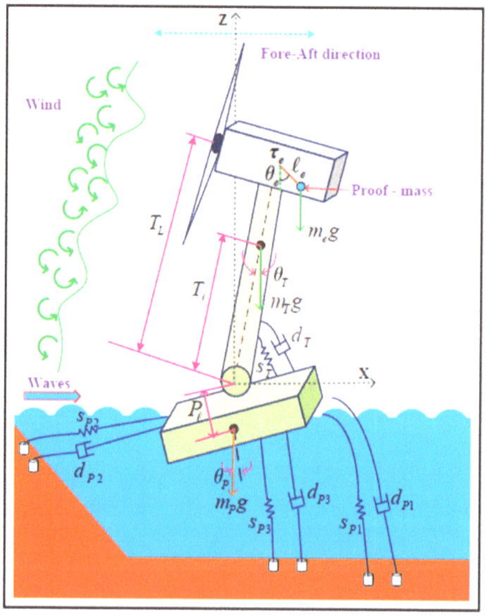

Figure 1. Barge-type TORA-based 5DOF OFWT model.

Figure 1 presents a pivoting linear rigid beam with the same point of origin XYZ reference frame, where the pivot point connects the platform with the tower. The size from the middle of the platform to the middle of the mutual pivot point is denoted by P_ℓ, and, respectively, the mass of the barge-type platform and its moment of inertia are denoted by m_P and I_P. The platform and tower's attached dampers coefficient, $d_{P1}, d_{P2}, d_{P3}, d_T$, and the tightness of springs, $S_{P1}, S_{P2}, S_{P3}, S_T$, have constant values. Likewise, m_T and I_T, respectively, represent the mass of the tower and its moment of inertia. T_L stands for the distance between the nacelle's mid-point and the mutual pivot point, and T_ℓ stands for the distance between the center of mass of the tower and the mutual pivot point. Gravity is denoted with g, and m_e is the mass composed of the rotary proof mass m_{e1} with the attached rod mass m_{e2}. The rotating proof mass has a moment of inertia I_e, where ℓ_e stands for the rod size from the applicable control input τ_e to the rotary proof mass. The angles $\theta_e, \theta_T, \theta_P$ stand for angular positions and $\dot{\theta}_e, \dot{\theta}_T, \dot{\theta}_P$ stand for the angular velocities of the rotary proof mass, tower, and platform. The translational positions X_P, Z_P, and respective velocities are represented by $\dot{X}_P, \dot{Z}_P$.

After thoroughly examining and grasping the working theory of the TORA active mass damper control approach and using the OFWT system, the mathematical model of the TORA floating wind turbine has been obtained using the Euler–Lagrange equation of motion [14]. The following are some necessary equations taken from the derived model [14]. The overall kinetic energy of the considered model is determined by finding the platform kinetic energy K_{m_P}, tower kinetic energy K_{m_T}, and rotary proof-mass kinetic energy K_{m_e}, as follows:

$$\begin{aligned} K_T &= K_{m_P} + K_{m_T} + K_{m_e} \\ &= \frac{1}{2} I_P \dot{\theta}_P^2 + \frac{1}{2} I_T \dot{\theta}_T^2 + \frac{1}{2} I_e \dot{\theta}_e^2 + \frac{1}{2} m_P \left(\dot{X}_P^2 + \dot{Z}_P^2 \right) + \frac{1}{2} m_T \left[\left(\dot{X}_P + \frac{d}{dt}(T_\ell \sin\theta_T) \right)^2 \right. \\ &\left. + \left(\dot{Z}_P + \frac{d}{dt}(T_\ell \cos\theta_T) \right)^2 \right] + \frac{1}{2} m_T \left[\left(\dot{X}_P + \frac{d}{dt}(T_L \sin\theta_T) + \frac{d}{dt}(\ell_e \sin\theta_e) \right)^2 \right. \\ &\left. + \left(\dot{Z}_P + \frac{d}{dt}(T_L \cos\theta_T) + \frac{d}{dt}(\ell_e \cos\theta_e) \right)^2 \right] \end{aligned} \tag{1}$$

In the same way, the overall potential energy of the considered model is determined by finding the platform potential energy U_{m_P}, tower potential energy U_{m_T}, and rotary proof-mass potential energy U_{m_e}, as follows:

$$\begin{aligned} U_T &= U_{m_P} + U_{m_T} + U_{m_e} \\ &= \tfrac{1}{2} s_{P1} \theta_P^2 + \tfrac{1}{2} s_{P2} X_P^2 + \tfrac{1}{2} s_{P3} Z_P^2 - m_P g P_\ell \cos\theta_P + \tfrac{1}{2} s_T (\theta_T - \theta_P)^2 \\ &+ m_T g T_\ell \cos\theta_T + \tfrac{1}{2} s_T (\theta_T - \theta_P)^2 + m_T g T_\ell \cos\theta_T + m_e g (T_L \cos\theta_T - \ell_e \cos\theta_e) \end{aligned} \tag{2}$$

The Lagrangian of the system is determined by the difference between K_T and U_T as follows:

$$\begin{aligned} L &= K_T - U_T \\ &= \frac{1}{2} \left(I_P \dot{\theta}_P^2 + I_T \dot{\theta}_T^2 + I_e \dot{\theta}_e^2 \right) + \frac{1}{2} m_P \left(\dot{X}_P^2 + \dot{Z}_P^2 \right) + \frac{1}{2} m_T \left(\dot{X}_P^2 + 2 \dot{X}_P T_\ell \cos\theta_T \dot{\theta}_T \right. \\ &\left. + T_\ell^2 \dot{\theta}_T^2 + \dot{Z}_P^2 - 2 \dot{Z}_P T_\ell \sin\theta_T \dot{\theta}_T \right) - \frac{1}{2} s_{P1} \theta_P^2 + m_P g P_\ell \cos\theta_P - \frac{1}{2} s_T (\theta_P^2 + \theta_T^2 \\ &- 2\theta_P \theta_T) - m_T g T_\ell \cos\theta_T + \frac{1}{2} m_e \left(\dot{X}_P^2 + \dot{Z}_P^2 + T_L^2 \dot{\theta}_T^2 + \ell_e^2 \dot{\theta}_e^2 + 2 T_L \dot{\theta}_T \left(\dot{X}_P \cos\theta_T \right. \right. \\ &\left. - \dot{Z}_e \sin\theta_T \right) + 2 \ell_e \dot{\theta}_e \left(\dot{X}_P \cos\theta_e - \dot{Z}_P \sin\theta_e \right) + \left. 2 T_L \ell_e \dot{\theta}_T \dot{\theta}_e \cos(\theta_T - \theta_e) \right) \\ &- m_e g (T_L \cos\theta_T - \ell_e \cos\theta_e) - \frac{1}{2} \left(s_{P2} X_P^2 + s_{P3} Z_P^2 \right) \end{aligned} \tag{3}$$

By using the Euler–Lagrange equations of motion, the respective dynamics of the barge-type TORA-based five degrees of freedom OFWT model are as follows:

$$\left\{\begin{array}{l}(m_P+m_T+m_e)\ddot{X}_P+m_e\ell_e\cos\theta_e\ddot{\theta}_e+d_{P2}\dot{X}_P\\+(m_TT_\ell\cos\theta_T+m_eT_L\cos\theta_T)\ddot{\theta}_T-m_e\ell_e\sin\theta_e\dot{\theta}_e^2\\-(m_TT_\ell\sin\theta_T+m_eT_L\sin\theta_T)\dot{\theta}_T^2+s_{P2}X_P=0\\(m_P+m_T+m_e)\ddot{Z}_P-m_e\ell_e\sin\theta_e\ddot{\theta}_e+d_{P3}\dot{Z}_P\\-(m_TT_\ell\sin\theta_T+m_eT_L\sin\theta_T)\ddot{\theta}_T-m_e\ell_e\cos\theta_e\dot{\theta}_e^2\\-(m_TT_\ell\cos\theta_T+m_eT_L\cos\theta_T)\dot{\theta}_T^2+s_{P3}Z_P=0\\I_P\ddot{\theta}_P+(d_T+d_{P1})\dot{\theta}_P-d_T\dot{\theta}_T+(s_T+s_{P1})\theta_P\\+m_PgP_\ell\sin\theta_P-s_T\theta_T=0\\(I_T+m_TT_\ell^2+m_eT_L^2)\ddot{\theta}_T+m_eT_L\ell_e\cos(\theta_T-\theta_e)\ddot{\theta}_e\\-d_T\dot{\theta}_P+(m_TT_\ell\cos\theta_T+m_eT_L\cos\theta_T)\ddot{X}_P+d_T\dot{\theta}_T\\-(m_TT_\ell\sin\theta_T+m_eT_L\sin\theta_T)\ddot{Z}_P-s_T\theta_P+s_T\theta_T\\+m_eT_L\ell_e\sin(\theta_T-\theta_e)\dot{\theta}_e^2-\sin\theta_T(m_TgT_\ell+m_egT_L)=0\\(I_e+m_e\ell_e^2)\ddot{\theta}_e+m_e\ell_e\cos\theta_e\ddot{X}_P-m_e\ell_e\sin\theta_e\ddot{Z}_P\\+m_eT_L\ell_e\cos(\theta_T-\theta_e)\ddot{\theta}_T+m_eg\ell_e\sin\theta_e\\-m_eT_L\ell_e\sin(\theta_T-\theta_e)\dot{\theta}_T^2=\boldsymbol{\tau_e}\end{array}\right. \tag{4}$$

where the platform mass is m_P = 6,150,000 kg, the tower mass is m_T = 347,460 kg, the nacelle mass is m_{nc} = 350,000 kg, the proof mass is m_{e1} = 84,000 kg and the proof-mass rod mass is m_{e2} = 42,000 kg. The tower and proof-mass rod length are respectively taken as T_L = 90 m and ℓ_e = 6 m. In general, $P_\ell = -0.28$ m, 0, T_ℓ = 64 m is the center of the mass. Consequently I_e = 3,528,000 kg$\times$m^2, I_T = 182,170,000 kg$\times$m^2, and I_P = 169,450,000 kg$\times$m^2 are the inertia of the rotary proof mass, tower, and platform, where g = 9.81 m/s^2 is the gravity constant. The stiffness value of the springs is s_{P1} = 14,171,000,000 N/m, $s_{P2} = s_{P3}$ = 1,417,100,000 N/m and s_T = 97,990,000,000 N/m, and d_T = 2,103,200,000 N$\times$s/m, d_{P1} = 3,637,400,000 N$\times$s/m, and $d_{P2} = d_{P3}$ = 363,740,000 N$\times$s/m are the values of the dampers coefficient. Due to the small angles of the tower and platform, we have considered the following approximation: $sin\theta_T = \theta_T$, $sin\theta_P = \theta_P$, $cos\theta_T = 1$, and $cos\theta_P = 1$. For simplicity, let the system's states be $x = \left[X_P\ \dot{X}_P\ Z_P\ \dot{Z}_P\ \theta_P\ \dot{\theta}_P\ \theta_T\ \dot{\theta}_T\ \theta_e\ \dot{\theta}_e\right]^{\mathrm{T}} = \left[x_j\ x_k\ x_l\ x_m\ x_n\ x_o\ x_p\ x_q\ x_r\ x_s\right]^{\mathrm{T}}$ hence, the barge-type 5DOF TORA-based OFWT model into a generalized affine form will be as follows:

$$\dot{x} = n(x) + o(x)u_e \tag{5}$$

Equation (5) contains the following data:

$$\left\{\begin{array}{l}n(x)=\left[\begin{array}{lllll}x_k & n_j(x) & x_m & n_l(x) & x_o\end{array}\right.\\ \qquad\left.\begin{array}{lllll}n_n(x) & x_q & n_p(x) & x_s & n_r(x)\end{array}\right]^{\mathrm{T}}\\o(x)=\left[\begin{array}{lllll}0 & o_j(x) & 0 & o_l(x) & 0\end{array}\right.\\ \qquad\left.\begin{array}{lllll}o_n(x) & 0 & o_p(x) & 0 & o_r(x)\end{array}\right]^{\mathrm{T}}\\u_e=\boldsymbol{\tau_e}\end{array}\right. \tag{6}$$

where

$$\begin{array}{r}\det(M)=c_5\left(c_0^2c_6c_9-c_0^2c_7^2-c_0c_1^2c_9+2c_0c_1c_3c_7-c_0c_2^2c_9-c_0c_3^2c_6\right.\\ \left.+2c_0c_2c_4c_7-c_0c_4^2c_6+c_1^2c_4^2-2c_1c_2c_3c_4+c_2^2c_3^2\right)>0\end{array} \tag{7}$$

$$\left\{\begin{array}{l}n_j(x)=\frac{n_j}{\det(M)},\ n_l(x)=\frac{n_l}{\det(M)},\ n_n(x)=\frac{n_n}{\det(M)},\\n_p(x)=\frac{n_p}{\det(M)},\ n_r(x)=\frac{n_r}{\det(M)},o_j(x)=\frac{o_j}{\det(M)},\\o_l(x)=\frac{o_l}{\det(M)},\ o_n(x)=\frac{o_n}{\det(M)},o_p(x)=\frac{o_p}{\det(M)},\\o_r(x)=\frac{o_r}{\det(M)}\end{array}\right. \tag{8}$$

$$\begin{cases} d_1 = -s_{P2}x_j - d_{P2}x_k + c_2x_q^2 + c_4x_s^2 \\ d_2 = -s_{P3}x_l - d_{P3}x_m + c_1x_q^2 + c_3x_s^2 \\ d_3 = -(s_T + s_{P1} + c_{12})x_n - (d_T + d_{P1})x_o + s_Tx_p + d_Tx_q \\ d_4 = s_Tx_n + d_Tx_o + (c_{13} - s_T)x_p - d_Tx_q - c_8x_s^2 \\ d_5 = c_8x_q^2 - c_{14}\sin x_r \end{cases} \tag{9}$$

$$\begin{cases} n_j = -c_5\big(d_1c_0c_7^2 - d_4c_1c_4^2 - d_5c_3c_2^2 + d_1c_6c_4^2 + d_1c_9c_2^2 + d_5c_1c_2c_4 \\ \quad +d_4c_2c_3c_4 - d_5c_0c_1c_7 - 2d_1c_2c_4c_7 + d_2c_1c_2c_9 - d_2c_1c_4c_7 \\ \quad -d_2c_2c_3c_7 + d_4c_0c_1c_9 - d_4c_0c_3c_7 + d_5c_0c_3c_6 + d_2c_3c_4c_6 - d_1c_0c_6c_9\big) \\ o_j = -c_5\big(-c_2^2c_3 + c_1c_2c_4 - c_0c_1c_7 + c_0c_3c_6\big) \end{cases} \tag{10}$$

$$\begin{cases} n_l = -c_5\big(d_2c_0c_7^2 + d_4c_2c_3^2 + d_5c_4c_1^2 + d_2c_6c_3^2 + d_2c_9c_1^2 \\ \quad -d_5c_1c_2c_3 - d_4c_1c_3c_4 + d_1c_1c_2c_9 - d_1c_1c_4c_7 - d_1c_2c_3c_7 \\ \quad -2d_2c_1c_3c_7 + d_1c_3c_4c_6 + d_5c_0c_2c_7 - d_4c_0c_2c_9 \\ \quad +d_4c_0c_4c_7 - d_5c_0c_4c_6 - d_2c_0c_6c_9\big) \\ o_l = -c_5\big(c_1^2c_4 - c_1c_2c_3 + c_0c_2c_7 - c_0c_4c_6\big) \end{cases} \tag{11}$$

$$\begin{aligned} n_n = -d_3\Big(&c_0^2c_7^2 - c_6c_9c_0^2 + c_0c_9c_1^2 - 2c_0c_1c_3c_7 + c_0c_9c_2^2 \\ &-2c_0c_2c_4c_7 + c_0c_6c_3^2 + c_0c_6c_4^2 - c_1^2c_4^2 + 2c_1c_2c_3c_4 - c_2^2c_3^2\Big) \end{aligned} \tag{12}$$

$$\begin{cases} n_p = c_5\big(d_1c_1c_4^2 - d_2c_2c_3^2 - d_4c_0c_3^2 - d_4c_0c_4^2 - d_5c_7c_0^2 + d_4c_9c_0^2 \\ \quad +d_5c_0c_1c_3 - d_1c_2c_3c_4 + d_2c_1c_3c_4 - d_1c_0c_1c_9 + d_1c_0c_3c_7 \\ \quad +d_5c_0c_2c_4 + d_2c_0c_2c_9 - d_2c_0c_4c_7\big) \\ o_p = d_5\big(-c_7c_0^2 + c_0c_1c_3 + c_0c_2c_4\big) \end{cases} \tag{13}$$

$$\begin{cases} n_r = -c_5\big(-d_1c_3c_2^2 + d_5c_0c_1^2 + d_2c_4c_1^2 + d_5c_0c_2^2 + d_4c_7c_0^2 \\ \quad -d_5c_6c_0^2 + d_1c_1c_2c_4 - d_2c_1c_2c_3 - d_4c_0c_1c_3 - d_1c_0c_1c_7 \\ \quad +d_1c_0c_3c_6 - d_4c_0c_2c_4 + d_2c_0c_2c_7 - d_2c_0c_4c_6\big) \\ o_r = -c_5\big(c_0c_1^2 + c_0c_2^2 - c_6c_0^2\big) \end{cases} \tag{14}$$

where the equilibrium points of the 5DOF barge-type OFWT model are as follows:

$$\begin{cases} x_{eq} = \left[0,0,0,0,0,0,0,0,\theta_{eq},0\right]^{\mathrm{T}}, \\ \theta_{eq} = i\pi, \quad i = 0,\ \pm1,\ \pm2,\ldots \end{cases} \tag{15}$$

We supposed that the nacelle's rotary proof mass could not reach 90° from the equilibrium position due to the small-angle approximation. Therefore, the equilibrium position is $\theta_{eq} = 0$ in Equation (15). The robust adaptive control laws for the output constraints of the 5DOF barge-type TORA-based OFWT are developed in the following parts.

3. Design of Robust Adaptive ISMC Algorithm

The design of a robust control algorithm for the TORA actuator in a nacelle is required to reduce the overall system's vibrations effectively. The dynamics of 5DOF barge-type TORA-based OFWT have nonlinearities that are subjected to a variety of perturbations caused by unknown disturbances and parametric uncertainties. We can take the following matched perturbation in a suggested generalized form of Equation (5), as follows:

$$\dot{x} = n(x) + o(x)u_e + \Upsilon_i(t) \tag{16}$$

where $\Upsilon_i(t)$ is the matched perturbation and $i = 1,2,3,4,5$. To implement the robust control technique [22], the general affine form of Equation (16) must be divided into second-order nonlinear systems. In this way, we will be able to achieve the following subsystems:

$$\begin{cases} \begin{cases} \dot{x}_j = x_k \\ \dot{x}_k = n_j(x) + o_j(x)u_e + \Upsilon_1(t) \end{cases} \\ \begin{cases} \dot{x}_l = x_m \\ \dot{x}_m = n_l(x) + o_l(x)u_e + \Upsilon_2(t) \end{cases} \\ \begin{cases} \dot{x}_n = x_o \\ \dot{x}_o = n_n(x) + \Upsilon_3(t) \end{cases} \\ \begin{cases} \dot{x}_p = x_q \\ \dot{x}_q = n_p(x) + o_p(x)u_e + \Upsilon_4(t) \end{cases} \\ \begin{cases} \dot{x}_r = x_s \\ \dot{x}_s = n_r(x) + o_r(x)u_e + \Upsilon_5(t) \end{cases} \end{cases} \tag{17}$$

where $(x_j, x_k), (x_l, x_m), (x_n, x_o), (x_p, x_q), (x_r, x_s)$ are the state variables. A three-step method will be used to implement a robust adaptive control scheme for a rotary actuator of the proof mass. Firstly, by using the backstepping methodology, a backstepped OFWT subsystem model will be achieved. After this, we will design an adaptive chattering suppression gain for robust backstepping integral sliding mode control laws. Finally, we will design a high-gain observer with stability analysis for the suggested model. The backstepping procedure [30] has been adapted to convert the proposed model (17) to another pattern. For this, a new state variable is selected for the first subsystem of Equation (17) as follows:

$$\mu_1 = x_j \tag{18}$$

Next, we choose the Lyapunov function and take its derivative, as follows:

$$V(\mu_1) = 0.5\mu_1^2 \tag{19}$$

$$\dot{V}(\mu_1) = \mu_1\dot{\mu}_1 = \mu_1 x_k \tag{20}$$

Select a fictitious control input value x_k to maintain the stability of μ_1, as in:

$$x_k = -\lambda_1\mu_1 \tag{21}$$

where $\lambda_1 > 0$. Equation (20) is modified by substituting the value of x_k, giving:

$$\dot{V}(\mu_1) = -\lambda_1\mu_1^2 \tag{22}$$

By selecting μ_1, stable λ_1 dynamics can be achieved. Equation (21) will be backstepped and the variable changed, such as:

$$\mu_2 = x_k + \lambda_1\mu_1 \tag{23}$$

Following the aforementioned method, the chosen state variables of backstepping in a subsystem form of Equation (17) will be:

$$\begin{cases} \begin{cases} \mu_1 = x_j \\ \mu_2 = x_k + \lambda_1\mu_1 \end{cases} \\ \begin{cases} \mu_3 = x_l \\ \mu_4 = x_m + \lambda_2\mu_3 \end{cases} \\ \begin{cases} \mu_5 = x_n \\ \mu_6 = x_o + \lambda_3\mu_5 \end{cases} \\ \begin{cases} \mu_7 = x_p \\ \mu_8 = x_q + \lambda_4\mu_7 \end{cases} \\ \begin{cases} \mu_9 = x_k \\ \mu_{10} = x_s + \lambda_5\mu_9 \end{cases} \end{cases} \tag{24}$$

where $\lambda_2, \lambda_3, \lambda_4, \lambda_5$ represent the positive constants. The remaining Lyapunov functions are defined as $V(\mu_3) = 0.5{\mu_3}^2, V(\mu_5) = 0.5{\mu_5}^2, V(\mu_7) = 0.5{\mu_7}^2, V(\mu_9) = 0.5{\mu_9}^2$. The rest of the fictitious control inputs have been taken as: $x_m = -\lambda_2\mu_3,\ x_o = -\lambda_3\mu_5, x_q = -\lambda_4\mu_7, x_s = -\lambda_5\mu_9$. Taking the Equation (24) derivative and further evaluating it, the following results will be achieved:

$$\begin{cases} \begin{cases} \dot{\mu}_1 = \mu_2 - \lambda_1\mu_1 \\ \dot{\mu}_2 = n_j(x) + o_j(x)u_e + \lambda_1(\mu_2 - \lambda_1\mu_1) + \Upsilon_1(t) \end{cases} \\ \begin{cases} \dot{\mu}_3 = \mu_4 - \lambda_2\mu_3 \\ \dot{\mu}_4 = n_l(x) + o_l(x)u_e + \lambda_2(\mu_4 - \lambda_2\mu_3) + \Upsilon_2(t) \end{cases} \\ \begin{cases} \dot{\mu}_5 = \mu_6 - \lambda_3\mu_5 \\ \dot{\mu}_6 = n_n(x) + \lambda_3(\mu_6 - \lambda_3\mu_5) + \Upsilon_3(t) \end{cases} \\ \begin{cases} \dot{\mu}_7 = \mu_8 - \lambda_4\mu_7 \\ \dot{\mu}_8 = n_p(x) + o_p(x)u_e + \lambda_4(\mu_8 - \lambda_4\mu_7) + \Upsilon_4(t) \end{cases} \\ \begin{cases} \dot{\mu}_9 = \mu_{10} - \lambda_5\mu_9 \\ \dot{\mu}_{10} = n_r(x) + o_r(x)u_e + \lambda_5(\mu_{10} - \lambda_5\mu_9) + \Upsilon_5(t) \end{cases} \end{cases} \quad (25)$$

The second-order backstepping-based subsystems of the presented OFWT system are represented by Equation (25). Furthermore, two different control algorithms are presented for the stabilization of the OFWT model. The first one is called an adaptive integral sliding mode control and the second is the adaptive integral sliding mode control using BLF. The sliding manifold for each subsystem of Equation (25) is used for designing proposed control laws, as follows:

$$\sigma_1 = \sigma_a + \int \sigma_a dt \quad (26)$$

$$\sigma_2 = \sigma_b + \int \sigma_b dt \quad (27)$$

$$\sigma_3 = \sigma_c + \int \sigma_c dt \quad (28)$$

$$\sigma_4 = \sigma_d + \int \sigma_d dt \quad (29)$$

$$\sigma_5 = \sigma_e + \int \sigma_e dt \quad (30)$$

where $\sigma_a = \alpha_1\mu_1 + \mu_2,\ \ \sigma_b = \alpha_2\mu_3 + \mu_4,\ \sigma_c = \alpha_3\mu_5 + \mu_6,\ \sigma_d = \alpha_4\mu_7 + \mu_8,\ \sigma_e = \alpha_5\mu_9 + \mu_{10}$, and $\alpha_1, \alpha_2, \alpha_3, \alpha_4, \alpha_5$ are the positive constants. The surfaces (26)–(30) sum and respective derivatives sum are as follows:

$$\sigma_T = \sigma_1 + \sigma_2 + \sigma_3 + \sigma_4 + \sigma_5 \quad (31)$$

where

$$\dot{\sigma}_1 = \alpha_1\mu_1(1 - \lambda_1) + \mu_2(\alpha_1 + 1) + \lambda_1(\mu_2 - \lambda_1\mu_1) + n_j(x) + o_j(x)u_e + \Upsilon_1(t) \quad (32)$$

$$\dot{\sigma}_2 = \alpha_2\mu_3(1 - \lambda_2) + \mu_4(\alpha_2 + 1) + \lambda_2(\mu_4 - \lambda_2\mu_3) + n_l(x) + o_l(x)u_e + \Upsilon_2(t) \quad (33)$$

$$\dot{\sigma}_3 = \alpha_3\mu_5(1 - \lambda_3) + \mu_6(\alpha_3 + 1) + \lambda_3(\mu_6 - \lambda_3\mu_5) + n_n(x) + \Upsilon_3(t) \quad (34)$$

$$\dot{\sigma}_4 = \alpha_4\mu_7(1 - \lambda_4) + \mu_8(\alpha_4 + 1) + \lambda_4(\mu_8 - \lambda_4\mu_7) + n_p(x) + o_p(x)u_e + \Upsilon_4(t) \quad (35)$$

$$\dot{\sigma}_5 = \alpha_5\mu_9(1 - \lambda_5) + \mu_{10}(\alpha_5 + 1) + \lambda_5(\mu_{10} - \lambda_5\mu_9) + n_r(x) + o_r(x)u_e + \Upsilon_5(t) \quad (36)$$

Theorem 1. *Consider the backstepping-based proposed model (25) and select the sliding manifold as in Equation (31) for the adaptive ISMC algorithm, as follows:*

$$\begin{aligned}u_e = &-\left(o_j(x)+o_l(x)+o_p(x)+o_r(x)\right)^{-1}[\alpha_1\mu_1(1-\lambda_1)+\mu_2(\alpha_1+1)\\&+\lambda_1(\mu_2-\lambda_1\mu_1)+n_j(x)+\alpha_2\mu_3(1-\lambda_2)+\mu_4(\alpha_2+1)+\lambda_2(\mu_4-\lambda_2\mu_3)\\&+n_l(x)+\alpha_3\mu_5(1-\lambda_3)+\mu_6(\alpha_3+1)+\lambda_3(\mu_6-\lambda_3\mu_5)+n_n(x)+\alpha_4\mu_7(1\\&-\lambda_4)+\mu_8(\alpha_4+1)+n_p(x)+\lambda_4(\mu_8-\lambda_4\mu_7)+\alpha_5\mu_9(1-\lambda_5)+\mu_{10}(\alpha_5\\&+1)+\lambda_5(\mu_{10}-\lambda_5\mu_9)+n_r(x)+\hat{\kappa}(t)\sigma_T+\Omega_T sign(\sigma_T)]\end{aligned} \tag{37}$$

where the adaptive gain is represented by $\hat{\kappa}_T(t)$ and switching gain is represented by $\Omega_T > 0$. The proposed control algorithm will effectively stabilize the OFWT, against all the matched unknown bounded perturbations.

Proof of Theorem 1. Stability analysis using the Lyapunov function is a very precise method of control design. We took a Lyapunov candidate function, such as:

$$V_T = 0.5(\sigma_T^2 + \eta^{-1}\tilde{\kappa}_T{}^2(t)) \tag{38}$$

where $\tilde{\kappa}_T(t) = \kappa_T - \hat{\kappa}_T(t)$, $\dot{\tilde{\kappa}}_T(t) = -\dot{\hat{\kappa}}_T(t)$, and κ_T, η are the bounded positive constants. Now, substituting the respective values after taking the derivative of the Lyapunov function (38) to time yields:

$$\begin{aligned}\dot{V}_T &= \sigma_T\dot{\sigma}_T + \eta^{-1}\tilde{\kappa}_T(t)\dot{\tilde{\kappa}}_T(t)\\&= \sigma_T[\alpha_1\mu_1(1-\lambda_1)+\mu_2(\alpha_1+1)+\lambda_1(\mu_2-\lambda_1\mu_1)+n_j(x)\\&\quad+\Upsilon_1(x,t)+\alpha_2\mu_3(1-\lambda_2)+\mu_4(\alpha_2+1)+\lambda_2(\mu_4-\lambda_2\mu_3)\\&\quad+n_l(x)+\Upsilon_2(x,t)+\alpha_3\mu_5(1-\lambda_3)+\mu_6(\alpha_3+1)+\lambda_3(\mu_6\\&\quad-\lambda_3\mu_5)+n_n(x)+\Upsilon_3(x,t)+\alpha_4\mu_7(1-\lambda_4)+\mu_8(\alpha_4+1)\\&\quad+n_p(x)+\Upsilon_4(x,t)+\lambda_4(\mu_8-\lambda_4\mu_7)+\alpha_5\mu_9(1-\lambda_5)\\&\quad+\mu_{10}(\alpha_5+1)+\lambda_5(\mu_{10}-\lambda_5\mu_9)+n_r(x)+\Upsilon_5(x,t)+(o_j(x)\\&\quad+o_l(x)+o_p(x)+o_r(x))\,u_e]-\eta^{-1}\tilde{\kappa}_T(t)\dot{\hat{\kappa}}_T(t)\end{aligned} \tag{39}$$

Now, substituting the control input u_e into Equation (39) yields:

$$\begin{aligned}\dot{V}_T &= \sigma_T\left[\Upsilon_1(t)+\Upsilon_2(t)+\Upsilon_3(t)+\Upsilon_4(t)+\Upsilon_5(t)\right]\\&\quad-\Omega_T|\sigma_T|-\hat{\kappa}_T(t)\sigma_T^2-\eta^{-1}\tilde{\kappa}_T(t)\dot{\hat{\kappa}}_T(t)+\kappa_T\sigma_T^2-\kappa_T\sigma_T^2\\&= \sigma_T\left[\Upsilon_1(t)+\Upsilon_2(t)+\Upsilon_3(t)+\Upsilon_4(t)+\Upsilon_5(t)\right]\\&\quad-\Omega_T|\sigma_T|+\tilde{\kappa}_T(t)\sigma_T^2-\eta^{-1}\tilde{\kappa}_T(t)\dot{\hat{\kappa}}_T(t)-\kappa_T\sigma_T^2\\&= \sigma_T\left[\Upsilon_1(t)+\Upsilon_2(t)+\Upsilon_3(t)+\Upsilon_4(t)+\Upsilon_5(t)\right]\\&\quad-\Omega_T|\sigma_T|-\kappa_T\sigma_T^2+\tilde{\kappa}_T(t)\left(\sigma_T^2-\eta^{-1}\dot{\hat{\kappa}}_T(t)\right)\end{aligned} \tag{40}$$

where the adaptive gain is designed from $\dot{\hat{\kappa}}_T(t) = \eta\sigma_T{}^2$. Let $\beta > 0$ and fulfils the bounded unknown perturbation constraint as $|\Upsilon_1(t)| \le \beta_1$, $|\Upsilon_2(t)| \le \beta_2$, $|\Upsilon_3(t)| \le \beta_3$, $|\Upsilon_4(t)| \le \beta_4$, $|\Upsilon_5(t)| \le \beta_5$ where $\beta = \beta_1+\beta_2+\beta_3+\beta_4+\beta_5$. After applying these parameter values, Equation (40) will be as follows:

$$\dot{V}_T \le -|\sigma_T|[\Omega_T - \beta] - \kappa_T\sigma_T^2 \tag{41}$$

where $\Omega_T = \beta + \psi$, $\psi > 0$, and further simplifying yields:

$$\dot{V}_T \le -\psi|\sigma_T| - \kappa_T\sigma_T{}^2 \tag{42}$$

The derivative of the Lyapunov function is negative definite, as shown in Equation (42). As a result, the suggested control algorithm stabilizes the system effectively and robustly, completing the verification. □

Theorem 2. *Consider the backstepping-based proposed model (25) and select the sliding manifold as in Equation (31) for adaptive ISMC algorithm with output constraint ($|\mu_5| \leq \omega_0$ and $|\mu_7| \leq \omega_1$), as follows:*

$$\begin{aligned} u_e = &-\left(o_j(x)+o_l(x)+o_p(x)+o_r(x)\right)^{-1}[\alpha_1\mu_1(1-\lambda_1)+\mu_2(\alpha_1+1) \\ &+\lambda_1(\mu_2-\lambda_1\mu_1)+n_j(x)+\alpha_2\mu_3(1-\lambda_2)+\mu_4(\alpha_2+1)+\lambda_2(\mu_4-\lambda_2\mu_3) \\ &+n_l(x)+\alpha_3\mu_5(1-\lambda_3)+\mu_6(\alpha_3+1)+\lambda_3(\mu_6-\lambda_3\mu_5)+n_n(x)+\alpha_4\mu_7(1 \\ &-\lambda_4)+\mu_8(\alpha_4+1)+n_p(x)+\lambda_4(\mu_8-\lambda_4\mu_7)+\alpha_5\mu_9(1-\lambda_5)+\mu_{10}(\alpha_5 \\ &+1)+\lambda_5(\mu_{10}-\lambda_5\mu_9)+n_r(x)+\hat{\kappa}(t)\sigma_T+\left(\omega_0^2-\mu_5^2\right)^{-1}(\nu_0\mu_5|\sigma_3|)+(\omega_1^2 \\ &-\mu_7^2)^{-1}(\nu_1\mu_7|\sigma_4|)+\Omega_T(\left(\omega_0^2-\mu_5^2\right)^{-0.5}(\mu_5)+\left(\omega_1^2-\mu_7^2\right)^{-0.5}(\mu_7) \\ &+(|2\sigma_T|)^{0.5})\,sign(\sigma_T)-\psi_1 sign(\sigma_T)-\psi_2 sign(\sigma_T)] \end{aligned} \tag{43}$$

where ν_0, ν_1, ω_0, ω_1 are the positive gains, the adaptive and switching gains are respectively as: $\hat{\kappa}_T(t)$, $\Omega_T > 0$. The proposed control algorithm will effectively stabilize the OFWT to the output constraint, against all the matched unknown bounded perturbations.

Proof of Theorem 2. Select a Lyapunov candidate function, as follows:

$$\begin{aligned} V_T = |\sigma_T| &+ 0.5(\nu_0 \ln\left(\left(\omega_0^2-\mu_5^2\right)^{-1}\left(\omega_0^2\right)\right) \\ &+\nu_1 \ln\left(\left(\omega_1^2-\mu_7^2\right)^{-1}\left(\omega_1^2\right)\right)+\eta^{-1}\tilde{\kappa}_T^2(t)) \end{aligned} \tag{44}$$

where $\tilde{\kappa}_T(t) = \kappa_T - \hat{\kappa}_T(t)$, $\dot{\tilde{\kappa}}_T(t) = -\dot{\hat{\kappa}}_T(t)$, $\kappa_T > 0$, and $\eta > 0$. Now, substituting the respective values after taking the derivative of the Lyapunov function (44) to time yields:

$$\begin{aligned} \dot{V}_T = \dot{\sigma}_T sign(\sigma_T) &+ \left(\left(\omega_0^2-\mu_5^2\right)^{-1}(\nu_0\mu_5\dot{\mu}_5)\right) \\ &+\left(\left(\omega_1^2-\mu_7^2\right)^{-1}(\nu_1\mu_7\dot{\mu}_7)\right)-\eta^{-1}\tilde{\kappa}_T(t)\dot{\hat{\kappa}}_T(t) \end{aligned} \tag{45}$$

$$\begin{aligned} \dot{V}_T = &\,sign(\sigma_T)\left[\alpha_1\mu_1(1-\lambda_1)+\mu_2(\alpha_1+1)+\lambda_1(\mu_2-\lambda_1\mu_1)+n_j(x)+\Upsilon_1(t)+\alpha_2\mu_3(1\right. \\ &-\lambda_2)+\mu_4(\alpha_2+1)+\lambda_2(\mu_4-\lambda_2\mu_3)+n_l(x)+\alpha_3\mu_5(1-\lambda_3)+\Upsilon_2(t)+\mu_6(\alpha_3+1) \\ &+\lambda_3(\mu_6-\lambda_3\mu_5)+n_n(x)+\Upsilon_3(t)+\alpha_4\mu_7(1-\lambda_4)+\mu_8(\alpha_4+1)+n_p(x)+\Upsilon_4(t) \\ &+\lambda_4(\mu_8-\lambda_4\mu_7)+\alpha_5\mu_9(1-\lambda_5)+\mu_{10}(\alpha_5+1)+\lambda_5(\mu_{10}-\lambda_5\mu_9)+n_r(x)+\Upsilon_5(t) \\ &+\left(\omega_0^2-\mu_5^2\right)^{-1}(\nu_0\mu_5|\sigma_3|)+\left(\omega_1^2-\mu_7^2\right)^{-1}(\nu_1\mu_7|\sigma_4|)+(o_j(x)+o_l(x)+o_p(x) \\ &+o_r(x))\,u_e]-\psi_1-\psi_2-\psi_3-\psi_4-\eta^{-1}\tilde{\kappa}_T(t)\dot{\hat{\kappa}}_T(t) \end{aligned} \tag{46}$$

where the following parameters are defined as $\psi_1 = \left((\omega_0^2-\mu_5^2)^{-1}(\nu_0\mu_5(\alpha_3\mu_5+\alpha_3\int\mu_5 dt+\int\mu_6 dt))\right)$, $\psi_3 = ((\omega_0^2-\mu_5^2)^{-1}(\nu_0\lambda_3\mu_5^2))$, $\psi_2 = \left(\left(\omega_1^2-\mu_7^2\right)^{-1}(\nu_1\mu_7(\alpha_4\mu_7+\alpha_4\int\mu_7 dt+\int\mu_8 dt))\right)$, and $\psi_4 = \left(\left(\omega_1^2-\mu_7^2\right)^{-1}(\nu_1\lambda_4\mu_7^2)\right)$. Let $\beta > 0$ and fulfill the bounded unknown perturbation constraint as $|\Upsilon_1(t)| \leq \beta_1$, $|\Upsilon_2(t)| \leq \beta_2$, $|\Upsilon_3(t)| \leq \beta_3$, $|\Upsilon_4(t)| \leq \beta_4$, $|\Upsilon_5(t)| \leq \beta_5$, where $\beta = \beta_1 + \beta_2 + \beta_3 + \beta_4 + \beta_5$. Now, substituting the control input u_e in Equation (46) yields:

$$\begin{aligned}
\dot{V}_T = & \left[\Upsilon_1(t) + \Upsilon_2(t) + \Upsilon_3(t) + \Upsilon_4(t)\right. \\
& \left.+\Upsilon_5(t)\right] sign(\sigma_T) - \Omega_T\left(\left(\omega_0^2 - \mu_5^2\right)^{-0.5}\left(\mu_5^2\right)^{0.5}\right. \\
& +\left(\omega_1^2 - \mu_7^2\right)^{-0.5}\left(\mu_7^2\right)^{0.5} + \left(|2\sigma_T|)^{0.5}\right) + \kappa_T(t)|\sigma_T| \\
& -\hat{\kappa}_T(t)|\sigma_T| - \eta^{-1}\tilde{\kappa}_T(t)\dot{\hat{\kappa}}_T(t) - \kappa_T(t)|\sigma_T| \\
\leq & -\kappa_T(t)|\sigma_T| + \beta sign(\sigma_T) - \Omega_T(\ln(\omega_0^2 \\
& -\mu_5^2)^{-0.5}\left(\omega_0^2\right)^{0.5} + \ln\left(\omega_1^2 - \mu_7^2\right)^{-0.5}\left(\omega_1^2\right)^{0.5} \\
& +(|2\sigma_T|)^{0.5}) + \tilde{\kappa}_T(t)|\sigma_T| - \eta^{-1}\tilde{\kappa}_T(t)\dot{\hat{\kappa}}_T(t) \\
\leq & -\Omega_T\sqrt{2V_T}
\end{aligned} \tag{47}$$

where the adaptive gain is designed from $\dot{\hat{\kappa}}_T(t) = \eta|\sigma_T|$. The derivative of the Lyapunov function is negative definite, as shown in Equation (47). As a result, the suggested control algorithm stabilizes the system effectively and robustly, and will never become unbounded due to the conditions mentioned in Theorem 2. It completes the verification. □

4. Formulation of HGO

With information on desired output states, the high-gain observer can generate the estimation of the unknown states in the presence of system uncertainties and disturbances. The HGO's state estimation property ensures that the controller has access to unknown states. Figure 2 represents the schematic diagram of the adaptive robust control scheme for the OFWT system. The HGO uses the actual system states information and provides all known and unknown estimated states to the control input of the closed-loop system, as shown in the schematic diagram.

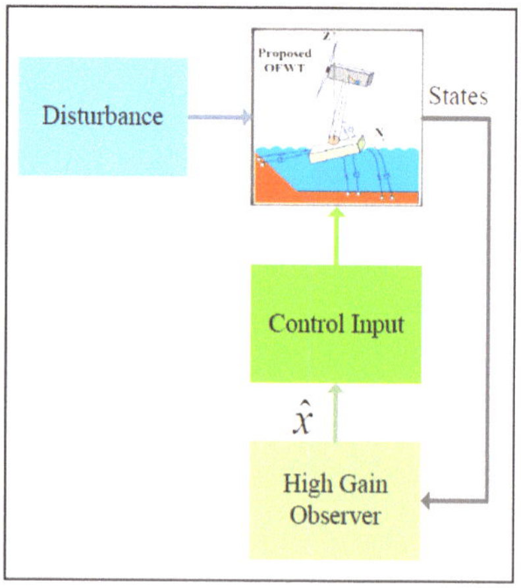

Figure 2. Schematic diagram of the adaptive robust control scheme for OFWT system.

In the form of a second order, we have presented five subsystems in this article, where each of them is defined in normal form. After that we developed HGOs for each subsystem for the nominal values of the system, as follows:

$$\begin{cases} \dot{\hat{x}}_j = \hat{x}_k + \gamma_1(x_j - \hat{x}_j) \\ \dot{\hat{x}}_k = \phi_j(\hat{x}, u) + \gamma_2(x_j - \hat{x}_j) \end{cases} \tag{48}$$

$$\begin{cases} \dot{\hat{x}}_l = \hat{x}_m + \gamma_3(x_l - \hat{x}_l) \\ \dot{\hat{x}}_m = \phi_l(\hat{x}, u) + \gamma_4(x_l - \hat{x}_l) \end{cases} \tag{49}$$

$$\begin{cases} \dot{\hat{x}}_n = \hat{x}_o + \gamma_5(x_n - \hat{x}_n) \\ \dot{\hat{x}}_o = \phi_n(\hat{x}, u) + \gamma_6(x_n - \hat{x}_n) \end{cases} \tag{50}$$

$$\begin{cases} \dot{\hat{x}}_p = \hat{x}_q + \gamma_7(x_p - \hat{x}_p) \\ \dot{\hat{x}}_q = \phi_p(\hat{x}, u) + \gamma_8(x_p - \hat{x}_p) \end{cases} \tag{51}$$

$$\begin{cases} \dot{\hat{x}}_r = \hat{x}_s + \gamma_9(x_r - \hat{x}_r) \\ \dot{\hat{x}}_s = \phi_r(\hat{x}, u) + \gamma_{10}(x_r - \hat{x}_r) \end{cases} \tag{52}$$

where $\phi_j(\hat{x}, u) = \hat{n}_j(x) + \hat{o}_j(x)u_e$, $\phi_l(\hat{x}, u) = \hat{n}_l(x) + \hat{o}_l(x)u_e$, $\phi_n(\hat{x}, u) = \hat{n}_n(x)$, $\phi_p(\hat{x}, u) = \hat{n}_p(x) + \hat{o}_p(x)u_e$, $\phi_r(\hat{x}, u) = \hat{n}_r(x) + \hat{o}_r(x)u_e$, and $\gamma_1 = a_1^1/b$, $\gamma_2 = a_2^1/b^2$, $\gamma_3 = a_1^2/b$, $\gamma_4 = a_2^2/b^2$, $\gamma_5 = a_1^3/b$, $\gamma_6 = a_2^3/b^2$, $\gamma_7 = a_1^4/b$, $\gamma_8 = a_2^4/b^2$, $\gamma_9 = a_1^5/b$, $\gamma_{10} = a_2^5/b^2$, while b has a small positive constant value, such as $b \ll 1$. Constant parameters a_1^1, a_2^1, a_1^2, a_2^2, a_1^3, a_2^3, a_1^4, a_2^4, a_1^5, a_2^5 are chosen from these polynomials $s^2 + a_1^1 s + a_2^1 = 0$, $s^2 + a_1^2 s + a_2^2 = 0$, $s^2 + a_1^3 s + a_2^3 = 0$, $s^2 + a_1^4 s + a_2^4 = 0$, and $s^2 + a_1^5 s + a_2^5 = 0$ in such a way that in the left half-plane, the polynomials' roots will exist.

Stability Analysis: The stability analysis of the high-gain observer of the suggested barge-type OFWT model is determined by rewriting Equation (16) into a generalized form as follows:

$$\begin{cases} \dot{x} = Ax + B\phi(x, u) \\ y = Cx \end{cases} \tag{53}$$

where $x \in R^\rho$, $\rho = 10$, $y \in R^m$, $m = 5$, $\varnothing : R^5 \times R \to R$ is a real-valued map, and $\varnothing(x, u)$ is the image of (x, u). Overall, the proposed OFWT model Equation (16) has a 10-degree order ($\rho = 10$) and each subsystem of the proposed OFWT model Equation (17) has a 2-degree order $\left(\rho = \sum_{i=1}^{5} \rho_i\right)$. According to the HGO design [35], the derivative of estimated states can be expressed as follows:

$$\dot{\hat{x}} = A\hat{x} + B\phi(\hat{x}, u) + HC(y - \hat{x}) \tag{54}$$

where

$$\tilde{x} = \begin{bmatrix} \tilde{x}_1 \\ \tilde{x}_2 \\ \tilde{x}_3 \\ \tilde{x}_4 \\ \tilde{x}_5 \\ \tilde{x}_6 \\ \tilde{x}_7 \\ \tilde{x}_8 \\ \tilde{x}_9 \\ \tilde{x}_{10} \end{bmatrix} = \begin{bmatrix} x_1 - \hat{x}_1 \\ x_2 - \hat{x}_2 \\ x_3 - \hat{x}_3 \\ x_4 - \hat{x}_4 \\ x_5 - \hat{x}_5 \\ x_6 - \hat{x}_6 \\ x_7 - \hat{x}_7 \\ x_8 - \hat{x}_8 \\ x_9 - \hat{x}_9 \\ x_{10} - \hat{x}_{10} \end{bmatrix} \tag{55}$$

here, $\tilde{x}$ is the estimation error of the observer. For simplicity, a notation can be defined as $\xi(x, \hat{x}) = \phi(x, u) - \phi(\hat{x}, u)$. Now, taking the derivative of Equation (55) yields:

$$\dot{\tilde{x}} = A_0 \tilde{x} + B_0 \xi(x, \hat{x}) \tag{56}$$

where $\xi(x, \hat{x}) = [\xi_1(x, \hat{x})\ \xi_2(x, \hat{x})\ \xi_3(x, \hat{x})\ \xi_4(x, \hat{x})\ \xi_5(x, \hat{x})]^{\mathrm{T}}$. Now, the scale estimation is defined as $\varpi_{ij} = (x_{ij} - \hat{x}_{ij})/b^{\rho_i - j}$, where $1 \le i \le 5$ and $1 \le j \le \rho_j$. After substituting the values, the scale estimation of each state is as follows: $\varpi_{11} = \tilde{x}_{11}/b$, $\varpi_{12} = \tilde{x}_{12}$. For simplicity, we assumed that $\varpi_{11} = \tilde{x}_{11}/b = \tilde{x}_1/b = \varpi_1$, $\varpi_{12} = \tilde{x}_{12} = \tilde{x}_2 = \varpi_2$ and similarly for the remaining parameters. From this, we can get $b\dot{\varpi}_1 = -a_1^1\varpi_1 + \varpi_2$, $b\dot{\varpi}_2 = -a_2^1\varpi_1 + b\xi_1(x, \hat{x})$, $\varpi_3 = \tilde{x}_3/b$, $\varpi_4 = \tilde{x}_4$, $\varpi_5 = \tilde{x}_5/b$, $\varpi_6 = \tilde{x}_6$, $\varpi_7 = \tilde{x}_7/b$, $\varpi_8 = \tilde{x}_8$, $\varpi_9 = \tilde{x}_9/b$, $\varpi_{10} = \tilde{x}_{10}$. From the scale estimation of each state, a generalized form can be written as follows:

$$D(b)\varpi = x - \hat{x} \tag{57}$$

taking the derivative of Equation (57) and substituting the values yields:

$$\dot{\varpi} = D^{-1}(b)(A - HC)D(b)\varpi + D^{-1}(b)B\xi(x, z, D(b)\varpi) \tag{58}$$

$$\dot{\varpi} = b^{-1}\zeta\varpi + B\xi(x, z, D(b)\varpi) \tag{59}$$

hence, Equation (59) can be written as:

$$b\dot{\varpi} = \zeta\varpi + bB\xi(x, z, D(b)\varpi) \tag{60}$$

since ζ is Hurwitz, thus it is concluded from Equation (60) that as the value b approaches zero, the uncertain term becomes zero, and the error converges to zero asymptotically. The derived matrices for the stability analysis are presented in Appendix A.

5. Simulation Results and Discussion

In this paper, MATLAB/SIMULINK software is utilized to simulate the mathematical model of the barge-type TORA-based 5DOF OFWT and as well as to verify the efficiency of the designed BLF-based adaptive backstepping ISMC compared with the adaptive backstepping ISMC law. As in the control design portion, we have selected a backstepping ISMC surface and designed two robust adaptive ISMC laws based on the high-gain observer. The objective of this study is to examine the OFWT model stability under the applied output constraints and to improve its performance in terms of stability by using the suggested control strategy. For this, the parametric values discussed in the suggested perturbed 5DOF OFWT model section have been used in simulation with the assumption that the system is in vibrating condition with these initial values: $x_j(0) = 0.2$, $x_k(0) = 0$, $x_l(0) = 0.15$, $x_m(0) = 0$, $x_n(0) = \pi/36$, $x_o(0) = 0$, $x_p(0) = \pi/36$, $x_q(0) = 0$, $x_r(0) = 0$, $x_s(0) = 0$, where the following are the initial design conditions for the HGO: $\hat{x}_j(0) = 0$, $\hat{x}_k(0) = 0$, $\hat{x}_l(0) = 0$, $\hat{x}_m(0) = 0$, $\hat{x}_n(0) = 0$, $\hat{x}_o(0) = 0$, $\hat{x}_p(0) = 0$, $\hat{x}_q(0) = 0$, $\hat{x}_r(0) = 0$, and $\hat{x}_s(0) = 0$.

We have been able to achieve the simulation results displayed in Figure 3 by utilizing the initial values rather than incorporating control input. Throughout every simulation outcome, X_P of the part (a) and Z_P of the part (b) represent the platform surge and heave linear displacement, while parts (c), (d), and (e) represent the platform pitch angle (θ_P), bending tower angle (θ_T), and rotary angle (θ_e) of the TORA actuator, respectively. Without applying control laws, the suggested model's natural response, including and excluding the actuated proof mass, has been studied using the system comparison analysis shown in Figure 3. Figure 3 shows that while achieving the equilibrium position, the effect of the vibration is smaller in the suggested proof-mass actuator model case, while Figure 3e shows the actuated proof-mass vibrating behavior. Similarly, without considering the actuated proof-mass case, Figure 3a,b go to their equilibrium position approximately in 60 s, while in the case of including the proof-mass actuator, and with no applying control input, the states go to the equilibrium position within 5 s. In the remaining subparts of Figure 3c–e, in both cases (excluding and including the proof-mass actuator model), the states go to their equilibrium position after 100 s, but the difference is in terms of vibrational amplitude. Furthermore, the barge-type OFWT is a fundamentally stable system, as has been demonstrated in the literature [6,9]. Our proposed model's simulated results demonstrate a similar pattern to that shown in previous articles. As the standard matched sinusoidal disturbance, $\Upsilon_i(t)$ is taken in our proposed model as a perturbation factor. The selection of disturbances $\Upsilon_1(t)$ and $\Upsilon_2(t)$ are used in such a way as to cause the drifting phenomena into the barge platform of the proposed model, where $\Upsilon_1(t) = 5sin(2\pi ft)[cos\varphi - sin\varphi]$, $\Upsilon_2(t) = 5sin(2\pi ft)[sin\varphi + cos\varphi]$, and φ is the drifting angle. Moreover, $\Upsilon_i(t) = 5sin(2\pi ft)$, $i = 3,4,5,\ f = 15$ Hz, and $\varphi = 30^\circ$.

The comparison study has been conducted by applying the proposed adaptive backstepping ISMC and BLF-based adaptive backstepping ISMC to the proof-mass actuator, as shown in Figure 4. Table 1 shows the positive constants parameters. For comparison purposes, we have chosen the same values for both tuning control laws. All the parts of Figure 4 show that the states go to their equilibrium position at the same time (within 10 s) by using the adaptive backstepping ISMC and the BLF-based adaptive backstepping ISMC. Parts (c) and (d) of Figure 4 show that the BLF-based control algorithm works accurately and bound the platform pitch angle and the tower bending angle within five degrees, while the adaptive backstepping ISMC violates the constraint. Furthermore, the proposed BLF-based adaptive backstepping ISMC better stabilizes the pitch and the bending movement of the barge-type OFWT tower within 10 s as compared with the active and passive control approaches mentioned in other articles [9,13]. Table 2 presents the root mean square error (RMSE) of Figure 4. For quantitative comparison purposes, Table 2 also presents the transient state and steady-state error response. From Table 2, it can be seen clearly that the error of controlled states (CS) using the adaptive backstepping ISMC and BLF-based adaptive backstepping ISMC shows a very small error. Overall, the RMSE of the controlled states using the adaptive backstepping ISMC has less error than the BLF-based adaptive backstepping ISMC.

Table 1. Proposed control tuning parameters.

Parameters	Value	Parameters	Value
α_1	3	λ_5	10
α_2	3	ν_0	1
α_3	3	ν_1	1
α_4	2	ω_0	5
α_5	1	ω_1	5
λ_1	1	η	10
λ_2	1	Ω_T	10
λ_3	1	a_1	11
λ_4	1	a_2	11

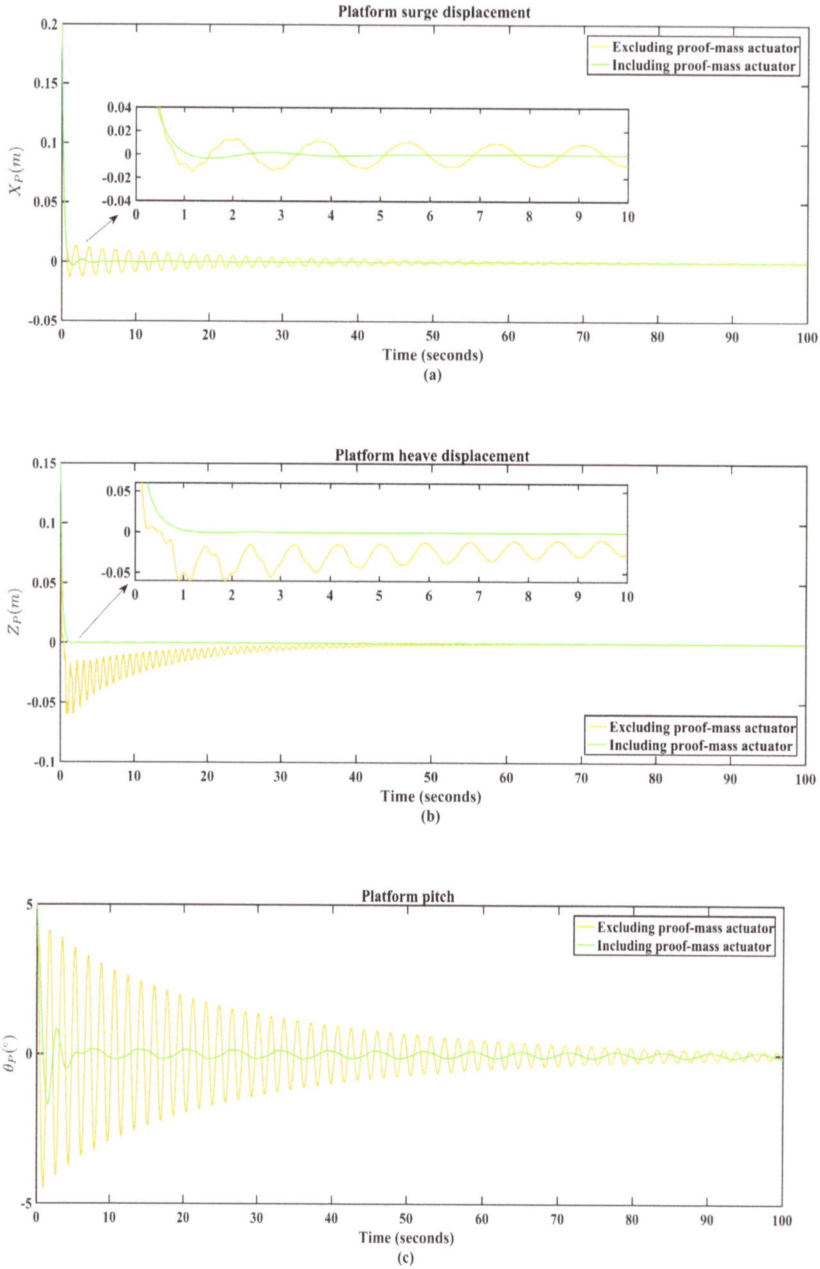

Figure 3. *Cont.*

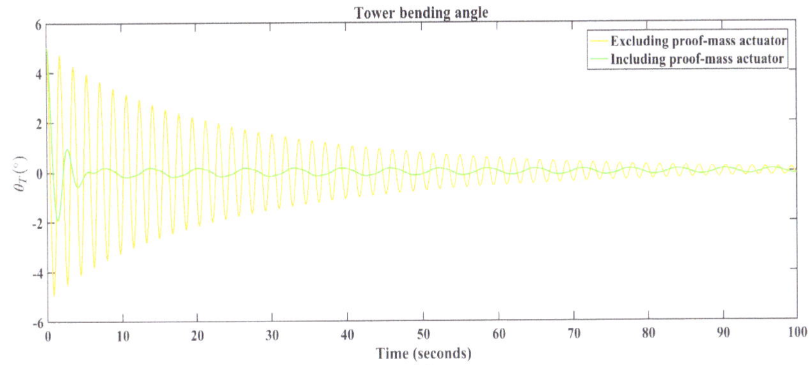

(d)

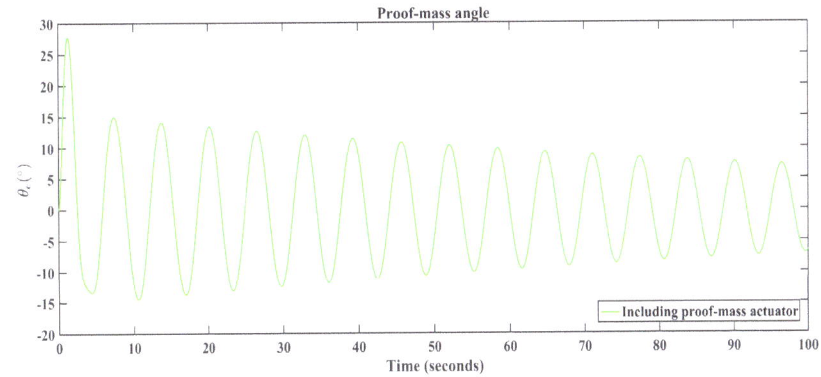

(e)

Figure 3. The response of barge−type 5DOF TORA−based OFWT without applying control laws, (**a**) X_P, (**b**) Z_P, (**c**) θ_P, (**d**) θ_T, and (**e**) θ_e, where brown results do not apply the proof-mass actuator in the OFWT model and green results do apply the proof-mass actuator in the OFWT model.

Table 2. Error analysis of the controlled states (CS) and estimated states (ES).

Signal under Observation	Error	X_P	Z_P	θ_P	θ_T	θ_e
CS using adaptive ISMC	Transient state	0.0291	0.0210	1.1648	1.2893	1.2992
	Steady state	0.0001	0.0001	0.0061	0.0070	0.0116
	RMSE	0.0158	0.0114	0.8208	0.9084	0.9155
CS using BLF-based adaptive ISMC	Transient state	0.0289	0.0211	1.1778	1.3108	3.4688
	Steady state	0.0001	0.0001	0.0063	0.0070	0.0102
	RMSE	0.0157	0.0114	0.8299	0.9236	2.4443
ES using HGO	RMSE	0.0051	0.0038	0.1279	0.1279	1.2221

Figure 5 represents the total backstepping ISMC surface response for designing HGO-based proposed control algorithms. It can be observed that the system trajectories reach the sliding manifold in 0.5 s i.e., $\sigma_T = 0 \,\forall\, t \geq 0.5$. The BLF-based adaptive backstepping ISMC input represents significantly less chattering than the adaptive backstepping ISMC input, as can be seen clearly in Figure 6. Using the HGO for the BLF−based adaptive backstepping ISMC, the original states and their estimated states with the respective error of the suggested model are shown in Figure 7, where the numerical difference between the original and their estimated states are very minor, as shown in Table 2. This model works for all of the hypotheses provided in the modeling section. The response of the adaptive gains used in the adaptive backstepping ISMC and BLF-based adaptive backstepping ISMC algorithms are shown in Figure 8, where the adaptive gain used in the adaptive backstepping ISMC settles at approximately 97 and the adaptive gain used in the BLF-based adaptive backstepping ISMC settles at approximately 43.

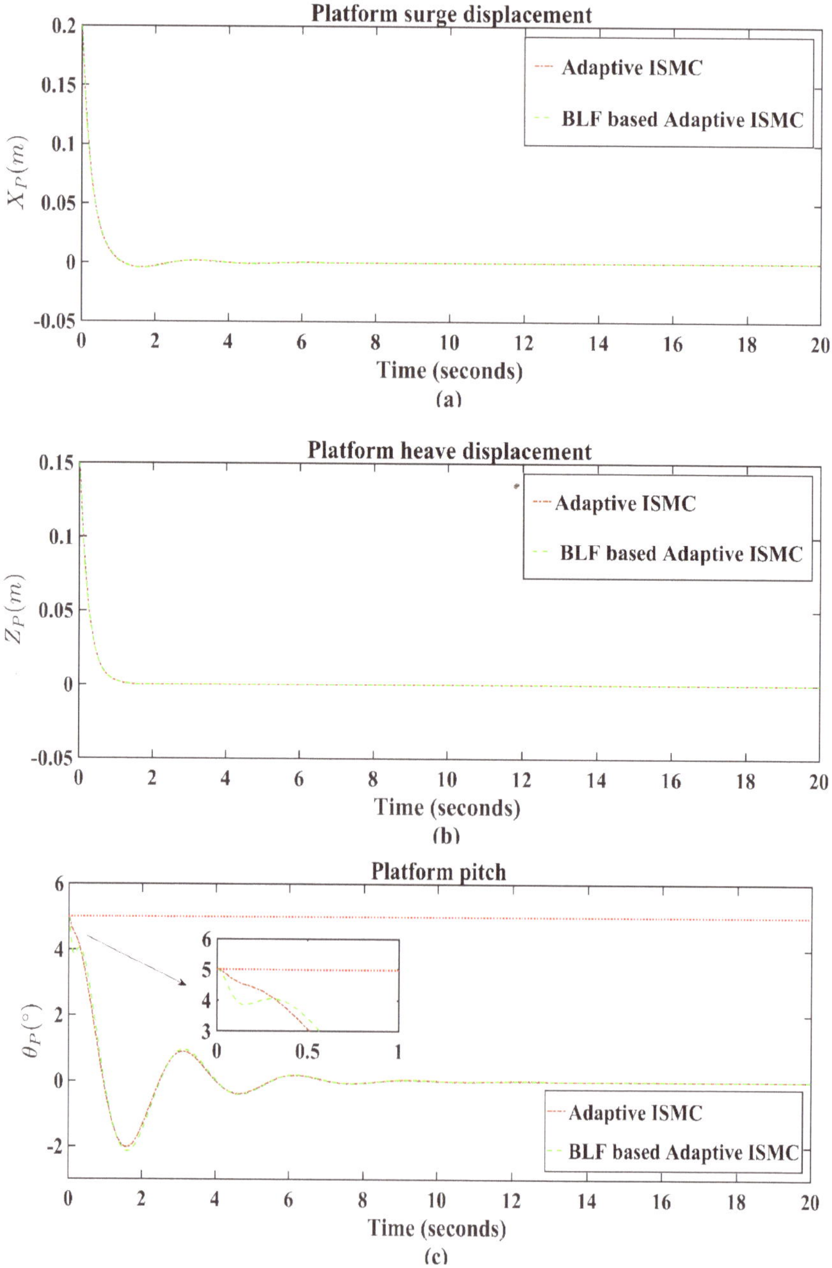

Figure 4. *Cont.*

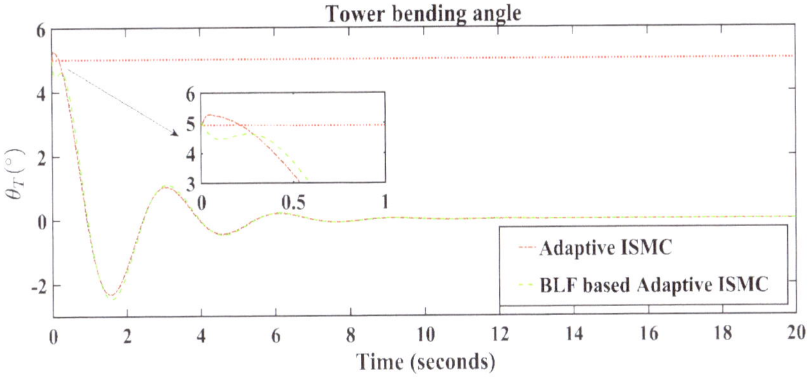

(d)

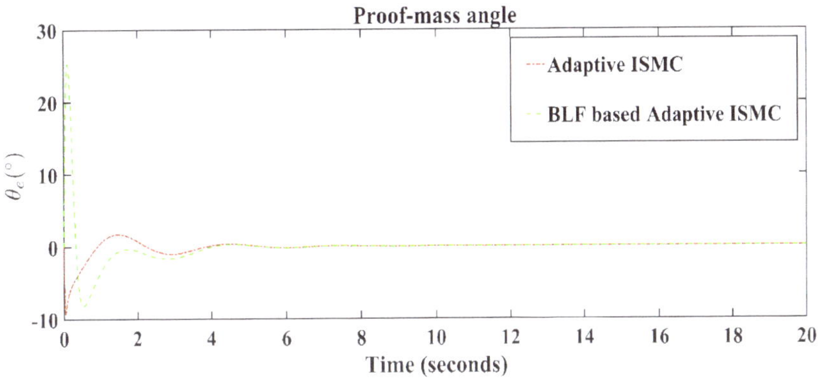

(e)

Figure 4. The response of barge−type 5DOF TORA−based OFWT when using adaptive backstepping ISMC and BLF-based adaptive backstepping ISMC, (**a**) X_P, (**b**) Z_P, (**c**) θ_P, (**d**) θ_T, and (**e**) θ_e, where brown results represent adaptive backstepping ISMC efficiency and green results represent BLF based adaptive backstepping ISMC efficiency.

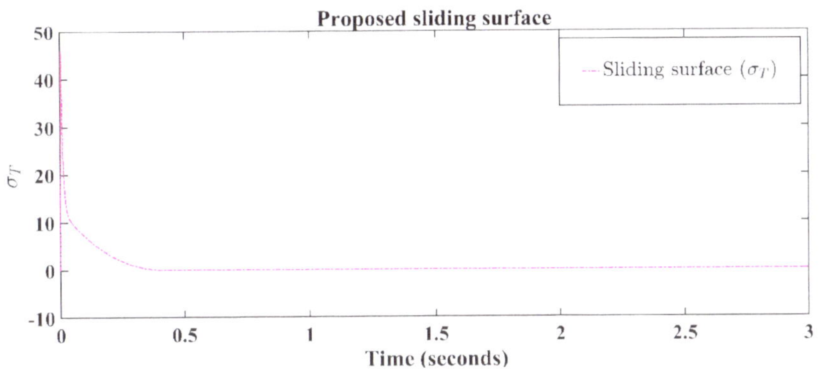

Figure 5. The response of the total backstepping ISMC surface (σ_T) for developing proposed control algorithms.

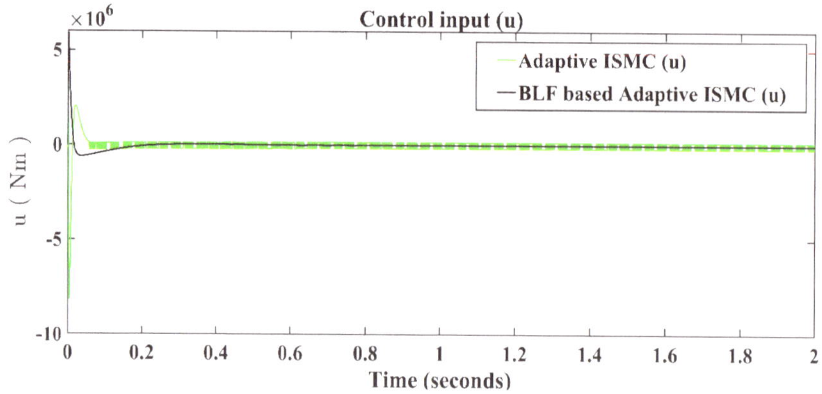

Figure 6. The response of the proposed control inputs, where green represents adaptive backstepping ISMC efficiency and black represents BLF—based adaptive backstepping ISMC efficiency.

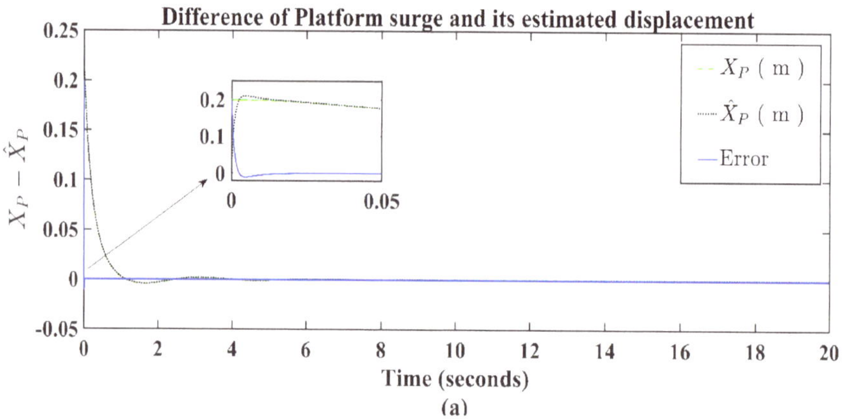

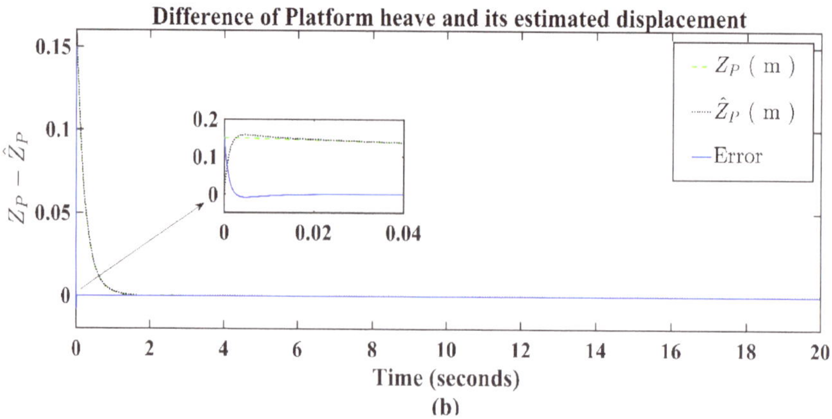

Figure 7. *Cont.*

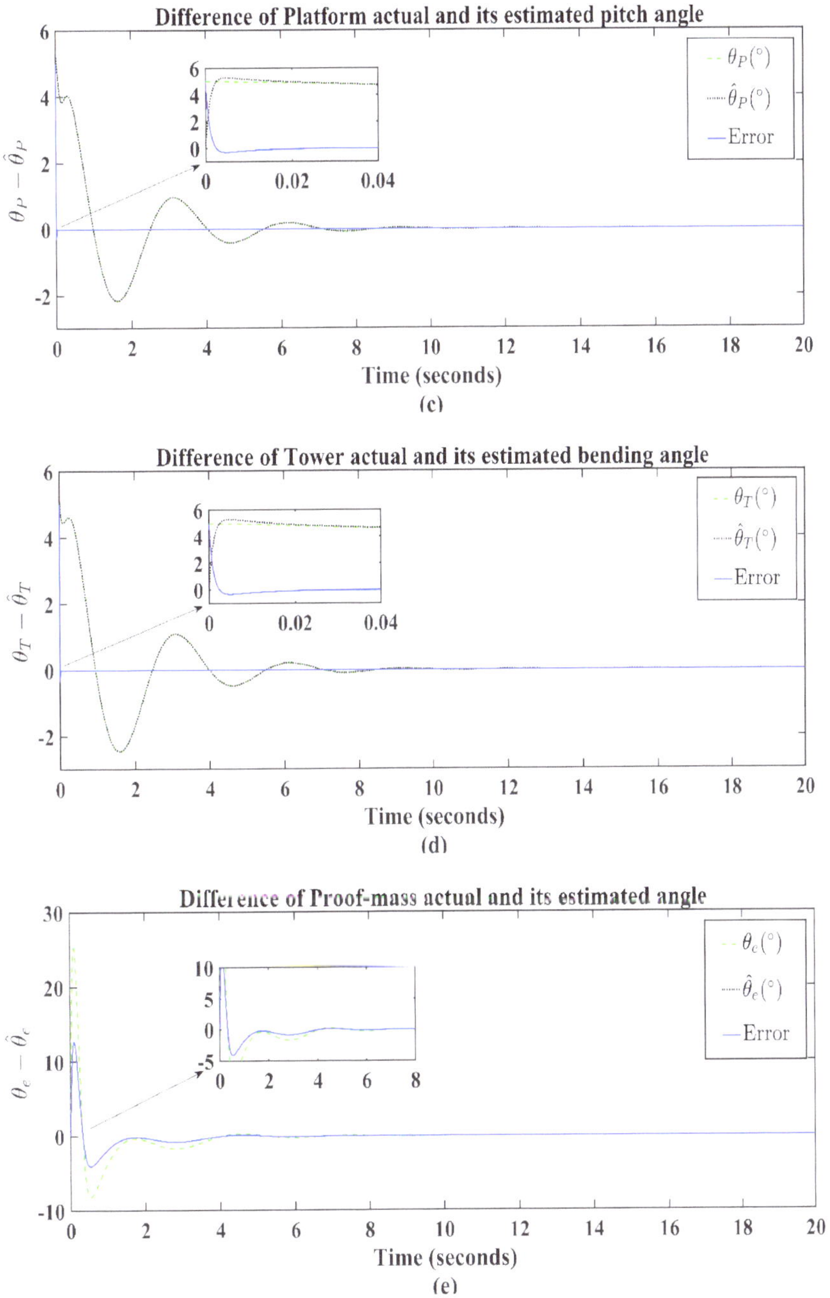

Figure 7. Simulation results of the controlled barge−type 5DOF TORA−based OFWT for the actual states (green) and their estimated states (black), and corresponding error (blue color): (**a**) Difference between platform surge and its estimated displacement, (**b**) Difference between platform heave and its estimated displacement, (**c**) Difference between platform's actual and its estimated pitch angle, (**d**) Difference between tower's actual and its estimated bending angle, (**e**) Difference between proof mass actual and its estimated angle.

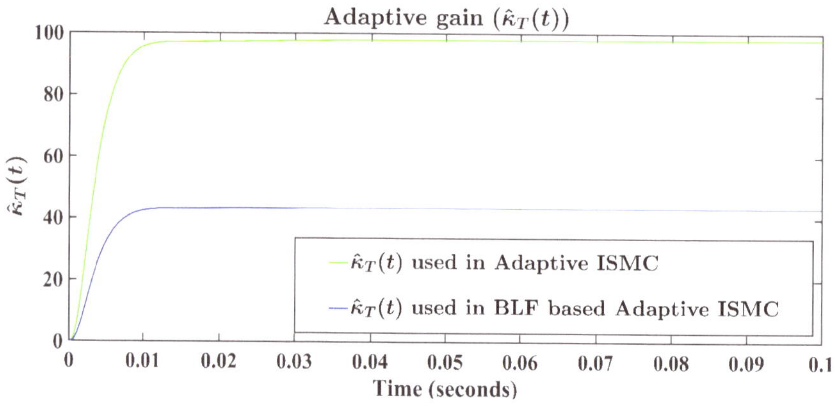

Figure 8. The response of the adaptive gain used in adaptive backstepping ISMC (green) and BLF-based adaptive backstepping ISMC (blue).

6. Conclusions

The TORA as an AMD control strategy for the stabilization of the barge-type 5DOF OFWT model is challenging and interesting. In this article, a novel control framework has been devised that consists of a high-gain observer-based adaptive integral sliding mode control with backstepping and barrier Lyapunov function. This control algorithm successfully solved the issue of the unavailability of the system states by providing all the proposed OFWT model estimated states. The numerical simulations and their numeric error table reveal that the proposed output constraint BLF-based adaptive backstepping ISMC can effectively handle vibrations as compared with the adaptive backstepping ISMC. In both control laws, the proposed model states go to their equilibrium position within 10 s, but the difference is that the BLF-based adaptive backstepping ISMC guarantees to satisfy the constraint on the states. Moreover, the proposed output constraint BLF-based adaptive backstepping ISMC shows less chattering than the adaptive backstepping ISMC. The results confirm the validity and efficiency of the suggested control approaches. The floating wind turbine structures are thus far at an early stage of development and various platforms are in a proof-of-concept study using a scale model incorporating a barge, spar, and tension leg platform. Therefore, more research on this topic is still required. Future work will consider model gearboxes and actuation components that produce the control input torque, like hydraulic or electric motors. Similarly, comparing the simulation results of this proposed model by using other simulation tools (FAST/SIMPACK) as well as experimental results can be conducted.

Author Contributions: Conceptualization, S.A.A.S. and B.G.; methodology, S.A.A.S. and B.G.; formal analysis, B.G., I.A. and H.U.; writing—original draft, S.A.A.S.; writing—review and editing, H.U., N.A. and A.S.; validation, S.A.A.S., B.G. and I.A.; visualization, N.A. and A.S.; supervision, B.G.; funding acquisition, B.G. All authors have read and agreed to the published version of the manuscript.

Funding: This research received no external funding.

Institutional Review Board Statement: Not applicable.

Informed Consent Statement: Not applicable.

Data Availability Statement: Not applicable.

Conflicts of Interest: The authors declare no conflict of interest.

Appendix A. Matrices

Following are the matrices derived in the stability analysis of high-gain observer for the suggested barge-type OFWT model.

$$A = \begin{bmatrix} 0&1&0&0&0&0&0&0&0&0\\ 0&0&0&0&0&0&0&0&0&0\\ 0&0&0&1&0&0&0&0&0&0\\ 0&0&0&0&0&0&0&0&0&0\\ 0&0&0&0&0&1&0&0&0&0\\ 0&0&0&0&0&0&0&0&0&0\\ 0&0&0&0&0&0&0&1&0&0\\ 0&0&0&0&0&0&0&0&0&0\\ 0&0&0&0&0&0&0&0&0&1\\ 0&0&0&0&0&0&0&0&0&0 \end{bmatrix} \tag{A1}$$

$$B = \begin{bmatrix} 0&1&0&0&0&0&0&0&0&0\\ 0&0&0&1&0&0&0&0&0&0\\ 0&0&0&0&0&0&0&0&0&0\\ 0&0&0&0&0&0&0&1&0&0\\ 0&0&0&0&0&0&0&0&0&1 \end{bmatrix}^{T} \tag{A2}$$

$$C = \begin{bmatrix} 1&0&0&0&0&0&0&0&0&0\\ 0&0&1&0&0&0&0&0&0&0\\ 0&0&0&0&1&0&0&0&0&0\\ 0&0&0&0&0&0&1&0&0&0\\ 0&0&0&0&0&0&0&0&1&0 \end{bmatrix} \tag{A3}$$

$$H = \begin{bmatrix} \gamma_1&\gamma_2&0&0&0&0&0&0&0&0\\ 0&0&\gamma_3&\gamma_4&0&0&0&0&0&0\\ 0&0&0&0&\gamma_5&\gamma_6&0&0&0&0\\ 0&0&0&0&0&0&\gamma_7&\gamma_8&0&0\\ 0&0&0&0&0&0&0&0&\gamma_9&\gamma_{10} \end{bmatrix}^{T} \tag{A4}$$

$$A_0 = \begin{bmatrix} -\gamma_1&1&0&0&0&0&0&0&0&0\\ -\gamma_2&0&0&0&0&0&0&0&0&0\\ 0&0&-\gamma_3&1&0&0&0&0&0&0\\ 0&0&-\gamma_4&0&0&0&0&0&0&0\\ 0&0&0&0&-\gamma_5&1&0&0&0&0\\ 0&0&0&0&-\gamma_6&0&0&0&0&0\\ 0&0&0&0&0&0&-\gamma_7&1&0&0\\ 0&0&0&0&0&0&-\gamma_8&0&0&0\\ 0&0&0&0&0&0&0&0&-\gamma_9&1\\ 0&0&0&0&0&0&0&0&-\gamma_{10}&0 \end{bmatrix} \tag{A5}$$

$$B_0 = \begin{bmatrix} 0&1&0&0&0&0&0&0&0&0\\ 0&0&0&1&0&0&0&0&0&0\\ 0&0&0&0&0&1&0&0&0&0\\ 0&0&0&0&0&0&0&1&0&0\\ 0&0&0&0&0&0&0&0&0&1 \end{bmatrix}^{T} \tag{A6}$$

$$D(b) = \begin{bmatrix} b & 0 & 0 & 0 & 0 & 0 & 0 & 0 & 0 & 0 \\ 0 & 1 & 0 & 0 & 0 & 0 & 0 & 0 & 0 & 0 \\ 0 & 0 & b & 0 & 0 & 0 & 0 & 0 & 0 & 0 \\ 0 & 0 & 0 & 1 & 0 & 0 & 0 & 0 & 0 & 0 \\ 0 & 0 & 0 & 0 & b & 0 & 0 & 0 & 0 & 0 \\ 0 & 0 & 0 & 0 & 0 & 1 & 0 & 0 & 0 & 0 \\ 0 & 0 & 0 & 0 & 0 & 0 & b & 0 & 0 & 0 \\ 0 & 0 & 0 & 0 & 0 & 0 & 0 & 1 & 0 & 0 \\ 0 & 0 & 0 & 0 & 0 & 0 & 0 & 0 & b & 0 \\ 0 & 0 & 0 & 0 & 0 & 0 & 0 & 0 & 0 & 1 \end{bmatrix} \quad \text{(A7)}$$

$$\zeta = \begin{bmatrix} -a_1^1 & 1 & 0 & 0 & 0 & 0 & 0 & 0 & 0 & 0 \\ -a_2^1 & 0 & 0 & 0 & 0 & 0 & 0 & 0 & 0 & 0 \\ 0 & 0 & -a_1^2 & 1 & 0 & 0 & 0 & 0 & 0 & 0 \\ 0 & 0 & -a_2^2 & 0 & 0 & 0 & 0 & 0 & 0 & 0 \\ 0 & 0 & 0 & 0 & -a_1^3 & 1 & 0 & 0 & 0 & 0 \\ 0 & 0 & 0 & 0 & -a_2^3 & 0 & 0 & 0 & 0 & 0 \\ 0 & 0 & 0 & 0 & 0 & 0 & -a_1^4 & 1 & 0 & 0 \\ 0 & 0 & 0 & 0 & 0 & 0 & -a_2^4 & 0 & 0 & 0 \\ 0 & 0 & 0 & 0 & 0 & 0 & 0 & 0 & -a_1^5 & 1 \\ 0 & 0 & 0 & 0 & 0 & 0 & 0 & 0 & -a_2^5 & 0 \end{bmatrix} \quad \text{(A8)}$$

References

1. Wang, S.; Wang, S. Impacts of wind energy on environment: A review. *Renew. Sustain. Energy Rev.* **2015**, *49*, 437–443. [CrossRef]
2. Liu, Y.; Xiao, Q.; Incecik, A.; Wan, D.C. Investigation of the effects of platform motion on the aerodynamics of a floating offshore wind turbine. *J. Hydrodyn.* **2016**, *28*, 95–101. [CrossRef]
3. Pérez-Collazo, C.; Greaves, D.; Iglesias, G. A review of combined wave and offshore wind energy. *Renew Sustain. Energy Rev.* **2015**, *42*, 141–153. [CrossRef]
4. Lackner, M.A.; Rotea, M.A. Structural control of floating wind turbines. *Mechatronics* **2011**, *21*, 704–719. [CrossRef]
5. Stewart, G.M. Load Reduction of Floating Wind Turbines Using Tuned Mass Dampers. 2012. Available online: https://scholarworks.umass.edu/theses/781/ (accessed on 20 February 2023).
6. Lalonde, E.R.; Dai, K.; Bitsuamlak, G.; Lu, W.; Zhao, Z. Comparison of semi-active and passive tuned mass damper systems for vibration control of a wind turbine. *Wind Struct.* **2020**, *30*, 663–678.
7. M'zoughi, F.; Garrido, I.; Garrido, A.J.; De La Sen, M. Fuzzy Airflow-Based Active Structural Control of Integrated Oscillating Water Columns for the Enhancement of Floating Offshore Wind Turbine Stabilization. *Int. J. Energy Res.* **2023**, *2023*, 4938451. [CrossRef]
8. Ahmad, I.; M'zoughi, F.; Aboutalebi, P.; Garrido, I.; Garrido, A.J. Fuzzy logic control of an artificial neural network-based floating offshore wind turbine model integrated with four oscillating water columns. *Ocean Eng.* **2023**, *269*, 113578. [CrossRef]
9. He, E.M.; Hu, Y.Q.; Zhang, Y. Optimization design of tuned mass damper for vibration suppression of a barge-type offshore floating wind turbine. *Proc. Inst. Mech. Eng. Part M J. Eng. Marit. Environ.* **2017**, *231*, 302–315. [CrossRef]
10. Hu, Y.; Wang, J.; Chen, M.Z.; Li, Z.; Sun, Y. Load mitigation for a barge-type floating offshore wind turbine via inerter-based passive structural control. *Eng. Struct.* **2018**, *177*, 198–209. [CrossRef]
11. Gavgani, S.A.M.; Jalali, H.H.; Farzam, M.F. Semi-active control of jacket platforms under wave loads considering fluid-structure interaction. *Appl. Ocean Res.* **2021**, *117*, 102939. [CrossRef]
12. Park, S.; Lackner, M.A.; Pourazarm, P.; Rodríguez, T.A.; Cross-Whiter, J. An investigation on the impacts of passive and semiactive structural control on a fixed bottom and a floating offshore wind turbine. *Wind Energy* **2019**, *22*, 1451–1471. [CrossRef]
13. Shah, S.A.A.; Gao, B.; Ahmed, N.; Liu, C. Advanced robust control techniques for the stabilization of translational oscillator with rotational actuator based barge-type OFWT. *Proc. Inst. Mech. Eng. Part M J. Eng. Marit. Environ.* **2021**, *235*, 327–343. [CrossRef]
14. Shah, S.A.A.; Gao, B.; Bhatia, A.K.; Liu, C.; Rauf, A. Anti-vibration control design for TORA based barge-type offshore floating wind turbine using extended order high gain observer. *Eng. Comput.* **2022**, *39*, 2705–2732. [CrossRef]
15. Frank, M.S.; Leylaz, G.; Sun, J.Q. Data-driven robust tracking control of underactuated mechanical systems using identified flat output and active disturbance rejection control. *Int. J. Control* **2021**, *95*, 3101–3117. [CrossRef]
16. Ahmed, N.; Raza, A.; Shah, S.A.A.; Khan, R. Robust composite-disturbance observer based flight control of quadrotor attitude. *J. Intell. Robot. Syst.* **2021**, *103*, 11. [CrossRef]
17. He, B.; Wang, S.; Liu, Y. Underactuated robotics: A review. *Int. J. Adv. Rob. Syst.* **2019**, *16*, 1–29. [CrossRef]
18. Ahmed, N.; Ali Shah, S.A. Adaptive output-feedback robust active disturbance rejection control for uncertain quadrotor with unknown disturbances. *Eng. Comput.* **2022**, *39*, 1473–1491. [CrossRef]

19. Liu, C.; Yu, C.; Gao, B.; Ali Shah, S.A.; Tapus, A. Towards a balancing safety against performance approach in human–robot co-manipulation for door-closing emergencies. *Complex Intell. Syst.* **2022**, *8*, 2859–2871. [CrossRef]
20. Yu, Z.; Guo, Y.-B.; Zhang, X.-H. Stability control of TORA with environmental disturbance. In Proceedings of the 27th IEEE Chinese Control and Decision Conference (2015 CCDC), Qingdao, China, 23–25 May 2015; pp. 6107–6110. [CrossRef]
21. Liu, C.; Gao, B.; Zhao, J.; Shah, S.A.A. Orbitally stabilizing control for the underactuated translational oscillator with rotational actuator system: Design and experimentation. *Proc. Inst. Mech. Eng. Part I J. Syst. Control. Eng.* **2019**, *233*, 491–500. [CrossRef]
22. Shah, S.A.A.; Gao, B.; Ahmed, N.; Liu, C.; Rauf, A. Disturbance observer-based sliding mode control of TORA system for floating wind turbines. In Proceedings of the IEEE 9th Annual International Conference on CYBER Technology in Automation, Control, and Intelligent Systems (CYBER), Suzhou, China, 29 July–2 August 2019; pp. 418–423. [CrossRef]
23. Saleh, B.; Yousef, A.M.; Ebeed, M.; Abo-Elyousr, F.K.; Elnozahy, A.; Mohamed, M.; Abdelwahab, S.A.M. Design of PID controller with grid connected hybrid renewable energy system using optimization algorithms. *J. Electr. Eng. Technol.* **2021**, *16*, 3219–3233. [CrossRef]
24. Saleh, B.; Yousef, A.M.; Abo-Elyousr, F.K.; Mohamed, M.; Abdelwahab, S.A.M.; Elnozahy, A. Performance analysis of maximum power point tracking for two techniques with direct control of photovoltaic grid-connected systems. *Energy Sources Part A Recovery Util. Environ. Eff.* **2022**, *44*, 413–434. [CrossRef]
25. Rauf, A.; Zafran, M.; Khan, M.S.A.; Khan, A.; Shah, S.A.A. Non-singular terminal sliding mode control of converter-fed DC motor system with mismatched disturbance compensation. *Int. J. Model. Identif. Control* **2021**, *39*, 285–292. [CrossRef]
26. Ahmed, N.; Bhatia, A.K.; Shah, S.A.A. Robust active disturbance attenuation control of an uncertain quadrotor. *Int. J. Intell. Unmanned Syst.* **2021**, *10*, 346–362. [CrossRef]
27. Zhang, Y.; Zhao, X.; Wei, X. Robust structural control of an underactuated floating wind turbine. *Wind Energy.* **2020**, *23*, 2166–2185. [CrossRef]
28. Shah, S.A.A.; Malik, F.M.; Raza, A.; Khan, R.; Ahmed, N. Sampled data gyroscope stabilization of two degree of freedom platform. In Proceedings of the 13th IEEE International Bhurban Conference on Applied Sciences and Technology (IBCAST), Islamabad, Pakistan, 12–16 January 2016; pp. 251–255.
29. Yoo, D.S. Design of integral sliding mode control for underactuated mechanical systems. *J. Korean Inst. Intell. Syst.* **2013**, *23*, 208–213.
30. Al-Dujaili, A.Q.; Humaidi, A.J.; Pereira, D.A.; Ibraheem, I.K. Adaptive backstepping control design for ball and beam system. *Int. Rev. Appl. Sci. Eng.* **2021**, *2*, 211–221. [CrossRef]
31. Mu, D.; Wang, G.; Fan, Y.; Qiu, B.; Sun, X. Adaptive trajectory tracking control for underactuated unmanned surface vehicle subject to unknown dynamics and time-varing disturbances. *Appl. Sci.* **2018**, *8*, 547. [CrossRef]
32. Bhatia, A.K.; Jiang, J.; Kumar, A.; Shah, S.A.A.; Rohra, A.; Zhen, Z. Adaptive preview control with deck motion compensation for autonomous carrier landing of an aircraft. *Int. J. Adapt. Control Signal Process.* **2021**, *35*, 769–785. [CrossRef]
33. Raza, A.; Malik, F.M.; Mazhar, N.; Khan, R. Two-time-scale robust output feedback control for aircraft longitudinal dynamics via sliding mode control and high-gain observer. *Alex. Eng. J.* **2021**, *61*, 4573–4583. [CrossRef]
34. Baala, Y.; Bri, S. DFIG-Based Wind Turbine Control Using High-Gain Observer. In Proceedings of the 1st International IEEE Conference on Innovative Research in Applied Science, Engineering and Technology (IRASET), Meknes, Morocco, 16–19 April 2020; pp. 1–5. [CrossRef]
35. Khalil, H.K. *Nonlinear Systems*, 3rd ed.; Patience Hall: Hoboken, NJ, USA, 2002; p. 115.

Journal of
Marine Science and Engineering

MDPI

Article

Stress Distribution on the Preliminary Structural Design of the CENTEC-TLP under Still Water and Wave-Induced Loads

Esmaeil Zavvar, Hossam S. Abdelwahab, Emre Uzunoglu, Bai-Qiao Chen and C. Guedes Soares *

Centre for Marine Technology and Ocean Engineering (CENTEC), Instituto Superior Técnico, Universidade de Lisboa, 1049-001 Lisboa, Portugal
* Correspondence: c.guedes.soares@centec.tecnico.ulisboa.pt

Abstract: An assessment is made of the stress distribution and the hydrodynamic response of the preliminary structural design of the tension leg platform of a 10 MW wind turbine. The platform supporting a 10 MW turbine is modelled and analysed by the finite element method. The stress distribution of the platform is determined in still water with the turbine at above-rated conditions, and the response of the tension leg platform is estimated in the time domain. The results of the time domain analysis show reasonable agreement between the present results and the available data. To check the design stiffener dimensions, span, and spacing against stress distribution, classification societies' recommendations are used. The results of the stress distribution analysis indicate that the critical locations of the platform are the interaction of the lower columns with the upper columns and the connection of the tower of the turbine.

Keywords: still water; CENTEC tension leg platform (CENTEC-TLP); hydrodynamic response; finite element models; time domain

Citation: Zavvar, E.; Abdelwahab, H.S.; Uzunoglu, E.; Chen, B.-Q.; Guedes Soares, C. Stress Distribution on the Preliminary Structural Design of the CENTEC-TLP under Still Water and Wave-Induced Loads. *J. Mar. Sci. Eng.* **2023**, *11*, 951. https://doi.org/10.3390/jmse11050951

Academic Editor: Kamal Djidjeli

Received: 21 February 2023
Revised: 13 April 2023
Accepted: 22 April 2023
Published: 28 April 2023

1. Introduction

In response to the ever-increasing need for electrical energy, several research projects have been conducted to identify strategies for the generation of a sizeable amount of power. Consequently, there has been a shift in attention towards the improvement of more environmentally friendly energy sources such as different forms of renewable energy. Wind turbines have received notice as a potential alternative among various other options as the economic prospects for their application are very good [1]. The usage of onshore wind has been prevalent for several decades, but offshore wind farms have been developing at a large pace [2]. Several studies have been made about possible locations for offshore wind farms and about the criteria to rank their economic potential [3]. Different kinds of support structures have been used [4], in particular the spar [5], semisubmersible [6], tension leg platform [7], and barge type [8]. Several of the present studies are being made for turbines of 10 MW, using as the reference model the one specified in [9].

Some researchers investigated the stress distribution on offshore wind turbines and support structures. For example, Reyno et al. [10] did research on the stress concentration in the door opening under design loading conditions. According to the results, the average value of the stress concentration is 1.45, while the amount of reduction achieved by reinforcement was 78%. Umut et al. [11] studied the effects that gravity, rotor forces, and wind loads/vortex loads have on the behaviour of square cross-sections of wind turbine towers with varying wall thicknesses. It was determined that square cross-sections might be a workable alternative to the more commonly used circular cross-sections.

Within the offshore wind turbines, there are two different categories of supporting structures which are fixed bottom and floating structures, both of which are subject to fatigue. For instance, Yeter et al. [12] investigated a fixed offshore wind turbine support structure subjected to combined wave and wind-induced loading. It was concluded that the brace component experienced the most severe fatigue damage, and the estimated fatigue life was above 1000 years. Yeter et al. [13] designed a jacket offshore wind turbine support structure installed in water depths of 40, 60, and 80 m. The final dimensions were verified by free vibration, structural strength, and stability assessments.

Regarding floating turbines, there are also studies investigating the different types of platforms. Teixeira et al. [14] performed a fatigue analysis for the tower of offshore wind turbines. Kim et al. [15] investigated a support structure of a floating wind turbine to evaluate the effect of wind speed on the stress transfer function. To this end, an artificial neural network (ANN) was used to minimize the number of simulations while increasing the accuracy of the findings. To increase the accuracy of the ANN model, a superposition model was developed. In another study on floating turbines, Bachynski and Moan [16] performed a study on the design and dynamic analysis of a single-column 5 MW tension leg platform wind turbine (TLPWT). It was concluded that for some TLPWTs, second-order sum-frequency wave forces are important for fatigue and extreme response calculations. Ringing forces of third order were shown to be essential for TLPWTs with large diameters (14–18 m), particularly when the turbine was idle or parked.

Oguz et al. [17] performed a numerical and experimental study on a SeaStar-type TLP floating offshore 5 MW wind turbine in regular and irregular waves with wind conditions. It was concluded that the surge, pitch, and heave motion of the experimental and the numerical results have a good agreement, but the roll, sway, and yaw motions were insignificant. A coupled analysis for a floating 5 MW wind turbine, which is fixed by mooring to the bed, was conducted by Sclavounos et al. [18]. The authors indicated that TLPs, due to their low root mean square response have the potential to offer benefits accelerations and negligible heave and pitch motions.

Nematbakhsh et al. [19] compared the typical potential flow calculations of wave-induced loads with CFD calculations demonstrating how detailed analysis can be performed. Jamalkia et al. [20], on the other hand, showed how the analysis of the dynamics response of the structure could be used for detecting damage.

While most of these studies were made with turbines of the order of 5 MW, more recent studies address platforms with 10 MW turbines [21–23].

The design of the tension leg platform which is the main subject of this paper has been used as a reference for the design of a wind farm in the north of Spain [24]. For this area, the environmental conditions have been assessed using specific models to determine the conditions of wind [25], waves [26], and currents [27].

Uzunoglu and Guedes Soares [7,28] explained the design process of the TLP platform and conducted a numerical investigation of the CENTEC-TLP with the 10 MW DTU wind turbine. The results showed minimal wave and wind responses in transport and operational conditions. The frequency domain design solution was conducted with in-house code, and NREL's FAST provided the time-domain solution verification. Tank testing was carried out within the ARCWIND project [29]. In the numerical design study, three environmental conditions consisting of rated, above-rated, and 50-year extreme, was published. The study concluded that the designed model had restrained responses in free-floating conditions, and installed conditions, and the structure of the TLP did not experience any problems regarding mooring breaking, slack, or excessive surge motions. Given the satisfactory dynamic response performance, the preliminary structural design of the CENTEC-TLP has been conducted and this study contributes to the understanding of stress distribution on this 4-column TLP hull.

This work aims to assess the preliminary structural design of the CENTEC-TLP 4-column design against the class society guidelines and to identify the locations exposed to higher stresses. Therefore, the platform is structurally analysed in still water conditions using ANSYS® APDL, with wind thrust applied at a constant rated wind speed. Then, hydrodynamic analysis is performed under wave and wind loads. In addition, a comparison with the results of the earlier work of Uzunoglu and Guedes Soares [7] is carried out. Finally, the stress distribution is evaluated on the sections of the platform to identify the critical sections that require further reinforcements.

2. Description of the System and Guidelines

2.1. Description of the Model

In this study, the CENTEC-TLP with the 10 MW DTU wind turbine is analysed to define the stress distribution under hydrostatic pressure. The characteristics of the platform and the wind turbine are presented in Figure 1, Tables 1 and 2. The overall length and breadth of the TLP are 49, while the wind turbine has a total mass (rotor, nacelle, and tower) of 1302 tonnes and a height of 129 m. The total mass of the whole structure (Rotor, Nacelle, Tower, TLP) is 23,388 tonnes. The installed draft is 20 m. Further information on the DTU 10 MW is available in Uzunoglu and Guedes Soares [7], and Bak et al. [9]. The CENTEC-TLP is moored by 12 mooring lines distributed in groups of three mooring lines at each corner. The properties of the mooring lines are shown in Table 3. The mass properties of the tower, turbine, and TLP are shown in Table 4. The properties of the stiffeners are shown in Figure 2 and Table 5. Young's modulus and Poisson's ratio are 210 GPa and 0.3, respectively. Steel NV-550 with 550 MPa yield stress is defined in this study.

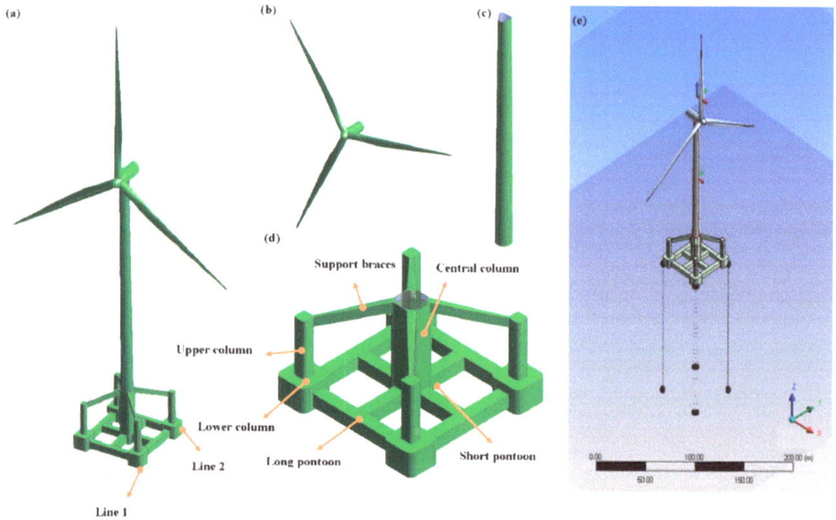

Figure 1. A model of the CENTEC-TLP with the DTU 10 MW. (**a**) A perspective view of the platform with the turbine in ANSYS APDL; (**b**) the blades of the wind turbine; (**c**) the tower; (**d**) the platform. (**e**) A perspective view of the platform with the turbine with mooring lines in ANSYS AQWA.

Table 1. Properties of the CENTEC-TLP.

Parameters	Value	Units
Pontoon diameters	4	(m)
Lower column height	7.5	(m)
Lower column diameter	10.5	(m)
Upper column height	17.5	(m)
Upper column side length	4	(m)
Height of the central column above the waterline	10	(m)
Steel thickness of components below installed draft	0.04	(m)
Steel thickness of components above-installed draft	0.04	(m)
Transport draft	3.85	(m)
Transport GM	27.76	(m)
Draft when installed	20	(m)
Volume	3422.4	(m)
Displacement (operational and installed)	3511	(ton)
Steel density	7850	(kg/m)

Table 2. The DTU 10 MW properties Hmedi et al. [29].

Parameter	Value	Units
Cut in wind speed	4	(m/s)
Cut out wind speed	25	(m/s)
Rated wind speed	11.4	(m/s)
Rated power	10	(MW)
Minimum/maximum rotor speeds	6–9.6	(rpm)
Rotor frequency (1P)	6.5–10	(s)
Blade passing frequency (3P)	2.08–3.33	(s)
Number of blades	3	(-)
Rotor diameter	178.3	(m)
Hub diameter	5.6	(m)
Hub height	129	(m)
Rotor mass	227,962	(kg)
Nacelle mass	446,036	(kg)
Tower mass	628,442	(kg)
Total mass of Rotor, Nacelle, Tower	1,302,440	(kg)

Table 3. Mooring line properties [30].

Parameter	Value	Units
Material	6 strand IWRC	(-)
Diameter	14	(cm)
Breaking strength	12,600	(kN)
EA	1.17×10^9	(N)
Number of lines	12	(-)
Pretension of lines	12,500	(kN)

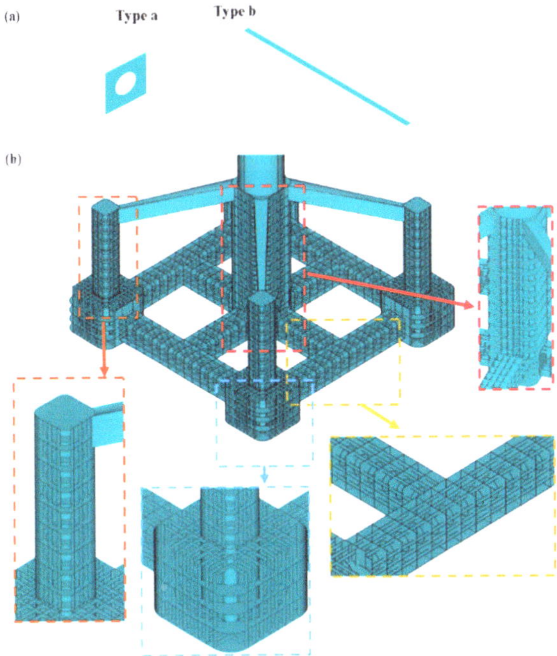

Figure 2. Designed stiffeners for the TLP; (**a**) stiffener types, (**b**) the stiffeners' arrangements inside the TLP.

Table 4. Mass details of the tower, turbine, and the platform.

Parameter	Tower	Turbine	Platform
Mass (kg)	628,442	673,998	2,208,600
Ixx (kg·m^2)	6.52×10^8	1.66×10^8	6.9×10^8
Iyy (kg·m^2)	6.52×10^8	1.06×10^8	6.9×10^8
Izz (kg·m^2)	7.84×10^6	1.17×10^{10}	1.04×10^9
CG (m)	[0, 0, 57.5]	[0.61, 0, 131.56]	[0, 0, 8.36]

Table 5. Properties of the stiffeners.

Part	Type	Units	Thickness (m)	Height (m)	Mesh Size (m)
TLP (the hull of the structure)	-	-	0.004	-	0.50
Long Pontoon	a	13	0.016	1.00	0.50
	b	3	0.016	0.50	0.50
Short Pontoon	a	6	0.016	1.00	0.50
	b	3	0.016	0.50	0.50
Central Column	a	14	0.016	2.00	0.50
	b	4	0.016	0.80	0.50
Upper Column	a	4	0.016	0.90	0.50
	b	3	0.016	-	0.50
Lower Column	a	3	0.016	-	0.50
	b	6	0.016	1.30	0.50

2.2. Frequency Domain Diffraction/Radiation Model

Before the time-domain simulations for the moored TLP, linear diffraction and radiation calculations are conducted in ANSYS® AQWA to determine the free-floating hydrodynamic coefficients, such as added mass and damping, as well as the first-order wave exciting loads. A fixed Cartesian coordinate system is defined with the vertical z-axis pointing upwards and zero incidences of the waves when they propagate along the direction of the positive x-axis. The fluid is assumed to be inviscid and incompressible, and the flow is irrotational so that the flow can be described by velocity potentials within the domain enclosed by the boundaries defined for the problem. These include the wetted body surface, the free surface, and the sea bottom. The velocity potentials are obtained using Green's function approach.

The submerged body surface is discretized into small panels, and a pulsating source is located on each panel. The combinations of source strengths required to diffract an incoming regular wave of a given frequency and to allow body oscillation in each degree of freedom are then calculated. Then, the incident forces, diffraction forces, added mass $a_{kj}(\omega)$, and radiation damping $b_{kj}(\omega)$ on the body are calculated. The simulations in the frequency domain for the free-floating platform were conducted for 31 frequencies. The range of frequencies is selected with care to have accurate time-domain calculations.

2.3. Time-Domain Simulation Model

The time-domain simulations are conducted using the ANSYS® AQWA time-domain module to calculate the motions and mooring load on the platform in regular waves. The equations of motion are solved at each time step, and the obtained results from the diffraction/radiation potential in the frequency domain are converted to the time domain using the retardation function approach for the hydrodynamic forces to include memory effects. According to Cummins et al. [31], the equations of motion are formulated as follows:

$$\sum_{j=1}^{6}\left[\left[M_{kj} + A_{kj}\right]\ddot{x}_j(t) + \int_{-\infty}^{t} K_{kj}(t-\tau)\dot{x}_j(\tau)d\tau + C_{kj}x_j(t)\right] = F_k^1(t) + F_k^m(t) \quad (1)$$

where $F_k^1(t)$ is the total first-order wave exciting forces including incident and diffraction forces. These forces can be obtained in the time domain based on the first-order wave loads obtained from the frequency domain solution. $F_k^m(t)$ are the mooring system forces for tethers, C_{kj} is the hydrostatic restoring matrix, A_{kj} is the added mass at an infinite frequency, and $K_{kj}(t)$ is the retardation function that describes the time of the generalized force k after an impulsive motion along each of the generalized coordinates j. The retardation functions $K(t)$ are the Fourier transform of the frequency-dependent damping coefficients $b_{kj}(\omega)$ obtained from the frequency domain solution as:

$$K_{kj}(t) = \frac{2}{\pi}\int_0^\infty b_{kj}(\omega)\cos(\omega t)d\omega \tag{2}$$

The frequency-independent added mass can be calculated based on the frequency-dependent added mass and retardation functions as follow:

$$A_{kj} = a_{kj} + \frac{1}{\omega}\int_0^\infty K_{kj}(t)\sin(\omega t)d\omega \tag{3}$$

While the sum frequency loads may be important for TLPs considering ringing and springing-type mooring responses, at this stage, the manuscript focuses on the TLP and considers only the first-order loads.

2.4. Mooring Simulation

Twelve mooring lines are represented in the numerical simulation with steel wires as shown in Figure 1e. The load elongation curve is defined in the numerical model according to Equation (4). the properties of the mooring lines are shown in Table 3. The applied pretension in each mooring element is tuned with the obtained pretension from the benchmark solution [7]. Before starting the time domain simulation, static equilibrium analysis is conducted to make sure that the tension leg platform starts the simulation at an equilibrium position similar to the benchmark solution.

$$Load = k_1(\delta L) \tag{4}$$

2.5. Guidelines and Standards for the TLP

In terms of analysis of offshore and floating wind turbine structures, documents such as IEC 61400-1 [32], DNV-OS-C105 [33], DNV-OS-J103 [34], DNV-OS-C101 [35], IEC 61400-3-2 [36], DNV-OS-J101 [37], and DNV-ST-0119 [38], DNV-RP-C201 [39], DNV-RP-C203 [40], and DNV-RP-C202 [41] provide some regulations and guidelines

To get some basic information about the properties of stiffeners, DNV-RP-C203 [40] and DNV-OS-C101 [35] recommendations are used. According to the DNV-OS-C101 [35], the minimum thickness should not be less than:

$$t = \frac{14.3\ t_0}{\sqrt{f_{yd}}}(\text{mm}) \tag{5}$$

where f_{yd} is design yield strength (N/mm^2) and t_0 is the thickness of the structure based on the application categories of components [33]. For plates subjected to lateral pressure, the minimum thickness is determined as follows:

$$t = \frac{15.8\ k_a s\sqrt{p_d}}{\sqrt{\sigma_{pd1}k_{pp}}}\ (\text{mm}) \tag{6}$$

where k_a, s, p_d, σ_{pd1}, and k_{pp} are defined as a correction factor for the aspect ratio of the plate field, stiffener spacing (m), design pressure (kN/m^2), design bending stress (N/mm^2), and fixation parameter for the plate, respectively. According to the DNV-OS-C101 [35], the minimum section module for stiffeners is defined as follows:

$$Z_s = \frac{l^2 s p_d}{k_m \sigma_{pd2} k_{ps}}, \ min\ 15 \times 10^3 \left(\text{mm}^3\right) \tag{7}$$

where $l, k_m, \ \sigma_{pd2},$ and k_{ps} are defined as stiffener span (m), bending moment factor, design bending stress (N/mm^2), and fixation parameter for stiffeners, respectively. The minimum shear area is defined as follows (DNV-OS-C101 [35]):

$$A_s = \frac{l s p_d}{2\tau_{pds}} 10^3 \left(\text{mm}^2\right) \tag{8}$$

where τ_{pds} is the design shear stress (N/mm^2). The height of the web is calculated as follows (DNV-RP-C202 [41]):

$$h \leq 0.4\ t_w \sqrt{E/f_y} \tag{9}$$

where t_w is the thickness of the web. Therefore, based on the above formulae, the minimum thickness for stiffeners with a spacing of 1 m and a span of 2 m is 16 mm, and the minimum height of the web should be 125 mm. However, to have less stress than f_y in the structure, the web height should be 420 mm. Checks with this size of stiffeners will be conducted in the next phase of the study, dealing with the detailed structural design.

3. Development of the FEM Model

To model the TLP, Shell 281 element with eight nodes in the ANSYS is used (Figure 3). The ANSYS contact capability is utilized to model the interaction of the stiffeners and TLP. In contact between two elements, the node-to-line method is established as the target and the contact. The hull of the TLP was defined as the target surface, and the edges of the stiffeners were introduced as the contact surface. The mesh size is 0.5 m (Yeter et al. [14]) for the TLP and 1.0 m for the tower, and the turbine is 0.3 m.

The mesh size is larger in the tower as the focus is on the stress distribution on the TLP. The meshed TLP with 10 MW wind turbine and stiffeners are presented in Figure 3. To analyze the stress distribution, the rotor, and the nacelle weight are defined as a point mass, and the wind pressure is defined as a force on the top of the tower (Figure 4a). The wind force and the moment of the nacelle and the rotor are 1500 kN and 40 MN·m, respectively. The TLP is under hydrostatic pressure as the still water, as shown in Figure 4b. The TLP's draft after installation is 20 m. Fixed boundary conditions are used in each corner of the model in the locations of tendons (Figure 4c).

To analyse the CENTEC-TLP under hydrostatic and hydrodynamic loads, after generating the model in APDL, the geometry of the model is transferred to AQWA. The simulations in the time domain are conducted on the full scale. The simulations performed in regular waves as well as the wind loads are shown in Table 6 for the above-rated conditions. The wind force and wind speed at the above-rated conditions are 800 kN and 22.4 m/s, respectively. The total simulation time is selected to be the 60 s after the transient be-haviour of the platform. The selected time step is 0.05 s. This study assumes that the TLP has mooring lines instead of steel tethers.

Table 6. The above-rated environmental conditions.

Waves		Wind (m/s)	Turbine
H (m)	**T (s)**		
8	11.5	22.4	Operational

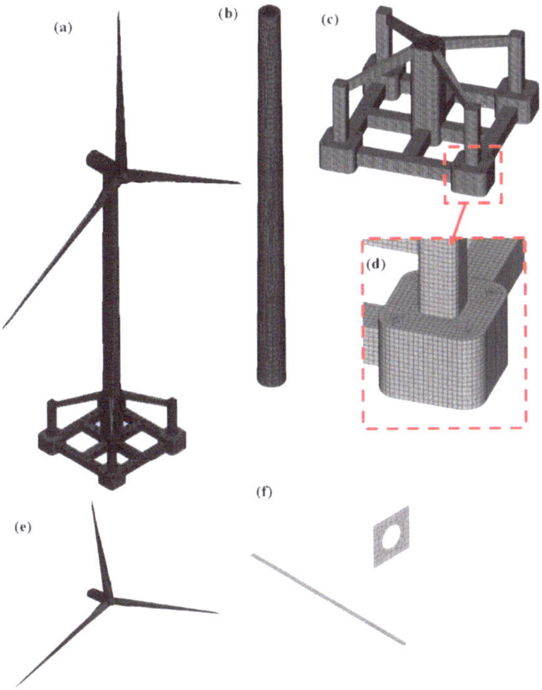

Figure 3. The meshed model of the CENTEC-TLP: (**a**) a perspective of the TLP with the tower and the 10 MW wind turbine; (**b**) the tower of the turbine; (**c**) the TLP; (**d**) the connection of the upper column with the lower column; (**e**) the meshed 10 MW turbine; (**f**) the meshed stiffeners.

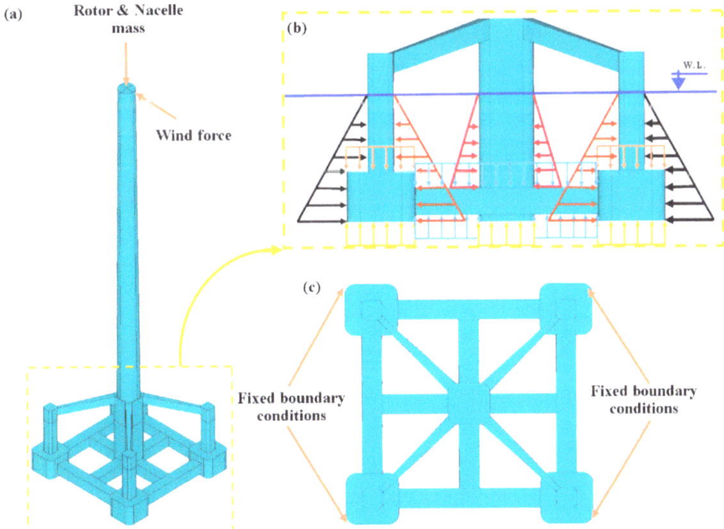

Figure 4. The loaded model: (**a**) location of the rotor and the nacelle mass and the wind pressure, (**b**) the concentrated force applied on the top of the CENTEC-TLP, (**c**) the locations of boundary conditions.

The benchmark values are validated by experimental models presented in [29]. Hence, to show the accuracy of the modeling, meshing, and method used, the results of the ANSYS AQWA are validated with the benchmark model.

4. Results and Discussion

In this part, the platform is first numerically analysed in the still water condition with a wind thrust applied at a constant rated wind speed. Then, hydrostatic results for the platform in both free-floating and operating conditions are validated against the benchmark solution available in the literature [7]. The study is limited to the analyses conducted in still water and above-rated conditions. At this stage, the primary objective is to investigate the highest regular waves that still have wind loads from the running turbine. In the future, the platform's strength can be investigated under extreme loads of irregular sea states and wind coming from various directions. Due to software limitations (AQWA), the hydrodynamic analysis does not include the coupling effects between the platform and the wind turbine.

4.1. Stress Distributions on the Platform

The platform is analysed with the minimum thickness and height of stiffeners proposed by DNV-OS-C101 [35] and DNV-RP-C202 Figure 5 shows the stress distribution, and Table 7 presents the stress of the reinforced TLP.

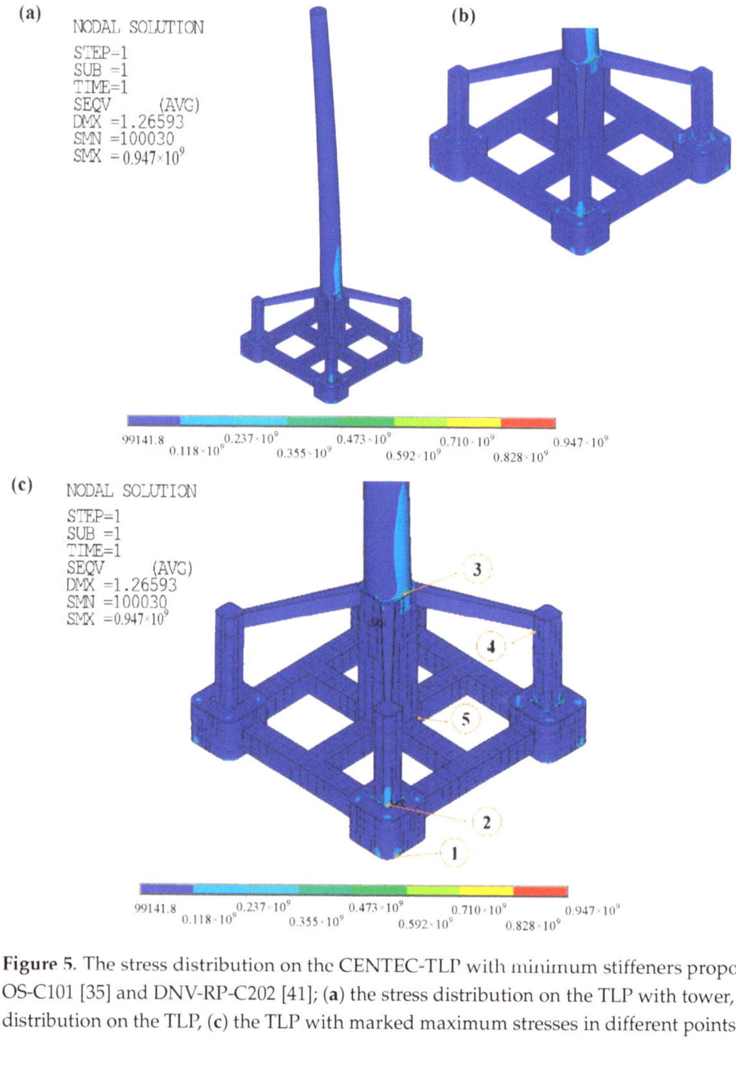

Figure 5. The stress distribution on the CENTEC-TLP with minimum stiffeners proposed by DNV-OS-C101 [35] and DNV-RP-C202 [41]; (**a**) the stress distribution on the TLP with tower, (**b**) the stress distribution on the TLP, (**c**) the TLP with marked maximum stresses in different points.

Table 7. The stress (MPa) in the CENTEC-TLP.

Location [1]	Von Mises Stress	X- Component of Stress	Y- Component of Stress	Z- Component of Stress	1st Principal Stress	2nd Principal Stress	3rd Principal Stress
1	601	460	498	554	659	333	0.123
2	947	541	477	537	607	300	1.39
3	762	492	545	470	550	237	5.27
4	456	490	509	459	509	387	4.78
5	681	289	465	319	506	173	0.175

[1] Locations are presented in Figure 5.

According to Figure 5 and Table 7, it can be concluded that the thickness, height, and number of stiffeners, which are determined by DNV-OS-C101 [35] and DNV-RP-C202 [41], are not enough for some parts of the TLP because the values of stress in some critical points, such as location 1, 2, and 3, are higher than the yield stress; thus, a yielding would occur in the TLP before those stress levels would be reached. As a result, these locations need to be reinforced.

These options could be stiffeners, fibre-reinforced polymer (FRP), bracket collar plates, doubler plates, or similar methods. Overall, the downwind of the TLP experienced higher stress than other parts. The increase of the thickness of the stiffeners would be one of the options to reinforce the TLP.

According to Table 8, the increase in the stiffeners' thickness in the lower, upper, and central columns leads to a decrease in the maximum stress in different locations. It can be concluded that instead of increasing the stiffeners' thickness in the mentioned three parts to 40 mm, it is better to increase the thickness of the stiffeners in the lower column to 40 mm, in the upper column to 30 mm, and in the central column to 35. Table 8 presents the effects of the combination of different thicknesses on the stress in the CENTEC-TLP.

Table 8. The effects of the stiffeners' thickness of the lower, the upper, and the central column on stress in the CENTEC-TLP.

Location [1]	Stress (MPa)				
	20 mm	25 mm	30 mm	35 mm	40 mm
1	576	567	560	555	550
2	924	778	676	599	539
3	673	595	528	474	429
4	429	419	413	406	401
5	561	467	402	354	317

[1] Locations are presented Locations are presented in Figure 5.

The results indicate that increasing the thickness of all stiffeners compared with the increase of the thickness of the lower, upper, and central columns has negligible effects on the maximum stresses of different locations. Thus, the thickness of the lower, upper, and central columns just increased. Table 9 presents the effects of the stiffeners' thickness on stress in the TLP.

Table 9. The stress (MPa) in the CENTEC-TLP by increased thicknesses.

Location [1]	Von Mises Stress	X- Component of Stress	Y- Component of Stress	Z- Component of Stress	1st Principal Stress	2nd Principal Stress	3rd Principal Stress
1	563	429	451	434	611	307	1.6
2	412	417	423	415	475	253	1.33
3	474	463	351	276	467	199	4.44
4	413	523	502	487	533	375	4.61
5	354	234	309	225	309	136	0.172

[1] Locations are presented Locations are presented in Figure 5.

The values are calculated in the current loading condition and the safety factor(s) need to be applied in WSD/LRFD approaches. In the following subsections, the stress distribution in each part of the TLP will be discussed.

4.1.1. The Pontoons

The stress distribution on the pontoons in the TLP is presented in Figure 6. According to Figure 6d,f the pontoons connected to the central column downwind of the TLP have more stress than other parts of the pontoons. It can be seen that the bottom of the pontoons, especially pontoons that are in the leeward of the TLP, have more stress be-cause of having both the hydrostatic pressure and the wind force together; therefore, that part should be reinforced by appropriate methods (Figure 6f).

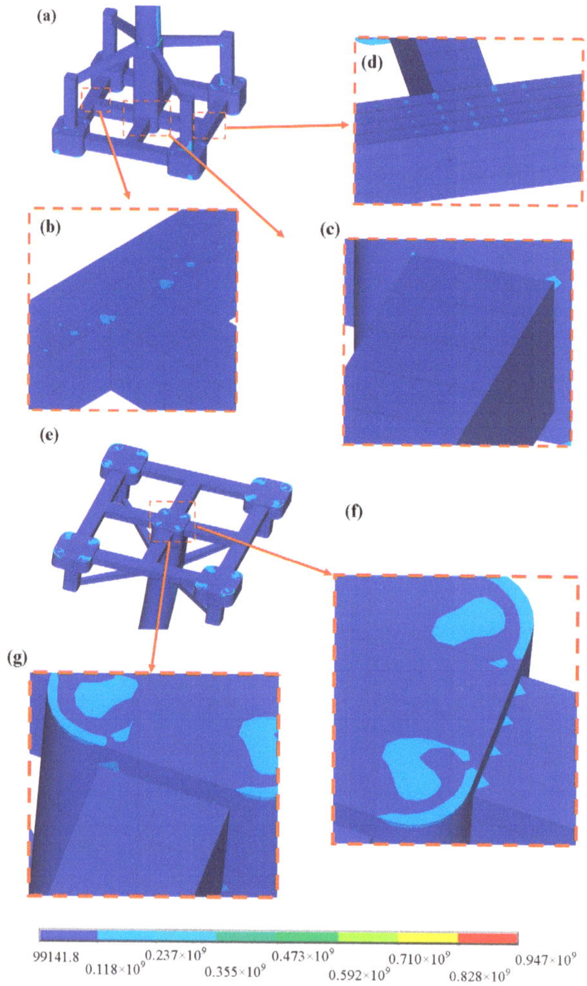

Figure 6. The stress distribution on the pontoons in the CENTEC-TLP; (**a**) a close view of the TLP, (**b**) the stress distribution on the pontoons on the opposite side of the TLP, (**c**) the stress distribution on the connection of pontoons with the central column, (**d**) the stress distribution on the downwind parts of the pontoon, (**e**) the stress distribution on the bottom of the TLP, (**f**) the stress distribution on the downwind parts of the pontoon in the bottom, (**g**) the stress distribution on the connection of the pontoon with the central column at the bottom.

4.1.2. The Lower Column and the Mooring Connections

The stress distribution on the lower column and the mooring connections in the TLP is presented in Figure 7. It can be seen that stress concentration appears in the mooring connections and the interaction of the lower column with the upper column, especially the corners that are close to the mooring lines. Figure 7d indicates that the inside parts of the lower column get less stress than the outside parts. All corners of the lower columns and the mooring connections need extra support by stiffeners or other reinforcing methods.

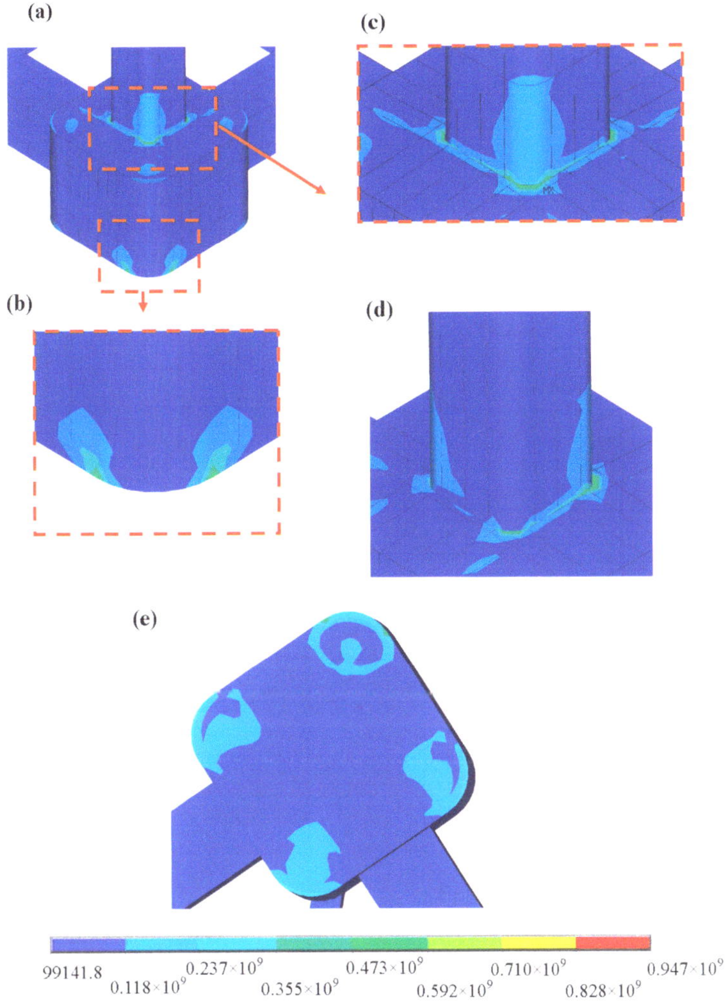

Figure 7. The stress distribution on the lower column and the mooring connections in the CENTEC-TLP; (**a**) a close view of the lower column, (**b**) the stress distribution on the mooring connections, (**c**) the stress distribution on the interaction of the lower column with the upper column (outside), (**d**) the stress distribution on the interaction of the lower column with the upper column (inner side), (**e**) the stress distribution on the bottom of the lower column.

4.1.3. The Central Column

Figure 8 indicates the connection between the tower of the DTU 10 MW and the central column. According to Figure 8c, the interaction of the tower with the central column and the bottom of the central column, all corners of the TLP should be reinforced.

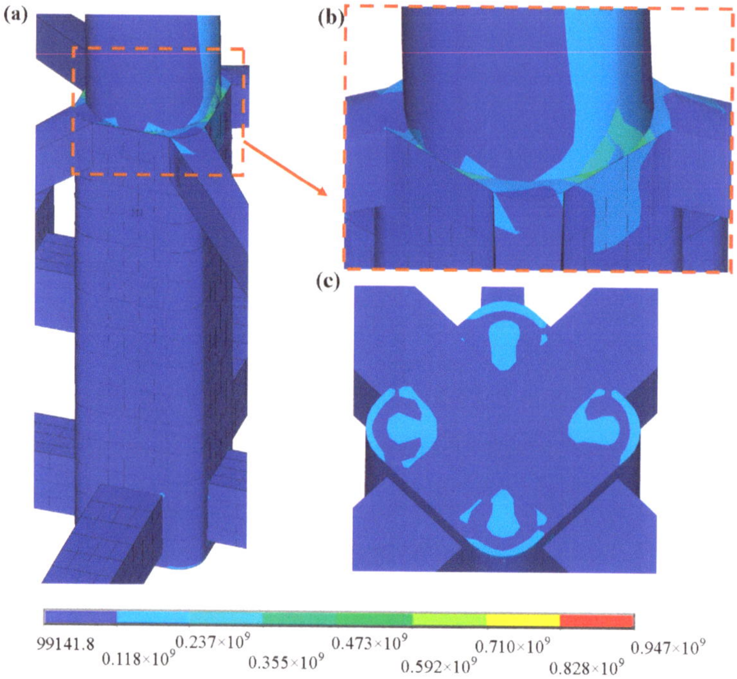

Figure 8. The stress distribution on the central column in the CENTEC-TLP; (**a**) a close view of the central column, (**b**) the stress distribution on the connection of the tower of the DTU 10 MW and central column, (**c**) the stress distribution on the bottom of the central column.

4.1.4. The Upper Column and the Support Brace

The stress distribution on the upper column and the support brace in the TLP is pre-sented in this section. Figure 9 shows that stress concentration could be seen in the inter-action of the lower column and support brace with the upper column on the leeward side of the TLP. According to Figure 9c, the bottom of the support brace requires more stiffener tan than the top of the support brace. Figure 9d indicates that other support braces which are not in downwind of the TLP have less stress concentration.

4.2. Hydrodynamic Response of the CENTEC-TLP

4.2.1. The Free-Floating Condition

The platform is designed to be assembled as a system with the platform, tower, and turbine at the shoreline. It can be transported in this condition with a draft of 3.85 m. AQWA calculates the metacentric height in the free-floating case as 26.41 m, which presents a 5% difference from the benchmark's metacentric height. This is because of some of the volumetric differences caused by mesh size differences and the mooring line modelling. By looking closely at Table 10, it can be seen that the results of the AQWA are similar to the results of the benchmark case. Therefore, the modelling, meshing size, and mass distribution are considered acceptable. The TLP validated experimentally is available in [29]. Figure 10 depicts the response amplitude operators of the platform's motions in the free-floating condition. The platform is transported in Galicia, Spain, where the majority of small waves have periods between 5 and 10 s, as shown in [7]. The results indicate that the platform's greatest pitch RAO between these wave periods is 0.38°/m, while the largest heave RAO is less than 1.4.

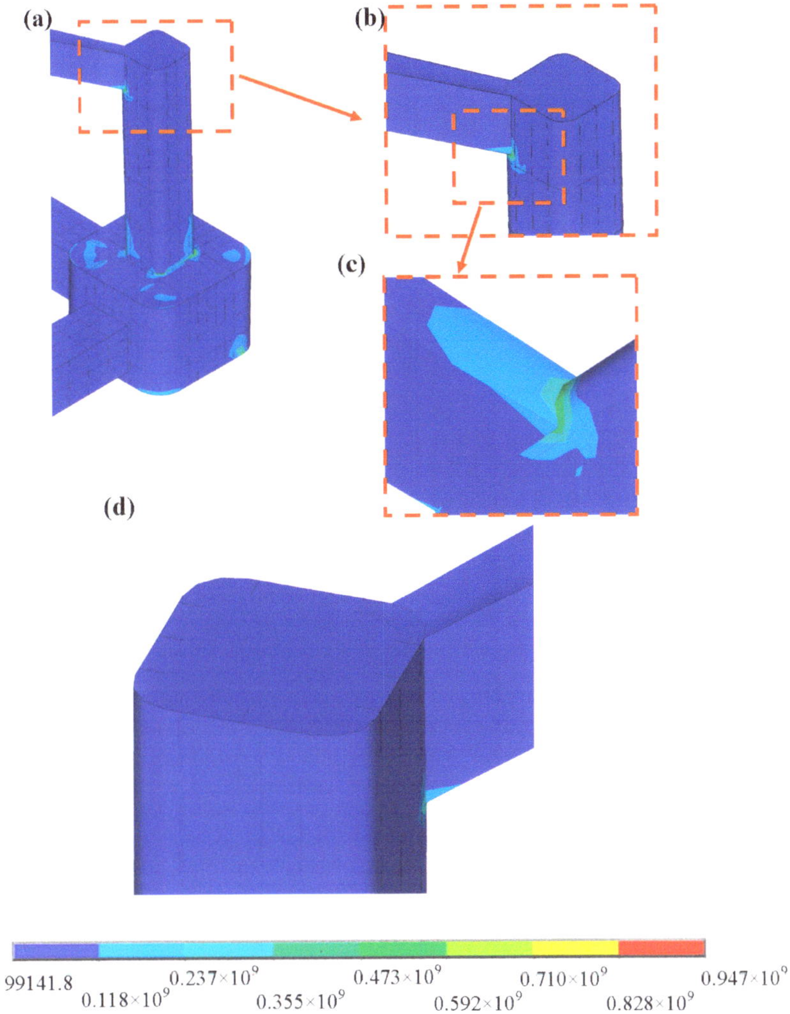

Figure 9. The stress distribution on the upper column and the support brace in the CENTEC-TLP; (**a**) a close view of the upper column, (**b**) the stress distribution on the connection of the support brace and upper column in the leeward, (**c**) the stress distribution on the bottom of support brace in the leeward, (**d**) the stress distribution on the bottom of support brace.

Table 10. Comparison of the characteristics of obtained in AQWA and Benchmark (BM) model in transport condition.

Parameter	BM	AQWA	Error %
Transport condition GM (m)	27.76	26.41	−5%
Transport draft (m)	3.85	3.85	0%
Displacement (ton)	3511	3511	0%
Static pitching angle (deg)	0.22	0.20	−9%
Heave natural period (s)	4.7	4.50	−4%
Roll and Pitch natural period (s)	22.3	24.38	9%

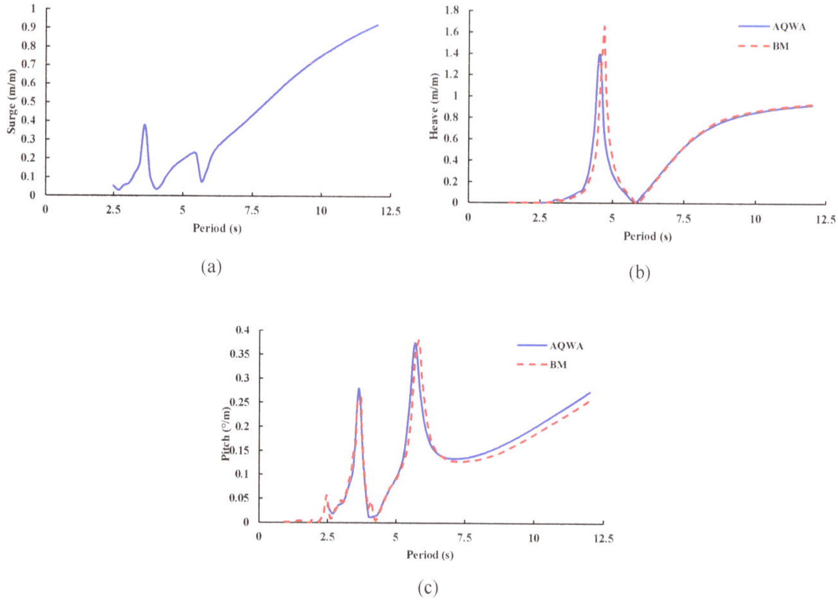

Figure 10. Response amplitude operator of the platform in its free-floating condition: (**a**) surge; (**b**) heave; (**c**) pitch.

4.2.2. Natural Frequencies in the Installed Condition

The platform is installed with 12 mooring lines at a 20-m draft. The characteristics of the platform in the installation position are shown in Table 10. The natural frequency of the benchmark model with the AQWA model is compared, and the results are provided in Table 11. Thus, it is concluded that, for this environmental setup, AQWA and the Benchmark results are in good agreement. The leading cause of the difference is modelling mooring lines as steel tethers.

Table 11. Comparison between the natural frequency obtained by AQWA and the benchmark (BM) model.

Parameter	Units	Surge	Pitch	Heave
BM	Hz	0.029	0.207	0.481
AQWA		0.030	0.280	0.500
BM	s	35.71	4.83	2.08
AQWA		33	4	2

4.2.3. The Above-Rated Operation Condition

The results of the isolated steady wind and regular wave loads are provided in this section. The data are provided for a single wave direction (head wave). The values of the wave and wind in the above-rated conditions are presented in Table 6. Figure 11 presents the time series of motions and Figure 12 mooring forces obtained in regular waves. Results of isolated steady wind and regular waves in the above-rated conditions are shown in Tables 12 and 13, respectively. The obtained results show that the difference in the surge motion is 5 to 7% and 10 to 20% in heave motion. In the pitch rotation, the difference is about 14 to 16%, and the maximum difference in the mooring lines is 9 to 19%. Under steady wind loads, the difference is about 0 to 2%. The differences in the isolated case are not too high. Thus, it can be concluded that there is a good agreement between the AQWA and the benchmark model.

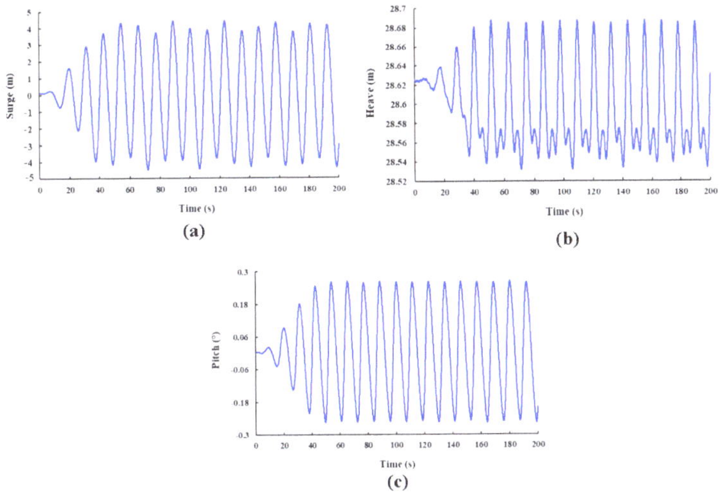

Figure 11. The time series of motions obtained in the regular wave: (**a**) surge (m); (**b**) heave (m); (**c**) pitch (degree).

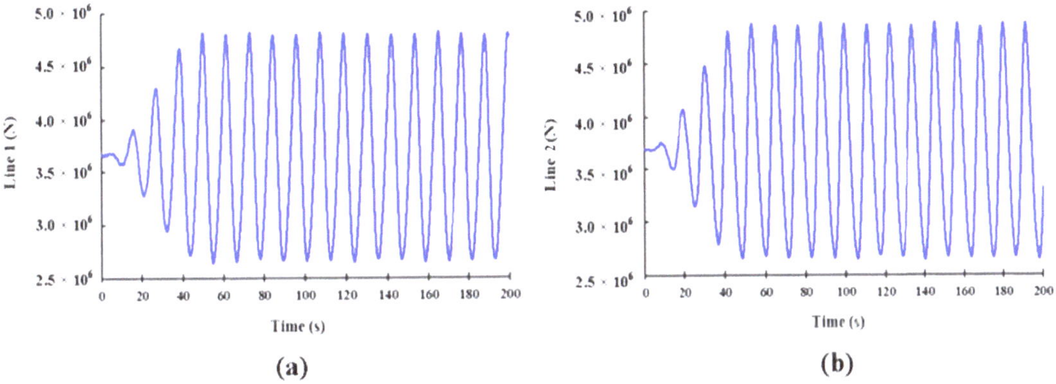

Figure 12. The mooring forces obtained in the regular wave. (**a**) Line 1; (**b**) Line 2.

Table 12. Platform responses to the regular waves in above-rated conditions.

Parameter		Benchmark (BM)	AQWA	Error %
Isolated Motions Type		**Wave**	**Wave**	
Surge (m)	min	−3.89	−4.16	−7%
	max	4.1	4.32	−5%
Heave (m)	min	−0.1	−0.09	10%
	max	0.05	0.06	−20%
Pitch (deg)	min	−0.29	−0.25	14%
	max	0.31	0.26	16%
Line 1 (kN)	min	2219	2640	−19%
	max	5058	4822	5%
Line 2 (kN)	min	2430	2656	−9%
	max	4894	4876	0%

Table 13. Results of the platform in the above-rated conditions at a steady wind.

Isolated Motions Type	Benchmark (BM)	AQWA	%
	Wind	Wind	
Surge (m)	1.93	1.9	2%
Heave (m)	0	0	0%
Pitch (deg)	0.09	0.09	0%
Line 1 (kN)	3339	3392	−2%
Line 2 (kN)	3993	4069	−2%

5. Conclusions

The stress distribution on the CENTEC-TLP at the design condition in still water and the response of the platform is determined in the time domain under wave and wind load conditions. The main conclusions are summarized as follows:

- Pontoons downwind of the TLP have high stress in the connection of the pontoons with the central column and lower columns.
- The stress concentration is seen in the interaction of the lower column and the support brace with the upper column on the leeward side of the TLP. The lower column, especially the corners close to the mooring lines, has higher stress values than other corners of the lower column. The mooring connections show high stress. Therefore, they should be reinforced with methods, such as an increased thickness of the stiffeners, number of stiffeners, or others.
- Overall, the leeward of the TLP needs more attention and precise design because that side had higher stress values.
- In the free-floating condition, the maximum difference between the benchmark results and the AQWA in GM is 5%.
- In natural frequencies, a difference of 8% in the roll and pitch, and 4% in the heave was shown.
- The mean difference between the results of the isolated wave of benchmark and AQWA model in surge, heave, and pitch is −7%, 20%, and 16%, respectively.
- The maximum mooring force of the isolated wave of the benchmark and AQWA model in line 1 is 4874, and in line 2 is 4893, which signifies a 4 and 0% difference.
- The differences of 2%, 0%, and 0% in the surge, heave, and pitch in the isolated wind are shown between the results of the benchmark and the AQWA model.
- The mooring line forces of the isolated wind of the benchmark and AQWA model in line 1 is 3392, and in line 2 is 4069, which signifies a −2% difference in both cases.
- This study focused on the above-rated condition which is a combination of wind and wave loading. Investigating the TLP against extreme loads and multivariate statistical methods or time-frequency domain analysis can be an important topic for future studies.

Author Contributions: Conceptualization, C.G.S., E.U. and B.-Q.C.; methodology, E.Z. and H.S.A.; formal analysis, E.Z. and H.S.A.; visualization, E.Z. writing—original draft preparation, E.Z. and H.S.A.; writing—review and editing, E.U., B.-Q.C. and C.G.S.; Supervision C.G.S. All authors have read and agreed to the published version of the manuscript.

Funding: This work was performed within the project ARCWIND—Adaptation and implementation of floating wind energy conversion technology for the Atlantic region, which is co-financed by the European Regional Development Fund through the Interreg Atlantic Area Programme under contract EAPA 344/2016. This work contributes scope of the Strategic Research Plan of the Centre for Marine Technology and Ocean Engineering (CENTEC), which is financed by the Portuguese Foundation for Science and Technology (Fundação para a Ciência e Tecnologia—FCT) under contract UIDB/UIDP/00134/2020. The second author is funded by FCT with a PhD grant under contract 2020.06969.BD.

Institutional Review Board Statement: Not applicable.

Informed Consent Statement: Not applicable.

Data Availability Statement: Not applicable.

Acknowledgments: Already under Funding.

Conflicts of Interest: The authors declare no conflict of interest.

References

1. Díaz, H.; Serna, J.; Nieto, J.; Guedes Soares, C. Market Needs, Opportunities and Barriers for the Floating Wind Industry. *J. Mar. Sci. Eng.* **2022**, *10*, 934. [CrossRef]
2. Díaz, H.; Guedes Soares, C. Review of the current status, technology and future trends of offshore wind farms. *Ocean Eng.* **2020**, *209*, 107381. [CrossRef]
3. Díaz, H.; Teixeira, A.; Guedes Soares, C. Application of Monte Carlo and Fuzzy Analytic Hierarchy Processes for ranking floating wind farm locations. *Ocean Eng.* **2022**, *245*, 110453. [CrossRef]
4. Bagbanci, H.; Karmakar, D.; Guedes Soares, C. Review of offshore floating wind turbines concepts. In *Maritime Engineering and Technology*; Guedes Soares, C., Garbatov, Y., Sutulo, S., Santos, T.A., Eds.; Taylor & Francis Group: London, UK, 2012; pp. 567–576. [CrossRef]
5. Leroy, V.; Delacroix, S.; Merrien, A.; Bachynski-Polić, E.; Gilloteaux, J.-C. Experimental investigation of the hydroelastic response of a spar-type floating offshore wind turbine. *Ocean Eng.* **2022**, *255*, 111430. [CrossRef]
6. Bagbanci, H.; Karmakar, D.; Guedes Soares, C. Comparison of Spar and Semisubmersible Floater Concepts of Offshore Wind Turbines Using Long-Term Analysis. *J. Offshore Mech. Arct. Eng.* **2015**, *137*, 061601. [CrossRef]
7. Uzunoglu, E.; Guedes Soares, C. Hydrodynamic design of a free-float capable tension leg platform for a 10 MW wind turbine. *Ocean Eng.* **2020**, *197*, 106888. [CrossRef]
8. Vijay, K.; Karmakar, D.; Uzunoglu, E.; Guedes Soares, C. Performance of barge-type floaters for floating wind turbine. In *Progress in Renewable Energies Offshore*; Guedes Soares, C., Ed.; Taylor & Francis Group: London, UK, 2016; pp. 637–645. [CrossRef]
9. Bak, C.; Zahle, F.; Bitsche, R.; Kim, T.; Yde, A.; Henriksen, L.C.; Hansen, M.H.; Blasques, J.P.A.A.; Gaunaa, M.; Natarajan, A. *The DTU 10-MW Reference Wind Turbine*; DTU Wind Energy, Technical University of Denmark: Lyngby, Denmark, 2013.
10. Reyno, H.; Park, J.-S.; Kang, S.-Y.; Kang, Y.-J. A Numerical Analysis for Stress Concentration of Openings in Offshore Tubular Steel Tower under Design Loading Condition. *J. Korea Acad. Coop. Soc.* **2015**, *16*, 1516–1523. [CrossRef]
11. Akbas, U.B.; Shen, J. Design issues of wind turbine towers. In Proceedings of the 8th International Conference on Structural Dynamics, Leuven, Belgium, 4–6 July 2011; pp. 1592–1598.
12. Yeter, B.; Garbatov, Y.; Guedes Soares, C. Fatigue damage assessment of fixed offshore wind turbine tripod support structures. *Eng. Struct.* **2015**, *101*, 518–528. [CrossRef]
13. Yeter, B.; Garbatov, Y.; Guedes Soares, C. Structural design of an adaptable jacket offshore wind turbine support structure for deeper waters. In *Maritime Technology and Engineering 3*; Guedes Soares, C., Santos, T.A., Eds.; Taylor & Francis Group: London, UK, 2016; pp. 583–594. [CrossRef]
14. Teixeira, R.; Nogal, M.; O'connor, A.; Nichols, J.; Dumas, A. Stress-cycle fatigue design with Kriging applied to offshore wind turbines. *Int. J. Fatigue* **2019**, *125*, 454–467. [CrossRef]
15. Kim, H.J.; Jang, B.S.; Park, C.K.; Bae, Y.H. Fatigue analysis of floating wind turbine support structure applying modified stress transfer function by artificial neural network. *Ocean Eng.* **2018**, *149*, 113–126. [CrossRef]
16. Bachynski, E.E.; Moan, T. Design considerations for tension leg platform wind turbines. *Mar. Struct.* **2012**, *29*, 89–114. [CrossRef]
17. Oguz, E.; Clelland, D.; Day, A.H.; Incecik, A.; López, J.A.; Sánchez, G.; Almeria, G.G. Experimental and numerical analysis of a TLP floating offshore wind turbine. *Ocean Eng.* **2018**, *147*, 591–605. [CrossRef]
18. Sclavounos, P.; Tracy, C.; Lee, S. Floating Offshore Wind Turbines: Responses in a Seastate Pareto Optimal Designs and Economic Assessment. In Proceedings of the International Conference on Offshore Mechanics and Arctic Engineering 2008, Estoril, Portugal, 15–20 June 2008; Paper OMAE2008-48234. pp. 31–41. [CrossRef]
19. Nematbakhsh, A.; Bachynski, E.E.; Gao, Z.; Moan, T. Comparison of wave load effects on a TLP wind turbine by using computational fluid dynamics and potential flow theory approaches. *Appl. Ocean Res.* **2015**, *53*, 142–154. [CrossRef]
20. Jamalkia, A.; Ettefagh, M.M.; Mojtahedi, A. Damage detection of TLP and Spar floating wind turbine using dynamic response of the structure. *Ocean Eng.* **2016**, *125*, 191–202. [CrossRef]
21. Mas-Soler, J.; Uzunoglu, E.; Bulian, G.; Guedes Soares, C.; Souto-Iglesias, A. An experimental study on transporting a free-float capable tension leg platform for a 10 MW wind turbine in waves. *Renew. Energy* **2021**, *179*, 2158–2173. [CrossRef]
22. Sreebhadra, M.; Rony, J.; Karmakar, D.; Guedes Soares, C. Extreme response analysis for TLP-type floating wind turbine using Environmental Contour Method. In *Trends in Maritime Technology and Engineering*; Guedes Soares, C., Santos, T.A., Eds.; Taylor and Francis: London, UK, 2022; pp. 315–324. [CrossRef]
23. Guo, X.; Zhang, Y.; Yan, J.; Zhou, Y.; Yan, S.; Shi, W.; Li, X. Integrated Dynamics Response Analysis for IEA 10-MW Spar Floating Offshore Wind Turbine. *J. Mar. Sci. Eng.* **2022**, *10*, 542. [CrossRef]
24. Diaz, H.M.; Silva, D.; Bernardo, C.; Guedes Soares, C. Micro siting of floating wind turbines in a wind farm using a multiple-criteria framework. *Renew. Energy* **2023**, *204*, 449–474. [CrossRef]

25. Salvação, N.; Guedes Soares, C. Wind resource assessment offshore the Atlantic Iberian coast with the WRF model. *Energy* **2017**, *145*, 276–287. [CrossRef]
26. Silva, D.; Bento, A.R.; Martinho, P.; Guedes Soares, C. High resolution local wave energy modelling in the Iberian Peninsula. *Energy* **2015**, *91*, 1099–1112. [CrossRef]
27. Costa, T.C.; Pereira, L.T.; Marta-Almeida, M.; Guedes Soares, C. Mapping of currents off the northwestern Iberian coast with the Regional Ocean Modelling System. *J. Oper. Oceanogr.* **2018**, *13*, 71–83. [CrossRef]
28. Uzunoglu, E.; Guedes Soares, C. Parametric modelling of marine structures for hydrodynamic calculations. *Ocean Eng.* **2018**, *160*, 181–196. [CrossRef]
29. Hmedi, M.; Uzunoglu, E.; Medina-Manuel, A.; Mas-Soler, J.; Vittori, F.; Pires, O.; Azcona, J.; Souto-Iglesias, A.; Guedes Soares, C. Experimental Analysis of CENTEC-TLP Self-Stable Platform with a 10 MW Turbine. *J. Mar. Sci. Eng.* **2022**, *10*, 1910. [CrossRef]
30. Chakrabarti, S. *Handbook of Offshore Engineering*; 2-Volume Set; Elsevier: Oxford, UK, 2005.
31. Cummins, W.E. The impulse response function and ship motions. *Schiffstechnik* **1962**, *9*, 101–109.
32. *IEC 61400-1*; Wind Turbines—Part 1: Design Requirements. International Electrotechnical Commission: Geneva, Switzerland, 2005; No. 2, p. 88. Available online: https://webstore.iec.ch/publication/26423 (accessed on 12 April 2023).
33. *DNVGL-OS-C105*; Structural Design of TLPS (LRFD Method). Det Norske Veritas: Høvik, Norway, 2011; Volume DNV-OS-C10, p. 38. Available online: https://rules.dnv.com/servicedocuments/dnvpm/ (accessed on 12 April 2023).
34. *DNV-OS-J103*; Design of Floating Wind Turbine Structures. Det Norske Veritas: Høvik, Norway, 2013. Available online: https://rules.dnv.com/servicedocuments/dnvpm/ (accessed on 12 April 2023).
35. *DNV-OS-C101*; Design of Offshore Steel Structures, General (LRFD Method). Det Norske Veritas: Høvik, Norway, 2011; p. 2015. Available online: https://rules.dnv.com/servicedocuments/dnvpm/ (accessed on 12 April 2023).
36. *IEC 61400-3-2*; Design Requirements for Floating Offshore Wind Turbines. International Electrotechnical Commission: Geneva, Switzerland, 2020. Available online: https://webstore.iec.ch/publication/29244 (accessed on 12 April 2023).
37. *DNV-OS-J101*; Design of Offshore Wind Turbine Structures. Det Norske Veritas: Høvik, Norway, 2014; pp. 212–214. Available online: https://rules.dnv.com/servicedocuments/dnvpm/ (accessed on 12 April 2023).
38. *DNVGL-ST-0119*; DNV GL Standard for Floating Wind Turbines. Det Norske Veritas: Høvik, Norway, 2018. [CrossRef]
39. *DNV-RP-C201*; Buckling Strength of Plated Structures. Det Norske Veritas: Høvik, Norway, 2002. [CrossRef]
40. *DNVGL-RP-C203*; Fatigue Design of Offshore Structures. Det Norske Veritas: Oslo, Norway, 2016. [CrossRef]
41. *DNV-RP-C202*; Buckling Strength of Shells. Det Norske Veritas: Høvik, Norway, 2013; p. 27. [CrossRef]

Journal of
Marine Science and Engineering

MDPI

Review

Experimental Challenges and Modelling Approaches of Floating Wind Turbines

Mohamad Hmedi [1,2], Emre Uzunoglu [1], Chen Zeng [1], J. F. Gaspar [1] and C. Guedes Soares [1,*]

1 Centre for Marine Technology and Ocean Engineering (CENTEC), Instituto Superior Técnico, Universidade de Lisboa, 1049-001 Lisbon, Portugal; mohamad.hmedi@ec-nantes.fr or mohamad.hmedi@centec.tecnico.ulisboa.pt (M.H.); emre.uzunoglu@centec.tecnico.ulisboa.pt (E.U.); chen.zeng@centec.tecnico.ulisboa.pt (C.Z.); jose.gaspar@centec.tecnico.ulisboa.pt (J.F.G.)

2 Laboratoire de recherche en Hydrodynamique, Energétique et Environnement Atmosphérique, CNRS UMR 6598, École Centrale Nantes, Nantes Université, 44321 Nantes, France

* Correspondence: c.guedes.soares@centec.tecnico.ulisboa.pt

Abstract: This paper reviews experimental methods for testing floating wind turbines. The techniques covered include early-stage and up-to-date approaches such as a porous disc method and hybrid model testing. First, the challenges induced by Froude and Reynolds similitudes and the importance of the various aerodynamic phenomena are discussed. The experimental methods are evaluated based on their cost, versatility, requirements, and limitations. The work primarily focuses on representing aerodynamic loads via hybrid and physical rotor testing, and a preliminary classification is proposed to facilitate the selection of the approaches. The work does not aim to identify an optimal method, but it provides insights into each method's distinctive features, serving as a roadmap for selecting the most appropriate methodology based on the specific testing goals and level of accuracy. Overall, this study offers a comprehensive resource for testing the coupled hydrodynamic and aerodynamic performance of floating wind turbines. The conclusions offer guidance for selecting an appropriate methodology based on the desired testing outcome.

Keywords: floating wind; wave tank testing; software in the loop; low Reynolds blades; drag disc; hybrid testing

Citation: Hmedi, M.; Uzunoglu, E.; Zeng, C.; Gaspar, J.F.; Guedes Soares, C. Experimental Challenges and Modelling Approaches of Floating Wind Turbines. *J. Mar. Sci. Eng.* **2023**, *11*, 2048. https://doi.org/10.3390/jmse11112048

Academic Editor: Constantine Michailides

Received: 25 June 2023
Revised: 15 October 2023
Accepted: 20 October 2023
Published: 25 October 2023

1. Introduction

Fossil fuel-based power plants are environmentally harmful and accentuate global warming. With the increase in energy demand and the intricate political status, the shift towards renewable resources becomes more crucial [1], which reflects the good economic prospects for the use of this type of energy [2]. To date, wind energy covers 35% of renewables in Europe, with the majority of wind turbines (WTs) situated on land [3]. While the conventional deployment mode is onshore, implementing WTs in offshore locations is more efficient. The move to offshore utilization is driven by favourable wind regimes at these locations, specifically higher wind speeds and reduced turbulence. In addition, the visual, environmental, and auditory impact of offshore WTs is lower compared to onshore installations, making them a reasonable choice for larger turbine deployment. Overall, the benefits of offshore WTs deployment make it a viable option for renewable energy generation [4].

Recent years have witnessed a marked trend towards the utilization of offshore locations for the deployment of WTs [1]. When water depth exceeds 50 m, floating structures are deemed the most appropriate option. Recent analyses have shown that bottom-fixed offshore wind turbines are financially infeasible above 50 m depth due to the high expenses incurred for installation [5,6]. By contrast, floating wind turbines (FWTs) have demonstrated better economic viability in deeper waters [7,8]. In this context, several designs have been developed, including tension leg platforms (TLPs), semi-submersibles, and spar-type structures [9]. These designs are inspired by the oil and gas industry and intend to support WTs under a coupled wind and wave excitation.

Given the complexity of the environment, the investigation of conceptual design dynamics is necessary to guarantee their operational capabilities. Various initiatives, both industrial and academic, have been undertaken to progress the state of the art in floating wind technology. Notable examples of implemented prototypes include Blue H TLP in Italy [10], Statoil's Hywind Spar in Norway [11], WindFloat semi-submersible in Portugal [12], and FLOATGEN barge in France [13]. These designs were mostly developed through a collaborative effort between industry and academia. LIFES50+ [14,15], INNWIND.EU [16], and ARCWIND [17] exemplify the interdisciplinary cooperative research between academia and industry in advancing floating wind technologies.

The collaborative work for floating wind projects follows a series of stages guided by the Technology Readiness Levels (TRLs) [18,19]. The first stages (i.e., TRLs 1–6) integrate both numerical simulations and experimental testing to validate the concept. Numerical tools involve mathematical models and computer algorithms to estimate the physical properties and foresee the behaviour of the floating turbine. Design tools (e.g., ParMod [20,21]) and simulation software (e.g., OpenFast [22], Qblade [23]) offer the advantage of exploring a wide range of parameters and testing various wind and wave scenarios. Meanwhile, experimental testing provides a higher degree of accuracy and reliability based on direct measurements of the system. Additionally, experiments can reveal unexpected phenomena or behaviours that may have been ignored by simulations, such as neglecting viscous effects in potential flow calculations [24] and the coupling between aerodynamics and hydrodynamics. Therefore, complementing the numerical simulations with experimental testing validates the mathematical predictions and provides a more comprehensive understanding of the system.

To validate the theoretical predictions and the design dynamics, experimental testing is necessary, particularly in areas where the theory falls short. Building scaled models for hydrodynamic testing in a wave tank requires satisfying Froude (Fr) similitude conditions [25–27]. Nevertheless, the Reynolds number (Re) is decreased, and thereby the aerodynamic performance of the turbine is altered [28]. Accurately modelling mass (gravity and inertia) and mooring loads introduce further challenges to the testing environment. Furthermore, given the interplay between the turbine's aerodynamics and the blade pitch controller [29], incorporating the controller is generally not overlooked. Blade pitch control strategies for onshore turbines induce negative aerodynamic damping effects when used for floating wind systems (i.e., the motion is amplified rather than dissipated) [30–34]. A noticeable effort is visible in testing and validating new control strategies for FWTs [35–37]. Therefore, testing the concept considering the coupling between all components is needed, where both Fr and Re quantities are conserved.

The need for a testing environment that respects both Fr and Re poses a challenge for FWT experimental campaigns. To overcome the Fr–Re scaling conflict, different methodologies have been adopted to represent the aerodynamics of the turbine. The earliest techniques involved static cables that represent steady wind loads [38] or a drag disc to replicate the rotor thrust [39–41]. Froude-scaled rotors (FSRs) were used with high wind speeds to match the turbine's thrust in the below-rated speed zone [42–44]. Given the necessity for higher wind speed, performance-scaled rotors (PSRs) (i.e., a redesigned rotor with an identical aerodynamic performance of the turbine at a lower Re number) were then utilized [45–52]. The requirement of a wind generating system above the wave tank, which is uncommon, paved the way to develop hybrid-testing approaches. The hybrid methodology combines physical testing and numerical simulation in real-time. An actuation system provides the coupling between the simulated WT and the physical floater in the wave basin [53–62] or inversely in the wind tunnel [63–65].

Considering the variety of approaches to replicate the turbine's aerodynamics, each has its advantages and limitations, with some being more adequate than others in some cases. This paper provides an overview of these methods, their strengths, and weaknesses. First, Froude and Reynolds scaling are reviewed, followed by the test's sensitivity to various aerodynamic phenomena. The experimental techniques are then detailed and evaluated based on their cost, versatility, requirements, and limitations. The comparison between the different methods mainly focuses on hybrid and physical rotor testing. The various methods are organized to achieve that goal, and a preliminary classification is proposed. The primary objective is to present an overview that compares the existing methods while providing insightful observations on their distinctive features. It should be noted that the ultimate aim is not to determine the "best" technique but rather to furnish a means of comparing the different methods. The conclusion provides a roadmap for selecting the suitable methodology based on the experiment's goal.

2. Froude and Reynolds Similitude

The Froude number is a dimensionless quantity used to indicate the influence of gravity on a fluid's motion:

$$Fr = \frac{U}{\sqrt{gL}} \tag{1}$$

while the *Re* number is the ratio of inertial forces to viscous forces:

$$Re = \frac{\rho UL}{\mu} = \frac{UL}{\nu} \tag{2}$$

where U is the wave (fluid) celerity, L is the characteristic length, g is the gravity acceleration, ρ is the fluid density, and μ and ν are the dynamic and kinematic viscosity, respectively.

The geometric scaling between the full (f) and model (m) scale environments is as follows:

$$L_f = \lambda L_m \tag{3}$$

where λ represents the scaling factor.

To ensure the hydrodynamic similitude (i.e., conserving the gravitational loads), the Fr number is kept constant between both scale environments:

$$Fr_f = Fr_m \tag{4}$$

Table 1 presents the Froude-scaling factors for some parameters for a constant fluid density. For a detailed understanding of the scaling methodology see [66].

Table 1. Scaling factors for some parameters using Fr similitude.

Parameter	Units	Factor
Length	m	λ
Time	s	$\sqrt{\lambda}$
Volume	m^3	λ^3
Force	N	λ^3
Moment	N·m	λ^4
Power	W	$\lambda^{7/2}$
Inertia	kg·m^2	λ^5
Angle	Rad	1

The velocity and time relationships according to Fr scaling are **as follows**:

$$U_f = \sqrt{\lambda} U_m \rightarrow t_f = \sqrt{\lambda}\, t_m \tag{5}$$

while for *Re* scaling, they are related as follows:

$$U_f = \frac{U_m}{\lambda} \rightarrow t_f = \lambda^2\, t_m \tag{6}$$

From Equation (5), if Fr similitude is applied, the relation between the full and model scale Re numbers is as follows:

$$Re_f = \lambda^{3/2}\, Re_m \quad (7)$$

Equation (7) demonstrates a lower Re number at the model scale where Fr similitude is applied. For instance, the Re number is reduced by a factor of 350, nearly three orders of magnitude, for a 1/50 Fr scaled model, as seen for the DeepCwind semi-submersible testing [67,68]. Applying Fr scaling to the floater and Re scaling to the rotor is limited, with the need for ferocious wind speeds at the model scale and difficulties with scaling down time. For instance, if the geometric scaling factor (λ) is 30, a wind speed of 300 m/s is required over the basin to represent a 10 m/s Re-scaled wind, and one second Re-scaled period represents 900 s at full scale (Equation (6)).

3. Significance of Aerodynamic Phenomena

FWTs experience environmental loads throughout their lifetime, classified into hydrodynamic, aerodynamic, and inertial loads (Figure 1). The equation of motion that governs the behaviour of an operational FWT, following Newton's second law, can be expressed as follows [69]:

$$[M + A]\vec{\ddot{x}} + B\vec{\dot{x}} + [C_h + C_m]\vec{x} = \overrightarrow{F_{ex}} + \overrightarrow{F_{aer}} + \overrightarrow{M_{gyr}} \quad (8)$$

where $\vec{x}$, $\vec{\dot{x}}$, and $\vec{\ddot{x}}$ are the system's motion, velocity, and acceleration tensors. M represents the mass and inertia matrix, A is the added mass matrix, B is the damping matrix, C_h is the hydrostatic stiffness matrix, C_m is the mooring stiffness. $\overrightarrow{F_{ex}}$ the incident wave excitation force, $\overrightarrow{F_{aer}}$ is the total aerodynamic tensor, and $\overrightarrow{M_{gyr}}$ are the gyroscopic moments. As the methods discussed in the paper focus on replicating the turbine's aerodynamics, the following presents the importance of different aerodynamic phenomena on the responses of an FWT.

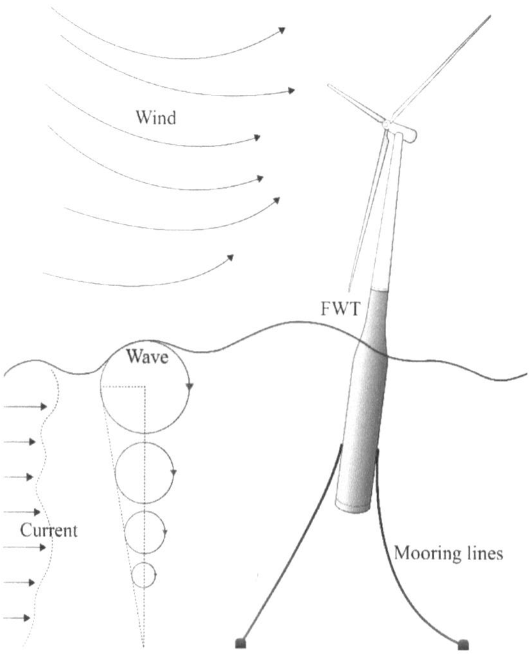

Figure 1. Floating wind turbine subjected to various loads.

3.1. Aerodynamic Loads

The aerodynamic conversion process at the rotor creates in the hub frame an aerodynamic tensor of six components [70]:

$$\overrightarrow{F_{aer}} = \left[F_x; F_y; F_z; M_x; M_y; M_z\right] \tag{9}$$

The thrust force F_x dominates the two force components F_y and F_z. The generator torque M_x also dominates the local aerodynamic pitch and yaw moments (i.e., M_y and M_z A load sensitivity analysis is made by eliminating the non-dominating aerodynamic forces and moments when simulating a semi-submersible FWT [70]. It is concluded that M_y affects the platform pitch motions and mooring line tension with 10–15% changes in the standard deviation are seen. M_z has significant effects on the yaw response (i.e., 2–10% in the parked case), and is usually small and excited at low frequency. If torsional responses is not aimed, M_z can be neglected for operational scenarios [70]. F_y influences the sway motion, roll motion, and tower side–side bending moment, which are typically smaller than the corresponding surge, pitch, or fore–aft bending moment. F_z has a negligeable effect on the platform motions in operational conditions. However, F_z affects by up to 15% the surge, pitch, and yaw motions for parked conditions and misaligned wind–wave conditions.

In addition, wind turbine controller settings modify the platform's dynamics. The controller adjusts the blade's pitch angle and the generator torque, which alters the aerodynamic loads [29]. Platform pitch instabilities may be induced if a land-based controller is used for FWT [30]. To prevent instabilities, FWT-specific control strategies are developed [32,71,72]. Thus, emulating the controller during the tests allows the replication of the coupling effect between the controller, aerodynamics, and the system dynamics.

3.2. Inertial Loads

In addition to the aerodynamic loads, inertial loads are produced due to the rotation of the blades. The centripetal acceleration of the rotor results in an inertial centrifugal force. Each blade element with a mass dm and flapping angle β, located at a distance r from the blade's root-end, is subjected to a centrifugal force dF_c along the chord direction [73]:

$$dF_c = r\,\omega_r^2\,dm\cos\beta \tag{10}$$

For perfectly symmetric blades, the total centrifugal force would be zero. However, as the blades are unbalanced, a small resultant force will be generated periodically at the rotor speed (i.e., 1P frequency).

Moreover, any rotational motion of the platform will produce a gyroscopic moment due to the rotor's rotation. For instance, a platform pitch motion induces a yaw moment for a horizontal axis WT. The gyroscopic moment is defined as follows [74]:

$$\overrightarrow{M_g} = J_r\,\overrightarrow{\omega_r} \wedge \overrightarrow{R_p} \tag{11}$$

with J_r is the rotor inertia, $\overrightarrow{\omega_r}$ is the rotor's rotational speed, and $\overrightarrow{R_p}$ corresponds to the platform rotational speed vector.

The sensitivity analysis conducted on a semi-submersible demonstrates that the gyroscopic moments and centrifugal forces primarily affect the sway, roll, and yaw motions up to 5% in operational conditions [70]. Disregarding the gyroscopic effects for a spar FWT overestimates the tower torsion up to 17% [75].

3.3. Loads Frequency Bandwidth

The excitation loads are distributed along a frequency bandwidth [40,53,76]. In experimental testing, the loads should be emulated with the correct amplitude and frequency content. Table 2 exemplifies the bandwidth for the OC3, OC4, and the triple spar models [53], explained as follows:

- The low frequency (LF) range may coincide with the natural frequencies of the FWT along the 6 degrees of freedom (DoF). It should be noted that TLPs have high natural frequencies in heave, roll and pitch as the moorings increase the system stiffness [5,21].
- The wave frequency (WF) range exciting the platform motions.
- The 1P and NP rotor frequencies range corresponding to the frequency of one full rotation of the rotor and the frequency at which one blade among N blades passes in front of the tower. NP would be referred to as the 3P frequency for a three-blade WT.
- The first fore–aft and side–side tower bending mode (Twr 1st).
- The high frequency (HF) range, including the blade modes (flapwise and edgewise).

Table 2. Limits of the frequency bandwidths at full scale.

	LF	WF	1P	3P and Twr 1st	HF
Low limit (Hz)	0.005	0.04	0.12	0.4	0.8
High limit (Hz)	0.05	0.25	0.2	0.8	2

4. Early-Stage Methods

Wave tank experiments were first performed to gauge the performance of FWTs. This section summarises the earliest testing campaigns, including the methodologies (i.e., the drag disc and static cables), strengths and weaknesses of each approach, and suggestions for improving their effectiveness.

4.1. Static Cables

The static cable approach was initially employed by Utsonomiya et al. [38] to evaluate the performance of a spar design. The main goal was to replicate the static tilt of the platform correctly under wind loads coupled with hydrodynamic loads from waves. The method involves using a cable (Figure 2) to imitate a static thrust obtained from the thrust curve of a given turbine. During Utsonomiya's campaign, decay, regular and irregular wave tests, and regular wave tests with a steady rotor thrust applied at the tower's top via the cable were conducted. However, only a steady thrust force was considered, neglecting other aerodynamic loads and interactions such as wind turbulence, aerodynamic damping, controller effects, and gyroscopic moments.

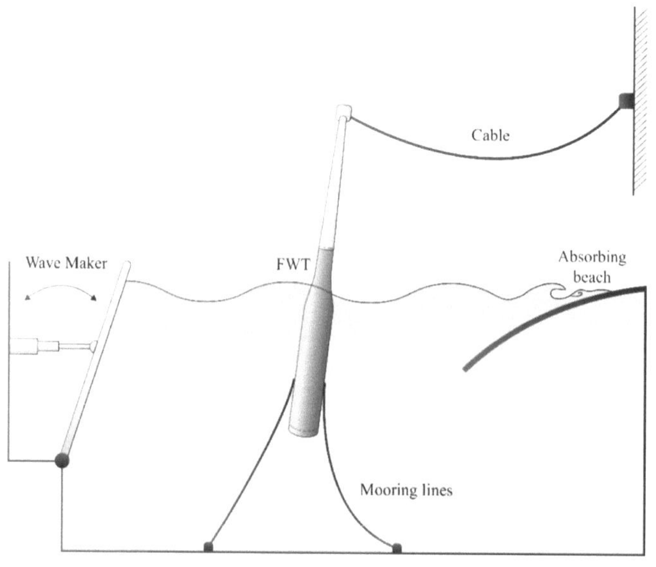

Figure 2. Static cable approach.

Although this approach is simple and quick, it lacks variability as it reproduces only a steady thrust force without considering the wind turbulence and the controller effects. Additionally, using the thrust curve of the WT does not account for the platform response in the calculation of the aerodynamic load, and the gyroscopic effects are not considered. Furthermore, this technique ignores other aerodynamic load components, such as aerodynamic moments and the coupling with the blade pitch controller. In other words, testing with a static cable is limited to steady wind cases where the rotor thrust is relatively constant.

One possible improvement to this method is to connect the cable to a dynamic winch that can vary the tension instantaneously to match the thrust force obtained in a real-time simulation of the turbine. This adaptation would convert the approach into a "Software in the Loop" (SIL) system described in Section 5. Another enhancement is using a rotating disc to reproduce gyroscopic moments [12]. Despite its limitations, this method remains a valuable and straightforward approach for investigating the hydrodynamic response of FWTs under steady wind loads [77].

4.2. Porous Disc

One possible method for reproducing the turbine thrust involves replacing the rotor with a drag/porous disc exposed to an airflow (Figure 3). For FWTs, the use of this technique has been documented in several research papers, including the 1:105 [12,39] and 1:67 [40] scale WindFloat semisubmersible, as well as combined FWT-wave energy converter (WEC) systems [41,78,79].

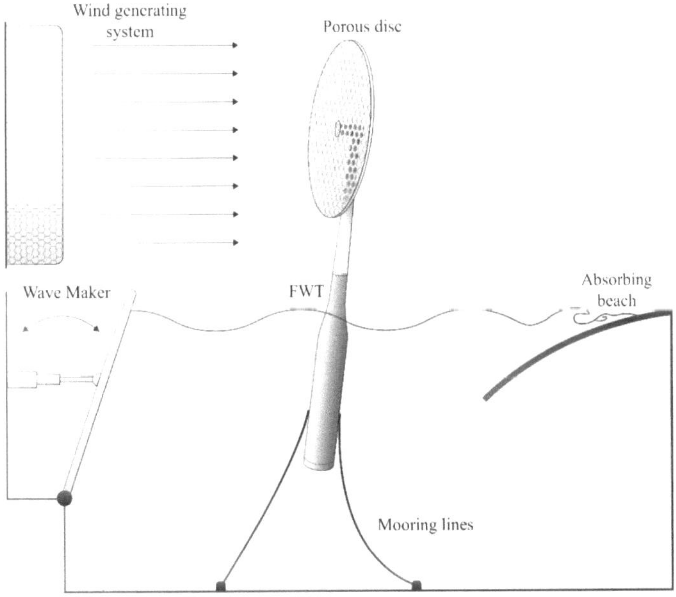

Figure 3. Porous disc approach.

The drag disc method boasts a high degree of versatility and ease of implementation for various turbines. The construction and calibration processes are relatively straightforward, highlighting the method's practicality. Additionally, the method yields a consistent thrust force compared to the thrust curve in the below-rated zone, which is beneficial when testing wind turbines in wave tanks. The ability to generate wind loads below the rated wind speed is particularly noteworthy, as it allows for comprehensive testing under controlled conditions. In general, the drag disc method offers a compelling option for FWT testing when a wind-generating system is available.

Despite its use for FWT testing, the porous disc approach has considerable drawbacks. One notable issue is the method's failure to account for the gyroscopic moments, the aerodynamic yaw and pitch forces and moments and the generator torque [41]. In addition, the vortex shedding generated by the disc can cause vibrations that add more uncertainty to the measurements. Furthermore, this technique is limited in applicability since it cannot accurately simulate wind loads in the above-rated wind speed zone. Once the wind speed exceeds the rated level, the generated thrust force will surpass the expected value from the thrust curve, resulting in an incorrect emulated thrust. Attempting to simulate a control system using the drag disc method is not feasible, so it fails to account for the coupling between aerodynamics and the controller.

Consequently, it is unsuitable for testing control strategies and prevents varying the thrust coefficient, leading to incorrect aerodynamic damping. The need for a wind generating system increases the test operating costs as such wind blowers are both expensive to operate and not commonly available over the basin, which creates a barrier to the widespread adoption of this method. Finally, a light disc is advisable to maintain a margin for calibrating the system's mass properties. While the WindFloat campaign has attempted to address some limitations (i.e., using a spinning disc to replicate gyroscopic effects [12,39]), the drag disc approach is still partially bounded, with several notable deficiencies.

5. Physical Rotor

5.1. Froude-Scaled Rotors

An obvious option for replicating the aerodynamics of FWTs is to use a Froude-scaled rotor (FSR) in the experiment (Figure 4). In 2007, Marintek ocean basin laboratory (currently SINTEF) evaluated a 1:67 scaled model of the Hywind Spar with an FSR. In their campaign, two DC motors served to set a constant blade pitch angle and the rotor's rotational speed [31,33]. Likewise, the AAUE-TLP fitted with an FSR of the NREL 5 MW WT and underwent testing at Aalborg University, where the generated wind speed was amplified to reproduce the thrust force correctly [80]. A similar approach was adopted by Ahn and Shin [81] to examine the dynamics of the Hywind Spar. In a study by Bahramiasl et al. [74], in which the wind speed was also increased, a 1:100 Sea-Star TLP model was examined to investigate the gyroscopic effects at different wind headings. In the same campaign, a motor initiated the rotor's rotation due to mechanical friction blocking the rotation from the blowing wind. In another investigation by Wen et al. [82], an FSR was utilized to evaluate the aerodynamic loading effects on the JTU-S4 spar model. Michailides et al. [43] investigated the performance of a combined wind–wave energy concept (SFC) in extreme conditions. They used an FSR calibrated in a wind tunnel to obtain the expected thrust for varying wind speeds and blade pitch angles.

A notable campaign is the DeepCwind in 2013 [42,44,83] at MARIN, where three FWT systems (Sea-Star type TLP [84], Hywind spar [85], DeepCwind semi-submersible [86]) were scrutinized. It brought the first floating offshore wind data available to the public and allowed others to learn from the issues that arose during their tests. The data collected from this campaign validated the numerical model of the three designs [68,87,88]. In DeepCwind tests, an FSR was used, and the wind speed was amplified to reproduce the rotor's thrust force accurately. For instance, a 21.8 m/s blowing wind was utilized to replicate the rated wind speed scenario of 11.4 m/s.

A significant shortcoming of increasing the wind speed is the generation of unwanted loads on the tower and the platform. Moreover, to prevent interaction with the waves, the fans were placed high enough with the nozzle tilted 2.16 degrees downward. The inclination of the nozzle introduced a vertical component of the wind velocity, resulting in a 20% reduction in the wind speed at the lower ends of the rotor. Furthermore, the generator torque was not accurately replicated, and the motors induced the blades' rotation. The blade pitch angle remained constant throughout all the tests as no control system was used. The lack of a controller and the need for higher wind speed made above-rated wind conditions infeasible for testing.

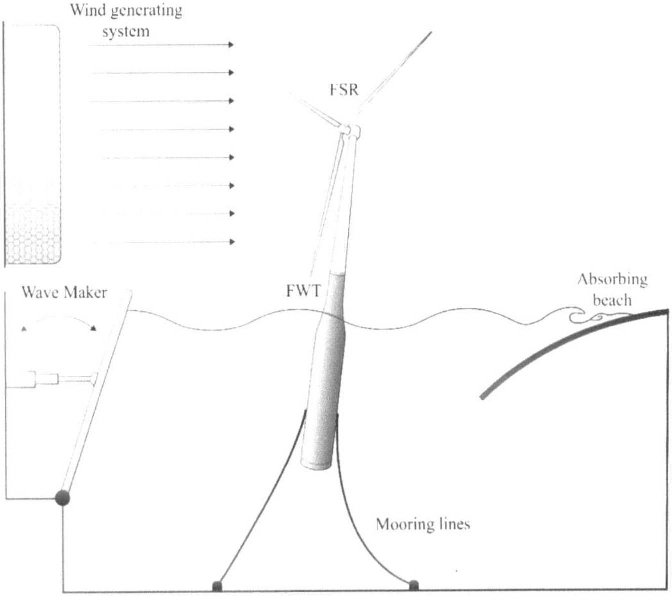

Figure 4. Froude-scaled rotor approach.

The results from these campaigns conclude the need for adjustments and modifications in future studies to replicate the aerodynamic loads accurately. It was suggested to use a performance-scaled rotor (PSR), where the blades are redesigned for a similar performance at low Re [42]. In addition to PSRs, attempts were made to tackle the limitations of this approach. In studies by Li et al. [89] and Duan et al. [47], a modified version of FSR was employed. The DeepCwind semi-submersible [88] and the Hywind spar [47] were the tested platforms, and the adjustment made was to let the rotor freely spin instead of using a motor to initiate the rotation. The motor was devoted as a generator to decelerate the rotor speed, and the tip speed ratio (TSR) was correctly kept to some extent. The improvement resulted a marginally increased wind speed compared to the initial version, such as using 12.8 m/s wind speed instead of 21.8 m/s for the rated scenario (i.e., 11.4 m/s).

5.2. Performance-Scaled Rotor

Following the conclusions of the DeepCwind campaign in 2013 [42], efforts were made to construct performance-scaled rotors (PSRs) [50,83], where the blades are redesigned for a low Re environment (Figure 5). Accordingly, the DeepCwind semi-submersible was retested with a PSR at MARIN [90,91]. The new testing campaign validated the results obtained in the first campaign, where the global performance was roughly identical. On the other hand, the new concept allowed the performance of new tests that were not feasible by the FSR. For instance, the above-rated wind speed tests were made, and the potential implementation of an active blade pitch controller was considered to investigate the controller's impact on the global performance of the floating turbine.

A MARIN PSR was used again to test the GustoMSC Tri-Floater, and the measured thrust matched the full-scale turbine predictions. The Windstar TLP [92] mounting the NREL 5 MW was tested using a PSR at the SKLOE laboratory (i.e., Shanghai Jiao Tong University) and demonstrated the effectiveness of this approach in resolving the scaling problem. PSR is utilized to test larger turbines such as the DTU 10 MW WT [93]. Among the tests found in the literature are the triple Spar model [94] and the sea-star TLP [95] at DHI Denmark. Using a PSR, an upscale of the DeepCwind semi-submersible with the DTU 10 MW underwent testing in the INNWIND.EU project at LHEEA—Ecole Centrale de Nantes (ECN) [16,96]. The University of Ulsan (UOU) 10 MW design [97], the KIER

TLP [98], and the EOLINK TLP [99] exemplify the wide use of the PSR method. Technical details and information on how the blades are designed and fabricated are out of the scope of this paper and can be found in references [28,51,100].

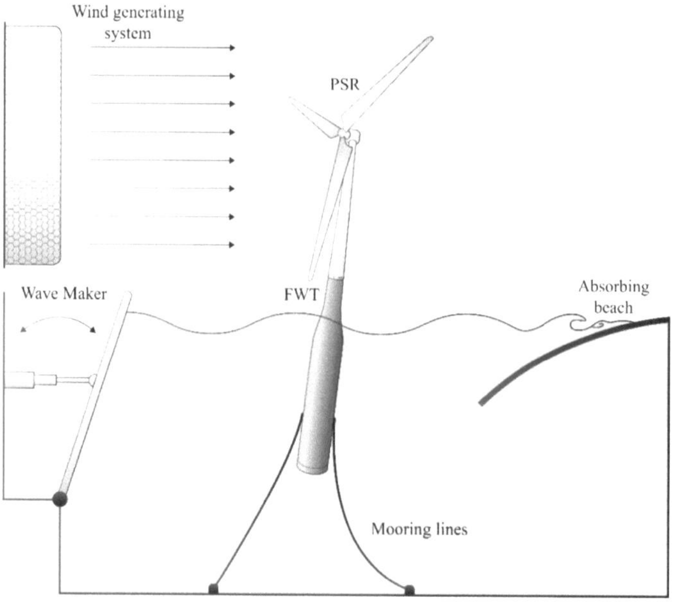

Figure 5. Performance-scaled rotor method.

The PSR approach brought several advantages, improving the fidelity of the experiments. A correct thrust force coefficient in a Froude-scaled environment could be maintained, as well as the TSR. In addition, the ability to implement and test wind turbine controllers is possible. The rotor torque, the aerodynamic damping, and gyroscopic moments can be replicated, along with other unexpected aerodynamic and hydrodynamic phenomena uncaptured by some numerical tools.

However, emulating physical wind over a wave tank requires a wind generation system, making PSR costly to adopt and operate [101]. Calibration of the wind speeds, wind gradient and turbulence increase the complexity of the experimental setup. The accuracy of the wind generation systems over the basin is much less than in wind tunnels. For instance, in most campaigns (e.g., [42,96]), multiple fans pumping the air through ducts and honeycomb mesh were used. The difficulty of controlling the boundaries of airflow after exiting the nozzles was noticed in most campaigns [46,98]. Thus, the ability to replicate the same test decreases, adding more uncertainty to the measurements. Moreover, the low Reynolds number at the model scale leads to inaccurate modelling of the WT's power and torque coefficients. The methodology has shown a good representation of the real scheme of a wind turbine. However, less versatility is present as each turbine has a specific blade geometry, and a redesigned blade for each turbine design and scaling factor is needed.

6. Hybrid Testing

Hybrid testing emerges as a sophisticated technique that synergistically combines physical testing and numerical simulation, fostering a holistic approach to experimentation. The inception of this approach originates back to the 1970 s in the field of earthquake engineering [102]. In that context, the technique was pioneered to investigate the dynamic response of structures subjected to earthquake conditions. The procedure involved computationally determining the structure's displacements and subsequently applying them in real-time to the physical model using multiple actuators.

FWTs tests borrowed the approach, drawing inspiration from the aforementioned concept. Similarly, the system's motions can be reproduced using a robot in a physical wind environment [64], or the aerodynamic loads can be emulated under physical wave conditions using actuators such as cables, propellers, or ducted fans [62,103]. The simplest version involves emulating a pre-recorded time series of motions or aerodynamic loads through the actuation system. However, synchronising the wave elevation with the emulated aerodynamic loads (or the blowing wind with the robot motions) poses a challenge. Furthermore, the real-time replication of blade pitch controller effects remains elusive. Additionally, the fidelity of these tests is constrained as the simulation fails to account for the platform's actual (i.e., the measured) motion in the underlying calculations.

A higher fidelity hybrid technique with increased complexity is developed to address these limitations. This approach entails the real-time integration of physical testing and numerical simulation. In the cases where an actuator replaces the rotor, the platform's motions are tracked and subsequently fed into a numerical tool. The simulation provides the aerodynamic loads considering the platform motions at each corresponding time step, which are then appropriately scaled and replicated by the actuator [54].

In a study conducted by Stewart and Muskulus [104], a comparison is made between utilizing pre-calculated time series of aerodynamic loads and real-time testing. The findings indicate that employing predefined loads results in significantly larger discrepancies in both loads and motions when comparing the simulation to the experimental data. This discrepancy arises from the absence of a feedback loop between the system's motions and the predefined aerodynamic loads. However, implementing real-time testing introduces challenges associated with compensating for delays [103].

This methodology is referred to by various names, such as "Software in the Loop" (SIL), "Hardware in the Loop" (HIL), and "Real-Time Hybrid Model" (ReaTHM) testing. All of these terms encompass tests that combine physical testing and numerical simulation. In general, this integration offers substantial benefits for testing complex structures like FWT. However, it necessitates meticulous compensation for delays to enable swift actuator execution within small time steps. Note that a similar approach is adopted in ocean engineering to test mooring line dynamics. The motions of a floating platform under wave excitation are measured experimentally and used in numerical simulations of the full-scale mooring lines [105–109]. The following section provides a comprehensive overview of the hybrid testing campaigns conducted on FWT and highlights the distinct advantages offered by each actuation system employed in these tests.

6.1. Propeller Actuators

When considering the selection of actuators for hybrid testing, researchers have utilized various options such as propellers, ducted fans, and cables. Among these choices, propellers and ducted fans have emerged as more commonly employed. Azcona et al. [54] utilized a ducted fan positioned at the hub height to replicate the rotor's thrust (Figure 6). The thrust force was calculated in real-time by the numerical tool FAST [110], which received the platform motions obtained from the basin measurement. The strategy was applied to test a 1:40 scale model of the OO-Star semi-submersible [15] equipped with a 6 MW wind turbine at the ECN wave basin. The successful implementation of this campaign demonstrates the SIL's capability to incorporate aerodynamic damping. The same principle was applied at the ECN basin to test the DeepCwind semi-submersible with the NREL 5 MW WT [61,111] and the DTU 10 MW WT [96]. Notably, an improvement was made by incorporating second-order hydrodynamics into the numerical model, enhancing the fidelity of the simulated load and the tests.

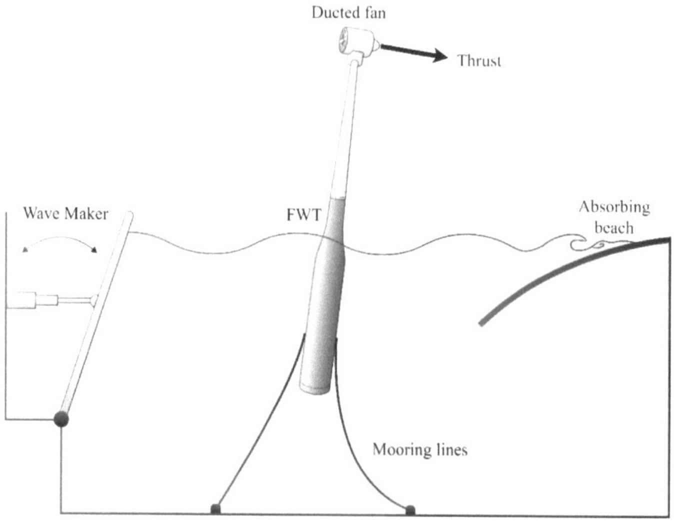

Figure 6. Hybrid testing using a ducted fan.

In addition to these SIL systems, numerous researchers have widely adopted the hybrid approach using ducted fans or propellers. Desmond et al. [112] employed a simplified version of the hybrid technique to evaluate the SCDnezzy2 semi-submersible, which supports two 7.5 MW WT. In that campaign, two ducted fans substituted the two rotors and generated steady wind loads of up to 10 N thrust on each tower. Similarly, Wright et al. [113] utilized a single ducted fan generating steady thrust to test a hexagonal TLP design in two scales. Andersen [114] employed a ducted fan to emulate a recorded time series of thrust to test a three-column semi-submersible.

However, emulating recorded loads is low fidelity compared to SIL since they do not replicate the blade pitch controller effects in real-time, and the instantaneous motions are not considered in the simulation of aerodynamic loads. A comparison between hybrid testing with and without SIL (i.e., real-time exchange between the simulation and the physical testing) was conducted by Matoug et al. [115]. The NAUTILUS semi-submersible [14] with the DTU 10 MW [93] and the WindQuest 10 MW [116] were tested, revealing a 14% improvement in cross-correlation between numerical simulation and experimental testing when SIL is employed.

Building upon these findings, implementing single propeller actuators with SIL has been extensively utilized in various campaigns. For instance, the Iberdrola TLP [117] was tested at the Kelvin Hydrodynamics Laboratory (KHL) at the University of Strathclyde [118] using a single ducted fan and following the same SIL concept described by Azcona et al. [54]. The approach was also applied to test the SOFTWIND Spar [53,76,119] at ECN. However, it should be noted that the single fan actuator aims to reproduce the thrust force exclusively, and the other aerodynamic loads and gyroscopic moments are not emulated with a single actuator. Therefore, efforts have been made to develop multi-actuator systems capable of reproducing the full aerodynamic tensor.

As the first version only emulated the thrust force, Azcona et al. developed a new multi-propeller system capable of reproducing the rotor moments [120]. This concept improves the SIL system by incorporating the moments generated by the aerodynamic and gyroscopic effects. Within the ARCWIND project, the multi-propellers system is utilized to test the CENTEC-TLP [5,17,62,121] and the SATH10 semi-submersible [122]. At the Instituto Hidraulica Cantabria (IHC), an actuator consisting of six drone propellers is developed by Urbán and Guanche [60] and utilized to test the TELWIND FWT [123]. The propellers are calibrated and tuned to account for the aerodynamic influence resulting from their proximity. The results demonstrate small deviations compared to the theoretical values of the loads. Another six-propeller actuator was developed by Otter et al. [59,124] at MaREI, Ireland, and exhibited good performance in emulating both thrust and torque. Additionally, a multi-propeller system was employed by Kanner et al. [125] to test a vertical-axis WT, expanding the application beyond horizontal-axis WTs.

6.2. Cable Winch Actuators

In addition to propellers and ducted fans, cable winches have been explored as actuators in hybrid testing. Early testing campaigns employed cables to replicate static thrust based on the turbine-specific thrust curves [38] (Section 4.1). However, to overcome the limitations of the static cable approach, researchers sought to improve the method by applying tensions instantaneously to the model based on real-time simulations. The underlying concept of using cable winches as actuators in hybrid testing is similar to the SIL system with propeller actuators.

One of the pioneering systems employing cable winches as actuators is developed at SINTEF in Norway. Their technique, known as ReaTHM testing, utilized six cables attached to a square frame at the hub height (Figure 7). Researchers at SINTEF employed this system to conduct experiments on the NOWITECH model, which featured the NREL 5 MW turbine [55–57]. To optimize the cable-based method, a sensitivity analysis was performed by Bachynski et al. [70], resulting in the emulation of a reduced number of aerodynamic loads based on the analysis outcomes. Further enhancements were introduced by Thys et al. [126] to encompass wind direction variability and tower loads, extending the frequency bandwidth to include the 3P frequency and the first tower bending frequency. The updated system was employed to test the NAUTILUS semi-submersible with the DTU 10 MW WT, with a comparative analysis against the previous system presented by Chabaud et al. [58]. The results indicate that the new system provides greater flexibility in replicating loads despite requiring higher cable tensions.

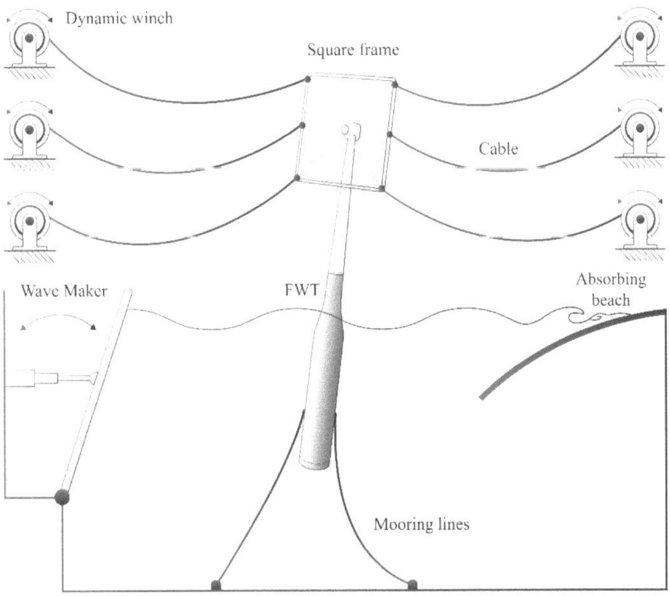

Figure 7. Hybrid testing using dynamic winches.

Another study by Antonutti et al. [127] utilized a similar hybrid system with cables to test a semi-submersible mounting of the Haliade 6 MW WT. Their system consists of five cables enabling actuation of the surge, pitch, and yaw degrees of freedom. The results demonstrate good agreement between the experimental and numerical models. Moreover, a hybrid system employing cable actuators was used by Hall et al. at the University of Maine, USA, to test a 1:50 scale model of the DeepCwind semi-submersible [67,128]. Unlike other cable-based systems, their approach replicates only the thrust force. Two opposing cables attached to the turbine at the hub height are used as the actuator. The results obtained from this hybrid testing are compared to those of complete physical testing with a PSR [50] and reveal a satisfactory agreement, particularly regarding motions and mooring tensions. It also demonstrates the capability to reproduce the effects of aerodynamic damping.

6.3. Wind Tunnel Tests

In the wave basin experiments, the primary objective is to scrutinize the hydrodynamic response of the platform combined with the emulation of the aerodynamic loads using actuators, thus bypassing the need for a scaled rotor subjected to wind flow. Conversely, wind tunnel testing aims to comprehensively examine the aerodynamic loads, wakes, and turbines' interactions. Within the context of hybrid testing, the motions of the FWT model are simulated using a numerical tool, enabling the replication of the FWT motions through actuators, while a physical rotor is subjected to airflow in the wind tunnel.

Pioneered by researchers from Politecnico de Milano in Italy, a robot of two degrees of freedom (DoF) was developed by Bayati et al. [63] to mimic surge and pitch motions effectively in real-time. The actuator's technical details are documented in [129,130]. The initial application of this robot involved exclusively imposing surge motions upon a scaled 1:75 model of the DTU 10 MW turbine during wind tunnel tests, facilitating a comprehensive exploration of wakes in both steady and unsteady wind conditions [131]. Subsequent enhancements to the robot resulted in the realization of additional degrees of freedom, thus evolving into a six-DoF mechanism (Figure 8) with the name "HexaFloat" [64]. This upgraded robot is then deployed in wind tunnel testing of the DTU 10 MW turbine to represent a floating support platform, as reported in the works of Bayati et al. [132] and Belloli et al. [133]. The presented robot has been utilized for various purposes within the field of wind energy research. Fontanella et al. [134] employed it to investigate the unsteady aerodynamics of the DTU 10 MW WT under imposed surge motions. Additionally, it is utilized in the COREWIND project to examine the aerodynamics of two 15 MW FWTs [135]. Furthermore, the same robot is used in phase III of the OC6 project to collect data for validating the aerodynamic loading on the rotor in motion caused by a floating support structure [136].

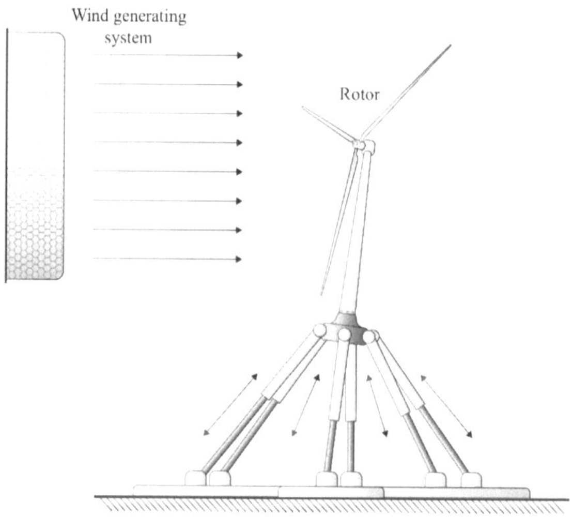

Figure 8. Hybrid testing in the wind tunnel with a 6 DoF robot.

In parallel work, Schliffke et al. [137] embarked upon a similar approach to scrutinize the repercussions of surge motions on wake characteristics. Their experimental campaign involved the utilization of the FLOATGEN barge, integrated with a 2 MW wind turbine, as the focal test model. Instead of deploying a scaled rotor, their concept entailed a porous disc. The imposition of surge displacements is achieved through a motor controlled by the manufacturer's software. Although a modest time delay between the assigned and emulated motion was noted, its effects were deemed negligible, primarily due to its independence from the imposed frequency and amplitude. Notably, their study focused predominantly on the ramifications of surge motion on wake characteristics without delving extensively into motion dynamics. Furthermore, Rockel et al. [138,139] conducted a series of hybrid tests in wind tunnels to ascertain the impact of pitch motions on FWT wakes and the intricate interplay between wakes and the FWT.

7. Methods Comparison and Applicability

7.1. Hybrid vs. Physical Rotor Testing

Hybrid testing conducted in wave basins offers the advantage of not requiring a physical wind generating system; alongside the measured motions, a wind input file is incorporated into the numerical simulation to introduce the wind turbulence during the tests. Additionally, it is possible to emulate the effects of the blade pitch controller and aerodynamic damping as they are included in the simulation results. For PSR, incorporating low-Reynolds modifications poses challenges in accurately modelling the turbine's power coefficient and torque characteristics. This limitation hinders the scalability of blade pitch control and impedes the precise capture of aerodynamic damping effects [101].

Unlike the limitation of physical rotors, the hybrid approach in the wave basin is not restricted to a specific turbine or scale. Utilizing actuators makes it feasible to replicate the aerodynamic loads of any turbine, making the hybrid technique more cost-effective and time-efficient compared to physical rotor testing. The latter necessitates the construction of a scaled rotor for each turbine design, along with the associated scaling coefficient. Hybrid testing campaigns have demonstrated favourable outcomes by emulating the thrust force accurately and replicating torques and gyroscopic moments. Moreover, hybrid testing grants complete control over the aerodynamics (or hydrodynamics) to better focus on one aspect (i.e., aero or hydro) of the study. Meanwhile, the PSR method showcased superior proficiency in catching unexpected, coupled effects. However, PSR may replicate global aerodynamic phenomena with inferior accuracy compared to the numerical models employed in hybrid testing. This discrepancy arises from partially including scaling effects, such as incorrect torque coefficient and absence of blade pitch control.

In addition, there are challenges associated with hybrid testing, primarily related to delays that can impact the fidelity of the tests. The employed actuators must possess sufficient bandwidth to emulate the assigned load or motions accurately within an appropriate time step [101]. The simulation tool implemented in hybrid testing must provide the calculated load within short time steps, justifying the use of BEM-based tools (i.e., OpenFast) rather than CFD. Moreover, the accuracy of the tests depends on several factors, including the precision of motion tracking measurements and the simulation tool employed. It is acknowledged that certain aerodynamic phenomena (in the case of hybrid testing in wave basins) and hydrodynamic responses (in the case of hybrid testing in wind tunnels) that are not accounted for in the numerical model will not be fully replicated in the experiments. Nevertheless, the existing numerical models may already include most known phenomena. So, the omission of phenomena poses a concern only when exhibiting significant coupling effects, which may happen but is unlikely to be known beforehand.

Furthermore, the calibration of the actuation system is challenging, especially for the one employing multiple emulators. While using multiple actuators allows the reproduction of a broader range of loads (in basin tests) or the mimicry of displacements with multiple degrees of freedom (in tunnel tests), it also introduces complexities and additional sources of uncertainty within the measurements [140]. In other words, more attention is required for the calibration process of the different actuators as well as the mass and inertia quantities, to effectively mitigate and minimize the error levels in both prototype modelling and load replication.

7.2. Actuators' Suitability in Hybrid Testing

It has been demonstrated that both propellers and cable actuators can effectively emulate the thrust force in hybrid testing. However, propeller systems encounter challenges in replicating the aerodynamic moments, as noted by Otter et al. [59], and incorporating the propellers' torques into force allocation is a complex task [140]. On the other hand, cable actuators excel in reproducing the moments, although the control of cables is comparatively more intricate than that of propellers. Additionally, the cable's frame requires a larger space on the top of the tower, unlike propellers, which occupy a smaller footprint. It is worth noting that the pretension of the cables may affect the results of free decay tests if not performed perfectly, as observed by Sauder et al. [55], whereas conducting free decay tests with turned-off propeller actuators is simpler.

In addition, the square frame required for cable actuators occupies a large area on the tower top. In contrast, propellers are lightweight and occupy a smaller footprint. Consequently, calibrating the tower top masses poses a greater challenge for cable systems. Furthermore, propellers are easily transferable between different basins. In contrast, cables require specialized infrastructure, limiting them to the facility for which they were initially designed.

Due to the high rotational speed of the propellers, undesirable vibrations are induced, whereas cables may excite the attaching frame's natural period. Careful attention must be given to the temperature of the active emulator to prevent burning the propeller's electrical motor or demagnetizing it. When dealing with multi-actuator systems, the interaction between propellers must be considered during the tuning process. The wakes of adjacent propellers can modify the air inflow of others, leading to performance changes of up to 10%, as noted by Urbán and Guanche [60] when tested on a fixed support.

Although, propellers and winches are distinct in how they handle load monitoring. Measuring the load generated by propellers is commonly performed via a load cell attached to the tower top. Inertial loads, which are proportional to acceleration, make real-time controlling of the propeller's thrust difficult. Consequently, propellers are often controlled in an open-loop manner, relying solely on fixed-support thrust calibration and empirical correction factors to address uncertainties such as propeller interactions. On the other hand, cable loads can be readily measured using a load cell for each cable, given their negligible mass. This measurement allows for closed-loop control with feedback algorithms that can promptly correct for uncertainties in real-time. Additionally, winch systems typically possess a higher bandwidth, resulting in shorter time delays, a critical factor for ensuring the accuracy of hybrid testing. Hence, winch systems are generally considered to be more accurate than propellers.

The choice between hybrid testing in wind tunnels or wave basins depends on the specifications of the phenomena under investigation. Hybrid testing in wind tunnels is suitable when studying wakes and turbine interactions, while hybrid testing in wave basins with propellers or cables is preferable when focusing on hydrodynamics. If the thrust loading is considered the dominant load for the tests, a single propeller (or a ducted fan) is recommended. On the other hand, a winch system is more appropriate when replicating the aerodynamic and gyroscopic moments. The high bandwidth of cable actuators makes them the preferred option for replicating 3p, tower and vertical loads.

7.3. Method Selection Roadmap

Evaluating and comparing the different methods based on various criteria can provide valuable insights into their performance and suitability for specific applications. Table 3 recaps this assessment based on the cost, accuracy, and other relevant factors. The PSR method exhibits superior performance in emulating the turbine's unexpected, coupled effects and surpasses the limitations of the porous disc and FSR methods, which are restricted to scenarios below rated wind speed. However, the PSR method may replicate global aerodynamics less accurately than numerical models used in hybrid testing due to scaling effects, including incorrect torque coefficients and the absence of blade pitch control. Therefore, hybrid testing stands as the sole approach that allows testing and evaluating control strategies.

Table 3. Overview of the experimental methodologies used for FWTs.

Criteri \ Method	Porous Disc	FSR	PSR	Hybrid
Aerodynamic loads	Limited	Limited	Effective	Effective
Blade pitch controller	Infeasible	Infeasible	Limited	Highly apt
Development cost	Medium	High	High	High
Operation cost	Medium	High	High	Low
Versatility	Flexible	Restrictive	Restrictive	Flexible
Scalability	Moderate	Limited	Limited	High

Furthermore, the hybrid approach demonstrates enhanced flexibility and cost-efficiency compared to other techniques. Unlike the PSR and FSR methods, which necessitate the construction of new scaled rotors for each turbine design and scaling factor, the actuators utilized in the hybrid approach can represent different turbine designs. These advantages allow for greater versatility and eliminate the need for costly scaled rotor modifications with each variation in design or scaling. Still, the material used for the hybrid framework may be expensive.

Overall, both the PSR and hybrid techniques serve as effective representations of turbine aerodynamics. The PSR method exhibits higher accuracy in emulating unpredicted aerodynamic characteristics, albeit at a higher operation cost. On the other hand, the hybrid technique offers a cost-effective alternative while still providing a reliable representation of turbine aerodynamics and coupled phenomena. The selection of the testing approach typically hinges on the specific objectives of the tests, as well as a careful balance between budget constraints and fidelity requirements. The choice will depend on a trade-off between financial considerations, the available equipment, and mainly the desired level of accuracy and detail in the results, as summarized in Figure 9 showing an agreement with the conclusion of Otter et al. [101].

In the presence of a wind-generating system in an ocean basin, the PSR method can allow for the investigation of unpredicted coupled effects. Otherwise, hybrid testing allows physical testing on the studied phenomenon, including known coupled effects. Hybrid testing in wind tunnels allows for the investigation of the turbines' interactions and wake effects. When studying the hydrodynamic response of FWTs coupled with the aerodynamics, propeller actuators present an option if the thrust force is the dominant aerodynamic effect in the test case. Propellers can be easily moved between different basins, allowing their use in facilities beyond their original development. On the other hand, when the aerodynamic moments are important and a higher bandwidth is sought, dynamic winches are more suitable actuators.

The classification and recommendations represent the current status of FWT testing campaigns. This review indicates a dominance for hybrid testing in the future with technological advancement and improvements in simulation tools. As larger turbines are targeted (i.e., 20 MW turbines and more), small scaling factors will be needed to construct physical rotors, leading to difficulties in building, calibrating, and fitting the system in the testing facility. Hybrid testing overcomes future scaling barriers from larger turbine sizes by representing the rotor with actuators. The improvements in simulation tools will result in an enhanced fidelity of the emulated loads, thus the capability of hybrid testing to replicate coupling effects.

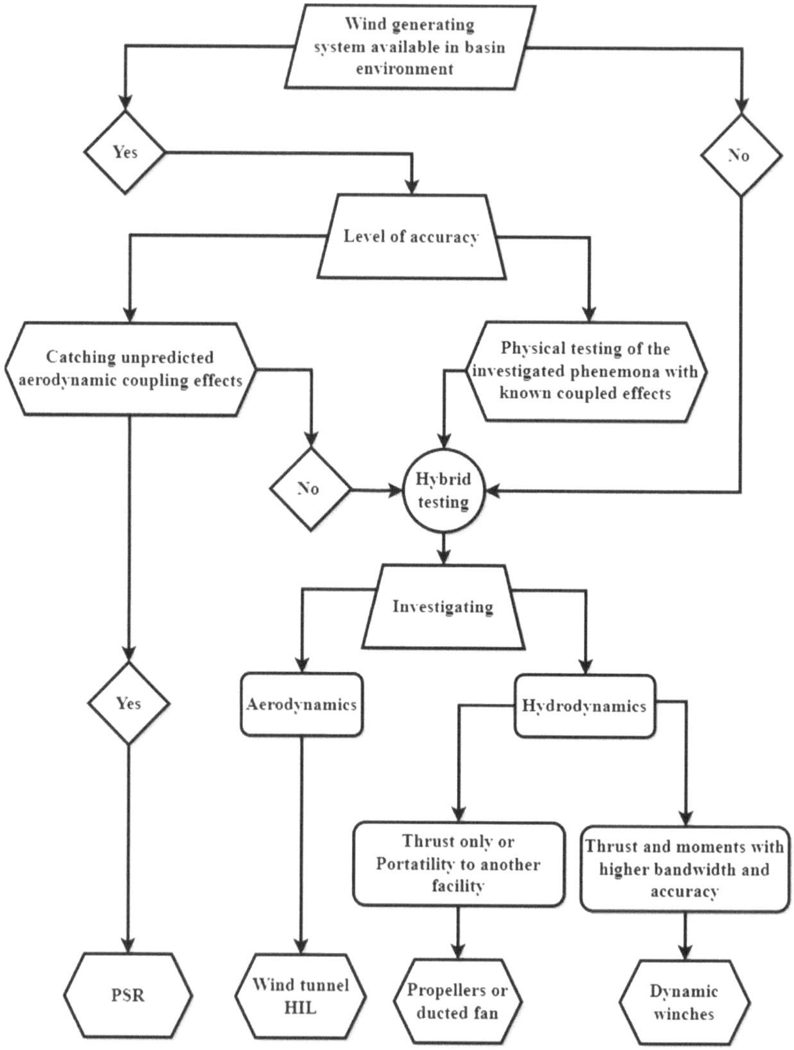

Figure 9. A recommended roadmap for selecting the testing approach.

8. Conclusions

This paper presents a comprehensive review of the experimental methodologies employed in studying FWTs. It begins by summarising the Froude and Reynolds similitude principles and the importance of various aerodynamic phenomena to underscore the associated experimental challenges. Subsequently, the methods are presented, spanning from early-stage techniques like the porous disc method to more contemporary approaches (i.e., the performance scaled rotor (PSR) method and the hybrid approach). Each method is described, providing insights into its specific characteristics, and is evaluated based on criteria such as accuracy, cost, and versatility.

The main emphasis of this review revolves around the assessment of the PSR method and the hybrid approach. The PSR method entails using a redesigned rotor that maintains identical performance characteristics in a lower Reynolds environment. This approach replicates the aerodynamic behaviour of the turbine, including unpredicted coupling effects. However, the introduction of low-Reynolds modifications presents challenges when attempting to model the power coefficient and torque characteristics of the turbine. Consequently, the scalability of blade pitch control becomes constrained in the PSR approach.

The hybrid approach combines physical experimentation and numerical simulation conducted in wave basins or wind tunnels, with actuators serving as the link between the two in real-time. The hybrid approach proves advantages by simplifying experimental tests in wave basins where a wind generation system is unavailable. Previous hybrid testing campaigns are assessed, specifically focusing on commonly utilized actuators such as cable winches and propellers. The actuation systems are evaluated, emphasizing the simplicity of propellers' calibration compared to cable systems, as well as the cable actuators' capability to reproduce the moments accurately at higher bandwidth. The selection of actuators is discussed, marking its dependence on the investigated phenomena and the targeted fidelity.

It is determined that hybrid testing is versatile, allowing for the evaluation of multiple turbine configurations at different scales. In addition, it provides a reliable representation of turbine aerodynamics and coupled phenomena. On the other hand, the PSR method is recognized for including unpredicted coupling effects while exhibiting lower versatility due to the requirement for costly modifications of scaled rotors with each design variation or scaling.

Based on the presented discussion, a roadmap is outlined for selecting the appropriate methodology, considering the investigated phenomena, level of accuracy, available equipment, and budget constraints. PSR is recommended when a wind-generating system over a wave basin is available, and including unpredicted coupling effects is of concern. Hybrid testing is advised when physically testing the model, including known coupling effects, while giving control on one aspect of the study (aero or hydro). For larger turbines, the hybrid approach is a more promising option in the foreseeable future.

Author Contributions: Conceptualization, M.H., E.U., C.Z., J.F.G. and C.G.S.; methodology, M.H., E.U., C.Z., J.F.G. and C.G.S.; software, M.H., E.U., C.Z., J.F.G. and C.G.S.; validation, M.H., E.U., C.Z., J.F.G. and C.G.S.; formal analysis, M.H., E.U., C.Z., J.F.G. and C.G.S.; investigation, M.H., E.U., C.Z., J.F.G. and C.G.S.; resources, M.H., E.U., C.Z., J.F.G. and C.G.S.; data curation, M.H., E.U., C.Z., J.F.G. and C.G.S.; writing—original draft preparation, M.H.; writing—review and editing, M.H., E.U., C.Z., J.F.G. and C.G.S.; visualization, M.H., E.U., C.Z., J.F.G. and C.G.S.; supervision, E.U. and C.G.S.; project administration, C.G.S.; funding acquisition, C.G.S. All authors have read and agreed to the published version of the manuscript.

Funding: M.H., C.Z. and J.F.G. received funding from the project Variable Geometry Wave Energy Conversion System for Floating Platforms (WEC4MUP), which is funded by the Portuguese Foundation for Science and Technology (Fundação para a Ciência e a Tecnologia—FCT) under contract WEC4MUP_PTDC/EME-REN/0242/2020. E.U. received funding from the project ARCWIND—Adaptation and Implementation of Floating Wind Energy Conversion Technology for the Atlantic Region—which is co-financed by the European Regional Development Fund through the Interreg Atlantic Area Programme under contract EAPA 344/2016. This work contributed to the Strategic Research Plan of the Centre for Marine Technology and Ocean Engineering (CENTEC), which is financed by the Portuguese Foundation for Science and Technology (Fundação para a Ciência e Tecnologia—FCT) under contract UIDB/UIDP/00134/2020.

Institutional Review Board Statement: Not applicable.

Informed Consent Statement: Not applicable.

Data Availability Statement: Not applicable.

Conflicts of Interest: The authors declare no conflict of interest.

References

1. Díaz, H.; Guedes Soares, C. Review of the Current Status, Technology and Future Trends of Offshore Wind Farms. *Ocean. Eng.* **2020**, *209*, 107381. [CrossRef]
2. Díaz, H.; Serna, J.; Nieto, J.; Guedes Soares, C. Market Needs, Opportunities and Barriers for the Floating Wind Industry. *J. Mar. Sci. Eng.* **2022**, *10*, 934. [CrossRef]
3. Wind Europe. *Wind Energy in Europe—Statistics and the Outlook for 2021–2025*; WindEurope: Brussels, Belgium, 2021; p. 36.
4. Bagbanci, H.; Karmakar, D.; Guedes Soares, C. Review of Offshore Floating Wind Turbines Concepts. In *Maritime Engineering and Technology*; Guedes Soares, C., Garbatov, Y., Sutulo, S., Santos, T.A., Eds.; Taylor & Francis Group: London, UK, 2012; pp. 553–562. [CrossRef]
5. Uzunoglu, E.; Guedes Soares, C. Hydrodynamic Design of a Free-Float Capable Tension Leg Platform for a 10 MW Wind Turbine. *Ocean. Eng.* **2020**, *197*, 106888. [CrossRef]
6. Damiani, R.; Dykes, K.; Scott, G. A Comparison Study of Offshore Wind Support Structures with Monopiles and Jackets for U.S. Waters. *J. Phys. Conf. Ser.* **2016**, *753*, 092003. [CrossRef]
7. Castro-Santos, L.; Silva, D.; Bento, A.R.; Salvação, N.; Guedes Soares, C. Economic Feasibility of Floating Offshore Wind Farms in Portugal. *Ocean. Eng.* **2020**, *207*, 107393. [CrossRef]
8. Castro-Santos, L.; Bento, A.R.; Silva, D.; Salvação, N.; Guedes Soares, C. Economic Feasibility of Floating Offshore Wind Farms in the North of Spain. *J. Mar. Sci. Eng.* **2020**, *8*, 58. [CrossRef]
9. Uzunoglu, E.; Karmakar, D.; Guedes Soares, C. Floating Offshore Wind Platforms. In *Floating Offshore Wind Farms*; Castro-Santos, L., Diaz-Casas, V., Eds.; Green Energy and Technology; Springer International Publishing: Cham, Switzerland, 2016; pp. 53–76. [CrossRef]
10. Collu, M.; Borg, M. Design of Floating Offshore Wind Turbines. In *Offshore Wind Farms*; Elsevier: Amsterdam, The Netherlands, 2016; pp. 359–385. ISBN 978-0-08-100779-2.
11. Skaare, B.; Nielsen, F.G.; Hanson, T.D.; Yttervik, R.; Havmøller, O.; Rekdal, A. Analysis of Measurements and Simulations from the Hywind Demo Floating Wind Turbine. *Wind Energy* **2015**, *18*, 1105–1122. [CrossRef]
12. Roddier, D.; Cermelli, C.; Aubault, A.; Weinstein, A. WindFloat: A Floating Foundation for Offshore Wind Turbines. *J. Renew. Sustain. Energy* **2010**, *2*, 033104. [CrossRef]
13. Reynaud, M.; Le Bouhris, E.; Soulard, T.; Yves, P. *Rapport de Suivi Environnemental de L'éolienne Flottante FLOATGEN, Site D'essais SEM-REV*; Zenodo: Genève, Switzerland, 2021. [CrossRef]
14. Galván, J.; Sánchez-Lara, M.J.; Mendikoa, I.; Pérez-Morán, G.; Nava, V.; Rodríguez-Arias, R. *NAUTILUS-DTU10* MW Floating Offshore Wind Turbine at Gulf of Maine: Public Numerical Models of an Actively Ballasted Semisubmersible. *J. Phys. Conf. Ser.* **2018**, *1102*, 012015. [CrossRef]
15. Pegalajar-Jurado, A.; Bredmose, H.; Borg, M.; Straume, J.G.; Landbø, T.; Andersen, H.S.; Yu, W.; Müller, K.; Lemmer, F. State-of-the-Art Model for the LIFES50+ OO-Star Wind Floater Semi 10 MW Floating Wind Turbine. *J. Phys. Conf. Ser.* **2018**, *1104*, 012024. [CrossRef]
16. Borisade, F.; Koch, C.; Lemmer, F.; Cheng, P.W.; Campagnolo, F.; Matha, D. Validation of INNWIND.EU Scaled Model Tests of a Semisubmersible Floating Wind Turbine. *Int. J. Offshore Polar Eng.* **2018**, *28*, 54–64. [CrossRef]
17. Hmedi, M.; Uzunoglu, E.; Medina-Manuel, A.; Mas-Soler, J.; Vittori, F.; Pires, O.; Azcona, J.; Souto-Iglesias, A.; Guedes Soares, C. Experimental Analysis of CENTEC-TLP Self-Stable Platform with a 10 MW Turbine. *J. Mar. Sci. Eng.* **2022**, *10*, 1910. [CrossRef]
18. Sanchez, R. *Technology Readiness Assessment Guide*; United States Department of Energy: Washington, DC, USA, 2011; p. 73.
19. ITTC. Recommended Procedures and Guidelines: Model Tests for Offshore Wind Turbines; Recommended Procedures and Guidelines 7.5-02-07-03.8. In Proceedings of the International Towing Tank Conference, Zürich, Switzerland, 13–18 June 2021; p. 19.
20. Uzunoglu, E.; Guedes Soares, C. Parametric Modelling of Marine Structures for Hydrodynamic Calculations. *Ocean. Eng.* **2018**, *160*, 181–196. [CrossRef]
21. Uzunoglu, E.; Guedes Soares, C. A System for the Hydrodynamic Design of Tension Leg Platforms of Floating Wind Turbines. *Ocean. Eng.* **2019**, *171*, 78–92. [CrossRef]
22. Jonkman, J. The New Modularization Framework for the FAST Wind Turbine CAE Tool. In Proceedings of the 51st AIAA Aerospace Sciences Meeting Including the New Horizons Forum and Aerospace Exposition, Grapevine (Dallas/Ft. Worth Region), TX, USA, 7–10 January 2013; American Institute of Aeronautics and Astronautics: Grapevine (Dallas/Ft. Worth Region), TX, USA, 2013. [CrossRef]
23. Perez-Becker, S.; Saverin, J.; Behrens de Luna, R.; Papi, F.; Combreau, C.; Ducasse, M.-L.; Marten, D.; Bianchini, A. *Deliverable 2.2—Validation Report of QBlade-Ocean*; Zenodo: Genève, Switzerland, 2022. [CrossRef]
24. Chen, P.; Chen, J.; Hu, Z. Review of Experimental-Numerical Methodologies and Challenges for Floating Offshore Wind Turbines. *J. Mar. Sci. Appl.* **2020**, *19*, 339–361. [CrossRef]
25. Sarpkaya, T. *Wave Forces on Offshore Structures*; Cambridge University Press: Cambridge, UK; New York, NY, USA, 2010; ISBN 978-0-521-89625-2.
26. Chakrabarti, S.K. Physical Modelling of Offshore Structures. In *Handbook of Offshore Engineering*; Elsevier: Amsterdam, The Netherlands, 2005; pp. 1001–1054. ISBN 978-0-08-044381-2.
27. Faltinsen, O.M. *Sea Loads on Ships and Offshore Structures*; Cambridge ocean technology series; Cambridge University Press: Cambridge, UK, 1999; ISBN 978-0-521-45870-2.

28. Müller, K.; Sandner, F.; Bredmose, H.; Azcona, J.; Manjock, A.; Pereira, R. Improved Tank Test Procedures for Scaled Floating Offshore Wind Turbines. In Proceedings of the International Wind Engineering Conference, IWEC 2014, Hannover, Germany, 3–5 September 2014. [CrossRef]
29. Pao, L.Y.; Johnson, K.E. Control of Wind Turbines. *IEEE Control Syst.* **2011**, *31*, 44–62. [CrossRef]
30. Jonkman, J. Influence of Control on the Pitch Damping of a Floating Wind Turbine. In Proceedings of the 46th AIAA Aerospace Sciences Meeting and Exhibit, Reno, NV, USA, 7–10 January 2008; American Institute of Aeronautics and Astronautics: Reno, NV, USA, 2008. [CrossRef]
31. Nielsen, F.G.; Hanson, T.D.; Skaare, B. Integrated Dynamic Analysis of Floating Offshore Wind Turbines. In Proceedings of the 25th International Conference on Offshore Mechanics and Arctic Engineering, Hamburg, Germany, 4–9 June 2006; ASMEDC: Hamburg, Germany, 2006; pp. 671–679. [CrossRef]
32. Larsen, T.J.; Hanson, T.D. A Method to Avoid Negative Damped Low Frequent Tower Vibrations for a Floating, Pitch Controlled Wind Turbine. *J. Phys. Conf. Ser.* **2007**, *75*, 012073. [CrossRef]
33. Skaare, B.; Hanson, T.D.; Nielsen, F.G.; Yttervik, R.; Hansen, A.M.; Thomsen, K.; Larsen, T.J. Integrated Dynamic Analysis of Floating Offshore Wind Turbines. In Proceedings of the European Wind Energy Conference and Exhibition, Milan, Italy, 7–10 May 2007.
34. Namik, H.; Stol, K. A Review of Floating Wind Turbine Controllers. In *Handbook of Wind Power Systems*; Pardalos, P.M., Rebennack, S., Pereira, M.V.F., Iliadis, N.A., Pappu, V., Eds.; Energy Systems; Springer: Berlin/Heidelberg, Germany, 2013; pp. 415–441. ISBN 978-3-642-41080-2.
35. Yu, W.; Lemmer, F.; Bredmose, H.; Borg, M.; Pegalajar-Jurado, A.; Mikkelsen, R.F.; Larsen, T.S.; Fjelstrup, T.; Lomholt, A.K.; Boehm, L.; et al. The Triple Spar Campaign: Implementation and Test of a Blade Pitch Controller on a Scaled Floating Wind Turbine Model. *Energy Procedia* **2017**, *137*, 323–338. [CrossRef]
36. Zhu, H.; Sueyoshi, M. Experimental Study and Attitude Control of a Floating Type Shrouded Wind Turbine. *Sustain. Energy Technol. Assess.* **2022**, *50*, 101811. [CrossRef]
37. Hu, R.; Le, C.; Gao, Z.; Ding, H.; Zhang, P. Implementation and Evaluation of Control Strategies Based on an Open Controller for a 10 MW Floating Wind Turbine. *Renew. Energy* **2021**, *179*, 1751–1766. [CrossRef]
38. Utsunomiya, T.; Sato, T.; Matsukuma, H.; Yago, K. Experimental Validation for Motion of a SPAR-Type Floating Offshore Wind Turbine Using 1/22.5 Scale Model. In Proceedings of the ASME 2009 28th International Conference on Ocean, Offshore and Arctic Engineering, Honolulu, HI, USA, 31 May–5 June 2009; ASMEDC: Honolulu, HI, USA, 2009; pp. 951–959. [CrossRef]
39. Cermelli, C.; Roddier, D.; Aubault, A. WindFloat: A Floating Foundation for Offshore Wind Turbines—Part II: Hydrodynamics Analysis. In Proceedings of the ASME 2009 28th International Conference on Ocean, Offshore and Arctic Engineering, Honolulu, HI, USA, 31 May–5 June 2009; ASMEDC: Honolulu, HI, USA, 2009; pp. 135–143. [CrossRef]
40. Cermelli, C.; Aubault, A.; Roddier, D.; McCoy, T. Qualification of a Semi-Submersible Floating Foundation for Multi-Megawatt Wind Turbines. In Proceedings of the Offshore Technology Conference, Houston, TX, USA, 3–6 May 2010; OTC-20674-MS. [CrossRef]
41. Wan, L.; Gao, Z.; Moan, T. Experimental and Numerical Study of Hydrodynamic Responses of a Combined Wind and Wave Energy Converter Concept in Survival Modes. *Coast. Eng.* **2015**, *104*, 151–169. [CrossRef]
42. Robertson, A.N.; Jonkman, J.M.; Goupee, A.J.; Coulling, A.J.; Prowell, I.; Browning, J.; Masciola, M.D.; Molta, P. Summary of Conclusions and Recommendations Drawn from the DeepCwind Scaled Floating Offshore Wind System Test Campaign. In Proceedings of the ASME 2013 32nd International Conference on Ocean, Offshore and Arctic Engineering, Nantes, France, 9–14 June 2013; American Society of Mechanical Engineers: Nantes, France, 2013; V008T09A053. [CrossRef]
43. Michailides, C.; Gao, Z.; Moan, T. Experimental and Numerical Study of the Response of the Offshore Combined Wind/Wave Energy Concept SFC in Extreme Environmental Conditions. *Mar. Struct.* **2016**, *50*, 35–54. [CrossRef]
44. Goupee, A.J.; Koo, B.J.; Kimball, R.W.; Lambrakos, K.F.; Dagher, H.J. Experimental Comparison of Three Floating Wind Turbine Concepts. *J. Offshore Mech. Arct. Eng.* **2014**, *136*, 020906. [CrossRef]
45. Pegalajar-Jurado, A.; Hansen, A.M.; Laugesen, R.; Mikkelsen, R.F.; Borg, M.; Kim, T.; Heilskov, N.F.; Bredmose, H. Experimental and Numerical Study of a 10MW TLP Wind Turbine in Waves and Wind. *J. Phys. Conf. Ser.* **2016**, *753*, 092007. [CrossRef]
46. de Ridder, E.-J.; Otto, W.; Zondervan, G.-J.; Huijs, F.; Vaz, G. Development of a Scaled-Down Floating Wind Turbine for Offshore Basin Testing. In Proceedings of the ASME 2014 33rd International Conference on Ocean, Offshore and Arctic Engineering, San Francisco, CA, USA, 8–13 June 2014; American Society of Mechanical Engineers: San Francisco, CA, USA, 2014; V09AT09A027. [CrossRef]
47. Duan, F.; Hu, Z.; Niedzwecki, J.M. Model Test Investigation of a Spar Floating Wind Turbine. *Mar. Struct.* **2016**, *49*, 76–96. [CrossRef]
48. Du, W.; Zhao, Y.; He, Y.; Liu, Y. Design, Analysis and Test of a Model Turbine Blade for a Wave Basin Test of Floating Wind Turbines. *Renew. Energy* **2016**, *97*, 414–421. [CrossRef]
49. Duan, F.; Hu, Z.; Liu, G.; Wang, J. Experimental Comparisons of Dynamic Properties of Floating Wind Turbine Systems Based on Two Different Rotor Concepts. *Appl. Ocean. Res.* **2016**, *58*, 266–280. [CrossRef]

50. Fowler, M.J.; Kimball, R.W.; Thomas, D.A.; Goupee, A.J. Design and Testing of Scale Model Wind Turbines for Use in Wind/Wave Basin Model Tests of Floating Offshore Wind Turbines. In Proceedings of the ASME 2013 32nd International Conference on Ocean, Offshore and Arctic Engineering, Nantes, France, 9–14 June 2013; American Society of Mechanical Engineers: Nantes, France, 2013; V008T09A00. [CrossRef]
51. Wen, B.; Tian, X.; Dong, X.; Li, Z.; Peng, Z.; Zhang, W.; Wei, K. Design Approaches of Performance-Scaled Rotor for Wave Basin Model Tests of Floating Wind Turbines. *Renew. Energy* **2020**, *148*, 573–584. [CrossRef]
52. Gueydon, S.; Lindeboom, R.; Van Kampen, W.; De Ridder, E.-J. Comparison of Two Wind Turbine Loading Emulation Techniques Based on Tests of a TLP-FOWT in Combined Wind, Waves and Current. In Proceedings of the ASME 2018 1st International Offshore Wind Technical Conference, San Francisco, CA, USA, 4–7 November 2018; American Society of Mechanical Engineers: San Francisco, CA, USA, 2018; V001T01A012. [CrossRef]
53. Arnal, V.; Bonnefoy, F.; Gilloteaux, J.-C.; Aubrun, S. Hybrid Model Testing of Floating Wind Turbines: Test Bench for System Identification and Performance Assessment. In Proceedings of the ASME 2019 38th International Conference on Ocean, Offshore and Arctic Engineering, Glasgow, UK, 9–14 June 2019; American Society of Mechanical Engineers: Glasgow, UK, 2019. [CrossRef]
54. Azcona, J.; Bouchotrouch, F.; González, M.; Garciandía, J.; Munduate, X.; Kelberlau, F.; Nygaard, T.A. Aerodynamic Thrust Modelling in Wave Tank Tests of Offshore Floating Wind Turbines Using a Ducted Fan. *J. Phys. Conf. Ser.* **2014**, *524*, 012089. [CrossRef]
55. Sauder, T.; Chabaud, V.; Thys, M.; Bachynski, E.E.; Sæther, L.O. Real-Time Hybrid Model Testing of a Braceless Semi-Submersible Wind Turbine: Part I—The Hybrid Approach. In Proceedings of the ASME 2016 35th International Conference on Ocean, Offshore and Arctic Engineering, Busan, Republic of Korea, 19–24 June 2016; American Society of Mechanical Engineers: Busan, Republic of Korea, 2016; V006T09A039. [CrossRef]
56. Bachynski, E.E.; Thys, M.; Sauder, T.; Chabaud, V.; Sæther, L.O. Real-Time Hybrid Model Testing of a Braceless Semi-Submersible Wind Turbine: Part II—Experimental Results. In Proceedings of the ASME 2016 35th International Conference on Ocean, Offshore and Arctic Engineering, Busan, Republic of Korea, 19–24 June 2016; American Society of Mechanical Engineers: Busan, Republic of Korea, 2016; V006T09A040. [CrossRef]
57. Berthelsen, P.A.; Bachynski, E.E.; Karimirad, M.; Thys, M. Real-Time Hybrid Model Tests of a Braceless Semi-Submersible Wind Turbine: Part III—Calibration of a Numerical Model. In Proceedings of the ASME 2016 35th International Conference on Ocean, Offshore and Arctic Engineering, Busan, Republic of Korea, 19–24 June 2016; American Society of Mechanical Engineers: Busan, Republic of Korea, 2016; V006T09A047. [CrossRef]
58. Chabaud, V.; Eliassen, L.; Thys, M.; Sauder, T. Multiple-Degree-of-Freedom Actuation of Rotor Loads in Model Testing of Floating Wind Turbines Using Cable-Driven Parallel Robots. *J. Phys. Conf. Ser.* **2018**, *1104*, 012021. [CrossRef]
59. Otter, A.; Murphy, J.; Desmond, C.J. Emulating Aerodynamic Forces and Moments for Hybrid Testing of Floating Wind Turbine Models. *J. Phys. Conf. Ser.* **2020**, *1618*, 032022. [CrossRef]
60. Urbán, A.M.; Guanche, R. Wind Turbine Aerodynamics Scale-Modeling for Floating Offshore Wind Platform Testing. *J. Wind. Eng. Ind. Aerodyn.* **2019**, *186*, 49–57. [CrossRef]
61. Vittori, F.; Bouchotrouch, F.; Lemmer, F.; Azcona, J. Hybrid Scaled Testing of a 5 MW Floating Wind Turbine Using the SiL Method Compared with Numerical Models. In Proceedings of the ASME 2018 37th International Conference on Ocean, Offshore and Arctic Engineering, Madrid, Spain, 17–22 June 2018; American Society of Mechanical Engineers: Madrid, Spain, 2018; V010T09A082. [CrossRef]
62. Vittori, F.; Pires, O.; Azcona, J.; Uzunoglu, E.; Guedes Soares, C.; Zamora Rodríguez, R.; Souto-Iglesias, A. Hybrid Scaled Testing of a 10 MW TLP Floating Wind Turbine Using the SiL Method to Integrate the Rotor Thrust and Moments. In *Developments in Renewable Energies Offshore*; Guedes Soares, C., Ed.; Taylor & Francis Group: London, UK, 2020; pp. 417–423. [CrossRef]
63. Bayati, I.; Belloli, M.; Facchinetti, A.; Giappino, S. Wind Tunnel Tests on Floating Offshore Wind Turbines: A Proposal for Hardware-in-the-Loop Approach to Validate Numerical Codes. *Wind. Eng.* **2013**, *37*, 557–568. [CrossRef]
64. Bayati, I.; Belloli, M.; Ferrari, D.; Fossati, F.; Giberti, H. Design of a 6-DoF Robotic Platform for Wind Tunnel Tests of Floating Wind Turbines. *Energy Procedia* **2014**, *53*, 313–323. [CrossRef]
65. Bayati, I.; Belloli, M.; Bernini, L.; Fiore, E.; Giberti, H.; Zasso, A. On the Functional Design of the DTU10 MW Wind Turbine Scale Model of LIFES50+ Project. *J. Phys. Conf. Ser.* **2016**, *753*, 052018. [CrossRef]
66. Bredmose, H.; Larsen, S.; Matha, D.; Rettenmeier, A.; Marino, E.; Saettran, L. *Collation of Offshore Wind Wave Dynamics: Marine Renewables Infrastructure Network for Emerging Energy Technologies D2.4*; DTU: Lyngby, Denmark, 2012.
67. Hall, M.; Moreno, J.; Thiagarajan, K. Performance Specifications for Real-Time Hybrid Testing of 1:50-Scale Floating Wind Turbine Models. In Proceedings of the ASME 2014 33rd International Conference on Ocean, Offshore and Arctic Engineering, San Francisco, CA, USA, 8–13 June 2014; American Society of Mechanical Engineers: San Francisco, CA, USA, 2014; V09BT09A047. [CrossRef]
68. Coulling, A.J.; Goupee, A.J.; Robertson, A.N.; Jonkman, J.M.; Dagher, H.J. Validation of a FAST Semi-Submersible Floating Wind Turbine Numerical Model with DeepCwind Test Data. *J. Renew. Sustain. Energy* **2013**, *5*, 023116. [CrossRef]
69. Blusseau, P.; Patel, M.H. Gyroscopic Effects on a Large Vertical Axis Wind Turbine Mounted on a Floating Structure. *Renew. Energy* **2012**, *46*, 31–42. [CrossRef]
70. Bachynski, E.E.; Chabaud, V.; Sauder, T. Real-Time Hybrid Model Testing of Floating Wind Turbines: Sensitivity to Limited Actuation. *Energy Procedia* **2015**, *80*, 2–12. [CrossRef]

71. Goupee, A.J.; Kimball, R.W.; Dagher, H.J. Experimental Observations of Active Blade Pitch and Generator Control Influence on Floating Wind Turbine Response. *Renew. Energy* **2017**, *104*, 9–19. [CrossRef]
72. Namik, H.; Stol, K. Individual Blade Pitch Control of Floating Offshore Wind Turbines. *Wind Energy* **2010**, *13*, 74–85. [CrossRef]
73. Manwell, J.F.; McGowan, J.G.; Rogers, A.L. *Wind Energy Explained: Theory, Design and Application*, 2nd ed.; Wiley: Chichester, UK, 2009; ISBN 978-0-470-01500-1.
74. Bahramiasl, S.; Abbaspour, M.; Karimirad, M. Experimental Study on Gyroscopic Effect of Rotating Rotor and Wind Heading Angle on Floating Wind Turbine Responses. *Int. J. Environ. Sci. Technol.* **2018**, *15*, 2531–2544. [CrossRef]
75. Høeg, C.E.; Zhang, Z. The Influence of Gyroscopic Effects on Dynamic Responses of Floating Offshore Wind Turbines in Idling and Operational Conditions. *Ocean. Eng.* **2021**, *227*, 108712. [CrossRef]
76. Arnal, V. Experimental Modelling of a Floating Wind Turbine Using a "Software-in-the-Loop" Approach. Ph.D. Thesis, Ecole Central Nantes, Nantes, France, 2020.
77. Tomasicchio, G.R.; D'Alessandro, F.; Avossa, A.M.; Riefolo, L.; Musci, E.; Ricciardelli, F.; Vicinanza, D. Experimental Modelling of the Dynamic Behaviour of a Spar Buoy Wind Turbine. *Renew. Energy* **2018**, *127*, 412–432. [CrossRef]
78. Wan, L.; Gao, Z.; Moan, T.; Lugni, C. Experimental and Numerical Comparisons of Hydrodynamic Responses for a Combined Wind and Wave Energy Converter Concept under Operational Conditions. *Renew. Energy* **2016**, *93*, 87–100. [CrossRef]
79. Wan, L.; Greco, M.; Lugni, C.; Gao, Z.; Moan, T. A Combined Wind and Wave Energy-Converter Concept in Survival Mode: Numerical and Experimental Study in Regular Waves with a Focus on Water Entry and Exit. *Appl. Ocean. Res.* **2017**, *63*, 200–216. [CrossRef]
80. Mortensen, S.M.; Laugesen, K.; Jensen, J.K.; Jessen, K.; Soltani, M. Experimental Verification of the Hydro-Elastic Model of a Scaled Floating Offshore Wind Turbine. In Proceedings of the 2018 IEEE Conference on Control Technology and Applications (CCTA), Copenhagen, Denmark, 21–24 August 2018; IEEE: Copenhagen, Denmark, 2018; pp. 1623–1630. [CrossRef]
81. Ahn, H.-J.; Shin, H. Model Test and Numerical Simulation of OC3 Spar Type Floating Offshore Wind Turbine. *Int. J. Nav. Archit. Ocean. Eng.* **2019**, *11*, 1–10. [CrossRef]
82. Wen, B.; Jiang, Z.; Li, Z.; Peng, Z.; Dong, X.; Tian, X. On the Aerodynamic Loading Effect of a Model Spar-Type Floating Wind Turbine: An Experimental Study. *Renew. Energy* **2022**, *184*, 306–319. [CrossRef]
83. Martin, H.R.; Kimball, R.W.; Viselli, A.M.; Goupee, A.J. Methodology for Wind/Wave Basin Testing of Floating Offshore Wind Turbines. *J. Offshore Mech. Arct. Eng.* **2014**, *136*, 020905. [CrossRef]
84. Tracy, C.H. Parametric Design of Floating Wind Turbines. Ph.D. Thesis, Massachusetts Institute of Technology, Cambridge, MA, USA, 2007.
85. Jonkman, J. *Definition of the Floating System for Phase IV of OC3*; NREL: Golden, CO, USA, 2010; NREL/TP-500-47535, 979456.
86. Robertson, A.; Jonkman, J.; Masciola, M.; Song, H.; Goupee, A.; Coulling, A.; Luan, C. *Definition of the Semisubmersible Floating System for Phase II of OC4*; NREL: Golden, CO, USA, 2014; NREL/TP-5000-60601, 1155123.
87. Jonkman, J.; Prowell, I.; Robertson, A.; Goupee, A.J.; Stewart, G.M. Numerical Prediction of Experimentally Observed Behavior of a Scale-Model of an Offshore Wind Turbine Supported by a Tension-Leg Platform. In Proceedings of the Offshore Technology Conference, Houston, TX, USA, 6–9 May 2013; OTC: Houston, TX, USA, 2013; OTC-24233-MS. [CrossRef]
88. Browning, J.R.; Jonkman, J.; Robertson, A.; Goupee, A.J. Calibration and Validation of a Spar-Type Floating Offshore Wind Turbine Model Using the FAST Dynamic Simulation Tool. *J. Phys. Conf. Ser.* **2014**, *555*, 012015. [CrossRef]
89. Li, L.; Gao, Y.; Hu, Z.; Yuan, Z.; Day, S.; Li, H. Model Test Research of a Semisubmersible Floating Wind Turbine with an Improved Deficient Thrust Force Correction Approach. *Renew. Energy* **2018**, *119*, 95–105. [CrossRef]
90. Lemmer, F.; Amann, F.; Matha, D.; Armendaríz, J.; Munduate, X.; Bottasso, C.; Campagnolo, F.; Montinari, P.; Manjock, A.; Pereira, R.; et al. Model Building and Scaled Testing of 5 MW and 10 MW Semi-Submersible Floating Wind Turbines. In Proceedings of the 12th Deep Sea Offshore Wind R&D Conference, EERA Deep Wind'2015, Trondheim, Norway, 4–6 February 2015.
91. Helder, J.; Pietersma, M. *UMaine-DeepCwind/OC4 Semi Floating Wind Turbine Repeat Tests*; Maritime Research Institute Netherlands: Wageningen, The Netherlands, 2013.
92. Zhao, Y.; She, X.; He, Y.; Yang, J.; Peng, T.; Kou, Y. Experimental Study on New Multi-Column Tension-Leg-Type Floating Wind Turbine. *China Ocean Eng.* **2018**, *32*, 123–131. [CrossRef]
93. Bak, C.; Zahle, F.; Bitsche, R.; Kim, T.; Yde, A.; Henriksen, L.C.; Hansen, M.H.; Blasques, J.P.A.A.; Gaunaa, M.; Natarajan, A. *The DTU 10-MW Reference Wind Turbine*; DTU: Lyngby, Denmark, 2013.
94. Bredmose, H.; Lemmer, F.; Borg, M.; Pegalajar-Jurado, A.; Mikkelsen, R.F.; Larsen, T.S.; Fjelstrup, T.; Yu, W.; Lomholt, A.K.; Boehm, L.; et al. The Triple Spar Campaign: Model Tests of a 10 MW Floating Wind Turbine with Waves, Wind and Pitch Control. *Energy Procedia* **2017**, *137*, 58–76. [CrossRef]
95. Bredmose, H.; Robert, M.; Anders Mandrup, H.; Robert, L.; Nicolai, H.; Bjarne, J.; Jens, K. *Experimental Study of the DTU 10 MW Wind Turbine on a TLP Floater in Waves and Wind*; EWEA: Brussels, Belgium, 2015.
96. Azcona, J.; Lemmer, F.; Matha, D.; Amann, F.; Bottasso, C.L.; Montinari, P.; Chassapoyannis, P.; Diakakis, K.; Voutsinas, S.; Pereira, R.; et al. *D4.2.4: Results of Wave Tank Tests*; Deliverable reports on INNWIND.EU; DTU: Delhi, India, 2016; p. 114.
97. Ahn, H.; Shin, H. Experimental and Numerical Analysis of a 10 MW Floating Offshore Wind Turbine in Regular Waves. *Energies* **2020**, *13*, 2608. [CrossRef]

98. Madsen, F.J.; Nielsen, T.R.L.; Kim, T.; Bredmose, H.; Pegalajar-Jurado, A.; Mikkelsen, R.F.; Lomholt, A.K.; Borg, M.; Mirzaei, M.; Shin, P. Experimental Analysis of the Scaled DTU10MW TLP Floating Wind Turbine with Different Control Strategies. *Renew. Energy* **2020**, *155*, 330–346. [CrossRef]
99. Connolly, A.; Guyot, M.; Le Boulluec, M.; Héry, L.; O'Connor, A. Fully Coupled Aero-Hydro-Structural Simulation of New Floating Wind Turbine Concept. In Proceedings of the ASME 2018 1st International Offshore Wind Technical Conference, San Francisco, CA, USA, 4–7 November 2018; American Society of Mechanical Engineers: San Francisco, CA, USA, 2018; p. V001T01A027. [CrossRef]
100. Shroff, S.; Makris, A.; Roukis, G.; Croce, A.; Karakalas, A.; Machairas, T.; Saravanos, D.; Chrysochoidis, N.; Theodosiou, T.; Berring, P.; et al. *D2.24: Manufactured and Laboratory Tested Scaled Blades and Parts of the Blade*; Deliverable reports on INNWIND.EU; DTU: Delhi, India, 2012; p. 150.
101. Otter, A.; Murphy, J.; Pakrashi, V.; Robertson, A.; Desmond, C. A Review of Modelling Techniques for Floating Offshore Wind Turbines. *Wind Energy* **2021**, *25*, 831–857. [CrossRef]
102. Carrion, J.E.; Spencer, B.F., Jr. *Model-Based Strategies for Real-Time Hybrid Testing*; Newmark Structural Engineering Laboratory Report Series 006; Newmark Structural Engineering Laboratory, University of Illinois at Urbana-Champaign: Champaign, IL, USA, 2007; ISSN 1940-9826.
103. Chabaud, V.; Steen, S.; Skjetne, R. Real-Time Hybrid Testing for Marine Structures: Challenges and Strategies. In Proceedings of the ASME 2013 32nd International Conference on Ocean, Offshore and Arctic Engineering, Nantes, France, 9–14 June 2013; American Society of Mechanical Engineers: Nantes, France, 2013. [CrossRef]
104. Stewart, G.; Muskulus, M. Aerodynamic Simulation of the MARINTEK Braceless Semisubmersible Wave Tank Tests. *J. Phys. Conf. Ser.* **2016**, *749*, 012012. [CrossRef]
105. Xu, S.; Ji, C.; Guedes Soares, C. Experimental Study on Taut and Hybrid Moorings Damping and Their Relation with System Dynamics. *Ocean. Eng.* **2018**, *154*, 322–340. [CrossRef]
106. Ji, C.; Xu, S. Verification of a Hybrid Model Test Method for a Deep Water Floating System with Large Truncation Factor. *Ocean. Eng.* **2014**, *92*, 245–254. [CrossRef]
107. Stansberg, C.T.; Øritsland, O.; Kleiven, G. VERIDEEP: Reliable Methods for Laboratory Verification of Mooring and Stationkeeping in Deep Water. In Proceedings of the Offshore Technology Conference, Houston, TX, USA, 1–4 May 2000; OTC: Houston, TX, USA, 2000; OTC-12087-MS. [CrossRef]
108. Stansberg, C.T.; Yttervik, R.; Oritsland, O.; Kleiven, G. Hydrodynamic Model Test Verification of a Floating Platform System in 3000m Water Depth. In Proceedings of the International Conference on Offshore Mechanics and Arctic Engineering, OMAE2000-OFT4145, New Orleans, LA, USA, 14–17 February 2000; Volume 2, pp. 325–334.
109. Ormberg, H.; Stansberg, C.T.; Yttervik, R.; Kleiven, G. Integrated Vessel Motion and Mooring Analysis Applied in Hybrid Model Testing. In Proceedings of the Ninth International Offshore and Polar Engineering Conference, Brest, France, 30 May–4 June 1999; ISOPE-I-99-051.
110. Jonkman, J.; Buhl, M.L., Jr. *FAST User's Guide*; National Renewable Energy Laboratory (NREL): Golden, CO, USA, 2005; p. 143.
111. Azcona, J.; Bouchotrouch, F.; Vittori, F. Low-frequency Dynamics of a Floating Wind Turbine in Wave Tank–Scaled Experiments with SiL Hybrid Method. *Wind Energy* **2019**, *22*, 1402–1413. [CrossRef]
112. Desmond, C.; Hinrichs, J.-C.; Murphy, J. Uncertainty in the Physical Testing of Floating Wind Energy Platforms' Accuracy versus Precision. *Energies* **2019**, *12*, 435. [CrossRef]
113. Wright, C.; O'Sullivan, K.; Murphy, J.; Pakrashi, V. Experimental Comparison of Dynamic Responses of a Tension Moored Floating Wind Turbine Platform with and without Spring Dampers. *J. Phys. Conf. Ser.* **2015**, *628*, 012056. [CrossRef]
114. Andersen, M.T. Floating Foundations for Offshore Wind Turbines. Ph.D. Thesis, Aalborg Universitetsforlag, Aalborg, Denmark, 2016.
115. Matoug, C.; Augier, B.; Paillard, B.; Maurice, G.; Sicot, C.; Barre, S. An Hybrid Approach for the Comparison of VAWT and HAWT Performances for Floating Offshore Wind Turbines. *J. Phys. Conf. Ser.* **2020**, *1618*, 032026. [CrossRef]
116. Achard, J.-L.; Maurice, G.; Balarac, G.; Barre, S. Floating Vertical Axis Wind Turbine—OWLWIND Project. In Proceedings of the 2017 International Conference on ENERGY and ENVIRONMENT (CIEM), Bucharest, Romania, 19–20 October 2017; IEEE: Bucharest, Romania, 2017; pp. 216–220. [CrossRef]
117. Zamora-Rodriguez, R.; Gomez-Alonso, P.; Amate-Lopez, J.; De-Diego-Martin, V.; Dinoi, P.; Simos, A.N.; Souto-Iglesias, A. Model Scale Analysis of a TLP Floating Offshore Wind Turbine. In Proceedings of the ASME 2014 33rd International Conference on Ocean, Offshore and Arctic Engineering, San Francisco, CA, USA, 8–13 June 2014; American Society of Mechanical Engineers: San Francisco, CA, USA, 2014; V09BT09A016. [CrossRef]
118. Oguz, E.; Clelland, D.; Day, A.H.; Incecik, A.; López, J.A.; Sánchez, G.; Almeria, G.G. Experimental and Numerical Analysis of a TLP Floating Offshore Wind Turbine. *Ocean. Eng.* **2018**, *147*, 591–605. [CrossRef]
119. Leroy, V.; Delacroix, S.; Merrien, A.; Bachynski-Polić, E.E.; Gilloteaux, J.-C. Experimental Investigation of the Hydro-Elastic Response of a Spar-Type Floating Offshore Wind Turbine. *Ocean. Eng.* **2022**, *255*, 111430. [CrossRef]
120. Pires, O.; Azcona, J.; Vittori, F.; Bayati, I.; Gueydon, S.; Fontanella, A.; Liu, Y.; de Ridder, E.J.; Belloli, M.; van Wingerden, J.W. Inclusion of Rotor Moments in Scaled Wave Tank Test of a Floating Wind Turbine Using SiL Hybrid Method. *J. Phys. Conf. Ser.* **2020**, *1618*, 032048. [CrossRef]

121. Hmedi, M.; Uzunoglu, E.; Guedes Soares, C.; Medina-Manuel, A.; Mas-Soler, J.; Abad-Gibert, V.; Souto-Iglesias, A.; Vittori, F.; Pires, O.; Azcona, J. Experimental Analysis of a Free-Float Capable Tension Leg Platform with a 10 MW Turbine. In *Trends in Renewable Energies Offshore*; Guedes Soares, C., Ed.; Taylor & Francis Group: London, UK, 2022; pp. 549–557. [CrossRef]
122. Vittori, F.; Azcona, J.; Eguinoa, I.; Pires, O.; Rodríguez, A.; Morató, Á.; Garrido, C.; Desmond, C. Model Tests of a 10 MW Semi-Submersible Floating Wind Turbine under Waves and Wind Using Hybrid Method to Integrate the Rotor Thrust and Moments. *Wind Energy Sci.* **2022**, *7*, 2149–2161. [CrossRef]
123. Armesto, J.A.; Jurado, A.; Guanche, R.; Couñago, B.; Urbano, J.; Serna, J. TELWIND: Numerical Analysis of a Floating Wind Turbine Supported by a Two Bodies Platform. In Proceedings of the ASME 2018 37th International Conference on Ocean, Offshore and Arctic Engineering, Madrid, Spain, 17–22 June 2018; American Society of Mechanical Engineers: Madrid, Spain, 2018; V010T09A073. [CrossRef]
124. Otter, A.; Flannery, B.; Murphy, J.; Desmond, C. Current Simulation with Software in the Loop for Floating Offshore Wind Turbines. *J. Phys. Conf. Ser.* **2022**, *2265*, 042028. [CrossRef]
125. Kanner, S.; Yeung, R.W.; Koukina, E. Hybrid Testing of Model-Scale Floating Wind Turbines Using Autonomous Actuation and Control. In Proceedings of the OCEANS 2016 MTS/IEEE Monterey, Monterey, CA, USA, 19–23 September 2016; IEEE: Monterey, CA, USA, 2016; pp. 1–6. [CrossRef]
126. Thys, M.; Chabaud, V.; Sauder, T.; Eliassen, L.; Sæther, L.; Magnussen, Ø. Real-Time Hybrid Model Testing of a Semi-Submersible 10 MW Floating Wind Turbine and Advances in the Test Method. In Proceedings of the ASME 2018 1st International Offshore Wind Technical Conference, San Francisco, CA, USA, 4–7 November 2018; American Society of Mechanical Engineers: San Francisco, CA, USA, 2018; V001T01A013. [CrossRef]
127. Antonutti, R.; Poirier, J.-C.; Gueydon, S. Coupled Testing of Floating Wind Turbines in Waves and Wind Using Winches and Software-in-the-Loop. In Proceedings of the Offshore Technology Conference, Houston, TX, USA, 4–7 May 2020; OTC: Houston, TX, USA, 2020; D021S019R001. [CrossRef]
128. Hall, M.; Goupee, A.J. Validation of a Hybrid Modeling Approach to Floating Wind Turbine Basin Testing: Validation of a Hybrid Modeling Approach to Floating Wind Turbine Basin Testing. *Wind Energy* **2018**, *21*, 391–408. [CrossRef]
129. Ambrosini, S.; Bayati, I.; Facchinetti, A.; Belloli, M. Methodological and Technical Aspects of a Two-Degrees-of-Freedom Hardware-In-the-Loop Setup for Wind Tunnel Tests of Floating Systems. *J. Dyn. Syst. Meas. Control* **2020**, *142*, 061002. [CrossRef]
130. Bayati, I.; Belloli, M.; Facchinetti, A. Wind Tunnel 2-DoF Hybrid/HIL Tests on the OC5 Floating Offshore Wind Turbine. In Proceedings of the ASME 2017 36th International Conference on Ocean, Offshore and Arctic Engineering, Trondheim, Norway, 25–30 June 2017; American Society of Mechanical Engineers: Trondheim, Norway, 2017; V010T09A076. [CrossRef]
131. Bayati, I.; Belloli, M.; Bernini, L.; Zasso, A. Wind Tunnel Wake Measurements of Floating Offshore Wind Turbines. *Energy Procedia* **2017**, *137*, 214–222. [CrossRef]
132. Bayati, I.; Facchinetti, A.; Fontanella, A.; Giberti, H.; Belloli, M. A Wind Tunnel/HIL Setup for Integrated Tests of Floating Offshore Wind Turbines. *J. Phys. Conf. Ser.* **2018**, *1037*, 052025. [CrossRef]
133. Belloli, M.; Bayati, I.; Facchinetti, A.; Fontanella, A.; Giberti, H.; La Mura, F.; Taruffi, F.; Zasso, A. A Hybrid Methodology for Wind Tunnel Testing of Floating Offshore Wind Turbines. *Ocean. Eng.* **2020**, *210*, 107592. [CrossRef]
134. Fontanella, A.; Bayati, I.; Mikkelsen, R.; Belloli, M.; Zasso, A. UNAFLOW: A Holistic Wind Tunnel Experiment about the Aerodynamic Response of Floating Wind Turbines under Imposed Surge Motion. *Wind. Energy Sci.* **2021**, *6*, 1169–1190. [CrossRef]
135. Fontanella, A.; Facchinetti, A.; Di Carlo, S.; Belloli, M. Wind Tunnel Investigation of the Aerodynamic Response of Two 15 MW Floating Wind Turbines. *Wind. Energy Sci.* **2022**, *7*, 1711–1729. [CrossRef]
136. Bergua, R.; Robertson, A.; Jonkman, J.; Branlard, E.; Fontanella, A.; Belloli, M.; Schito, P.; Zasso, A.; Persico, G.; Sanvito, A.; et al. OC6 Project Phase III: Validation of the Aerodynamic Loading on a Wind Turbine Rotor Undergoing Large Motion Caused by a Floating Support Structure. *Wind. Energy Sci.* **2023**, *8*, 465–485. [CrossRef]
137. Schliffke, B.; Aubrun, S.; Conan, B. Wind Tunnel Study of a "Floating" Wind Turbine's Wake in an Atmospheric Boundary Layer with Imposed Characteristic Surge Motion. *J. Phys. Conf. Ser.* **2020**, *1618*, 062015. [CrossRef]
138. Rockel, S.; Camp, E.; Schmidt, J.; Peinke, J.; Cal, R.; Hölling, M. Experimental Study on Influence of Pitch Motion on the Wake of a Floating Wind Turbine Model. *Energies* **2014**, *7*, 1954–1985. [CrossRef]
139. Rockel, S.; Peinke, J.; Hölling, M.; Cal, R.B. Wake to Wake Interaction of Floating Wind Turbine Models in Free Pitch Motion: An Eddy Viscosity and Mixing Length Approach. *Renew. Energy* **2016**, *85*, 666–676. [CrossRef]
140. Hmedi, M.; Uzunoglu, E.; Guedes Soares, C. Review of Hybrid Model Testing Approaches for Floating Wind Turbines. In *Trends in Maritime Technology and Engineering*; Guedes Soares, C., Santos, T.A., Eds.; Taylor & Francis Group: London, UK, 2022; Volume 2, pp. 421–428. [CrossRef]

Journal of
Marine Science and Engineering

MDPI

Article

Experimental Analysis of CENTEC-TLP Self-Stable Platform with a 10 MW Turbine

Mohamad Hmedi [1], Emre Uzunoglu [1], Antonio Medina-Manuel [2], Jordi Mas-Soler [2,3], Felipe Vittori [4], Oscar Pires [4], José Azcona [4], Antonio Souto-Iglesias [5] and C. Guedes Soares [1,*]

1 Centre for Marine Technology and Ocean Engineering (CENTEC), Instituto Superior Técnico, Universidade de Lisboa, 1049-001 Lisbon, Portugal
2 CEHINAV, ETSIN, Universidad Politecnica de Madrid (UPM), 28040 Madrid, Spain
3 TPN, Dept. of Naval Arch. & Ocean Eng. Escola Politecnica, University of Sao Paulo, Sao Paulo 05508-010, Brazil
4 Wind Energy Department, National Renewable Energy Center (CENER), 31621 Sarriguren, Spain
5 CEHINAV, DACSON, ETSIN, UPM, 28006 Madrid, Spain
* Correspondence: c.guedes.soares@centec.tecnico.ulisboa.pt

Abstract: This work evaluates the experimental test results regarding the operational performance of a free-float capable tension leg platform with a 10 MW wind turbine. It covers the platform dynamics in the selected installation area: Ribadeo, Spain. The model and the facility are initially presented, along with the experimental setup and the load cases. The testing campaign includes a software-in-the-loop method to emulate the rotor thrust and the aerodynamic and gyroscopic moments in pitch and yaw. The result sets are structured to start from basic information from system identification cases and continue with responses against regular and irregular waves accompanied by steady and stochastic wind scenarios. The performance in operational and extreme conditions is assessed as well as fault scenarios. The experiments demonstrate auspicious motion dynamics and mooring line behavior when examined against class society rules.

Keywords: wave tank testing; tension leg platform; floating wind; software in the loop; offshore wind

Citation: Hmedi, M.; Uzunoglu, E.; Medina-Manuel, A.; Mas-Soler, J.; Vittori, F.; Pires, O.; Azcona, J.; Souto-Iglesias, A.; Guedes Soares, C. Experimental Analysis of CENTEC-TLP Self-Stable Platform with a 10 MW Turbine. *J. Mar. Sci. Eng.* **2022**, *10*, 1910. https://doi.org/10.3390/jmse10121910

Academic Editor: Spyros A. Mavrakos

Received: 1 November 2022
Accepted: 23 November 2022
Published: 5 December 2022

1. Introduction

Environmental and political status expand the shift to renewable resources as energy dependence is ever-increasing [1]. Among renewable resources, wind energy is one of the fastest-growing sectors and covers 35% of renewables [2]. While the total wind capacity is increasing [3], offshore wind has some advantages over onshore wind. The high offshore wind potential and the long distance from the coast allow higher energy extraction with a reduced impact on residential areas [4]. Thus, offshore wind is advantageous with the possibility of implementing larger turbines [5].

In this regard, the move offshore has been conspicuous, even more so in deeper water and floating turbines [6]. Several projects evaluated alternative platforms to support multi-megawatt wind turbines [7–10]. ARCWIND (Adaptation and Implementation of Floating Wind Energy Conversion Technology for the Atlantic Region) is one project seeking to develop capable 10 MW floating wind turbines (FWTs). In ARCWIND, the wind energy potential in the Atlantic area is assessed [11–13], the best locations for wind farms are chosen [14], the feasibility of different FWT technologies is studied [15–17], and the logistics and economics of deploying these concepts in a full-scale farm environment are considered [18]. Within the project, Three FWT models are designed: a spar [19], a barge semisubmersible [20], and a tension leg platform (TLP) [21].

Regarding performance, TLPs have a minimal dynamic response in the restricted modes (i.e., heave, roll, pitch). Their stability relies on taut mooring lines, and the platform responds in a way similar to a rigid structure in terms of motion amplitudes and resonance frequencies. The minimal pitching response is an advantage for power production as it affects the aerodynamic performance of wind turbines [22–24]. Thus, compared to other floaters, the TLP's generated power is less affected by motion amplitudes, for instance, larger pithing angles that can change the turbine aerodynamics.

Over the years, alternative TLP designs with various hull geometries, such as one-column [25–27] and four-column [28,29] designs, have been considered as possible hull forms. The single-column hull form originates from the SeaStar platform [26,30]. It was suggested as a means of reducing the hull weight compared to the conventional four-column form, which ConocoPhillips first installed in the 1980s. However, a smaller waterplane area signifies that, without ballasting, the hull form is limited in providing stability during installation cases and will likely require specialized vessels, such as installation barges, to transport the system to the installation site [31]. Since installation in larger numbers (i.e., wind farms) becomes economically unfavorable, this hull form has not been used even in demo projects to date. Hence, developing alternatives to keep the advantages of the TLP dynamics while addressing the disadvantages becomes necessary. The CENTEC-TLP was developed with these constraints in mind.

Unlike SeaStar-type TLP designs [27,32], the CENTEC-TLP has sufficient stability in free-floating conditions, as it follows the principles of a barge for transportation purposes. It makes use of a large waterplane area, thus benefitting from a low transport draft, making it accessible to low-depth ports and shipyards. The platform's design development followed a series of stages as described by renewable industries in terms of Technology Readiness Levels (TRLs) [33,34]. The proof of concept (i.e., TRLs 1–3) involved the design procedure explained in [21,35,36], development of the hull form [15] and evaluation of the numerical performance [37–39]. The validation stage (i.e., TRLs 4–6) included testing a scaled model [40], as experimental testing allows the investigation of the system dynamics in a realistic environment [41]. This phase aimed to confirm the system dynamics, verify the mooring system, and collect sufficient data for future optimization of the numerical model if needed.

In this scope, the CENTEC-TLP was tested at the Universidad Politécnica de Madrid (UPM) towing tank. Initial results regarding the transportation dynamics and wave resistance are published in [42]. The results show a maximum heeling of 2 degrees in 3 m significant wave height. Given that a significant wave height above 1.5 meters is not usually considered favorable for installation, it can be stated that the design shows favorable motion dynamics for transport. The studies continued with the evaluation of the system dynamics and the mooring lines in operational, extreme, and fault conditions. The tests were conducted with the same prototype used in the transportation tests. They included adding a software-in-the-loop (SiL) system for wind load generation [43]. The aerodynamic system improvement comprises the emulation of the rotor thrust and the aerodynamic and gyroscopic pitch and yaw moments explained in [44], unlike the initial systems representing the thrust only [45]. The inclusion of the SiL avoids the Froude–Reynolds scaling conflict [46] and allows the testing of wind turbine pitch control strategies [47].

This work expands the initial set of data published in the experimental results from the CENTEC-TLP testing campaign [40]. It focuses on the installed platform and evaluates the operational performance. The paper starts by describing the design, the prototype, and the towing tank. Identification cases of free decays are then presented. They are followed by the response amplitude operators (RAOs) in regular waves and steady wind. Then, irregular waves and turbulent wind cases are evaluated, representing operational, extreme, and faulty conditions. The mooring line breaking and slacking are examined according to class society rules. All results are presented at full scale.

2. Model and Prototype Description

2.1. The Design

2.1.1. The Wind Turbine and Tower

The CENTEC-TLP (Figure 1—left) was designed for the DTU 10 MW turbine and tower [48]. The turbine is upscaled from the 5 MW NREL reference turbine [49], with a shorter tower and larger hub diameter compared to direct upscaling. Its specifications and mass properties are listed in Table 1. In addition, a clearance (i.e., the height of the platform-to-tower connection) of 10 m is set to prevent the waves from affecting the blades and the tower. This value matches the DeepCwind design [50] at 10 m above the mean sea level (MSL), regardless of the environmental conditions.

Figure 1. The CENTEC-TLP design (**left**) and prototype (**right**).

Table 1. Specifications of the DTU 10MW WT.

Parameter	Value	Units
Cut-in wind speed	4	[m/s]
Cut-out wind speed	25	[m/s]
Rated wind speed	11.4	[m/s]
Rated power	10	[MW]
Min–max rotor speeds	6–9.6	[rpm]
Rotor diameter	178.3	[m]
Rotor mass	227,962	[kg]
Nacelle mass	446,036	[kg]
Tower mass	628,442	[kg]
First tower mode frequency	4.02	[s]

The tower's mass and inertial data are in Table 2, and those for the nacelle and the rotor are in Table 3. The tower and the rotor nacelle assembly (RNA) combination details are provided in Table 4. Note that the 10-m base height locates the structure's centre of gravity (CoG) at 95.9 m above MSL. The eccentricity in the longitudinal direction results from the offset of the centre of mass of the turbine.

Table 2. Mass details of the tower model and the moments of inertia about its center of gravity.

Parameter	Value	Units
Tower mass	628,442	[kg]
I_{xx}	6.52×10^8	[kg m^2]
I_{yy}	6.52×10^8	[kg m^2]
I_{zz}	7.84×10^6	[kg m^2]
CoG (x,y,z)	0, 0, 57.6	[m]

Table 3. Mass details of the nacelle and rotor assembly.

Parameter	Value	Units
Total mass	673,998	[kg]
Center of gravity	[0.61, 0, 131.56]	[m, m, m]

Table 4. Mass details of the tower, rotor, and the nacelle combination.

Parameter	Value	Units
Total mass	1,302,440	[kg]
CoG (x, y, z)	[0.32, 0, 95.9]	[m, m, m]
I_{xx}	2.43×10^9	[kg m^2]
I_{yy}	2.45×10^9	[kg m^2]
I_{zz}	2.24×10^7	[kg m^2]
First fore–aft natural period	4.02	[s]

2.1.2. The CENTEC-TLP Platform

The CENTEC-TLP (Figure 1) is designed in the ARCWIND project to support the DTU 10 MW wind turbine. The physical properties and the numerical performance can be found in [15,21,37,51]. It is symmetrically designed with a square-shaped ring pontoon and four rectangular stability columns. A central column supports the tower at the centre of a cross-shaped inner pontoon. The design involves four diagonal support braces for better structural integrity. The sides length is 49 m and the draft is 20 m. The primary material is steel with no ballasting, which keeps the mass low as possible (i.e., 2200 tons) while adding the free-float capabilities. The system floats on the pontoons during the transportation phase as a barge. Compared to cylindrical designs, rectangular pontoons are easier and cheaper to fabricate [52]. They also maintain the waterplane area when the platform heaves, preserving its stability.

2.2. The Prototype

2.2.1. Towing Tank and Data Acquisition

The CEHINAV research group carried out the experimental tests. They manage the Universidad Politécnica de Madrid (UPM) towing tank, whose length and breadth are 100 m and 3.8 m, respectively. The facility is equipped with a single paddle wave generator capable of generating up to 0.3 m waves. It also includes a passive wave absorption area. During the tests, the depth was set to 127.2 m (2.12 m model scale). This value was set to avoid exceeding the wave generator limits. The CENTEC-TLP design is adapted to the Ribadeo (Spain) site, where the water depth is 132 m [53,54].

The data acquisition is made with a sampling frequency of 100 Hz. This value is sufficient since the typical frequencies of the investigated phenomena are in the order of 1.2 Hz. The collected data include the mooring line tensions, the wave elevation, and the recorded motions. An optical tracking system is used for recording the six DOF motions. The system consists of four infrared cameras fixed on the tank's walls, tracking the displacements and rotations for at least five markers. The in-house system "HarbourDuino Mini" manages the storage of the tracked motions and the synchronization with other measurements.

2.2.2. Prototype Properties

The experiment involves a 1:60 Froude-scaled prototype. The scaling factor targets the installation depth and fulfils the 20% Chakrabarti's limit [55] of the tank width for model testing. In addition, the scale satisfies the ITTC recommendations [56,57] to fit the prototype in the basin dimensions, avoid interactions with the reflected waves from the side walls, uphold the wave maker capacity, and correctly scale the mooring lines. A silicone woven fibreglass and epoxy resin composite material are used to build the prototype. The material selection considers the geometrical complexity of the hull to guarantee a low mass, leaving a proper margin for calibrations. Table 5 summarizes the design and prototype properties.

Table 5. Design and prototype properties (full scale).

Parameter	Design	Prototype	Deviation
Draft [m]	20	20	0%
Total mass [t]	3490.6	3499.2	0.24%
Z_{CoG} [m]	28.47	25.82	9.3%
I_{xx} [kg m^2]	1.25×10^{10}	1.42×10^{10}	13.6%
I_{yy} [kg m^2]	1.25×10^{10}	1.42×10^{10}	13.6%
I_{zz} [kg m^2]	1.05×10^{9}	1.68×10^{9}	60%

For a TLP, setting the mooring line pretensions correctly is a priority to replicate its dynamics. In this regard, the buoyancy excess plays a major role. Accordingly, the mass and the displaced volume are set as the main targets for the scaling. The total mass (including the tower and RNA) scaling presents a minimal deviation of 0.24%. The draft, and therefore the displaced volume, matches the design value. Regarding the inertia, the lower columns were reinforced for better structural integrity, as shown in Figure 2. The reinforcements lead to a higher mass in these zones. The current calibration is the chosen compromise to match the roll, pitch, and yaw inertias.

Figure 2. Lower columns' reinforcements.

2.2.3. Modeling of the Mooring Lines

In the design case, the CENTEC-TLP is moored with 12 tendons of 14 cm diameter. The 6-strand wire rope material is used with three lines attached to each corner. In the experiments, four tendons were used to represent three designed mooring lines each. Accordingly, the mooring stiffness and pretensions were conserved on each corner. This truncation is set according to the ITTC recommendations [58], where the total mass, the total stiffness, and restoring moments of the floating structure are conserved. Each mooring line is modelled with a spring connected to the platform via a stainless-steel wire and a screw (Figure 3). The mooring properties on each side are presented in Tables 6 and 7. The slight difference in the mooring pretensions results from the inclusion of the lines' weight. The load cell hanging in the lines is heavier than their actual mass scale and causes a difference in the pretensions. The tension adjustment is made through the connecting bolts while maintaining the 20 m draft. In the nomenclature, the lines T3 and T4 are located upwind, while T1 and T2 are downwind (Figure 1—right).

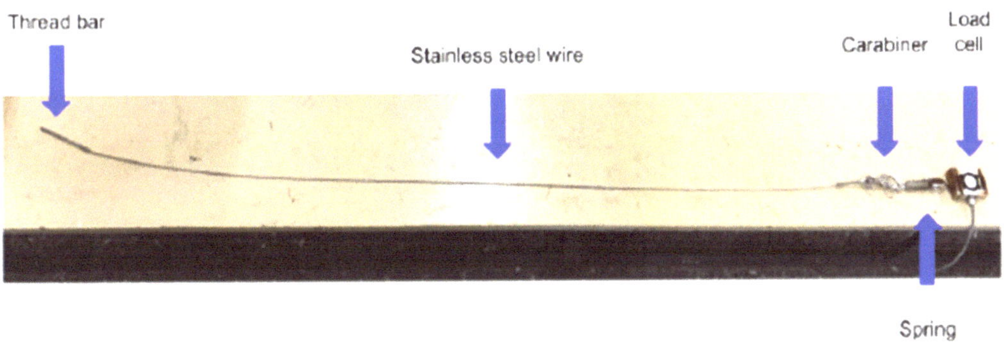

Figure 3. The mooring line used in the experiment.

Table 6. Design and prototype mooring pretensions.

Tendon	Design [kN]	Measured [kN]	Deviation
T1	10,910	10,920	0.09%
T2	10,910	10,770	1.28%
T3	10,910	11,070	1.46%
T4	10,910	10,970	0.5%

Table 7. Design and prototype mooring stiffness.

Tendon	Design [kN/m]	Measured [kN/m]	Deviation
T1	9.47×10^4	9.44×10^4	0.32%
T2	9.47×10^4	9.32×10^4	1.58%
T3	9.47×10^4	9.57×10^4	1.06%
T4	9.47×10^4	9.51×10^4	0.42%

2.2.4. Emulation of Aerodynamic Loads

Froude-scaled rotors under an air flow produce out-of-scale aerodynamic loads due to a lower Reynolds number compared to the full-scale rotor. In the experiments, an actuator emulates the aerodynamic loads from the wind turbine at the hub height. The technique is referred to as the software-in-the-loop method (SiL), where the aerodynamic loads are Froude-scaled and included via actuators. SiL is a hybrid approach that combines physical testing and numerical simulation [59,60]. The communication between the actuators and the simulation is done in real time through a set of sensors and a motion-tracking system [46,47]. ITTC categorizes this technique as a functional practice to investigate the dynamic response of FWTs in operational and extreme conditions [33].

The National Renewable Energy Center (CENER) provided the SiL application in the experiment. Unlike initial SiL systems limited to the thrust force [45], an improved version of SiL is adopted, emulating, in addition to the rotor thrust, the aerodynamic and gyroscopic yaw and pitch moments. The emulator is composed of four drone propellers equally distributed around a drone chassis located at the hub height, as shown in Figure 1 (right). The calibration and the performance of the SiL system on the CENTEC-TLP are detailed in [43]. This configuration produces an approximate force range of 0-24N.

A schematic of the system is illustrated in Figure 4. Each propeller is powered by a brushless motor and fed with an AC/DC power supply. The thrust forces are controlled by the motor rotational speed set by the electronic speed controller (ESC), which is regulated by a pulse width modulation (PWM) signal generated by the LabVIEW control software, using servo libraries for Arduino. FAST [61] is used to simulate the turbine aerodynamics in real-time taking into account the measured motions. FAST's output contains all loads from the rotor, including the applied aerodynamic, gravitational, and inertial loads. The loads are fed to the propellers via a calibration curve relating the requested force to the required PWM signal.

Concerning the tower, the first bending mode is set as a goal for the scaling. The first set of tests is conducted when the tower is cantilevered on a stable base (Figure 5), with the mass on its top representing the RNA. Bending tests are conducted, and the fore-aft and side-side periods are 4.25 s, roughly equal to the actual design period of 4 s. Tower identification test when fixed on the platform, is presented in Section 4.1.

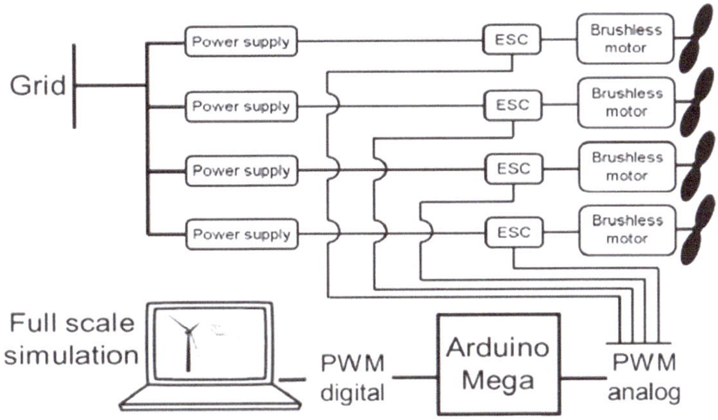

Figure 4. Schematic of the software in the loop technique.

Figure 5. Tower bending test.

3. Operational Criteria and Experimental Matrix

3.1. Criteria

The experimental campaign focuses on operational, extreme, and fault conditions. The wind and wave conditions represent the installation environment in Ribadeo, Spain, selected within the ARCWIND project [11,62]. The results are examined and compared to the criteria discussed by DNV [63–65] and API [66,67] and summarized in Table 8.

Table 8. Operational and design criteria for the CENTEC-TLP.

Requirement	Target
Surge, sway, and yaw period	>25 s
Roll, pitch, and heave period	<5 s
Anchor angle limit	5°
Max surge displacement	9 m
Mooring line breaking	$F_{Line} < F_{break}$ = 36,000 kN
Mooring line slack criteria	F_{Line} > 20% Pretension = 2182 kN
Pitch	RMS < 2°

Ideally, tension leg platforms should limit the surge motion to stay below a five-degree mooring angle limit at the anchors [68]. This condition is given for steel tethers with limited flexibility. However, CENTEC-TLP utilizes flexible wire rope lines. Still, this recommendation was considered in the design process. Hence, the 5-degree anchor angle safety limit is equivalent to a 9 m surge displacement.

On the other hand, breaking the mooring lines is an ultimate limit state for the TLP. Hence, this value needs to be observed in all conditions. The line-breaking tension (F_{break}) is 12,000 kN for each of the 12 tendons [15]. The experimental prototype has four lines, each mimicking the behavior of three tendons, as described in Section 2.2.3. Thus, the breaking tension in this assessment is set to 36,000 kN.

Finally, the line tension should stay above 20% of the pretension (i.e., 2182 kN) to avoid a case considered slack [64]. Once the mooring line tensions drop below a certain value, they enforce high dynamic loads when they come back into tension, causing a series of failures.

3.2. Experimental Matrix

Several tests were conducted at UPM to evaluate the operational and survivability criteria. The experimental test matrix is summarized in Table 9. The evaluated cases are classified into two categories, including controlled and stochastic environments. Controlled scenarios involve identification cases of free decays, regular waves, and steady wind conditions. The stochastic conditions serve to evaluate the operational performance in irregular waves and turbulent wind environments, in addition to the survivability in extreme and fault scenarios. The system is designed for Ribadeo, in the north of Spain, with the scatter diagram shown in Table 10. Accordingly, the tested conditions are selected from the scatter diagram using the waves with the highest probability in the installation site.

Table 9. Experimental test matrix.

Controlled Environments	Stochastic Environments
Free decays	Irregular waves and turbulent wind
Regular waves	50-year extreme
Regular waves and steady wind	Emergency shutdown
Steady wind	

3.2.1. Decay Tests

Identification tests as free decays are made to evaluate the system's natural periods. This testing is conducted by displacing the model from its static equilibrium position in a particular DoF, and then releasing it to oscillate in free motion. The system is observed to swing around the equilibrium position with a smaller amplitude after every oscillation because the damping forces remove mechanical energy from the system, whereas there is no excitation force. Theoretically, the angular frequency of this oscillatory motion may differ from the natural frequency; however, the values will be close to identical when damping forces are relatively weak. Accordingly, the natural period is obtained by averaging the results of the Fourier analysis applied to the motion signals from different runs. Another set of surge and pitch decays, including steady wind loads, is made to assess the aerodynamic damping effects on the system. The wind scenarios represent the below-rated (8.65 m/s), rated (11.4 m/s), and the above-rated (20.17 m/s) wind speeds.

Table 10. Scatter diagram of Ribadeo showing the probabilities of wave height and period combinations [12,62].

Hs (m) \ Tp (s)	<5	5–6	6–7	7–8	8–9	9–10	10–11	11–12	12–13	13–14	14–15	15–16	16–17	17–18	>18
0.5	0.00	0.00	0.00	0.00	0.01	0.00	0.00	0.00	0.00	0.00	0.00	0.00	0.00	0.00	0.00
1	0.26	0.59	0.71	0.85	1.62	1.93	1.03	0.37	0.15	0.04	0.01	0.00	0.01	0.00	0.00
1.5	0.58	1.99	1.61	1.08	1.92	4.07	4.85	3.89	1.97	0.54	0.17	0.00	0.03	0.01	0.00
2	0.06	1.44	2.19	0.84	1.01	1.67	3.14	4.81	4.86	1.93	0.51	0.00	0.10	0.01	0.00
2.5	0.00	0.16	1.66	0.72	0.77	0.81	1.14	2.47	4.52	3.35	0.99	0.00	0.18	0.02	0.00
3	0.00	0.00	0.40	0.66	0.57	0.47	0.60	1.16	2.40	2.91	1.55	0.00	0.22	0.01	0.00
3.5	0.00	0.00	0.04	0.30	0.43	0.33	0.36	0.65	1.46	1.83	1.51	0.00	0.34	0.03	0.00
4	0.00	0.00	0.00	0.04	0.17	0.23	0.23	0.41	0.82	1.19	1.19	0.00	0.43	0.03	0.00
4.5	0.00	0.00	0.00	0.00	0.04	0.13	0.11	0.23	0.37	0.74	0.78	0.00	0.42	0.04	0.00
5	0.00	0.00	0.00	0.00	0.01	0.04	0.08	0.10	0.26	0.45	0.52	0.00	0.31	0.04	0.00
5.5	0.00	0.00	0.00	0.00	0.00	0.02	0.04	0.07	0.12	0.25	0.38	0.00	0.27	0.04	0.00
6	0.00	0.00	0.00	0.00	0.00	0.00	0.01	0.03	0.08	0.14	0.24	0.00	0.18	0.04	0.00
6.5	0.00	0.00	0.00	0.00	0.00	0.00	0.00	0.02	0.03	0.06	0.11	0.00	0.13	0.03	0.00
7	0.00	0.00	0.00	0.00	0.00	0.00	0.00	0.01	0.01	0.02	0.06	0.00	0.12	0.03	0.00
7.5	0.00	0.00	0.00	0.00	0.00	0.00	0.00	0.00	0.00	0.01	0.03	0.00	0.05	0.03	0.00
8	0.00	0.00	0.00	0.00	0.00	0.00	0.00	0.00	0.00	0.01	0.01	0.00	0.02	0.01	0.00
8.5	0.00	0.00	0.00	0.00	0.00	0.00	0.00	0.00	0.00	0.00	0.01	0.00	0.01	0.00	0.00

3.2.2. Steady Wind and Regular Waves

The second set of tests includes steady wind conditions are summarized in Table 11. The steady wind loads were emulated via CENER's SiL system. The offsets in the motions and mooring line tensions are assessed according to operational criteria (Table 8), highlighting the contribution of wind loads to the system dynamics.

Table 11. Steady wind conditions.

Condition	U_w [m/s]
Cut-in	4
Below Rated	8.65
Rated	11.4
Above rated	20.17
Limit	23.06
Cut-out	25
50-year	50

In the third set of tests, the system was subjected to only regular head waves and regular head waves combined with steady wind loads. The regular waves combined with steady wind conditions are summarized in Table 12. The first wave was selected to identify the problems that may occur close to the resonance conditions. The limit and forced-limit tests represent the operational scenario bounds in the installation site, and the 50-years test refers to the extreme sea state. The analysis was then carried out according to the ITTC recommendations [69]: the response amplitude operators (RAOs) are estimated using the Fourier analysis of the motions, mooring tensions, and wave signals. Only the steady part of the signal, excluding the time when the refracted waves reach the model, was analysed, and a minimum of 10–20 cycles were taken to obtain the RAOs. The example in Figure 6 shows the TLP unique behavior in heave when excited by regular waves. Due to the taut tendons, both positive and negative surge induce a downward heave motion (i.e., a set down). That is, the heave motion has spikes at twice the surge motion frequency. In this case, only the spikes occurring at the wave frequency are considered for the RAOs.

Table 12. Regular wave and steady wind conditions.

Condition	H [m]	T [s]	U_w [m/s]
Resonance	1.5	5	4
Below rated	1.5	10.5	8.65
Rated	2	12.5	11.4
Above rated	3.5	13.5	20.17
Limit	4	14.5	23.06
Forced-limit	8	16.5	23.06
50-year	10.81	15.38	50

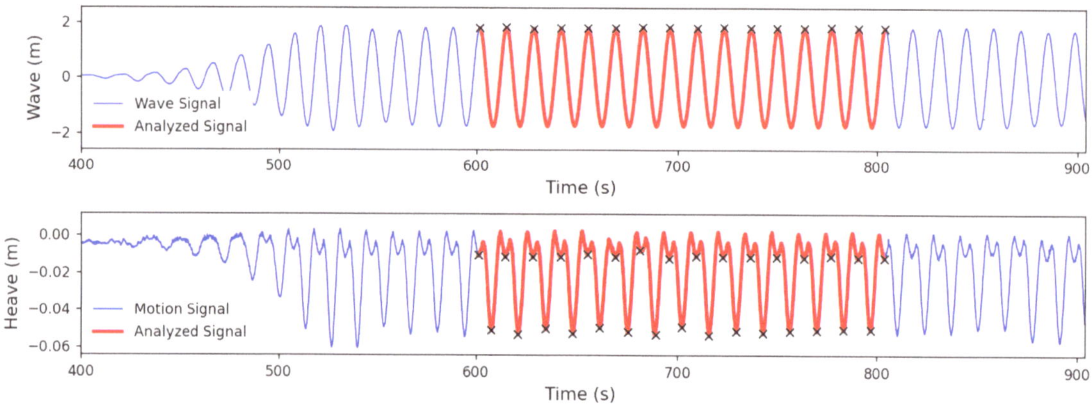

Figure 6. Example of heave and wave signals in regular wave conditions.

3.2.3. Stochastic Environments

The operational performance in stochastic wind and wave environments and the survivability in extreme conditions are also evaluated. Wind speed scenarios covering all the turbine states were selected for this analysis (Table 13). The wind and waves are aligned and have a 180-degree heading (i.e., head waves). The JONSWAP spectrum with a peak parameter of 1 is used for the irregular waves in operational conditions and 1.45 in extreme scenarios. Statistical analysis of the motions and mooring lines was made to assess the system dynamics and examine them against class society rules.

Table 13. Stochastic and extreme wind and wave conditions.

Condition	Hs [m]	Tp [s]	U_w [m/s]
Below rated	1.5	10.5	8.65
Rated	2	12.5	11.4
Above rated	3.5	13.5	20.17
Limit	4	14.5	23.06
50-year	10.81	15.38	50

The last test sets assess the dynamics in fault conditions. DNV recommends testing emergency shutdown scenarios since mooring slack does not necessarily occur in extreme events but may occur in emergency stop cases [64]. In addition, controller error (i.e., one blade pitch angle set incorrectly) may cause platform yawing in operational conditions [70]. Accordingly, emergency shutdown tests in four conditions (Table 14) in the rated and above-rated steady wind conditions (i.e., 11.4 and 20.17 m/s) are examined.

Table 14. Wind and wave conditions for emergency shutdown tests.

Wave Type	Wind Type	Condition	Hs [m]	Tp [s]	U_W [m/s]
Regular	Steady	Rated	2	12.5	11.4
	Steady	Above rated	3.5	13.5	20.17
Irregular	Turbulent	Rated	2	12.5	11.4
	Turbulent	Above rated	3.5	13.5	20.17

4. Results in Controlled Environment Settings

4.1. Decay Tests

The system's natural periods are identified with decay tests. The measured values are recapped in Table 15. The surge and sway periods are within the design target surpassing 25 s (Table 8) and read 27.4 s and 27.8 s, respectively. The roll and pitch also respect the design criteria and reside below the 5 s limit. The deviations compared to the design are less than 5% for most measured periods.

Table 15. Natural Period of the system.

	Surge [s]	Sway [s]	Roll [s]	Pitch [s]	Yaw [s]	Tower [s]
Exp	27.4	27.8	4.9	4.9	15.3	4.7
Ref	29	29	4.8	4.8	16.8	4
deviation	4.1%	4.1%	2%	2%	8.9%	17.5%

Due to the high stiffness of the mooring lines, heave decay tests were difficult to carry out. Displacing the model in heave resulted in tendon tension loss or rapid decays with limited oscillations. In both cases, the signal has noise due to the mooring responses. While the white noise wave spectrum test can provide an estimate of the heave period, the physical limitation of the wave maker did not allow for testing waves below a 5 s period. In this case, the analysis from the white noise test did not provide a clean signal to identify the heave natural period. Since the mass and the line stiffness were correctly scaled and other modes, such as surge, had deviations under 5 percent, it was considered close to the design value (i.e., approximately 2 s). Compared to pitching and rolling periods, this value is farther away from the first-order wave period. Hence, heave is unlikely to be excited without the second-order loads.

Table 16 shows the results of the surge and pitch decays in steady wind. The aerodynamic damping effect is seen by the slight change in the periods. The SiL system used for the wind load generation is connected to the wind turbine simulator "FAST". All phenomena captured by FAST are reproduced by the SiL actuators, including the aerodynamic damping.

Table 16. Natural periods of the system with steady wind loads.

	Wind Speed [m/s]	Surge [s]	Pitch [s]
Below rated	8.65	28.17	5.11
Rated	11.4	28.65	5.09
Above rated	20.17	27.85	5.03

Regarding the tower, the first-mode period was measured with bending tests. The period in the cantilevered configuration (i.e., 4.25 s in Section 2.1) increased to 4.7 s when the tower was clamped to the floater. This variation is also seen in other testing campaigns [10,71,72]. The increase in the tower's first-mode periods is related to the floater flexibility and the rigidity of the clamping between the tower and the floater. The value of 4.7 s is accepted as it lays outside the first-order wave excitation range of 5 to 25 s [65].

4.2. Steady Wind Responses

Steady wind tests (Table 11) are made to check the system response against the wind in operational and extreme conditions. The primary influence of wind is on surge and pitch since only 180-heading wind is used. Tables 17 and 18 show the mean offsets of motions and line tensions, respectively. The results fulfil the operational and survivability criteria (Table 8) and leave a sufficient margin for wave loads. The maximum surge noted at the rated wind scenario is 3.65 m; the value is below the maximum allowed surge displacement (9 m). The rotational motions are also negligible, as expected from a TLP. The tendon tensions are situated in the allowed range (i.e., above 2182 kN and below 36,000 kN) and avoid the breaking and slacking phenomenon.

Table 17. Motions in steady wind conditions.

	Cut-In	Below Rated	Rated	Above Rated	Limit	Cut-Out	50-Years
Surge [m]	0.45	1.73	3.65	1.19	1.08	1.06	0.21
Sway [m]	−0.02	−0.038	−0.06	−0.02	−0.024	−0.026	−0.004
Heave [m]	−0.001	−0.017	−0.067	−0.01	−0.007	−0.005	−0.001
Roll [deg]	0.001	0.007	0.005	0.001	0.0	0.003	0
Pitch [deg]	−0.013	−0.058	−0.113	−0.036	−0.034	−0.029	−0.006
Yaw [deg]	0.0	0.03	0.06	0.013	0.01	0.01	−0.005

Table 18. Line tensions in steady wind conditions.

	Cut-In	Below Rated	Rated	Above Rated	Limit	Cut-Out	50-Years
Line T1 [kN]	10,262	9246	7777	9641	9745	9823	10,470
Line T2 [kN]	10,195	9188	7691	9590	9695	9734	10,359
Line T3 [kN]	11,171	12,190	13,695	11,771	11,666	11,622	10,991
Line T4 [kN]	11,256	12,292	13,826	11,892	11,790	11,715	11,044

4.3. Regular Wave and Steady Wind Responses

The tested steady wind and regular waves are summarized in Table 12. The turbine is parked when only subjected to waves. It operates when the wind is introduced. Figures 7 and 8 illustrate the motions and the line tensions RAOs, respectively. As a TLP, the system responds minimally to waves in most degrees of freedom. The amplitudes stay below 0.02 deg/m in roll, under 0.045 deg/m for pitch, and less than 0.05 deg/m in yaw. In surge, a longer wave period induces a higher response. While heaving is small in term of motion dynamics, it mainly affects the mooring response of the TLP.

As for the moorings, the tendons T3 and T4 are attached to the upwind columns, while T1 and T2 are the downwind ones. T3 and T4 behave identically in most of the tested waves. The case is the same for T1 and T2. However, the resonance test case (i.e., 5 s wave period) shows different behaviour. T1 and T3 have a higher response compared to T2 and T4. The roll motion seen in Figure 7 (i.e., 0.014 deg/m) may be causing this difference. In addition, the mooring RAOs are roughly identical in both regular waves and the combined regular waves and steady wind conditions. While the heave RAOs show some discrepancy due to wind (i.e., at 8.65 and 11.4 m/s), the discrepancy almost vanishes in the mooring response.

The 50-year extreme condition result is highlighted in green (i.e., at 15.38 s). The response is nearly linear with respect to the operational conditions. The system responds minimally to waves and wind, as reflected in the mooring lines. In this case, the rotational motions are more affected, recording the highest peaks. The tests with wave and wind are roughly identical to those with waves only in most scenarios. These results show that the RAOs depend mainly on the wave frequency and direction.

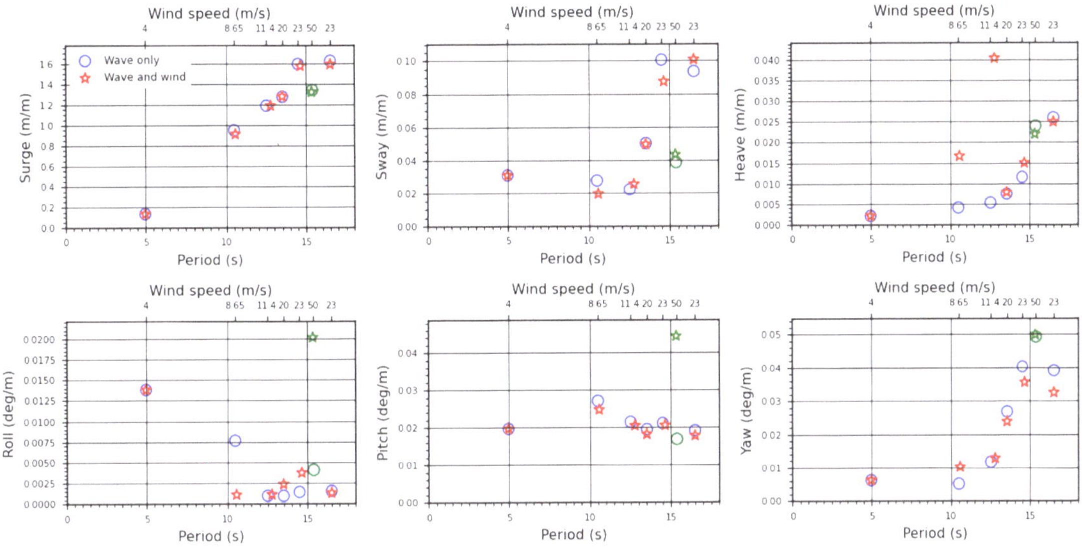

Figure 7. Response amplitude operators of the motions in regular wave conditions. The green markers (at T = 15.38 s) represent the 50-year extreme condition.

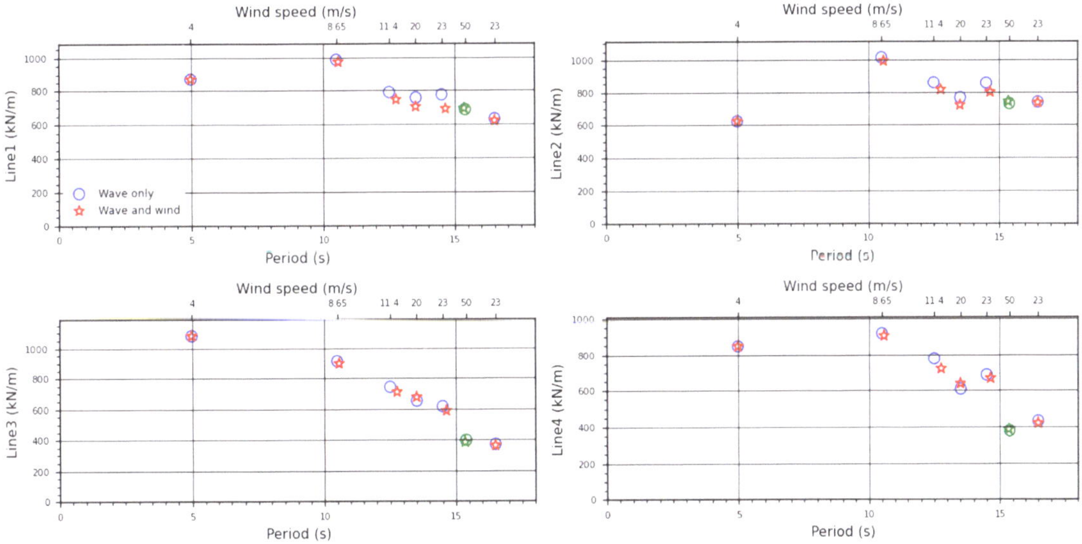

Figure 8. Response amplitude operators of the moorings in regular wave conditions. The green markers (at T = 15.38 s) represent the 50-year extreme condition.

5. Results in Stochastic Environment

5.1. Stochastic and Extreme Wind and Wave Conditions

The operational and extreme conditions are recapped in Table 13. In all scenarios, the wind and waves are unidirectional. The JONSWAP spectrum was used for the waves. The peak parameter is 1 and 1.45 for the operational and 50-years cases, respectively. The wind loads are emulated using CENER's SiL system. The match between the wave height and the wind speed is done using the scatter diagram of the Ribadeo installation area.

Figures 9 and 10 show the maximum, minimum, and standard deviation of the motions and mooring line tensions. The maximum surge motion is 6.3 m, occurring in the above-rated condition. The maxima are close for the rated, above-rated, and limit conditions. They are lower than 9 m, keeping the anchor angle below the 5-degree safety limit. Compared to the above-rated wind case, the rated wind case introduces a higher turbine trust. At the same time, wave forces increase, given the wind speed and wave height correlation. These results show that the change in wind and wave forces are similar.

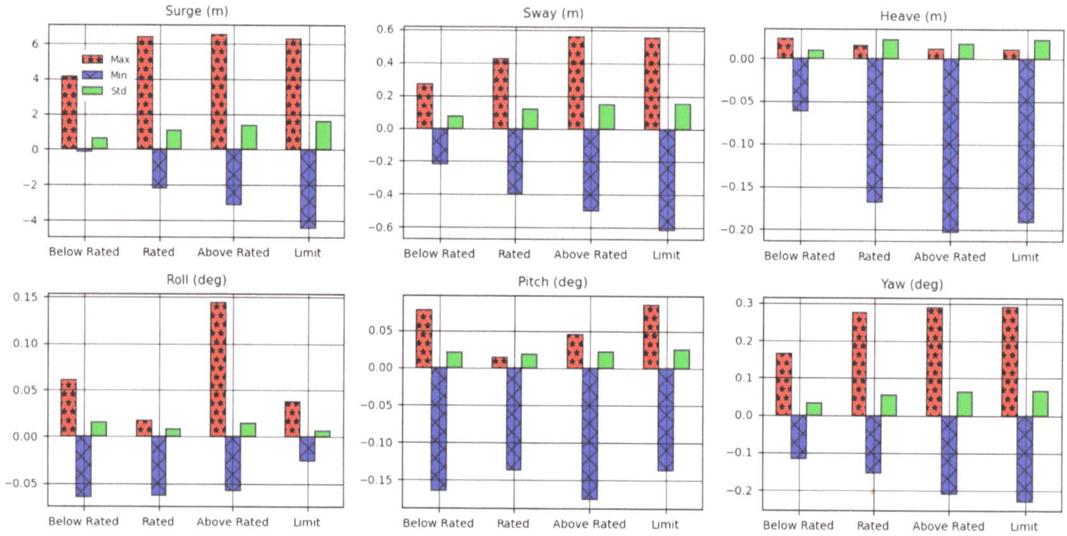

Figure 9. Maximum, minimum, and standard deviation of the motions in operational conditions.

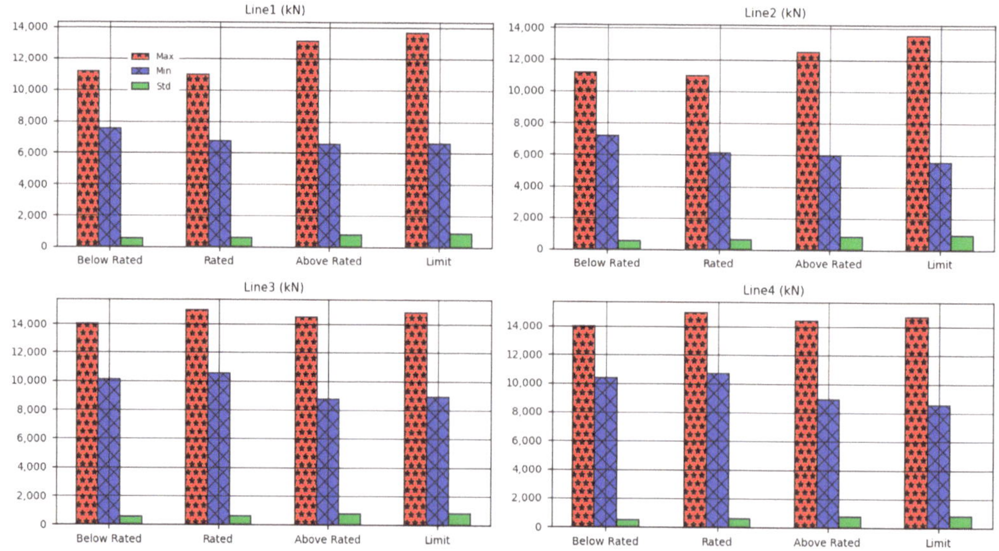

Figure 10. Maximum, minimum, and standard deviation of the mooring tensions in operational conditions.

The maxima for heave and sway motions may be considered small. However, the minimum for heave can reach 20 centimeters. While the absolute values may look negligible, they are not direct indicators of the TLP's performance. They will reflect as changes in mooring line tensions. This relation is further discussed when presenting the mooring line tensions.

The rolling and pitching motions stay below approximately 0.05 and 0.16 degrees, respectively. The existence of rolling motion signifies that the model is not entirely symmetrical regarding its setup, which can be confirmed by checking the initial mooring line tensions. In all cases, the motions are limited, as expected from a tension leg platform.

The primary advantage of these smaller motions is that the reflection on the aerodynamics is limited as the platform never pitches to a point where it can alter the aerodynamic response of the blades. Higher platform pitch eventually affects the blade geometry that encounters the wind. A study by Tran & Kim [22] shows that platform pitching significantly affects power generation. Also, Li et al. [24] showed that any increase in the platform pitching increases the fluctuations in the generated power and main shaft bearing forces. At the given angles, these cases are not factors that should be considered for a TLP.

Despite their low absolute values, it is understood from Figure 10 that motions still reflect in the mooring line performance. While the TLP provides a behaviour close to a fixed platform, it requires matching two criteria to ensure structural integrity: no loss of mooring line tension and no mooring line breaking should occur. The mooring line-breaking tension is 36,000 kN, while slack tension is 2182 kN.

In Figure 10, the maximum line tension (15,000 kN) reaches below half of the breaking limit. The minimum line tension is 5800 kN, proving that no slacking has occurred. Thus, the line tensions stay within the criteria range and respect the DNV and API rules [63,66] for mooring line safety. The behavior of individual lines clarifies that tension relies on the mooring line location. The turbine thrust causes the tensions in lines T3 and T4 to be higher than lines T1 and T2. The difference between the values is approximately 2000 kN, showing that this margin cannot be overlooked.

As for the 50-year scenario, the motions and mooring tensions statistics are summarized in Tables 19 and 20. In 3 h of testing, 98% of the surge displacement belongs to the safety margin of 9 m (Figure 11). While the surge safety limit is surpassed, the maximum value (i.e., 11.45 m) is still acceptable since wire rope has higher flexibility compared to steel tendons, for which the 5-degree recommendation was given.

Table 19. Maximum, minimum, and standard deviation of the motions in 50-year conditions.

	Surge [m]	Sway [m]	Heave [m]	Roll [deg]	Pitch [deg]	Yaw [deg]
Max	11.45	1.20	0.16	0.34	1.1	2.3
Min	−8.40	−0.81	−0.63	−0.8	−0.53	−0.59
Std	3.15	0.26	0.08	0.17	0.23	0.20

Table 20. Maximum, minimum, and standard deviation of the line tensions in 50-year conditions.

	T1 [kN]	T2 [kN]	T3 [kN]	T4 [kN]
Max	17,494	17,227	22,227	23,372
Min	4977	4057	3440	2544
Std	1537	1756	2127	2352

As for yawing, 98% remains below a half degree, and the maximum is 2.3 degrees. Pitch reaches a maximum of 1.1 degrees while 97% stays lower than 0.5 degrees. The platform heaves mostly between −30 and 20 cm and attains 63 cm at some point. While heaving directly influences the moorings, no slacking or breaking occurs. The peak in the heave (i.e., −63 cm) effect is seen in the mooring line tensions with a minimum of 2544 kN in line T4. The lines T3 and T4, respectively T1 and T2, present roughly identical behavior. The slight discrepancy in their behavior can be related to sway and roll, attaining a maximum of 1.2 m and 0.8 degrees, respectively. Mooring statistics (Figure 12) show that the tensions remain in the acceptable range, and the maximums (i.e., 23,000 kN) are rarely attained.

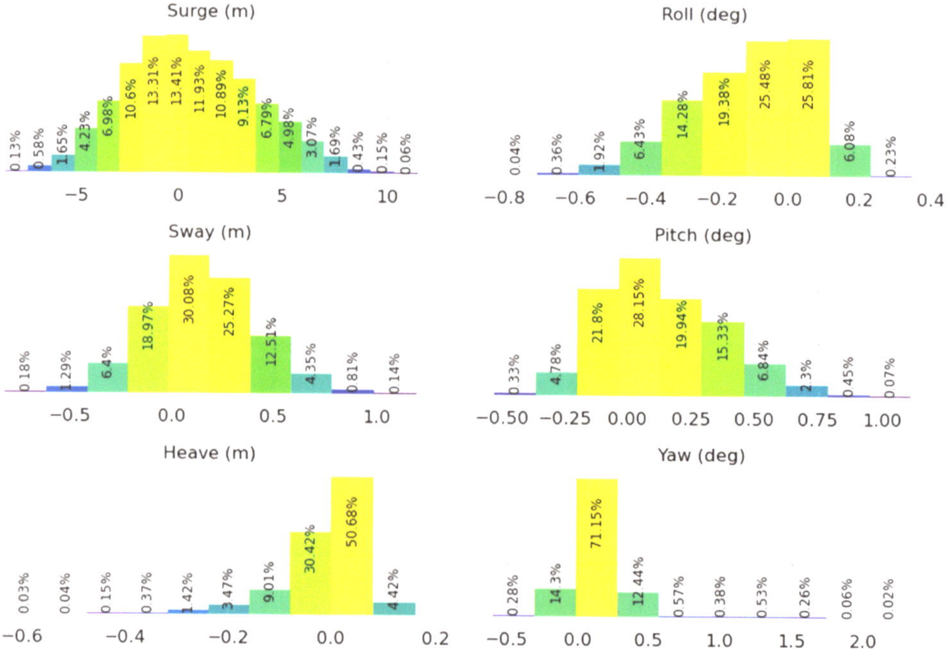

Figure 11. Histogram of the motions in the 50-year condition.

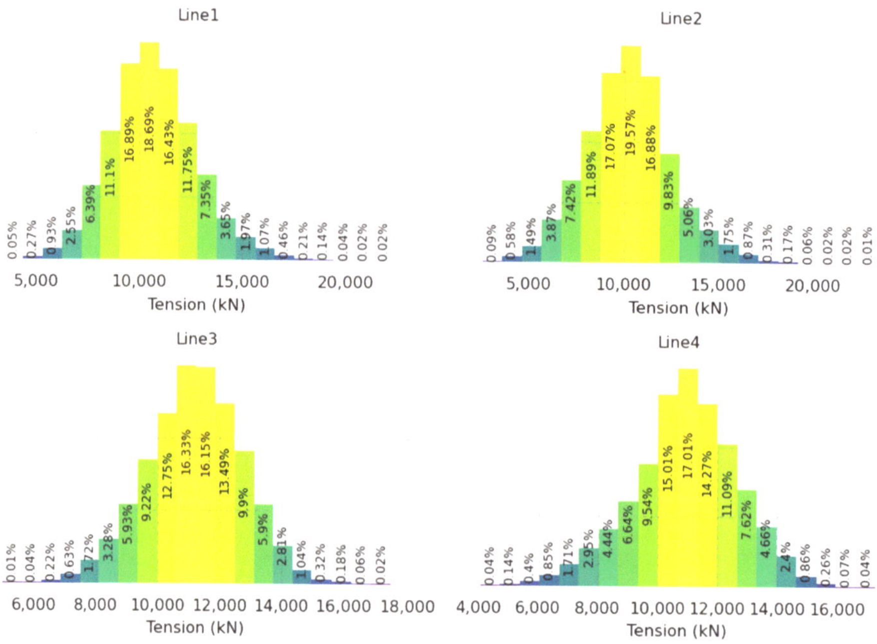

Figure 12. Histogram of the mooring tensions in the 50-year condition.

5.2. Emergency Shutdown

The shutdown scenarios involve both regular and irregular waves with steady and turbulent wind, respectively. The tests comprise rated and above-rated wind speeds detailed in Table 14. The motion data at the shutdown, including the line tensions, are summarized in Tables 21 and 22, respectively. The results show that the surge remains below 9 m, so the anchor angle is below 5 degrees. The heave and rotational motions are also restricted to a few centimeters and 0.15 degrees. Regarding the mooring lines, no slacking nor breaking occur. The minimum tension is 6460 kN far enough from the tension limit of a slack line (i.e., 2182 kN). The maximum tension seen in tendon T4 (i.e., 14,670 kN) has a sufficient margin (22,000 kN) till the breaking limit

Table 21. Motion statistics in emergency shutdown scenario.

	Regular Wave–Steady Wind				Irregular Wave–Turbulent Wind			
	Rated		Above Rated		Rated		Above Rated	
	Max	Min	Max	Min	Max	Min	Max	Min
Surge [m]	5.5	−3.4	2.8	−2.7	4.6	−2.33	3.9	−1.95
Sway [m]	0.09	−0.1	0.15	−0.22	0.2	−0.19	0.14	−0.13
Heave [m]	0.002	−0.14	0.002	−0.04	0.006	−0.1	0.001	−0.7
Roll [deg]	0.018	−0.016	0.012	−0.01	0.016	−0.017	0.013	−0.014
Pitch [deg]	0.04	−0.15	0.056	−0.058	0.044	−0.13	0.03	−0.1
Yaw [deg]	0.12	−0.06	0.08	−0.048	0.13	−0.06	0.12	−0.05

Table 22. Line tension statistics in emergency shutdown scenario.

	Regular Wave–Steady Wind				Irregular Wave–Turbulent Wind			
	Rated		Above Rated		Rated		Above Rated	
	Max	Min	Max	Min	Max	Min	Max	Min
Line T1 [kN]	11,880	6720	12,140	9040	12,050	7170	12,130	7840
Line T2 [kN]	11,650	6460	12,140	8810	11,850	7130	11,620	7060
Line T3 [kN]	14,450	9560	12,170	9335	14,510	9300	13,470	9900
Line T4 [kN]	14,500	9230	12,530	9020	14,670	9550	13,450	9560

The emergency shutdown causes a maximum surge of 5 m (Figure 13) which leaves the anchor angle below 5 degrees. While the fault controller may result in high yawing for semis [70], the yaw for the CENTEC-TLP remains limited to 0.2 degrees with the defective controller. In addition, the line tensions (Figure 14) remain in the margin where no slacking nor breaking occur.

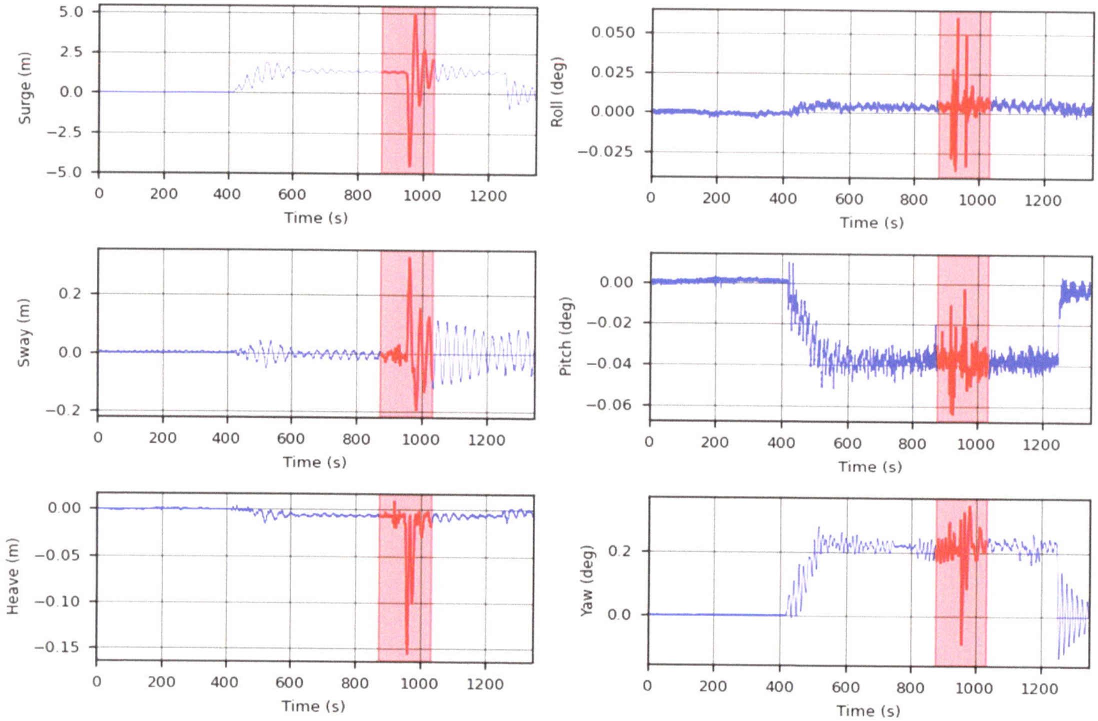

Figure 13. Motion signals in emergency shutdown case.

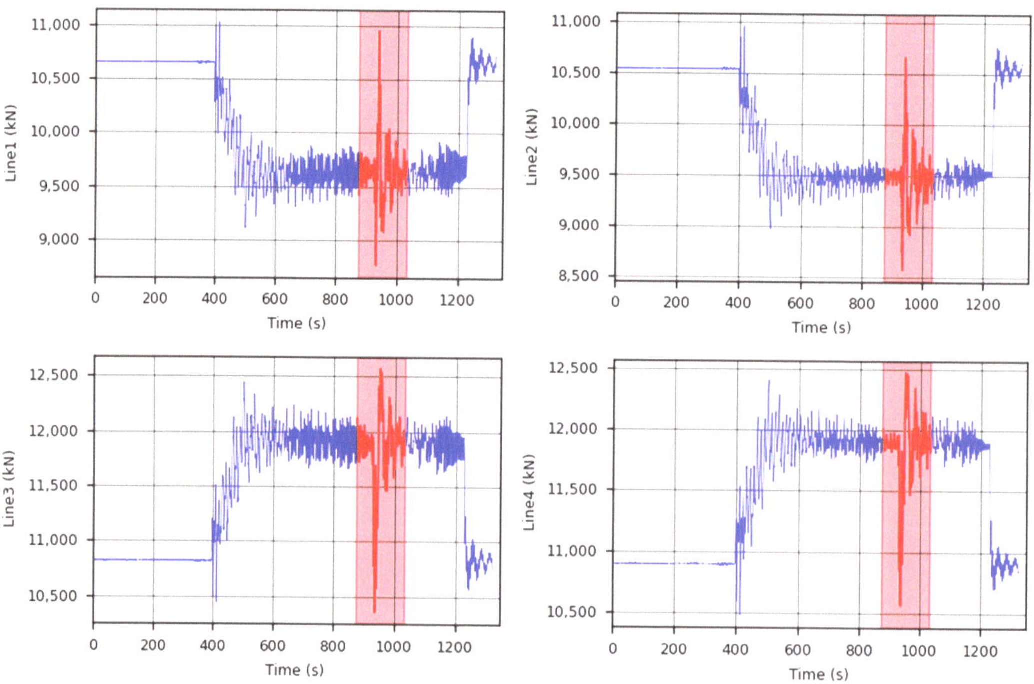

Figure 14. Mooring tensions signals in the emergency shutdown case.

6. Conclusions

This work presents the hydrodynamic performance of the CENTEC-TLP in operational, extreme, and fault conditions. The objective is achieved by experimental testing of a scaled prototype. The natural frequency of the system validates that the system is replicated with high accuracy. Response amplitude operators obtained from regular wave tests present a minimal system response when subjected to waves. The surging due to steady wind loads remains within acceptable limits.

Stochastic wind and wave tests in operational conditions prove the model's robustness in these scenarios. The minimal response in pitching emphasizes the advantages of the TLP in capturing the aerodynamic power. In addition, the motions remain restricted in the 50-year extreme condition when examined against the survivability criteria.

The mooring lines' tensions are also verified as part of the operational and survivability criteria. The tensions remain within the acceptable range, leaving sufficient margins to slack mooring and breaking limits. The same conclusion applies to the emergency shutdown and controller error cases where the motions remain restrained, and the moorings tensions stay within limits.

Author Contributions: Conceptualization, M.H., E.U., A.S.-I. and C.G.S.; methodology, M.H., E.U., A.M.-M., J.M.-S. and A.S.-I.; software, M.H., E.U., F.V., O.P. and J.A.; validation, M.H., A.M.-M., F.V., O.P. and J.A.; formal analysis, M.H. and A.M.-M.; investigation, M.H. and A.M.-M.; resources, A.S.-I. and C.G.S.; data curation, M.H. and A.M.-M.; writing—original draft preparation, M.H.; writing—review and editing, M.H., E.U., A.M.-M., J.A. and C.G.S.; visualization, M.H. and E.U.; supervision, E.U., A.S.-I. and C.G.S.; project administration, C.G.S.; funding acquisition, C.G.S. All authors have read and agreed to the published version of the manuscript.

Funding: This work has been funded by the project ARCWIND—Adaptation and Implementation of Floating Wind Energy Conversion Technology for the Atlantic Region, which is co-financed by the European Regional Development Fund through the Interreg Atlantic Area Programme under contract EAPA 344/2016. The first author has been funded by the project Variable geometry Wave Energy Conversion system for floating platforms (WEC4MUP), which is funded by the Portuguese Foundation for Science and Technology (Fundação para a Ciência e a Tecnologia—FCT) under contract (WEC4MUP_PTDC/EME-REN/0242/2020). The research was also supported by the Spanish Ministry for Science, Innovation and Universities (MCIU) under grant RTI2018-096791-B-C21 "Hidrodinámica de elementos de amortiguamiento del movimiento de aerogeneradores flotantes". This work contributes to the Strategic Research Plan of the Centre for Marine Technology and Ocean Engineering (CENTEC), which is financed by the Portuguese Foundation for Science and Technology (Fundação para a Ciência e Tecnologia—FCT) under contract UIDB/UIDP/00134/2020.

Institutional Review Board Statement: Not applicable.

Informed Consent Statement: Not applicable.

Data Availability Statement: Not applicable.

Conflicts of Interest: The authors declare no conflict of interest.

References

1. Díaz, H.; Guedes Soares, C. Review of the Current Status, Technology and Future Trends of Offshore Wind Farms. *Ocean Eng.* **2020**, *209*, 107381. [CrossRef]
2. *Energy Data-2020 Edition*; Statistical books; Publications Office: Luxembourg, 2020; ISBN 978-92-76-20629-3.
3. WindEurope. *Wind Energy in Europe-Statistics and the Outlook for 2021–2025*; WindEurope: Brussels, Belgium, 2021; p. 36.
4. Díaz, H.; Serna, J.; Nieto, J.; Guedes Soares, C. Market Needs, Opportunities and Barriers for the Floating Wind Industry. *JMSE* **2022**, *10*, 934. [CrossRef]
5. Bagbanci, H.; Karmakar, D.; Guedes Soares, C. Review of Offshore Floating Wind Turbines Concepts. In *Maritime Engineering and Technology*; Guedes Soares, C., Garbatov, Y., Sutulo, S., Santos, T.A., Eds.; Taylor & Francis Group: London, UK, 2012; pp. 553–562. [CrossRef]
6. Uzunoglu, E.; Karmakar, D.; Guedes Soares, C. Floating Offshore Wind Platforms. In *Floating Offshore Wind Farms*; Castro-Santos, L., Diaz-Casas, V., Eds.; Green Energy and Technology; Springer International Publishing: Cham, Germany, 2016; pp. 53–76. [CrossRef]

7. Pegalajar-Jurado, A.; Bredmose, H.; Borg, M.; Straume, J.G.; Landbø, T.; Andersen, H.S.; Yu, W.; Müller, K.; Lemmer, F. State-of-the-Art Model for the LIFES50+ OO-Star Wind Floater Semi 10MW Floating Wind Turbine. *J. Phys. Conf. Ser.* **2018**, *1104*, 012024. [CrossRef]
8. Galván, J.; Sánchez-Lara, M.J.; Mendikoa, I.; Pérez-Morán, G.; Nava, V.; Rodríguez-Arias, R. *NAUTILUS-DTU10* MW Floating Offshore Wind Turbine at Gulf of Maine: Public Numerical Models of an Actively Ballasted Semisubmersible. *J. Phys. Conf. Ser.* **2018**, *1102*, 012015. [CrossRef]
9. Leroy, V.; Delacroix, S.; Merrien, A.; Bachynski-Polić, E.E.; Gilloteaux, J.-C. Experimental Investigation of the Hydro-Elastic Response of a Spar-Type Floating Offshore Wind Turbine. *Ocean Eng.* **2022**, *255*, 111430. [CrossRef]
10. Robertson, A.N.; Jonkman, J.M.; Goupee, A.J.; Coulling, A.J.; Prowell, I.; Browning, J.; Masciola, M.D.; Molta, P. Summary of Conclusions and Recommendations Drawn From the DeepCwind Scaled Floating Offshore Wind System Test Campaign. In *International Conference on Offshore Mechanics and Arctic Engineering*; American Society of Mechanical Engineers: New York, NY, USA, 2013.
11. Salvação, N.; Guedes Soares, C. Wind Resource Assessment Offshore the Atlantic Iberian Coast with the WRF Model. *Energy* **2018**, *145*, 276–287. [CrossRef]
12. Silva, D.; Martinho, P.; Guedes Soares, C. Wave Energy Distribution along the Portuguese Continental Coast Based on a Thirty Three Years Hindcast. *Renew. Energy* **2018**, *127*, 1064–1075. [CrossRef]
13. Salvação, N.; Bentamy, A.; Guedes Soares, C. Developing a New Wind Dataset by Blending Satellite Data and WRF Model Wind Predictions. *Renew. Energy* **2022**, *198*, 283–295. [CrossRef]
14. Díaz, H.; Guedes Soares, C. A Novel Multi-Criteria Decision-Making Model to Evaluate Floating Wind Farm Locations. *Renew. Energy* **2022**, *185*, 431–454. [CrossRef]
15. Uzunoglu, E.; Guedes Soares, C. Hydrodynamic Design of a Free-Float Capable Tension Leg Platform for a 10 MW Wind Turbine. *Ocean Eng.* **2020**, *197*, 106888. [CrossRef]
16. Yang, Y.; Bashir, M.; Wang, J.; Michailides, C.; Loughney, S.; Armin, M.; Hernández, S.; Urbano, J.; Li, C. Wind-Wave Coupling Effects on the Fatigue Damage of Tendons for a 10 MW Multi-Body Floating Wind Turbine. *Ocean Eng.* **2020**, *217*, 107909. [CrossRef]
17. Baita-Saavedra, E.; Cordal-Iglesias, D.; Filgueira-Vizoso, A.; Morató, À.; Lamas-Galdo, I.; Álvarez-Feal, C.; Carral, L.; Castro-Santos, L. An Economic Analysis of An Innovative Floating Offshore Wind Platform Built with Concrete: The SATH®Platform. *Appl. Sci.* **2020**, *10*, 3678. [CrossRef]
18. Filgueira-Vizoso, A.; Castro-Santos, L.; Iglesias, D.C.; Puime-Guillén, F.; Lamas-Galdo, I.; García-Diez, A.I.; Uzunoglu, E.; Díaz, H.; Guedes Soares, C. The Technical and Economic Feasibility of the CENTEC Floating Offshore Wind Platform. *J. Mar. Sci. Eng.* **2022**, *10*, 1344. [CrossRef]
19. Yang, Y.; Bashir, M.; Michailides, C.; Mei, X.; Wang, J.; Li, C. Coupled Analysis of a 10 MW Multi-Body Floating Offshore Wind Turbine Subjected to Tendon Failures. *Renew. Energy* **2021**, *176*, 89–105. [CrossRef]
20. Vittori, F.; Azcona, J.; Eguinoa, I.; Pires, O.; Rodríguez, A.; Morató, Á.; Garrido, C.; Desmond, C. Model Tests of a 10 MW Semi-Submersible Floating Wind Turbine under Waves and Wind Using Hybrid Method to Integrate the Rotor Thrust and Moments. *Wind Energy Sci.* **2022**, *7*, 2149–2161. [CrossRef]
21. Uzunoglu, E.; Guedes Soares, C. A System for the Hydrodynamic Design of Tension Leg Platforms of Floating Wind Turbines. *Ocean Eng.* **2019**, *171*, 78–92. [CrossRef]
22. Tran, T.-T.; Kim, D.-H. The Platform Pitching Motion of Floating Offshore Wind Turbine: A Preliminary Unsteady Aerodynamic Analysis. *J. Wind Eng. Ind. Aerodyn.* **2015**, *142*, 65–81. [CrossRef]
23. Jeon, M.; Lee, S.; Lee, S. Unsteady Aerodynamics of Offshore Floating Wind Turbines in Platform Pitching Motion Using Vortex Lattice Method. *Renew. Energy* **2014**, *65*, 207–212. [CrossRef]
24. Li, Z.; Wen, B.; Dong, X.; Long, X.; Peng, Z. Effect of Blade Pitch Control on Dynamic Characteristics of a Floating Offshore Wind Turbine under Platform Pitching Motion. *Ocean Eng.* **2021**, *232*, 109109. [CrossRef]
25. Tracy, C.H. Parametric Design of Floating Wind Turbines. Ph.D.Thesis, Massachusetts Institute of Technology, Cambridge, MA, USA, 2007.
26. Kibbee, S.E.; Chianis, J.; Davies, K.B.; Sarwono, B.A. The Seastar Tension-Leg Platform. In Proceedings of the All Days, OTC, Houston, TX, USA, 2 May 1994; p. OTC-7535-MS.
27. Bachynski, E.E. *Design and Dynamic Analysis of Tension Leg Platform Wind Turbines*; Norges Teknisk-Naturvitenskapelige Universitet, Fakultet for Ingeniørvitenskap og Teknologi, Institutt for Marin Teknikk: Trondheim, Norway, 2014; ISBN 978-82-326-0097-7.
28. Han, Y.; Le, C.; Ding, H.; Cheng, Z.; Zhang, P. Stability and Dynamic Response Analysis of a Submerged Tension Leg Platform for Offshore Wind Turbines. *Ocean Eng.* **2017**, *129*, 68–82. [CrossRef]
29. Adam, F.; Myland, T.; Dahlhaus, F.; Großmann, J. GICON®-TLP for Wind Turbines–the Path of Development. In *Renewable Energies Offshore*; Guedes Soares, C., Ed.; CRC Press: London, UK, 2015; pp. 651–656.
30. Kibbee, S.; Chianis, J.; Davies, K.; Sarwono, B. *A Mini-Platform for Deepwater-the SeaStar TLP*; The Society of Naval Architects and Marine Engineers: Houston, TX, USA, 1995.
31. Oguz, E.; Clelland, D.; Day, A.H.; Incecik, A.; López, J.A.; Sánchez, G.; Almeria, G.G. Experimental and Numerical Analysis of a TLP Floating Offshore Wind Turbine. *Ocean Eng.* **2018**, *147*, 591–605. [CrossRef]

32. Matha, D. *Model Development and Loads Analysis of an Offshore Wind Turbine on a Tension Leg Platform with a Comparison to Other Floating Turbine Concepts: April 2009*; National Renewable Energy Laboratory (NREL): Golden, CO, USA, 2010; p. 793961, NREL/SR-500-45891. [CrossRef]
33. ITTC. Recommended Procedures and Guidelines: Model Tests for Offshore Wind Turbines. Recommended Procedures and Guidelines 7.5-02-07-03.8. In Proceedings of the International Towing Tank Conference, Zürich, Switzerland, 13–18 June 2021; p. 19.
34. Sanchez, R. *Technology Readiness Assessment Guide*; United States Department of Energy: Washington, DC, USA, 2011; p. 73.
35. Uzunoglu, E.; Guedes Soares, C. Supervisory System for the Automation of Model Building and Simulations with the Wind Turbine Code FAST. In *Progress in Renewable Energies Offshore*; Guedes Soares, C., Ed.; Taylor & Francis Group: London, UK, 2016; pp. 627–636. ISBN 978-1-138-62627-0.
36. Uzunoglu, E.; Guedes Soares, C. Parametric Modelling of Marine Structures for Hydrodynamic Calculations. *Ocean Eng.* **2018**, *160*, 181–196. [CrossRef]
37. Uzunoglu, E.; Guedes Soares, C. Response Dynamics of a Free-Float Capable Tension Leg Platform for a 10 MW Wind Turbine at the Northern Iberian Peninsula. In *Developments in Renewable Energies Offshore*; Guedes Soares, C., Ed.; Taylor & Francis Group: London, UK, 2020; pp. 408–416. [CrossRef]
38. Zavvar, E.; Chen, B.Q.; Uzunoglu, E.; Guedes Soares, C. Stress Distribution on the CENTEC-TLP in Still Water and Rated Wind Speed. In *Trends in Maritime Technology and Engineering Volume 2*; Guedes Soares, C., Santos, T.A., Eds.; Taylor & Francis Group: London, UK, 2022; pp. 519–526. [CrossRef]
39. Zavvar, E.; Abdelwahab, H.S.; Uzunoglu, E.; Chen, B.Q.; Guedes Soares, C. Numerical Study of the Wave Induced Motions and Loads on the CENTEC-TLP Floating Wind Turbine. In *Trends in Renewable Energies Offshore*; Guedes Soares, C., Ed.; Taylor & Francis Group: London, UK, 2022; pp. 567–573. [CrossRef]
40. Hmedi, M.; Uzunoglu, E.; Guedes Soares, C.; Medina-Manuel, A.; Mas-Soler, J.; Abad-Gibert, V.; Souto-Iglesias, A.; Vittori, F.; Pires, O.; Azcona, J. Experimental Analysis of a Free-Float Capable Tension Leg Platform with a 10 MW Turbine. In *Trends in Renewable Energies Offshore*; Guedes Soares, C., Ed.; Taylor & Francis Group: London, UK, 2022; pp. 549–557. [CrossRef]
41. Faltinsen, O.M. *Sea Loads on Ships and Offshore Structures*, Cambridge ocean technology series; 1. paperback ed.; repr.transferred to digital printing; Cambridge University Press: Cambridge, UK, 1999; ISBN 978-0-521-45870-2.
42. Mas-Soler, J.; Uzunoglu, E.; Bulian, G.; Guedes Soares, C.; Souto-Iglesias, A. An Experimental Study on Transporting a Free-Float Capable Tension Leg Platform for a 10 MW Wind Turbine in Waves. *Renew. Energy* **2021**, *179*, 2158–2173. [CrossRef]
43. Vittori, F.; Pires, O.; Azcona, J.; Uzunoglu, E.; Guedes Soares, C.; Zamora Rodríguez, R.; Souto-Iglesias, A. Hybrid Scaled Testing of a 10MW TLP Floating Wind Turbine Using the SiL Method to Integrate the Rotor Thrust and Moments. In *Developments in Renewable Energies Offshore*; Guedes Soares, C., Ed.; Taylor & Francis Group: London, UK, 2020; pp. 417–423. [CrossRef]
44. Pires, O.; Azcona, J.; Vittori, F.; Bayati, I.; Gueydon, S.; Fontanella, A.; Liu, Y.; de Ridder, E.J.; Belloli, M.; van Wingerden, J.W. Inclusion of Rotor Moments in Scaled Wave Tank Test of a Floating Wind Turbine Using SiL Hybrid Method. *J. Phys. Conf. Ser.* **2020**, *1618*, 032048. [CrossRef]
45. Azcona, J.; Bouchotrouch, F.; González, M.; Garciandía, J.; Munduate, X.; Kelberlau, F.; Nygaard, T.A. Aerodynamic Thrust Modelling in Wave Tank Tests of Offshore Floating Wind Turbines Using a Ducted Fan. *J. Phys. Conf. Ser.* **2014**, *524*, 012089. [CrossRef]
46. Hmedi, M.; Uzunoglu, E.; Guedes Soares, C. Review of Hybrid Model Testing Approaches for Floating Wind Turbines. In *Trends in Maritime Technology and Engineering Volume 2*; Guedes Soares, C., Santos, T.A., Eds.; Taylor & Francis Group: London, UK, 2022; pp. 421–428. [CrossRef]
47. Otter, A.; Murphy, J.; Pakrashi, V.; Robertson, A.; Desmond, C. A Review of Modelling Techniques for Floating Offshore Wind Turbines. *Wind Energy* **2021**, *25*, 831–857. [CrossRef]
48. Bak, C.; Zahle, F.; Bitsche, R.; Kim, T.; Yde, A.; Henriksen, L.C.; Hansen, M.H.; Blasques., J.P.A.A.; Gaunaa, M.; Natarajan, A. The DTU 10-MW Reference Wind Turbine. In *Danish Wind Power Research 2013*; 2013. Available online: https://backend.orbit.dtu.dk/ws/portalfiles/portal/55645274/The_DTU_10MW_Reference_Turbine_Christian_Bak.pdf (accessed on 30 October 2022).
49. Jonkman, J.; Butterfield, S.; Musial, W.; Scott, G. *Definition of a 5-MW Reference Wind Turbine for Offshore System Development*; National Renewable Energy Laboratory (NREL): Golden, CO, USA, 2009; p. 75, NREL/TP-500-38060.
50. Robertson, A.; Jonkman, J.; Masciola, M.; Song, H.; Goupee, A.; Coulling, A.; Luan, C. *Definition of the Semisubmersible Floating System for Phase II of OC4*; National Renewable Energy Laboratory (NREL): Golden, CO, USA, 2014; p. 1155123, NREL/TP-5000-60601.
51. Uzunoglu, E.; Guedes Soares, C. An Integrated Design Approach for a Self-Float Capable Tension Leg Platform for Wind Energy. In *Developments in Maritime Technology and Engineering*; Guedes Soares, C., Santos, T.A., Eds.; Taylor & Francis Group: London, UK, 2021; pp. 673–681. [CrossRef]
52. Hmedi, M.; Uzunoglu, E.; Guedes Soares, C. Influence of Platform Configuration on the Hydrodynamic Performance of Semi-Submersibles for Offshore Wind Energy. In *Trends in Maritime Technology and Engineering Volume 2*; Guedes Soares, C., Santos, T.A., Eds.; Taylor & Francis Group: London, UK, 2022; pp. 411–420. [CrossRef]
53. Silva, D.; Gonçalves, M.; Bentamy, A.; Guedes Soares, C. Assessment of the Use of Scatterometer Wind Data to Force Wave Models in the North Atlantic Ocean. *Ocean Eng.* **2022**, *266*, 112803. [CrossRef]

54. Hmedi, M.; Uzunoglu, E.; Guedes Soares, C. Effect of Geometry Modifications on the Dynamics of a Free-Float Capable Tension Leg Platform. In *Trends in Renewable Energies Offshore*; Guedes Soares, C., Ed.; Taylor & Francis Group: London, UK, 2022; pp. 575–584. [CrossRef]
55. Chakrabarti, S.K. Physical Modelling of Offshore Structures. In *Handbook of Offshore Engineering*; Elsevier: Amsterdam, The Netherlands, 2005; pp. 1001–1054. ISBN 978-0-08-044381-2.
56. ITTC. Recommended Procedures and Guidelines: Floating Offshore Platform Experiments. Recommended Procedures and Guidelines 7.5-02-07-03.1. In Proceedings of the International Towing Tank Conference, Zürich, Switzerland, 13–18 June 2021; p. 12.
57. ITTC. Recommended Procedures and Guidelines: Seakeeping Experiments. Recommended Procedures and Guidelines 7.5-02-07-02.1. In Proceedings of the International Towing Tank Conference, Zürich, Switzerland, 13–18 June 2021; p. 33.
58. ITTC. Recommended Procedures and Guidelines: Passive Hybrid Model Tests of Floating Offshore Structures with Mooring Lines. Recommended Procedures and Guidelines 7.5-02-07-03.5. In Proceedings of the International Towing Tank Conference, Zürich, Switzerland, 13–18 June 2021; p. 8.
59. Sauder, T.; Chabaud, V.; Thys, M.; Bachynski, E.E.; Sæther, L.O. Real-Time Hybrid Model Testing of a Braceless Semisubmersible Wind Turbine: Part I—The Hybrid Approach. In Proceedings of the Volume 6: Ocean Space Utilization; Ocean Renewable Energy; American Society of Mechanical Engineers: Busan, Republic of Korea, 2016; p. V006T09A039.
60. Chabaud, V.; Steen, S.; Skjetne, R. Real-Time Hybrid Testing for Marine Structures: Challenges and Strategies. In *Proceedings of the Volume 5: Ocean Engineering*; American Society of Mechanical Engineers: Nantes, France, 2013. [CrossRef]
61. Jonkman, J. The New Modularization Framework for the FAST Wind Turbine CAE Tool. In *Proceedings of the 51st AIAA Aerospace Sciences Meeting including the New Horizons Forum and Aerospace Exposition, Grapevine, TX, USA, 7–11 January 2013*; American Institute of Aeronautics and Astronautics: Grapevine, TX, USA, 2013.
62. Silva, D.; Bento, A.R.; Martinho, P.; Guedes Soares, C. High Resolution Local Wave Energy Modelling in the Iberian Peninsula. *Energy* **2015**, *91*, 1099–1112. [CrossRef]
63. DNV. *Marine Operations and Marine Warranty*; DNVGL-STN-001; DNV: Berum, Norway, 2018; p. 676.
64. DNV. *Floating Wind Turbine Structures*; DNVGL-ST-F205; DNV: Berum, Norway, 2018; p. 162.
65. DNV. *Global Performance Analysis of Deepwater Floating Structures*; DNVGL-RP-F205; DNV: Berum, Norway, 2019.
66. PI Recommended Practice for Planning, Designing, and Constructing Fixed Offshore Platforms. [American Petroleum In-stitute]. RECOMMENDED PRACTICE 2A-WSD (RP 2A-WSD) TWENTY-FIRST EDITION, DECEMBER. 2000.
67. API RP 2T: Planning, Designing, and Constructing Tension Leg Platforms. 2010.
68. Chakrabarti, S.K. *Handbook of Offshore Engineering*; Elsevier: Amsterdam, The Netherlands, 2005; ISBN 978-0-08-044381-2.
69. ITTC. Recommended Procedures and Guidelines: Analysis Procedure for Model Tests in Regular Waves. Recommended Procedures and Guidelines 7.5-02-07-03.2. In Proceedings of the International Towing Tank Conference, Zürich, Switzerland, 13–18 June 2021; p. 5.
70. Uzunoglu, E.; Guedes Soares, C. Yaw Motion of Floating Wind Turbine Platforms Induced by Pitch Actuator Fault in Storm Conditions. *Renew. Energy* **2019**, *134*, 1056–1070. [CrossRef]
71. Arnal, V.; Bonnefoy, F.; Gilloteaux, J.-C.; Aubrun, S. Hybrid Model Testing of Floating Wind Turbines: Test Bench for System Identification and Performance Assessment. In *Proceedings of the Volume 10: Ocean Renewable Energy, Glasgow, UK, 9–14 June 2019*; American Society of Mechanical Engineers: Glasgow, UK, 2019. [CrossRef]
72. Roddier, D.; Cermelli, C.; Aubault, A.; Weinstein, A. WindFloat: A Floating Foundation for Offshore Wind Turbines. *J. Renew. Sustain. Energy* **2010**, *2*, 033104. [CrossRef]

Journal of
Marine Science
and Engineering

MDPI

Article

Hydrodynamic Analysis of the WIND-Bos Spar Floating Offshore Wind Turbine

Thiago S. Hallak [1], C. Guedes Soares [1,*], Oscar Sainz [2], Sergio Hernández [2] and Alfonso Arévalo [2]

1 Centre for Marine Technology and Ocean Engineering (CENTEC), Instituto Superior Técnico, Universidade de Lisboa, 1049-001 Lisboa, Portugal
2 Bluenewables, Guia de Isora, 38689 Tenerife, Spain
* Correspondence: c.guedes.soares@centec.tecnico.ulisboa.pt

Abstract: The WIND-bos spar Floating Offshore Wind Turbine is studied both experimentally and numerically. The experimental model of the moored WIND-bos platform is presented, and the different numerical models that have been developed to analyze the hydrodynamics of the platform are described. The results provide a detailed comparison of numerical and experimental motion responses of the floating structure in regular and irregular waves. The numerical study includes frequency domain results from spectral analysis, weakly nonlinear time-domain results from a validated in-house code, and coupled time-domain results from commercial software. The importance of damping calibration is put in evidence, whereas damping ratios are calculated iteratively in the coupled time-domain simulations, and nonlinear damping force is considered within the developed numerical scheme. The results compare well and also show that the novel concept has a good motion performance in general.

Keywords: wind energy; spar; floating offshore wind turbine; coupled analysis; numerical-experimental comparison

Citation: Hallak, T.S.; Guedes Soares, C.; Sainz, O.; Hernández, S.; Arévalo, A. Hydrodynamic Analysis of the WIND-Bos Spar Floating Offshore Wind Turbine. *J. Mar. Sci. Eng.* 2022, *10*, 1824. https://doi.org/10.3390/jmse10121824

Academic Editor: Eva Loukogeorgaki

Received: 3 October 2022
Accepted: 22 November 2022
Published: 28 November 2022

1. Introduction

Floating Offshore Wind Turbines (FOWTs) are floating structures that enable wind power absorption in locations offshore where fixed wind turbines are not feasible, mainly due to the considerably great water depths in such locations. Although most of the present wind farms are in shallow water with fixed platforms, the present trend is to move away from the coast using FOWTs [1]. Prospective operation sites for FOWTs are, for instance, the coasts of North Atlantic European countries, such as Portugal, Norway, Ireland, and Scotland, as well as the USA, China, and Japan, among others [2].

Even though the majority of the installed wind power capacity still lies onshore, the offshore wind energy sector is currently undergoing greater expansion: as of 2020, the global installed power capacity of offshore wind was 35.3 GW, more than ten times what was ten years before. However, around 80% of this had been installed in only three countries, namely the UK, China, and Germany, which accounted for, respectively, 10.2 GW, 9.9 GW, and 7.8 GW of installed capacity offshore by the end of 2020 [3]. In 2021, the increase in wind power capacity offshore was led by China: around 17 GW of offshore wind was installed in 2021 only in that country, which accounts for 80% of the new installations in that year, followed by the UK with 11% and Vietnam with 4%. All of that took the global installations of offshore wind to a full power capacity of 57.2 GW [4], meaning that almost 40% of the installations offshore were accomplished last year. Remarkably, the parcel of floating wind is also rather small, for less than 200 MW of the offshore wind turbines are mounted on FOWTs, which indicates enormous amounts of wind resources that are yet to be exploited in deeper waters [5].

Historically, the technology for FOWTs also benefited from the knowledge previously acquired by the Oil & Gas (O&G) industry. However, as pointed out in [6], transferring the offshore O&G technology directly to the offshore wind industry would not have been economical and, in some cases, inappropriate, as happens with the moorings, which for wind and other renewable energy converters need to be designed for much smaller water depths [7]. The main technological challenges have been addressed, and different conceptual designs emerged seeking to develop cost-effective solutions for the offshore wind industry. The current trend of such concepts is toward bigger turbines and greater output to reduce overall CAPEX and OPEX and the commissioning of floating structures leading to an expansion of wind energy farms based on FOWTs [1].

The qualitative valuation of three floating concepts—barge, Tension Leg Platform (TLP), and spar, was introduced in [8] in relative terms of stability, motions, costs, and others. Then, ref. [9] analyzed three designs representing the same concepts (the ITI Energy barge, the MIT/NREL TLP, and the OC3-Hywind spar) using the tool FAST [10]. Those comparative studies did not include semi-submersible platforms, which is unfortunate. It was identified that barge platforms may not be suitable for FOWTs mainly due to the considerably high fatigue loads observed in the barge-supported turbine. These loads include, for instance, side-by-side and fore-aft tower base bending moments and may also compromise the nacelle with high inertial loads. On the other hand, the MIT/NREL TLP and the OC3-Hywind showed similar and relatively improved fatigue responses. Even though no Tension Leg Platform (TLP) has been through real-scale prototype testing, the single column with four spikes TLP has been considered in various studies, e.g., [11,12]. More recently, the CENTEC TLP platform is an example of an innovative concept currently under development that allows significant cost reduction in installation costs for TLPs [13–15]. The concept is designed to accommodate the DTU 10.0 MW Reference Wind Turbine [16].

Two of the types, however, can be illustrated with concepts that went through prototype testing at sea, namely the Hywind spar platform and the WindFloat semi-submersible platform. The development of these concepts led to the commissioning of the first two wind farms ever, in 2017 and 2019, respectively. The Hywind and WindFloat wind farms operate in Scotland at 30.0 MW and in Portugal at 25.0 MW, respectively. The development of the Hywind technology can be followed in [17–20], while the development of the WindFloat system can be consulted in [21,22].

Because semi-submersible platforms have relatively low installation costs and good site adaptability, they are often the FOWT solution considered in the full-scale analysis [23]. Two examples of semi-submersible FOWT concepts that have been thoroughly analyzed and deserve some detail are the OC5-DeepCWind platform and the braceless CSC semi-submersible platform. Both of them were designed to accommodate the NREL 5.0 MW wind turbine [24]. The CSC platform has been tested in wave basins [25,26] and later studied by different authors to develop and calibrate a fully coupled model of the oceanic system, as provided by [27–29]. The latest contributions offered accurate global motion prediction in various load cases and proved that the concept has good motion performance in general. The DeepCWind platform is an object of study for authors of various institutions because necessary data are available open-source.

Model uncertainty analysis of the DeepCWind platform was performed in [30], which presented hydrodynamic calculations by different codes, which in turn have been used in a validation study by [31]. Then, ref. [32] assessed the experimental uncertainty of the DeepCWind also in a test campaign, this time under hydrodynamic loading. The effect of wave nonlinearity on extreme responses and fatigue of the DeepCWind platform was studied in [33]. It was concluded that fatigue damage at the tower base is dominated by wind effects, as well as mooring lines fatigue; the surge motion, on the other hand, is dominated by wave effects when extreme responses are observed. The coupling of the DeepCWind platform with conical point-absorbers Wave Energy Converters (WECs) was studied by [34]. The results show an increase in initial stability and power absorption, but the motion response in irregular waves is worsened due to the mechanical coupling, which affects especially for the pitch motion. The results are presented for three different WEC diameters, also showing that the greater the WECs diameter, the stronger those trends. A similar concept was later studied by [35], which obtained similar results even though the conclusions drawn were mainly qualitative.

While several of the mentioned studies have been made for 5.0 MW wind turbines, recent efforts of adaptation or development have addressed 10.0 MW wind turbines, such as in [36–39].

Hybrid wind-wave floating solutions, such as the ones discussed above, are examples of Multi-Use Platforms (MUPs). These concepts have been reviewed by [40,41], with structure classifications and design challenges described. The DeepCWind platform is one among various configurations of FOWT that have been considered in MUP studies, such as the Mermaid semi-submersible with Oscillating Water Columns (OWCs) [42]. More recent conceptual designs are the spar platform with surge flaps and a tidal current converter [43], the concentric WEC-array adapted to semi-submersible wind platforms [44], and the FOWT proposed with a desalination plant and a solar deck [45], cite a few.

The design considered in the present paper is a novel spar FOWT: the WIND-bos platform. Spars, in general, have a considerably small waterplane area and a considerably large draft. Moreover, ballast weight is included deep down in the platform. Thus, due to these two reasons, the natural periods in the heave, roll, and pitch of spars are usually much higher in comparison to common sea wave periods. Because of the ballast weight, spar platforms also tend to be very stable. However, the WIND-bos platform has a particular geometry, which is different from the conventional spars, such as the Hywind concept, which is also a spar. The hull of the WIND-bos platform is composed of two main bodies built of steel and concrete, which has several advantages that allow significant cost reduction and also allow smooth, safe, and cost-efficient transport and installation processes. These two bodies are interconnected by means of three vertical steel legs, as shown in Figure 1. The platform is designed to accommodate the DTU 10.0 MW Reference Wind Turbine.

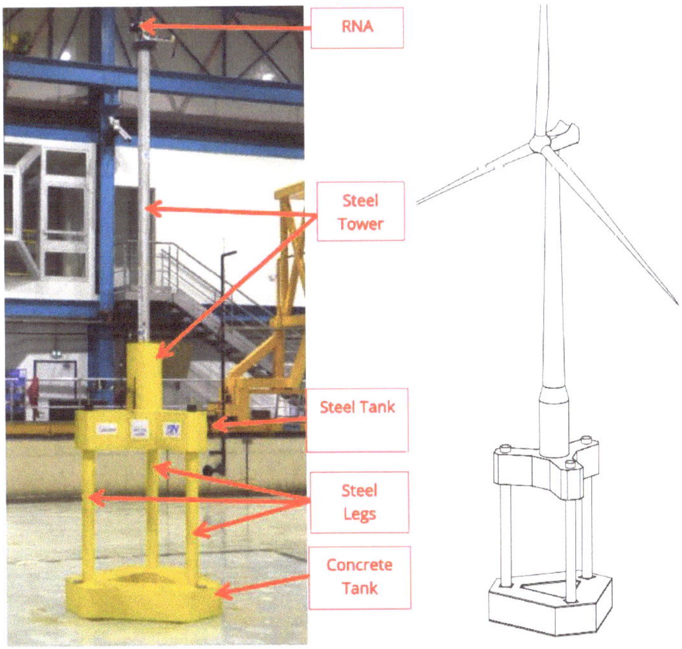

Figure 1. Experimental model of the WIND-bos platform (**left**) and its view in perspective (**right**).

The hydrodynamic behavior of the WIND-bos platform is studied in detail. First, the scaled model of the platform is analyzed based on the tests performed in the experimental campaign conducted under the scope of the MaRINET2 program. Then, numerical models are used to simulate the dynamic behavior of the platform according to three different methods, namely, frequency domain, weakly nonlinear time domain, and coupled time-domain models. The weakly nonlinear time-domain solver is developed in-house and predicts the 6 degrees of freedom (DoF) motion responses of the platform in the time domain under regular and irregular waves. Because of the many models developed, the study provides a deep numerical-experimental comparison for the hydrodynamic analysis of the novel FOWT concept.

2. Experimental Campaign

The experimental campaign was accomplished in the BGO First Facility operated by Oceanide, in La Seyne-sur-mer, France. The 1/40 scale model of the WIND-bos platform tested in the wave basin is shown in Figure 1, when outside of the tank, besides a perspective view of the platform accommodating the wind turbine. In Table 1, the environmental conditions tested are presented and include both regular and irregular wave scenarios, whereas the latter are generated based on JONSWAP spectra. Regular and irregular waves are considered, but the results are presented for head wave incidence only. The environmental parameters result from the Froude scaling of real sea conditions, e.g., $H' = \alpha H$ and $T' = \sqrt{\alpha}T$, where $\alpha = 1/40$ is the scale parameter, and the apostrophe stands for values in the model scale.

Table 1. Wave conditions (real scale).

Regular Wave Conditions					
Sea State	**T [s]**	**H [m]**	**Sea State**	**T [s]**	**H [m]**
RW01	4.0	1.0	RW11	13.0	2.0
RW02	4.0	2.0	RW12	13.0	4.0
RW03	5.0	2.0	RW13	15.0	2.0
RW04	5.0	4.0	RW14	15.0	4.0
RW05	7.0	2.0	RW15	17.0	2.0
RW06	7.0	4.0	RW16	17.0	4.0
RW07	9.0	2.0	RW17	19.0	2.0
RW08	9.0	4.0	RW18	19.0	4.0
RW09	11.0	2.0	RW19	21.0	2.0
RW10	11.0	4.0	RW20	21.0	4.0

Irregular Wave Conditions			
Sea State	**Tp [s]**	**Hs [m]**	**γ [-]**
IW1	6.0	2.0	3.3
IW2	9.0	4.0	3.3
IW3	11.0	6.0	3.3
IW4	12.0	10.0	3.3
IW5	8.0	2.0	3.3
IW6	11.0	4.0	3.3
IW7	13.0	6.0	3.3
IW8	14.0	10.0	3.3

A linear mooring system consisting of three horizontal dry steel wires is used to keep the model heading and to provide station-keeping. The mooring lines' diameter is only 40 mm (or 1 mm in the model scale), whereas the length of the lines is 337.72 m. The mooring is much lighter in comparison to the catenary mooring that is usually considered in real-scale prototypes (not detailed in this paper); also, a set of springs was used to reach the target pretension values; thus, the restoring forces related to the mooring do play an important role in the dynamic behavior of the system and should essentially behave as linear springs. The mooring lines parameters are presented in Table 2, whereas the configuration of the lines is shown in detail in Figure 2, and the fairleads and anchors positions are presented in Table 3.

Table 2. Mooring parameters.

Parameter	Description	Value	Unit
N	Number of lines	3	[-]
α	Angle between lines	120	[deg]
L	Line length	337.72	[m]
W_D	Dry weight	9.86	[kg/m]
EA	Stiffness	78.9	[kN/m]
Γ_0	Pre-tension	34.21	[kN]

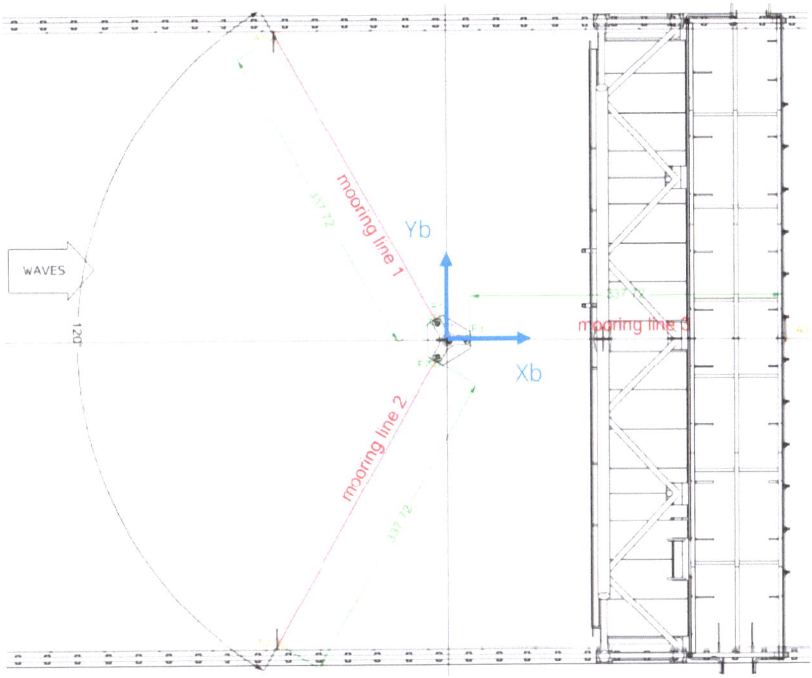

Figure 2. Scheme of the experimental facilities with the mooring lines configuration.

Table 3. Fairleads and anchors positions.

Line	x [m]	y [m]	z [m]
Fairleads			
1	−12.34	21.37	−17.04
2	−12.34	21.37	−17.04
3	24.68	0.00	−17.04
Anchors			
Line	**x [m]**	**y [m]**	**z [m]**
1	−180.59	314.20	15.00
2	−180.59	314.20	15.00
3	362.42	0.00	15.00

The sampling frequency was set at 15.8 Hz (or 100 Hz in the model scale) for the instrumentation. A low-pass filter [0.0; 0.5 Hz] has been applied to remove the high-frequency noise, and a band-pass filter [0.029; 0.5 Hz] has been applied on motion time series to extract wave frequency signals during regular wave tests. The Response Amplitude

Operators (RAOs) are evaluated considering the wave amplitudes at the basin center, i.e., future model location, during wave calibration. Waves were also calibrated by 20 runs for regular waves and 10 runs for irregular waves. These runs were performed without the scale model in the basin. Free decay tests of the free-floating platform were performed in the vertical and X and Y rotation modes of motion and in 6 DoF for the moored platform. Pulling tests were also performed to obtain the initial restoring coefficients in surge and sway due to the effects of mooring lines. Tilt tests were also performed in roll and pitch to obtain the metacentric heights. The following values were measured: GM_T = 11.97 m and GM_L = 11.89 m.

Additionally, in this tank test campaign, several combined wave and wind runs were launched using an equivalent thrust force at the hub height using a drone. The combined wave and wind results are not included in this paper.

3. Numerical Modelling

3.1. Frequency Domain

An in-house code is written in Matlab® to obtain results in the frequency domain. The code considers the hydrodynamic coefficients of the platform, namely, the wave radiation (added mass and potential damping) and wave excitation force coefficients and phases, also the hydrostatic coefficients, the rigid body mass matrix, and an added damping matrix that calibrates the damping forces with the damping ratios observed in the tank test campaign. The numerical evaluation generates the linear system in complex variables and evaluates the complex Response Amplitude Operators (RAOs) in 6 DoF for each frequency. Basic spectral analysis techniques are then considered to obtain the statistics of motion for each sea state.

3.2. Weakly Nonlinear Time Domain

The weakly nonlinear time-domain model is written in Matlab® and solves the hydrodynamic problem by performing time integration. The code is under development and has already been validated in regular waves [46]. Nonlinear hydrodynamic force models may be accounted for in time-domain codes rather straightforwardly [47–49]. To simulate structures with tubular elements, such as the WIND-bos, nonlinear Morison models are usually required to account for viscous forces. That same approach is considered in this paper to validate the in-house time-domain code with new irregular wave results.

The vertical tubular elements of the WIND-bos and previous analysis of motion responses for the same wave scenarios pointed to the surge drag as the only strong nonlinear hydrodynamic force in action, and so it is the only nonlinear force added to the code. The code is, therefore, said to be a weakly nonlinear time-domain code.

The equation of motion in 6 DoF of the platform may be written in the form of Newton's 2nd law:

$$[\mathrm{M}]\cdot\ddot{\mathbf{x}}(t) = \mathbf{F_e}(t) + \mathbf{F_r}(t) + \mathbf{F_m}(t) + \mathbf{F_{hyds}}(t) + \mathbf{F_{add}}(t), \quad (1)$$

where M is the mass matrix of the rigid body; $\ddot{\mathbf{x}}$ is the acceleration vector; $\mathbf{F_e}$ is the wave excitation force; $\mathbf{F_{rad}}$ is the wave radiation force; $\mathbf{F_m}$ is the mooring force; $\mathbf{F_{hyds}}$ is the hydrostatic force; and $\mathbf{F_{add}}$ is the added force. The force formulation is based on [46,50], whereas the wave radiation force is evaluated according to the Cummins Equation [51]:

$$\mathbf{F_{hyds}}(t) = -\left[\mathrm{C_{hyds}}\right]\cdot\mathbf{x}(t), \quad (2)$$

$$\mathbf{F_m}(t) = -[\mathrm{C_m}]\cdot\mathbf{x}(t), \quad (3)$$

$$\mathbf{F_r}(t) = -[\mathrm{M_{A\infty}}]\cdot\ddot{\mathbf{x}}(t) - \int_0^{\infty}[\kappa(\tau)]\cdot\dot{\mathbf{x}}(t-\tau)d\tau, \quad (4)$$

where:

$$\kappa_{ij}(\tau) = \frac{2}{\pi}\int_0^{\infty} B_{ij}(\omega)\cos(\omega\tau)d\omega. \quad (5)$$

In the weakly nonlinear time-domain model, the added force term is added to the system to account for viscous effects in 6 DoF. The evaluation is performed taking into account the expected damping ratios as observed in the wave tank test campaign in 6 DoF and also includes a nonlinear force term that represents the surge drag force acting in the three long vertical steel legs of the platform and the two main underwater bodies:

$$\mathbf{F_{add}}(t) = -[B_{add}]\cdot\dot{\mathbf{x}}(t) - \frac{1}{2}\rho C_D A_X \dot{x}_1(t)|\dot{x}_1(t)|\hat{\mathbf{i}}, \quad (6)$$

where ρ is the water density; C_D is the weighted drag force coefficients of the underwater bodies; and A_X is the total projected area of the underwater bodies in the x direction. Substituting Equations (2)–(6) back into Equation (1):

$$\begin{aligned}&([M] + [M_{A\infty}])\cdot\ddot{\mathbf{x}}(t) + \textstyle\int_0^{\infty}[\kappa(\tau)]\cdot\dot{\mathbf{x}}(t - \tau)d\tau + [B_{add}]\cdot\dot{\mathbf{x}}(t)\\ &+\tfrac{1}{2}\rho C_D A_X \dot{x}_1(t)|\dot{x}_1(t)|\hat{\mathbf{i}} + \left([C_m] + \left[C_{hyds}\right]\right)\cdot\mathbf{x}(t) = \mathbf{F_e}(t).\end{aligned} \quad (7)$$

Remarkably, it is possible to consider any input force on the right-hand side. In this paper, the wave excitation force in irregular waves is evaluated by the code together with the wave elevation series, whereas this calculation is based on an arbitrarily large number of Airy components. For each one out of the I frequency bands, the code considers N Airy waves distributed within, such that the energy balance is satisfied, and the phases are random. The code also considers the 1.0 min ramp duration for the wave elevation and wave force series:

$$\psi_{i,j} \sim U[0\,,\,2\pi[, \quad (8)$$

$$\omega_{i,j} \sim U\left[\omega_i - \frac{d\omega_i}{2}\,,\,\omega_i + \frac{d\omega_i}{2}\right[, \quad (9)$$

$$A_{i,j} \sim \mathcal{N}\left(0\,,\,\frac{\sqrt{2S(\omega_i)d\omega_i}}{N}\right), \quad (10)$$

$$\zeta_w(t) = \textstyle\sum_{j=1}^{N}\sum_{i=i}^{I} A_{i,j}\sin\left(\omega_{i,j}t + \psi_{i,j}\right), \quad (11)$$

$$\mathbf{F_e}(t) = \textstyle\sum_{j=1}^{N}\sum_{i=i}^{I} f_e(\omega_i)A_{i,j}\sin\left(\omega_{i,j}t + \psi_{i,j} + \xi(\omega_i)\right), \quad (12)$$

where i,j is the index of the j-th wave within the i-th frequency band; ψ is the phase of the wave; ω is the frequency of the wave; A is the amplitude of the wave; $\mathbf{f_e}$ are the 6 DoF wave excitation force coefficients with correspondent phase ξ; $S(\omega)$ is the sea spectrum; U is the uniform distribution; and $\mathcal{N}(\mu,\,\sigma^2)$ is the normal distribution with mean μ, and variance σ^2. Note that Equation 9 applied to an even discretization of frequency bands $d\omega_i$ leads to an exact amount of $N \times I$ Airy waves uniformly distributed within the interval $\left[\omega_1 - \frac{d\omega}{2},\, \omega_N + \frac{d\omega}{2}\right[$. It is also important to remark that Equations 11 and 12 hold true for an arbitrarily large N; however, the discretization of waves cannot be overcounted, for the amplitude of each individual wave must be represented by the numerical tolerance of the processing machine.

Last but not least, time integration is performed using the Runge–Kutta 4th-Order Method in 6 DoF, with the forces and motions being evaluated every 0.04 s. The computational time required for each 3.0 h sea state is around a few hours in a regular computer.

3.3. Coupled Time Domain

The coupled time-domain simulations are performed by SESAM DeepC® and include a model of the coupling between the platform and mooring system. The mooring forces arise from diverse physical phenomena, for instance, due to the elasticity of the lines, the inertia of the lines, hydrodynamic damping acting on the lines, and wave excitation forces acting on the lines, just to cite some of the phenomena that may have indeed affect somehow the dynamics of the system in the basin. The software allows the user to input all relevant parameters ruling the dynamic of mooring lines; however, attention must also be paid to the measurement of values in the basin, as this can only be performed during the period of the wave tank test campaign. That said, Figure 3 presents the coupled model in SESAM DeepC®, with the lines modeled just as measured in the basin. Moreover, the coupled time-domain simulations are performed with a linear added damping matrix to match the significant amplitude of motion of simulations with the expected significant amplitude of motion as observed in the tank test campaign.

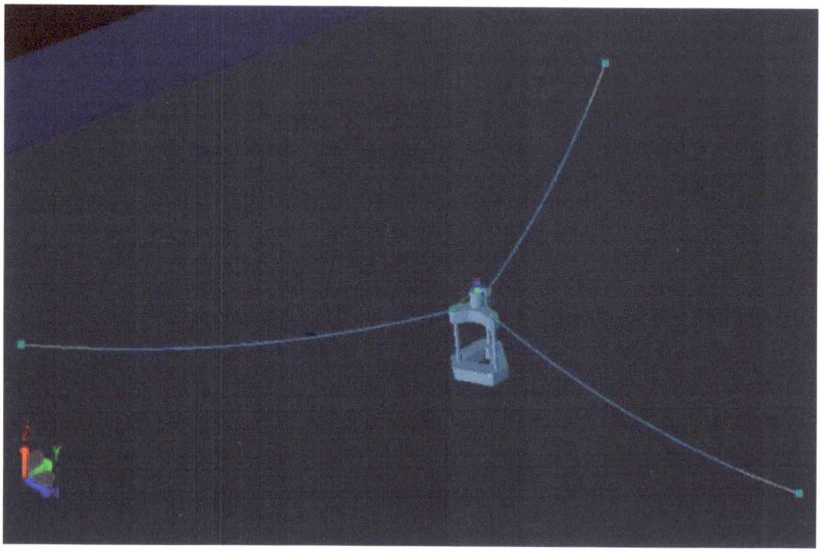

Figure 3. Coupled model in SESAM DeepC®.

By now, the numerical approaches do not include second-order effects such as wave mean drift and difference-of-frequency wave excitation forces. However, this should now be required to assess the first-order motions of the present platform, as measured in the experimental program.

4. Results

4.1. Free Decay and Iterative Damping Results

Free decay tests are performed to obtain basic characteristics of the platform, such as the natural periods at the different modes of motion, as well as the damping ratios. The WIND-bos platform is a spar FOWT, i.e., it has a small waterplane area in regard to its underwater sections. Thus, it is expected that most of the damping acting on the platform comes from viscous damping on the submerged elements but not from the potential damping related to wave radiation on the free surface. Viscous forces deviate significantly from linearity.

For the coupled time-domain simulations, to simulate the platform with a coherent damping matrix, an iterative method was applied: for each wave scenario, an added damping matrix was evaluated iteratively and added to the linear system till the numerical damping ratios correspond to the experimental damping ratios at the particular amplitudes of motion. There is a significant increase in computational time when doing so; however, the simulations are performed considering from 1.0 to 12.0 m significant wave height sea states, showing that the results might indeed be undermined if a single damping matrix is considered.

For underdamped oscillations, the damping ratio may be evaluated based on the rule of logarithmic decrement:

$$\delta = \frac{1}{k} \ln \frac{X_i}{X_{i+k}}, \tag{13}$$

where δ is the damping ratio; k is the number of full oscillations considered when performing the evaluation; and X_i is the i-th peak value. When using Equation 13, it is important to check that the motion is centered around the x-axis, i.e., $X_i \rightarrow 0$ as $i \rightarrow +\infty$.

The damping ratios obtained from experiments are presented in Table 4 for various amplitude ranges, illustrating that the damping ratios vary considerably according to the motion amplitudes of the platform. Based on the experimental free decay time series, the natural periods for heave, roll, and pitch read, respectively, 44.71, 27.33, and 27.35 s for the free-floating platform. Then, for the moored platform, the same natural periods read 44.38, 25.93, and 25.95 s for heave, roll, and pitch, respectively. Moreover, for the moored platform, the natural periods in surge, sway, and yaw read 123.54, 120.48, and 40.14 s, respectively. The free decay tests are presented in Figure 4 for the moored platform.

Table 4. Damping ratios obtained from experimental free decay tests.

Mode	Initial Peak	# of Cycles	Avg. amp. [m]	Damping Ratio
Surge	1	1	3.362	67.5%
	1	2	2.739	54.7%
	1	6	1.599	33.3%
	3	4	0.894	22.5%
	5	5	0.551	15.6%
	7	3	0.455	13.5%
Mode	**Initial Peak**	**# of cycles**	**Avg. amp. [m]**	**Damping Ratio**
Heave	1	1	4.369	22.2%
	1	2	4.034	18.3%
	1	6	3.004	16.3%
	3	4	2.458	15.3%
	5	5	1.816	10.8%
	7	3	1.606	5.7%
Mode	**Initial Peak**	**# of Cycles**	**Avg. amp. [°]**	**Damping Ratio**
Pitch	1	1	4.448	23.4%
	1	2	4.106	18.6%
	1	6	3.024	17.1%
	3	4	2.454	16.4%
	5	5	1.777	11.5%
	7	3	1.559	6.0%

The results presented in Table 4 show that the platform has considerably high damping, especially on the surge mode, whereas the damping on the pitch is fairly lower, considering that the pitch angle platform must be kept small, and so will be the damping ratio. Thus, it is reasonable to consider a constant damping ratio for pitch and heave in the small regular wave scenarios, and the values compute 16% in heave and 8.0% pitch. A similar trend to Table 4 is observed when plotting the final damping ratios obtained iteratively during the coupled time-domain simulations, as shown in Table 5.

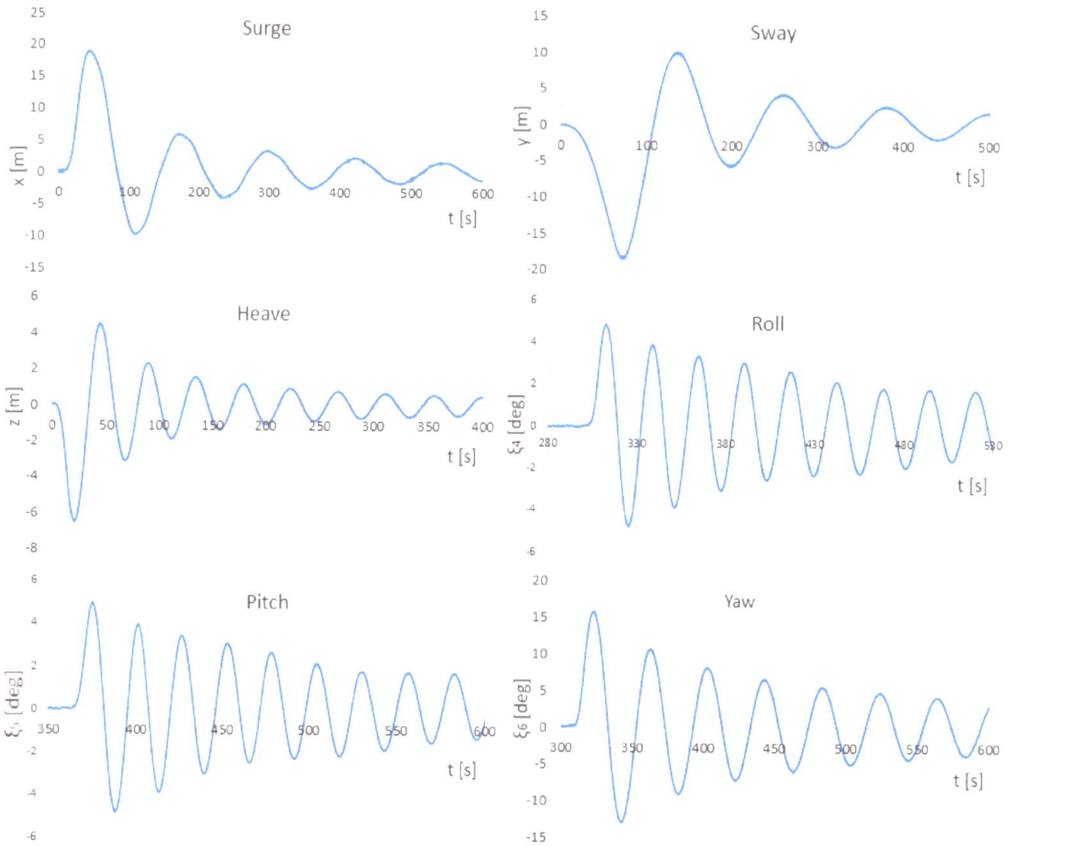

Figure 4. Free decay tests of the moored WIND-bos platform (experimental).

Table 5. Damping ratios obtained iteratively for the coupled time-domain simulations.

IW1, IW5		**IW2, IW6**	
Heave	10.0%	Heave	16.0%
Roll/Pitch	6.0%	Roll/Pitch	8.0%
IW3, IW7		**8.0 m Height IWs**	
Heave	20.0%	Heave	24.0%
Roll/Pitch	8.5%	Roll/Pitch	9.5%
IW4, IW8		**12.0 m Height IWs**	
Heave	31.0%	Heave	33.0%
Roll/Pitch	12.%	Roll/Pitch	14.0%

Because the surge response is affected by drift, a constant damping ratio was considered in the surge mode for the coupled time-domain simulations. When comparing the results between the free-floating and the moored platform, it is observed that the damping ratios are rather slightly affected; thus, the damping forces acting on the lines may be neglected in practice, which was expected from the type of mooring used, whereas the major effect of mooring is indeed stiffness.

By comparing the natural period results, it is observed that the heave natural period is decreased by only 1.0% after the inclusion of mooring lines. In the case of roll and pitch, the natural periods are decreased more considerably, by around 5.0%, most likely due to the elasticity of the lines. Because the natural periods of horizontal motions (surge, sway, and yaw) can only be evaluated for moored platforms, it is not possible to compare the results with a free-floating platform.

4.2. Regular Waves Results

The post-processed RAOs are plotted for different wave incidences in Figure 5.

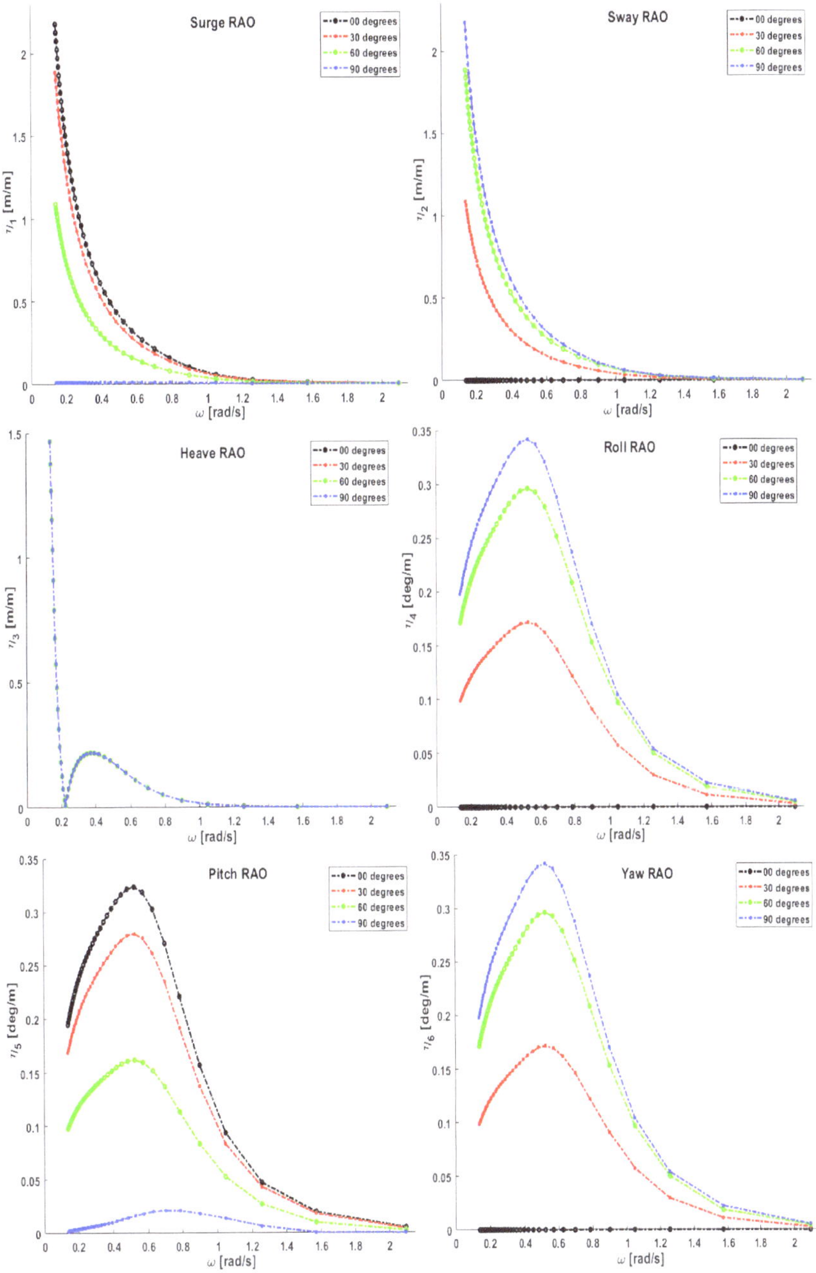

Figure 5. Frequency domain RAOs for several wave incidences.

In Figure 5, it is observed that the platform has a symmetric behavior when responding to sea waves, for the heave motion does not depend on the wave direction, and the roll and pitch motions compensate for each other. The natural frequency of heave is around 0.14 rad/s, where a peak is observed in Figure 5. The heave amplitude of motion reaches almost 1.5 m per meter of wave for long waves. However, the heave natural frequency is outside the range of common sea waves. In the case of roll and pitch, the peak of response is observed at 0.50 rad/s (~12.5 s), which is actually about half of the natural periods of those modes.

Figures 6–8 present the RAOs according to the different methods, mostly obtained from the regular wave scenarios, both experimentally and numerically. The amplitudes of motion are obtained based on the averaged distance between maxima and minima after the stationary regime has been achieved. Clearly, in those figures, the weakly nonlinear time-domain code is the method that achieves results that are closer to the experimental values, whereas the frequency domain results and the coupled time-domain results deviate in some cases and for different particular reasons. Still, the results obtained by all methods compare well.

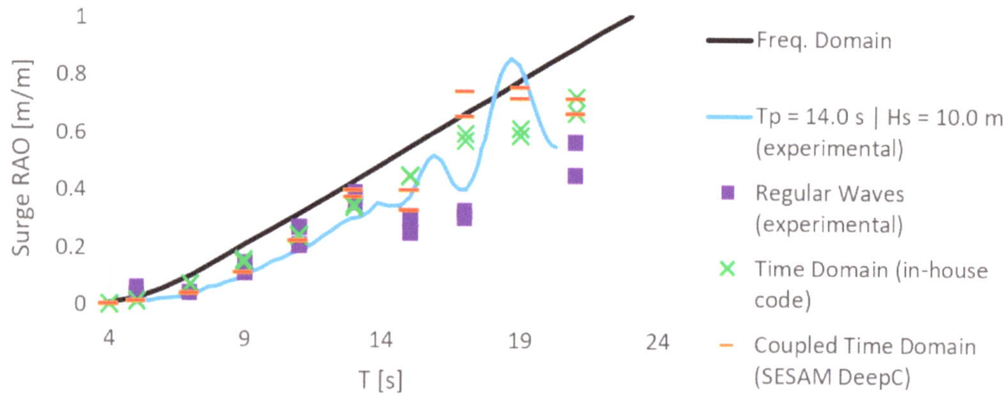

Figure 6. Surge RAOs according to different methods.

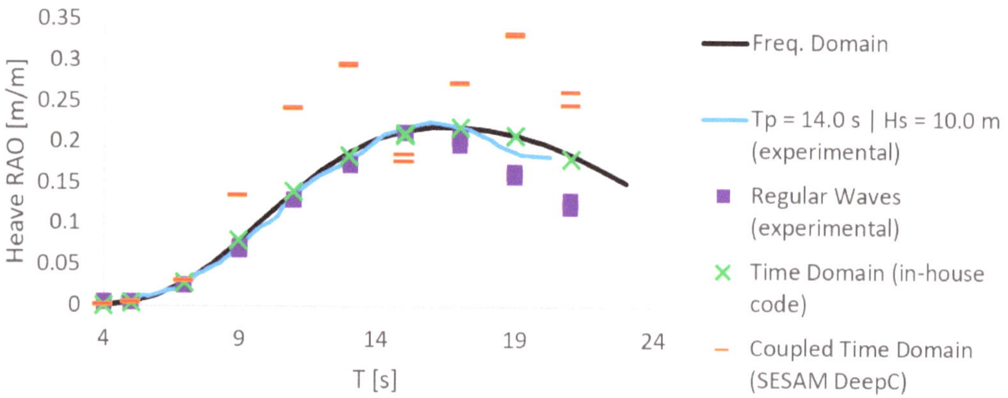

Figure 7. Heave RAOs according to different methods.

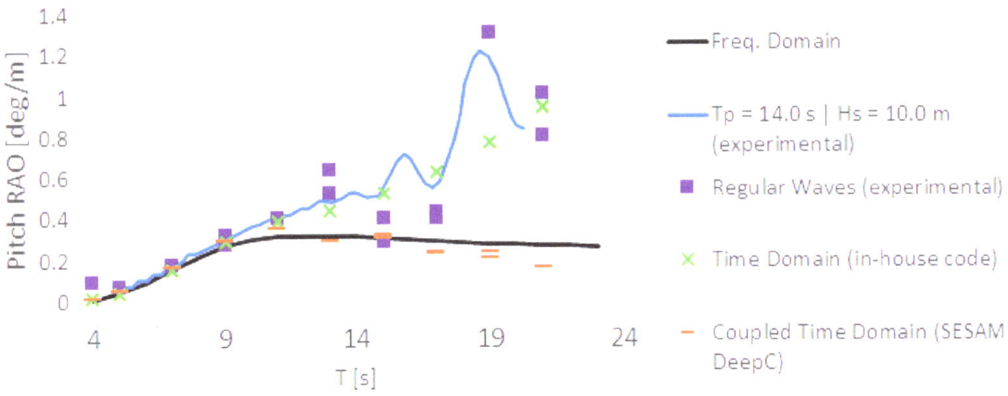

Figure 8. Pitch RAOs according to different methods.

The coupled time-domain model is the only one that predicts an increase in surge motion in wave periods between 17.0 and 21.0 s, for instance, whereas this increase in motion response has indeed been observed in the basin in RW17 and RW18 (19.0 s wave period), and may be related with mooring effects. On the other hand, the coupled time-domain model fails to predict the heave motion with accuracy. The comparison between different methods shows that the heave of the platform in regular waves is ruled mainly by first-order forces. Thus, the coupled time-domain simulations are probably failing to predict the heave mode because the addition of nonlinear mooring effects is not the ideal approach to increase the accuracy of the heave motion prediction.

In the case of pitch motion, the coupled time-domain simulations show that this mode is very sensitive to the pitch damping ratio assigned, especially at high wave periods, and therefore calibration is required to match values. In Figure 8, the coupled time-domain results are obtained after the significant amplitudes of motion matched with the expected ones for the same damping ratio. Because of this, an underestimation of response is being observed, for the damping ratio is calibrated with values obtained around the resonant frequency. The frequency domain model also fails to predict the pitch response to irregular waves, whereas the weakly nonlinear time-domain model is fairly good, capturing the trend of pitch response, especially when compared with the experimental RAOs obtained in the high energy scenario IW8.

4.3. Irregular Waves Results

The primary results obtained from the irregular wave simulations are the full-time series of motion, from which several parameters regarding hydrodynamic performance may be further evaluated. Figures 9–12 present the wave elevation, surge motion, heave motion, and pitch motion, respectively, in test IW8 as measured in the wave basin. These plots illustrate the response of the platform for an environmental scenario of interest, whereas it would be unpractical to present the whole set of time series obtained. In Figures 13–16, the wave elevation and motion response are plotted for the first 10.0 min of simulation during the weakly nonlinear time-domain simulation. Then, in Figures 17–19, the motion response of the platform is plotted for comparison regarding sea state IW8 and according to the coupled time-domain solver.

The comparison provided by the time series points out that experimental responses tend to have more accentuated maxima and minima in comparison to the numerical predictions. That may be further proven with the measures of motion statistics. For instance, in IW8, the platform heaved up to 2.0–2.5 m and reached more than 5.0 deg in pitch. The weakly nonlinear code shows responses with a considerably regular zero-crossing wave period.

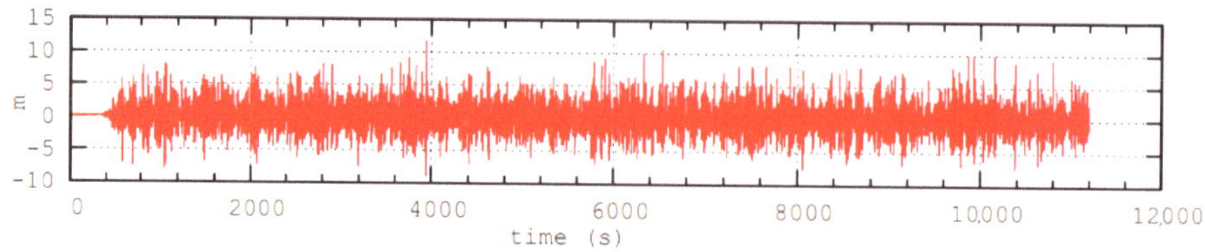

Figure 9. IW8 wave series according to experiments (Hs = 10.0 m, Tp = 14.0 s).

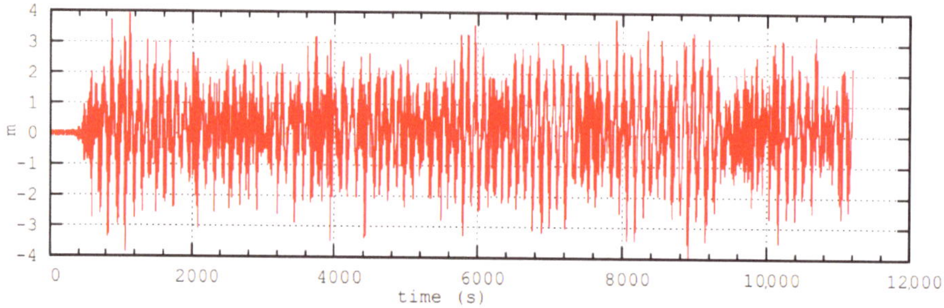

Figure 10. Surge response to IW8 according to experiments.

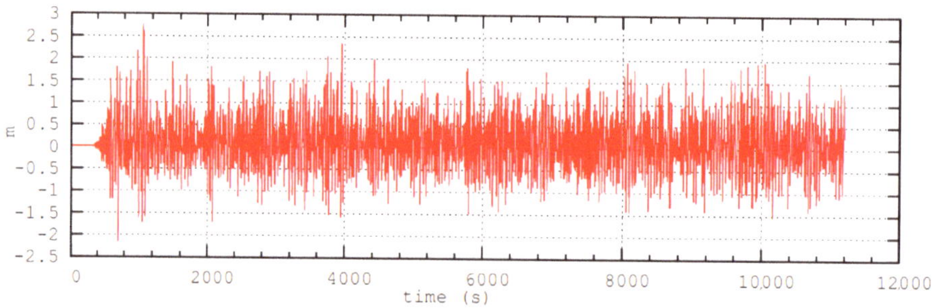

Figure 11. Heave response to IW8 according to experiments.

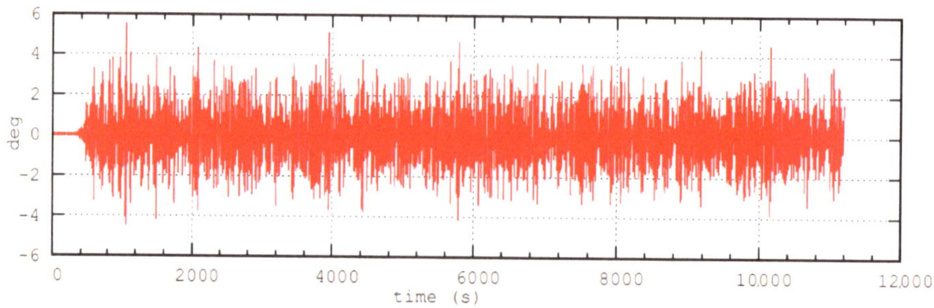

Figure 12. Pitch response to IW8 according to experiments.

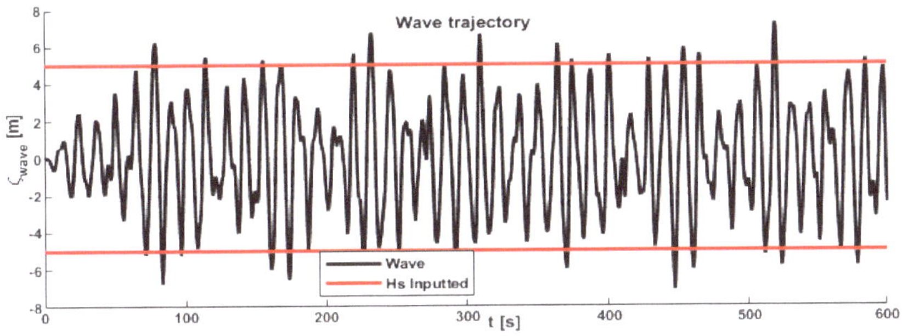

Figure 13. IW8 wave series according to weakly nonlinear time−domain solver (Hs = 10.0 m, Tp = 14.0 s).

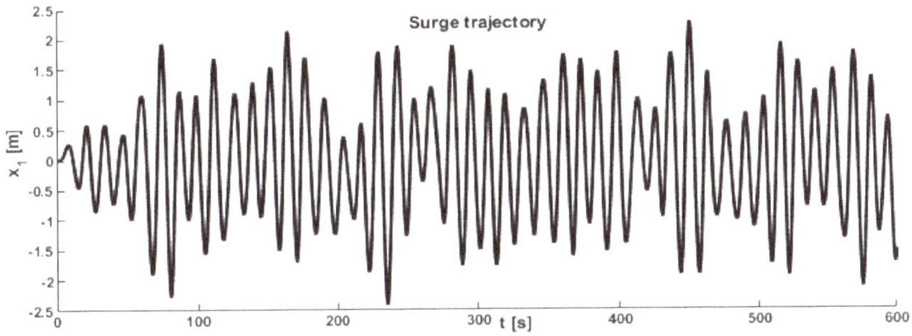

Figure 14. Surge response to IW8 according to weakly nonlinear time−domain solver.

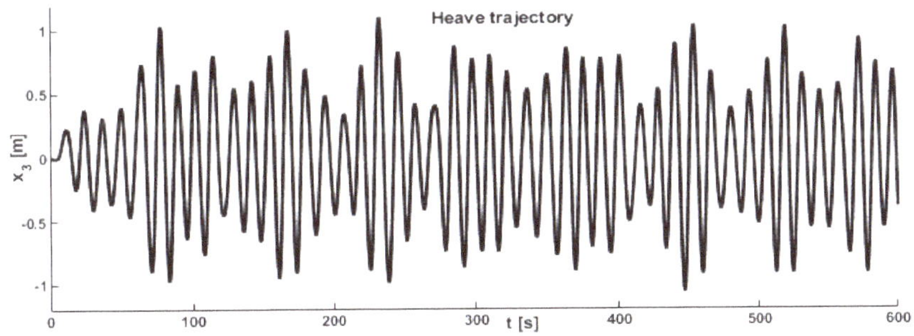

Figure 15. Heave response to IW8 according to weakly nonlinear time−domain solver.

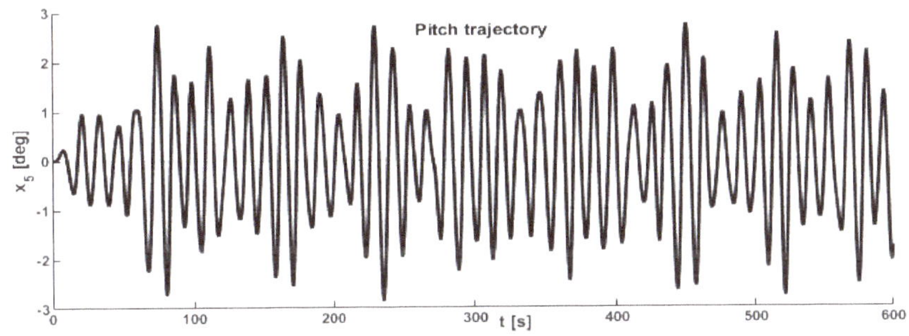

Figure 16. Pitch response to IW8 according to weakly nonlinear time−domain solver.

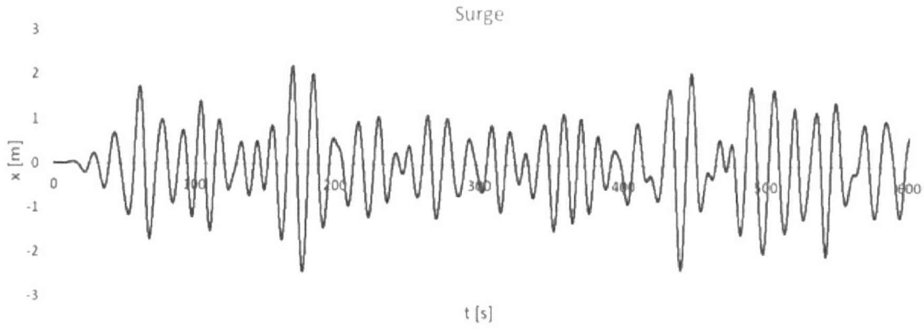

Figure 17. Surge response to IW8 according to the coupled time–domain solver SESAM DeepC®.

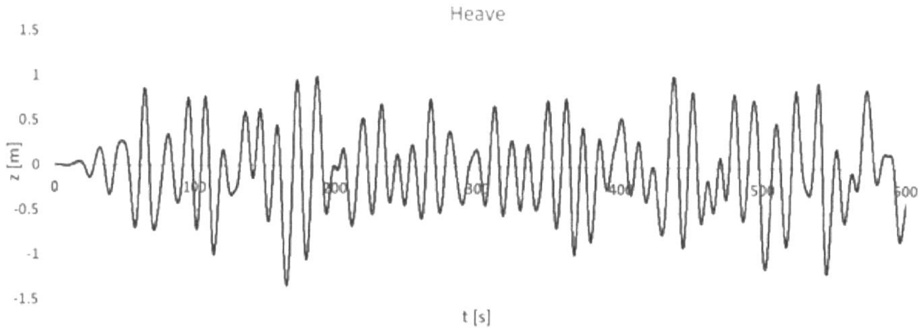

Figure 18. Heave response to IW8 according to the coupled time–domain solver SESAM DeepC®.

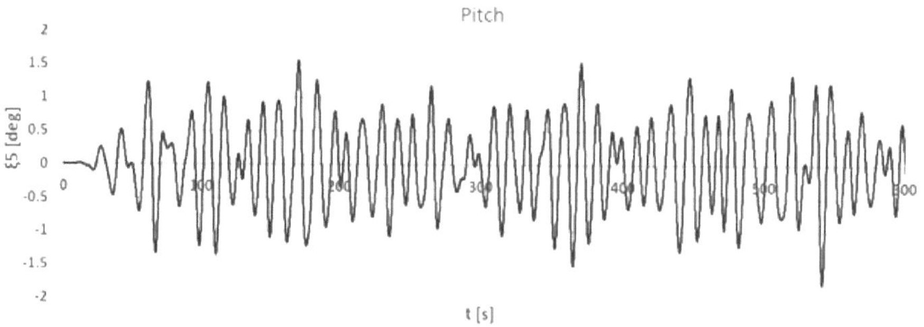

Figure 19. Pitch response to IW8 according to the coupled time–domain solver SESAM DeepC®.

Tables 6–8 present the wave and motion statistics results obtained during the irregular wave scenarios both numerically and experimentally. The results show that, between the three distinct numerical models considered, the coupled time-domain code is the one that predicts significant amplitudes of response closer to the experimental values, as can be observed from the standard deviation values. The absolute maxima and minima, however, tend to be lower in comparison to the experimental data, especially for the most energetic sea states. The frequency domain results, on the other hand, do not have a good agreement with the experimental data—the significant amplitude of motion is considerably overestimated in sea states IW5, IW6, and IW7, especially for the pitch motion, whereas in the less energetic sea states the trend is of underestimation. The in-house code also shows good agreement with the data, especially regarding the surge and pitch motions that were effectively calibrated in terms of nonlinear surge drag force.

Table 6. Surge motion statistics in irregular wave conditions according to different methods.

		Surge Motion					
Sea State	**Model** [1]	**Mean [m]**	**Stdd [m]**	**Tz [s]**	**Tc [s]**	**Max [m]**	**Min [m]**
IW1	FD	0.000	0.018	5.925	5.693	0.043	−0.043
	WNTD	0.000	0.023	6.311	5.803	0.057	−0.056
	CTD	−0.021	0.019	10.103	-	0.043	−0.086
	Exp.	0.128	0.188	8.015	-	0.916	−0.751
IW2	FD	0.000	0.062	8.548	7.173	0.169	−0.176
	WNTD	−0.001	0.125	8.129	8.425	0.329	−0.343
	CTD	−0.022	0.106	9.782	-	0.419	−0.440
	Exp.	0.130	0.304	15.369	-	1.171	−1.066
IW3	FD	0.000	0.470	10.993	10.940	1.176	−1.162
	WNTD	0.000	0.352	13.311	10.157	0.947	−0.957
	CTD	−0.028	0.283	11.701	-	0.887	−1.040
	Exp.	0.042	0.469	15.388	-	1.518	−2.196
IW4	FD	0.000	1.007	11.890	11.053	2.512	−2.530
	WNTD	0.005	0.502	13.075	8.752	1.390	−1.311
	CTD	−0.057	0.619	12.949	-	2.264	−2.696
	Exp.	0.320	1.040	21.902	-	3.569	−3.928
IW5	FD	0.000	1.018	11.899	11.053	2.550	−2.564
	WNTD	0.000	0.055	8.375	8.585	0.128	−0.133
	CTD	−0.020	0.040	9.326	-	0.141	−0.166
	Exp.	−0.103	0.116	5.744	-	0.398	−0.668
IW6	FD	0.000	1.120	11.983	11.030	2.757	−2.832
	WNTD	0.000	0.166	10.947	10.268	0.435	−0.405
	CTD	−0.022	0.187	11.638	-	0.688	−0.810
	Exp.	−0.091	0.279	13.024	-	0.803	−1.383
IW7	FD	0.000	1.058	11.880	12.073	2.826	−2.744
	WNTD	0.002	0.630	11.778	12.207	2.158	−2.285
	CTD	−0.026	0.439	13.399	-	1.849	−1.891
	Exp.	0.000	0.597	16.519	-	2.232	−2.488
IW8	FD	0.000	1.100	11.944	12.004	2.667	−2.660
	WNTD	−0.005	1.017	11.794	11.905	2.985	−3.188
	CTD	−0.050	0.887	14.734	-	3.178	−3.445
	Exp.	0.162	1.170	20.010	-	3.922	−3.987

[1] Model abbreviations: Frequency domain (FD), weakly nonlinear time domain (WNTD), coupled time domain (CTD), and experimental (Exp.).

The comparison between zero-crossing periods provided by Tables 6–8 shows that the different methods tend to estimate similar values, with two exceptions. First, the heave motion in the less energetic sea states, and second, the surge motion for the most energetic sea states. Both conditions point out that some strong second-order difference-of-frequency excitation forces may be affecting the heaving and surging of the platform, depending upon the wave conditions. This happens because both the heave and surge natural periods are much higher than the sea waves periods. This is further confirmed by the significant amplitude of motion evaluated in the basin, which is not captured by any of the numerical methods when heave motion is observed in IW1 and IW5, for instance. The coupled time-domain model is also not able to predict the heave zero-crossing period, pointing out that this issue must not be related to mooring effects.

The mean values presented in Tables 6–8 show that both the frequency domain and weakly nonlinear time codes predict responses that are zero-averaged. That is expected due to the underlying hypotheses of these methods, whereas the same statement should never hold true regarding experimental data. In the case of coupled time-domain simulations, the nonlinear mooring effects also induce the platform to move around a different position other than the free-floating static equilibrium position. The mean values of the experimental data sometimes lie far from that position, so attention must be paid when comparing the maxima and minima values observed in the basin.

The motion response statistics presented in Tables 6–8 show that the platform has, in general, a good hydrodynamic performance in irregular waves. The pitch angle is kept small for sea states up to Hs = 4.0 m, at least, considering only the dynamics of the platform and simple mooring system. The inclusion of aerodynamic and wind turbine effects will add damping to the pitch mode of the platform; thus, it can be expected that, in reality, the platform will behave well even in more energetic sea states. However, to achieve that, the design of the real mooring system, not analyzed so far, must also be smart in order to not add mooring effects that worsen the pitch response of the FOWT.

Table 7. Heave motion statistics in irregular wave conditions according to different methods.

		Heave Motion					
Sea State	**Model** [1]	**Mean [m]**	**Stdd [m]**	**Tz [s]**	**Tc [s]**	**Max [m]**	**Min [m]**
IW1	FD	0.000	0.007	6.033	5.677	0.018	−0.017
	WNTD	0.000	0.008	6.098	6.020	0.018	−0.019
	CTD	0.001	0.009	8.257	-	0.033	−0.033
	Exp.	−0.012	0.103	34.066	-	0.329	−0.335
IW2	FD	0.000	0.032	8.476	8.520	0.087	−0.090
	WNTD	0.000	0.062	8.367	8.393	0.150	−0.158
	CTD	−0.004	0.098	12.329	-	0.352	−0.438
	Exp.	0.021	0.223	28.776	-	0.778	−0.760
IW3	FD	0.000	0.263	11.016	11.000	0.063	−0.647
	WNTD	0.000	0.177	9.984	10.046	0.429	−0.443
	CTD	−0.014	0.323	17.822	-	1.031	−1.297
	Exp.	0.052	0.344	22.191	-	1.287	−1.245
IW4	FD	0.000	0.533	12.006	10.923	1.353	−1.354
	WNTD	0.000	0.247	10.023	10.145	0.691	−0.673
	CTD	−0.046	0.719	22.268	-	2.101	−2.641
	Exp.	0.156	0.591	19.687	-	2.487	−1.838
IW5	FD	0.000	0.540	12.016	10.927	1.381	−1.382
	WNTD	0.000	0.027	8.401	8.485	0.055	−0.057
	CTD	0.204	0.031	10.475	-	0.114	−0.107
	Exp.	0.009	0.088	27.642	-	0.303	−0.279
IW6	FD	0.000	0.595	11.986	11.872	1.478	−1.485
	WNTD	0.000	0.090	9.970	10.206	0.339	−0.214
	CTD	−0.006	0.151	13.602	-	0.216	−0.205
	Exp.	0.040	0.201	18.318	-	0.822	−0.712
IW7	FD	0.000	0.553	11.985	11.880	1.381	−1.438
	WNTD	0.000	0.306	11.808	12.080	0.984	−1.060
	CTD	−0.017	0.379	19.285	-	1.445	−1.782
	Exp.	0.055	0.321	17.674	-	1.396	−1.273
IW8	FD	0.000	0.581	11.861	11.028	1.363	−1.354
	WNTD	0.000	0.508	12.142	11.876	1.431	−1.428
	CTD	−0.050	0.935	27.135	-	2.619	−3.279
	Exp.	0.133	0.579	17.419	-	2.752	−2.163

[1] Model abbreviations: Frequency domain (FD), weakly nonlinear time domain (WNTD), coupled time domain (CTD), and experimental (Exp.).

Table 8. Pitch motion statistics in irregular wave conditions according to different methods.

		Pitch Motion					
Sea State	**Model** [1]	**Mean [°]**	**Stdd [°]**	**Tz [s]**	**Tc [s]**	**Max [m]**	**Min [m]**
IW1	FD	0.000	0.046	5.685	5.698	0.104	−0.110
	WNTD	0.000	0.052	6.033	5.802	0.122	−0.125
	CTD	−0.003	0.046	6.345	-	0.163	−0.170
	Exp.	−0.014	0.122	6.315	-	0.535	−0.528
IW2	FD	0.000	0.122	7.178	6.480	0.331	−0.353
	WNTD	0.000	0.238	8.365	7.927	0.608	−0.627
	CTD	−0.003	0.217	8.874	-	0.824	−0.811
	Exp.	−0.009	0.287	9.053	-	1.111	−0.980
IW3	FD	0.000	0.689	10.997	10.114	1.717	−1.695
	WNTD	0.000	0.492	10.899	10.143	1.268	−1.302
	CTD	−0.002	0.390	10.455	-	1.386	−1.522
	Exp.	0.044	0.543	10.687	-	2.313	−1.857
IW4	FD	0.000	1.358	10.940	11.077	3.565	−3.600
	WNTD	0.000	0.703	10.167	8.250	1.966	−1.830
	CTD	−0.002	0.671	11.157	-	2.423	−2.429
	Exp.	−0.034	1.039	12.162	-	3.750	−3.712
IW5	FD	0.000	1.383	10.953	11.077	3.597	−3.608
	WNTD	0.000	0.105	8.418	8.625	0.220	−0.234
	CTD	−0.002	0.092	8.096	-	0.356	−0.362
	Exp.	0.115	0.126	6.207	-	0.744	−0.359
IW6	FD	0.000	1.499	11.880	11.057	3.846	−3.952
	WNTD	0.000	0.251	10.223	10.240	0.632	−0.606
	CTD	−0.002	0.259	10.335	-	1.074	−1.076
	Exp.	0.117	0.366	10.073	-	1.559	−1.212
IW7	FD	0.000	1.409	11.874	10.234	3.835	−3.725
	WNTD	0.000	0.778	12.000	12.102	2.719	−2.834
	CTD	−0.002	0.402	11.613	-	1.658	−1.569
	Exp.	0.060	0.685	13.088	-	2.661	−2.495
IW8	FD	0.000	1.467	11.944	11.040	3.642	−3.688
	WNTD	0.000	1.285	12.334	11.953	3.829	−3.879
	CTD	−0.001	0.654	12.148	-	2.522	−2.322
	Exp.	0.021	1.293	14.023	-	5.524	−4.488

[1] Model abbreviations: Frequency domain (FD), weakly nonlinear time domain (WNTD), coupled time domain (CTD), and experimental (Exp.).

In regular waves, the only case where the pitch response may be severe is for a 19.0 s period regular wave; thus, ideally, the mooring system should also be designed in order to mitigate the response at this particular frequency. The motion performance in heave and pitch in both regular and irregular waves is within reasonable limits for a FOWT.

Last but not least, the values of skewness and kurtosis of the motion time series have been calculated during the numerical schemes. The values are not presented in Tables 6–8 because the observed skewness is, as a matter of fact, always around 0, and the observed kurtosis is always around 3, meaning that, within the numerical models developed, the predicted responses can be considered Gaussian processes.

5. Conclusions

In this paper, the hydrodynamic performance of the WIND-bos platform is studied in detail, both experimentally and numerically. The structure under development is a novel concept of spar-type FOWT, which presents good motion performance, for the significant amplitudes of motion are kept small for sea states of interest. Within the numerical models developed, the predicted motion responses can be considered zero-mean Gaussian processes. The further addition of aerodynamics and wind turbine effects shall lead to an even better pitch response. The design of the real mooring system has not been analyzed so far and must be smart in order to not add coupling effects that could worsen pitch response.

The results provided by the in-house weakly nonlinear time-domain code offer good prediction capabilities both in regular and irregular waves. In regular waves, the results match with the experiments in heave, in pitch (with the exception of a single frequency) and, basically, underestimate the surge response slightly. In irregular waves, the results are mostly underestimated because second-order effects have not been considered so far. The next step within the research lines is to consider QTFs, especially the difference-of-frequency second-order excitation forces, that are known to affect the dynamics of many FOWTs and ocean systems consisting of different underwater bodies, for instance, transmitting wave energy to the resonant modes of the platform, which usually lie in the low-frequency range of sea waves.

The results provided by the coupled time-domain solver in regular waves deviate from experiments in some cases, even though the prediction of surge amplitude of motion is fairly good. On the other hand, the in-house solver agrees with most of the experimental data in regular waves. However, in irregular waves, the simulations performed by the coupled time-domain solver presented results that compare well with the experiments and are relatively better than the weakly nonlinear solver, mainly because the coupling with the mooring system is simulated at each time step. The coupled time-domain code is also the only one that is reliable among the models considered to perform, for instance, mooring fatigue analysis, which shall be considered in the future when the real design of the mooring system is accounted for.

Author Contributions: Conceptualization: T.S.H., C.G.S.; software T.S.H.; formal analysis, T.S.H.; experiments: O.S., S.H. and A.A.; data curation: O.S., S.H. and A.A.; validation: T.S.H.; visualization, T.S.H.; supervision: C.G.S.; writing—original draft T.S.H.; writing—review and editing: C.G.S., O.S., S.H. and A.A. All authors have read and agreed to the published version of the manuscript.

Funding: The Experiments have been done within the MARINET2 project, funded by the EU.

Institutional Review Board Statement: Not applicable.

Informed Consent Statement: Not applicable.

Data Availability Statement: Not applicable.

Acknowledgments: The first author has a Ph.D. Scholarship by the Portuguese Foundation for Science and Technology (Fundação para a Ciência e Tecnologia—FCT), under contract No. SFRH/BD/145602/2019. This work contributes to the Strategic Research Plan of the Centre for Marine Technology and Ocean Engineering (CENTEC), which is financed by the Portuguese Foundation for Science and Technology (Fundação para a Ciência e Tecnologia—FCT) under contract UIDB/UIDP/00134/2020.

Conflicts of Interest: The authors declare no conflict of interest.

References

1. Diaz, H.M.; Guedes Soares, C. Review of the current status, technology and future trends of offshore wind farms. *Ocean Eng.* **2020**, *209*, 107381. [CrossRef]
2. ESMAP (Energy Sector Management Assistance Programme). Offshore Wind Technical Potential by Country. Available online: https://esmap.org/publications (accessed on 29 April 2022).
3. GWEC (Global Wind Energy Council). Global Wind Report 2021. Available online: https://gwec.net/global-wind-report-2021/ (accessed on 29 April 2022).
4. GWEC (Global Wind Energy Council). Global Wind Report 2022. Available online: https://gwec.net/global-wind-report-2022/ (accessed on 29 April 2022).
5. Diaz, H.M.; Serna, J.; Nieto, J.; Guedes Soares, C. Market needs, opportunities and barriers for the floating wind industry. *J. Mar. Sci. Eng.* **2022**, *10*, 934. [CrossRef]
6. Jonkman, J.M. Dynamics of offshore floating wind turbines–model development and verification. *Wind Energy* **2009**, *12*, 459–492. [CrossRef]
7. Xu, S.; Wang, S.; Guedes Soares, C. Review of the mooring design for floating wave energy converters. *Ren. Sust. Energy Rev.* **2019**, *111*, 595–621. [CrossRef]
8. Butterfield, S.; Musial, W.; Jonkman, J.M.; Sclavounos, P.; Wayman, L. Engineering challenges for floating offshore wind turbines. In Proceedings of the Copenhagen Offshore Wind 2005 Conference and Expedition, Copenhagen, Denmark, 26–28 October 2005.

9. Jonkman, J.M.; Matha, D. Dynamics of offshore floating wind turbines–analysis of three concepts. *Wind Energy* **2011**, *14*, 557–569. [CrossRef]
10. Jonkman, J.M.; Buhl, M.L., Jr. *FAST User's Guide*; NREL/EL-500-38230; National Renewable Energy Laboratory (NREL): Golden, CO, USA, 2005.
11. Ren, N.; Li, Y.; Ou, J. The effect of additional mooring chains on the motion performance of a floating offshore wind turbine with a tension leg platform. *Energies* **2012**, *5*, 1135–1149. [CrossRef]
12. Bahramiasl, S.; Abbaspour, M.; Karimirad, M. Experimental study on gyroscopic effect of rotating rotor and wind heading angle on floating wind turbine responses. *Int. J. Environ. Sci. Technol.* **2018**, *15*, 2531–2544. [CrossRef]
13. Uzunoglu, E.; Guedes Soares, C. Hydrodynamic design of a free-float capable tension leg platform for a 10.0 MW wind turbine. *Ocean Eng.* **2020**, *197*, 106888. [CrossRef]
14. Mas-Soler, J.; Uzunoglu, E.; Bulian, G.; Guedes Soares, C.; Souto-Iglesias, A. An experimental study on transporting a free-float capable tension leg platform for a 10 MW wind turbine in waves. *Renew. Energy* **2021**, *179*, 2158–2173. [CrossRef]
15. Filgueira-Vizoso, A.; Castro-Santos, L.; Iglesias, D.C.; Puime-Guillén, F.; Lamas-Galdo, I.; Garcia-Diez, A.I.; Uzunoglu, E.; Diaz, H.M.; Guedes Soares, C. The technological and economic feasibility of the CENTEC floating offshore wind turbine. *J. Mar. Sci. Eng.* **2022**, *10*, 1344. [CrossRef]
16. Bak, C.; Zahle, F.; Bitsche, R.; Taesong, K.; Yde, A.; Henriksen, L.C.; Natarajan, A.; Hansen, M.H. *Description of the DTU 10.0 MW Reference Wind Turbine*; Report-I-0092; Denmark Technical University: Copenhagen, Denmark, 2013.
17. Hanson, T.D.; Skaare, B.; Yttervik, R.; Nielsen, F.G.; Havmøller, O. Comparison of measured and simulated responses at the first full scale floating wind turbine Hywind. In Proceedings of the EWEA 2011, Brussels, Belgium, 14–17 March 2011.
18. Skaare, B.; Hanson, T.D.; Yttervik, R.; Nielsen, F.G. Dynamic response and control of the Hywind Demo floating wind turbine. In Proceedings of the EWEA 2011, Brussels, Belgium, 14–17 March 2011.
19. Skaare, B.; Nielsen, F.G.; Hanson, T.D.; Yttervik, R.; Havmøller, O.; Rekdal, A. Analysis of measurements and simulations from the Hywind Demo floating wind turbine. *Wind Energy* **2015**, *18*, 1105–1122. [CrossRef]
20. Skaare, B. Development of the Hywind concept. In Proceedings of the 36th International Conference on Ocean Offshore and Arctic Engineering, OMAE2017-62710, Trondheim, Norway, 25–30 June 2017. [CrossRef]
21. Cermelli, C.; Roddier, D.; Aubault, A. WindFloat: A floating foundation for offshore wind turbines—Part II: Hydrodynamic analysis. In Proceedings of the 28th International Conference on Ocean Offshore and Arctic Engineering, OMAE2009-79231, Honolulu, HI, USA, 31 May–5 June 2009. [CrossRef]
22. Roddier, D.; Cermelli, C.; Aubault, A.; Weinstein, A. WindFloat: A floating foundation for offshore wind turbines. *J. Renew. Sustain. Energy* **2010**, *2*, 033104. [CrossRef]
23. Li, C.; Zhou, S.; Shan, B.; Hu, G.; Song, X.; Liu, Y.; Hu, Y.; Yiqing, X. Dynamics of a Y-shaped semi-submersible floating wind turbine: A comparison of concrete and steel support structures. *Ships Offshore Struct.* **2022**, *17*, 1663–1683. [CrossRef]
24. Jonkman, J.M.; Butterfield, S.; Musial, W.; Scott, G. *Definition of a 5 MW Reference Wind Turbine for Offshore System Development*; NREL (National Renewable Energy Laboratory): Golden, CO, USA, 2005.
25. Sauder, T.; Chabaud, V.; Thys, M.; Bachynski, E.E.; Saether, L.O. Real-time hybrid model testing of a braceless semi-submersible wind turbine. Part I: The hybrid approach. In Proceedings of the 35th International Conference on Ocean Offshore and Arctic Engineering, OMAE2016-54435, Busan, Republic of Korea, 19–24 June 2016. [CrossRef]
26. Bachynski, E.E.; Thys, M.; Sauder, T.; Chabaud, V.; Saether, L.O. Real-time hybrid model testing of a braceless semi-submersible wind turbine. Part II: Experimental results. In Proceedings of the 35th International Conference on Ocean Offshore and Arctic Engineering, OMAE2016-54437, Busan, Republic of Korea, 19–24 June 2016. [CrossRef]
27. Berthelsen, P.A.; Bachynski, E.E.; Saether, L.O. Real-time hybrid model testing of a braceless semi-submersible wind turbine. Part III: Calibration of a numerical model. In Proceedings of the 35th International Conference on Ocean Offshore and Arctic Engineering, OMAE2016-54640, Busan, Republic of Korea, 19–24 June 2016. [CrossRef]
28. Karimirad, M.; Bachynski, E.E.; Berthelsen, P.A.; Ormberg, H. Comparison of real-time hybrid model testing of a braceless semi-submersible wind turbine and numerical simulations. In Proceedings of the 36th International Conference on Ocean Offshore and Arctic Engineering, OMAE2017-61121, Trondheim, Norway, 25–30 June 2017. [CrossRef]
29. Hegseth, J.M.; Bachynski, E.E.; Karimirad, M. Comparison and validation of hydrodynamic load models for a semi-submersible floating wind turbine. In Proceedings of the 37th International Conference on Ocean Offshore and Arctic Engineering, OMAE2018-77676, Madrid, Spain, 17–22 June 2018. [CrossRef]
30. Uzunoglu, E.; Guedes Soares, C. On the model uncertainty of wave induced platform motions and mooring loads of a semi-submersible based wind turbine. *Ocean Eng.* **2018**, *148*, 277–285. [CrossRef]
31. Robertson, A.N.; Wendt, F.; Jonkman, J.M.; Popko, W.; Dagher, H.; Gueydon, S.; Qvist, J.; Vittori, F.; Azcona, J.; Uzunoglu, E.; et al. OC5 Project Phase II: Validation of global loads of the DeepCWind floating semisubmersible wind turbine. *Energy Procedia* **2017**, *137*, 38–57. [CrossRef]
32. Robertson, A.N.; Bachynski, E.E.; Gueydon, S.; Wendt, F.; Schünemann, P.; Jonkman, J.M. Assessment of experimental uncertainty for a floating wind semisubmersible under hydrodynamic loading. In Proceedings of the 37th International Conference on Ocean Offshore and Arctic Engineering, OMAE2018-77703, Madrid, Spain, 17–22 June 2018. [CrossRef]
33. Xu, K.; Zhang, M.; Shao, Y.; Gao, Z.; Moan, T. Effect of wave nonlinearity on fatigue damage and extreme response of a semi-submersible floating wind turbine. *Appl. Ocean Res.* **2019**, *91*, 101879. [CrossRef]

34. Hallak, T.S.; Karmakar, D.; Guedes Soares, C. Hydrodynamic performance of semi-submersible FOWT combined with point-absorber WECs. In *Developments in Maritime Technology and Engineering*; Guedes Soares, C., Santos, T.A., Eds.; Taylor & Francis Group: London, UK, 2021; Volume 2, pp. 577–585. [CrossRef]
35. Da Silva, L.S.P.; Sergiienko, N.Y.; Cazzolato, B.; Ding, B. Dynamics of hybrid offshore renewable energies platforms: Heaving point-absorbers connected to a semi-submersible floating offshore wind turbine. *Renew. Energy* **2022**, *199*, 1424–1439. [CrossRef]
36. Armesto, J.A.; Jurado, A.; Guanche, R.; Couñago, B.; Urbano, J.; Serna, J. Telwind: Numerical analysis of a floating wind turbine supported by a two bodies platform. In Proceedings of the 37th International Conference on Ocean Offshore and Arctic Engineering, OMAE2018-77587, Madrid, Spain, 17–22 June 2018. [CrossRef]
37. Baita-Saavedra, A.; Cordal-Iglesias, D.; Filgueira-Vizoso, A.; Morató, À.; Lamas-Galdo, I.; Álvarez-Feal, C.; Carral, L.; Castro-Santos, L. An economic analysis of an innovative floating offshore wind platform built in concrete: The SATH platform. *Appl. Sci.* **2020**, *10*, 3678. [CrossRef]
38. Yang, Y.; Bashir, M.; Michailides, C.; Mei, X.; Wang, J.; Li, C. Coupled analysis of a 10 MW multi-body floating offshore wind turbine subjected to tendon failures. *Renew. Energy* **2021**, *176*, 89–105. [CrossRef]
39. Guo, X.; Zhang, Y.; Yan, J.; Zhou, Y.; Yan, S.; Shi, W.; Li, X. Integrated dynamic response analysis for IEA 10-MW spar floating offshore wind turbine. *J. Mar. Sci. Eng.* **2022**, *10*, 542. [CrossRef]
40. Karmakar, D.; Guedes Soares, C. Review of the present concepts of multi-use offshore platforms. In *Renewable Energies Offshore*; Guedes Soares, C., Ed.; Taylor & Francis Group: London, UK, 2015; pp. 867–875.
41. Nassar, W.M.; Anay-Lara, O.; Ahmed, K.H.; Campos-Gaona, D.; Elgenedy, M. Assessment of multi-use offshore platforms: Structure classification and design challenges. *Sustainability* **2020**, *12*, 1860. [CrossRef]
42. Armesto, J.A.; Sarmiento, J.; Ayllón, V.; Iturrioz, A.; Jurado, A.; Guanche, R.; Losada, I.J. Numerical and experimental study of a multi-use platform. In Proceedings of the 35th International Conference on Ocean Offshore and Arctic Engineering, OMAE2016-54427, Busan, Republic of Korea, 19–24 June 2016. [CrossRef]
43. Li, L.; Gao, Y.; Yuan, Z.; Day, S.; Hu, Z. Dynamic response and power production of a floating integrated wind, wave and tidal energy system. *Renew. Energy* **2018**, *116*, 412–422. [CrossRef]
44. Kamarlouei, M.; Gaspar, J.F.; Calvário, M.; Hallak, T.S.; Mendes, M.J.G.C.; Thiebaut, F.; Guedes Soares, C. Experimental study of wave energy converter arrays adapted to a semi-submersible wind platform. *Renew. Energy* **2022**, *188*, 145–163. [CrossRef]
45. Amin, I.; Dai, S.; Day, S.; Ali, M.E.A.; Balah, A.; Shawky, H.; Oterkus, S.; Oterkus, E. Experimental study on the motion response of an integrated floating desalination and offshore wind turbine on a non-ship platform. *Ocean Eng.* **2021**, *234*, 109275. [CrossRef]
46. Hallak, T.S.; Guedes Soares, C.; Sainz, O.; Hernández, S.; Arévalo, A. *Time Domain Analysis of the WIND-Bos Spar in Regular Waves. Trends in Renewable Energies Offshore*; Taylor & Francis Group: London, UK, 2022; pp. 559–566. [CrossRef]
47. Eliassen, L.; Bachynski, E.E. The effect of turbulence model on the response of a large floating wind turbine. In Proceedings of the 36th International Conference on Ocean Offshore and Arctic Engineering, OMAE2017-61179, Trondheim, Norway, 25–30 June 2017. [CrossRef]
48. Rivera-Arreba, I.; Bruinsma, N.; Bachynski, E.E.; Viré, A.; Paulsen, B.T.; Jacobsen, N.G. Modeling of a semisubmersible floating wind platform in severe waves. In Proceedings of the 37th International Conference on Ocean Offshore and Arctic Engineering, OMAE2018-77680, Madrid, Spain, 17–22 June 2018. [CrossRef]
49. Han, K.; Vada, T.; Wang, Q.Q. Time domain coupled analysis and load transfer for floating wind turbine structures. In Proceedings of the 37th International Conference on Ocean Offshore and Arctic Engineering, OMAE2018-77880, Madrid, Spain, 17–22 June 2018. [CrossRef]
50. Hallak, T.S.; Kamarlouei, M.; Gaspar, J.F.; Guedes Soares, C. Time domain analysis of a conical point-absorber moving around a hinge. In *Trends in Maritime Technology and Engineering*; Guedes Soares, C., Santos, T.A., Eds.; Taylor & Francis Group: London, UK, 2022; Volume 2, pp. 401–409. [CrossRef]
51. Cummins, W.E. The impulse response function and ship motions. *Schiffstechnik* **1962**, *47*, 101–109.

Journal of
Marine Science and Engineering

MDPI

Article

Experimental and Numerical Study of the Influence of Clumped Weights on a Scaled Mooring Line

Tomas Lopez-Olocco [1,2,*], Leo M. González-Gutiérrez [1,2], Javier Calderon-Sanchez [1,2], Adolfo Marón Loureiro [3], Leandro Saavedra Ynocente [3], Ana Bezunartea Barrio [3] and Nicolás Vivar Valdés [4]

1 Universidad Politécnica de Madrid (UPM), 28040 Madrid, Spain; leo.gonzalez@upm.es (L.M.G.-G.); javier.calderon@upm.es (J.C.-S.)
2 Canal de Ensayos Hidrodinámicos de la ETSI Navales (CEHINAV), Universidad Politécnica de Madrid, 28040 Madrid, Spain
3 Laboratorio de Dinámica del Buque, Canal de Ensayos Hidrodinámicos el Pardo (INTA-CEHIPAR), 28048 Madrid, Spain; maronla@inta.es (A.M.L.); saavedrayl@inta.es (L.S.Y.); bezunarteaba@inta.es (A.B.B.)
4 Offshore Engineering Department, SEAPLACE S.L., 28016 Madrid, Spain; n.vivar@alumnos.upm.es
* Correspondence: tomas.lopez@upm.es

Citation: Lopez-Olocco, Tomas; González-Gutiérrez, L.M.; Calderon-Sanchez, J.; Marón-Loureiro, A.; Saavedra-Ynocente, L.; Ana Bezunartea-Barrio, A.; Vivar-Valdés, N. Experimental and Numerical Study of the Influence of Clumped Weights on a Scaled Mooring Line. *J. Mar. Sci. Eng.* **2022**, *10*, 676. https://doi.org/10.3390/jmse10050676

Academic Editors: Emre Uzunoglu, Antonio Souto-Iglesias and Carlos Guedes Soares

Received: 30 March 2022
Accepted: 8 May 2022
Published: 16 May 2022

Abstract: Recently, several experimental and numerical studies have underlined the advantages of adding clumped weights at discrete positions of mooring lines. To confirm the influence of these weights, an experimental study was performed for a 1:30 scale model of a mooring line. In this study, the clumped weight is modeled as a scaled disc placed at different positions along the mooring line. The series of experiments has been carried out at the CEHIPAR towing tank using a submerged studless chain both with and without clumped weights. The experiments consist of the excitation of the suspension point with horizontal periodic motions using different amplitudes and periods, where the mooring line's tension at the fairlead is measured using a load cell and a dynamometer, and the motion of a part of the line is recorded using low-cost submerged cameras. Similarly to previous experiments, the fairlead tensions increase with higher amplitudes and lower periods, and a clear pattern in the motions of the line at different depths is found. The dissipated energy and the fairlead tension is also increased by the addition of the clumped weight, and the variation of this energy with its position along the line is monitored. The presence of clumped weights is also implemented into a finite element numerical code, previously validated without clumped weights, where all the previous experiments with clumped weights are replicated with remarkable accuracy. This double experimental and computational approach to the problem provides an important dataset for numerical code validations and opens future discussions about the impact of clumped weights on floating plattorms.

Keywords: mooring line; clump weights; hybrid mooring system; experiments; floating offshore wind turbines

1. Introduction

Floating wind power is considered one of the most promising sources of sustainable energy today. The international framework is clearly favorable for offshore wind as a key technology in meeting decarbonization objectives. The offshore sector is largely being explored due to its higher wind potential when compared to onshore cases. An important factor in motivating offshore locations for wind turbines is the problem that onshore locations are often designated for purposes other than energy production. Classical advantages justify the installation of offshore wind turbines with respect to onshore locations. The energy yield of a wind turbine installed in open sea is, in general, above the production of an onshore location due to higher and steadier winds. The visual and audible impact of a wind farm is less restrictive for the design for offshore than for onshore wind turbines. Finally, most of the world's population is located close to the coastline and transmission losses are therefore low. The growing evolution of offshore wind installations in Europe is quantified in many reports [1], where a tendency toward growth is clearly observed and the important role that it will play in the future is increasingly likely.

The offshore wind energy industry is as economically promising as it is technical challenging, due to the aforementioned *intention to place Floating Offshore Wind Turbines (FOWT) in deeper waters. The reason for this interest is that 61% of US coastal areas, nearly all of Japan's, and various European locations, such as the coast of Norway and Spain and Portugal in the Mediterranean, require floating foundation technology due to the great depth of the water. Today, the majority of European offshore wind turbines are placed in shallow waters, where according to experimental research [2] snap loading may occur more frequently. However, in order to install more offshore farms where better atmospheric conditions are found, highly reliable mooring solutions are required. The mooring system designed for these facilities must be ready to withstand the loads applied by the environment. This imposes the necessity of longer, and therefore heavier and more expensive mooring lines. Due to this necessity, the cost reduction of this growing offshore technology demands the optimization of the mooring line system, where the mechanical fatigue and the extreme and impulsive loads are the most aggressive phenomena for the lines. In order to prevent accidents and long and expensive repairs, a thorough study of the mooring line dynamics is necessary, where extreme loads and fatigue loads are adequately modeled.

We can assert that in the context of offshore wind energy, the reduction of motions allows the turbine to produce power in a wider range of sea states, and lower motions and accelerations reduce the tension of the lines and related costs, see [3]. Therefore, the dual goal is to achieve, when possible, a load reduction of both the mooring lines and the platform motions.

Due to the complex dynamics that govern the mooring line, the presence of snap loads and tension peaks in the mooring lines have been studied in multiple publications, i.e., [4,5]. The evolution of the line tension in such an aggressive and chaotic environment, combined with the fatigue phenomenon and the footprint limitation, makes it necessary to perform a safe and appropriate mooring design in offshore applications. Today, several new concepts and alternative designs for FOWT mooring lines are being studied at water depths greater than 50 m [6]. These modern improvements for mooring systems are to include point masses (Clump Weights) that provide drag and weight to increase the elastic range of the line and enhance the damping of the floating system.

2. State of the Art

The research of mooring systems admits several classifications, being one of them the methodology employed. The most common methodologies are the experimental ones [2,7] that use scaled models, and the numerical one [8,9] where full-scale models are able to be employed. Additionally, there is a vast number of published articles where both approaches are used [10,11]. For each of these methodologies, two types of experiments/simulations have been found in the context of mooring systems, either moored floating objects/structures are placed in a towing tank and are excited by currents or waves [12–20], or isolated mooring lines are forced to move the fairlead with prescribed motion. These last ones can include clumped weights [6,21] or otherwise only the mooring line [2,7–9,11,22–25]. In this experimental and computational study, we follow the second approach and we focus on the isolated mooring dynamics with prescribed motion for the fairlead. The objective of the present study is to extend current research to include the presence of clumped weights, previously addressed by [6] from a quasi-static perspective, but using a dynamic finite element model and comparing its results to the experimental measurements also performed.

The role of the mooring system has a significant contribution to the motion of floating structures, causing about 80% of the total damping of the structure at 200 m depth [26]. This energy dissipation is basically caused by three terms: drag contribution, friction with the seabed, and the internal mechanical damping due to the line deformations. Limiting the scope of research to that found on forced mooring lines, another interesting classification is observable depending on whether the authors are more focused on the damping quantification of the mooring system [26–28] or the mooring dynamics and the different regimes found for different kinds of excitation [10]. The computation of the mooring damping has been computed by quasi-static methods when the mooring line motion is very slow, see [26]. A simple calculation procedure for the practical estimation of damping due to mooring lines is presented and verified experimentally in [29]. The most commonly used method of estimating the total mooring damping is the indicator diagram presented in [28], which has the advantage of including all dynamic responses of the mooring line.

As most of the experiments performed in moored systems are normally scaled models of full-scale prototypes, the importance of the scaling laws and the dimensional analysis of the problem is clear. In [28], one of the first dimensional analyses of the mooring dynamics was performed and, afterwards, the analysis was applied to the scaling of the damped energy. A slightly more simplified dimensional analysis was performed in [25], where the final differential equation that models the phenomenon contains five non-dimensional parameters. It is also worth mentioning that when the experiment must represent a full-scale design, not only should the Froude number associated to gravity remain constant between scales, but also other non-dimensional numbers such as the one associated with line stiffness should be considered. Some experiments [7,30] added a spring at the anchor position to try to keep the stiffness of the line in the proper scaling. One important conclusion of [25] is that even when using a spring connected to the mooring line in order to have a scaled stiffness, a correct scaling is only obtained at quasi-static level. If perfect geometrical and dynamic scaling in vacuum is addressed, the chain must be scaled to have an adequate propagation celerity for longitudinal elastic waves.

Another possible classification comes from the completeness of the differential equation that models the mooring dynamics. In [11], a numerical and experimental study for five different scales of chain mooring systems was performed. They compared the different theoretical approaches to the problem showing the difference to the static case, where no terms including velocities or accelerations were part of the model; the quasi-static case where no terms based on accelerations were considered; and finally, the dynamic approach where accelerations and added mass forces were included.

For the numerical computation of the mooring line, the OPASS [10] code based on the finite element method is used. Other numerical alternatives such as the Open-Moor software, similar to OPASS, is explained in [8] and validated against the forced isolated mooring line described in [25]. An interesting advantage of this latter one is that the time integration is performed using the generalized α-method that permits larger time steps.

Some experiments are able to measure the trajectories and velocities of different points of the mooring line. Different techniques are applied to this process: either image processing methods are applied to snapshots from the camera [2] or sophisticated optical devices such as Qualysis to directly obtain the point position [7]. During the experiments performed in [2] in shallow water, the importance of line stretching for damping was demonstrated also showing that the most critical factors for snap conditions are high pretensions and low amplitudes and periods. A similar study was performed by [22] but also adding a uniform current to the surge and heave fairlead oscillations. A sharp increase of the maximum tension force was observed at large pretension for the mooring line with the surge- and heave-imposed motions.

The dynamic behavior of different mooring lines and the impact of weight and sea bottom friction was analyzed in [7]. An original part of this experimental work is the quantification of the friction coefficients and the effect of the mooring line on the seabed. This work was complemented by the numerical study performed in [9] where an exhaustive sensitivity analysis was performed to evaluate the variability of mooring loads because of inaccuracies in the definition of model inputs.

In [24], the slack-taut regime was searched by numerically changing the initial pretension, amplitude and frequency of the top mooring line excitation were made up of three different parts. The results showed that the slack-taut phenomenon, characterized by maximum tensions several times that of the pretension, would occur beyond a certain threshold of amplitude, frequency, and pretension.

A nonlinear dynamic model of the mooring-line based on FEM simulations was formulated in [23]. An isolated mooring line was tested for a large combination of parameters: pretension, amplitude and frequency, diameter and elasticity modulus. They concluded that the mooring-lines' stiffness presents a hysteresis behavior which becomes more obvious as the suspension point frequency/amplitude gets larger.

We should remark that in this kind of research where an isolated mooring line is studied under forced movement, just a few papers mention the presence of a clumped weight, especially combining the experimental and computational perspectives. Some computational models analyzed implemented the clumped weight as part of the lumped-mass approach, where the total mass of the element (chain and clumped weight) is concentrated in the element's adjacent nodes, see, for example, [18,31]. A numerical-experimental comparison was performed in [21] where a lumped mass computational method was described and compared to the experiments performed in a towing tank. Both approaches were performed using a single clumped weight and limiting the study to a single frequency and amplitude.

The improvement given by the clumped weights was demonstrated in [6] where the key design parameters of clump weights and their effects on mooring system performance were investigated from a quasi-static point of view and assuming that dynamical effects can be neglected. The paper described a methodology of determining the optimum values of these parameters to improve mooring performance.

Table 1 compares the recent research literature of excited mooring lines and the innovation of this study.

As can be observed, most of the papers where the forced mooring lines are studied did not include the clumped weight in the experiment or dynamic numerical simulation. In this paper, we try to fill this gap and analyze experimentally and computationally what effects the presence of a clumped weight has on the mooring line. Another innovation developed for this study is the system designed to track the mooring line kinematics, which is a low-cost system based on an image processing code and submerged cameras. Additionally, the simulation tool OPASS, specifically designed for mooring line dynamics now includes the possibility of clumped weight implementation. Other important conclusions about the evolution of the geometric stiffness of the line and the energy dissipation when clumped weights are included, are also important results obtained from this work.

Table 1. Comparison table of the existing research literature of mooring lines forced oscillations and the present work.

	Methodology	Results	Use of Clumped Weights
Current work	Experimental and numerical	Dynamic analysis of tension response for different positions of clumped weights, force-displacement curves, PSD analysis, dissipated energy, trajectories, numerical validations with experiments, non-dimensional analysis.	Yes
Nakajima [21]	Experimental and numerical	Fairlead tension time series, frequency response curves of tension. Single frequency and amplitude with clumped weights.	Yes
Luo [6]	Numerical	Dynamic analysis of tension response for different positions of clumped weights.	Yes
Azcona [10]	Experimental and numerical	Fairlead tension, force-displacement curves, trajectories, and numerical validation with experiments.	No

Table 1. *Cont.*

	Methodology	Results	Use of Clumped Weights
Li [23]	Experimental and numerical	Time series of fairlead tension, trajectories, force-displacement curve, dissipated energy, comparison of maximum fairlead tension with the static.	No
Barrera [9]	Experimental and numerical	Experimental fairlead tension time series comparison with numerics, experimental time series of the displacement for different points of the line compared with numerics, sensitivity analysis of the number of elements and computational parameters.	No
Bergdhal [25]	Experimental and numerical	Fairlead tension time series compared with numerics.	No
Yan [27]	Experimental and numerical	force-displacement curve and energy dissipation.	No
Barrera [7]	Experimental	Time series of fairlead tension, fairlead tension-displacement curves, trajectories, accelerations, dissipated energy, sand footprints, hydrodynamic response with different sea states	No
Kitney [30]	Experimental	Force-displacement curves and energy dissipation, non-dimensional analysis.	No
Gao [22]	Experimental	Force-displacement curve, fairlead tension, and PSD analysis, trajectories	No
Hsu [2]	Experimental	Force-displacement curve, energy dissipation and mooring line velocity vectors of image processing.	No
Chen [8]	Numerical	Fairlead tension time series, validation with experiments in the literature.	No
Qiao [24]	Numerical	Fairlead tension time series and PSD analysis, trajectories.	No
Webster [28]	Numerical	Force displacement curves and energy dissipation, non-dimensional analysis.	No

This paper is organized as follows: in Section 3, the case of study, the dimensional analysis, and the experimental setup are presented; in Section 4, the numerical tool used is described and the modeling and implementation of Clumped Weights on it is explained; in Section 5, the experimental results are shown; in Section 6, the numerical and experimental validations are conducted.

3. Experimental Methodology

Among the variety of FOWT designs, the semisubmersible one is receiving particular attention due to various advantages [32]. The two most remarkable ones are: first, these platforms can be fully assembled onshore and deployed to their final operational destination and, second, the mooring systems are widely employed and cost competitive. Recently, different projects have developed proof of concept designs, such as the OC4-DeepCWind shown in Figure 1, in order to generate test data to be used in FOWT modeling tools' validation [33].

Figure 1 shows the OC4-DeepCWind FOWT mooring line arrangement based on three chains separated by 120°. We study a typical deep water configuration where the platform is placed at an approximate sea depth of 200 m.

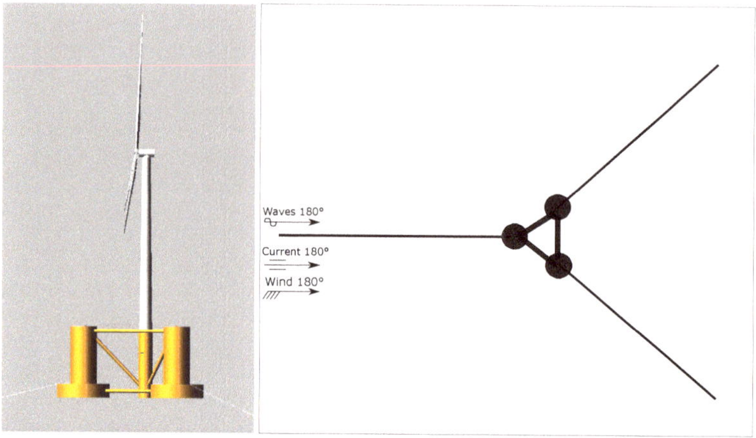

Figure 1. (**Left**) OC4-DeepCWind FOWT design. (**Right**) Top view of the mooring system of the OC4-DeepCWind FOWT [33].

Considering the worst case scenario where waves and wind are aligned in the 180 degree direction as shown in the right panel of Figure 1. This situation, where the wave motion and wind are aligned with the upwind mooring line, provokes the same tension for the two downwind cables, while the upwind line becomes the most loaded one. It is well known, see [26], that in this scenario, surge motions of the moored platform will become very important, and therefore need to be analyzed for the upwind mooring line.

In this paper, three different configurations of mooring line reproducing the upwind mooring line are analyzed: an isolated scaled mooring line and the same line with the addition of a single clumped weight at two different positions. The experiments consist of the excitation of the suspension point, hereafter called fairlead, with horizontal periodic movements reproducing the platform surge motion.

Depending on the installation site, see [10], when surge motion is considered in normal operation conditions, these types of platforms typically have periods ranging from 10 to 30 s and amplitudes from 3 to 12 m.

3.1. Dimensional Analysis

In this subsection, we will try to extend the dimensional analysis performed in [28] to our case where the clumped weight is attached to the mooring line using a line scale factor λ.

Let us start with the analysis of the mooring line equation of motion formulated in global coordinates. This equation will include the elastic deformation of the chain, internal mechanical damping of the chain, and external hydrodynamic forces (weight in water, drag, added mass, and clumped weight). Although the numerical code includes the external action of the sea bottom and structural damping in this dimensional analysis, neither the role of the sea bottom nor the internal damping will be considered. Structural damping is negligible in this kind of mooring dynamics and sea bottom effects are very hard to scale. Similarly to the analysis performed by [25], no bending is considered and we also assume that the experiment is performed in calm water, and therefore, no water current is included in the analysis. When only chains are used as mooring lines, good accuracy is obtained when bending effects are considered negligible, see [10]. The differential equation of motion that supports the dimensional analysis for a mooring line element, is:

$$\gamma\ddot{\mathbf{R}} - \frac{(\partial EA\varepsilon\mathbf{t})}{\partial l_0} - \mathbf{F}_D - \mathbf{F}_W - \mathbf{F}_{CW} - \mathbf{F}_{AM} = 0, \tag{1}$$

where $\mathbf{F}_D$, $\mathbf{F}_W$, $\mathbf{F}_{CW}$, $\mathbf{F}_{AM}$ are the drag, weight in water, clumped weight, and added mass forces to the mooring line, respectively, γ is the mass per unit length of the mooring line, $\ddot{\mathbf{R}}$ is the acceleration of the line element and the partial differential term is the tension of the line element, where ε and $\mathbf{t}$ are the deformation of the line element and the vector tangential to the line element, respectively. Bold text denotes vectorial notation.

The list of dimensional parameters involved in the analysis are:

- F_f Fairlead tension.
- L Total unstretched length of the mooring line.
- H Water depth.
- D Equivalent diameter of the fictitious mooring line that would result if the structural cross-section of the chain were to be formed into a solid, cylindrical rod.
- EA Product of Young's modulus of the chain by the structural cross sectional area of the chain.
- ρ_0 Water density.
- w Chain weight in water per unit volume.
- g Gravity acceleration.
- F_D Mooring line drag force magnitude.
- F_{AM} Mooring line added mass force magnitude.
- ω Fairlead oscillating angular frequency.
- A Fairlead oscillating amplitude.
- F_{DCW} Clumped weight drag force.
- F_{AMCW} Clumped weight added mass force.
- D_{CW} Clumped weight diameter.
- L_{CW} Clumped weight length.
- W_{CW} Weight in water of the clumped weight.
- t Time.

Performing the typical non-dimensional analysis with the previous list of parameters, and using H, D, and D_{CW} as three characteristic lengths corresponding to the water depth, mooring line diameter, and clumped weight diameter, and $\sqrt{H/g}$ and wHD^2/g as characteristic time and mass, respectively. The list of non-dimensional numbers is very similar to the one obtained by [25,28] but adding three new parameters due to the clumped weight's presence. The non-dimensional numbers are classified into three groups according to the situation; they can be summarized as:

Related to the mooring line

- $\Pi_1 = L/H$ Scope of the mooring line.
- $\Pi_2 = \rho_0 g/w$ Ratio between water and line densities.
- $\Pi_3 = F_D/wHD^2$ Drag force coefficient.
- $\Pi_4 = F_{AM}/wHD^2$ Added mass coefficient.
- $\Pi_5 = \omega\sqrt{H/g}$ Ratio between the oscillating frequency and gravitational time.
- $\Pi_6 = A/H$ Ratio between the oscillation amplitude and water depth.
- $\Pi_7 = t\sqrt{g/H}$ Non-dimensional time
- $\Pi_8 = F_f/wHD^2$ Non-dimensional fairlead tension

Related to the mooring line structure

- $\Pi_9 = EA/wHD^2$ Relative stiffness of the mooring line, but also according to [25], the ratio between the propagation celerity of longitudinal elastic waves and the characteristic velocity.

Related to the clumped weight

- $\Pi_{10} = F_{DCW}/wD_{CW}^3$ Clumped weight drag force.
- $\Pi_{11} = F_{AMCW}/wD_{CW}^3$ Clumped weight added mass force.
- $\Pi_{12} = W_{CW}/wD_{CW}^2 L_{CW}$ Weight in water of the clumped weight.

A complete dynamic similarity would imply 12 equal non-dimensional numbers, nine coming from the mooring line plus three from the clumped weight. The first nine non-dimensional numbers include the five used by [25] and are also part of the list of Webster [28] who made a similar analysis but also included three more effects due to bending, water current, and pretension.

In reference [25], it is assumed that two different length scales should be used for the mooring line, one for the water depth λ and another for the mooring line diameter η. Additionally, due to the presence of the clumped weights, two more length scales ψ and α are used. Considering a prototype full-scale size denoted with subindex p and a small-scale model denoted with subindex m, we can express the scaling ratios mathematically as:

$$\frac{L_m}{L_p} = \frac{H_m}{H_p} = \lambda \tag{2}$$

$$\frac{D_m}{D_p} = \eta \tag{3}$$

$$\frac{D_{CWm}}{D_{CWp}} = \psi \tag{4}$$

$$\frac{L_{CWm}}{L_{CWp}} = \alpha \tag{5}$$

We assume first that the water density, ρ_0, the chain density ρ_c, and gravity, g, are the same in the model and prototype. Conserving the non-dimensional numbers Π_5, Π_7, and Π_2:

$$\frac{\omega_p}{\omega_m} = \frac{t_m}{t_p} = \sqrt{\frac{H_m}{H_p}} = \sqrt{\lambda} \tag{6}$$

$$\frac{w_m}{w_p} = 1 \tag{7}$$

Similarly, conserving the non-dimensional numbers Π_3, Π_4, Π_8, Π_9, and Π_6:

$$\frac{F_{fm}}{F_{fp}} = \frac{F_{AMm}}{F_{AMp}} = \frac{F_{Dm}}{F_{Dp}} = \frac{EA_m}{EA_p} = \frac{H_m}{H_p}\frac{D_m^2}{D_p^2} = \lambda\eta^2 \tag{8}$$

$$\frac{A_m}{A_p} = \lambda \tag{9}$$

Finally, conserving the non-dimensional numbers related to the clumped weight Π_{10}, Π_{11}, and Π_{12}:

$$\frac{F_{AMCWm}}{F_{AMCWp}} = \frac{F_{DCWm}}{F_{DCWp}} = \psi^3 \tag{10}$$

$$\frac{(W_{CW})_p}{(W_{CW})_m} = \frac{(D_{CW}^2)_p}{(D_{CW}^2)_m}\frac{(L_{CW})_p}{(L_{CW})_m} = \psi^2\alpha \tag{11}$$

We note that according to Equations (6) and (9), the Froude number defined as $Fr = \frac{A\omega}{\sqrt{gH}}$ is conserved at both scales. Amplitudes and periods of the fairlead motions are scaled according to Equations (6) and (9).

Following the general statements of the dimensional analysis, we should try to scale the maximum number of terms in the differential equations that represent the mooring line. This implies scaling the dominant hydrodynamic loads apart from the weight of the mooring line and the inertial loads. However, as the Reynolds number in the experiments is far from the full-scale value, the hydrodynamic loads due to drag are not scaled according to the scaling laws.

As the OC4-DeepCWind is placed at a 200 m water depth and the CEHIPAR calm water tank has a 6.5 m water depth, a scale factor $\lambda = 1/30$ is chosen. Table 2 summarizes the most important properties of the full-scale mooring system.

Table 2. OC4-DeepCWind mooring system properties [33].

	Value	Units
Number of mooring lines	3	–
Angle between adjacent lines	120	deg
Depth to Anchors below SWL	200	m
Depth to fairlead below SWL	14	m
Radius to Anchors from Platform Centerline	837.6	m
Radius to fairleads from Platform Centerline	40.868	m
Unstretched mooring line length	835.5	m
Mooring line diameter	0.0766	m
Equivalent mooring line mass density	113.35	Kg/m
Equivalent mooring line mass in water	108.63	Kg/m
Equivalent mooring line extensional stiffness	753.6×10^6	N
Hydrodynamic Drag coefficient for mooring lines	1.1	–
Hydrodynamic added-mass coefficient for mooring lines	1.0	–
Seabed drag coefficient for mooring lines	1.0	–
Structural damping of mooring lines	2.0	%

For the scaled experiments, a 27 m long DIN5685A studless chain was selected as its properties are close to the OC4-DeepCWind platform's mooring system when the non-dimensional ratios of $\lambda = 1/30$ and $\eta = 1/23$ are used. The chain used for the dynamic load test is made of stainless steel, see Figure 2 where the main dimensions are indicated. In our case, the diameter D_w of the chain is 7 mm and the length L is 27 m. The length of the link t_w is 16 mm and the diameter of the chain link d_w is 2 mm. The mass per unit length γ of the mooring line is 67.8 g/m.

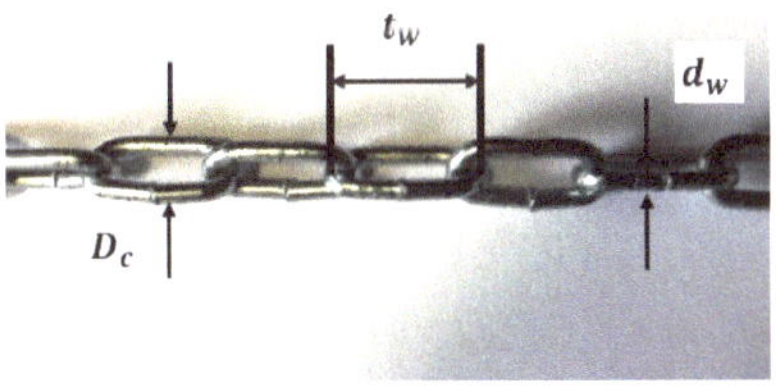

Figure 2. Principal dimensions of the selected chain.

Conserving the non-dimensional number Π_{12} as expressed by Equation (11), the dimensions of the model scale clumped weight are defined as follows. Assuming typical industrial values of a clumped weight, this is a weight in water of 1Tn, we approximate a cylindrical shape volume $V_{CW} = 2.4$ m^3 ($D_{CW} = 1.5$ m, $L_{CW} = 1.33$ m). An example of an industrial clumped weight is shown in Figure 3.

A scale ratio $\psi = 1/18$ is adopted for the clumped weight, not far from the values used to scale the mooring line length ($\lambda = 1/30$) and diameter ($\eta = 1/23$). Instead of using a block shape, in our case, the clumped weight design was modeled as a rigid circular disc characterized by a length L_{CW} and a diameter D_{CW}, in order to maximize drag. To accomplish this, a length scale factor is fixed to the value $\alpha = 1/80$, which is slightly above the other dimensional ratios.

Figure 3. Typical clumped weight used in the offshore industry.

The chosen values for the scaled clumped weight are collected in Table 3 and the manufactured device is shown in Figure 4.

Table 3. Involved variables in the design of the scaled clumped weight.

	Symbol	Value	Units
Mass	$(M_{cw})_m$	122.0	g
Volume of displaced water	$(V_{cw})_m$	80.0	cm^3
Diameter	$(D_{cw})_m$	82.3	mm
Width	$(L_{cw})_m$	16.5	mm

The clumped weight was directly attached to the chain, as shown in Figure 4. The same line has been used for both configurations of the experimental campaign. Consequently, to insert the devices, the chain was cut twice, one for each position of the clumped weight. Special care has been taken to preserve the total length of the line for each configuration. After the aforementioned procedure, the static tension of the line at the fairlead was measured before prescribing the imposed motions.

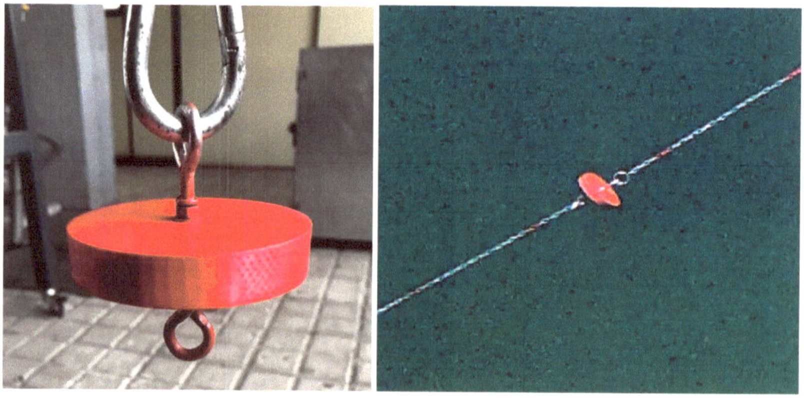

Figure 4. Designed and manufactured scaled clumped weight (**left**) and snapshot of the clumped weight installed in the mooring line obtained with the submerged camera (**right**).

3.2. Experimental Setup

The experimental campaign was conducted in the calm water tank at the *Canal de Ensayos Hidrodinámicos de el Pardo* (CEHIPAR), see Figure 5a. The main dimensions of the tank were 320 m in length, 12.5 m in width, and water depth is kept at 6.5 m. The tank is equipped with a towing carriage, see Figure 5b, which supports the actuator and the instrumentation for the data acquisition.

(a) (b)

Figure 5. CEHIPAR facilities. (**a**) CEHIPAR towing tank. (**b**) CEHIPAR towing tank carriage.

When the chain is submerged, it acquires a typical catenary shape, see Figure 6. The fairlead is connected to the moving and measurement devices and the bottom end of the chain is anchored to the floor. The anchor consisted of a heavy steel plate, 20 times the weight of the total length of the chain to avoid possible displacements during the tests. Special care has been taken to place the anchor at 25 m in the horizontal direction from the initial fairlead position to have the desired pretension for all the configurations.

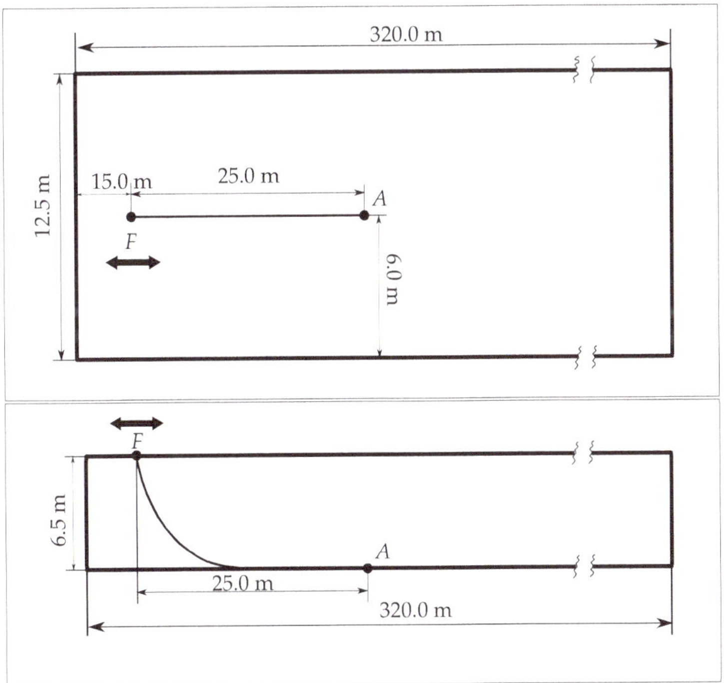

Figure 6. Schematic layout of the top and side views of the experimental setup.

The suspension point was connected to both a load cell and a dynamometer through a metal ring. The measuring devices where fixed to an aluminum beam. Figure 7 shows the described connection. The auxiliary aluminum structure was connected to a linear guide designed to impose surge movements on the water tank when connected to the actuator.

As mentioned, the movement of the suspension point was generated by a mechanical actuator driven by an electrical motor fixed in the structure above the towing carriage shown in Figure 5. The model of the electrical motor is SGMG-44V and its principal technical characteristics are: 4.4 kW of maximum power, 28.4 Nm of maximum torque, and 25 rps of maximum velocity. The maximum acceleration for the motor–actuator system was 1.2 m/s^2, which is higher than the maximum acceleration required in the test matrix for this experimental campaign.

To the measure of the axial tension load of the mooring line, the one-component load cell HBM with an SG full bridge was used. The measured range of the load cell was 0–60 N and the accuracy of the load cell was ±0.179 N. The sampling rate of the load cell was 100 Hz. The time duration of the force measurement for each case was about 180 s to ensure an acceptable number of cycles in the periodic regime. The connection of the load cell to the beam attached to the actuator allowed the load cell to freely rotate, as can be seen in Figure 7. The load cell used permitted measurement of the tension regardless of the chain rotation. The load cell was covered with a special material to prevent the ingress of humidity into the device during the experiments. The load calibration and determination of uncertainties are described in Appendix A.1.

Figure 7. The load cell and dynamometer connection to the aluminum beam attached to the linear guide.

As it is necessary to simultaneously measure both the horizontal and vertical components of the fairlead tension, a 6 DOF dynamometer was also used. The measured range of the dynamometer was 0–250 N for the x-direction and 0–1000 N for the z-direction. The resolution of the dynamometer was 0.042 N for the x-direction and 0.125 N for the z-direction, and the sampling rate was 100 Hz. To measure the fairlead displacement of the mooring line, a three-component accelerometer HBM with an SG full bridge was used. The measured range of the accelerometer was −2.5 G to 2.5 G and the accuracy of the accelerometer was ±0.0131 m/s^2, and the sampling rate was 100 Hz. The displacement was obtained by a double numerical integration of the signal using the trapezoid rule. The accelerometer's calibration and determination of uncertainties are described in Appendix A.2.

In this facility, neither the side walls nor the bottom of the water tank are made of transparent material, so a low-cost submerged camera was used to record the mooring movements. An underwater BARLUS camera to record the evolution of the mooring line was used. The technical characteristics were a sampling rate of 24 fps, a 3.6 mm lens, and an image resolution of 5 MP. Special care was taken to place the camera's line of sight perpendicular to the plane of the catenary-shaped line. The camera was fixed to an auxiliary structure to guarantee the same position during the experimental campaign. The distance between the camera and the line was selected to record the largest area possible in order to obtain the best image quality with minimal optical error. A first calibration process, not described here for the sake of brevity, was addressed to determine the equivalent distance of a pixel image.

In order to represent the kinematics of a discrete number of points of the mooring line, equally spaced red marks were distributed along the mooring line. The distance between two adjacent marks along the mooring line is approximately 27 cm and the exact position of these markers is provided in Table 4. The trajectories of the marks were obtained by image processing. This kind of experiments can be found in the literature, see [10] with isolated chains, but are not frequent when clumped weights are added to the line.

Table 4. Position of the red marks used for capturing the line movements.

Marker Number	Position along the Line Measured from the Suspension Point (m)
1	1.62
2	1.89
3	2.16
4	2.43
5	2.70
6	2.97
7	3.24
8	3.51
9	3.78
10	4.05
11	4.32

3.3. Test Matrix

The surge motions of the top of the mooring line are driven by an electrical motor placed on the top of the carriage. The imposed top-end surge motions are sinusoidal motions described by Equation (12).

$$\delta = A \sin \frac{2\pi t}{T}, \tag{12}$$

where δ represents the fairlead horizontal displacement with respect to the position where the pretension is measured, A and T represent the imposed amplitude and period, respectively. The amplitude values used are in the range [0.125,0.225], which correspond to [3.875,6.975] at real scale, and are presented in Table 5. Model-scale and the corresponding full-scale periods are presented in Table 6.

Table 5. Amplitudes of the fairlead motions for $\lambda = 1/30$.

ID	Model Scale (m)	Full Scale (m)
A 1	0.225	6.975
A 2	0.200	6.200
A 3	0.175	5.425
A 4	0.150	4.650
A 5	0.125	3.875

A total number of 105 dynamic tests were conducted for the surge motion, and each test had a duration of 180 s to ensure an acceptable number of cycles. For each amplitude and period, three configurations have been tested: in the first, no clumped weight is used and, in the other two, the clumped weight is placed at $L_s/3$ (Configuration 1) and $L_s/2$ (Configuration 2) from the fairlead, where L_s is the suspended length of the line. Figure 8 shows the catenary-shaped mooring line in the three configurations.

Table 6. Fairlead motions periods in seconds for $\lambda = 1/30$.

ID	Model Scale	Full Scale
T 1	2.80	15.590
T 2	3.00	16.703
T 3	3.50	19.487
T 4	4.00	22.271
T 5	4.50	25.055
T 6	5.00	27.839
T 7	5.50	30.623

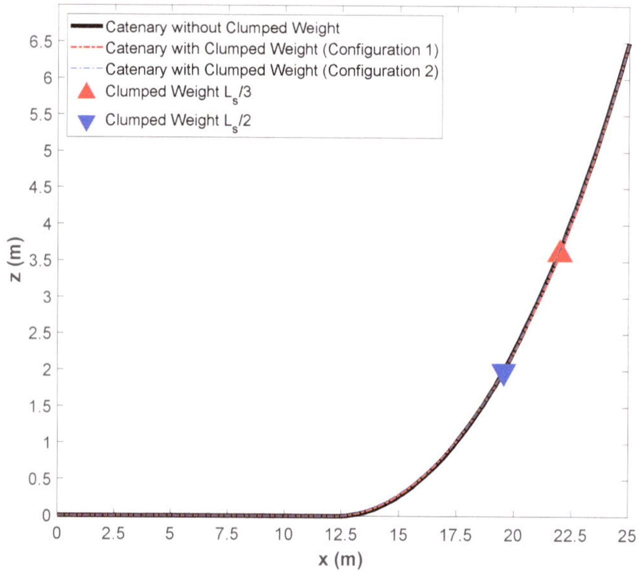

Figure 8. All three tested configurations. Note: the three lines are almost overlapping, the catenary shape is slightly changed in each configuration.

4. Computational Model

4.1. Quasi-Static Modeling

For the quasi-static analysis, the theory developed by [34] is used. The analytical equations for a line, with a portion of the line resting on the seabed, are:

$$x_F - x_A = L - \frac{V_F}{\gamma_w} + \frac{H_F}{\gamma_w}\ln\left[\frac{V_F}{H_F} + \sqrt{1 + \left(\frac{V_F}{H_F}\right)^2}\right] + \frac{H_F L}{EA} + \frac{C_B \gamma_w}{2EA}\left[-\left(L - \frac{V_F}{\gamma_w}\right)^2 + \left(L - \frac{V_F}{\gamma_w} - \frac{H_F}{C_B \gamma_w}\right) MAX\left(L - \frac{V_F}{\gamma_w} - \frac{H_F}{C_B \gamma_w}, 0\right)\right], \tag{13}$$

$$z_F - z_A = \frac{H_F}{\gamma_w}\left[\sqrt{1 + \left(\frac{V_F}{H_F}\right)^2} - 1\right] + \frac{V_F^2}{2EA\gamma_w}, \tag{14}$$

and

$$\gamma_w = \frac{\rho_c - \rho_0}{\rho_c}\gamma g, \tag{15}$$

where L is the total unstretched length of the line, EA is the extensional stiffness, γ_w is the apparent weight in fluid per unit length, C_B is the seabed friction coefficient, H_F and V_F are the horizontal and vertical forces applied at the fairlead, respectively, ρ_c is the density of the chain, ρ_0 is the density of the water, γ apparent mass per unit length of the chain, and g is the acceleration due to gravity.

A scheme of the mooring line where a clumped weight has been added is illustrated in Figure 9. We can divide the line into two segments, the first segment going from the anchor to the position of the point mass, while the second one starts at the point mass location of the fairlead. The equations of the catenary are applied to each segment before and after the clumped weight was added and solved together as a nonlinear system of six equations: four catenary Equations (13), (14), (16) and (17) and two extra equations for the local force equilibrium at the position of the clumped weight. A Newton–Raphson iterative method is used to solve the resulting system of 6 nonlinear equations with 6 unknowns: H_A, V_A, H_F, V_F, T_{Hs2}, T_{Hs1}, T_{Hs1}, T_{Hs2} , T_{Vs1}, T_{Vs2}, x_P, z_P.

$$x_F - x_A = \frac{H_F}{\gamma_w}\left\{\ln\left[\frac{V_F}{H_F} + \sqrt{1+\left(\frac{V_F}{H_F}\right)^2}\right] - \ln\left[\frac{V_F - \gamma_w L}{H_F} + \sqrt{1+\left(\frac{V_F-\gamma_w L}{H_F}\right)^2}\right]\right\} + \frac{H_F L}{EA} \tag{16}$$

$$z_F - z_A = \frac{H_F}{\gamma_w}\left[\sqrt{1+\left(\frac{V_F}{H_F}\right)^2} - \sqrt{1+\left(\frac{V_F-\gamma_w L}{H_F}\right)^2}\right] + \frac{1}{EA}\left(V_F L + \frac{\gamma_w L^2}{2}\right) \tag{17}$$

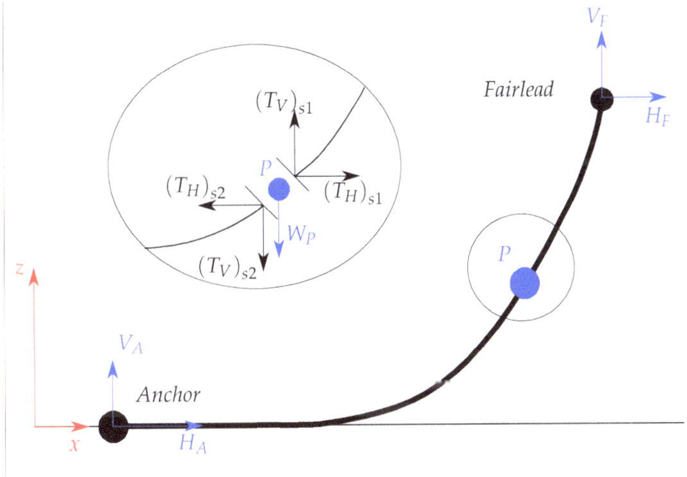

Figure 9. Free body diagram of a mooring line with a clumped weight positioned at point P.

Note that the equations are presented slightly differently than in [34] because of the change in the reference system and the assumption that the chain is modeled as a Morison's slender cylinder.

4.2. Basic Dynamic Equations

Using the finite element method, the partial differential Equation (PDE) (18) which models the mooring line elements is transformed into a nodal system of ordinary differential equations (ODE). This ODE system is discretized in time and solved by the OPASS code [10].

$$\gamma\ddot{\mathbf{R}} + C_4(1+\varepsilon)\left[\ddot{\mathbf{R}} - (\ddot{\mathbf{R}}\cdot\mathbf{t})\mathbf{t}\right] - \frac{(\partial EA\varepsilon\mathbf{t})}{\partial l_0} - \frac{(\partial\beta EA\dot{\varepsilon}\mathbf{t})}{\partial l_0} - \mathbf{F_1} - \mathbf{F_2} - \mathbf{F_3} = 0, \tag{18}$$

where γ is the line mass per unit of mooring line unstretched length, $\mathbf{R}$ is the position of the line element analyzed, $\mathbf{t}$ is the vector tangential to the line at the point analyzed, ε is the deformation, EA is the extensional stiffness, C_4 is constant for the calculation of the hydrodynamic inertial force per unit of unstretched line length, β is a coefficient for the calculation of the structural damping force, and l_0 is the location of the point analyzed along the line.

The external forces $\mathbf{F_1}$, $\mathbf{F_2}$, $\mathbf{F_3}$ are the resultant forces from weight in water, the tangential and normal drag components, respectively, per unit of unstretched length. These hydrodynamic forces are model-led by Equations (19)–(21).

$$\mathbf{F_1} = -\gamma_r g \mathbf{k} = -\frac{\rho_c - \rho_0}{\rho_c}\gamma g \mathbf{k}, \tag{19}$$

where $\mathbf{k}$ is the unit vector parallel to the z axis.

The normal and tangential drag forces are modeled by the Morison equation ([35]), as follows:

$$\mathbf{F_2} = \frac{1}{2}C_{dt} D \rho_0 (1+\varepsilon) V_t^2 \mathbf{t} \tag{20}$$

$$\mathbf{F_3} = \frac{1}{2}C_{dn} D \rho_0 (1+\varepsilon) V_n^2 \mathbf{n}, \tag{21}$$

where the vector $\mathbf{V} = V_t\mathbf{t} + V_n\mathbf{n}$ denotes the relative velocity between the water and the mooring line, expressed by its tangential and normal components. Moreover, C_{dt} and C_{dn} are the tangential and normal drag coefficients of the bare line.

Using bar elements for the spatial discretization of the mooring line, linear polynomials for the elements' internal interpolation of the line motions, and 0-order polynomials for the mass and added mass terms, the global equation of motion is condensed into Equation (22)

$$M \cdot \ddot{\mathbf{X}} + C \cdot \dot{\mathbf{X}} + K \cdot \mathbf{X} + \mathbf{F} = 0, \tag{22}$$

where M is the mass matrix of the system, C is the damping matrix, K is the stiffness matrix, $\mathbf{F}$ is the vector of external forces, and $\ddot{\mathbf{X}}$, $\dot{\mathbf{X}}$, and $\mathbf{X}$ are, respectively, the vectors of acceleration, velocities, and positions of the line DOF defined at the nodes.

4.3. Clumped Weight Implementation

The presence of a clumped weight modifies the system of Equation (22), where an additional weight in water, drag, inertia, and added mass due to the clumped weight are added to the matrix system. We assume a rigid union between the clumped weight and the mooring line, where all local forces due to the clumped weight are distributed per unit length. The external forces acting on the mooring line element i are weight in water $\mathbf{F_{1i}}$, tangential drag $\mathbf{F_{2i}}$ and normal drag $\mathbf{F_{3i}}$, and the added mass $\mathbf{F_{4i}}$.

The weight of the mooring line element is increased by the weight of the clump weight, resulting in an external force for that element described in Equation (23), where W_{CW} is the weight of the Clump Weight in water and L_i is the length of the chain element.

$$\mathbf{F_{1i}} = -\left(\gamma_r g + \frac{W_{CW}}{L_i}\right)\mathbf{k} \tag{23}$$

As described in Section 3.1, the shape of the clump weight is a circular disk represented as schematic view in Figure 10.

For those elements that include a clumped weight, the tangential drag is modified according to Equation (24), where the second term is the tangential drag force per unit length added by the clumped weight. The tangential drag coefficient C_{dtCW} is interpolated for every time step according to its local clumped weight Reynolds Number. The selected C_D-Re curve is illustrated in [36].

$$C_2^{CW} = \frac{1}{2}\rho_0 C_{dt} D + \frac{1}{2}\rho_0 C_{dtCW}\frac{\pi D_{CW}^2}{4}\frac{1}{L_i} \tag{24}$$

Regarding the inertial forces, the implementation of the clumped weight impacts the mass matrix. The body mass of the element that includes the clump weight is increased, and the mass per unit length of line γ is augmented by the mass of the device per unit length of the line in a quantity of $\frac{m_{CW}}{L_i}$.

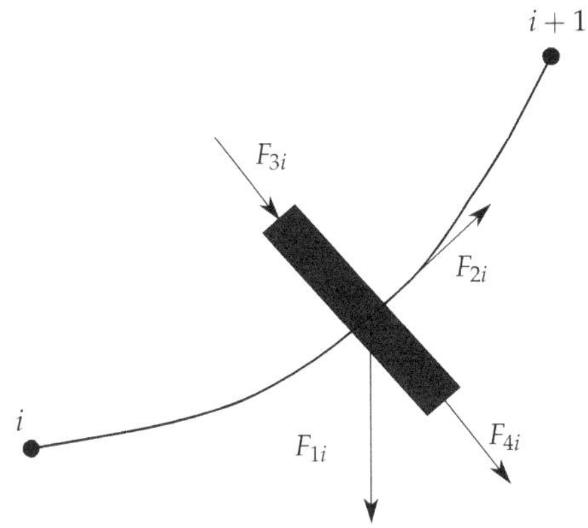

Figure 10. External forces acting on an element of line.

The added mass force per unit of unstretched line length was calculated in OPASS based on a normal added mass coefficient C_{mn} for the Morison approach.

$$C_4^{CW} = C_{mn}\rho_0\frac{\pi D^2}{4} + \frac{1}{3}\rho_0\frac{D_{CW}^3}{L_i} \tag{25}$$

The second term of Equation (25) is the well-known theoretical ideal fluid added mass per unit length of the mooring line element for a circular disk described by [37]. This term is only considered for those elements that include a clump weight.

4.4. Computational Parameters Used in the Simulations

The dimensions of the chain where described in Section 3.1. A mass per unit length $\gamma = 0.0678$ Kg/m and a wire diameter of the chain link $d_w = 2$ mm were measured. Equation (26) provided by [38] determines the axial stiffness in kN when d_w is in meters of the chain used in the simulations.

$$EA = 0.854 \cdot 10^8 d_w \tag{26}$$

As the computational model considers the chain to be a line with a constant circular section (Morison approach), an equivalent hydrodynamic diameter has to be determined. Consequently, the diameter of the equivalent circular section was calculated assuming the same mass per unit length of the chain, see Equation (27). This procedure is based on the assumption that in catenary mooring lines, the weight in water has a dominant effect on the motions, loads, and restoring forces.

$$D = \sqrt{\frac{4\gamma}{\pi\rho_c}} \tag{27}$$

The added mass coefficient C_{mn} of 1.0 is determined following the standards of [39] and the normal and tangential drag coefficients C_{dn} and C_{dt} were obtained following the indications of DNV [40]. For a studless chain, this guideline provides a value for C_{dn} of 2.4 and, for C_{dt}, a value of 1.15; these values correspond to the wire diameter of the chain link. For the numerical implementation, the values have to be referred to the equivalent hydrodynamic diameter, and in order to obtain the equivalent drag coefficient, we assume the same drag force per unit length and proceed as follows:

$$C_{Dn} = \frac{C_{dn} d_w}{D} \quad ; \quad C_{Dt} = \frac{C_{tn} d_w}{D} \tag{28}$$

The contact model implemented in OPASS uses bi-linear springs. When a node is in contact with the seabed, a spring with stiffness K_{sc} provides the floor reaction force per indentation depth and per unit of line length. A damping D_{sc} is also included in the model. The resulting parameters are summarized in Table 7.

Table 7. Parameters of the computational model.

	Value	Units
Equivalent hydrodynamic diameter (D)	0.0033	m
Unstretched Length (L)	27.0	m
Line mass density (ρ_c)	7850	Kg/m^3
Mass per unit length (γ)	0.0678	Kg/m
Axial stiffness (EA)	3.416×10^5	N
Coefficient for the structural Rayleigh damping (β)	0.0007	–
Added mass coefficient (C_{mn})	1.0	–
Normal drag coefficient (C_{Dn})	1.445	–
Tangential drag coefficient (C_{Dt})	0.697	–
Vertical seabed stiffness (K_{sc})	20.0	N/m^2
Vertical seabed damping (D_{sc})	0.1	Ns/m^2
Tangential friction coefficient (C_{ft})	0.0	–
Normal friction coefficient (C_{fn})	0.0	–

5. Experimental Results

Once the shape and position of the catenary have been verified, the static tension measured by the load cell is compared to the theoretical pretension predicted with the formulation described in Section 4.1 as a function of the offset value, see Figure 11. We define the offset distance as the horizontal distance between the top end of the mooring line and the anchor position. We observe that varying the offset value from a typical catenary shape to fully taut, the pretension of the mooring line also increases. The experimental measures in each configuration are also included (offset = 25 m), and the largest difference from the numerical predictions is approximately 0.21%. Before starting the fairlead oscillations according to the test matrix, a static test was performed to ensure the accuracy and repeatability of these values.

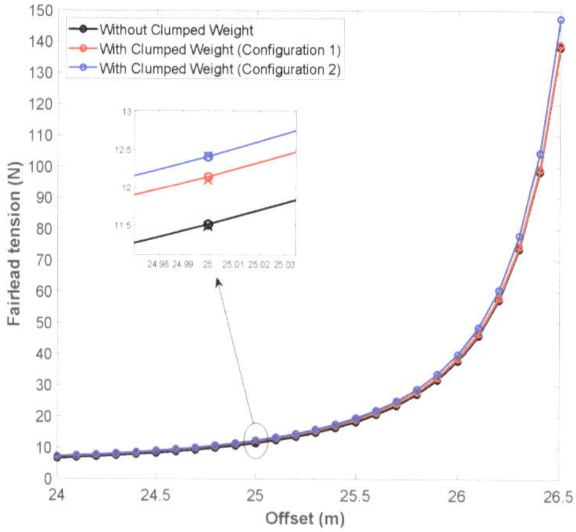

Figure 11. Variation of the tension with offset distance in all configurations. Experimental values are added to the figure with crossed symbols 'x' in the detailed view.

For the selected offset in the experiments (offset = 25 m), the maximum difference in the initial pretension is obtained when the clumped weight is placed in Configuration 2, resulting an increment of 10% with respect to the mooring line without clumped weight.

5.1. Mooring Line Dynamics

The fairlead tension measured by the load cell is represented and compared to the fairlead displacement for different values of the period $T = 2.8,\ 3.5,\ 5.5$ s and amplitudes $A = 0.125,\ 0.175,\ 0.225$ m in Figure 12. After reaching a periodic regime and assuming that the measurements satisfy the ergodic theorem, the tension values are the time average during 20 periods and the vertical bars represent its standard deviation.

As can be observed, the maximum and minimum values of the fairlead tension during the cycle are obtained for the minimum displacement with respect to the initial position. As expected, we observe that for constant amplitudes, smaller periods imply larger accelerations and tensions. Similarly, for constant periods, larger amplitudes also imply larger accelerations and fairlead tensions. We clearly observe that the effect of adding a clumped weight in the first configuration (Configuration 1) to the isolated mooring is to increase the fairlead tension for all amplitudes and periods.

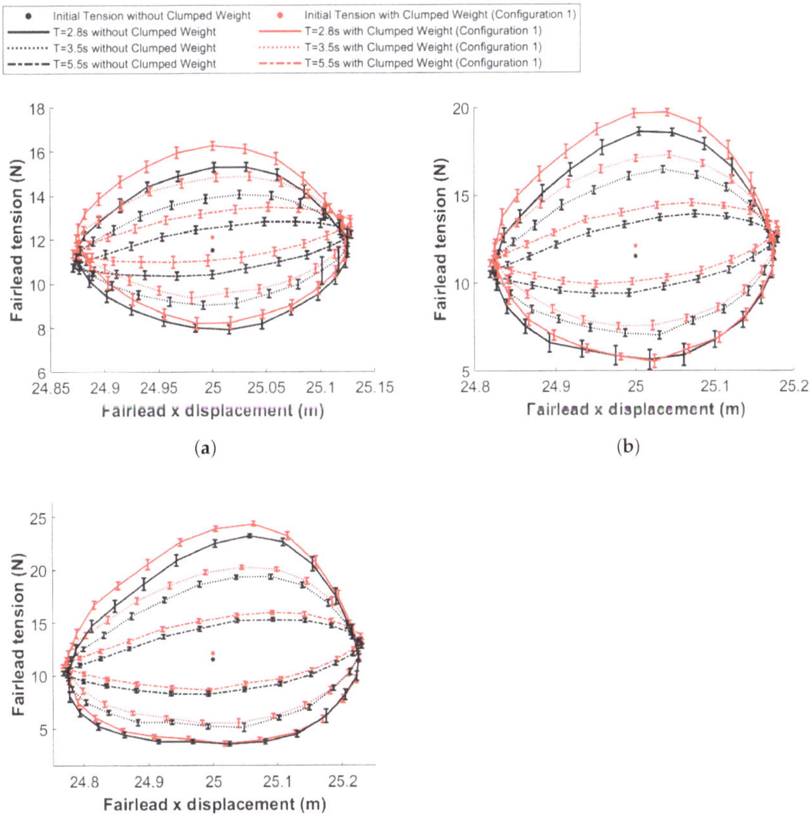

Figure 12. Fairlead tension during the fairlead displacement with and without clumped weight for three different fairlead amplitude values and Configuration 1. (**a**) $A = 0.125$ m. (**b**) $A = 0.175$ m. (**c**) $A = 0.225$ m.

Similarly, in Figure 13, the fairlead tension for the isolated mooring is also compared to the clumped weight (Configuration 2) for the same set of amplitudes and periods. We can observe that the change of the clumped weight position has an important influence on the magnitude of the increase of the fairlead tension. The maximum fairlead tension grows approximately in the same magnitude of the clumped weight in Configuration 1, while the maximum fairlead tension is increased by almost double of this weight in Configuration 2. In other words, when clumped weights are present, the maximum fairlead tension is increased by just 6% in Configuration 1 and by 9% in Configuration 2, with respect to the isolated mooring line.

Additionally, the maximum and minimum fairlead tension of the complete test matrix cases is shown in Tables 8 and 9.

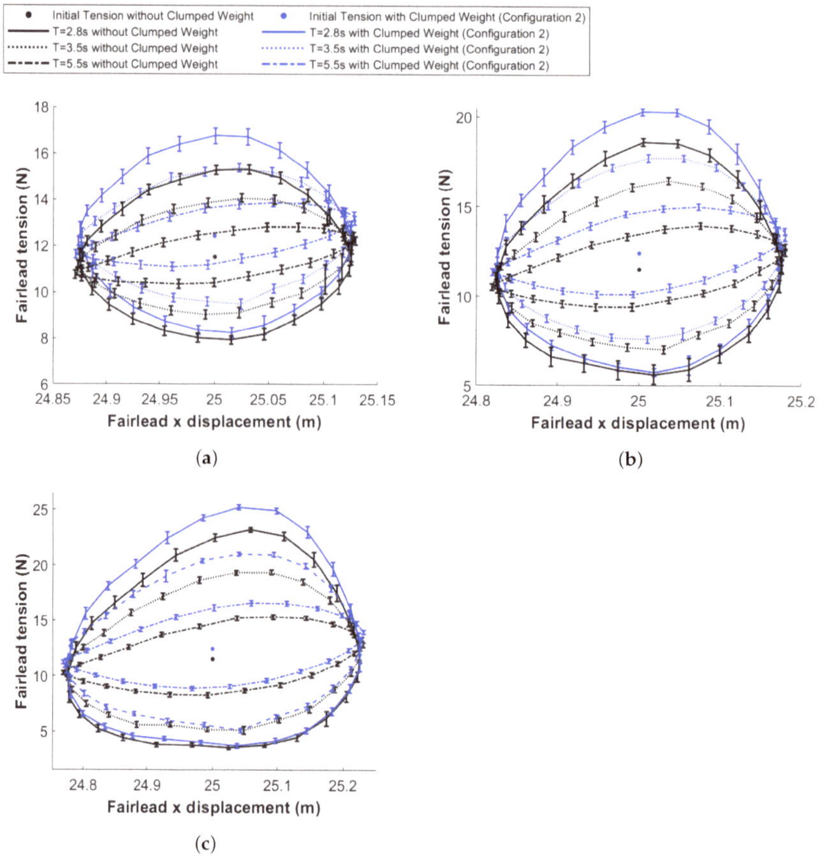

Figure 13. Fairlead tension during the fairlead displacement with and without clumped weight for three different fairlead amplitude values and Configuration 2. (**a**) $A = 0.125$ m. (**b**) $A = 0.175$ m. (**c**) $A = 0.225$ m.

Table 8. Maximum fairlead tension in *N* for all the test matrix and configurations. Notation *WO CW*, *CW1*, and *CW2* denotes mooring line without clumped weight, mooring line with clumped weight (Configuration 1) and mooring line with clumped weight (Configuration 2), respectively.

Amplitudes (m)		Periods (s)							Configurations
		2.8	3.0	3.5	4.0	4.5	5.0	5.5	
0.125		15.35	14.88	14.07	13.55	13.2	12.96	12.80	WO CW
		16.30	15.78	14.99	14.31	13.92	13.65	13.47	CW1
		16.83	16.31	15.37	14.77	14.35	14.06	13.87	CW2
0.150		16.94	16.32	15.19	14.45	13.93	13.57	13.32	WO CW
		18.23	17.32	16.07	15.22	14.67	14.40	14.00	CW1
		18.44	17.74	16.56	15.71	15.13	15.72	14.42	CW2
0.175		18.68	17.84	16.40	15.66	14.71	14.24	13.89	WO CW
		19.86	18.93	17.29	16.18	15.44	14.92	14.54	CW1
		20.41	19.50	17.83	16.72	15.94	15.41	15.00	CW2
0.200		20.73	19.68	17.80	16.50	15.63	15.00	14.55	WO CW
		21.88	20.74	18.72	17.31	16.35	15.69	15.20	CW1
		22.37	21.33	19.60	18.12	17.10	16.28	15.78	CW2
0.225		23.20	21.87	19.46	17.86	16.72	15.95	15.33	WO CW
		24.31	22.83	20.35	18.59	17.44	16.58	15.95	CW1
		25.24	23.96	21.11	19.38	18.36	17.33	17.05	CW2

Table 9. Minimum fairlead tension in *N* for all the test matrix and configurations. Notation *WO CW*, *CW1*, and *CW2* denotes mooring line without clumped weight, mooring line with clumped weight (Configuration 1) and mooring line with clumped weight (Configuration 2), respectively.

Amplitudes (m)		Periods (s)							Configurations
		2.8	3.0	3.5	4.0	4.5	5.0	5.5	
0.125		7.826	8.189	8.959	9.558	9.876	10.14	10.34	WO CW
		7.758	8.247	9.453	10.04	10.44	10.64	10.90	CW1
		8.151	8.626	9.561	10.18	10.59	10.88	11.09	CW2
0.150		6.601	7.116	8.057	8.693	9.289	9.662	9.917	WO CW
		6.305	7.064	8.282	9.283	9.832	10.21	10.50	CW1
		6.941	7.413	8.55	9.359	9.874	10.33	10.63	CW2
0.175		5.48	5.935	6.999	7.931	8.571	9.04	9.379	WO CW
		5.56	6.147	7.036	8.004	8.914	9.578	9.995	CW1
		5.797	6.311	7.486	8.419	9.10	9.507	10.07	CW2
0.200		4.388	4.921	6.074	6.895	7.779	8.358	8.805	WO CW
		4.626	5.181	6.384	7.163	7.961	8.923	9.402	CW1
		4.79	5.40	6.424	7.467	8.297	8.911	9.39	CW2
0.225		3.463	3.93	5.115	6.116	6.912	7.655	8.194	WO CW
		3.38	4.222	5.478	6.496	7.326	7.881	8.586	CW1
		3.857	4.203	5.449	6.519	7.452	8.117	8.608	CW2

A fast Fourier transform (fft) was performed to obtain the power spectra for the mooring line under the imposed motions. The power spectra of the fairlead tension have been analyzed for the complete test matrix. As the fft does not change significantly in our test matrix, a single representative case is presented in Figure 14 for an amplitude of $A = 0.175$ m and $T = 3.5$ s. The 0 Hz frequency value corresponds to the quasi-static case as mentioned in [41]. In all cases, they are mostly dominated by the first harmonic followed by the third and second harmonics. As Figure 14 shows, the presence of the clumped weight along the line does not significantly change the analysis: the power amplitude and frequency of the dominant harmonics being almost the same in all configurations.

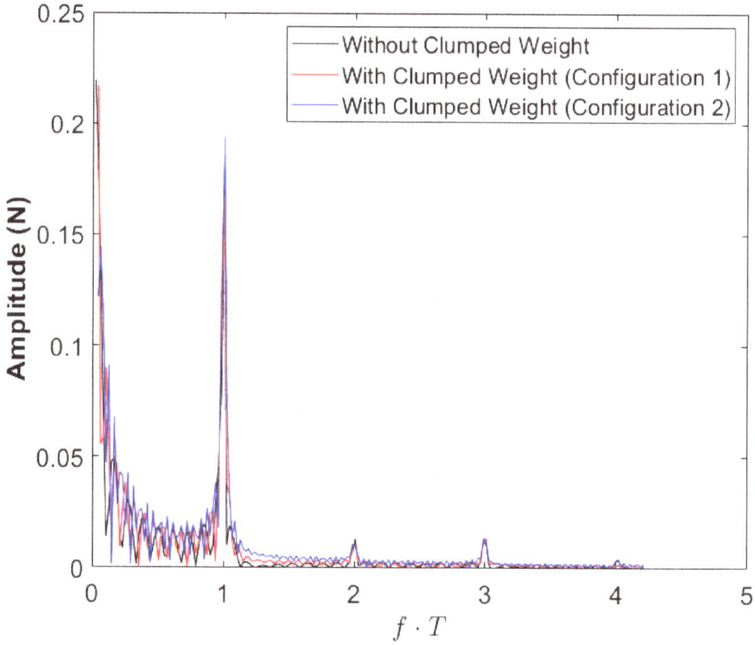

Figure 14. Spectrum plot of the top-end tensions for an amplitude of $A = 0.175$ s and period $T = 3.5$ s for the three tested configurations.

The energy dissipation is computed using the diagram method detailed in [28], where the dissipated energy of the mooring line is the area enclosed in one cycle by the hysteresis loop:

$$E = \oint F_d \cdot d\delta, \tag{29}$$

where F_d and δ are the damping force and the fairlead displacement, respectively.

The dissipation of energy in a mooring line is mainly caused by structural damping, drag with possible VIV, and friction with the seabed [42]. In this study, a simplification is made and the structural damping and friction with the seabed are considered negligible with respect to drag and are modeled as:

$$F_d = F_D = \frac{1}{2}\rho_0 C_D D L_s \dot{\delta}^2 \tag{30}$$

Since the displacement is periodic, $\delta = A \cdot sin(\omega t)$, the dissipation energy can be calculated as:

$$E = 4\int_0^{\delta_{max}} F_d \cdot d\delta = \frac{4}{3}\rho_0 D L C_D A^3 \omega^2 \tag{31}$$

The dissipated energy depends on several factors, such as the frequency and amplitude, as shown in Figures 12 and 13 where the enclosed areas are larger when the frequencies and amplitudes are increased.

The dissipated energy of the complete test matrix computed in dimensional and non-dimensional forms according to Equation (29) are shown in Figure 15. The non-dimensional dissipated energy is obtained by dividing by a reference potential energy, $A \cdot w \cdot H$, as performed by [28]. Both the dimensional dissipated energy and the non-dimensional version present the same tendency, see Figure 15. It can be appreciated that larger amplitudes implicate larger dissipated energy. This pattern is presented in several publications dealing with isolated mooring lines, e.g., [7,23,27]. In the same way, smaller periods increase the dissipated energy. In addition, when clumped weights are present, the pattern is consistent and the damping in all imposed motions is higher, regardless of the location along the line.

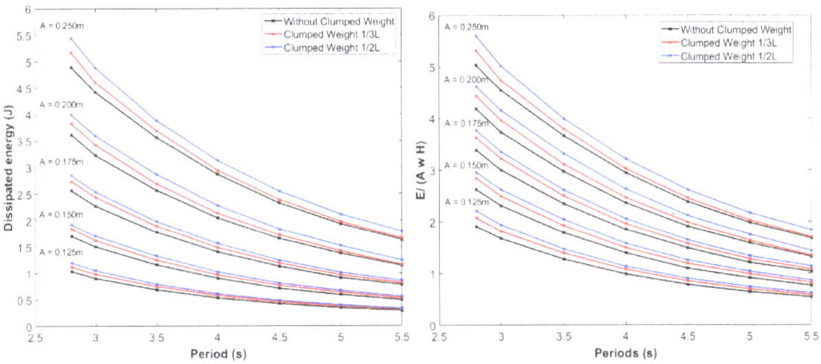

Figure 15. Dissipated energy for all the test matrix. Dimensional form (**left**) and non-dimensional form (**right**).

The dissipated energy for the line without clumped weight is plotted against $\omega^2 A^3$ in Figure 16 for all the motions of the test matrix. Similarly to [23], a good linear fit is accomplished for the experimental results. The same linear tendency is also observed when clumped weights are added in both configurations.

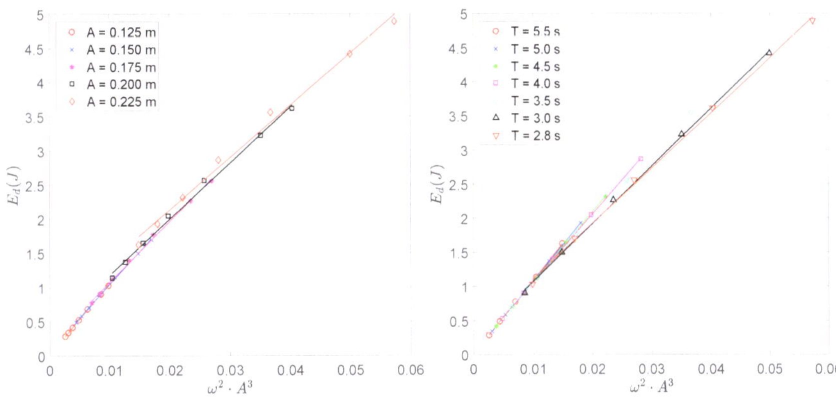

Figure 16. Dissipated energy under the imposed top-end motions for the mooring line without clumped weight.

Assuming that the drag coefficient C_D can be obtained from the slope of the fitted curves, which is $\frac{4}{3}\rho_0 DLC_D$, the mean value obtained for the complete test matrix was $C_D = 1.75$ for the isolated mooring line, which is a little higher than the standard value used for the simulations $C_{Dn} = 1.455$, see Section 4.4. The mean values of C_D obtained when clumped weights are present were 1.82 and 1.90 for Configuration 1 and Configuration 2, respectively. This increase of the damping coefficient explains why the line dissipates more energy when clumped weights are added.

The non-dimensional increase of dissipated energy due to the clumped weight is expressed by Equation (32).

$$\frac{\Delta E}{E_{CW}} = \frac{\Delta E}{\rho_0 A^3 \omega^2 D_{CW}^2} \tag{32}$$

The dimensionless increase of dissipated energy for both clump weight configurations with respect to the line without clump weight, for each of the different amplitudes and periods tested, is shown in Figure 17. The amplitude of the motion has more influence than the period in the increase of dissipated energy, as seen in Figure 17. Configuration 2 has an evident increase of the dissipated energy, almost two times higher than Configuration 1 for all cases. The presence of the clumped weights shows a more dominant effect for smaller amplitudes in terms of dissipated energy.

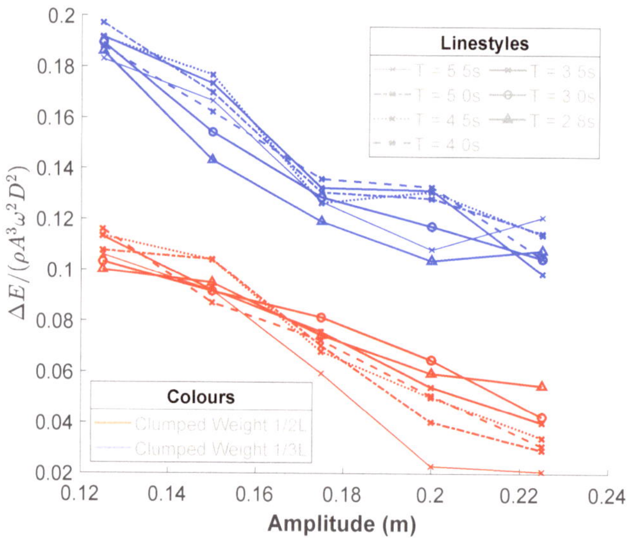

Figure 17. Dimensionless increase of dissipated energy due to the Clumped weights.

5.2. Mooring Line Kinematics

In this subsection, the results of the image processing of the camera recordings will be presented. The postprocessing of the visual data was performed with MATLAB tools and consisted of applying a moving method for detecting local outliers according to a specified window. A Piecewise Cubic Hermite Interpolating Polynomial has been used for replacing the outliers. Finally, a classical filter for frequencies higher than the third harmonic has been applied and optical errors have been excluded.

Figure 18 shows the trajectories of markers 2 to 10 for the mooring line without clumped weight and with clumped weight (Configuration 1). The results show the same pattern described in previous publications [7,10], where the points near the top of the line move very similarly to the fairlead, while deeper points move more vertically. Despite the slightly changed trajectories of the points, the pattern barely changes with the addition of clumped weights in both configurations. Trajectories of the mooring line with clumped weight (Configuration 2) has not been added because the similarity in the results overlaps with the other two configurations with a slight change of the position because of the different shape of the catenary.

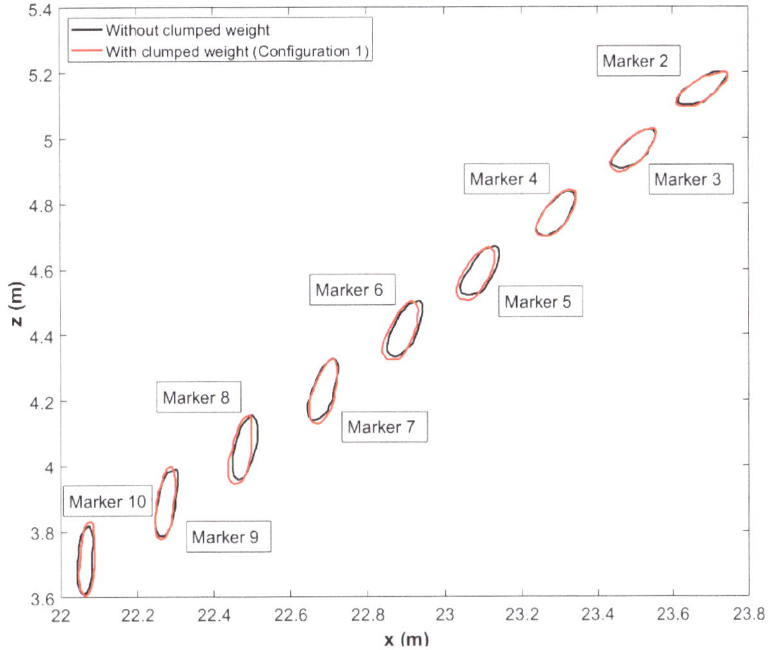

Figure 18. Trajectories of the tracked points of the mooring line for a top-end motion of an Amplitude of 0.175 m and periods 3.5 s. Markers 2 to 10 are indicated in the figure.

Figure 19 presents the tracked motions of three mooring line points with and without clumped weight in Configuration 1, for an amplitude of 0.125 m and three periods of 2.8, 3.5, 5.5 s. For all points, the ranges of motion in the x- and z-directions of the reference system in Figure 9 are practically the same, with some changes observed for the shortest periods, and a slight change in the initial position of the marker due to the different catenary curves caused by the presence of the clumped weight.

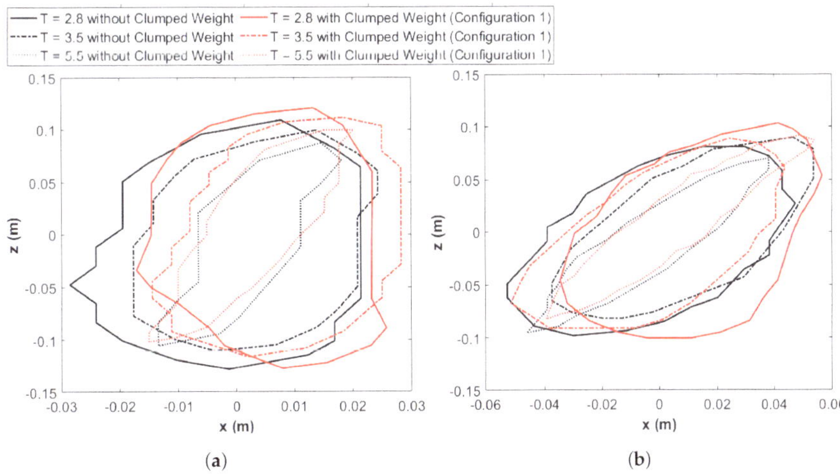

Figure 19. *Cont.*

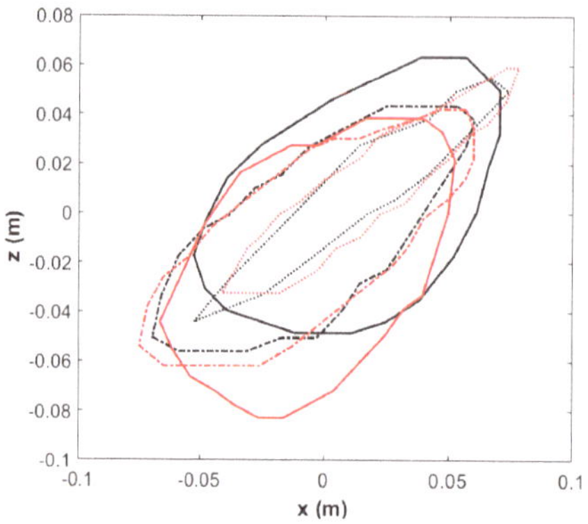

(c)

Figure 19. Tracked motions of three points of the mooring line, for a top-end motion of an Amplitude of 0.125 m and periods 2.8, 3.5, 5.5 s. (**a**) Trajectories of the mooring line lower point of Figure 18 (Marker 9). (**b**) Trajectories of the mooring line middle point of Figure 18 (Marker 5). (**c**) Trajectories of the mooring line upper point of Figure 18 (Marker 2).

6. Numerical Results

The numerical code OPASS has been validated with these new experimental results, where clumped weights are part of the mooring line. First, the fairlead tension in the absence of clumped weights has been computed and compared with the experimental results for three oscillation periods $T = 2.8,\ 3.5,\ 5.5$ s and two amplitudes $A = 0.125,\ 0.225$ m in Figure 20, and for two oscillations periods $T = 2.8,\ 5.5$ s and five amplitudes values $A = 0.225,\ 0.220,\ 0.175,\ 0.150, 0.125$ m in Figure 21. As can be observed, the prediction of the numerical tool approximates the values measured in the experiments for all periods and amplitudes tested very well. When the clumped weight is added in both configurations, the prediction of the numerical code is still valid, see Figures 22 and 23 for Configuration 1 and Figures 24 and 25 for Configuration 2. From these figures, a slightly overpredicted fairlead tension at the maximum tension condition can be observed. This tendency appears in all the cases tested. In addition, the numerical code can capture the increase of the fairlead tension in approximately the same proportion as the experimental results shown in Figures 12 and 13.

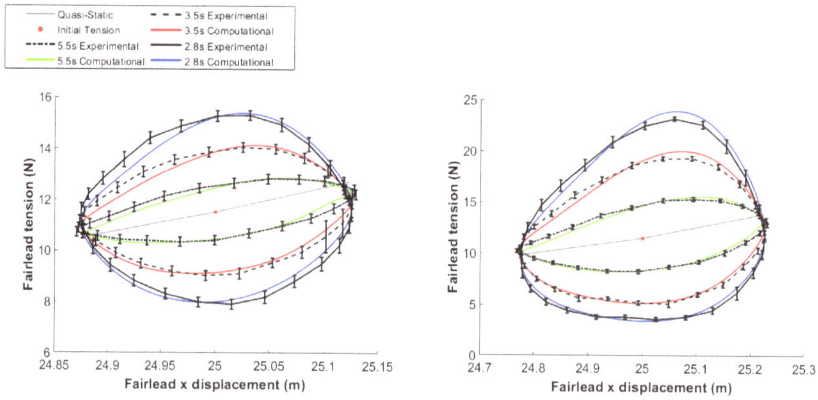

(**a**) (**b**)

Figure 20. Comparison between the computed fairlead tension values and the experimental measurements of the isolated mooring line for two different oscillation amplitudes $A = 0.125,\ 0.225$ m and three different periods $T = 2.8,\ 3.5,\ 5.5$ s for the isolated mooring line. (**a**) $A = 0.125$ m. (**b**) $A = 0.225$ m.

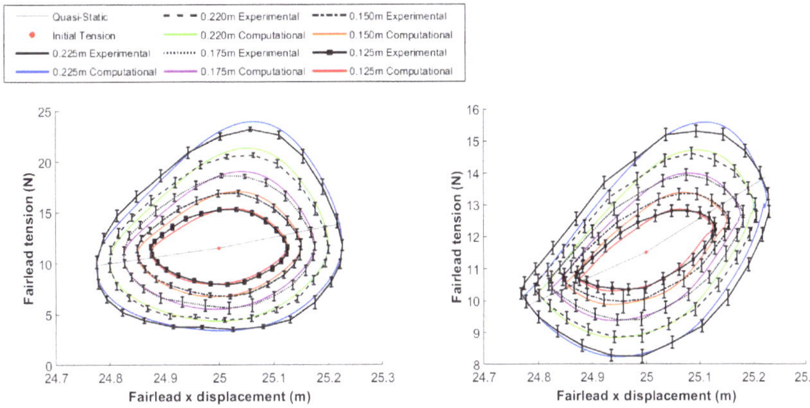

(a) (b)

Figure 21. Comparison between the computed fairlead tension values and the experimental measurements of the isolated mooring line for two different oscillation amplitudes $A = 0.225, 0.220, 0.175, 0.150, 0.125$ m and two different periods $T = 2.8,\ 5.5$ s for the isolated mooring line. (**a**) $T = 2.8$ s. (**b**) $T = 5.5$ s.

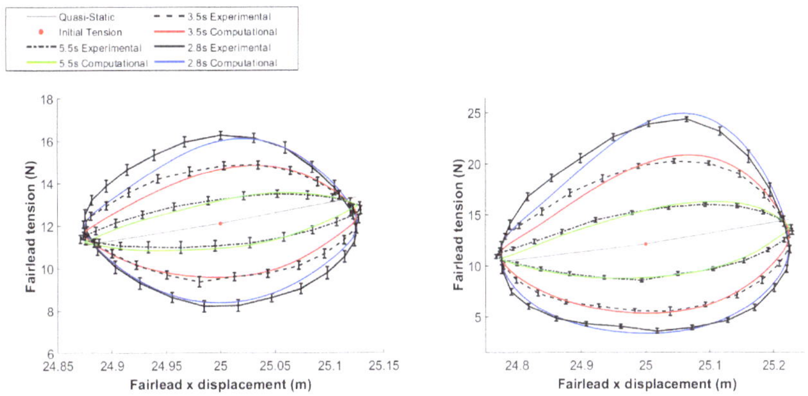

(a) (b)

Figure 22. Comparison between the computed fairlead tension values and the experimental measurements of the mooring line with the clumped weight in configuration 1 for two different oscillation amplitudes $A = 0.125,\ 0.225$ m and three different periods $T = 2.8,\ 3.5,\ 5.5$ s for the isolated mooring line. (**a**) $A = 0.125$ m. (**b**) $A = 0.225$ m.

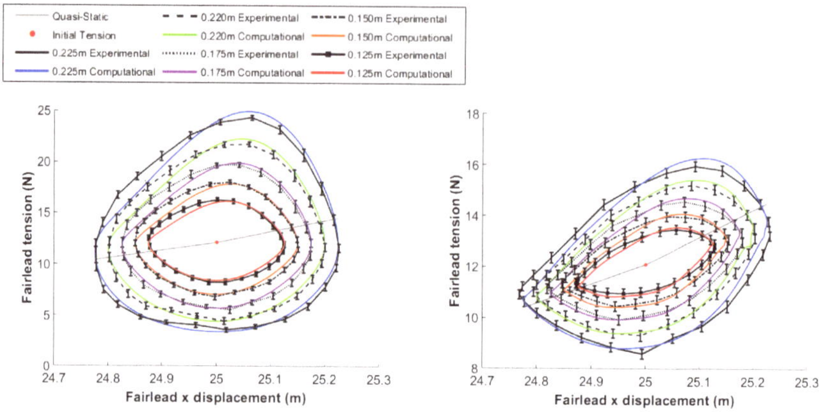

(a) (b)

Figure 23. Comparison between the computed fairlead tension values and the experimental measurements with clumped weight in configuration 1 for different oscillation amplitudes $A = 0.225$, 0.220, 0.175, 0.150, 0.125 m and two different periods $T = 2.8$, 5.5 s for the isolated mooring line. (**a**) $T = 2.8$ s. (**b**) $T = 5.5$ s.

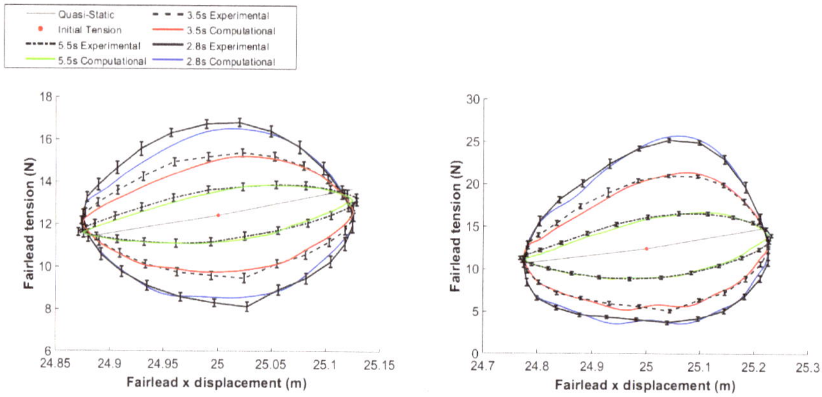

(a) (b)

Figure 24. Comparison between the computed fairlead tension values and the experimental measurements of the mooring line with the clumped weight in configuration 2 for two different oscillation amplitudes $A = 0.125$, 0.225 m and three different periods $T = 2.8$, 3.5, 5.5 s for the isolated mooring line. (**a**) $A = 0.125$ m. (**b**) $A = 0.225$ m.

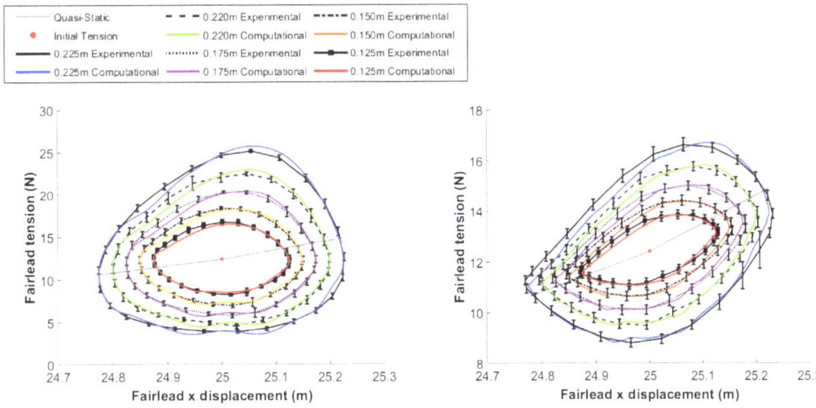

(**a**) (**b**)

Figure 25. Comparison between the computed fairlead tension values and the experimental measurements with clumped weight in configuration 2 for different oscillation amplitudes $A = 0.225$, 0.220, 0.175, 0.150, 0.125 m and two different periods $T = 2.8,\ 5.5$ s for the isolated mooring line. (**a**) $T = 2.8$ s. (**b**) $T = 5.5$ s.

The trajectory of part of the mooring line in the absence of clumped weights has been computed numerically for three oscillation periods $T = 2.8,\ 3.5,\ 5.5$ s and one amplitude $A = 0.125$ m, see Figure 26. It can be observed that the prediction of the numerical tool approximates the values measured in the experiments for all periods and amplitudes tested very well. When the clumped weight is added, the prediction of the numerical code is still valid, see, for example, Figure 27 for Configuration 1.

The optical errors have not been corrected in the postprocessing and their effects are present for the nodes placed at the ends: markers 1 to 3 and 8 to 11. This could be the explanation for why the trajectories numerically obtained have better agreement with the experimental results for the central nodes (Markers 4 to 7). This could be improved in the near future using several cameras instead of only a single one.

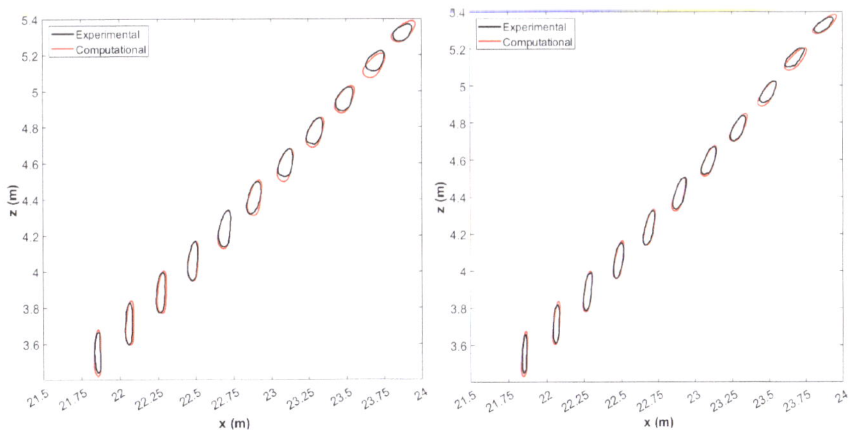

(**a**) (**b**)

Figure 26. *Cont.*

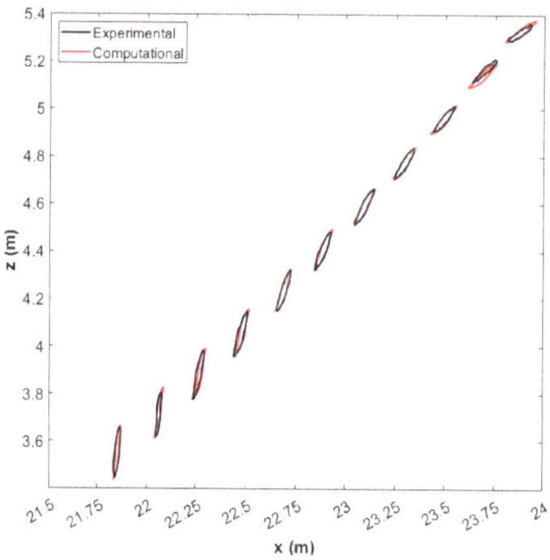

(c)

Figure 26. Comparison between computed trajectories of catenary nodes and experimental tracked points for different oscillation periods $T = 2.8,\ 3.5,\ 5.5$ s and amplitude $A = 0.125$ m for the isolated mooring line without clumped weight. (**a**) $T = 2.8$ s. (**b**) $T = 3.5$ s. (**c**) $T = 5.5$ s.

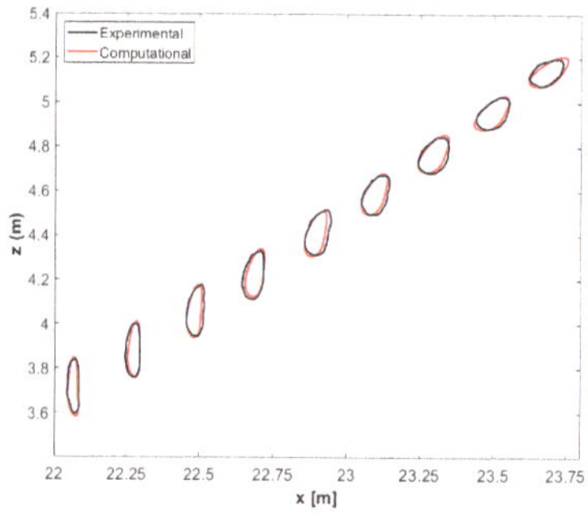

(**a**)

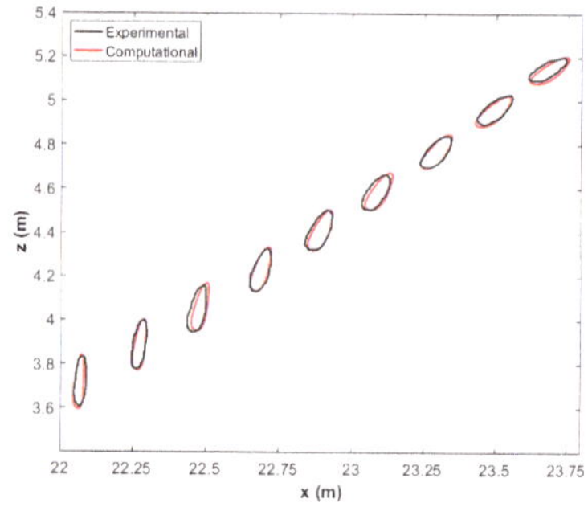

(**b**)

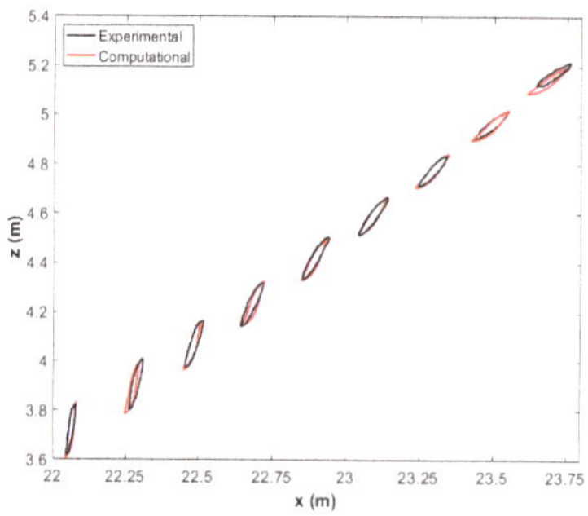

(c)

Figure 27. Comparison between computed trajectories of catenary nodes and experimental tracked points for different oscillation periods $T = 2.8,\ 3.5,\ 5.5$ s and amplitude $A = 0.125$ m for the isolated mooring line with clumped weight (Configuration 1). (**a**) $T = 2.8$ s. (**b**) $T = 3.5$ s. (**c**) $T = 5.5$ s.

7. Conclusions

The influence of the clumped weights when added to mooring line dynamics is studied from a dual point of view: experimentally, using scaled experiments in a towing tank, and numerically, using a finite element numerical tool where the action of the clumped weights has been modeled. The performance of the numerical tool is notably accurate when its predictions are compared to the experiments for both an isolated line and for the presence of clumped weights, in terms of fairlead tension and line kinematics.

The first evidence observed is that the geometric stiffness of the mooring line in the horizontal plane was increased when a clumped weight was added. Consequently, the disadvantage of this addition is that more rigid systems are more loaded and therefore require more expensive lines with larger breaking loads. The advantage of the stiffness increment obtained with the addition of the clumped weights is that the fairlead tension is increased for the same excursion, and therefore, we should expect a surge motion reduction of the moored platform in practical applications.

As expected, the experimental results show an increase in the fairlead tension when clumped weights are present, approximately increasing its value by the weight of the clumped weight in absolute terms and, in relative terms, increasing it between 6% and 9% of the maximum fairlead tension with respect to the isolated mooring line.

When the clumped weight is closer to the fairlead, the line dissipates less energy than when it is placed further away. Therefore, setting the clumped weight farther from the fairlead with the obvious limitation of seabed contact seems to be an adequate practice to reduce the surge motion of the moored platform. Keeping in mind that the role of the clumped weights is also to increase the dissipated energy, the clumped weight's presence is more effective in terms of non-dimensional energy dissipation when low amplitude motions of the fairlead are present.

To track the mooring line kinematics, a low cost system based on an image processing code and submerged cameras has been developed to search and locate the mooring line movements. Good agreement between the trajectories obtained after the image postprocessing and numerical counterparts is shown for the central nodes tracked, while the nodes placed at the ends seem to be affected by optical errors.

Previously validated for an isolated line OPASS code, a simulation tool specifically designed for mooring line dynamics now includes the possibility of clumped weight implementation and has been validated against experiments for two different clumped weight configurations with good accuracy. However, when the fairlead accelerations grow, and the mooring line excitation is more violent, the numerical approach tends to overpredict the maximum tension in the fairlead. This could be due to the difficulty of accurately determining the line drag and added mass coefficients.

Author Contributions: T.L.-O.: conceptualization, investigation, software, formal analysis, data curation, writing—original draft, writing—review and editing; L.M.G.-G.: conceptualization, methodology, investigation, writing—original draft, and writing—review and editing; J.C.-S.: writing—review and editing; A.M.L.: investigation, experimental setup, experiments; L.S.Y.: investigation, experimental setup, experiments, and data curation; A.B.B.: investigation, experimental setup, experiments; N.V.V.: experiments. All authors have read and agreed to the published version of the manuscript.

Funding: This research was founded by the Spanish Ministry of Economy and Competitiveness (MEC) under grant RTI2018-096791-B-C21 and RTI2018-096791-B-C22 *Hidrodinámica de elementos de amortiguamiento del movimiento de aerogeneradores flotantes*.

Institutional Review Board Statement: Not applicable.

Informed Consent Statement: Not applicable.

Acknowledgments: All the authors would like to thank to Jon Martinez for his help in the development of the Matlab script for tracking motion of the mooring line. The authors would like to thank Ciaran Stone for his valuable assistance during the preparation of this manuscript.

Conflicts of Interest: All authors have participated in the conception, analysis, and interpretation of the data, revising the article critically and approving the final version. The authors have no affiliation with any organization with a direct or indirect financial interest in the subject matter discussed in the manuscript. The authors confirm that this work is original and has not been published elsewhere.

Appendix A

Appendix A.1. Load Cell Uncertainties

As explained in Section 3.2, before performing the experimental campaign, the load cell was calibrated with different weights ranging from 0 to 6 Kg. A fitted curve of the measured data using a linear interpolation $val(x) = 0.08992x - 0.0001097$, and an R^2 of 0.9999 was performed, as represented in Figure A1.

The uncertainties were estimated by adding the standard deviation of the residuals σ to the combined standard uncertainty of the instrument $U_c = 0.001528$ mV/V that takes into account the temperature, deformation, and weight uncertainties.

$$error = \sqrt{\sigma^2 + U_c^2} = \sqrt{(0.000196\ \mathrm{mV/V})^2 + (0.001528\ \mathrm{mV/V})^2} \quad \text{(A1)}$$

The error in the measurements was estimated according to Equation (A1), resulting in an error of ± 0.179 N after correcting with the calibration line, represented in Figure A1.

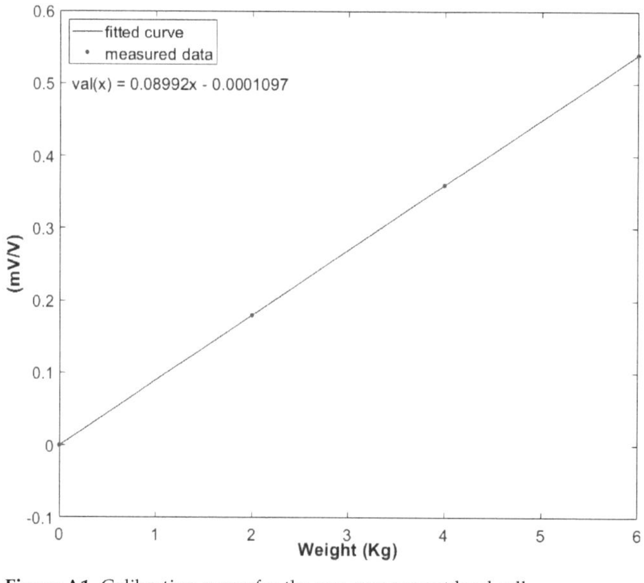

Figure A1. Calibration curve for the one-component load cell.

Appendix A.2. Accelerometer Uncertainties

As explained in Section 3.2, before performing the experimental campaign the accelerometer was calibrated with different accelerations ranging from -9.81 to 9.81 m/s^2 in the x-direction. A fitted curve of the measured data using a linear interpolation $val(x) = -0.09355x + 0.9178$ and an R^2 of 0.9999 was performed, as represented in Figure A2.

The uncertainties were estimated by adding the standard deviation of the residuals σ to the combined standard uncertainty of the instrument $U_c = 0.002153$ mV/V.

$$error = \sqrt{\sigma^2 + U_c^2} = \sqrt{(0.000683\ \mathrm{mV/V})^2 + (0.002153\ \mathrm{mV/V})^2} \quad \text{(A2)}$$

The error in the measurements was estimated according to Equation (A2), resulting in an error of ± 0.0131 m/s^2 from the acceleration due to gravity after correcting with the calibration line, represented in Figure A2.

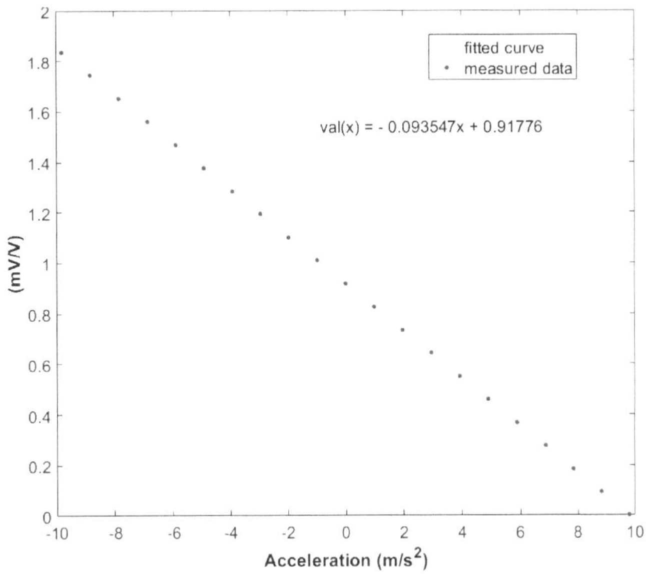

Figure A2. Calibration curve for the three-component accelerometer in the x-direction.

References

1. Ramírez, L.; Fraile, D.; Brindley, G. Offshore Wind in Europe—Key Trends and Statistics 2020. 2020. Available online: https://windeurope.org/intelligence-platform/product/offshore-wind-in-europe-key-trends-and-statistics-2020/ (accessed on 8 May 2021).
2. Hsu, W.Y.; Chuang, T.C.; Yang, R.Y.; Hsu, W.T.; Thiagarajan, K.P. An Experimental Study of Mooring Line Damping and Snap Load in Shallow Water. *J. Offshore Mech. Arct. Eng.* **2019**, *141*, 051603. [CrossRef]
3. Davidson, J.; Ringwood, J.V. Mathematical modelling of mooring systems for wave energy converters—A review. *Energies* **2017**, *10*, 666. [CrossRef]
4. Hsu, W.t.; Thiagarajan, K.P.; MacNicoll, M.; Akers, R. Prediction of extreme tensions in mooring lines of a floating offshore wind turbine in a 100-year storm. In *International Conference on Offshore Mechanics and Arctic Engineering*; American Society of Mechanical Engineers: New York, NY, USA, 2015; Volume 56574, p. V009T09A050.
5. Shah, A.; Umar, A.; Siddiqui, N. A methodology for assessing the reliability of taut and slack mooring systems against instability. *Ocean Eng.* **2005**, *32*, 1216–1234. [CrossRef]
6. Luo, Y. Optimum Design Of Clump Weights For Offshore Mooring Systems. In Proceedings of the Second International Offshore and Polar Engineering Conference, San Francisco, CA, USA, 14 June 1992.
7. Barrera, C.; Guanche, R.; Losada, I.J. Experimental modelling of mooring systems for floating marine energy concepts. *Mar. Struct.* **2019**, *63*, 153–180. [CrossRef]
8. Chen, L.; Basu, B. Development of an open-source simulation tool for mooring systems. In Proceedings of the 2018 Civil Engineering Research in Ireland Conference, Dublin, Ireland, 29–30 August 2018; pp. 823–828.
9. Barrera, C.; Guanche, R.; Rodríguez, Á.; Armesto, J.A.; Losada, I.J. On the importance of mooring system parametrisation for accurate floating structure designs. *Mar. Struct.* **2020**, *72*, 102765. [CrossRef]
10. Azcona, J.; Munduate, X.; González, L.; Nygaard, T.A. Experimental validation of a dynamic mooring lines code with tension and motion measurements of a submerged chain. *Ocean Eng.* **2017**, *129*, 415–427. [CrossRef]
11. Crudu, L.; Obreja, D.; Marcu, O. Moored offshore structures-evaluation of forces in elastic mooring lines. In *IOP Conference Series: Materials Science and Engineering*; IOP Publishing: Bristol, UK, 2016; Volume 147, p. 012096.
12. Liu, Z.; Tu, Y.; Wang, W.; Qian, G. Numerical analysis of a catenary mooring system attached by clump masses for improving the wave-resistance ability of a spar buoy-type floating offshore wind turbine. *Appl. Sci.* **2019**, *9*, 1075. [CrossRef]
13. Yuan, Z.M.; Incecik, A.; Ji, C. Numerical study on a hybrid mooring system with clump weights and buoys. *Ocean Eng.* **2014**, *88*, 1–11. [CrossRef]
14. Ji, C.; Yuan, Z. Experimental study of a hybrid mooring system. *J. Mar. Sci. Technol.* **2015**, *20*, 213–225. [CrossRef]
15. Barbanti, G.; Marino, E.; Borri, C. Mooring system optimization for a spar-buoy wind turbine in rough wind and sea conditions. In *Conference of the Italian Association for Wind Engineering*; Springer: Berlin/Heidelberg, Germany, 2018; pp. 87–98.
16. Xu, K.; Larsen, K.; Shao, Y.; Zhang, M.; Gao, Z.; Moan, T. Design and comparative analysis of alternative mooring systems for floating wind turbines in shallow water with emphasis on ultimate limit state design. *Ocean Eng.* **2021**, *219*, 108377. [CrossRef]

17. Sirigu, S.A.; Bonfanti, M.; Begovic, E.; Bertorello, C.; Dafnakis, P.; Giorgi, G.; Bracco, G.; Mattiazzo, G. Experimental investigation of the mooring system of a wave energy converter in operating and extreme wave conditions. *J. Mar. Sci. Eng.* **2020**, *8*, 180. [CrossRef]
18. Hermawan, Y.A.; Furukawa, Y. Coupled three-dimensional dynamics model of multi-component mooring line for motion analysis of floating offshore structure. *Ocean Eng.* **2020**, *200*, 106928. [CrossRef]
19. Bruschi, N.; Ferri, G.; Marino, E.; Borri, C. Influence of Clumps-Weighted Moorings on a Spar Buoy Offshore Wind Turbine. *Energies* **2020**, *13*, 6407. [CrossRef]
20. Cabrerizo-Morales, M.; Molina-Sanchez, R.; Pérez-Rojas, L. Small-Scale Study of Mooring Line Tension Thresholds Based on Impulsive Load Analysis during Big Floating Structure Operation and Commissioning. *Water* **2021**, *13*, 1056. [CrossRef]
21. Nakajima, T.; Motora, S.; Fujino, M. On the dynamic analysis of multi-component mooring lines. In Proceedings of the Offshore Technology Conference, Houston, TX, USA, 3–6 May 1982.
22. Gao, X.; Liu, X.; Xue, X.; Chen, N.Z. Fracture mechanics-based mooring system fatigue analysis for a spar-based floating offshore wind turbine. *Ocean Eng.* **2021**, *223*, 108618. [CrossRef]
23. Li, Y.; Guo, S.; Chen, W.; Yan, D.; Song, J. Analysis on restoring stiffness and its hysteresis behavior of slender catenary mooring-line. *Ocean Eng.* **2020**, *209*, 107521. [CrossRef]
24. Qiao, D.; Yan, J.; Liang, H.; Ning, D.; Li, B.; Ou, J. Analysis on snap load characteristics of mooring line in slack-taut process. *Ocean Eng.* **2020**, *196*, 106807. [CrossRef]
25. Bergdahl, L.; Palm, J.; Eskilsson, C.; Lindahl, J. Dynamically scaled model experiment of a mooring cable. *J. Mar. Sci. Eng.* **2016**, *4*, 5. [CrossRef]
26. Huse, E. Influence of mooring line damping upon rig motions. In Proceedings of the Offshore Technology Conference, Houston, TX, USA, 5–8 May 1986.
27. Yan, J.; Qiao, D.; Li, B.; Wang, B.; Liang, H.; Ning, D.; Ou, J. An improved method of mooring damping estimation considering mooring line segments contribution. *Ocean Eng.* **2021**, *239*, 109887. [CrossRef]
28. Webster, W.C. Mooring-induced damping. *Ocean Eng.* **1995**, *22*, 571–591. [CrossRef]
29. Huse, E.; Matsumoto, K. Practical estimation of mooring line damping. In Proceedings of the Offshore Technology Conference, Houston, TX, USA, 2–5 May 1988.
30. Kitney, N.; Brown, D.T. Experimental investigation of mooring line loading using large and small-scale models. *J. Offshore Mech. Arct. Eng.* **2001**, *123*, 1–9. [CrossRef]
31. Hall, M. MoorDyn V2: New Capabilities in Mooring System Components and Load Cases. In *International Conference on Offshore Mechanics and Arctic Engineering*; American Society of Mechanical Engineers: New York, NY, USA, 2020; Volume 84416, p. V009T09A078.
32. Roddier, D.; Cermelli, C.; Aubault, A.; Weinstein, A. WindFloat: A floating foundation for offshore wind turbines. *J. Renew. Sustain. Energy* **2010**, *2*, 033104. [CrossRef]
33. Robertson, A.; Jonkman, J.; Masciola, M.; Song, H.; Goupee, A.; Coulling, A.; Luan, C. *Definition of the Semisubmersible Floating System for Phase II of OC4*; Technical Report; National Renewable Energy Lab. (NREL): Golden, CO, USA, 2014.
34. Jonkman, J.M. *Dynamics Modeling and Loads Analysis of an Offshore Floating Wind Turbine*; University of Colorado at Boulder: Boulder, NT, USA, 2007.
35. Morison, J.; Johnson, J.; Schaaf, S. The force exerted by surface waves on piles. *J. Pet. Technol.* **1950**, *2*, 149–154. [CrossRef]
36. White, F.M. *Fluid Mechanics*; Tata McGraw-Hill Education: New York, NY, USA, 1979.
37. Patton, K.T. *An Experimental Determination of Hydrodynamic Masses and Mechanical Impedances*; Technical Report; Navy Underwater Sound Lab.: New London, CT, USA, 1965.
38. OrcaFlex, M. *Version 9.8 b*; Orcina Ltd.: Daltongate, LA, USA, 2014.
39. Veritas, B. NR 493 R02 E. *Classif. Mooring Syst. Perm. Offshore Units* **2002**, *583*.
40. Veritas, D.N. *Offshore Standard DNV-OS-E301 Position Mooring*; Det Norske Veritas: Høvik, Norway, 2010.
41. Guo, S.; Chen, W.; Fu, Y. Non-linearly restoring performance of SFT's catenary mooring-lines under consideration of its dynamic behaviors. *Procedia Eng.* **2016**, *166*, 202–211. [CrossRef]
42. Qiao, D.; Ou, J. Mooring line damping estimation for a floating wind turbine. *Sci. World J.* **2014**, *2014*, 840283. [CrossRef]

Journal of
Marine Science and Engineering

MDPI

Article

Approach for Installation and Logistics of a Floating Offshore Wind Farm

Hugo Díaz and C. Guedes Soares *

Centre for Marine Technology and Ocean Engineering (CENTEC), Instituto Superior Tecnico, Universidade de Lisboa, 1049-001 Lisbon, Portugal
* Correspondence: c.guedes.soares@centec.tecnico.ulisboa.pt

Abstract: An approach is developed for floating wind farm installation by creating a model that allows the planning of the main logistics and installation aspects of a floating wind farm. The method aims to preserve the coherency between logistic methods and project performance, as floating wind farm installation is a complex problem, due to the number of components that impose specific constraints in areas such as transportation and manufacturing. The planning tool is developed based on the knowledge about how to use the various approaches and on the analysis of different logistical solutions. A techno-economic view of the logistics and installation involved in the floating wind farm design is the main objective of this study. A case study for the CENTEC TLP concept implemented in Spanish and Irish waters is performed. Case study findings reveal the major aspects that affect the design and logistics factors. The tool is developed for the main types of floating wind platforms and is applicable for the preliminary design of wind farms worldwide. The tool suggests a strategy of design and logistic aspects dependent on a few inputs such as the number of components for installation, distances and the maximum number of turbines. These findings are especially important for the new floating wind farms being positioned further offshore.

Keywords: floating wind farm design; offshore wind farm installation; offshore wind logistics; techno-economic analysis

Citation: Díaz, H.; Guedes Soares, C. Approach for Installation and Logistics of a Floating Offshore Wind Farm. *J. Mar. Sci. Eng.* **2023**, *11*, 53. https://doi.org/10.3390/jmse11010053

Academic Editor: Bang-Fuh Chen

Received: 11 November 2022
Revised: 21 December 2022
Accepted: 23 December 2022
Published: 30 December 2022

1. Introduction

Floating offshore wind is a promising industry. Compared to the fixed offshore wind industry, the space available at sea is enormous. Moreover, the floating turbines are considered less intrusive from a visual and noise point of view.

The main challenge for the floating wind industry is the Levelized Cost of Electricity (LCOE). LCOE can be seen as the lifetime cost of an energy source per unit of energy generated [1]. LCOE for floating offshore wind could vary in the next decades from 125 EUR/MWh to 50 EUR /MWh [2]. Floating wind energy must improve its LCOE for better acceptance. The industry is already showing some improvements, and the latest auction results (in Spain or France) [3] suggest that floating offshore wind LCOE might decrease based on the new technology improvements [4].

Reducing investment expenditures is a way to achieve the LCOE target. The investment expenditures in the installation are estimated to represent between 12 and 22% of the total costs over a lifetime. Conducting the construction properly can minimize investment expenditures.

The installation and logistics phases are where significant cost reductions can be achieved [2,5]. Catapult Offshore Renewable Energy Group finds that many LCOE reduction initiatives are related to logistics. According to Poulsen and Bay Hasager [6], efficient logistics plans are expected to reduce costs significantly.

Currently, several floating wind farms are under planning worldwide [7]. There is a need for researchers and practitioners to improve logistics during offshore wind farm manufacturing, especially during the construction and installation phase. Increasing distances from the coast and rising sizes of the wind turbines [8] along with the development of ever more refined floating platforms [9,10] are also making the subject even more relevant due to more difficult logistical challenges. The improvement in the installation and logistics could reduce construction costs and contribute to a better LCOE.

In this context, this study develops a model for wind farm design and offshore logistics identifying the main costs. The method evaluates and compares possible offshore logistics spread to support the wind turbine construction phase. Different installation scenarios and the resulting requirements for the installation are presented.

Chapter 2 explore the available knowledge on floating wind logistics and identifies research areas of improvement. A description of the methodology adopted is presented in chapter 3. A project's descriptive parameters and structure are explained in chapter 4. Chapter 5 identifies the main logistical aspects related to the case study evaluated. A discussion about the main elements identified is presented in chapter 6. Chapter 7 performs a sensitivity analysis of model variables. Chapter 8 identifies the main challenges for the EU market. This paper concludes with the findings of this research.

2. State-of-the-Art

The studies on modelling logistic requirements and installing an offshore wind farm have increased over the last few years. This modelling and analysis allow practitioners to analyze the installation of a floating wind farm in advance so that developers can prepare for specific outcomes in terms of cost or delay. Several authors focus on construction operations and weather risk analyses.

Poulsen and Hasager [6] propose a logistics model related to modules, components, people, tools, and onshore and offshore transportation equipment. Kaiser and Snyder [11] analyze wind turbine logistics during construction. Main installation vessels, foundations, turbines, substations and install inner-array or export cable are described.

Thomsen [12] describes the installation process with detailed installation vessel characteristics. Skiba [13] analyses the challenge of installation works at sea. Moreover, other studies have been conducted to improve the use of installation vessels, which have been identified as the bottleneck of the installation process [14].

The European Wind Energy Association [15] presented a method allowing a dynamical installation planning of shorter periods based on up-to-date weather forecasts. Scholz-Reiter et al. [16] applied a mixed integer linear programming (MILP) model to determine the installation schedule in different weather conditions and loading operations. Lange et al. [17] developed a discrete simulation tool of the supply chain based on weather windows to install units efficiently. Ait-Alla et al. [18] presented a MILP model considering the weather in a deterministic manner. They reviewed the outcome of two installation scenarios. The model minimizes the installation costs by considering vessel utilization and fixed costs that span the length of the installation period. Barlow et al. [19] review the susceptibility of vessels and operations to weather constraints during installation. Irawan et al. [20] look to address offshore wind construction scheduling issues through an integer linear programming method. The authors combine weather data and vessel availability to identify the optimal installation with the lowest costs and shortest schedules.

Muhabie et al. [21] consider weather restrictions, distances, vessel capabilities and assembly scenarios in a discrete event simulation tool. Sarker and Faiz [22] present a cost model developed for wind turbine installation and transportation. The study also analyses the impact of decision variables on the total cost. Barlow et al. [23] explore the impact of critical logistical decisions on the cost and duration of the installation. Boulougouris [24] explained the appropriate reactions to disrupted installation schedules using a rolling-horizon optimization tool. Backe [25] developed the logistical planning of offshore wind farm installation through mathematical optimization.

Paterson et al. [26] present a software tool using Monte Carlo simulation and embedded forecasting and logistical models. Maples et al. [27] propose a specialized turbine assembly procedure. Navigant Consulting [28] review the evolution of installation concepts in the previous ten years. Vis and Ursavas [29] suggest a preassembly strategy based on the number of components for installation onsite and the maximum number of turbines to be loaded on a vessel.

The literature described above mainly focuses on installation vessel planning and optimization based on weather conditions. Besides, the documents are related to fixed offshore wind farms. Therefore, there is no intention in the present paper to push it further, even in the case of floating wind farms, where the offshore installation aspects are simple. Nevertheless, further academic contributions could be made in the logistical aspects (wind farm design, storage, main components, costs) of the floating wind farm construction.

To tackle this objective, a model is constructed based on offshore wind literature and data applied to the wind farms developed in the Arcwind project. However, dedicated literature was not found on proper planning and improving logistics during the construction phase. Considering the lack of knowledge, the present paper aims to develop a floating farm design logistics tool. The tool can work separately or be included in the previous tools developed by Díaz and Guedes Soares and related to the site selection and wind farm design (see Figure 1) [30–35]. This tool offers adaptability to planners and investors as required. The tool can simulate the complete installation of a floating offshore wind farm, handled in phases, and considers the constraints of the operations, materials and vessels.

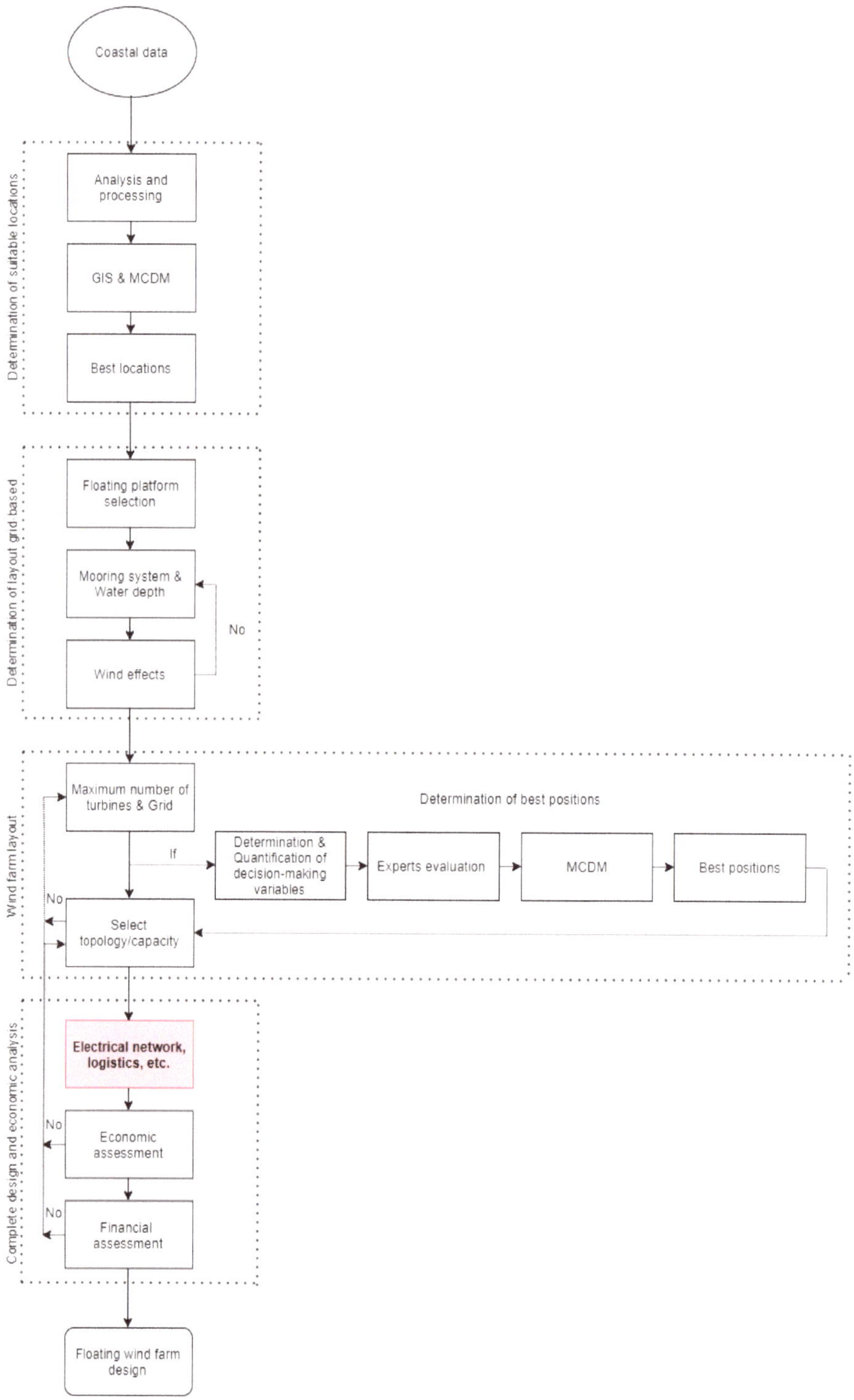

Figure 1. Floating wind farms design flowchart. The red colour block represents the step presented in this paper.

3. Methodology to Improve Floating Offshore Wind Farm Installation Planning

This paper employs a logistics simulation tool to determine the main installation aspects of a floating wind farm in advance. Moreover, the installation processes are analyzed to identify the durations and to prepare a complete economic evaluation.

The proposed model is based on literature and in-house knowledge related to floating wind technologies. The model can precisely measure the cost to a more targeted level, enabling companies to analyze the costs further. Several parameters are required for each wind farm concept: the cost per time unit of supplying resource capacity (technicians, vessels, staff) and the unit times of consumption of resource capacity by products and services, among others related to the technology, deployed.

The method can incorporate real operations by including process duration and weather conditions. The model calculates the cost of an hour worked in the wind farm depending on the total hours used. These indicators help plan improvements.

The main steps of this work are shown in Figure 2. The system represents a specific logistic setup for a floating farm construction. The objective is to determine the main logistic aspects of a floating farm as efficiently as possible. Based on the method application, some patterns are identified, and some recommendations are drawn to improve wind farm planning. A secondary objective is to identify the fundamental steps in a construction phase of a wind farm construction to reduce the execution time.

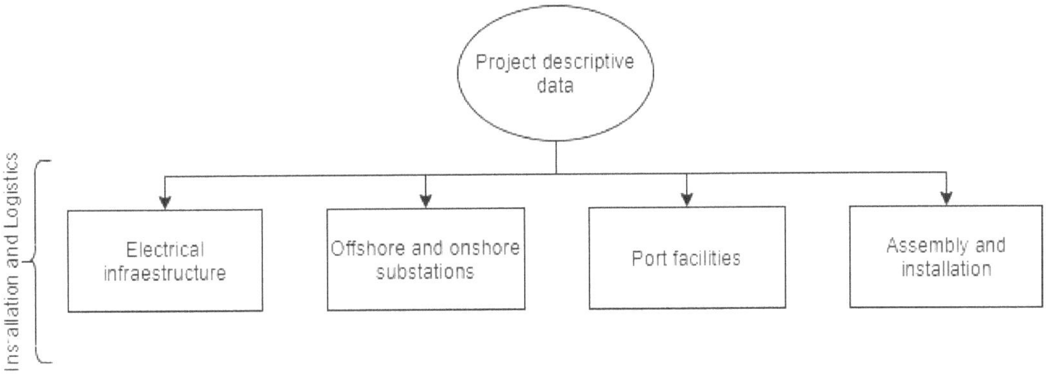

Figure 2. Main aspects addressed with this tool.

When determining the logistic aspects of floating wind farms in any given region, four main issues can be addressed more or less independently of each other: determining the electrical infrastructure (array cable, export cable and substation), offshore and onshore substations, port or shipyard characteristics and, finally, installation time. Methodological approaches for addressing these independent, fundamental aspects are developed in this work and then combined into the economic and financial module of floating wind farm design.

Creating the Model

Analyzing the logistic impacts of constructing floating farms requires a large amount of specific data such as capital costs, labor rates, region-specific inputs, personal expenditure patterns, and price deflators.

The first step of the model is to define a floating wind farm project. This is critical, particularly for the floating wind, because the main components (e.g., floating foundation type) are susceptible to project parameters. For this first version of the methodology, representative projects are derived based on a review of the multiple project proposals and concepts under consideration at the time of the model's development [36].

The model contains default values for all inputs necessary to perform an analysis. However, the user can also modify values or select between different options (e.g., floating foundation type, mooring system, etc.). To develop the input data, the Consortium of the Arcwind project estimated the data associated with individual line items for the representative project described in previous studies [36–38]. Some data, such as logistic times, required space, and costs, are derived from different sources (floating wind professionals, renewable energy studies, project-related case studies, etc.). These sources helped to establish a baseline for analysis (Appendix A). Once specific data were estimated from different study parameters, other external resources were used (see Table 1). Then, the estimates were refined based on the data available.

Table 1. External documentation used for the model design.

Institution	Topic	Reference
U. Bremen	Improving logistics scheduling and operations to support offshore wind construction phase	[39]
U. Groningen	Assessment approaches to logistics for offshore wind energy installation	[29]
NREL	Installation, operation, and maintenance strategies to reduce the cost of offshore wind energy	[27]
Garrad Hassan	Optimization of installation, operation and maintenance at offshore wind projects in the U.S.	[40]
Catapult ORE	Macroeconomic benefits of floating offshore wind in the UK	[41]
DNVGL	Transport and installation of wind power plants	[42]
Garrad Hassan	Assessment of ports for offshore wind development in the United States	[43]
NREL	Offshore wind jobs and economic development impacts in the United States: Four regional scenarios	[44]
U. Cranfield	A lifecycle techno-economic model of offshore wind energy for different entry and exit instances	[45]
U. Aalborg	How expensive is expensive enough? Opportunities for cost reductions in offshore wind energy logistics	[6]
Navigant Consulting, Inc.	Offshore wind market and economic analysis	[46]
U. Aalborg	Is the supply chain ready for the green transformation? The case of offshore wind logistics	[47]
Renewable UK	Future employment and skills in the UK wind and marine energy industries	[48]
NREL	2010 cost of wind energy review	[49]
IEA Wind	Offshore wind farm baseline documentation	[50]
Food and Resource Economic Institute	Economic valuation of the visual externalities of offshore wind farms	[51]
Douglas Westwood	Assessment of vessel requirements for the U.S. offshore wind sector	[52]
NREL	A spatial-economic cost-reduction pathway analysis for U.S. offshore wind energy development from 2015–2030	[53]
Panticon	Offshore wind logistics brief report 7	[54]
BVG associates	Guide to an offshore wind farm	[55]
BVG associates	UK offshore wind supply chain: capabilities and opportunities	[56]
BVG associates	Approaches to cost-reduction in offshore wind	[57]
BVG associates	Future renewable energy costs: Offshore wind	[58]
BVG associates	Ocean power innovation network value chain study: Summary report	[59]
Renewable UK	Offshore wind: Forecasts of future costs and benefits	[60]
The Crown Estate	A guide to an offshore wind farm	[61]
Strathclyde	Advanced logistics planning for offshore wind farm operation and maintenance activities	[62]
Deloitte	Analysis on the furthering of competition in relation to the establishment of large offshore wind farms in Denmark	[63]
U. Clemson	The potential economic impact of an offshore wind farm to the state of South Carolina	[64]
S. U. New York	Offshore wind development research	[65]
Carbon Trust	Floating wind joint industry project: Phase III summary report	[66]
Leanwind project	Logistic efficiencies and naval architecture for wind installations with novel developments	[67]
Green port Hull and BVG associates	Job roles in offshore wind	[68]

4. Project Descriptive Data

4.1. Locations

In the scope of the Arcwind project, several locations were determined through the study of the ZEE of the European Atlantic coast and the application of the methodologies proposed by Diaz et al. [32] and Loughney et al. [69]. More than one hundred locations could host a floating wind farm guaranteeing an energy potential capable of covering a large part of the electricity demand of the Atlantic countries.

After the application of the MCDM methodologies by Diaz et al. and Lounghley et al. [30,69] and based on the results obtained two locations by floating platform were selected to complete floating wind farm designs.

Ribadeo site is located on the Galician Coast (see Figure 3) in the Northwest of Spain (latitude, longitude: 43.837, −7.326). The location is characterized by a wind velocity of 9.3 m/s and a wind potential of 4923 h/yr. F15 site is a location situated on the West of Ireland coast in the region of Clare (latitude, longitude: 52.7, −10.5). The area is characterized by excellent wind resources and good conditions for the positioning of floating wind turbines. Table 2 shows other details that emphasize the potential of the areas proposed.

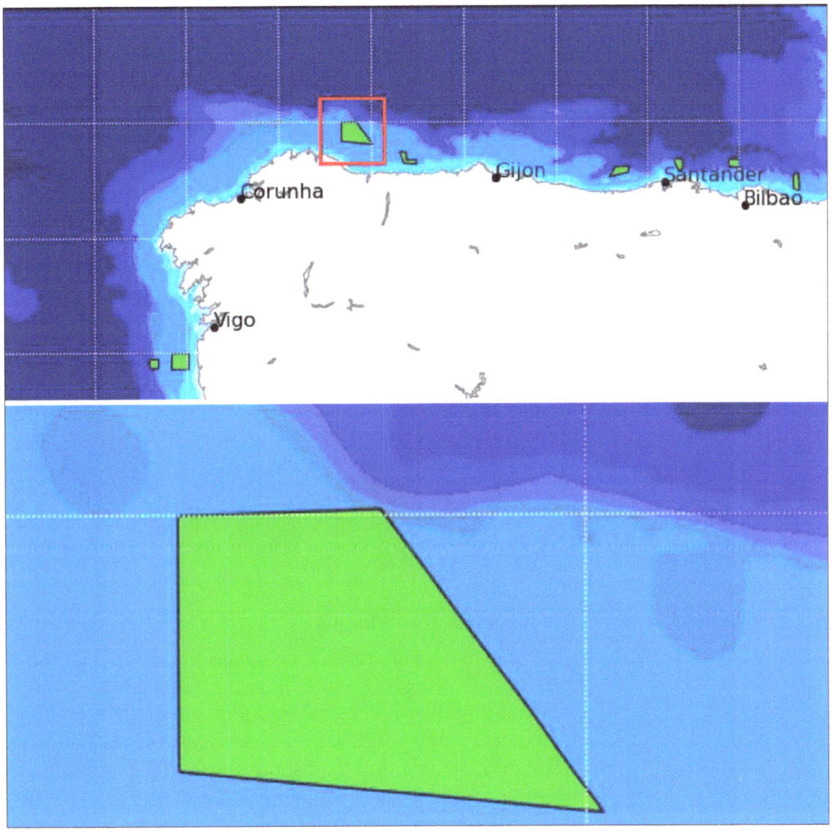

Figure 3. Example of a floating farm used in this study (Ribadeo offshore wind farm).

Table 2. Main characteristics of the Ribadeo (ES) and F15 (IE) locations.

Item	Ribadeo (ES)	F15 (IE)	Units
Wind velocity	9.33	10.81	m/s
Water depth	150	210	m
Wave conditions	2.15	3.02	m
Marine currents	0.51	0.12	m/s
Distance to local electrical grid	18.60	115.17	km
Distance from coastal facilities	89.50	117.36	km
Distance from shore	17.61	41.23	km
Distance from maritime routes	1.50	20.62	km
Distance from protected areas	10.92	32.02	km
Area of the territory	405	30.25	km^2

4.2. Floating Platform Selection

Floating support structures have emerged in recent years and are in a phase of rapid growth. The type of platform selected depends, among others, on water depth, ground conditions, turbine size, and wave loading. The fabrication and installation capabilities will also impact the foundation construction and design process.

Tension leg platforms (TLP) are floating foundations vertically moored to the seafloor by tendons. Given the structure dimensions, it is possible to manufacture and transport this to staging harbors using heavy-lift transport vessels. Compared with other foundation solutions, this concept will significantly increase the costs associated with this operation. Once mobilized at the staging port, suitable installation vessels transport the foundations to the wind farm site. TLP will be generally manufactured and stored in a location near the wind farm site, but transition pieces are generally manufactured and transported to store in the same place that the foundations. TLPs are installed by embedding the suction piles mooring systems into the ground using large hammers and, where necessary, drills. A steel transition piece typically is used to provide the interface between the WTG and the foundation. Transition pieces are connected to the foundation either through a grouped or bolted connection.

TLP foundations for offshore WTGs are usually manufactured and delivered directly to the wind farm site using offshore tugs. Still, the requirement for many units at low-cost places significant focus on standardization and process optimization for manufacturing and installation. The use of concrete helps to reduce costs.

5. Installation and Logistics

Floating offshore wind energy installations show significant differences, ranging from the installation process to different cost structures within the life cycle of a wind turbine. The connection of offshore wind farms to the energy grid, weather conditions, installation processes, and operation and maintenance of offshore wind turbines influence the energy cost. The accessibility to offshore locations represents a considerable difference in the cost structure during the installation process [70–73].

According to estimates, the total logistics costs for a floating offshore wind farm are about 5–10% of the total investment costs, 10% less than the regular price for a fixed offshore wind farm. To understand the installation process and the connected logistic effort, it is necessary to know the main components of a floating wind turbine that also have to be dealt with as transport units.

The upper section composed by a hub, usually three rotor blades and a nacelle is supported by a tower (modular structure) that is fixed on the floating structure (see Table 3). References [74–76] describe the three types of foundation structures developed in the Arcwind project, CENTEC-TLP, Sath and Telwind. In this case, the TLP concept was selected.

Table 3. General data and assumptions.

Item	Ribadeo	F15	Units
Number of turbines	88	25	-
Turbine rating	10	10	MW
Rotor diameter	178.3	178.3	m
Hub height	129	129	m
Water depth	150	210	m
Export cable longitude	17.6	41.2	km
Distance to installation port	89.5	117	km
Array row spacing	12	6.5	rotors
Array turbine spacing	12	6.5	rotors
Cable burial depth	2	2	m
Substructure and foundation type	TLP	TLP	-
Floating substructure anchor type	Suction pile anchors	Suction pile anchors	-
Number of anchors per floating substructure	12	12	-
Tower installation method	Two pieces	Two pieces	-
Turbine installation method	Individual components	Individual components	-
Installation vessel strategy	Tug vessel installation	Tug vessel installation	-
Vessel day rate cost range	Mean	Mean	-
Number of installation seasons	1	1	-

There needs to be a standardized setup of a logistics network to install a floating offshore wind farm. The specific offshore wind farm projects define the number of wind energy turbines and the location. Another decisive factor for the logistics network is the vessel concept for the installation [77].

The complexity of floating wind farm installation depends on the collaboration of the entire logistics network. The involved network partners, the installation method at sea, port characteristics, and the type and number of available construction spaces are key parameters.

In this scenario, the installation process starts with the component manufacturers that produce the main components of the offshore wind energy turbine. The production of the main components is divided into foundation structures (in this case, TLPs), the two upper tower segments, the set of rotor blades, the nacelle, the hub, the lower tower segment (transition piece), and the mooring system. The following partner in the network is the logistics service provider that arranges the transports between the component manufacturers and the port of shipment. The port operator is responsible for storing and handling the components at the port of shipment. Finally, the components are taken over by the construction company that operates the installation process and tugs. These vessels transport and install the floating wind energy turbines and related equipment.

This methodology highlights critical challenges related to supply chain management for floating wind projects, focusing on the installation and transportation infrastructure, equipment, and logistics. This study concerning the previous step of wind farm design covers the following topics:

- Electrical infrastructure
- Offshore and onshore substations
- Port facilities
- Assembly and installation

- **Key components**

The key components of the supply chain can be broken down into four major components for floating wind farms, namely wind turbines, support structures, cables (including array cables and export cables), and offshore/onshore substations. These components represent the major physical offshore components that comprise the floating wind farm. There are undoubtedly other minor components as well as onshore infrastructure, all of which should not be overlooked. For each of these key components, the supply chain also includes the equipment and infrastructure associated with the transport, storage, and installation of the components and the related service providers. Contractually, the supply transportation, storage, and installation of these components may be provided individually by firms or combined work packages (e.g., supply, transportation, and installation cables). Typically, a single project director will oversee the entire project and the various contractors delivering the work packages. A critical aspect of the project director's role is supply chain management. To facilitate the discussion of some of the requirements and challenges associated with the installation and transportation of major offshore wind components, it is helpful to provide an overview of these components and the defining characteristics critical to the transportation and installation aspects of supply chain management.

- **Wind Turbine generators**
  Wind turbine generators include:
  - The nacelle, which contains the primary components for power generation and controls.
  - The rotor, including the rotor hub, blades, and blade pitch system.
  - The tower and transition piece, which supports the rotor and nacelle and interfaces with the foundation.

The primary components for Wind turbine generators (WTGs) include hundreds of subcomponents and parts which must be assembled before transport to the project installation. Assembly and finishing of these major components are typically performed at specialist coastal facilities by the WTG original equipment manufacturer. For example, the nacelle must be assembled with the drive train, generator, and control systems. These assembly sites are crucial, both in terms of investment and the development of supply chain hubs in the nearby area. Given the number of parts and subcomponents, these sites represent significant opportunities for clustering the supply chain.

After the major components are assembled, the WGT can be transported and installed in several different ways. The objective of pre-assembly onshore is to minimize the number of lifts and amount of assembly required offshore. The total pre-assembly, however, results in heavier, more complex transport that is subject to more restrictive weather limitations.

- **Foundations**

The floating foundation type and the associated mooring system represent fundamental aspects of the floating farm design. In this model, three full-scale floating wind turbines are considered for tool application (Semisub, Spar and TLP), but the TLP concept is selected to be the option for intermediate depth waters of Ribadeo and Ireland locations as case studies. There are currently several widely different concepts under development. Therefore, a generic approach to floating wind farm designs is difficult. However, in this study specific aspects of these three types of platforms are considered for the farm design to some degree.

A free-float capable TLP (FTLP) has been extensively studied by Uzunoglu and Guedes Soares [74] as a floating wind platform (see Figure 4). This technology presents advantages such as relying on ordinary tugs instead of specialized vessels for transport. This approach is convenient because wind farms are planned for a high number of units. The dynamically stable design capable of free-floating on a shallow draft performs well as a common TLP [78]. The structure can be towed without being concerned with depth limits [79].

Figure 4. TLP concept developed by the Centre of Marine Technology and Ocean Engineering (CENTEC).

5.1. Electrical Infrastructure

Electrical cables are typically made up of three copper cores set into an XLPE (cross-linked polyethene) base, surrounded by steel wire armoring. Fiberoptic cables provide a communication channel for the wind farm. For array cables, 33 kV has been used to date, although there is significant R&D qualification work and future projects using large capacity WTGs that will utilize 66 kV lines. Export cables typically operate at 132 kV, but increasingly, 320 kV is being used.

Cables are typically manufactured at coastal manufacturing facilities where cables are loaded onto a vessel. For export cables, load out generally is in a single length of possibly tens of kilometers. This is due to the challenge of installing field joints and the desire to fabricate and load the export cable in a single length. Load-out of export cables can take several days, and due to the size and weight of these cables, special equipment (extremely large carousels) is required for transport and storage. Dedicated cable laying vessels are generally used to transport the export cables to the project site and install the export cables, typically using a subsea cabling plough to bury the cable in the seabed.

Array cables are generally lighter and shorter than export cables and thus typically do not have the same logistics constraints. Array cables can be transported in shorter lengths and are more flexible and therefore less onerous to handle than export cables. Array cables are often pre-cut to the necessary lengths, transported on individual drums, and stored at a staging port. The installation approach for array cables is similar to export cables in that a cabling plough is typically used to bury the cables, however cables, but the range of vessel options is generally larger. Array cables are often installed by offshore service vessels. Table 4 shows the main electrical system characteristics determined through the proposed tool for the floating wind farms under study.

Table 4. Electrical system characteristics.

Item	Ribadeo	F15	Units
Array cable voltage	33	33	kV
Array cable size 1	95	95	mm^2
Array cable size 2	500	500	mm^2
Strings			
Full strings	29	8	-
Turbines per partial string	1	1	-
Turbines per cable size			
95 mm^2	1	1	-
500 mm^2	3	3	-
Turbine interfaces per cable size			
95 mm^2	58	16	-
500 mm^2	88	25	-
Substation interfaces per cable size			
500 mm^2	30	9	-
Cable dimensions			
System angle	18.1	17.8	degrees
Free cable length	354.1	419.3	m
Fixed cable length	1902	884	m
Array cable length			
95 mm^2	40,845.8	12,093.4	m
500 mm^2	358,506	73,735	m
Array cable section mass			
95 mm^2	28.7	28.7	tonne
500 mm^2	59.6	59.6	tonne
Cable selections per vessel trip			
95 mm^2	205	205	-
500 mm^2	98	98	-
Export cable			
Export cable voltage	220	220	kV
Export cable size	1200	800	mm^2
Number of export cables	3	1	-
Export cable length	62,218	46,111	m
Export cable section mass	2296.8	4418.5	tonne
Cable sections per vessel trip	2	1	-
Cable and ancillary supply cost			
95 mm^2	13,322,332	3,935,768	EUR
500 mm^2	258,944,132	53,381,607	EUR
Cable and ancillary supply cost (Export cable)	71,694,638	38,699,082	EUR

5.2. *Offshore and Onshore Substation*

The offshore substation steps the voltage from the array cable operating voltage to the export system voltage. This plant includes transformers, reactors, switchgear, control, fire protection, and low-voltage auxiliary systems. All this equipment is contained in a sizeable fabricated topside structure which usually includes two or more stories and is installed upon a support structure (usually a jacket). Depending on the size of the project, there may be more than one offshore substation.

Several different concepts have been developed for offshore substations. The most commonly used concept is a lifted substructure and topside transported to the site (typically on a barge) and installed with a heavy lift crane vessel. Other concepts include:

- Self-installing substructures and topsides.
- Subsea base frames with floating jack-up topsides.
- Topside float-over designs.

The design, transportation, and installation of offshore substations for offshore wind projects have many similarities to offshore oil and gas installations. It draws heavily from the infrastructure and equipment used in that industry. The main components of substations are presented in Table 5.

Table 5. Offshore and onshore substations.

Item	Ribadeo	F15	Units
Number of substations	2	1	-
Topside			
Topside mass	4135	1401.5	tonne
Land-based topside assembly			
Jacket mass	1654	560.6	tonne
Pile mass	497.9	272.4	tonne
Topside cost	54,788,875	21,098,488	EUR
Switchgear	4,930,000	1,232,500	EUR
Shunt reactors	14,875,000	4,313,750	EUR
Ancillary system	5,100,000	5,100,000	EUR
Main power transformers (MPTs)			
Single MPT rating	250	290	MVA
Number of MPTs	4	1	-
MPTs total cost	10,625,000	3,081,250	EUR
Land-based topside assembly	2,282,250	647,063	EUR
Jacket			
Jacket mass	1654	560.6	tonne
Jacket cost	8,786,875	2,978,188	EUR
Pile mass	497.9	272.4	tonne
Pile cost	952,145	520,944	EUR
Onshore substation			
Onshore substation and transformer	13,152,645	6,912,999	EUR
Spare parts, capacitor bank.	408,702	368,645	EUR
Onshore transmission infrastructure			
Overhead transmission line	1,415,763	1,203,399	EUR
Switchyard	5,453,276	4,635,285	EUR
Subsea cable cost	362,210,011	124,400,838	EUR
Offshore substation, substructure and foundation cost	204,680,290	38,972,181	EUR
Onshore transmission infrastructure costs	20,430,388	14,150,684	EUR
Total electrical infrastructure cost	587,320,689	177,523,704	EUR

5.3. Port Facilities

Port facilities can be divided into manufacturing and mobilization ports. A manufacturing port is typically located in the vicinity of OWF manufacturing facilities and is used to deploy the manufacturers' support structure and WTG components before deploying offshore. Mobilization ports are used when temporary parking of equipment before deployment offshore is necessary, particularly in developing areas without an existing nearby manufacturing port. In the case of floating concepts, the port of manufacturing and mobilization could be the same since the concrete platforms do not need a special place as in the case of steel where the shipyards represent a key place. If there is a manufacturing port nearby, it generally has its staging area, which makes a separate mobilization port unnecessary. However, equipment is necessary, thus requiring a marshalling or staging port. Ports also need adequate water depth to allow the floating platforms and tug vessels to move the equipment and a waterway width typically over 200 m to transport the rotor-hub installed blades.

Port facilities play an essential role in the development of reliable supply chain management systems (see Table 6). Preferably they have enough area for future expansion and are in the vicinity of existing major port facilities for joint sharing of the facilities should there be a need. Hence, it is necessary to have dedicated FOWF port facilities. Another important aspect is the location of the port facility of the project locations. FOWF supply chain manufacturers will be located in ports they deem best located for their needs in terms of proximity to their target markets.

Table 6. Port characteristics.

Item	Ribadeo	F15	Units
Entrance and exit	42,583	12,097	EUR
Quayside docking	3,619,880	1,067,943	EUR
Wharfage	863,931	246,483	EUR
Turbine			
Area	1864	1864	m^2
Cost	57,359	16,295	EUR
Cranage			
600 ton	6,033,133	1,779,903	EUR
1000 ton	9,687,013	2,861,446	EUR
Crawler crane mobilization and demobilization	127,500	127,500	EUR
Total port cost	4,526,395	1,326,523	EUR
Total staging cost	15,905,005	4,785,145	EUR
Total port and staging cost	20,431,400	6,111,667	EUR

Port facilities are expected to have a significant amount of land to store the WTG and support structure components waiting to be assembled out at the quayside. Considering the ever-growing size of WTG components and the preference for ports to be utilized to supply multiple FOWF projects simultaneously, one can easily understand the amount of land required in these operations. WTG foundation components are generally produced by service providers other than the main WTG supplier; nevertheless, they operate from the same port facilities, where the manufacturing facility is located. In many cases, when the foundation is manufactured in steel the TLPs and Transition pieces are transported from two separate manufacturer ports to the marshalling port facility or another manufacturing port with load-out capacity, where these components are assembled at the later stage of the project.

Another important factor determining the location of supply port facilities, apart from the apparent logistics reasons, is the availability of the in-port transportation and load-out personnel experienced in delicate load-out operations to sea-going installation vessels.

Installation Vessels and Equipment

The selection of vessels and equipment for floating offshore wind farm construction is a critical process. It must balance the potentially competing objectives of lowering costs, minimizing durations, and ensuring safe operations with minimal environmental impact. FOWFs are composed of large, prefabricated, specialized components pre-assembled in the coastal facilities. Over time, these components have increased in size as more giant turbines have been utilized to bring the Levelized Cost of Energy (LCOE) down to levels competitive with other energy sources in the coming decades. Due to the size and distinct shapes of these components and the serial nature of the installations, the offshore wind industry was quick to develop and chart floating concepts once the applications moved from experimental development to the commercial deployment phase. There are many similarities between offshore wind and oil and gas marine installation operations, especially concerning Health, Safety, and Environment (HSE) practices and installation methods for structural components. However, purpose-built floating foundations are designed to address the unique challenges associated with the installation of OWF components "easily".

Towing capacity and marine maneuvers to minimize repeated operations using carrying the optimum amount of equipment during each deployment will be a bottleneck. Optimized transporting plans are essential to ensure that the number of deployments corresponds with the port logistics towards completing the installation phase with as few deployments as possible. The coordination of port logistics with the manufacturing and delivery of other components, in order not to keep any group in the supply chain idle for an extended amount of time, is important to ensure the best optimization and efficiency of the supply chain system and to minimize storage requirements. Depending on the proximity of the wind farm to shore, additional offshore tugs may be necessary for the optimized supply chain management.

In the current market, the installation vessels are either wholly owned by the OWF development Engineering Procurement Construction and Installation companies or international marine construction service companies. Heavy lift Vessels (HLV), which dominate the substation installations, are generally owned by maritime construction service companies that have historically serviced the offshore oil and gas industry's construction needs. Substation installation, given its sizes, design, and generally singular units, is more similar to oil and gas installation than other components of FOWF installations.

Dredging and onshore marine construction companies are other contributors to the current fleet. As the offshore oil and gas industry, installation vessels in the OWF industry also operate under a day rate system. A more predictable summer timeframe and a fixed rate in the less predictable winter operations fluctuate up and down based on market conditions. This statement is particularly true for the oil and gas industry, as observed in the years. However, the day rates for the FOWF installation vessels have a direct correlation with the available ships on the market, which again is affected by the announced installation targets of individual governments and their actions to support these targets.

The increased number of floating wind projects announced in Europe alone will push up the existing installation vessel market, even before including the projects announced outside Europe. Implementation of large-scale projects outside Europe, without the necessary infrastructure towards a sustainable and self-sufficient supply chain, will intensify international competition for the available vessels and result in upward pressure on prices similar to the current state of the fix offshore wind farms.

5.4. Assembly and Installation

One of the critical aspects of supply chain management is coordinating the supply of components with the installation process. When there is a mismatch in installation and supply rate, this can lead to either significant construction delays or increased storage requirements that can significantly impact the cost. For most critical offshore wind farm components, particularly foundations and turbines, manufacturing each unit is considerably longer than the time to install each unit. This generally requires a sufficient supply of components to be stored at the manufacturing port or marshalling port (if applicable) to ensure that the installation process is not delayed due to waiting on parts. The installation process is subject to seasonal variation in the rate at which units can be installed, generally with higher rates during the summer than the winter.

On the other hand, the manufacturing process is generally not subject to such seasonal variation (or at least not to the same degree). This difference is essential when assessing the delivery approach and the port requirements for a project, particularly storage. Having a sufficient cushion of components in storage is critical to managing delay risks in construction. Any installation delays will have knock-on effects that can lead to greater than day-for-day delays in completion. Therefore, time control is crucial for the proper functioning of the project and the correct installation (see Table 7) [80,81].

One of the key aspects of managing the supply and installation is selecting a delivery method appropriate for a given project. There are several methods for delivering wind farm components from the operational areas to the offshore wind farm site. These include the following:

1. Loading at a manufacturing port and transporting to and off-loading at a staging port for storage. Loading of components onto a transport vessel (or barge) at the manufacturers' facility and off-loading into a floating barge in a sheltered harbor near the offshore site, to be stored, awaiting assembled process.

2. Towing of the components with an offshore tug vessel at the manufacturers' premises, and positioning at the offshore site.
3. It loads the mooring components directly onto the installation vessel (also offshore tug) at the manufacturers' premises and installation at the offshore site.

Table 7. Main components installation time.

Item	Ribadeo	F15	Units
Anchor and mooring installation	973.9	286.6	days
Substructure load-out and prep for turbine installation	44	12.5	days
Turbine installation	144.8	41.1	days
Substructure and turbine tow-out and installation	256.3	78.6	days
Electrical installation			
Array cable installation	251	58	days
Export cable installation	37	35	days
Substation installation	4	4	days
Total floating bottom installation time	935.5	291.9	days

The selected approach for any project must consider the distances between manufacturing locations and the project site availability of various types of vessels, and the capabilities of the relevant port facilities, including potential manufacturing facilities and marshalling ports.

6. Discussion

The floating offshore wind logistics and installation has never been technically evaluated due to the lack of knowledge and technology barriers. This work tackles a new option to determine the logistic aspects related to a floating farm design. The tool is applied in two locations on the European Atlantic coast which contribute to understanding the new alternatives to assimilate and contribute to the diversification of the European energy matrix.

In this section, the floating wind project "Ribadeo and F15", located on the European Atlantic coast, was evaluated from a logistical point of view. The offshore knowledge is implemented in an in-house tool that allows for determining the critical aspects of a floating farm installation.

The project comprises a total of 113 wind turbines of the type DTU-10MW [82]. The areas are located 18 and 41 km from the shore in Ribadeo and F15 locations, respectively. The water depth at the installation site is up to 150 m in both cases.

The TLP concept requires a port capacity of 1864 m^2 per turbine and has a total cost considering the staging of 20.4 MEUR and 6.1 MEUR for each project. The port should have cranes up to capacities of 600 t and 1000 t.

In the conceptualization of this tool, several vessels are considered. Because of the floating technology, the conventional tug and offshore tug are considered for the installation phase. As mentioned previously, only a boat that can tug a floating turbine is not considered the specific pre-assembly. The vessels will be leased.

The tool considered a complete pre-assembly strategy of the floating platform, as defined in the previous section, was deemed to be possible. Table 4 shows the given distribution of processing time in some of the assembly aspects.

The results showed that the installation strategy would need, on average, 936 and 292 days for the Spanish and Irish floating farms, respectively. This process improves the currently used technique in a fixed offshore wind, with a time reduction improvement that could represent more than 30%.

The values of the floating wind energy projects evaluated in this study are estimated based on international market prices/regulations and installation assumptions based on the technology used. Some of this information is determined in previous modules of the Arcwind floating farm design (see Figure 1).

7. Comparison and Sensitivity Analysis

Due to the relatively immature stage of development of the floating wind industry, there is a limited knowledge of how installation and logistics will be undertaken. Options include a new installation method and infrastructure required or using new methods; the length of time for different tasks; and the strategies (operational space, capabilities, etc.). It is also difficult to get accurate values and specific conditions, for example, different materials, port costs, vessel day rates, etc. This is partially because only three floating farms have been installed but also because this information is commercially sensitive. Some values are confidential, and others are highly dependent on the market.

Therefore, validation of the installation and logistics tool comparing results with figures in the current literature. In summary, it was found that the main economic and time values are in line with the documents cited in Table 1. This indicates that the outputs from the present model are reasonable, although a bit higher end of the DNV GL [42] estimates and in line with NREL [27,44,53], even when the other studies focus more on other types of floating foundations.

However, it is essential to remember that in the literature, figures are for projects in different regions based on different approaches and models. The structure and scope of several models are unavailable, so it is impossible to identify where potential differences in the assumptions and functionality of the models could account for the variance in results. It is anticipated that as empirical data become available from the future floating projects of actual wind farms, the installation and logistics tool can be further validated and calibrated based on these data.

Given the difficulties validating values, sensitivity analysis (see Appendix B) was also conducted to confirm that the impact of variations is as expected. The analysis was performed on the items presented in Section 5. The model's parameters varied, including the number of turbines, water depth, distance to the electrical grid, distance to the port and weather conditions (schedule variability). The expected increases and decreases were found, validating that the model works as intended.

Moreover, the not linear variation in some variables, under the increase or decrease in one of the main parameters, exemplified the robustness of the model and the strengths derived from the knowledge of industry experts and the data considered.

8. Challenges for the EU Market

The EU floating wind market is in its infancy, and the first projects will face several supply-chain challenges that will impact logistics [83], including the following:

- There is currently a lack of manufacturing capability for the key components in the vicinity of major development areas (i.e., the Atlantic coast). Manufacturers of key components, particularly WTGs, will likely not invest in domestic production facilities until there is a sufficient pipeline of advanced projects. Therefore, initial projects will likely rely heavily on manufacturing far from project sites.
- The EU Atlantic ports generally do not currently have sufficient capacity to support large-scale floating offshore wind projects. With a few exceptions, existing ports usually are lacking in one or more critical areas such as load-bearing capacity, crane capacity, and storage capability.
- EU installation contractors and service providers generally have limited experience building floating offshore wind projects, thus representing a significant knowledge and experience gap.

Given these factors, supply chain management will be critical for EU floating wind projects. These projects will need to develop creative solutions to balance storage needs effectively, delivery rates, and installation rates considering the constraints mentioned above and challenges. The use of concrete or steel could create significant variations in the performance of the logistics chain and its associated costs. The development of floating wind substations could be also a critical aspect for the support the floating wind farms in deeper waters. With a heavy reliance on imports and domestic manufacturing far from project sites, marshalling or staging ports will likely be necessary; however, many EU ports do not have sufficient quayside storage, thus requiring storage at another location, for example, at a WTG manufacturing facility or floating foundation shipyard. This may result in extra handling and associated cost.

The floating option may help manage storage space constraints and handle the unitization of project cargo, i.e., making many pieces of cargo (e.g., WTG components) into single-unit transportation. Such units can then be towed with tug vessels.

One of the key selling points for floating offshore wind is the economic development potential. Maximize this potential for early projects and facilitate knowledge transfer; opportunities for joint venture work should be fully explored. Such arrangements can bring expertise from the North Europe region but locate as much manufacturing work as possible locally within the vicinity of the project and take advantage of local resources, including manufacturing capabilities and other skills from the naval industry and civil construction industry.

9. Conclusions

The floating offshore wind industry is growing and is expected to contribute to the global energy supply mix in the coming decades. The construction of floating wind projects has many similarities to the construction of offshore fixed infrastructure. Still, the serial nature of the manufacturing and installation of offshore wind components and the intense cost reduction pressure drives a need for robust supply chain management that is unique to the floating wind. Effective supply chain management for offshore wind requires sufficient planning early in the project development process to identify appropriate ports and vessels and select supply, transportation, and installation approaches optimized for a given project. Supply chain management approaches have evolved with the industry, and large-scale commercial projects in the EU will significantly benefit from these developments. However, the first EU projects will face significant supply chain challenges and novel, and innovative approaches may be required to address some of these challenges. Leveraging European experience in the fixed offshore wind while maximizing local content will help reduce cost and manage risk while helping to build a domestic industry, ensuring that floating wind can fulfil its promise of clean, reliable, and secure electrical power.

Author Contributions: Conceptualization, H.D.; formal analysis, H.D.; writing—original draft preparation, H.D. and C.G.S.; writing—review and editing, C.G.S.; project administration, C.G.S. All authors have read and agreed to the published version of the manuscript.

Funding: This work was conducted within the ARCWIND project–Adaptation and implementation of floating wind energy conversion technology for the Atlantic region, which is co-financed by the European Regional Development Fund through the Interreg Atlantic Area Programme under contract EAPA 344/2016. This work contributes to the Strategic Research Plan of the Centre for Marine Technology and Ocean Engineering (CENTEC), which is financed by the Portuguese Foundation for Science and Technology (Fundação para a Ciência e Tecnologia–FCT) under contract UIDB/UIDP/00134/2020.

Institutional Review Board Statement: Not applicable.

Informed Consent Statement: Not applicable.

Data Availability Statement: Not applicable.

Conflicts of Interest: The authors declare no conflict of interest.

Appendix A. Simplification of Some of the Formulas Used to Determine the Logistic Values

Full Strings = (Number of turbines/((1.73 × Array cable size × Array cable voltage × Power capacity factor × (1 − (Cable burial depth − 1) × Burial depth factor))/1000)/Turbine rating)

Turbines per Partial String = Number of turbines/Full strings

Turbines per cable size 95 mm^2 = (((1.73 × Array cable size 1 × Array cable voltage*Power factor × (1 − (Cable burial depth − 1) × Burial depth factor))/1000)/Turbine rating)

Turbines per cable size 500 mm^2 = (((1.73 × Array cable size 2 × Array cable voltage × Power factor*(1 − (Cable burial depth − 1) × Burial depth factor))/1000)/Turbine rating)

Turbine interfaces per cable size 95 mm^2 = (Turbine per cable size 95 mm × Full strings + (Turbines per partial string)) × 2

Turbine interfaces per cable size 500 mm^2 = (Turbines per cable size 500 mm-Turbines per cable size 95 mm^{-1}) × Full strings + (Turbines per partial string-Turbine per cable size 95 mm^{-1}) × 2 + Full strings

Substation Interfaces per Cable Size = Full strings + 1

System angle = −0.005 × (Water depth + 18.7)

Free cable length = (Water depth/Cos(System angle)) × (1.04) + 190

Fixed cable length = (Turbine spacing × Rotor diameter) − (2 × ((Tan(System angle) × Water depth) + 70)

Array cable length 95 mm^2 = (Turbine space × Rotor diameter + Water depth × 2) × (Turbine interfaces per cable size 95 mm/2) × 1.1

Array cable length 500 mm^2 = 1.1 × (((Full strings + 1)/Number of substations) × ((2 × Free cable length + Fixed cable length) + Sqrt(((((Full strings + 1)/Number of substations) − 1) × ((2 * Free cable length) + (Array spacing row × Rotor diameter) − (2 × ((Tan(System angle)) × Water depth) + 70))))^2 + (2 × Free cable length + Fixed cable length)^2)))/2))

Cable and ancillary supply cost 95 mm^2 = Array cable size cost 33 kV × Array cable length

Cable and ancillary supply cost 500 mm^2 = Array cable size cost 33 kV × Array cable length

Number of Export Cables = (Turbine rating × Number of turbines)/((1.73 × (Export cable size) × Export cable voltage × Power factor × (1 − (Cable burial depth − 1) × Burial depth factor))/1000)

Cable length = (Distance to shore × 1000 + Water depth) × Number of export cables × 1.1

Topside mass = 4 × (Single MPT rating × Number of MPTs) + 285

Topside cost = Topside mass × 14,500 + 4,500,000

Switchgear = Number of MPTs × 1,500,000

Shunt reactors = Single MPT rating × Number of MPTs × 35,000 × 0.5

Single MPT Rating = (Number of turbines × Turbine rating × 1.15)/Number of MPTs

Number of MPTs = (Number of turbines × Turbine rating/250)

MPTs total cost = Single MPT rating × 12,500 × Number of MPTs

Land-based topside assembly = (Switchgear + Shunt reactors + MPTs total cost) × 0.08

Jacket mass = 0.4 × Topside mass

Jacket cost = Jacket mass × 6000

Pile mass = 8 × Jacket mass^0.5574

Pile cost = Pile mass × 2250

Onshore substation and transformer = 11,652 × (Interconnect voltage + Turbine rating × Number of turbines) + 1,200,000

Spare parts, capacitor bank, etc. = 11,795 × (Turbine rating × Number of turbines)^0.36 + 350,000

Quayside Docking = (Anchor and mooring installation time + Preparation inshore area + Preparation for turbine installation + Turbine installation time + Substructure and turbine installation) * Quayside docking rate

Wharfage = (Turbine mass + Tower mass + Substructure and foundation mass) × Number of turbines + (Electrical topside mass + Electrical jacket mass + Electrical pile mass) × 2.75

Turbine area = Deck area per turbine × Turbine per vessel trip

Turbine cost = Turbine area × Turbine installation time × Storage area rate

Cranage1 = 5000 × (Anchor and mooring installation time + Onshore assembly area + Turbine installation area + Turbine installation time + Substructure installation time)

Cranage2 = 8000 × (Anchor and mooring installation time + Onshore assembly area + Turbine installation area + Turbine installation time + Substructure installation time)

Anchor and mooring installation = (Loadout anchor + Survey anchor position) × Anchors per foundation + (Anchor and mooring water depth-time factor) × Water depth × Anchors per substructure + Distance to installation port × 1000 × 2/(Transit Speed + Medium anchor handling tug × 1852) × Number_of_Turbines/24 × (1 + Support structure weather contingency index)

Substructure load-out and preparation for installation = (Prepare TLP for towing to assembly space + Prepare ballast system) + (Tow to project site) × Number of turbines/24

Appendix B. Sensitivity Analysis

Table A1. Variability associated with the number of turbines.

Item		Units	Scenario					
			−20%	−10%	−5%	5%	10%	20%
Electrical system characteristics								
Array cable length								
95 mm^2	83,259.99	m	−20.7%	−10.3%	−3.4%	6.9%	10.3%	20.7%
500 mm^2	730,778.8	m	−34.2%	−18.0%	−11.8%	6.4%	19.7%	41.2%
Export cable length	62,218	m	0.0%	0.0%	0.0%	0.0%	0.0%	0.0%
Cost								
95 mm^2	31,442,013	EUR	−20.7%	−10.3%	−3.4%	6.9%	10.3%	20.7%
500 mm^2	6.19×10^8	EUR	−34.1%	−18.0%	−11.8%	6.4%	19.7%	41.2%
Cable and ancillary supply cost (Export cable)	84,346,633	EUR	2.6%	25.5%	0.0%	0.0%	25.5%	52.8%
Offshore and onshore substations								
Onshore substation								
Onshore substation and transformer	15,473,700	EUR	−13.6%	−6.8%	−3.0%	3.0%	6.8%	13.6%
Spare parts, capacitor bank.	480,826.2	EUR	−2.1%	−1.0%	−0.4%	0.4%	1.0%	1.9%
Onshore transmission infrastructure								
Overhead transmission line	1,665,604	EUR	0.0%	0.0%	0.0%	0.0%	0.0%	0.0%
Switchyard	6,415,619	EUR	0.0%	0.0%	0.0%	0.0%	0.0%	0.0%
Port characteristics								
Entrance and exit	50,098.36	EUR	−20.5%	−10.2%	−4.5%	4.5%	10.2%	20.5%
Quayside docking	4,258,682	EUR	−20.5%	−10.2%	−4.5%	4.5%	10.2%	20.5%
Wharfage	1,016,389	EUR	−20.4%	−10.2%	−4.5%	4.5%	10.2%	20.4%
Turbine								
Area	1863.7	m^2	0.0%	0.0%	0.0%	0.0%	0.0%	0.0%
Cost	67,481.47	EUR	−20.5%	−10.2%	−4.5%	4.5%	10.2%	20.5%
Cranage								
600 ton	7,097,804	EUR	−20.5%	−10.2%	−4.5%	4.5%	10.2%	20.5%
1000 ton	11,396,486	EUR	−20.5%	−10.3%	−4.6%	4.5%	10.2%	20.4%
Crawler crane mobilization and demobilization	150,000	EUR	0.0%	0.0%	0.0%	0.0%	0.0%	0.0%

Table A1. *Cont.*

Item		Units	Scenario					
			−20%	−10%	−5%	5%	10%	20%
Main components installation time								
Anchor and mooring installation	974.0607	days	−20.5%	−10.2%	−4.5%	4.5%	10.2%	20.5%
Substructure load-out and prep for turbine installation	44	days	−20.5%	−10.2%	−4.5%	4.5%	10.2%	20.5%
Turbine installation	144.8333	days	−20.5%	−10.2%	−4.5%	4.5%	10.2%	20.5%
Substructure and turbine tow-out and installation	256.6667	days	−20.5%	−10.2%	−4.5%	4.5%	10.2%	20.5%
Electrical installation								
Array cable installation	439	days	−30.8%	−15.9%	−9.8%	6.2%	17.5%	36.2%
Export cable installation	37	days	−16.2%	−16.2%	0.0%	0.0%	0.0%	2.7%
Substation installation	4	days	0.0%	0.0%	0.0%	0.0%	0.0%	0.0%
Total floating bottom installation time	1899.561	days	−22.4%	−11.3%	−5.7%	4.8%	11.7%	24.0%

Table A2. Variability associated with the water depth.

Item		Units	Scenario					
			−20%	−10%	−5%	5%	10%	20%
Electrical system characteristics								
Array cable length								
95 mm^2	83,259.99	m	−1.8%	−0.9%	−0.4%	0.4%	0.9%	1.8%
500 mm^2	730,778.8	m	−1.8%	−0.9%	−0.4%	0.4%	0.9%	1.8%
Export cable length	62,218	m	−0.2%	−0.1%	0.0%	0.0%	0.1%	0.2%
Cost								
95 mm^2	31,442,013	EUR	−1.8%	−0.9%	−0.4%	0.4%	0.9%	1.8%
500 mm^2	6.19×10^8	EUR	−1.8%	−0.9%	−0.4%	0.4%	0.9%	1.8%
Cable and ancillary supply cost (Export cable)	84,346,633	EUR	−0.2%	−0.1%	−0.1%	0.1%	0.1%	0.2%
Offshore and onshore substations								
Onshore substation								
Onshore substation and transformer	15,473,700	EUR	0.0%	0.0%	0.0%	0.0%	0.0%	0.0%
Spare parts, capacitor bank	480,826.2	EUR	0.0%	0.0%	0.0%	0.0%	0.0%	0.0%
Onshore transmission infrastructure								
Overhead transmission line	1,665,604	EUR	0.0%	0.0%	0.0%	0.0%	0.0%	0.0%
Switchyard	6,415,619	EUR	0.0%	0.0%	0.0%	0.0%	0.0%	0.0%
Port characteristics								
Entrance and exit	50,098.36	EUR	0.0%	0.0%	0.0%	0.0%	0.0%	0.0%
Quayside docking	4,258,682	EUR	0.0%	0.0%	0.0%	0.0%	0.0%	0.0%
Wharfage	1,016,389	EUR	0.0%	0.0%	0.0%	0.0%	0.0%	0.0%
Turbine								
Area	1863.7	m^2	0.0%	0.0%	0.0%	0.0%	0.0%	0.0%
Cost	67,481.47	EUR	0.0%	0.0%	0.0%	0.0%	0.0%	0.0%
Cranage								
600 ton	7,097,804	EUR	−0.9%	−0.5%	−0.2%	0.2%	0.5%	0.9%
1000 ton	11,396,486	EUR	−0.9%	−0.5%	−0.2%	0.2%	0.5%	0.9%
Crawler crane mobilization and demobilization	150,000	EUR	0.0%	0.0%	0.0%	0.0%	0.0%	0.0%
Main components installation time								
Anchor and mooring installation	974.0607	days	−1.4%	−0.7%	−0.3%	0.3%	0.7%	1.4%
Substructure load-out and prep for turbine installation	44	days	0.0%	0.0%	0.0%	0.0%	0.0%	0.0%
Turbine installation	144.8333	days	0.0%	0.0%	0.0%	0.0%	0.0%	0.0%
Substructure and turbine tow-out and installation	256.6667	days	0.0%	0.0%	0.0%	0.0%	0.0%	0.0%
Electrical installation								
Array cable installation	439	days	−0.7%	−0.5%	−0.2%	0.2%	0.5%	0.9%
Export cable installation	37	days	0.0%	0.0%	0.0%	0.0%	0.0%	0.0%
Substation installation	4	days	0.0%	0.0%	0.0%	0.0%	0.0%	0.0%
Total floating bottom installation time	1899.561	days	−0.9%	−0.5%	−0.2%	0.2%	0.5%	0.9%

Table A3. Variability associated with the distance to the electrical grid.

Item		Units	Scenario					
			−20%	−10%	−5%	5%	10%	20%
Electrical system characteristics								
Array cable length								
95 mm^2	83,259.99	m	1.8%	1.8%	1.8%	1.8%	1.8%	1.8%
500 mm^2	730,778.8	m	1.8%	1.8%	1.8%	1.8%	1.8%	1.8%
Export cable length	62,218	m	−16.4%	−6.5%	−1.6%	8.3%	13.2%	23.1%
Cost								
95 mm^2	31,442,013	EUR	1.8%	1.8%	1.8%	1.8%	1.8%	1.8%
500 mm^2	6.19×10^8	EUR	1.8%	1.8%	1.8%	1.8%	1.8%	1.8%
Cable and ancillary supply cost (Export cable)	84,346,633	EUR	−16.1%	−6.4%	−1.5%	8.2%	13.1%	22.8%
Offshore and onshore substations								
Onshore substation								
Onshore substation and transformer	15,473,700	EUR	0.0%	0.0%	0.0%	0.0%	0.0%	0.0%
Spare parts, capacitor bank.	480,826.2	EUR	0.0%	0.0%	0.0%	0.0%	0.0%	0.0%
Onshore transmission infrastructure								
Overhead transmission line	1,665,604	EUR	0.0%	0.0%	0.0%	0.0%	0.0%	0.0%
Switchyard	6,415,619	EUR	0.0%	0.0%	0.0%	0.0%	0.0%	0.0%
Port characteristics								
Entrance and exit	50,098.36	EUR	0.0%	0.0%	0.0%	0.0%	0.0%	0.0%
Quayside docking	4,258,682	EUR	0.0%	0.0%	0.0%	0.0%	0.0%	0.0%
Wharfage	1,016,389	EUR	0.0%	0.0%	0.0%	0.0%	0.0%	0.0%
Turbine								
Area	1863.7	m^2	0.0%	0.0%	0.0%	0.0%	0.0%	0.0%
Cost	67,481.47	EUR	0.0%	0.0%	0.0%	0.0%	0.0%	0.0%
Cranage								
600 ton	7,097,804	EUR	0.0%	0.0%	0.0%	0.0%	0.0%	0.0%
1000 ton	11,396,486	EUR	0.0%	0.0%	0.0%	0.0%	0.0%	0.0%
Crawler crane mobilization and demobilization	150,000	EUR	0.0%	0.0%	0.0%	0.0%	0.0%	0.0%
Main components installation time								
Anchor and mooring installation	974.0607	days	0.0%	0.0%	0.0%	0.0%	0.0%	0.0%
Substructure load-out and prep for turbine installation	44	days	0.0%	0.0%	0.0%	0.0%	0.0%	0.0%
Turbine installation	144.8333	days	0.0%	0.0%	0.0%	0.0%	0.0%	0.0%
Substructure and turbine tow-out and installation	256.6667	days	0.0%	0.0%	0.0%	0.0%	0.0%	0.0%
Electrical installation								
Array cable installation	439	days	0.0%	0.0%	0.0%	0.0%	0.0%	0.0%
Export cable installation	37	days	−8.1%	0.0%	0.0%	2.7%	2.7%	5.4%
Substation installation	4	days	0.0%	0.0%	0.0%	0.0%	0.0%	0.0%
Total floating bottom installation time	1899.561	days	−0.1%	0.0%	0.0%	0.1%	0.2%	0.2%

Table A4. Variability associated with the distance to the port.

Item		Units	Scenario					
			−20%	−10%	−5%	5%	10%	20%
Electrical system characteristics								
Array cable length								
95 mm^2	83,259.99	m	0.0%	0.0%	0.0%	0.0%	0.0%	0.0%
500 mm^2	730,778.8	m	0.0%	0.0%	0.0%	0.0%	0.0%	0.0%
Export cable length	62,218	m	3.2%	3.2%	3.2%	3.2%	3.2%	3.2%
Cost								
95 mm^2	31,442,013	EUR	0.0%	0.0%	0.0%	0.0%	0.0%	0.0%
500 mm^2	6.19×10^8	EUR	0.0%	0.0%	0.0%	0.0%	0.0%	0.0%
Cable and ancillary supply cost (Export cable)	84,346,633	EUR	3.1%	3.1%	3.1%	3.1%	3.1%	3.1%
Offshore and onshore substations								
Onshore substation								
Onshore substation and transformer	15,473,700	EUR	0.0%	0.0%	0.0%	0.0%	0.0%	0.0%
Spare parts, capacitor bank.	480,826.2	EUR	0.0%	0.0%	0.0%	0.0%	0.0%	0.0%
Onshore transmission infrastructure								
Overhead transmission line	1,665,604	EUR	0.0%	0.0%	0.0%	0.0%	0.0%	0.0%
Switchyard	6,415,619	EUR	0.0%	0.0%	0.0%	0.0%	0.0%	0.0%
Port characteristics								
Entrance and exit	50,098.36	EUR	0.0%	0.0%	0.0%	0.0%	0.0%	0.0%
Quayside docking	4,258,682	EUR	0.0%	0.0%	0.0%	0.0%	0.0%	0.0%
Wharfage	1,016,389	EUR	0.0%	0.0%	0.0%	0.0%	0.0%	0.0%
Turbine								
Area	1863.7	m^2	0.0%	0.0%	0.0%	0.0%	0.0%	0.0%
Cost	67,481.47	EUR	0.0%	0.0%	0.0%	0.0%	0.0%	0.0%
Cranage								
600 ton	7,097,804	EUR	0.0%	0.0%	0.0%	0.0%	0.0%	0.0%
1000 ton	11,396,486	EUR	0.0%	0.0%	0.0%	0.0%	0.0%	0.0%
Crawler crane mobilization and demobilization	150,000	EUR	0.0%	0.0%	0.0%	0.0%	0.0%	0.0%
Main components installation time								
Anchor and mooring installation	974.0607	days	−0.6%	−0.3%	−0.1%	0.1%	0.3%	0.6%
Substructure load-out and prep for turbine installation	44	days	0.0%	0.0%	0.0%	0.0%	0.0%	0.0%
Turbine installation	144.8333	days	0.0%	0.0%	0.0%	0.0%	0.0%	0.0%
Substructure and turbine tow-out and installation	256.6667	days	−5.1%	−2.6%	−1.3%	1.3%	2.6%	5.1%
Electrical installation								
Array cable installation	439	days	0.0%	0.0%	0.0%	0.2%	0.2%	0.2%
Export cable installation	37	days	0.0%	0.0%	0.0%	2.7%	2.7%	2.7%
Substation installation	4	days	0.0%	0.0%	0.0%	0.0%	0.0%	0.0%
Total floating bottom installation time	1899.561	days	−1.0%	−0.5%	−0.2%	0.4%	0.6%	1.1%

Table A5. Variability associated with the initial work schedule of 320 days (as a function of weather conditions).

Item		Units	Scenario					
			−20%	**−10%**	**−5%**	**5%**	**10%**	**20%**
Electrical system characteristics								
Array cable length								
95 mm^2	83,259.99	m	0.0%	0.0%	0.0%	0.0%	0.0%	0.0%
500 mm^2	730,778.8	m	0.0%	0.0%	0.0%	0.0%	0.0%	0.0%
Export cable length	62,218	m	0.0%	0.0%	0.0%	0.0%	0.0%	0.0%
Cost								
95 mm^2	31,442,013	EUR	0.0%	0.0%	0.0%	0.0%	0.0%	0.0%
500 mm^2	6.19×10^8	EUR	0.0%	0.0%	0.0%	0.0%	0.0%	0.0%
Cable and ancillary supply cost (Export cable)	84,346,633	EUR	0.0%	0.0%	0.0%	0.0%	0.0%	0.0%
Offshore and onshore substations								
Onshore substation								
Onshore substation and transformer	15,473,700	EUR	0.0%	0.0%	0.0%	0.0%	0.0%	0.0%
Spare parts, capacitor bank.	480,826.2	EUR	0.0%	0.0%	0.0%	0.0%	0.0%	0.0%
Onshore transmission infrastructure								
Overhead transmission line	1,665,604	EUR	0.0%	0.0%	0.0%	0.0%	0.0%	0.0%
Switchyard	6,415,619	EUR	0.0%	0.0%	0.0%	0.0%	0.0%	0.0%
Port characteristics								
Entrance and exit	50,098.36	EUR	14.8%	6.8%	4.5%	−4.5%	−6.8%	−14.8%
Quayside docking	4,258,682	EUR	12.7%	4.9%	2.7%	−6.3%	−8.5%	−16.3%
Wharfage	1,016,389	EUR	14.8%	6.8%	4.5%	−4.5%	−6.8%	−14.7%
Turbine								
Area	1863.7	m^2	0.0%	0.0%	0.0%	0.0%	0.0%	0.0%
Cost	67,481.47	EUR	17.8%	13.5%	5.2%	−1.7%	−13.3%	−19.0%
Cranage								
600 ton	7,097,804	EUR	12.7%	4.9%	2.7%	−6.3%	−8.5%	−16.3%
1000 ton	11,396,486	EUR	12.7%	4.9%	2.6%	−6.3%	−8.5%	−16.3%
Crawler crane mobilization and demobilization	150,000	EUR	0.0%	0.0%	0.0%	0.0%	0.0%	0.0%
Main components installation time								
Anchor and mooring installation	974.0607	days	14.8%	6.8%	4.5%	−4.5%	−6.8%	−14.8%
Substructure load-out and prep for turbine installation	44	days	14.8%	6.8%	4.5%	−4.5%	−6.8%	−14.8%
Turbine installation	144.8333	days	25.6%	12.1%	4.0%	−15.5%	−23.3%	−29.9%
Substructure and turbine tow-out and installation	256.6667	days	14.8%	6.8%	4.5%	−4.5%	−6.8%	−14.8%
Electrical installation								
Array cable installation	439	days	24.1%	11.4%	4.8%	−9.8%	−10.7%	−25.1%
Export cable installation	37	days	2.7%	0.0%	0.0%	0.0%	0.0%	−16.2%
Substation installation	4	days	0.0%	0.0%	0.0%	0.0%	0.0%	0.0%
Total floating bottom installation time	1899.561	days	15.4%	6.3%	2.0%	−7.0%	−8.8%	−18.0%

References

1. US Energy Information Administration (EIA). *Levelized Cost of New Generation Resources in the Annual Energy Outlook 2021*; US Energy Information Administration: Washington, DC, USA, 2021.
2. Offshore Renewable Energy (ORE). *Catapult Floating Offshore Wind: Cost Reduction Pathways to Subsidy Free*; Offshore Renewable Energy: Glasgow, UK, 2021.
3. Offshore Renewable Energy (ORE). *Catapult Floating Wind: Technology Assessment*; Offshore Renewable Energy: Glasgow, UK, 2015.

4. Quest Floating Wind Energy. *Global Floating Wind Market and Forecast Report 2021–2034*; Quest Floating Wind Energy: Houston, TX, USA, 2021.
5. Bulder, B.H.; Swamy, S.K.; Warnaar, P.M.J.; van den Brink, M.L.M.; de la Vieter, M.L. *Pathways to Potential Cost Reductions for Offshore Wind Energy*; TU Delft: Delft, The Netherlands, 2021.
6. Poulsen, T.; Hasager, C.B. How expensive is expensive enough? Opportunities for cost reductions in offshoreWind energy logistics. *Energies* **2016**, *9*, 437. [CrossRef]
7. Wind Europe. *Floating Offshore Wind Energy–A Policy Blueprint for Europe*; Wind Europe: Bruxelles, Belgium, 2019.
8. Díaz, H.; Guedes Soares, C. Review of the current status, technology and future trends of offshore wind farms. *Ocean Eng.* **2020**, *209*, 107381. [CrossRef]
9. Wang, C.M.; Utsunomiya, T.; Wee, S.C.; Choo, Y.S. Research on floating wind turbines: A literature survey. *IES J. Part A Civ. Struct. Eng.* **2010**, *3*, 267–277. [CrossRef]
10. Uzunoglu, E.; Karmakar, D.; Guedes Soares, C. Floating Offshore Wind Platforms. In *Floating Offshore Wind Farms*; Castro-Santos, L., Díaz-Casas, V., Eds.; Springer International Publishing: Berlin/Heidelberg, Germany, 2016; pp. 53–76.
11. Kaiser, M.J.; Snyder, B. *Offshore Wind Energy Installation and Decommissioning Cost Estimation in the U.S. Outer Continental Shelf*; Energy Research Group, LLC.: Baton Rouge, LO, USA, 2010.
12. Thomsen, K.E. *Offshore Wind–A Comprehensive Guide to Successful Offshore Wind Farm Installation*; Academic Press: Cambridge, MA, USA, 2014; ISBN 978-0-12-410422-8.
13. Skiba, M. *Offshore Wind Energy—The Installation Challenge*; RWE: Essen, Germany, 2010.
14. Scholz-Reiter, B.; Heger, J.; Lütjen, M.; Schweizer, A. A MILP for installation scheduling of offshore wind farms. *Int. J. Math. Model. Methods Appl. Sci.* **2011**, *5*, 371–378.
15. European Wind Energy Association (EWEA). *Oceans of Opportunity*; European Wind Energy Association: Bruxelles, Belgium, 2009.
16. Scholz-Reiter, B.; Reza Karimi, H.; Lütjen, M.; Heger, J.; Schweizer, A. Towards a Heuristic for Scheduling Offshore Installation. In Proceedings of the 24th International Congress on Condition Monitoring and and Diagnostics Engineering Management, Stavanger, Norway, 30 May–1 June 2011; pp. 1–10.
17. Lange, K.; Rinne, A.; Haasis, H.D. Planning maritime logistics concepts for offshore wind farms: A newly developed decision support system. *Lect. Notes Comput. Sci.* **2012**, *7555*, 142–158. [CrossRef]
18. Ait-Alla, A.; Quandt, M.; Lütjen, M. Simulation-based aggregate Installation Planning of Offshore Wind Farms. *Int. J. Energy* **2013**, *7*, 23–30.
19. Barlow, E.; Öztürk, D.T.; Revie, M.; Boulougouris, E.; Day, A.H.; Akartunali, K. Exploring the impact of innovative developments to the installation process for an offshore wind farm. *Ocean Eng.* **2015**, *109*, 623–634. [CrossRef]
20. Irawan, C.A.; Ouelhadj, D.; Jones, D.; Stålhane, M.; Sperstad, I.B. Optimisation of maintenance routing and scheduling for offshore wind farms. *Eur. J. Oper. Res.* **2017**, *256*, 76–89. [CrossRef]
21. Muhabie, Y.T.; Petcu, C.; Rigo, P.; Caprace, J.D. Weather Down Time Analysis for Offshore Wind Farm Installations. *PIANC Yearb.* **2016**, 162–169.
22. Sarker, B.R.; Faiz, T.I. Minimizing transportation and installation costs for turbines in offshore wind farms. *Renew. Energy* **2017**, *101*, 667–679. [CrossRef]
23. Barlow, E.; Tezcaner Ozturk, D.; Revie, M.; Akartunali, K.; Day, A.; Boulougouris, E. *On Using Simulation to Model the Installation Process Logistics for an Offshore Wind Farm*; University of Strathclyde: Glasgow, UK, 2017.
24. Boulougouris, E. Simulation and Optimization for Offshore Windfarm Installation Logistics. In Proceedings of the NORCOWE 2016, Bergen, Norway, 14–15 September 2016; pp. 1–30.
25. Backe, S.; Haugland, D. Strategic optimization of offshore wind farm installation. *Lect. Notes Comput. Sci.* **2017**, *10572*, 285–299. [CrossRef]
26. Paterson, J.; D'Amico, F.; Thies, P.R.; Kurt, R.E.; Harrison, G. Offshore wind installation vessels—A comparative assessment for UK offshore rounds 1 and 2. *Ocean Eng.* **2018**, *148*, 637–649. [CrossRef]
27. Maples, B.; Saur, G.; Hand, M.; van Pietermen, R.; Obdam, T. *Installation, Operation, and Maintenance Strategies to Reduce the Cost of Offshore Wind Energy*; National Renewable Energy Lab.: Golden, CO, USA, 2013.
28. Navigant Consulting. *Navigant Global Evaluation of Offshore Wind Shipping Opportunity*; Navigant Consultin, Inc.: Chicago, IL, USA, 2013.
29. Vis, I.F.A.; Ursavas, E. Assessment approaches to logistics for offshore wind energy installation. *Sustain. Energy Technol. Assess.* **2016**, *14*, 80–91. [CrossRef]
30. Díaz, H.; Guedes Soares, C. A novel multi-criteria decision-making model to evaluate floating wind farm locations. *Renew. Energy* **2022**, *185*, 431–454. [CrossRef]
31. Díaz, H.; Guedes Soares, C. Micro-siting of floating turbines through multiple-criteria decision-making. *Proc. Int. Conf. Offshore Mech. Arct. Eng.OMAE* **2021**, *5*, 1–9. [CrossRef]
32. Díaz, H.; Guedes Soares, C. An integrated GIS approach for site selection of floating offshore wind farms in the Atlantic continental European coastline. *Renew. Sustain. Energy Rev.* **2020**, *134*, 110328. [CrossRef]
33. Martinez, H.D.; Guedes Soares, C. Multicriteria Decision Approach to the Design of Floating Wind Farm Export Cables. *Energies* **2022**, *15*, 6593. [CrossRef]

34. Diaz, H.; Teixeira, A.P.; Guedes Soares, C. Application of Fuzzy and Monte Carlo Analytic Hierarchy Processes for ranking floating wind farm locations. *Ocean Eng.* **2022**, *245*, 110453. [CrossRef]
35. Díaz, H.; Guedes Soares, C. A Multi-Criteria Approach to Evaluate Floating Offshore Wind Farms Siting in the Canary Islands (Spain). *Energies* **2021**, *14*, 865. [CrossRef]
36. ARCWIND. Project Arcwind. Available online: http://www.arcwind.eu/ (accessed on 5 November 2021).
37. Castro-Santos, L.; Silva, D.; Bento, A.R.; Salvação, N.; Guedes Soares, C. Economic feasibility of floating offshore wind farms in Portugal. *Ocean Eng.* **2020**, *207*, 107393. [CrossRef]
38. Amaro, N.; Egorov, A.; Glória, G. Fostering Offshore Wind Integration in Europe through Grid Connection Impact Assessment. *J. Mar. Sci. Eng.* **2022**, *10*, 463. [CrossRef]
39. Chartron, S. Improving logistics scheduling and operations to support offshore wind construction phase. *Logist. Res.* **2019**, *12*, 1–14. [CrossRef]
40. Sevilla, F.; Redfern, R.; Storey, A.; Baldock, N. *Optimization of Installation Operation and Maintenance At Offshore Wind Projects in the U.S.*; U.S. Department of Energy: Oak Ridge, TN, USA, 2014.
41. Crown Estate Scotland; ORE Catapult. *Macroeconomic Benefits of Floating Offshore Wind in the UK*; Crown Estate Scotland: Edinburg, UK, 2018.
42. *DNVGL-ST-0054*; Transport and Installation of Wind Power Plants. DNV GL AS: Bærum, Norway, 2017.
43. Garrad Hassan America Inc. *Assessment of Ports for Offshore Wind Development in the United States*; Garrad Hassan America Inc: San Diego, CA, USA, 2014.
44. Tegen, S.; Keyser, D.; Flores-Espino, F.; Miles, J.; Zammit, D.; Loomis, D. *Offshore Wind Jobs and Economic Development Impacts in the United States: Four Regional Scenarios Offshore Wind Jobs and Economic Development Impacts in the United States: Four Regional Scenarios*; National Renewable Energy Laboratory: Golden, CO, USA, 2015.
45. Ioannou, A.; Angus, A.; Brennan, F. A lifecycle techno-economic model of offshore wind energy for different entry and exit instances. *Appl. Energy* **2018**, *221*, 406–424. [CrossRef]
46. Hahn, M.; Gilman, P. *Offshore Wind Market and Economic Analysis: Annual Market Assessment*; Navigant Consultinc, Inc.: Burlington, MA, USA, 2013.
47. Poulsen, T.; Lema, R. Is the supply chain ready for the green transformation? The case of offshore wind logistics. *Renew. Sustain. Energy Rev.* **2017**, *73*, 758–771. [CrossRef]
48. Renewable UK. *Working for a Green Britain*; Renewable UK: London, UK, 2011.
49. Tegen, S.; Hand, M.; Maples, B.; Lantz, E.; Schwabe, P.; Smith, A. *2010 Cost of Wind Energy Review*; National Renewable Energy Lab.: Golden, CO, USA, 2012; Volume 111.
50. Smart, G.; Smith, A.; Warner, E.; Sperstad, I.B.; Prinsen, B.; Lacal-Arantegui, R. *IEA Wind Task 26: Offshore Wind Farm Baseline Documentation*; National Renewable Energy Lab.: Golden, CO, USA, 2016.
51. Ladenburg, J.; Dubgaard, A.; Martinsen, L.; Tranberg, J. *Economic Valuation of the Visual Externalities of Offshore Wind Farms*; U.S. Department of Energy: Washington, DC, USA, 2005.
52. Douglas-Westwood. *Assessment of Vessel Requirements for the U.S. Offshore Wind Sector*; Westwood Global Energy Group: Aberdeen, UK, 2013.
53. Beiter, P.; Musial, W.; Smith, A.; Kilcher, L.; Damiani, R.; Maness, M.; Sirnivas, S.; Stehly, T.; Gevorgian, V.; Mooney, M.; et al. *A Spatial-Economic Cost—Reduction Pathway Analysis for U.S. Offshore Wind Energy Development from 2015–2030*; National Renewable Energy Lab.: Golden, CO, USA, 2016.
54. Musuku, V.; Poulsen, T. *Offshore Wind Logistics Brief Report 7—Consolidation in the Offshore Wind Industry with a Focus on Logistics Related Mergers and Acquisitions*; Panticon: Faxe, Denmark, 2020.
55. The Crown Estate. *Guide to an Offshore Wind Farm*; The Crown Estate: London, UK, 2019.
56. BVG Associates. *UK Offshore Wind Supply Chain: Capabilities and Opportunities*; BVG Associates: Swindon, UK, 2014.
57. BVG Associates. *Approaches to Cost-Reduction in Offshore Wind*; BVG Associates: Swindon, UK, 2015.
58. Valpy, B.; English, P. *Future Renewable Energy Costs: Offshore Wind*; KIC-Innoenergy: Eindhoven, The Netherlands, 2014.
59. BVG Associates. *Ocean Power Innovation Network Value Chain Study: Summary Report*; BVG Associates: Swindon, UK, 2019.
60. Willow, C.; Valpy, B. *Offshore Wind—Forecasts of Future Costs and Benefits*; Renewables UK: London, UK, 2011.
61. The Crown Estate. *A Guide to an Offshore Wind Farm*; The Crown Estate: London, UK, 2010.
62. Dalgic, Y.; Lazakis, I.; Dinwoodie, I.; McMillan, D.; Revie, M. Advanced logistics planning for offshore wind farm operation and maintenance activities. *Ocean Eng.* **2015**, *101*, 211–226. [CrossRef]
63. Deloitte. *Analysis on the Furthering of Competition in Relation to the Establishment of Large Offshore Wind Farms in Denmark*; Deloitte: London, UK, 2011.
64. Flynn, R.J.; Carey, R.T. *The Potential Economic Impact of an OffShore Wind Farm to the State of South Carolina*; Strom Thurmond Institute: Clemson, SC, USA, 2007.
65. Yahalom, S.; Johansson, E.; Fink, E.; Chang, G.; Kopits, S.; Losz, A.; Singer, J.; Choi, J.; Obzay, K. *Offshore Wind Development Research*; U.S. Department of Energy: Washington, DC, USA, 2014.
66. Carbon Trust. *Phase III Summary Report—Floating Wind Joint Industry Project*; Carbon Trust: London, UK, 2021.
67. Livaniou, S.; Iordanis, S.; Anaxagorou, P.; Mocanu, B.; Sykes, R.; Goormachtigh, J.; Christensen, J.M.; Kristensen, J.T.; Antrobus, M. *Logistic Efficiencies and Naval Architecture for Wind Installations with Novel Developments*; Leanwind: Cork, Ireland, 2015.

68. Green Port Hull; BVG Associates. *Job Roles in Offshore Wind*; Green Port Hull: Hull, UK, 2017.
69. Loughney, S.; Wang, J.; Bashir, M.; Armin, M.; Yang, Y. Development and application of a multiple-attribute decision-analysis methodology for site selection of floating offshore wind farms on the UK Continental Shelf. *Sustain. Energy Technol. Assess.* **2021**, *47*, 101440. [CrossRef]
70. Guedes Soares, C.; Bento, A.R.; Gonçalves, M.; Silva, D.; Martinho, P. Numerical evaluation of the wave energy resource along the Atlantic European coast. *Comput. Geosci.* **2014**, *71*, 37–49. [CrossRef]
71. Bento, A.R.; Martinho, P.; Guedes Soares, C. Wave energy assessement for Northern Spain from a 33-year hindcast. *Renew. Energy* **2018**, *127*, 322–333. [CrossRef]
72. Bentamy, A.; Croize-Fillon, D. Spatial and temporal characteristics of wind and wind power off the coasts of Brittany. *Renew. Energy* **2014**, *66*, 670–679. [CrossRef]
73. Salvação, N.; Guedes Soares, C. Wind resource assessment offshore the Atlantic Iberian coast with the WRF model. *Energy* **2018**, *145*, 276–287. [CrossRef]
74. Uzunoglu, E.; Guedes Soares, C. Hydrodynamic design of a free-float capable tension leg platform for a 10 MW wind turbine. *Ocean Eng.* **2020**, *197*, 106888. [CrossRef]
75. Baita-Saavedra, E.; Cordal-Iglesias, D.; Filgueira-Vizoso, A.; Castro-Santos, L. Economic aspects of a concrete floating offshore wind platform in the Atlantic Arc of Europe. *Int. J. Environ. Res. Public Health* **2019**, *16*, 4122. [CrossRef] [PubMed]
76. Baita-Saavedra, E.; Cordal-Iglesias, D.; Filgueira-Vizoso, A.; Morató, À.; Lamas-Galdo, I.; Álvarez-Feal, C.; Carral, L.; Castro-Santos, L. An economic analysis of an innovative floating offshore wind platform built with concrete: The SATH® platform. *Appl. Sci.* **2020**, *10*, 3678. [CrossRef]
77. *DNV DNVGL-ST-N001*; Marine Operations and Marine Warranty. DNV GL AS: Bærum, Norway, 2016.
78. Uzunoglu, E.; Guedes Soares, C. Response dynamics of a free-float capable tension leg platform for a 10 MW wind turbine at the Northern Iberian Peninsula. In *Developments in Renewable Energies Offshore*; Guedes Soares, C., Ed.; Taylor and Francis: London, UK, 2021; pp. 408–416.
79. Mas-Soler, J.; Uzunoglu, E.; Bulian, G.; Guedes Soares, C. Souto-Iglesias, A. An experimental study on transporting a free-float capable tension leg platform for a 10 MW wind turbine in waves. *Renew. Energy* **2021**, *179*, 2158–2173. [CrossRef]
80. Martins, D.; Muraleedharan, G.; Guedes Soares, C. Analysis on weather windows defined by significant wave height and wind speed. In *Renewable Energies Offshore*; Guedes Soares, C., Ed.; Taylor & Francis: London, UK, 2015; pp. 91–98. ISBN 9781138028715.
81. Walker, R.T.; van Nieuwkoop-McCall, J.; Johanning, L.; Parkinson, R.J. Calculating weather windows: Application to transit, installation and the implications on deployment success. *Ocean Eng.* **2013**, *68*, 88–101. [CrossRef]
82. Bak, C.; Zahle, F.; Bitsche, R.; Taeseong, K.; Yde, A.; Henriksen, L.C.; Natarajan, A.; Hansen, M.H. Description of the DTU 10 MW Reference Wind Turbine. In *Danish Wind Power Research 2013*; Report-I-0092; DTU Library: Copenhagen, Denmark, 2013.
83. Diaz, H.; Serna, J.; Nieto, J.; Guedes Soares, C. Market Needs, Opportunities and Barriers for the Floating Wind Industry. *J. Mar. Sci. Eng.* **2022**, *10*, 934. [CrossRef]

Journal of
Marine Science and Engineering

MDPI

Article

Dynamic Analysis of a Novel Installation Method of Floating Spar Wind Turbines

Mohamed Hassan * and C. Guedes Soares *

Centre for Marine Technology and Ocean Engineering (CENTEC), Instituto Superior Técnico, Universidade de Lisboa, 1049-001 Lisboa, Portugal
* Correspondence: mohamed.hassan@centec.tecnico.ulisboa.pt (M.H.); c.guedes.soares@centec.tecnico.ulisboa.pt (C.G.S.)

Abstract: This paper presents the performance of a new, floating, mono-hull wind turbine installation vessel (Nordic Wind) in the installation process. The vessel can transport pre-assembled wind turbines from the marshalling port to the offshore installation site. Each assembled turbine will be positioned over the pre-installed floating spar structure. The primary difficulty lies in examining the multibody system's reactions when subjected to combined wind, current, and wave forces. Time-domain simulations are utilized to model the interconnected system, incorporating mechanical coupling between components, the mooring system for the spar, and the installation vessel. The primary objective is to focus on the monitoring and connection stages preceding the mating operations between the turbine and the floating spar. Additionally, it involves examining the impacts of wind, current, and wave conditions on the motion responses of the installation vessel and the spar, as well as the relative motions at the mating point, gripper forces, and mooring forces. The simulations show that the resulting gripper forces are reasonable to compensate. The relative motion at the mating point is not significantly affected by the orientations of the turbine blades, but it is influenced by the prevailing wave conditions. In addition, vessel heading optimization can minimize the relative motions at the mating point and gripper forces. Given the examined environmental conditions, the presented installation concept exhibits a commendable performance.

Keywords: floating wind turbine; offshore installation; float-over; weather monitoring; spar

Citation: Hassan, M.; Guedes Soares, C. Dynamic Analysis of a Novel Installation Method of Floating Spar Wind Turbines. *J. Mar. Sci. Eng.* **2023**, *11*, 1373. https://doi.org/10.3390/jmse11071373

Academic Editor: Constantine Michailides

Received: 17 May 2023
Revised: 25 June 2023
Accepted: 27 June 2023
Published: 5 July 2023

1. Introduction

Renewable energy from wind power is expanding globally, establishing itself as a prominent and increasingly significant source of sustainable energy, as the economic conditions are favorable [1,2]. To address the need for alternative energy sources, a considerable number of wind power generators are being deployed in wind farms, both on land and offshore. European offshore wind capacity grew by adding 3.4 GW of offshore capacity during 2021, reaching a total installed capacity of 17.4 GW [3]. Due to the varying quality of wind resources and geographical constraints, numerous countries are now contemplating the utilization of deep-water offshore areas for future wind power facility development. As of the end of 2020, Europe's floating wind fleet reached a cumulative capacity of 62 MW, representing 83% of the global floating wind capacity. Floating offshore wind (FOW) is seen as a crucial factor in unlocking Europe's vast and virtually limitless wind energy potential. Around 80% of the offshore wind resources in European seas are located in waters with depths of 60 m or greater, where traditional bottom-fixed offshore wind (BFOW) installations are not economically viable or attractive.

Figure 1 illustrates the three primary designs for floating offshore wind: tension leg platform, semi-submersible, and spar buoy. The semi-submersible and spar buoy structures, unlike the tension leg platform, are loosely moored to the seabed. This loose mooring configuration facilitates simpler installation processes. In contrast, the tension leg platform is firmly connected to the seabed [4–7].

Figure 1. Floating wind turbine substructure designs (DNV, [8]).

The world's first floating wind turbine farm, utilizing Equinor's Hywind technology from Norway, was commissioned in Scotland in late 2017. The floating wind turbine farm comprises five 6 MW turbines, provided by Siemens. Prior to installation, the turbine blades, nacelle, and tower are pre-assembled onshore. The installation process involves using the SAIPEM 7000, one of the largest semi-submersible crane vessels, to lift and place the components onto the spar floaters in the deep fjord of Norway [9]. Subsequently, the pre-assembled turbines on the spar floaters are then transported via wet towing to offshore Scotland. Once in position, they are connected with mooring systems to secure their stability. The Hywind Tampen project, with a capacity of 88 MW, attained the Financial Investment Decision in 2019 and is currently in the pre-construction phase. The objective of the project is to achieve a cost reduction of 40% compared to Hywind Scotland, which served as the initial demonstration of the spar-buoy technology [10]. Nevertheless, the current installation method still involves assembling the turbines in a sheltered location and then towing each complete assembly individually to the installation site for final commissioning.

To continue cost reduction and development of floating offshore wind projects, the following key areas shall be considered:

- Industrialization: optimize design of substructure and mooring and reduce the cost per ton towards mass production levels.
- Economics of scale: larger parks will drive down costs for infrastructure and logistics.
- Upscaling: follow the bottom-fixed industry towards turbine sizes of 10–15 MW.
- Technology development: establish new and more efficient methods for installation, operation, and maintenance. Offshore wind developers have recognized the preference for new installation methods that rely on cost-effective large crane vessels when developing commercial wind farms. This shift is driven by the aim to reduce expenses associated with installation operations. This will enhance the development of mass production towards potential future long-term markets. In the context of the Hywind installation challenge, there is a preference for innovative installation concepts that prioritize the use of novel installation vessels and facilities. The goal is to minimize offshore lifts and reduce the overall operation time involved in the installation process. The assembly and installation cost for offshore wind turbines (OWT) is relatively high and can account for approximately 20% of the total capital expenditures incurred in the development of offshore wind farms [11–14]. It is, therefore, crucial to employ cost-effective installation methods in order to mitigate the high assembly and installation costs associated with offshore wind turbines. Implementing efficient and economical

installation techniques is vital for the overall financial viability and success of offshore wind farm projects.

2. Current Installation Vessels

Two common types of vessels used for offshore wind turbine installation are jack-up vessels and floating crane vessels. Jack-up vessels offer a stable working platform for the installation of foundations, allowing for safe and efficient operations. These vessels are equipped with extendable legs that can be lowered to the seabed, providing stability during installation activities. However, the process of installing and retrieving the jack-up legs is time-consuming and highly dependent on weather conditions. Additionally, the stability of the jack-up unit relies heavily on the specific characteristics of the seabed, which can introduce uncertainties. Furthermore, jack-up vessels have limitations regarding the water depth in which they can operate effectively. In contrast to jack-up vessels, floating crane vessels offer greater flexibility for offshore installations and are particularly effective in the mass installation of wind farms. These vessels can swiftly move between foundations, facilitating efficient and rapid deployment. Their enhanced mobility makes them well-suited for large-scale wind farm projects. However, during the installation process, the motions of the floating installation vessel and the lifted objects are intricately coupled and are highly responsive to environmental conditions. This interdependence introduces challenges, as the movement of one affects the stability and control of the other. Therefore, careful consideration of environmental factors is crucial to ensure safe and successful installations. It is important to note that the installation cost associated with floating crane vessels is typically higher compared to jack-up crane vessels. Therefore, when making the selection of a crane vessel for practical operations, a well-balanced consideration between technical feasibility and cost is necessary. Finding the right balance ensures that the chosen vessel meets the technical requirements, while keeping the overall project cost within a feasible range.

Currently, there are ongoing efforts to utilize floating crane vessels for various offshore wind turbine installation tasks. These include the installation of monopile foundations for bottom-fixed wind turbines, the assembly of the tower-rotor-nacelle components for floating wind turbines, as well as the installation of tower-nacelle assemblies and rotors for floating wind turbines. These attempts demonstrate the exploration of alternative installation methods to meet the specific requirements of different types of wind turbine structures. The installation of blades for offshore wind turbines presents greater challenges compared to other components, such as foundations and transition pieces. This is primarily due to the need for precise installation during the final blade mating phase, which demands a high degree of accuracy. Additionally, at significant lifting heights, there is relatively large motion between the turbine hub and the blade root, further complicating the installation process. The combination of high installation precision and substantial motion necessitates careful planning and specialized techniques to ensure successful blade installation. Enhancing the weather window and mitigating unforeseen delays emerge as crucial challenges in reducing the cost of offshore installations. To accomplish this, it is essential to conduct precise evaluations of the installation vessels and methods to gauge their performance accurately. Furthermore, there is a need to develop numerical methods and models that can estimate the dynamic responses of systems during the installation process. These efforts aim to enhance the understanding and prediction of system behavior, allowing for informed decision-making and optimization of offshore installation operations [15–24].

Due to its substantial capacity and cost-effectiveness, the float-over method has gained significant popularity in the oil and gas industry for installing large decks onto offshore platforms. The float-over method offers several notable advantages for installing integrated topsides onto a floating substructure, especially when dealing with large topsides that surpass the lifting capacity of existing heavy-lift fleets. These advantages encompass time and cost savings by allowing the integration and commissioning of modules to take place on land, instead of at sea. Additionally, decoupling the fabrication schedules of the topside from the availability of heavy-lift vessels is another significant benefit [25–30].

The innovative installation concept, named by the current author as Nordic Wind, draws inspiration from the float-over method. The primary objective of this concept is to eliminate weather-dependent high lifts conducted from a floating vessel and reduce offshore installation time. This concept has the versatility to be employed for both fixed and floating offshore wind turbine installations.

3. The Installation Concept

3.1. Concept Overview

The design of the installation vessel allows for the transportation of at least three pre-assembled wind turbines. The actual number of turbines that can be transported will vary, based on their individual capacity and weight. When compared to the previously mentioned methods, this innovative concept will significantly reduce the need for offshore lifts and installations. As a result, it will lead to a decreased operational time and cost. To ensure a safe installation operation, the vessel is equipped with precisely designed complex mechanisms. The concept includes a motion compensation gripper positioned at the stern of the vessel, along with two lifting grippers and a skidding truss structure system. The connection between the vessel and the floating spar will be established using the motion compensation gripper. The primary purpose of the motion compensation gripper is to minimize the horizontal plane motions that occur between the vessel and the floating spar. The truss structure system is designed with the capability to move in both forward and backward directions. This allows for the selection and lifting of the turbine assembly using both the upper and lower lifting grippers. The design of the lifting grippers should enable them to securely hold turbine assemblies weighing a minimum of 1200 tons. Additionally, the lifting grippers will be designed with heave compensation capabilities to reduce the impact forces experienced during the mating operation. To secure the floating spar to the seabed, three mooring chain lines will be utilized. On the other hand, the installation vessel will rely on its own thrusters and employ a dynamic positioning system (DP). Figures 2 and 3 illustrate the key components of the installation concept. The stability of the vessel has been assessed under different loading conditions, considering both the transit and installation phases.

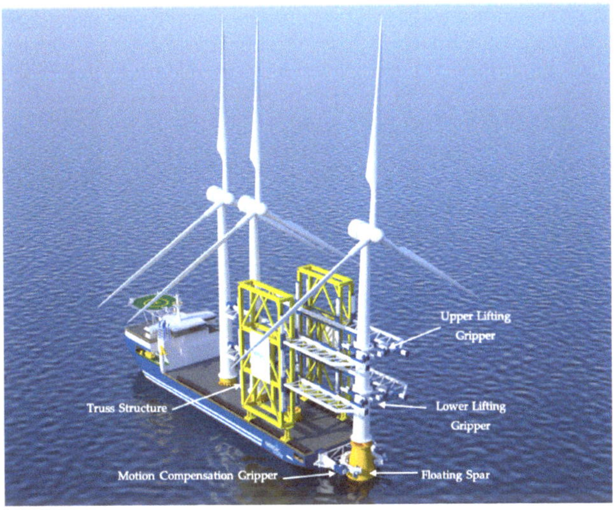

Figure 2. Nordic Wind concept's main components overview (artistic impression).

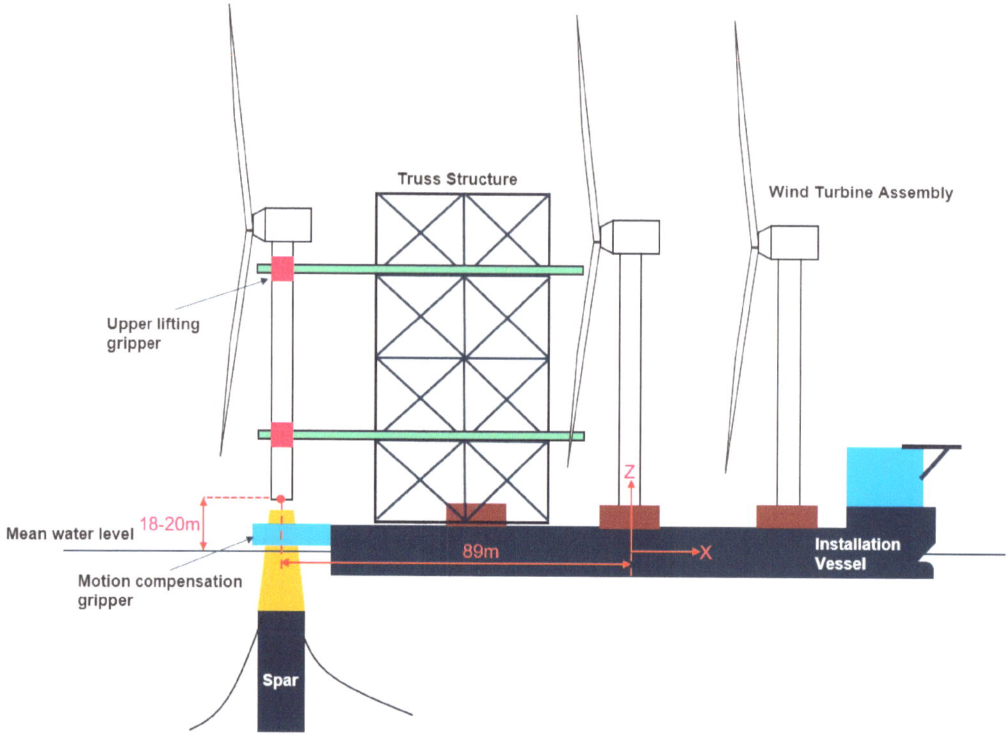

Figure 3. Illustration of installation concept's main components (not to scale).

3.2. Installation Procedure

The installation method consists of several key steps, which are as follows:

- Fully assembling the turbine onshore using a crane before the arrival of the installation vessel.
- Loading the turbine assembly onto the installation vessel.
- Transporting the turbine assembly to the installation site.
- Approaching the floating spar with the installation vessel and connecting it to the spar using the motion compensation gripper.
- Transferring the turbine assembly from the installation vessel to the floating spar.
- Mating the turbine assembly in place.

In Figure 4, the installation steps were illustrated starting from assembly, loadout, transportation, and, finally, the mating operation. The installation process will be started once the installation vessel connects the motion compensation gripper to the floating spar. The marine and operational crew will monitor the weather and the forecasted motion and response in order to allow for safe lifting and mating operations. The weather and vessel motions can be monitored by any decision support tool onboard, such as the Octopus system or any other system of the same type. This kind of system can accurately and clearly show how the incoming weather will impact the operation execution. The forecasted motions will be presented in a simple way to aid the crew of the vessel to make decisions based on the available data [31–33].

Onshore turbine assembly

Turbine assembly loadout to the installation vessel

Transportation to installation site

Connect floating spar to the installation vessel

Lifting and transfer the turbine assembly

Mating the turbine assembly

Figure 4. Nordic Wind concept's main installation steps.

Once the vertical relative motion is within the limits, the mating operation between the turbine and the spar can start. The mating process will be considered complete when the guiding pins located at the bottom of the tower successfully enter the docking device situated inside the floating spar. Once the mating operation is completed, the spar draft will be increased by 5 m–10 m, depending on the tower weight. Table 1 summarizes the installation vessel, floating spar, and turbine main particulars used during the dynamic study. The vessel hydrostatic stability has been checked during several stages of the operation and a fully loaded condition has been considered for this study.

Table 1. Installation vessel, floating spar, and turbine main parameters [20].

Installation Vessel		Floating Spar		Turbine	
Length overall	165 m	Length	92 m	Weight	1200 MT
Breadth	38 m	Beam WL	9.5 m	Dia_Bottom	7.5 m
Draft	7 m	Draught	76 m	Dia_Top	4.0 m
Displacement	30,500 MT	Displacement	12,600 MT	Height	115 m

This paper primarily focuses on the monitoring and positioning phases that precede the mating operations. The lifting grippers are not included in the modelling process, since they do not impact the relative motions in the horizontal plane or the forces exerted by the grippers. The design of the lifting grippers should prioritize the minimization of impact forces. These modelling features will be subject to future investigations. Furthermore, an analysis of the operational limits will be taken into consideration.

4. Dynamic Model

4.1. General

The primary objective of this study is to examine the hydrodynamic performance of the new installation concept during wind turbine installation. To achieve this, accurate and realistic modelling is essential in order to accurately quantify the dynamic responses during various installation phases and critical stages. This detailed analysis will facilitate improved operational planning. There are several key challenges to model the entire installation process, as explained in Figure 5.

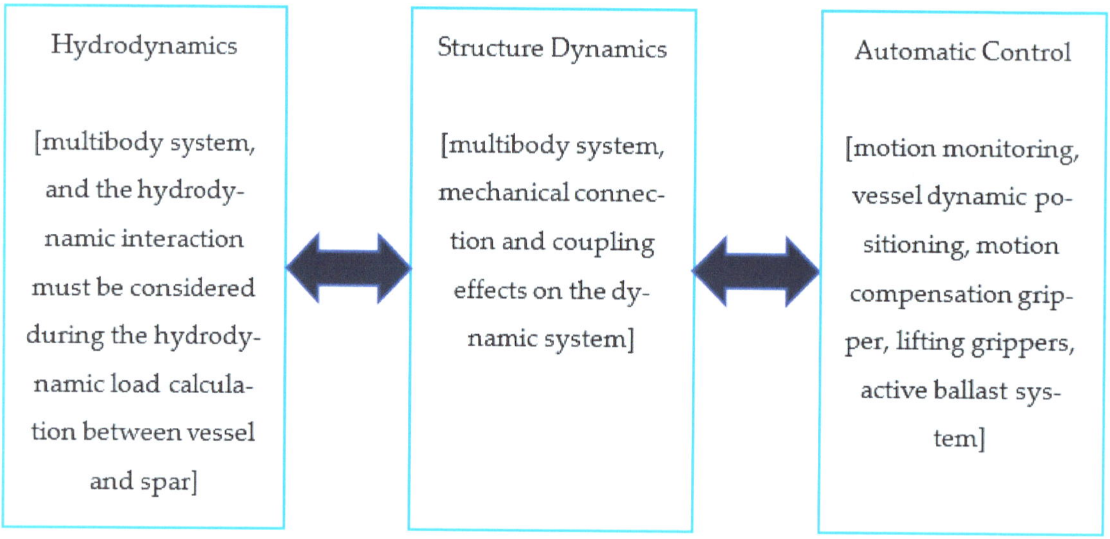

Figure 5. Key challenges in dynamic model (adopted from Jiang, [34]).

To overcome the above challenges, the dynamic model has been divided into several key stages in Figure 6, and are explained as follows:

- Loading condition 1: which corresponds to the monitoring stage, where structure responses are monitored prior to gripper connection.
- Loading condition 2: which corresponds to the connection stage between the floating spar and the installation vessel.
- Loading condition 3: which corresponds to the installation phase for lowering the turbine to the floating spar (mating operation).

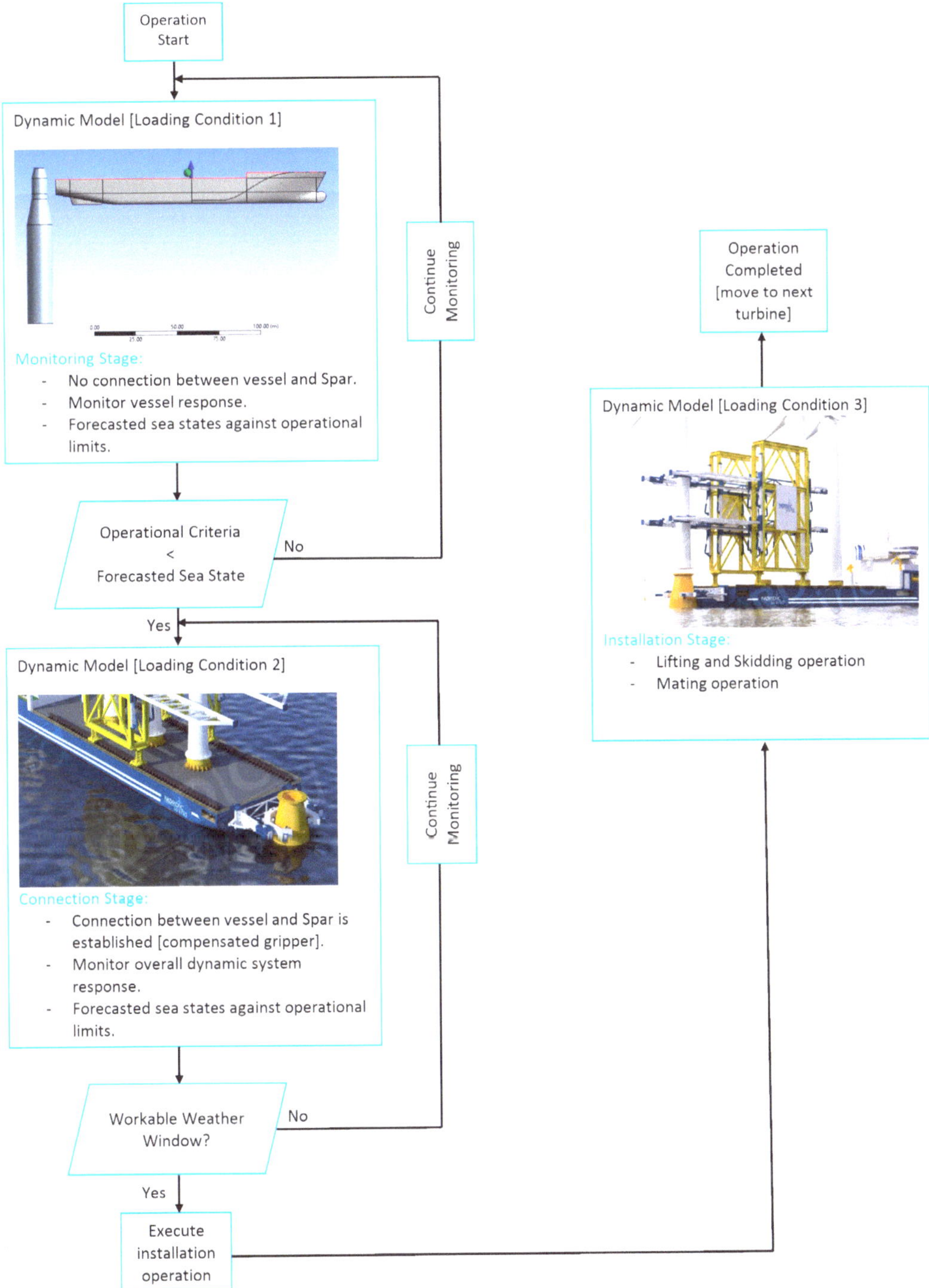

Figure 6. Dynamic system stages for the installation phase.

4.2. Numerical Model

The coupled dynamic system of the installation vessel and the floating spar was simulated and modelled using ANSYS AQWA software for numerical analysis. ANSYS AQWA offers a comprehensive set of tools specifically designed to analyze and assess the impact of environmental loads on various types of offshore and marine structures, including both floating and fixed structures [35].

The global coordinate system (GCS) employed in this context is a right-handed coordinate system. Its orientation is defined as follows: the X-axis points towards the bow, the Y-axis towards the port side, and the Z-axis points upwards. The origin of the GCS is situated at the mid-ship, center line, and still waterline position when the vessel is at rest. Each floating body exhibits six degrees of freedom. Assuming a linear and harmonic response, the coupled differential equations of motion for the floating bodies can be expressed in the following format:

$$\sum_{J=1}^{n}\left[-\omega^2\left(M_{ij}+A_{ij}\right)-i\omega B_{ij}+C_{ij}\right]\xi_j=F_i \quad (1)$$

where:

ω = Wave frequency.
M_{ij} = the generalized mass matrix for the floating bodies.
A_{ij} and B_{ij} = the added mass and potential damping matrix, respectively.
C_{ij} = the restoring force matrix.
ζ_j = the response motion in each of the six degrees of freedom for each floating body.
F_i = the complex amplitude of the wave exciting force acting on the floaters.

The utilization of the generalized mode concept proves to be effective in addressing multi-body hydrodynamic interaction problems. By employing this concept, a multi-body system consisting of NB body units can be represented by 6 multiplied by the number of bodies (6XNB) degrees of freedom, assuming that each body behaves as a rigid body. In the case of multi-bodies that are hydrodynamically and mechanically coupled, the equation mentioned above needs to be solved in a coupled matrix equation. In the model, the installation vessel and the spar foundation are represented as two rigid bodies that are interconnected through mechanical and hydrodynamic couplings at their interfaces. Both bodies experience wave-induced forces, hydrodynamic reaction forces, and mechanical coupling effects [36–38].

The equation of multi-body motion is given below:

$$(M+A(\infty))\ddot{x}+D_1\dot{x}+D_2f(\dot{x})+Kx+\int_0^t h(t-\tau)\dot{x}(\tau)d\tau=q(t,\ x,\ \dot{x}) \quad (2)$$

where:

M = the total mass matrix of the vessel and the spar.
x = the vector of rigid body motion with 12 degrees of freedom (DOFs).
A = the frequency-dependent added mass matrix.
D_1 = the linear damping matrix.
D_2 = the quadratic damping matrix.
K = the coupled stiffness matrix.
h = the coupled retardation function of the vessel and the spar, calculated based on the frequency-dependent added mass or potential damping.
q = the vector of external forces, which includes wave exciting forces, wave drift forces, or any other external force.

In this initial study, all wind turbine assemblies present on the installation vessel are connected rigidly. Mooring systems were separately incorporated for both the spar and the installation vessel.

4.3. Frequency Domain

In the present study, the added mass and potential damping were computed in the frequency domain and subsequently utilized in time domain simulations. These calculations were performed to analyze the coupled motion of two bodies through the implementation of retardation functions. The analysis also considers the viscous damping effects on both the vessel and spar. Figure 7 illustrates the hydrodynamic panel model employed for the installation vessel and the floating spar. The calculations incorporate first- and second-order hydrodynamics, based on the potential theory. Figure 8 depicts the Response Amplitude Operators (RAOs) for heave and pitch motions of both the installation vessel and the spar. These RAOs were evaluated under specific sea conditions, known as following sea conditions.

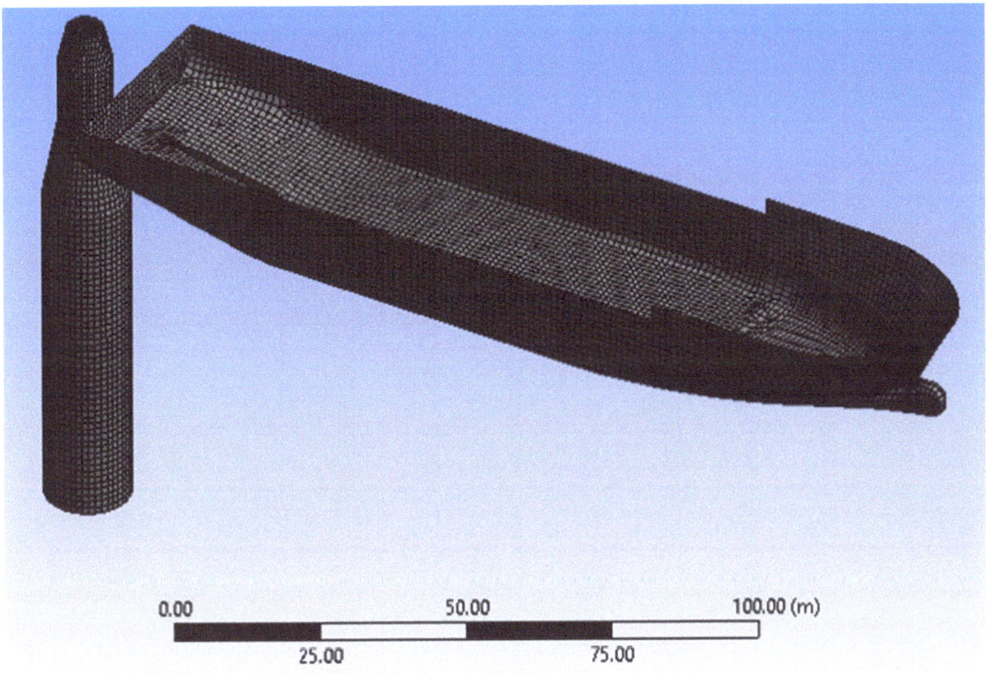

Figure 7. Hydrodynamic panel model for the vessel and the floating spar [20].

4.4. Wind Loads

Within ANSYS AQWA, wind loads can be simulated using various spectrum options or by specifying a constant velocity. This allows for flexible modelling of wind-induced loads in the analysis. The wind field in the analysis is assumed to propagate parallel to the horizontal plane. The fluctuating component of the wind velocity in the mean direction is described using the NPD spectrum, as defined in API RP2A. This spectrum provides a characterization of the varying wind velocity for the simulation. The average wind speed profile at a specific elevation z, denoted as U_z, can be approximated by the following equation:

$$U_z = U_{10} \times \left[1 + C\left(\ln\left(\frac{z}{10}\right)\right)\right] \quad (3)$$

where:

U_{10} = wind speed averaged over 1 h at reference elevation of 10 m.
$C = 0.0573 \times \sqrt{(1 + 0.15 \times U_{10})}$.

The wind spectrum is defined by the following equation:

$$S(f) = \frac{320 \times \left(\frac{U_{10}}{10}\right)^2 \times \left(\frac{z}{10}\right)^{0.45}}{\left(1 + f^{0.468}\right)^{3.561}} \tag{4}$$

where:

$S(f)$ = spectral energy density at frequency f [$(m/2)^2$: Hz].
f = frequency (Hz).

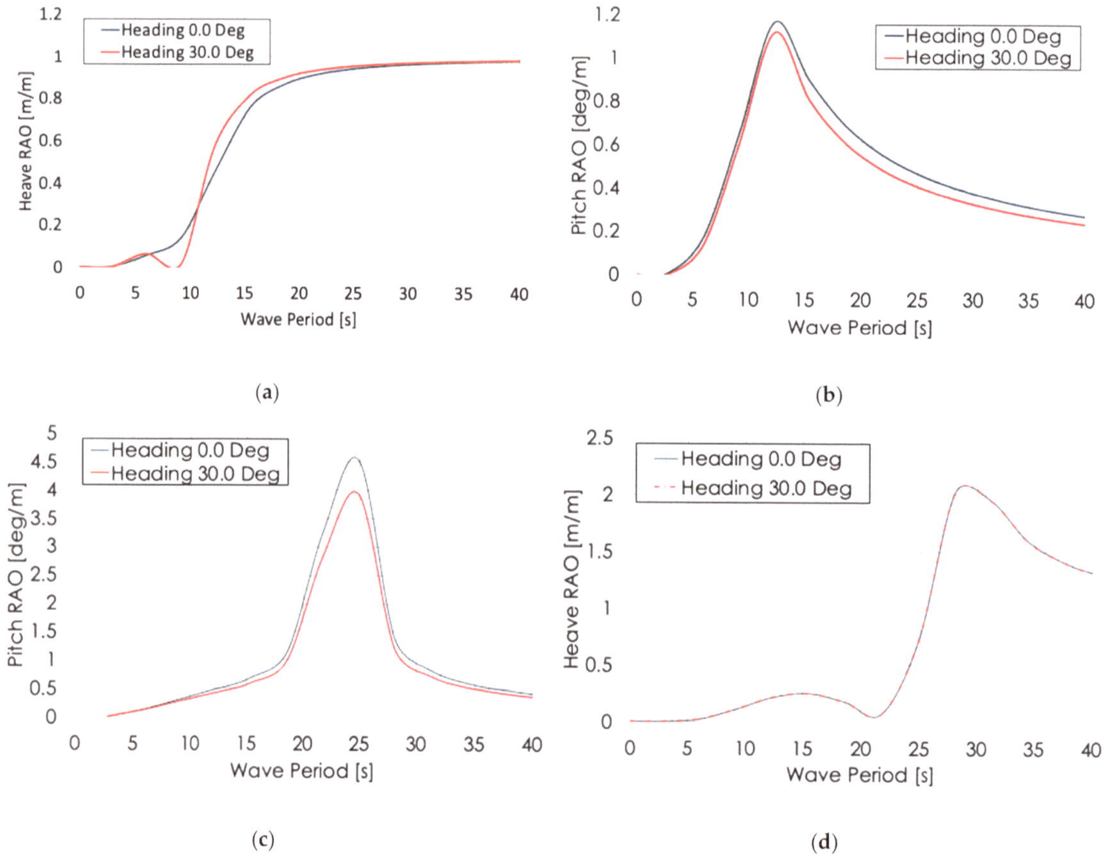

Figure 8. (**a**) Heave RAO for the installation vessel; (**b**) Pitch RAO for the installation vessel; (**c**) Heave RAO for the spar; (**d**) Pitch RAO for the spar.

In the present study, the wind forces generated by the wind turbines installed on the vessel deck were considered. The wind coefficients required for the calculations were determined using HAWC2 software, as detailed in Reference [34]. The wind coefficients for the installation vessel were obtained through model tests. Figure 9 illustrates the changes in force and moment coefficients, with respect to wind direction for the wind turbine. Specifically, two blade orientations were taken into consideration: blade pitch angles of 0 degrees and 90 degrees.

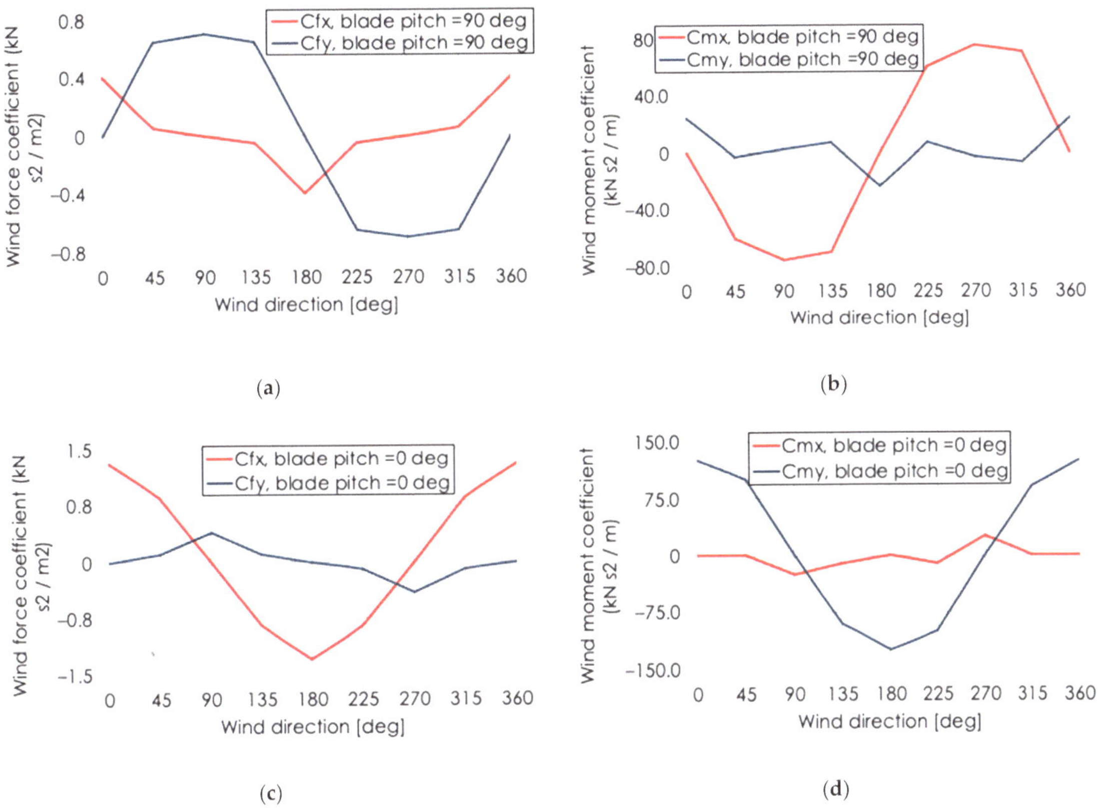

Figure 9. (**a**) Wind force coefficients for blade pitch 90 deg; (**b**) Wind moment coefficients for blade pitch 90 deg; (**c**) Wind force coefficients for blade pitch 0 deg; (**d**) Wind moment coefficients for blade pitch 0 deg.

4.5. Current Loads

In this study, the current was modeled using the power law method, where the current direction remains fixed and does not vary with depth. The current speed, denoted as S, varies with position (X, Y, Z), according to the following formula:

$$S = S_b + \left(S_f - S_b\right)\left[\frac{Z - Z_b}{Z_f - Z_b}\right]^{\frac{1}{p}} \quad (5)$$

where:

S_f and S_b are the current speeds at the surface and the seabed, respectively.
p is the power law exponent.
Z_f is the Z coordinate of the still water level.
Z_b is the Z coordinate of the seabed directly below (X, Y).

4.6. Mechanical Coupling

Understanding the effects of modelling mechanical couplings on the dynamic characteristics of a system is important. In this study, the mechanical coupling between the ship and the floating spar substructure was achieved using a motion compensation gripper. In the dynamic model, the motion compensation gripper is simplified using a mechanical joint. This mechanical joint enables independent rotation of both bodies, allowing them to move freely, with respect to each other. However, it should be noted that, despite the independent rotation allowed by the mechanical joint, the two bodies will remain rigidly connected in the longitudinal and transverse directions. The primary objective of the motion compensation gripper frame is to facilitate the installation process by compensating for the relative motions in the horizontal plane between the floating spar and the installation vessel [39,40]. By mitigating these relative motions, the motion compensation gripper frame helps to ensure smoother and more efficient installation operations. Figure 10 shows the mechanical joint between the installation vessel and the floating spar, as modelled in ANSYS AQWA.

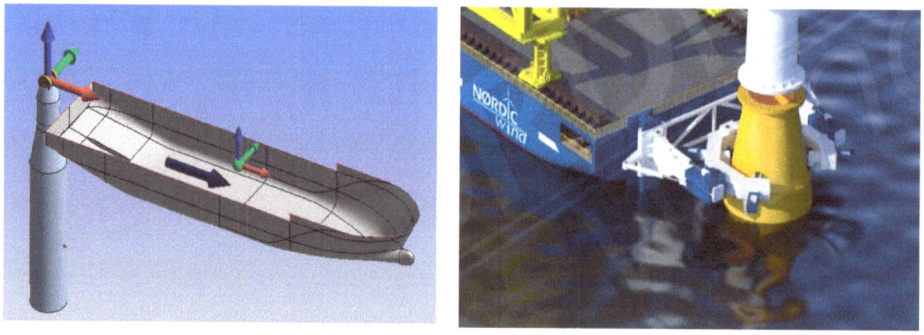

Figure 10. Mechanical connection model (ANSYS AQWA—**Left**), motion compensation gripper (artist impression—**Right**), [20].

4.7. Mooring System

The mooring system for the spar is composed of three separate lines that are connected to the substructure through bridles. The pretension and configuration of the bridles provide sufficient yaw stiffness, which helps prevent slack in the mooring lines during normal operation. This ensures the stability and proper functioning of the spar structure. Figure 11 illustrates the arrangement of the catenary mooring lines for the spar-type floating wind turbine. For further information on the specific characteristics of the mooring lines, please refer to Table 2.

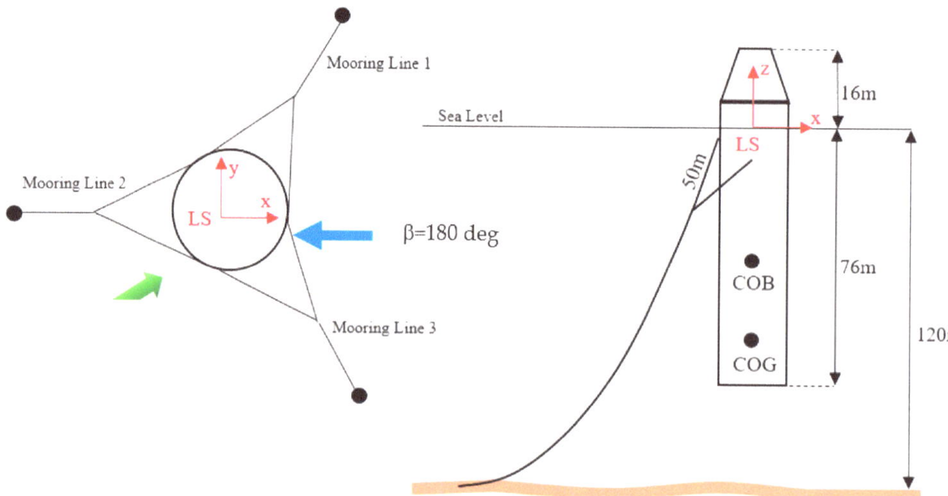

Figure 11. Mooring system layout.

Table 2. Mooring lines' properties, [20].

Parameter	Spar	Unit
Type	Chain grade R4S	-
Mooring line length	644	m
Diameter	0.23	m
Mass/unit length	315.36	kg/m
Pre-tension	1167	kN
Stiffness EA	1.45×10^6	kN
Maximum tension	13,570	kN

4.8. Hydrodynamic Interaction

The computation of linearized hydrodynamic wave-loading on floating bodies is performed using AQWA-LINE, which utilizes three-dimensional radiation/diffraction theory. The concept of fluid interactive loading between bodies is commonly discussed in the context of radiation/diffraction theory. Radiation forces, in essence, pertain to the impact of one body's flow field on another. The significance of such interaction depends on factors such as the separation distance between the bodies and the relative sizes of the bodies involved. During the connecting phase between the installation vessel and the floating spar, the importance of this interaction becomes apparent as the two bodies come into close proximity. In addition to the radiation coupling, the hydrodynamic interaction also includes the shielding effect, which refers to the change in the flow field around a body due to the presence of another nearby body. This effect can have a significant impact on the wave loads and motions of the interacting bodies [41]. A series of time domain simulations were initially conducted without incorporating the mechanical gripper connection between the installation vessel and the spar. These simulations aimed to assess the dynamic behavior and response of the system under various operating conditions and environmental loads. By conducting these simulations, it was possible to establish a baseline understanding of the system's behavior before introducing the mechanical gripper connection. Moreover, it was feasible to assess how the hydrodynamic interaction between the two bodies affected the resultant movements of the motion reference point (MRP).

5. Results and Discussion

5.1. Simulations Framework

The primary goal of the study is to assess the performance of the installation concept under varying wind, wave, and current conditions. To achieve this, multiple time domain simulations were conducted, considering different environmental conditions specific to the Hywind Tampen location. The selected environmental conditions for the study are provided in Table 3. Here, H_S is the significant wave height. T_P is the wave peak period. V_W is the 1-h average wind speed at 10 m height. β is the wave heading. α is the blade pitch angle. V_C is the current speed. For simulating irregular waves, the JONSWAP spectrum, with a peak factor γ of 3.3, was utilized [42,43]. In the time domain analysis, each combination of environmental conditions was simulated with intervals of 0.1 s, allowing for a detailed assessment of the system's behavior. The total duration of each simulation for every environmental condition combination was set to 10,800 s. It is worth noting that, in all simulation cases, the wave, wind, and current were assumed to be collinear, ensuring a consistent alignment of these factors throughout the analyses.

Table 3. Environmental conditions.

Condition	H_S [m]	T_P [s]	Β [deg]	V_W [m/s]	V_C [m/s]	α [deg]
1	1.5	6, 7, 8, 9, 10, 11, 12	0, 30, 150, 180, 210, 330	7.0	0.5	0, 90
2	2.0			9.0		
3	2.5			12.0		

For time-domain simulations, some reference points are considered to give insight into vessel and spar motions. The following points are considered and shown in Figure 12:

- Vessel center of gravity (Vessel COG)
- Spar center of gravity (Spar COG)
- Motion reference point (MRP)

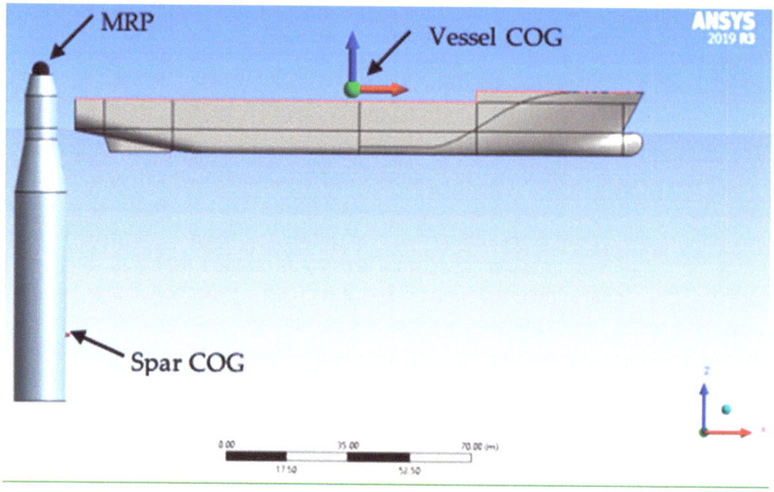

Figure 12. Reference points for time domain analysis.

The motion reference point (MRP) is located at the top of the spar. The mean value for the vertical motion at the MRP is 16.0 m from the water level.

5.2. Multibody Interactions Effects

As can be seen in Figure 13, the hydrodynamic interaction effects are variable and can be seen over all wave peak periods. When considering a wave period of 9 s, as an example, the wave surface elevation at the spar is less when considering multibody interaction (vessel and spar) and spar only, as shown in Figures 14 and 15. Figure 15 show the maximum and the minimum relative motions at MRP, with and without multibody interaction, for various wave peak periods ranging from 6–12 s, with a maximum significant wave height of 2.5 m and wave heading of 180°. This explains the reduction in the longitudinal and vertical motions at the motion reference point (MRP).

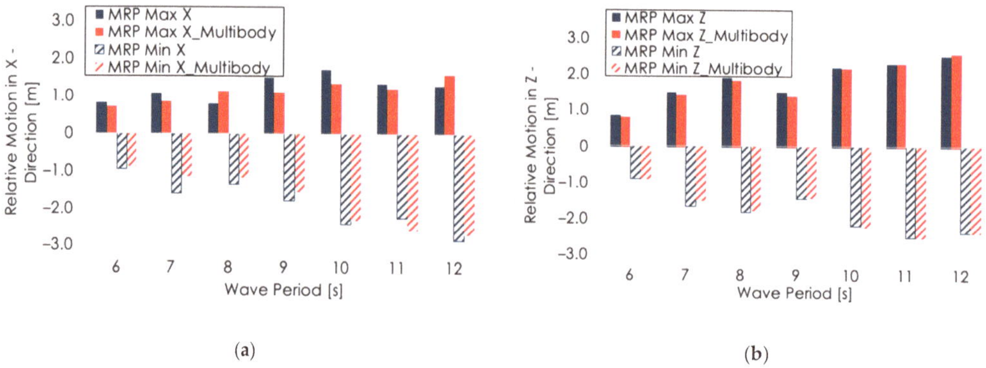

Figure 13. (**a**) Minimum and maximum longitudinal motion at MRP, with and without multibody interaction, for Hs 2.5 m and wave heading 180°—LC1; (**b**) Minimum and maximum vertical motion at MRP, with and without multibody interaction, for Hs 2.5 m and wave heading 180°—LC1.

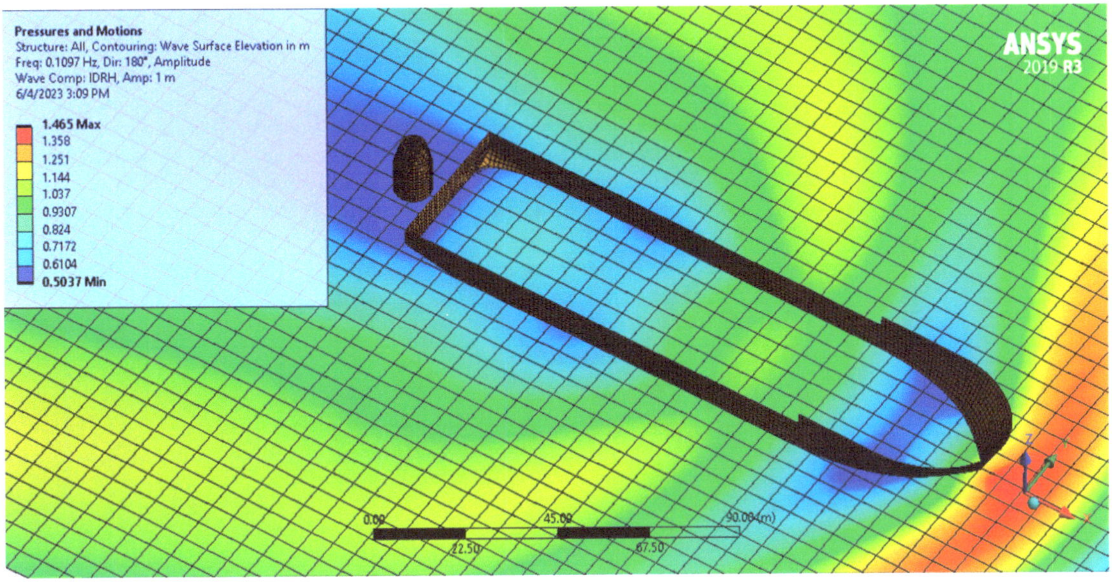

Figure 14. Wave surface elevation including multibody interactions, modelled in AQWA.

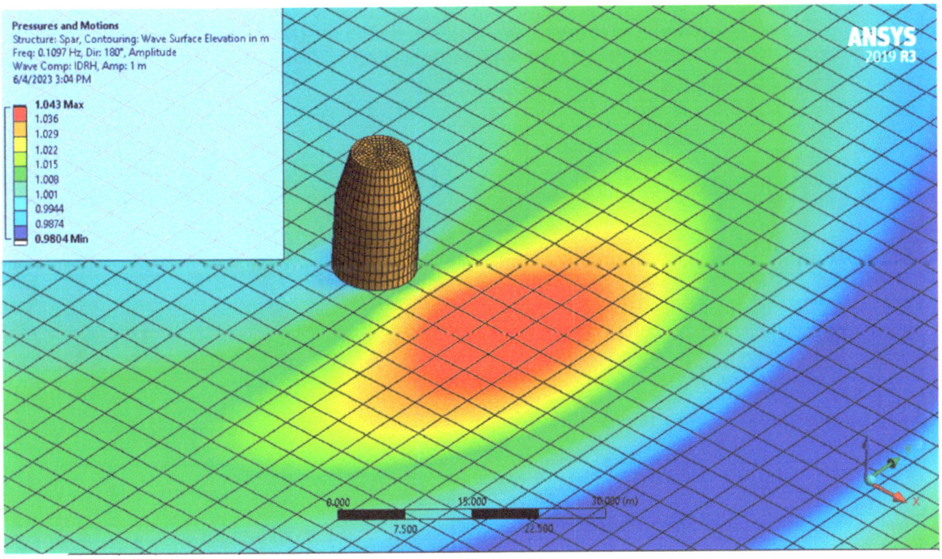

Figure 15. Wave surface elevation spar only, modelled in AQWA.

Figure 13 illustrates a comparison between the maximum and minimum relative motions at the motion reference point (MRP) for both the installation vessel and the spar. The mechanical gripper will function to compensate these motions to facilitate and ease the turbine mating operation. In addition to compensating for the relative motions between the installation vessel and the spar during the connection phase, it is also important to address the heave motion of the installation vessel during the lifting and mating phase of the wind turbine assembly. This can be achieved by implementing a control system for the lifting grippers that compensates for the heave motion of the installation vessel, ensuring a safe and efficient mating operation. Since the lifting grippers do not affect the relative motions in the horizontal plane, they were not modelled in the simulations evaluating the performance of the motion compensation gripper.

5.3. Gripper Forces

Figure 16 shows the gripper forces in the x and y directions. The forces are presented for Hs 2.5 m and wave heading of 180°. As can be seen, the multibody interaction has more effect on reducing gripper force in the x direction compared to the y direction. Figure 17 shows the standard deviation of gripper forces in x and y directions, respectively.

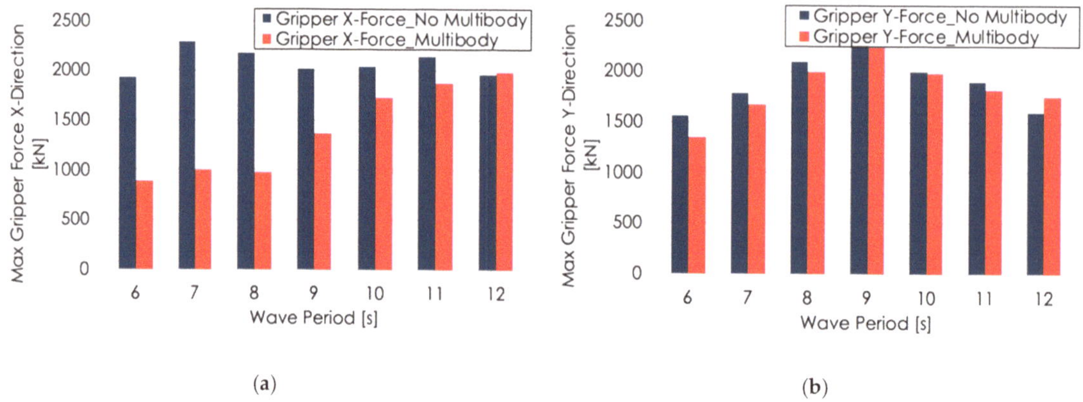

Figure 16. (**a**) Maximum gripper forces at the x direction, with and without multibody interaction; (**b**) Maximum gripper forces at the y-direction, with and without multibody interaction.

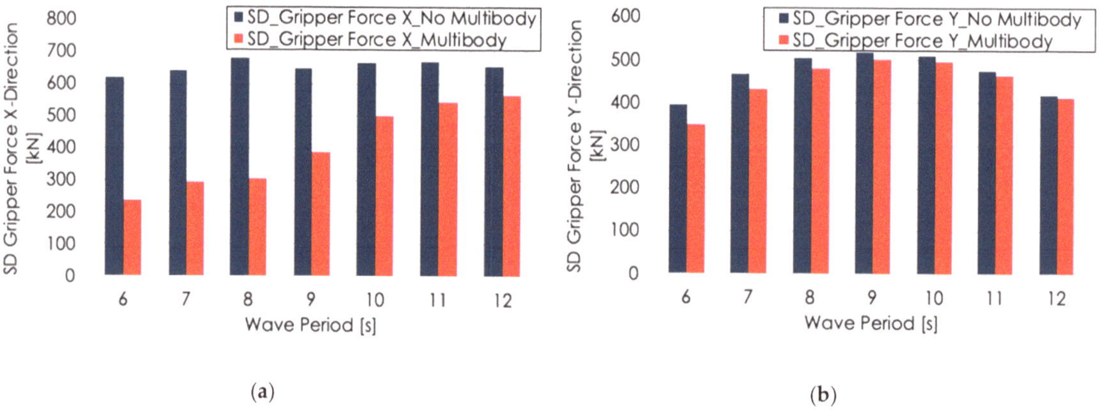

Figure 17. (**a**) Standard deviation gripper forces at the x direction, with and without multibody interaction; (**b**) Standard deviation gripper forces at the y direction, with and without multibody interaction.

Figure 18 shows results for the gripper forces in the x and y directions for different wave headings. The results show that the force in the x direction is more sensitive to the wave heading compared to the force in the y direction. Figure 19 presents the results of the vertical relative motion at MRP for different wave headings and significant wave heights. Figure 20 shows the results of the gripper forces for different significant wave heights. The results presented in Figures 16–20 shall be used together with the onboard weather forecasted decision tool to further optimize the mating operation.

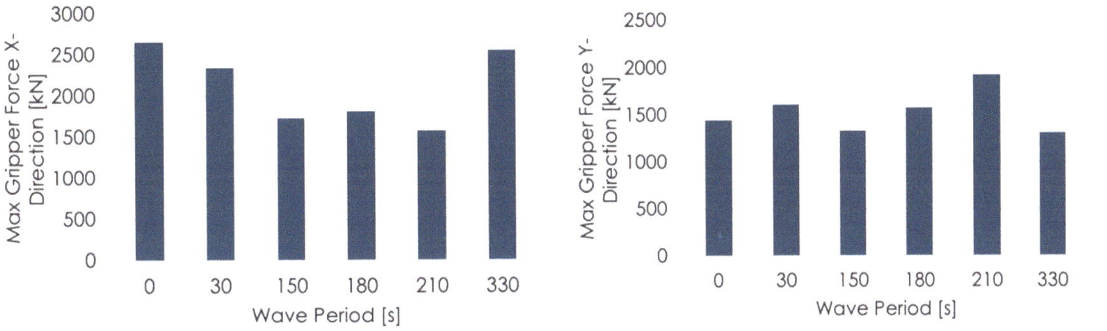

(a) (b)

Figure 18. (**a**) Maximum gripper forces at the x direction for different wave headings; (**b**) Maximum gripper forces at y direction for different wave headings (Hs 2.5 m and Tp 12 s).

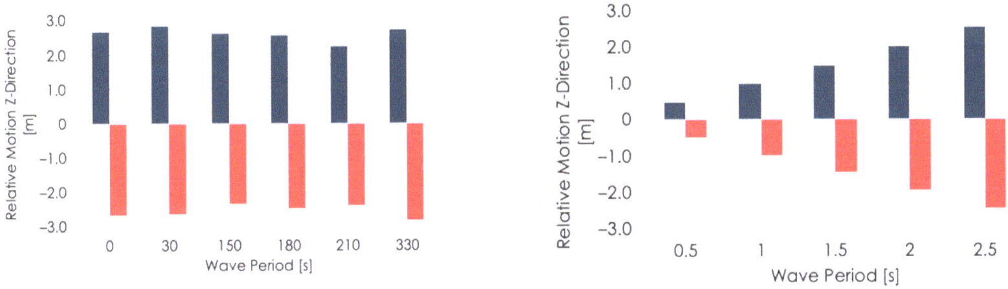

(a) (b)

Figure 19. (**a**) Minimum and maximum Z-relative motion at MRP (Hs 2.5 m and TP 12 s); (**b**) Minimum and maximum relative motion for different Hs and Tp 12 s.

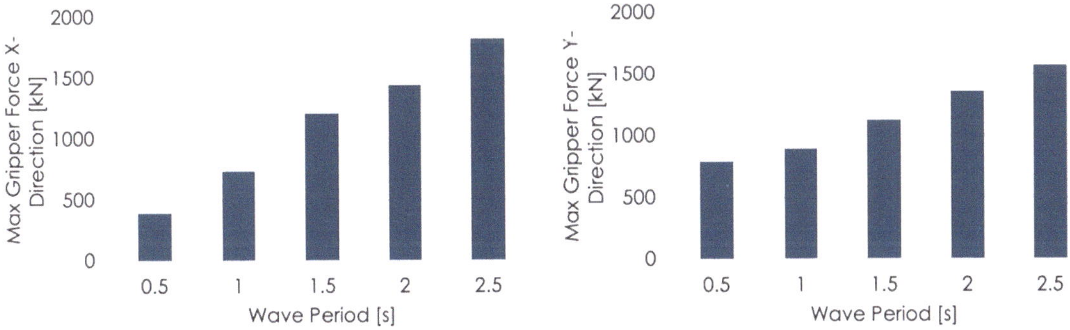

(a) (b)

Figure 20. (**a**) Maximum gripper forces at x direction for different wave headings; (**b**) Maximum gripper forces at y direction for different wave headings (Heading 180 deg and Tp 12 s).

Figure 21 shows the effect of the wind and current on the gripper forces. Two blade pitch angles, 0 and 90 degrees, respectively, were considered. As can be seen, an increase in the gripper forces in the x and y directions were observed. The wind and current has more effect on the gripper force in the x direction compared to the gripper force in the y direction. Moreover, the effect of wind on the gripper forces is higher when the blade pitch angle is 0 degrees, compared to a blade pitch angle of 90 degrees.

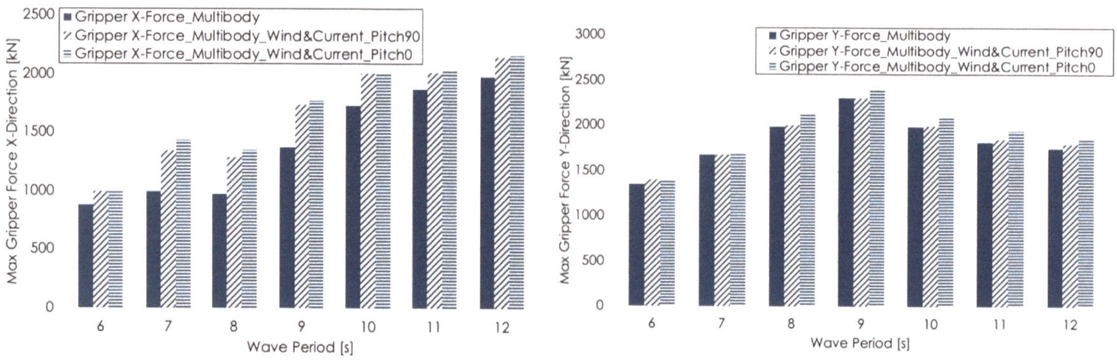

(**a**) (**b**)

Figure 21. (**a**) Maximum gripper forces at x direction, including wind and current for blade pitch angles of 0 and 90 degrees, respectively; (**b**) Maximum gripper forces at y direction, including wind and current for blade pitch angles of 0 and 90 degrees, respectively.

5.4. Mooring Line Tension

Considering the mooring layout and wave headings illustrated in Figure 11, it is anticipated that lines 1 and 3 will endure higher loads compared to line 2. The top tension in line 1 at β = 180 deg, Hs of 2.5 m, and Tp of 12 s, is studied here. Figure 22 presents the dynamic top tension for three conditions: wave only; wave, wind and current for blade pitch angle of 0 degrees; and wave, wind, and current for blade pitch angle of 90 degrees. As can be seen in Figure 22, the mooring line tension has increased due to wind and current effects. Furthermore, it is worth noting that the mooring line tension experiences a significant increase when the blade pitch angle is set to 0 degrees, as compared to when it is set to 90 degrees.

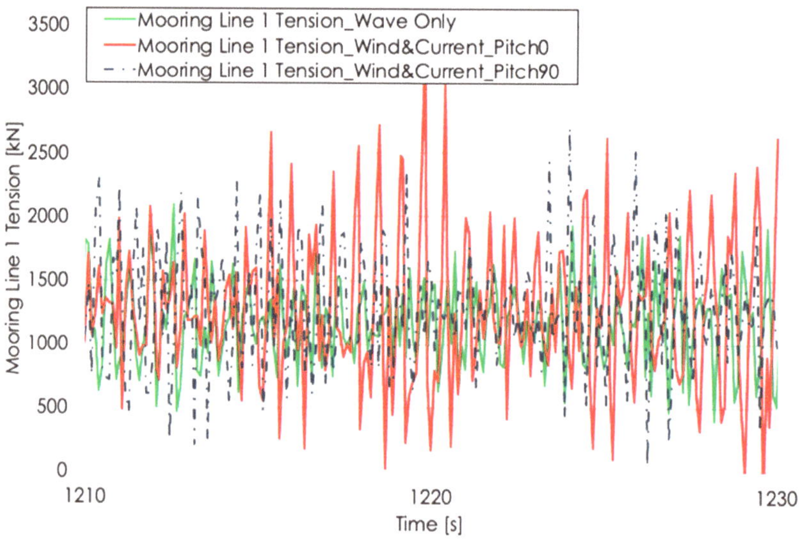

Figure 22. Time history of mooring line 1 dynamic tension, Hs = 2.5 m, Tp = 12 s, and β = 180 deg.

6. Conclusions

The main findings of this study can be summarized as follows:

- Offshore wind turbine installation concept: The study proposes a new method for installing offshore wind turbines using a floating vessel. The concept involves the use of a numerical model that includes a floating vessel, a spar foundation, a mechanical gripper, and mooring lines.
- Monitoring and positioning phases: The study focuses on the monitoring and positioning phases that precede the mating of the wind turbine assembly. These phases are crucial for ensuring a successful installation process.
- Multibody interactions: The study explores the impact of multibody interactions on the installation process. It demonstrates that these interactions have a substantial effect on reducing relative motions and gripper forces within a specific range of wave periods (6–9 s). However, their influence becomes negligible as the wave period increases.
- Relative motion: The relative motion between the wind turbine and the spar at the mating point is identified as a critical factor in the success of the mating operation. The study determines that this relative motion is primarily influenced by the first-order motions and is particularly sensitive to the significant wave height and wave peak period. However, it is found to be less affected by wind and current conditions.
- Blade orientations: Two blade orientations, with blade pitches of 90 degrees and 0 degrees, respectively, are considered in the study. It is observed that the choice of blade pitch has an impact on low-frequency motions and mooring line tensions. Specifically, the 0-degree blade pitch leads to a greater increase in the tension forces compared to the 90-degree blade pitch. Gripper forces, on the other hand, show less sensitivity to blade orientation angles.
- Gripper forces and hydraulic cylinder capacity: The study finds that the resulting gripper forces are within reasonable limits for compensation. These forces can be utilized to determine the appropriate hydraulic cylinder capacity for relative motion compensation during the installation process.
- Vessel heading optimization: The study highlights the importance of vessel heading optimization in minimizing relative motions and gripper forces between the installation vessel and the floating spar. For example, the gripper forces in the x direction are more sensitive to wave heading compared to the y direction. This optimization can facilitate the mating operation and reduce the impact force. Decision support tools, such as the Octopus system, are suggested for monitoring weather and vessel motions, providing accurate information on how incoming weather conditions may affect the operation execution. These tools can guide offshore crew in the optimization process.

Author Contributions: M.H.: Methodology, formal analysis; visualization, write—original draft. C.G.S.: write—review; Supervision. All authors have read and agreed to the published version of the manuscript.

Funding: This work contributes to the strategic research plan of the Centre for Marine Technology and Ocean Engineering (CENTEC), which is financed by the Portuguese Foundation for Science and Technology (Fundação para a Ciência e Tecnologia—FCT) under contract UIDB/UIDP/00134/2020.

Institutional Review Board Statement: Not applicable.

Informed Consent Statement: Not applicable.

Data Availability Statement: Not applicable.

Acknowledgments: The authors are grateful to FCT for funding.

Conflicts of Interest: The authors declare no conflict of interest.

Nomenclature, Abbreviations, and Symbols

ANSYS AQWA	Dynamic Analysis Software
α	Blade pitch angle
COG	Center of gravity
DP	Dynamic Positioning
HAWC2	Horizontal Axis Wind turbine simulation Code 2nd generation
GCS	Global coordinate system
MRP	Motion Reference Point
NB	Number of bodies
NPD	Norwegian Petroleum Directorate
OWT	Offshore wind turbine
Vc	Current speed
Octopus	Offshore motion forecasting and decision support tool
BFOW	Bottom Fixed Offshore Wind
β	Wave heading
Dia	Diameter
FOW	Floating Offshore Wind
Hs	Significant wave height
γ	Peak factor
MW	Megawatt
Nordic Wind	Current installation method name
RAOs	Response amplitude operators
Tp	Wave peak period
Vw	Wind Speed
WL	Waterline

References

1. Diaz, H.M.; Serna, J.; Nieto, J.; Guedes Soares, C. Market needs, opportunities, and barriers for the floating wind industry. *J. Mar. Sci. Eng.* **2022**, *10*, 934. [CrossRef]
2. Diaz, H.M.; Guedes Soares, C. Review of the current status, technology and future trends of offshore wind farms. *Ocean Eng.* **2020**, *209*, 107381. [CrossRef]
3. Offshore Wind in Europe, Key Trends, and Statistics. 2021. Available online: https://windeurope.org/ (accessed on 10 April 2022).
4. Uzunoglu, U.; Karmakar, D.; Guedes Soares, C. Floating Offshore Wind Platforms. In *Floating Offshore Wind Farms*; Castro-Santos, L., Diaz-Casas, V., Eds.; Springer International Publishing: Cham, Switzerland, 2016; pp. 53–76.
5. Guedes Soares, C.; Bhattacharjee, J.; Karmakar, D. Overview and prospects for development of wave and offshore wind energy. *Brodogradnja* **2014**, *65*, 87–109.
6. Bagbanci, H.; Karmakar, D.; Guedes Soares, C. Comparison of spar and semisubmersible floater concepts of offshore wind turbines using long-term analysis. *J. Offshore Mech. Arct. Eng.* **2015**, *137*, 061601. [CrossRef]
7. Sclavounos, P.; Lee, S.; DiPietro, J.; Potenza, G.; De Michele, G. Floating offshore wind turbines: Tension leg platform and taught leg buoy concepts supporting 3–5 MW wind turbines. In Proceedings of the European Wind Energy Conference EWEC, Warsaw, Poland, 20–23 April 2010.
8. DNV GL Launches Revised Design Standards and New Certification Guideline for Floating Wind Turbines. Available online: https://www.DNV.com (accessed on 20 October 2022).
9. Available online: https://www.saipem.com/en/projects/hywind (accessed on 23 February 2023).
10. Gao, Z.; Verma, A. A summary of the recent work at NTNU on marine operations related to installation of offshore wind turbines In Proceedings of the ASME 2018 37th International Conference on Ocean, Offshore and Arctic Engineering OMAE 2018, Madrid, Spain, 17–22 June 2018.
11. James, R.; Costa Ros, M. *Floating Offshore Wind: Market and Technology Review*; Carbon Trust: London, UK, 2015; pp. 79–93.
12. Herman, S.A. *Offshore Wind Farms—Analysis of Transport and Installation Costs*; Report No. ECN-I-02-002; Energy Research Centre of The Netherlands: Petten, The Netherlands, 2002.
13. Castro-Santos, L.; Silva, D.; Bento, A.R.; Salvacao, N.; Guedes Soares, C. Economic feasibility of floating offshore wind farms in Portugal. *Ocean Eng.* **2020**, *207*, 107393. [CrossRef]
14. Mone, C.; Stehly, T. Cost of Wind Energy Review Report. 2014. Available online: http://www.nrel.gov (accessed on 3 March 2023).
15. Verma, A.S.; Jiang, Z.; Vedvik, N.P.; Gao, Z.; Ren, Z. Impact assessment of a wind turbine blade root during an offshore mating process. *Eng. Struct.* **2019**, *180*, 205–222. [CrossRef]
16. Numerical assessment of wind turbine blade damage due to contact/impact with tower during installation. In *IOP Conference Series: Materials Science and Engineering*; IOP Publishing: Bristol, UK, 2017; p. 012025.

17. Zhao, Y.; Cheng, Z.; Sandvik, P.C.; Gao, Z.; Moan, T. An integrated dynamic analysis method for simulating installation of single blades for wind turbines. *Ocean Eng.* **2018**, *152*, 72–88. [CrossRef]
18. Butterfield, S.; Musial, W.; Jonkman, J.; Sclavounos, P. *Engineering Challenges for Floating Offshore Wind Turbines*; Technology Report; National Renewable Energy Lab (NREL): Golden, CO, USA, 2007.
19. Zhao, Y.; Cheng, Z.; Sandvik, P.C.; Gao, Z.; Moan, T.; Van Buren, E. Numerical modelling and analysis of the dynamic motion response of an offshore wind turbine blade during installation by a jack-up crane vessel. *Ocean Eng.* **2018**, *165*, 353–364. [CrossRef]
20. Hassan, M.A.A.A.; Guedes Soares, C. Installation of pre-Assembled offshore floating wind turbine using a floating vessel. In *Developments in Renewable Energies Offshore*; Guedes Soares, C., Ed.; Taylor and Francis: London, UK, 2021; pp. 461–468.
21. Hassan, M.A.A.A.; Guedes Soares, C. Multibody dynamics of floaters during installation of a spar floating wind turbine. In *Maritime Transportation and Harvesting of Sea Resources*; Guedes Soares, C., Teixeira, A.P., Eds.; Taylor & Francis Group: London, UK, 2018; pp. 371–380.
22. Hassan, M.A.A.A.; Guedes Soares, C. Analysis of vessel shielding effects during installation of spar floating wind turbine. In *Advances in Renewable Energies Offshore*; Guedes Soares, C., Ed.; Taylor & Francis: London, UK, 2019; pp. 693–702.
23. Acero, W.G.; Li, L.; Gao, Z.; Moan, T. Methodology for assessment of the operational limits and operability of marine operations. *Ocean Eng.* **2016**, *125*, 308–327. [CrossRef]
24. Ahn, D.; Shin, S.C.; Kim, H.C. Comparative evaluation of different offshore wind turbine installation vessels for Korean west south wind farm. *Int. J. Nav. Archit. Ocean Eng.* **2017**, *9*, 45–54. [CrossRef]
25. Collu, M.; Maggi, A.; Brennan, F. Stability Requirements for Floating Offshore Wind Turbine (FOWT) during Assembly and Temporary Phases: Overview and Application. *Ocean Eng.* **2014**, *84*, 164–175. [CrossRef]
26. Wang, S.; Li, X. Dynamic analysis of three adjacent bodies in twin-barge float over installation. *Ocean Syst. Eng.* **2014**, *4*, 39. [CrossRef]
27. Geba, K.; Welaya, Y.; Leheta, H.; Abdel-Nasser, Y. Motion, structural analysis of floaters float-over during the mating operation. In Proceedings of the Twenty-Fourth International Ocean and Polar Engineering Conference (ISOPE), Busan, Republic of Korea, 15–20 June 2014.
28. Seij, M.; Grook, J. State of the Art in Float-Overs, OTC-19072. In Proceedings of the Offshore Technology Conference, Houston, TX, USA, 30 April–3 May 2007.
29. Tribout, C.; Emery, D.; Weber, P.; Kaper, R. Float-Overs Offshore West Africa, OTC-19073. In Proceedings of the Offshore Technology Conference, Houston, TX, USA, 30 April–3 May 2007.
30. Wang, A.; Jiang, X.; Yu, C.; Zhu, S.; Li, H.; Wei, Y. Latest progress in float-over technologies for offshore installations and decommissioning. In Proceedings of the International Offshore and Polar Engineering Conference, Beijing, China, 20–25 June 2010.
31. Martins, D.; Muraleedharan, G.; Guedes Soares, C. Weather window analysis of a site off Portugal. In *Maritime Technology and Engineering*; Guedes Soares, C., Santos, T.A., Eds.; Taylor & Francis Group: London, UK, 2015; pp. 1329–1338.
32. Perera, L.P.; Rodrigues, J.M.; Pascoal, R.; Guedes Soares, C. Development of an onboard decision support system for ship navigation under rough weather conditions. In *Sustainable Maritime Transportation and Exploitation of Sea Resources*; Rizzuto, E., Guedes Soares, C., Eds., Taylor and Francis Group: London, UK, 2012; pp. 837–844.
33. Octopus. *The Octopus Reference Manual—Version 8.0*; ABB: Ede, The Netherlands, 2018.
34. Jiang, Z.; Gao, Z. Dynamic response analysis of a catamaran installation vessel during the positioning of a wind turbine assembly onto a spar foundation. *Mar. Struct.* **2018**, *61*, 1–24. [CrossRef]
35. ANSYS. *The AQWA Reference Manual—Version 19.1*; ANSYS: Canonsburg, PA, USA, 2019.
36. Faltinsen, O.M. *Sea Loads on Ships and Ocean Structures*; Cambridge University Press: Cambridge, UK, 1990.
37. Chakrabarti, S.K. *Hydrodynamics of Offshore Structures*; WIT Press: Torbay, UK, 1987.
38. Journée, J.M.J.; Massie, W.W. *Offshore Hydromechanics Book*, 1st ed.; Delft University of Technology: Delft, The Netherlands, 2001.
39. Hatledal, L.; Zhang, H.; Hilder, H. Numerical simulation of novel gripper mechanism between catamaran and turbine foundation for offshore wind turbine installation. In Proceedings of the 36th International Conference on Ocean, Offshore and Arctic Engineering, Trondheim, Norway, 25–30 June 2017. ASME paper OMAE2017–62342.
40. Jin, J.; Jiang, Z.; Gao, Z. Installation of pre-assembled offshore wind turbines using a catamaran vessel and an active gripper motion control method. In *Grand Renewable Energy Conference*; Japan Council for Renewable Energy: Tokyo, Japan, 2018.
41. Lee, D.H.; Choi, H.S. The motion behavior of shuttle tanker connected to a turret-moored FPSO. In Proceedings of the Third International Conference on Hydrodynamics, Seoul, Republic of Korea, 12–15 October 1998; pp. 173–178.
42. Veritas, D.N. *Recommended Practice DNV-RP-H103, Modelling and Analysis of Marine Operations*; DNV: Bærum, Norway, 2014.
43. Veritas, D.N. *Recommended Practice DNV-RP-C205, Environmental Conditions and Environmental Loads*; DNV: Bærum, Norway, 2014.

Journal of
Marine Science and Engineering

MDPI

Article

Mooring System Transport and Installation Logistics for a Floating Offshore Wind Farm in Lannion, France

Jorge Altuzarra [1,*], Alberto Herrera [2], Onintze Matías [1], Joaquín Urbano [3], Cristina Romero [3], Shan Wang [4] and C. Guedes Soares [4,*]

1 Vicinay Marine Innovación, Plaza Ibaiondo 1, 1010-106, 48940 Leioa, Spain
2 Core Marine Solutions S.L., Plaza Ibaiondo 1, 107 Mod2, 48940 Leioa, Spain
3 Esteyco, Avda. Burgos, 12B-Bajo, 28036 Madrid, Spain
4 Centre for Marine Technology and Ocean Engineering (CENTEC), Instituto Superior Técnico, Universidade de Lisboa,1049-001 Lisboa, Portugal
* Correspondence: jaltuzarra@vicinayinnovacion.com (J.A.); c.guedes.soares@centec.tecnico.ulisboa.pt (C.G.S.)

Abstract: This study addresses the planning procedures for the installation of the mooring systems that support the floating offshore wind turbines in a wind farm. It considers the logistics of the installation process and discusses the important role of the weather windows in the planning of those operations at a preliminary stage of the project. The case study is based on a wind farm array of 47 Telwind floating wind turbine platforms, to be located in Lannion (France), with a potential of 470 MW. The study includes the transport and logistics requirements of different mooring components, such as chains, connectors and drag anchors; the description of the installation operations considering the typology of vessels that are necessary in these manoeuvres; as well as the planning and costs associated with the transport and installation. Given the diversity of elements and operations involved in the installation procedure, it is demonstrated that the research results of duration and costs of this type of operations are only possible to obtain using a simulation tool.

Keywords: floating offshore wind turbine; mooring; installation; transport; operation; hook-up; weather window

Citation: Altuzarra, J.; Herrera, A.; Matías, O.; Urbano, J.; Romero, C.; Wang, S.; Guedes Soares, C. Mooring System Transport and Installation Logistics for a Floating Offshore Wind Farm in Lannion, France. *J. Mar. Sci. Eng.* **2022**, *10*, 1354. https://doi.org/10.3390/jmse10101354

Academic Editor: Alessandro Antonini

Received: 10 August 2022
Accepted: 19 September 2022
Published: 22 September 2022

1. Introduction

Nowadays, renewable energies offshore, such as wind energy, wave energy and tidal energy, have become important due to environmental issues and energy crises [1,2]. Wind energy offshore, in particular, emerges as a promising renewable clean alternative, mainly due to its availability and price competitiveness. The potential of marine renewables and, in particular, floating wind for the future of power generation and the energy transition to sustainable sources is known to the scientific community, technologists, and all energy industry [3,4].

The quality of the wind at greater distances from the coast is higher, and the resource itself increases, as well as the challenges associated with obtaining this energy. The wind resource assessment in [5] showed that the Iberian coast had an average annual energy density reaching up to 971, 549 and 398 W/m^2 in the north, centre and southern regions, respectively, values that are representative of what can be found on other areas of the European Atlantic coast. Another study shows typical average values of wave energy of about 40 kW/m in the summer and 10 kW/m in the winter for the corresponding areas [6].

The European Union has directed resources to develop this type of energy so that it becomes reliable, safe, and profitable [1]. The study of Wind Europe indicates that the North Sea will have the highest quantity of offshore wind farms by 2030, with a central value closed to 50 GW, higher than the 10 GW of 2016 [7].

Over recent years, the floating wind industry has suffered an important expansion. Floating wind turbines present an economical and technically feasible approach to access the deeper water sites to obtain the rich resource of wind power. Therefore, they have the potential to be the next generation of wind technology [8]. Moreover, the saturation of maritime space, the energy demand, and the geographical constraints require new offshore installation areas. All these requirements have created the need to identify locations for floating wind farm installation [9].

In the design of offshore wind farms, the first stage is often the choice of sites [10]. Site selection also plays a crucial role in the financial returns of the project, ease of construction, ongoing operations and maintenance, and overall safety. Geographical Information System (GIS)-assisted wind farm location criteria [11] were typically used, such as [9,12–15]. Díaz and Guedes Soares [9] identified the most suitable locations for floating offshore wind turbines within the European Atlantic Area (Portugal, Spain and France), considering the available wind resource, existing uses of maritime space and environmental constraints, and the operational needs of a floating wind farm. Díaz and Guedes Soares [14] developed a decision tool for planning offshore wind farm locations, combining multi-criteria decision analysis and GIS. Díaz et al. [15] showed that the floating wind farm locations suitability map developed with the AHP and MADA methods integrated into a GIS for the maritime surroundings of the European Atlantic coast is a substantial aid in the land-use management of these waters.

Platforms are an essential element of floating offshore wind turbines. Regarding the offshore wind platforms installed on offshore farms, there are several types of platforms within the main types of bottom-fixed and floating. Uzunoglu et al. [16] reviewed the main floating wind turbine platform concepts and the design standards for floating platforms. The first types of platforms to reach the commercial stage are the spar [17] and the semisubmersible type [18,19].

The connection of the platforms to the grid is another important component, as discussed in [20], and substations are often important components of this system. Two types of floating substation configurations were compared in [21] with respect to technical and cost performance for a 200 MW Wind Farm for the Northeast U.S., showing that the semi-type substation platform cost is lower than the TLP-type cost for the case where each tendon has a dedicated anchor, whereas the cost for the TLP-type with two tendons sharing an anchor is highly comparable to, if not less than, the semi-type platform.

One of the challenges in floating offshore wind turbines continued being the high costs associated [22]. The investment cost per MW in offshore wind is approximately 50% more expensive than in onshore wind, mostly due to the costs of additional components, underwater equipment, construction, and installation processes [4]. One of the most important issues in terms of an offshore wind farm is to be competitive in economic terms. Thus, a fundamental aim is to maximize energy production, minimize capital and operating costs, and stay within the constraints imposed by the site. Economic assessment studies have been made for locations in the Atlantic coast showing the relation of the location to the economic feasibility [23,24].

The other way of reducing production costs is to scale up the turbines, moving to larger turbines than the 5 MW ones that are more commonly used at present. So, recently, the design of a 10 MW floating offshore wind turbine (FOWTs) has been the main focus of research in order to further reduce the levelized cost of electricity (LCOE) of wind turbines to a more competitive level [25]. With regards to this, there are several European Union (EU)-funded research projects. For instance, a semi-submersible platform was developed in the INNWIND project [26]. Two semi-submersible concepts, a barge and a Tension Leg Platform (TLP) were developed in the LIFES50+ project [27]. Within the ARCWIND project, three novel concepts are being developed, a multi-body floating platform, the so-called TELWIND [28], a barge type [29] and a TLP [30]. Other concepts are also being studied, such as those described in [27] and [31].

One of the most important differences between fixed and floating substructures is mooring and anchoring systems. For floating structures, the station-keeping systems based on mooring lines and anchors are crucial to guarantee structure survivability and its components under different met ocean conditions. Traditionally, a successful mooring design considers several limit states (LS) (DNV-OS-E301, 2021 [32]), such as ultimate (ULS), accidental (ALS), fatigue (FLS) and service (SLS) [33]. The structural forces into the offshore wind anchoring and mooring systems are calculated either by using the quasi-static method [34,35] or dynamic analysis [36–39]. Masciola et al. [36] compared the response of the DeepCwind semisubmersible design in coupled simulations using FAST and the lumped-mass mooring model OrcaFlex, as well as the default quasi-static mooring model, against the 1:50-scale test data. They found that platform motions were influenced by mooring dynamics only in extreme sea states but that mooring dynamics are important to the prediction of mooring line tensions in all load cases. Bae et al. [37] performed numerical simulations on the performance of a Floating Wind Turbine (FWT) with a broken mooring line using an aero-hydro-servo-elastic-mooring coupled dynamic analysis in the time domain considering the OC4 DeepCwind semi-submersible wind turbine. Dynamic coupling analysis in frequency domain (FD) and time domain (TD) using ANSYS AQWA and Orcaflex were presented by [39] for two different mooring configurations on the hull of a Paired Column Semisubmersible (PCSemi). The Chain-Polyester-Chain (CPC) mooring concept performed better in deep waters and is more reliable for the PCSemi.

FOWT has several advantages compared with onshore ones, however, the complex and varied marine environment has brought great challenges to the transportation, installation, and operation of the equipment. It was mentioned in [40] that, currently, installation, operation, and maintenance (IO&M) costs contribute approximately 30% to the LCOE of offshore wind plants. The cost of the mooring and anchoring system is also included. There have been some reported attempts in the literature to minimize the operation and maintenance (O&M) costs of offshore wind farms [41–44]. The O&M costs are composed of labour costs (technician costs), material costs (component costs), transportation costs (vessels and associated costs), fixed costs (port, insurance, bidding) and potential revenue losses [44]. The LEANWIND project studied how to reduce the costs of farms by analysing their installation, operation and maintenance, and logistics and supply-chain [45].

To have cost-effective installation methods, the relevant numerical simulations of the installation of FOWTs need to be conducted as in [46–48]. A novel installation concept using a floating vessel was investigated in Hassan and Guedes Soares [47] and the coupled dynamic system of the installation vessel and the floating spar were performed using Ansys AQWA software. The numerical simulation of a three-body system, including the assembly, the catamaran vessel and the foundation was performed using SIMO and RIFLEX under the SIMA environment in [48]. The WindFloat Atlantic project [19] avoided the use of large offshore heavy-lift vessels by using an onshore crane, also allowing most of the commissioning works to be completed onshore. The offshore activities were greatly simplified, being designed to allow a tow-to-port O&M strategy for large component replacement. A similar strategy is described in [49].

For the installation of offshore platforms, the site has to be accessible for a certain period called a weather window [50,51], in which weather conditions are suitable for the specific work at sea. Deriving detailed information on suitable installation weather windows that will be available for a specific site will be also beneficial for the reduction in installation costs [52]. To determine it, a limiting operational environmental criterion [53] is given by the maximum values of wind speed and wave height for safe working and/or transfer conditions of personnel. In addition, there are some additional restrictions on specific cases. The weather window has been investigated for specific sites, such as, the Irish west coast [54], North Atlantic Ocean [51] and the south west of England [55].

Although the development of floating offshore wind turbine technology has increased greatly and the O&M aspects have been discussed for different projects, the installation logistics and cost of mooring components have been rarely discussed. This work addresses that problem by using a methodology based on simulations to assess the duration, costs and risks associated with the installation of an offshore wind farm in weather windows. This method is applied to a case study based on a wind farm in Lannion, located on the northwest coast of France. The selected technology for this study is the TELWIND platform developed by Esteyco.

Section 2 presents the characteristics of the floater, the installation site, the mooring system and the wind farm array design. Section 3 describes the required installation operations and their limitations and presents the simulation method. Section 4 presents the results and discussions of the simulation results. Two different installation strategies are studied and compared. Section 5 provides a summary and draws some conclusions from the present study.

2. Description of the Platform and Mooring System

2.1. TELWIND Platform

The TELWIND platform is illustrated in Figure 1 and consists of two parts (upper structure and lower tank), which are joined rigidly by tendons. The draft of the structure can be adjusted according to the installation site by modifying the length of the tendons joining its two main sections, making it possible to operate at depths of less than 100 m. The upper structure provides buoyancy in excess while the lower tank provides weight, ensuring the stability of the system and the counteraction of the forces applied in the structure to reduce the wind turbine induced motions.

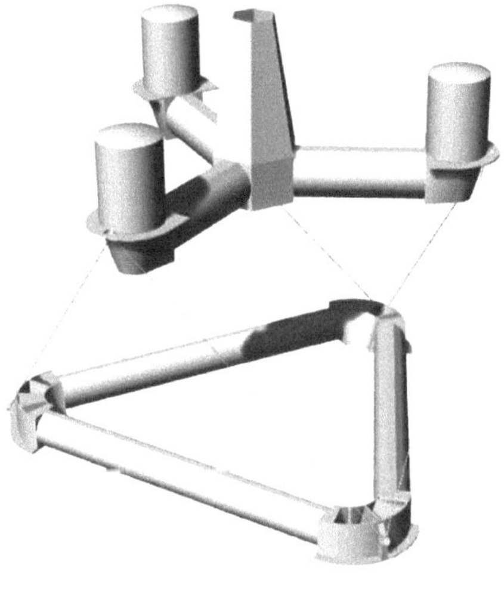

Figure 1. TELWIND platform.

2.2. Station Keeping System

The TELWIND platform station keeping system designed within the ARCWIND project is a spread mooring system in which the mooring lines are attached to each of the upper structure columns. Figures 2 and 3 show the proposed mooring system based on three catenary mooring lines of Ø133 mm R3 chain, and the assembly drawing of the mooring system including main components and connectors. The connection to the structure columns is performed using a dual-axis rotational connection, which absorbs the platform rotations to avoid out of plane loads in the first links of the mooring line. The mooring line is divided into two sections to facilitate the installation. One section is approximately 414 m of bottom chain with a 15-ton drag anchor connected at one end. The other section is a 60 m top chain segment preinstalled on the structure. These sections of the chain are connected on two of the lines using an H-Link connector, while an in-line tensioner is used for the connection of the third line and tensioning the whole mooring system to the target pre-tension during the hook-up operation.

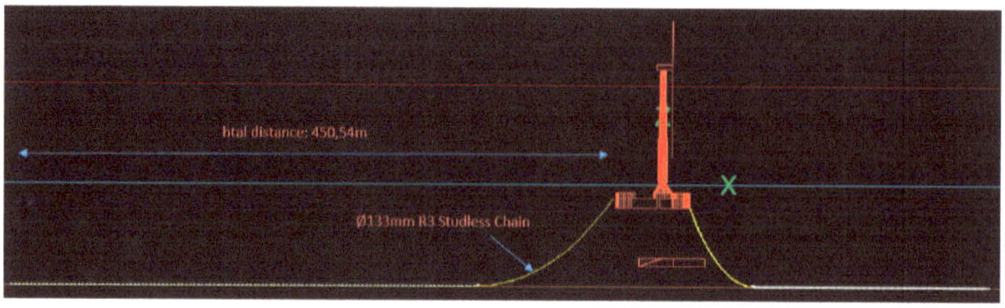

Figure 2. TELWIND station keeping system.

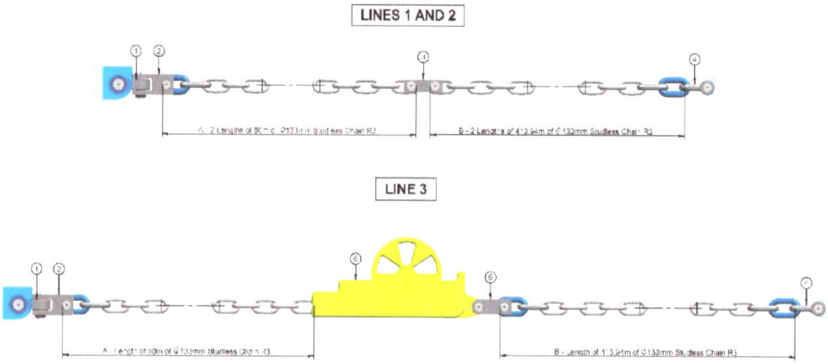

Figure 3. Station keeping system assembly drawing.

2.3. Wind Farm Array Design

The total capacity of the Lannion Floating Wind Farm is 470 MW, which is composed of 47 floating offshore wind turbines (FOWTs) of 10 MW, oriented towards N-NE. Figure 4 illustrates the design and dimensions of the wind farm array developed within the AR-CWIND project, which is divided into two clusters, one of 24 FOWTs and the other with 23 FOWTs, distributed in 4 rows per cluster with 5 FOWTs per row and an additional row per cluster with 4 and 3 FOWTs, respectively.

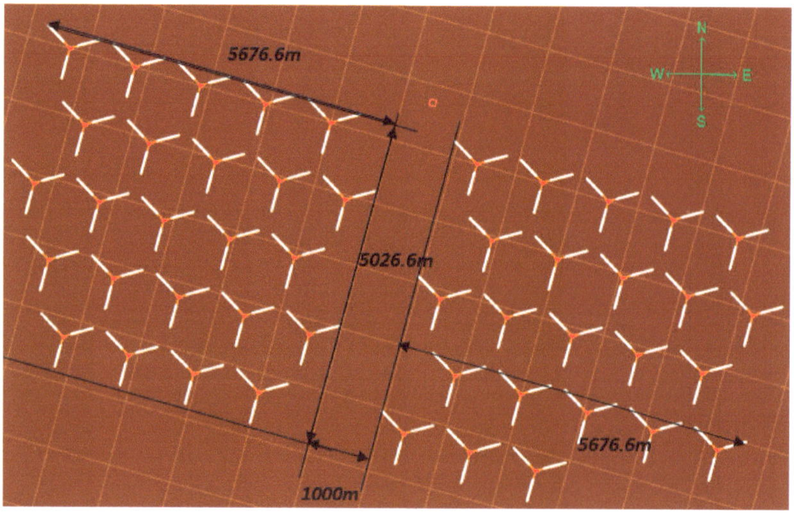

Figure 4. Wind farm array design and dimensions.

The substation is in the central path of the two clusters. The FOWTs of each cluster are connected to each other by inter-array cables. A radial topology is defined for the distribution of the inter-array cables and their connections to the substation as shown in Figure 5. The inter-array cables use a lazy-wave configuration at each end of the cable to be attached to the floating structures.

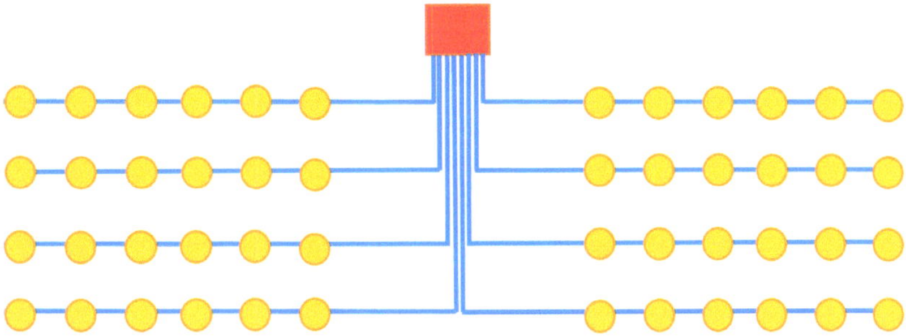

Figure 5. Wind farm radial topology.

It is to be noted that this study focuses on the installation of the mooring systems only and it does not include the power cable installation methods and timings, which would be necessary for a complete installation of the floating wind farm. This includes cable transport, deployment through seabed trenching and cable burying, installation of buoyancy modules for lazy waves, array, FOWTs and substation connections.

3. Transport and Installation Strategies

3.1. Mission Planner Simulations Method

Marine operations are planned and designed using classification societies' standards. These standards require that a contingency time is added to the planned operation time, the sum of both being the operation reference period. The weather limitations of the operation are reduced by a factor to account for uncertainty in the weather forecasts [53]. This approach is not suitable to estimate the risks associated with the weather windows in the early stages of the project when the available data are scarce.

The Mission Planner developed by CoreMarine [56] is adopted for the simulation. This tool takes the sequence of tasks to be performed during an offshore operation, each one with a duration and a set of limitations. Using a weather hindcast data base, the operation is simulated. The first task starts at the specified time; if the weather conditions surpass the allowed ones, the vessel waits until there is a suitable weather window. The result of each individual simulation is the time that was required to perform the operation. All the individual simulations are then analysed, and statistical data are obtained. For this case study, a 10-year hindcast weather data base [57] has been used.

For example, Figures 6 and 7 present part of the output of a simulation. There are different options depending on the nature of the operation. For instance, Figure 6 shows a case where the tasks do not need to be performed immediately one after the other. On the other hand, Figure 7 presents a case where the 1st three tasks must be performed immediately one after the other and, thus, a certain amount of time must pass to have an appropriate weather window. In this last figure, it can also be appreciated that each task can have a different weather limitation.

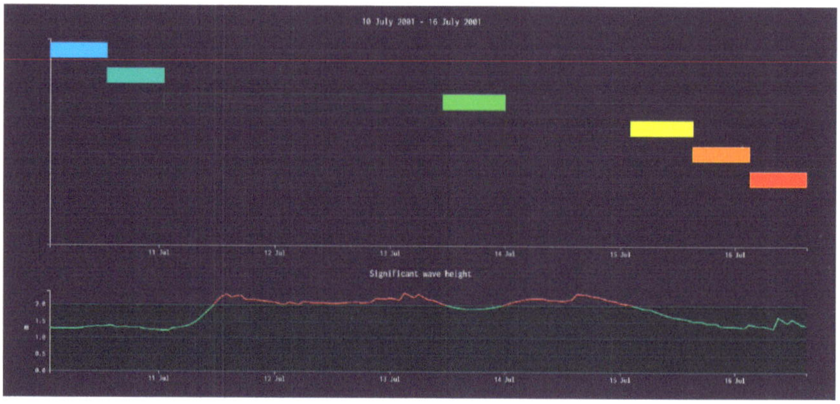

Figure 6. Mission Planner input example.

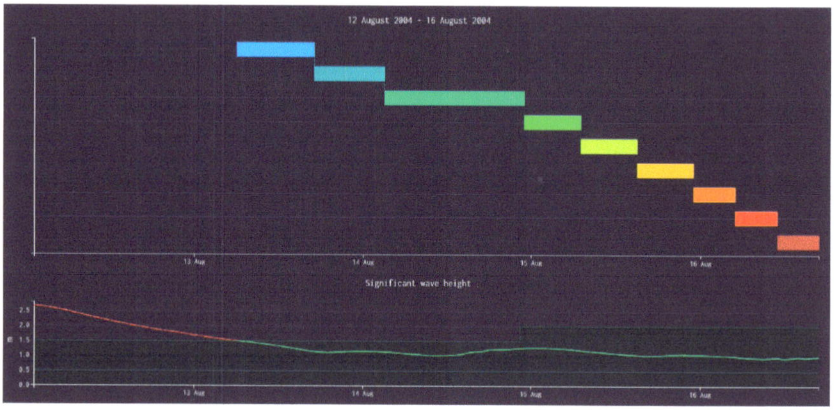

Figure 7. Mission Planner input example.

The overall result, presented in Figure 8, gives a visual representation of the appropriate time frames to perform the operations. Apart from this window, the data are extracted and analysed.

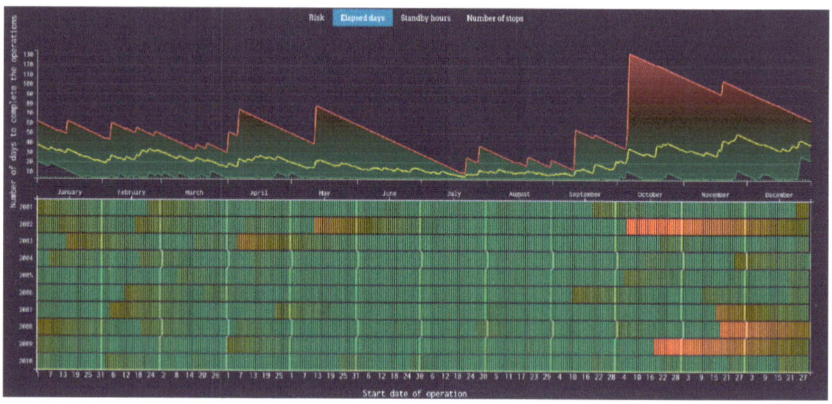

Figure 8. Mission Planner output graph.

3.2. Location

The Lannion (France) site, located at 49.23° N/3.68° W, is the proposed area for the construction of a floating wind farm of 470 MW. The area, with a water depth of 75 m and being located approximately 33 km off the coast of France, has been selected through an assessment of feasible areas for offshore floating wind farms, considering utilization, economic, technical, and social constraints [14]. Figure 9 shows the proposed installation site and the port selected as the base port for the storage of structures and moorings for the construction of the wind farm; the Cherbourg port is located at 154 km [9]. The general wave conditions in the area are given in [58–60].

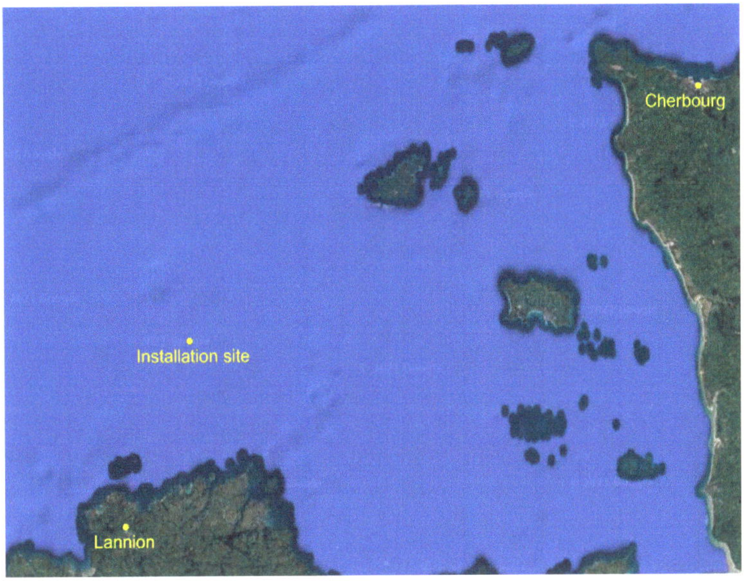

Figure 9. Lannion site. Map from © 2021 Google [61].

3.3. Total Mooring Components for the Wind Farm

Considering the proposed station keeping system and the number of FOWTs comprising the Lannion floating wind farm, Tables 1–3 show the total number of offshore mooring chain lengths, mooring connectors and drag anchors that the mooring systems of the entire wind farm array consist of.

Table 1. Floating wind farm offshore mooring chain.

Component	Quantity	Weight (Tons)
Chain Length of 60 m of Ø133 mm Studless Chain	141	2978
Chain Length of 414 m of Ø133 mm Studless Chain	141	20545

Table 2. Floating wind farm mooring connectors.

Component	Quantity	Weight (Tons)
Unijoint Connection for Ø133 mm Chain	141	423
Inverted H-Link for Ø133 mm Chain	141	183
H-Link for Ø133 mm Chain	94	99
In-Line Tensioner for Ø133 mm Chain	47	329
H-Link for Ø133 mm Chain and In-Line Tensioner	47	49
Anchor Shackle for Ø133 mm Chain	141	102

Table 3. Floating wind farm drag anchors.

Component	Quantity
Drag anchor (15 T)	141

3.4. Mooring Components Shipping Ports

The different components that constitute the mooring systems are assumed to be shipped out from the closest ports to their respective manufacturing facilities.

- Mooring chains and connectors are assumed to be manufactured in Spain. The port of shipment for the mooring chain and connectors is the Vicinay Sestao factory, which has a direct loading and unloading area at its facilities (43.3° N/−2.98° W).
- Drag Anchors are assumed to be manufactured in the Netherlands. The port of shipment for the drag anchors is the port of Rotterdam (51.56° N/4.73° E).

Table 4 shows the distances and transport durations of the shipments to the selected base port of Cherbourg. The speed of the ship during transport considered is 10 knots.

Table 4. Transport distance and duration.

Component	Distance (nm)	Transport Duration (h)
Mooring chain Mooring connectors	504	50
Drag anchors	278	28

3.5. Mooring Installation Operations

The installation of the mooring systems of the wind farm is composed of three different campaigns: mooring pre-lay, platform towing, and mooring and cable hook-up. The following subsections describe the procedure for carrying out these operations, including the duration and weather limitations for carrying out the operations in terms of maximum Hs. The wind speed limitation for all operations has been set to 20 knots.

3.5.1. Mooring Pre-Lay Campaign

This operation refers to the deployment of the bottom chains and drag anchors, which are pre-installed and wet stored at the seabed. The 141 mooring lines are pre-laid in 7 campaigns, 20 lines in each of the first 6 campaigns, and 21 lines in the last one. This requires the use of two Large Class Anchor Handling Tug Supply (AHTS) vessels, with more than 250 t bollard pull capacity, Dynamic Positioning 2 (DP2) and 2 Work Class Remotely Operated Vehicles (WROVs) working in parallel. More vessels could be used simultaneously, but always in pairs since 2 are needed for proof loading the anchors. The procedure for a pre-lay campaign is presented as follows:

1. AHTS drag anchor installation (2 AHTS)
    a. Mobilization
        i. AHTS arrival at the mobilization site
        ii. Load chain (10 segments of 414 m)
        iii. Load project equipment (Only on the 1st campaign)
        iv. Load 10 drag anchors (15t) and sea fastening
        v. Depart to the offshore location site
            1. 86 nautical miles at 12 knots
        vi. Drag anchor installation (10 times)
            1. As found survey of the lay route
            2. Overboard the drag anchor
            3. Deploy to seabed
            4. Land on the seabed and drag in

5. Pretension to the vessel maximum bollard pull
6. Lay chain and wet storage
7. As left survey and prepare the next anchor on deck
8. Relocate to the next anchor installation location

2. Drag anchor proof tensioning (20 times)
    a. AHTS #1 and #2 at mooring line location
    b. AHTS # recovers mooring line
    c. AHTS #2 passes the winch wire to AHTS #1
    d. AHTS #1 connects AHTS#2 wire to tensioning bridle
    e. AHTS #1 and AHTS #2 position on the field ready to tension
    f. Take tension up to calculated BP
    g. Proof tension anchor
    h. Take off tension
    i. AHTS #1 recovers tensioning bridle
    j. AHTS #1 disconnects AHTS #2
    k. AHTS deploys and wet stores the mooring line
    l. AHTS #1 and #2 relocate to the next mooring line

3. Demobilization
    a. Demobilize vessels

Table 5 shows the duration and weather limitations of the operations to be carried out in this campaign. The drag anchor and mooring line installation operation run in parallel, while the drag anchor tensioning operations need to be performed one after the other using both AHTS. The duration of a single mooring line installation (M) is approximately 22 h per AHTS. The drag anchor tensioning (DAT) operation duration is 12 h. The maximum significant wave considered for safe operation is 2.5 m for both installation and tensioning.

Table 5. Mooring Pre-Laying weather limitations and duration.

Task	Max Hs (m)	Duration (h)
M 1 and 2	2.5	22
M 3 and 4	2.5	22
M 5 and 6	2.5	22
M 7 and 8	2.5	22
M 9 and 10	2.5	22
M 11 and 12	2.5	22
M 13 and 14	2.5	22
M 15 and 16	2.5	22
M 17 and 18	2.5	22
M 19 and 20	2.5	22
DA T 1	2.5	12
DA T 2	2.5	12
DA T 3	2.5	12
DA T 4	2.5	12
DA T 5	2.5	12
DA T 6	2.5	12
DA T 7	2.5	12
DA T 8	2.5	12
DA T 9	2.5	12
DA T 10	2.5	12
DA T 11	2.5	12
DA T 12	2.5	12
DA T 13	2.5	12
DA T 14	2.5	12
DA T 15	2.5	12
DA T 16	2.5	12
DA T 17	2.5	12
DA T 18	2.5	12
DA T 19	2.5	12
DA T 20	2.5	12

3.5.2. FOWT Towing Campaign

The towing campaign runs parallel to the mooring and power cable hook-up. Two fleets of tugs are needed, each one composed of two tugs. The tugs assist the main AHTS vessel during the hook-up until the floater is storm safe. At that point, they return to the mobilization port. This campaign happens 23 times per fleet plus an additional one for one fleet, a total of 47 times. The procedure steps are presented below.

1. Mobilization
    a. The tug fleet arrives at the mobilization base
    b. Floater connection
2. Towing and installation
    a. Transit to the offshore site
        i. 86 nautical miles at 3 knots
    b. Arrive at the offshore site
    c. Assist AHTS with FOTW hook-up until it is storm safe
    d. Return to mobilization base
3. Demobilization
    a. Tug fleet demobilized

Although it has been considered for the cost analysis, the towing operation has not been simulated since it depends on a more restrictive one, the hook-up. This is because, as there are two fleets of tugs per AHTS, in charge of both towing the structures and assisting the AHTS during hook-up operations, this hook-up campaign is assumed to be the most restrictive in terms of operating times. Thus, the standby ratio of the tugs, a parameter that measures the time during which vessels are not operating due to weather constraints, will depend on this hook-up operation.

3.5.3. Mooring and Power Cable Hook-Up Campaign

This campaign includes the hook-up and tensioning operations of the mooring system. For this study, it has been assumed for the Hook-up that the connection of the first two lines of each mooring system is performed on the deck of the AHTS, while the third line is connected utilizing an In-Line Tensioner (ILT), which is used for the final tensioning of the complete system. The floater will be held in position by the holding tugs until the last mooring line is connected and the floater is deemed to be storm safe.

This operation is especially sensitive to weather conditions, so it has been designed to be completed in as little time as possible. Contrary to the rest of the campaigns, which involve transit from the port to the installation site and vice versa, an AHTS vessel will always stay on site, performing the hook-up. It will only return once to resupply ILTs. The floater tank must be ballasted during the operation, which takes between 3 and 4 h. A specific study is needed to determine the best moment to perform this deballasting, but for this study, it is not strictly necessary. The 47 hook-ups are carried out using large class AHTS with more than 250 t bollard pull capacity, DP2 and 2 WROVs. The operation procedure steps are shown below.

1. Mobilization (two times, 1st and 25th campaign)
    a. AHTS arrival to the mobilization site
    b. Load 23 ILTs and sea fasten them
    c. Load project equipment (Only on the 1st campaign)
    d. Depart to the mobilization site
    e. Transit to the Offshore location
        i. 86 nautical miles at 12 knots
2. FOWT mooring hook-up
    a. FOWT tow arrives at the offshore installation site
    b. AHTS arrives to the offshore installation site (1st and 25th campaign only)

c. DP trials
d. SIMOPS RA with FOWT tow master
e. As found survey
f. FOWT moved to the hook-up position
g. Tank deballasting
h. Mooring line #1 connection
    i. AHTS connects the wet stored mooring #1 to the recovery winch
    ii. The mooring is picked up and re-laid towards the floater
    iii. AHTS recovers to the back deck the chain tail from the FOWT
    iv. AHTS recovers the subsea mooring line to the back deck
    v. The mooring line is connected to the FOWT chain tail
    vi. Overboard and deploy the connected mooring line
    vii. Mooring line #1 connection completed
i. Mooring line #2 connection
    i. AHTS connects the wet stored mooring #2 to the recovery winch
    ii. The mooring is picked up and re-laid towards the floater
    iii. AHTS recovers to the back deck the chain tail from the FOWT
    iv. AHTS recovers the subsea mooring line to the back deck
    v. The mooring line is connected to the FOWT chain tail
    vi. Overboard and deploy the connected mooring line
    vii. Mooring line #2 connection completed
j. Mooring line #3 connection and pre-tensioning
    i. AHTS connects the wet stored mooring #3 to the recovery winch
    ii. The mooring is picked up and re-laid towards the floater
    iii. AHTS recovers to the back deck the chain tail from the FOWT
    iv. AHTS recovers the subsea mooring line to the back deck
    v. The mooring line is connected to the FOWT chain tail
    vi. Overboard and deploy the connected mooring line
    vii. Pre-tension the mooring line with the ILT
    viii. Mooring line #3 connection completed and the FOWT is storm safe
k. Demobilize the tow vessels

3. Demobilization FOWT power cable hook-up (two times)
    a. Recover the power cable from the seabed
    b. Recover the messenger line from the FOWT
    c. Connect the messenger line and the wet handshake to the FOWT
4. Hook-up completion
    a. As built survey
    b. AHTS transit to the next installation location
5. Demobilization (Only in the 47th campaign)
    a. Demobilize project equipment

Table 6 shows the duration and weather limitations of the operations carried out for each FOWT. M1, M2 and M3 + ILT is the hook-up of the mooring lines until the FOWT is storm safe. These tasks must be performed consecutively, and the maximum safe significant wave is 1.5 m. Each task takes 10 h except the last, which takes 20 h. C1 and C2 are the cable hook-up operations, which take approximately 6 h and have a maximum allowable wave height of 2 m.

Table 6. Mooring and power cable hook-up weather limitations and duration.

Task	Max Hs (m)	Duration (h)
M 1	1.5	10
M 2	1.5	10
M 3 + ILT	1.5	20
C 1	2	6
C 2	2	6

3.5.4. Dependencies and Contingency Time Frames

Tables 7 and 8 present the number of campaigns and elements installed or connected per campaign and the time dependencies between operations, which are especially important in combination with the optimal installation seasons. Since it is usually better to concentrate the operations in the summer months, the dependencies between the operations can push the most restrictive campaigns (FWT mooring and cable Hook-up) into months with few weather windows.

Table 7. Element and campaign summary.

Operation	No. of Elements per Campaign	Total Elements	Number of Campaigns
Mooring Pre-Lay campaign	20	141	7
FTW Mooring and Power Cable Hook-up	1	47	47

Table 8. Main operation dependencies.

Campaign Name	Time Dependencies
Mooring Pre-Lay campaign	No dependencies on other T and I campaigns
FTW Mooring and Power Cable Hook-up	Start after three mooring installation campaigns (21 days approximately)

3.6. *Estimation of Costs of Transport and Installation Vessels*

3.6.1. Transport Vessel

The type of vessel considered for the transport of the mooring components is a general cargo vessel with a deadweight of 4500 tons. In the case of drag anchors, the vessel has a capacity for the transportation of 14–15 anchors. Table 9 shows the required time for loading and unloading the mooring components per transport. The cost per vessel is approximately EUR 75,000.00.

Table 9. Transport vessel loading and unloading duration.

Component	Loading Duration (h)	Unloading Duration (h)
Mooring chain Mooring connectors	96	96
Drag anchors	48	48

3.6.2. Installation Vessels

Tables 10–12 present the main vessels used in the installation operations, an approximate day rate for the charter and the approximate costs of the installation equipment and fuel. Port costs are not included in the assessment.

Table 10. Main vessel types and their characteristics.

Vessel	Characteristics	Operation	Day Rate (EUR)
AHTS	BP > 250 t DP2 2xWROV	Mooring line installation Hook-up operations	40,000.00

- Anchor handler supply tug vessel (AHTS)

Table 11. AHTS costs.

Element	Characteristics	Operation	Day Rate (EUR)	Lump Sum (EUR)
AHTS	BP > 250t DP2 2xWROV	Mooring line installation Hook-up operations	40,000.00	-
Installation aids			-	500,000.00

- Fuel costs

Table 12. Fuel costs.

Vessel	Operation (EUR/Day)	Standby (EUR/Day)
AHTs	8000.00	2000.00
Tug	4000.00	1000.00

4. Results and Discussion

4.1. Installation Weather Windows Analysis Results

In the following subsections, the results of the simulations are presented. The parameter used to analyse the results are:

- Weather window assessment and associated risk. For each operation, the results of the simulations are represented by the average time to complete the operation and the standard deviation of the values.
- Duration and vessels are required for each strategy. For mooring and cable hook-up operation, another strategy is studied to reduce the average time for installation, the risk and the cost.
- Standby ratio. This metric measures the time when the vessels are not operating due to weather constraints. It is calculated by comparing the total hours required to perform an entire operation without considering the weather, and the total time that the vessels need to be chartered. The standby ratio is directly related to installation costs, so the optimization of installation sequences aims to reduce this parameter.

The summary of the results is presented in Table 13. For each operation, the duration in days using the averages of the optimal installation months is presented.

Table 13. Weather window analysis results summary.

Operation	Optimal Installation Months	Duration (Days)	Vessel Required	Units per Month
Mooring Pre-Lay campaign	April-August	127	2xAHTS	29
FTW Mooring and Power Cable Hook-up. Option A	All year	455	1xAHTS	3
FTW Mooring and Power Cable Hook-up. Option B	April-August	153	2xAHTS	9

4.1.1. Mooring Pre-Lay

The minimum duration of a single mooring pre-lay campaign, which includes the installation of 20 drag anchors and mooring lines, as well as the tensioning of these 20 anchors, is 19 days (460 h), as specified in Table 5.

Table 14 and Figure 10 show the average, maximum and minimum time needed to complete the same operation in each month of the year. This information is used to obtain the optimal time frames for installation and the risk associated to each month, which is represented by the standard deviation. The most suitable months to perform the drag anchor and mooring installation are the summer months. While it is possible to have good weather windows during the winter and autumn months, the risk is significantly higher.

Table 14. Mooring pre-laying campaign weather windows.

Month	Average (Days)	Maximum (Days)	Minimum (Days)	σ (Days)
January	41	72	22	11.44
February	33	55	20	9.45
March	28	46	20	6.08
April	24	39	20	3.91
May	22	32	20	3.22
June	21	26	20	1.37
July	21	25	20	0.98
August	21	27	20	1.62
September	24	38	20	4.23
October	31	57	20	9.42
November	32	60	20	10.71
December	39	69	20	13.01

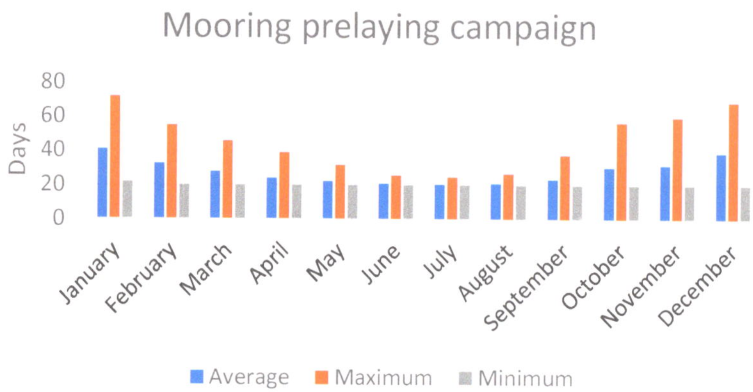

Figure 10. Mooring pre-laying campaign duration.

A total of 7 campaigns are necessary to install the 141 mooring lines Table 15 shows the number of campaigns that could be carried out in each month as well as the percentage completed out of the 7 total campaigns. It is observed that the installation can be performed between April and August with two AHTS vessels. The total duration could be approximately 127 days.

Table 15. Mooring Pre-Laying campaign completion % per month.

Month	No. Campaigns	Completion (% of Total)
January	0.75	10.7%
February	0.94	13.4%
March	1.1	15.7%
April	1.27	18.2%
May	1.37	19.5%
June	1.46	20.8%
July	1.49	21.3%
August	1.45	20.7%
September	1.3	18.6%
October	1.01	14.4%
November	0.96	13.7%
December	0.76	11.2%
TOTAL	13.88	198.3%

4.1.2. Mooring and Power Cable Hook-up Campaign

The minimum duration of the mooring and power cable hook-up campaign is approximately 2 days (52 h), as specified in Table 6. Each campaign includes the hook-up of the three mooring lines per floating wind turbine, the tensioning of the mooring system utilizing the In-Line tensioner, and the power cable hook-up.

Table 16 and Figure 11 shows the average, maximum and minimum amount of time it takes to complete the same operation in each month of the year. This information is used to obtain the optimal time frames for installation and the risk associated with each month, which is represented by the standard deviation. Same as in mooring pre-lay campaigns, the summer months are optimal to perform the hook-up campaigns. Compared with the previous operation, the average installation time and the standard deviation is larger due to the longer operation duration and additional constraints.

Table 16. Hook-up campaign weather windows.

Month	Average (Days)	Maximum (Days)	Minimum (Days)	σ (Days)
January	26	74	3	18.40
February	14	47	3	12.19
March	14	47	3	10.36
April	9	34	3	6.35
May	9	36	3	7.12
June	5	22	3	3.69
July	6	24	3	3.96
August	5	25	3	3.76
September	7	39	3	6.40
October	11	58	3	11.25
November	14	45	3	11.31
December	17	89	3	19.30

Table 17 shows the number of campaigns that could be carried out each month as well as the percentage completed out of the 47 total campaigns needed for the hook-up of the structures of the wind farm.

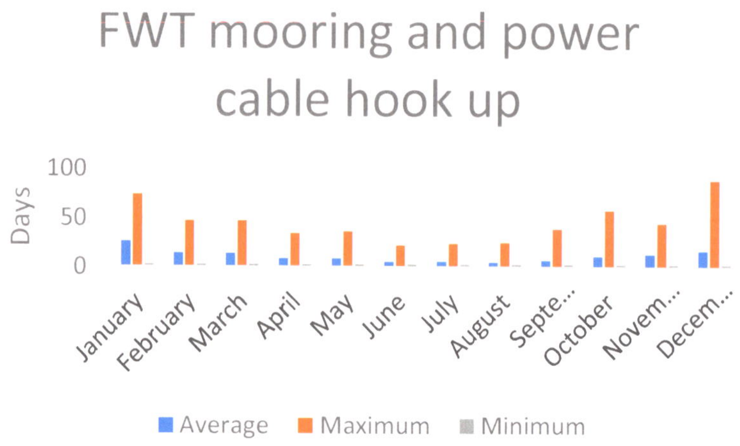

Figure 11. Hook-up campaign duration.

Table 17. Hook-up campaign completion % per month.

Month	No. Campaigns	Completion (% of Total)
January	1.22	2.6%
February	2.21	4.7%
March	2.26	4.8%
April	3.67	7.8%
May	3.62	7.7%
June	5.88	12.5%
July	5.45	11.6%
August	6.11	13.0%
September	4.56	9.7%
October	2.73	5.8%
November	2.30	4.9%
December	1.83	3.9%
TOTAL	41.83	89.0%

In this case, it is observed that the hook-up manoeuvres cannot be performed within a year with a single AHTS vessel. With a single AHTS vessel, it would take approximately 1.25 years to do the hook-up of all the FWTs that comprised the wind farm. Therefore, the use of another AHTS is proposed to optimize the time taken to carry out these campaigns, and with it, its cost.

Two strategies are considered:

- Option A: One AHTS vessel working all year round for 1 year and 3 months. For the 47 FWTs, it would take approximately 455 days (the first full year and second year until the end of march)
- Option B: Two AHTS vessels working in the months from April to August for a single year. For the 47 FWT, it would take approximately 153 days of operation and it would end at the end of August.

4.2. *Preliminary Schedule and Cost Estimation Results*

4.2.1. Installation Preliminary Schedule and Cost Estimation

A preliminary schedule for the installation of the moorings is presented in Table 18 considering the strategies proposed in the previous section. The numbers in each operation represent the number of campaigns performed in that month. There are 7 Mooring pre-lay campaigns and 47 FWT towing and hook-up campaigns.

Table 18. Installation preliminary schedule.

Year	Month	Option A			Option B		
		Mooring PRE-Lay	FWT Towing	Hook-Up	Mooring Pre-Lay	FWT Towing	Hook-Up
1	January	1	1	1			
	February	1	2	2			
	March	1	2	2	1		
	April	1	4	4	1	7	7
	May	1	4	4	2	7	7
	June	1	6	6	2	11	11
	July	1	5	5	1	11	11
	August		6	6		11	11
	September		4	4			
	October		3	3			
	November		2	2			
	December		2	2			
2	January		1	1			
	February		2	2			
	March		2	2			
	April		1	1			

Two alternatives are studied:

- Option A: The towing and hook-up operation happens continuously throughout the year using one AHTS and two tugs.
- Option B: The towing and hook-up operation happens from April to August and use two AHTS and eight tugs. To minimize the cost, these campaigns are scheduled to happen in the most productive months.

The total estimated duration and costs per operation considering both strategies are presented in Table 19. The standby ratio is the time that the ship is chartered but is not operating due to the weather. It can be used as an efficiency indicator. The fuel costs are based on this ratio since they are higher while operating than while on standby.

Table 19. Total installation cost estimation.

Operation		Vessel Required	Estimated Duration (Days)	Standby (%)	Vessel Cost (EUR)	Fuel Cost (EUR)	Total Cost (EUR)
Option A	Mooring Pre-lay	2xAHTS	212	71.9%	17,960,000	1,562,000	19,522,000
	FWT towing	1xAHTS	485	76.2%	23,280,000	3,326,500	26,606,500
	Hook-up	4xtugs (2 fleets)	485	78.7%	19,900,000	1,589,000	21,489,000
Option B	Mooring Pre-lay	2xAHTS	153	61.1%	13,240,000	1,326,000	14,566,000
	FWT towing	2xAHTS	153	62.2%	14,688,000	2,610,500	17,298,500
	Hook-up	8xtugs (4 fleets)	153	66.3%	13,240,000	1,231,000	14,471,000

Table 20 presents the total estimated duration and cost for the installation of the whole wind farm. The strategy labelled as Option B proves to be more economical since it avoids the months with low productivity and has a lower standby ratio. It has a total duration of 6 months.

Table 20. Installation cost per FWT and MW.

Option	Duration (Months)	From-to	Estimated Cost (EUR)
A	16	January–April	67,617,500
B	6	March–August	46,335,500

4.2.2. Transport Preliminary Schedule and Cost Estimation

On the premise that the necessary components will be ready at the commissioning port at the start of each of the scheduled installation phases, Table 21. presents the number of mooring components shipped per month considering the more economical installation option obtained in the previous section (option B).

Table 21. Transport preliminary schedule.

Year	Month	Option B								
		Chain Length of 60 m	Chain Length of 414 m	Unijoint	Inverted H-Link	H-Link	ILT	H-Link-ILT	Anchor Shackle	Drag Anchor
1	January									29
	February	141	30	90	90	60	27	27	141	28
	March		60							28
	April		31							28
	May		20	51	51	34	20	20		14
	June									14

The transport of the mooring chains and connectors is carried out between February and May using two vessels per month for the first two months and a single vessel for the other two. Considering the large number of anchors in the wind farm, it was decided to ship them from January to June utilizing two vessels per month for the first four months and another vessel per month for the remaining two.

Based on the total amount of each component required for the construction wind farm, the number of ships required to transport each component and their cost are defined in Table 22.

Table 22. Transport costs.

Component	Quantity	Number of Vessels	Cost (EUR)
Mooring chain	23523 tons	6	450.000
Mooring connectors	1185 tons		
Drag anchors	141 uds	10	750.000

5. Conclusions

The mooring system transport and installation logistics for a floating offshore wind platform, TELWIND, in Lannion, located on the northwest coast of France, are discussed in this paper. The discussed station keeping system is a spread mooring system designed within the ARCWIND project.

The procedures for carrying out the mooring pre-lay, platform towing, and mooring and cable hook-up campaigns, including the duration and weather limitations for carrying out the operations in terms of maximum Hs, are described and discussed. Based on different transport and installation strategies, the preliminary schedule and cost are estimated.

The most suitable months to perform the drag anchor and mooring installation are the summer months. While it is possible to have good weather windows during the winter and autumn months, the risk is significantly higher. Summer months are also optimal to perform the hook-up campaigns. Compared with the mooring pre-lay, the average installation time and the standard deviation is larger due to the longer operation duration and additional constraints, for the mooring and power cable hook-up campaign.

The cost estimation results for installation show that the strategy with two AHTS vessels working in the months from April to August for a single year proves to be more economical since it avoids the months with low productivity and has a lower standby ratio. For the 47 FWT, it would take approximately 153 days of operation and it would end at the end of August.

Regarding the transports, it was decided to ship them from January to June, employing two vessels per month for the first four months and another vessel per month for the remaining two.

The case study demonstrated that the Mission Planner simulation method is efficient for analysing the marine operations; however, it is suitable for the case when sufficient project data are available. The limitations of this study are related to the scarce data available in the initial stages of the design of a wind farm. For example, the mooring components, location, and the individual tasks are specified for the present study, while the rest of the input data are based on industry knowledge on installation time frames, weather limits and vessel costs. Further optimization can be made when planning the simultaneous offshore operations with dependences between them. The presented costs are highly dependent on the current state of the market.

The simulation study allowed for the assessment of the representative duration for the installation procedure as well as its overall cost, allowing for the conclusion that the use of a simulation tool is the only feasible way to produce these research results that are indispensable for proper costing and planning these operations.

The results of this model could be used on early tenders of Requests For Information (RFIs). The weather risk is significant, and it is recommended that it is considered on the initial stages of the projects through models such as the one presented in this work.

Author Contributions: J.A. designed the mooring systems, performed the hook-up analyses, conducted the investigation, and prepared the draft manuscript. A.H. worked on the methodology concept and the installation methods and contributed actively to the review of the draft. O.M. also helped in the review of the draft and supervised the work. J.U. and C.R. developed the TELWIND platform and contributed to the review of the draft. S.W. and C.G.S. contributed to the writing of the paper. All authors have read and agreed to the published version of the manuscript.

Funding: This research was funded by Interreg Atlantic Area European Regional Development Fund; "Adaptation and Implementation of Floating Wind Energy Conversion Technology for the Atlantic Region", ARCWIND project—grant number EAPA_344/2016. This work contributes to the Strategic Research Plan of the Centre for Marine Technology and Ocean Engineering (CENTEC), which is financed by the Portuguese Foundation for Science and Technology (Fundação para a Ciência e Tecnologia—FCT) under contract UIDB/UIDP/00134/2020.

Institutional Review Board Statement: Not applicable.

Informed Consent Statement: Not applicable.

Data Availability Statement: All the data presented in this work are based in the ARCWIND project internal deliverables. They are recorded in the project website data base with restricted access.

Acknowledgments: The valuable information provided by Esteyco concerning TELWIND technology is sincerely acknowledged. The information provided by Vicinay Mooring Connectors, Vicinay Sestao and Vicinay Marine regarding the mooring chain and connectors logistics and transport is acknowledged as well. Finally, all the ARCWIND project consortium must be recognized for their intensive work and collaboration.

Conflicts of Interest: The authors declare no conflict of interest.

References

1. *An EU Strategy to Harness the Potential of Offshore Renewable Energy for a Climate Neutral Future*; European Commission: Brussel, Belgium, 2020; p. 11.
2. DNV Energy Transition Outlook 2021. In *A Global and Regional Forecast to 2050*; DNV: Bærum, Norway, 2021; pp. 88–91.
3. Diaz, H.; Serna, J.; Nieto, J.; Guedes Soares, C. Market needs, opportunities and barriers for the floating wind industry. *J. Mar. Sci. Eng.* **2022**, *10*, 934. [CrossRef]
4. Díaz, H.; Guedes Soares, C. Review of the current status, technology and future trends of offshore wind farms. *Ocean. Eng.* **2020**, *209*, 107381. [CrossRef]
5. Salvação, N.; Guedes Soares, C. Wind resource assessment offshore the Atlantic Iberian coast with the WRF model. *Energy* **2018**, *145*, 276–287.

6. Silva, D.; Bento, A.R.; Martinho, P.; Guedes Soares, C. High resolution local wave energy modelling in the Iberian Peninsula. *Energy* **2015**, *91*, 1099–1112. [CrossRef]
7. Wind Europe. *Wind Energy in Europe: Scenarios for 2030*; WindEurope: Brussels, Belgium, 2017.
8. Sharma, J.S.; Dinesh, V.; Arul, A. Review on Floating Offshore Wind Turbines. In *Offshore Technology Conference Asia*; Virtual and Kuala: Lumpur, Malaysia, 2022; Paper OTC-31391-MS.
9. Díaz, H.; Guedes Soares, C. An integrated GIS approach for site selection of floating offshore wind farms in the Atlantic continental European coastline. Renew. Sustain. *Energy Rev.* **2020**, *134*, 110328.
10. Giebel, G.; Hasager, C.B. An overview of offshore wind farm design. *MARE-WINT* **2016**, 337–346.
11. Baban, S.M.; Parry, T. Developing and applying a GIS-assisted approach to locating wind farms in the UK. *Renew. Energy* **2001**, *24*, 59–71. [CrossRef]
12. Aydin, N.Y.; Kentel, E.; Duzgun, H.S. GIS-based site selection methodology for hybrid renewable energy systems: A case study from western Turkey. *Energy Convers. Manag.* **2013**, *70*, 90–106. [CrossRef]
13. Vasileiou, M.; Loukogeorgaki, E.; Vagiona, D.G. GIS-based multi-criteria decision analysis for site selection of hybrid offshore wind and wave energy systems in Greece. *Renew. Sustain. Energy Rev.* **2017**, *73*, 745–757. [CrossRef]
14. Díaz, H.; Guedes Soares, C. A novel multi-criteria decision-making model to evaluate floating wind farm locations. *Renew. Energy* **2022**, *185*, 431–454. [CrossRef]
15. Díaz, H.; Loughney, S.; Wang, J.; Guedes Soares, C. Comparison of multicriteria analysis techniques for decision making on floating offshore wind farms site selection. *Ocean. Eng.* **2022**, *248*, 110751.
16. Uzunoglu, E.; Karmakar, D.; Guedes Soares, C. Floating offshore wind platforms. In *Floating Offshore Wind Farms*; Castro-Santos, L., Diaz-Casas, V., Eds.; Springer International Publishing: Cham, Switzerland, 2016; pp. 53–76.
17. Skaare, B.; Nielsen, F.G.; Hanson, T.D.; Yttervik, R.; Havmøller, O.; Rekdal, A. Analysis of measurements and simulations from the Hywind Demo floating wind turbine: Dynamic analysis of the Hywind Demo floating wind turbine. *Wind Energy* **2015**, *18*, 1105–1122. [CrossRef]
18. Roddier, D.; Cermelli, C.; Aubault, A.; Weinstein, A. WindFloat: A floating foundation for offshore wind turbines. *J. Renew. Sustain. Energy* **2010**, *2*, 033104. [CrossRef]
19. Duarte, T.; Price, S.; Peiffer, A.; Pinheiro, J.M. WindFloat Atlantic Project: Technology Development Towards Commercial Wind Farms. In *Offshore Technology Conference*; OnePetro: Houston, TX, USA, 2022; Paper OTC-32058-MS.
20. Amaro, N.; Egorov, A.; Gloria, G. Fostering Offshore Wind Integration through Grid Connection Impact Assessment. *J. Mar. Sci. Eng.* **2022**, *10*, 463. [CrossRef]
21. Shelley, S.A.; Boo, S.Y.; Kim, D.; Luyties, W.H. Concept Design of Floating Substation for a 200 MW Wind Farm for the Northeast US. In *Offshore Technology Conference*; OnePetro: Houston, TX, USA, 2020; Paper OTC-30543-MS.
22. Castro-Santos, L.; Martins, E.; Guedes Soares, C. Methodology to calculate the costs of a floating offshore renewable energy farm. *Energies* **2016**, *9*, 324–350. [CrossRef]
23. Castro-Santos, L.; Silva, D.; Bento, A.R.; Salvacao, N.; Guedes Soares, C. Economic feasibility of floating offshore wind farms in Portugal. *Ocean Eng.* **2020**, *207*, 107393. [CrossRef]
24. Castro-Santos, L.; Bento, A.R.; Silva, D.; Salvacao, N.; Guedes Soares, C. Economic feasibility of floating offshore wind farms in the north of Spain. *J. Mar. Sci. Eng.* **2020**, *8*, 58–76. [CrossRef]
25. Yang, Y.; Bashir, M.; Michailides, C.; Mei, X.; Wang, J.; Li, C. Coupled analysis of a 10 MW multi-body floating offshore wind turbine subjected to tendon failures. *Renew. Energy* **2021**, *176*, 89–105. [CrossRef]
26. Azcona, J.; Vittori, F.; Schmidt, U.; Svanije, F.; Kapogiannis, G.; Karvelas, X.; Manolas, D.; Voutsinas, S.; Amann, F.; Faerron-Guzman, R.; et al. Design Solutions for 10 MW Floating Offshore Wind Turbines. *INNWIND. EU Deliv. D* **2017**, *4*, 37.
27. Yu, W.; Müller, K.; Lemmer, F.; Bredmose, H.; Borg, M.; Sanchez, G.; Landbo, T. Public definition of the two LIFES50+ 10MW floater concepts. *LIFES50+ Deliv.* **2017**, *4*.
28. Armesto, J.A.; Jurado, A.; Guanche, R.; Couñago, B.; Urbano, J.; Serna, J. Telwind: Numerical analysis of a floating wind turbine supported by a two bodies platform. In *International Conference on Offshore Mechanics and Arctic Engineering*; American Society of Mechanical Engineers: New York, NY, USA, 2018; Volume 51319, p. V010T09A073.
29. Baita-Saavedra, E.; Cordal-Iglesias, D.; Filgueira-Vizoso, A.; Morató, À.; Lamas-Galdo, I.; Álvarez-Feal, C.; Carral, L.; Castro-Santos, L. An economic analysis of an innovative floating offshore wind platform built with concrete: The SATH® platform. *Appl. Sci.* **2020**, *10*, 3678. [CrossRef]
30. Uzunoglu, E.; Guedes Soares, C. Hydrodynamic design of a free-float capable tension leg platform for a 10 MW wind turbine. *Ocean Eng.* **2020**, *197*, 106888. [CrossRef]
31. Guo, X.; Zhang, Y.; Yan, J.; Zhou, Y.; Yan, S.; Shi, W.; Li, X. Integrated Dynamics Response Analysis for IEA 10-MW Spar Floating Offshore Wind Turbine. *J. Mar. Sci. Eng.* **2022**, *10*, 542. [CrossRef]
32. DNV Group. *DNV-OS-E301 Position Mooring*; DNV GL: Oslo, Norway, 2015.
33. Xu, S.; Wang, S.; Guedes Soares, C. Review of mooring design for floating wave energy converters. *Renew. Sustain. Energy Rev.* **2019**, *111*, 595–621. [CrossRef]
34. Masciola, M.; Jonkman, J.; Robertson, A. Implementation of a multisegmented, quasi-static cable model. In Proceedings of the Twenty-Third International Offshore and Polar Engineering Conference, Anchorage, AK, USA, 30 June–5 July 2013.

35. Depalo, F.; Wang, S.; Xu, S.; Guedes Soares, C. Design and Analysis of a Mooring System for a Wave Energy Converter. *J. Mar. Sci. Eng.* **2021**, *9*, 782. [CrossRef]
36. Masciola, M.; Robertson, A.; Jonkman, J.; Coulling, A.; Goupee, A. Assessment of the importance of mooring dynamics on the global response of the DeepCwind floating semisubmersible offshore wind turbine. In Proceedings of the Twenty-Third International Offshore and Polar Engineering Conference, Anchorage, AK, USA, 30 June–5 July 2013.
37. Bae, Y.H.; Kim, M.H.; Kim, H.C. Performance changes of a floating offshore wind turbine with broken mooring line. *Renew. Energy* **2017**, *101*, 364–375. [CrossRef]
38. Depalo, F.; Wang, S.; Xu, S.; Guedes Soares, C.; Yang, S.H.; Ringsberg, J.W. Effects of dynamic axial stiffness of elastic moorings for a wave energy converter. *Ocean Eng.* **2022**, *251*, 111132. [CrossRef]
39. Amaechi, C.V.; Odijie, A.C.; Wang, F.; Ye, J. Numerical investigation on mooring line configurations of a Paired Column Semisubmersible for its global performance in deep water condition. *Ocean Eng.* **2022**, *250*, 110572. [CrossRef]
40. Maples, B.; Saur, G.; Hand, M.; Van De Pietermen, R.; Obdam, T. *Installation, Operation, and Maintenance Strategies to Reduce the Cost of Offshore Wind Energy (No. NREL/TP-5000-57403)*; National Renewable Energy Laboratory (NREL): Golden, CO, USA, 2013.
41. Sarker, B.R.; Faiz, T.I. Minimizing maintenance cost for offshore wind turbines following multi-level opportunistic preventive strategy. *Renew. Energy* **2016**, *85*, 104–113. [CrossRef]
42. Santos, F.P.; Teixeira, A.P.; Guedes Soares, C. Influence of Logistic Strategies on the Availability and Maintenance Costs of an Offshore Wind Turbine. In *Safety, Reliability and Risk Analysis: Beyond the Horizon*; Steenbergen, R.D.J.M., van Gelder, P.H.A.J.M., Miraglia, S., Vrouwenvelder, A.C.W.M.T., Eds.; CRC Press: Boca Raton, FL, USA, 2014; pp. 791–799.
43. Santos, F.P.; Teixeira, A.P.; Soares, C.G. Modelling and simulation of the operation and maintenance of offshore wind turbines. *J. Risk Reliab.* **2015**, *229*, 385–393. [CrossRef]
44. Dalgic, Y.; Lazakis, I.; Dinwoodie, I.; McMillan, D.; Revie, M. Advanced logistics planning for offshore wind farm operation and maintenance activities. *Ocean. Eng.* **2015**, *101*, 211–226. [CrossRef]
45. Akbari, N.; Irawan, C.A.; Jones, D.F.; Menachof, D. A multi-criteria port suitability assessment for developments in the offshore wind industry. *Renew. Energy* **2017**, *102*, 118–133. [CrossRef]
46. Hassan, M.A.A.A.; Guedes Soares, C. Analysis of vessel shielding effects during installation of spar floating wind turbine. In *Advances in Renewable Energies Offshore*; Guedes Soares, C., Ed.; Taylor & Francis Group: London, UK, 2019; pp. 693–702.
47. Hassan, M.A.A.A.; Guedes Soares, C. Installation of pre-assembled offshore floating wind turbine using a floating vessel. In *Developments in Renewable Energies Offshore*; Guedes Soares, C., Ed.; Taylor and Francis: London, UK, 2021; pp. 461–468.
48. Jin, J.; Jiang, Z.; Vatne, S.R.; Ren, Z.; Zhao, Y.; Gao, Z. Installation of pre-assemsbled offshore wind turbines using a catamaran vessel and an active gripper motion control method. In *Grand Renewable Energy proceedings*; Japan Council for Renewable Energy: Tokyo, Japan, 2018; p. 156.
49. Mas-Soler, J.; Uzunoglu, E.; Guedes Soares, C.; Bulian, G.; Souto-Iglesias, A. An experimental study on transporting a free-float capable tension leg platform for a 10 MW wind turbine in waves. *Renew. Energy* **2021**, *179*, 2158–2173. [CrossRef]
50. Santos, F.P.; Teixeira, A.P.; Guedes Soares, C. Maintenance Planning of an Offshore Wind Tur bine using Stochastic Petri Nets with Predicates. *J. Offshore Mech. Arct. Eng.* **2018**, *140*, 021904.
51. Martins, D.; Muraleedharan, G.; Guedes Soares, C. Analysis on weather windows defined by significant wave height and wind speed. In *Renewable Energies Offshore*; Guedes Soares, C., Ed.; Taylor & Francis Group: London, UK, 2015; pp. 91–98.
52. Walker, R.T.; van Nieuwkoop-McCall, J.; Johanning, L.; Parkinson, R.J. Calculating weather windows: Application to transit, installation and the implications on deployment success. *Ocean Eng.* **2013**, *68*, 88–101. [CrossRef]
53. DNV Offshore Standard DNV GL-ST-N001. In *Marine Operations and Marine Warranty*; DNV: Bærum, Norway, 2018.
54. O'Connor, M.; Lewis, T.; Dalton, G. Weather window analysis of Irish west coast wave data with relevance to operations & maintenance of marine renewables, Renew. *Energy* **2013**, *52*, 57–66.
55. Walker, R.T.; Johanning, L.; Parkinson, R. Weather windows for device deployment at UK test sites: Availability and cost implications. In Proceedings of the 9th European Wave and Tidal Energy Conference, Southampton, UK, 5–9 September 2011.
56. CoreMarine Mission Planner. Available online: https://missionplanner.core-marine.com/ (accessed on 30 March 2021).
57. NKUA. High Resolution Wave and Wind Hindcast Database. In *Europe 2000–2010*; NKUA: Athens, Greece.
58. Guedes Soares, C.; Bento, A.R.; Goncalves, M.; Silva, D.; Martinho, P. Numerical evaluation of the wave energy resource along the Atlantic European coast. *Comput. Geosci.* **2014**, *71*, 37–49. [CrossRef]
59. Goncalves, M.; Martinho, P.; Guedes Soares, C. A hindcast study on wave energy variability and trends in Le Croisic, France. In *Progress in Renewable Energies Offshore*; Guedes Soares, C., Ed.; Taylor & Francis Group: London, UK, 2016; pp. 3–9.
60. Goncalves, M.; Martinho, P.; Guedes Soares, C. A 33-year hindcast on wave energy assessment in the western French coast. *Energy* **2018**, *165*, 790–801. [CrossRef]
61. Map data. © 2021 Google, Imagery: © 2021 TerraMetrics, LLC.

Journal of
Marine Science and Engineering

MDPI

Article

Failure Rate Assessment for Onshore and Floating Offshore Wind Turbines

He Li [1], Weiwen Peng [2], Cheng-Geng Huang [2] and C. Guedes Soares [1,*]

1 Centre for Marine Technology and Ocean Engineering (CENTEC), Instituto Superior Técnico, Universidade de Lisboa, 1049-001 Lisbon, Portugal
2 School of Intelligent Systems Engineering, Sun Yat-Sen University, Guangzhou 510275, China
* Correspondence: c.guedes.soares@centec.tecnico.ulisboa.pt

Abstract: A detailed analysis is performed on a dataset of failure and maintenance records from various onshore wind farms located in different geographical areas for the safety, risk, reliability, availability, and maintainability characterization of wind turbines. Specifically, characteristics related to failures, including the criticality of failure modes, failure frequencies, failure rates, and lifetime distributions of components, are analyzed to support the failure identification and failure prevention of wind turbines. Additionally, characteristics of maintenance, including typical maintenance measures of failures, policies for spare components, delayed maintenance, as well as related times such as reaction time, travelling time, and mean time to repair, are provided to support the maintenance management of wind farms. Based on the operational data analysis results, a reliability influence factor-based failure data correction approach is presented to transfer the onshore data to floating offshore turbines by modeling the differences in failure occurrences based on experts' judgment. A comprehensive comparison with existing studies validates the performance of the proposed approach.

Keywords: wind energy; wind turbine; failure rate correction; failure data; maintenance data

Citation: Li, H.; Peng, W.; Huang, C.-G.; Guedes Soares, C. Failure Rate Assessment for Onshore and Floating Offshore Wind Turbines. *J. Mar. Sci. Eng.* **2022**, *10*, 1965. https://doi.org/10.3390/jmse10121965

Academic Editor: Vengatesan Venugopal

Received: 2 November 2022
Accepted: 7 December 2022
Published: 10 December 2022

1. Introduction

Wind energy, including floating offshore wind, is taking over the market share compared to traditional energies [1–4]. The ever-decreasing levelized cost of energy [5] and progressive advancement of modern technologies [5–7], together with rapidly accumulating domain knowledge and experiences [8–10], have contributed to speeding up the development of wind energy in the last several decades [11,12]. At present, the economic performance of onshore wind energy competes with traditional energy sources despite the drawbacks of seasonally variable output and environmental problems such as bird crashes and sound pollution [2,3,13]. However, the electricity produced from offshore wind is still expensive [14,15]. For instance, the power generation cost of a 2 MW offshore wind turbine operating at 50% efficiency and for 60% of the available time is about 135 GBP/MWh, which is obviously higher than that of onshore wind turbines at a power generation cost of 100 GBP/MWh [16].

The Operation and Maintenance (O&M) cost of wind turbines takes up to 20% of the overall energy cost, and such a proportion becomes bigger when referring to floating offshore wind [17,18]. Wind farm developers, operators, and users are trying to reduce the O&M costs of wind projects by understanding the failure features of wind turbines, improving the capability of maintenance resource management, and finding feasible measures to prevent the equipment from undergoing unexpected failures. However, the above aspects have been restricted by insufficient O&M data, primarily due to confidential issues among wind farms, operators, manufacturers, and other stakeholders [19–21].

To enrich the data on the failure, risk, reliability, availability, maintainability, and economic applications of wind turbines, the industry has established several databases, as listed in Table 1.

Table 1. Databases of wind turbines.

Country	Names of Databases	Number of WTs	Years
Germany	WindStats Germany, LWK, WMEP	6428	1989–2006
Denmark	WindStats Denmark	2345	1994–2004
Finland	VTT	72	1996–2008
Sweden	Sweden	723	1997–2005
Spain	CIRCE	4300	~2013
USA	CREW	800–900	2011–2015
India	India	15	2000–2004
China	Huadian, East China, SE China	1555	2009–2013
UK	Round 1 UK *, SPARTA *, Strathclyde	1516	2004–2016
Netherlands	NoordzeeWind	36	2007–2009
——	LGS-Onshore **	76	2018–2020

*: offshore wind; **: this paper; LWK: Landwirtschaftskammer Schleswig-Holstein; WMEP: Wissenschaftliches Mess- und Evaluierungsprogramm; VTT: Technical Research Centre of Finland; CIRCE: Universidad de Zaragoza; CREW: Continuous Reliability Enhancement for Wind; SPARTA: System Performance, Availability and Reliability Trend Analysis.

The mentioned datasets include the following main parameters:

- Failure rates from the WindStats Germany [22] and the Swedish [23] databases;
- Downtimes from the WMEP [24] and the VTT [25] databases;
- Productivity based on the VTT database [25];
- Stop rates (not limited to failures) from the NoordzeeWind database [26];
- The monthly number of repairs based on the SPARTA database [17];
- Failure and repair classifications based on the Strathclyde database [27].

However, the feasibility of such published information depends on the completeness of collected data and the comprehensiveness of analyses. Integrated analysis of multiple databases benefits the comprehensiveness of the analysis and the robustness of the results extracted. Failure rates of wind turbines based on a combination of several databases have been reported in [19,20,28].

The released databases of wind turbines installed from 1986 to 2016 indicate that the most recent populations are more than five years old and have, thus, passed the warranty period. However, the advancements in materials, manufacturing, design concepts, and O&M tools have led to a technical revolution in wind energy, especially in wind turbine applications [29–31]. The above indicates that the old datasets represent the properties of the previous wind turbines but may not apply to the recent ones. Hence, reporting new and more recent operation data with a comprehensive analysis can provide stakeholders and practitioners in the wind energy sector with urgently needed information on wind turbines' failure and maintenance features.

Floating offshore wind turbines, representing the next step in the wind energy market, are new concepts with limited installations [3,4]. Failure, risk, reliability, availability, and maintainability investigations of such equipment are restricted by unavailable failure and operation data. Reported onshore-data-based analyses such as reliability analysis [18,32], failure rate assessment [33], mean time to failure prediction [18,34], and maintenance strategy planning [30,35] cannot be applied to floating offshore wind facilities without pre-treatments. To this end, the construction and application of the failure rate correction approach to transfer the fruitful accumulated operation data of onshore wind turbines to the corresponding components of floating offshore structures would enrich the database of floating offshore wind turbines at an early stage of operation under the situation of unavailable operation data.

Stress factor-based and reliability influence factor-based (RIF-based) methods are applied to the failure rate correction/prediction of similar products [33]. Stress factor-based approaches reflect differences between the failure rates of new components of a product and those of similar ones using environmental and power utilization stress factors. For instance, Santos et al. [36] predicted failure rates of offshore wind turbines' components according to the operation data of onshore structures with the assistance of a stress factor model. Theoretically, however, the method requires extensive environmental and operation data collection, restricted to model applications.

RIF-based failure rate correction methods map the differences between the RIFs of components so that the failure rates of new (sometimes unknown) elements can be corrected/predicted. For instance, based on a similar topside system, Rahimi et al. [37] assigned weights to the RIFs of components of a subsea system to infer the failure properties of the system with the assistance of the scaling factor and boundary values designed. To upgrade the above deterministic model, Bhardwaj et al. [38] proposed a failure rate prediction model using Bayesian networks to represent the uncertainty of RIFs. The Bayesian network-based failure correction model was validated by analyzing a subsea processing system. Brissaud et al. [39] constructed an influencing reference coefficient of RIFs for failure rate assessment, which predicts the failure rates of components as an interval (considering uncertainties of RIFs) rather than a single value as existing methods did.

However, the approaches we have already mentioned are complicated, and they either require extensive data collection or additional indicators (or both simultaneously). With these restrictions, more recently, Li et al. [33] proposed a risk-RIF-based failure rate correction model to transform the failure data of onshore and bottom-fixed offshore wind turbines to a comparable level of floating equipment. The method requires fewer inputs but has better accuracy. However, it is pointed out that the risk-RIF-based failure rate correction model calculates a general indicator to reflect the overall differences of RIFs between various types of wind turbines and cannot map the discrepancies between each component. From a failure mechanism point of view, the harsh and complicated sea conditions introduce more fragility and initiate more failures in components exposed to them, such as floating foundations, mooring, towers, and blades, compared with those installed in nacelles, such as gearboxes, generators, and convertors. Hence, the failure rate correction should reflect the specific failure properties of each component.

Accordingly, following the risk-RIF-based failure rate correction model in [33], this paper proposes a failure rate correction approach for mapping the failure data of new products in their infantile stage, such as floating offshore wind turbines, from similar maturely operated equipment such as onshore wind turbines. The novel contributions of this paper are as follows:

(i) A new operation dataset of wind turbines is opened, which is the most up-to-date and can better reflect the failure and maintenance properties of recent wind turbines and strengthen the database of the wind energy sector.
(ii) A failure rate correction approach is proposed to transfer well-accumulated onshore wind turbines to supplement the insufficient operation data of the floating offshore wind sector. This extends the model in [33], in which the failure rate correction calculates a general indicator to reflect the overall differences between onshore and floating offshore wind turbines, and cannot reflect the discrepancies between each component. Meanwhile, in the present paper, the failure rate correction model is constructed using a bottom-up approach from elements to components until the wind turbine level is reached, to examine the failure rate differences of all the components.

The proposed failure rate correction approach can be applied to all scenarios of approximately inferring the failure rates of new products from maturely operated ones with sufficient operation data or knowledge, and it is not limited to the wind energy sector.

The rest of this paper is arranged as follows. Section 2 presents the particulars of the dataset. Section 3 analyzes the failure data of wind turbines. Section 4 provides maintenance information. Section 5 introduces the failure rate correction approach. The results, comparisons, validations, and discussions are listed in Section 6. The conclusions are provided in Section 7.

2. Features of the Dataset

This analysis includes 423 failures of 76 multi-MW (Megawatt) collected from wind turbines operating in four wind farms in China, see Figure 1. The observation consists of about 1.44 million operation hours or more than 164 turbine years. Different manufacturers make wind turbines with five configurations. For confidentiality reasons, the types of wind turbines and the power, rotor diameter, blade size, manufacturers, and operators of the wind turbines are not mentioned here. The wind turbines are less than three years old and are operated by users, but the manufacturers carry out maintenance as the wind turbines are in their warranty period.

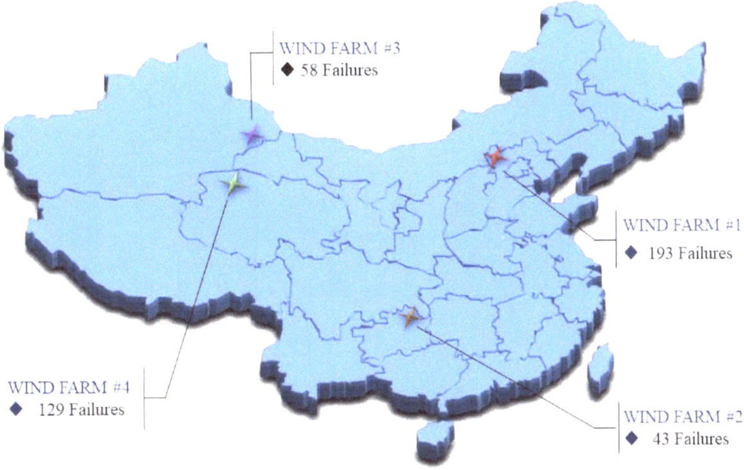

Figure 1. Wind farm locations in China.

The four wind farms are located in different areas with significant geographical and climatic discrepancies. Specifically, wind farms #1, #3, and #4 are located in flatlands with a more than 40 °C temperature difference between summer and winter (below 0 °C). However, wind farm #2 is situated on a mountain with a comparatively small temperature difference between winter and summer. The wind speed in wind farms #3 and #4 is higher than in the other wind farms, but sand content in the air is higher than in wind farm #1 and significantly higher than in wind farm #2.

3. Failure Data Analysis

3.1. Failure Definition

There is no standardized document issued to define wind turbine failures. Hence, researchers have defined wind turbine failures from different viewpoints based on the available data. Based on the downtime data, Wilkinson et al. [40] defined a failure as an event with downtime longer than one hour and requiring at least a manual restart of the wind turbines. Based on the operational data, Carroll et al. [27] defined a failure as an unscheduled visit to a wind turbine in which materials (elements or components) are consumed to enable the wind turbines to function.

This analysis based on maintenance records, including corrective and preventive maintenance, defines a failure as (i) an event that causes wind turbine stoppage; (ii) the triggering of Supervisory Control and Data Acquisition System (SCADA) or Condition Monitoring System (CMS) alarms due to a dangerous operational state; (iii) harmful states of components identified in predetermined inspections, such as the iron element (Fe) content of lubricating oil exceeding 200 mg/kg; or (iv) the predetermined periodical replacement of components.

3.2. Failure Properties

Overall, the observed wind turbines comprise 22 elements. A representative configuration of wind turbines is displayed in Figure 2. This analysis categorizes the 22 elements into seven units [12,41–43], including the Rotor (Blades, Hub, Main Bearing, and Main Shift), Generator, Gearbox, Electrical Facilities (Converter, Transformer, Monitoring and SCADA, Weather Unit, Power and Controller, and Other Electrical Facilities), Pitch & Yaw (Pitch System and Yaw System), Cooling & Hydraulics (Cooling System and Hydraulics), and Auxiliary (Crane, Climbing Aid, Brake, and Nacelle).

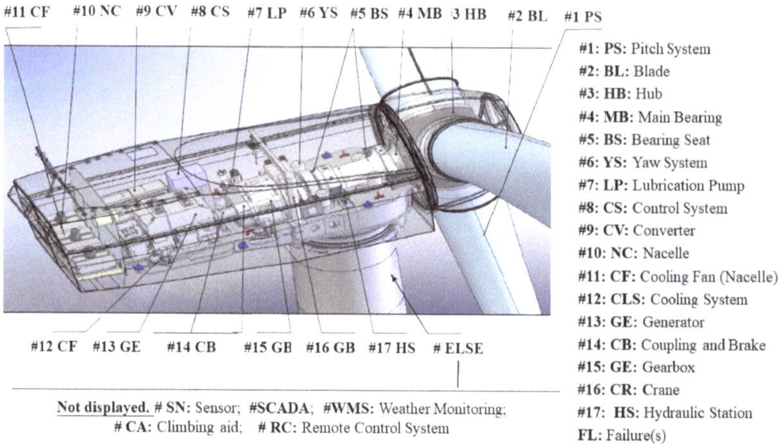

Figure 2. A representative configuration of a wind turbine.

3.2.1. Failure Mode and Failure Criticality

Failure modes are observable fault presentations, representing unhealthy working states of wind turbines. The collected 423 failures involved 112 failure modes, indicated in Appendix A Table A1. Electrical Facilities, Pitch & Yaw, and Cooling & Hydraulics process more than 71% (80 out of 112) failure modes of all the wind turbines. From a failure frequency point of view, the mentioned components contribute to over 67% of failures. The majority are functional failures rather than the structural failures of the Rotor, Generator, Gearbox, and Auxiliary (see Figure 3).

Electrical Facilities and Cooling & Hydraulics failures are common failures in wind farms #3 and #4. There is no evidence to illustrate the exact reasons, but the weather conditions may be a contributor. Specifically, air sand content and temperature differences (day and night, winter and summer) in these two wind farms are higher than in the others, which may introduce vulnerabilities to the wind turbines.

Failures mostly happen inside Nacelles, and a few are related to the SCADA, weather monitoring, and remote control systems. In addition, the majority of functional failures (79% of the total) require restarting or adjustment, while hard failures that require material consumption are a minority. Generally, most hard failures, such as anemometer damage, and several functional failures, such as wrong pitch angle, lead to severe consequences such as showdowns. For instance, 38% of failures of Rotors, 30% of failures of Electrical Facilities, and 24% of failures of Gearboxes made the wind turbines shut down, which is higher than for Cooling & Hydraulics (20%), Pitch & Yaw (15%), Generators (12%), and Auxiliary components (0%).

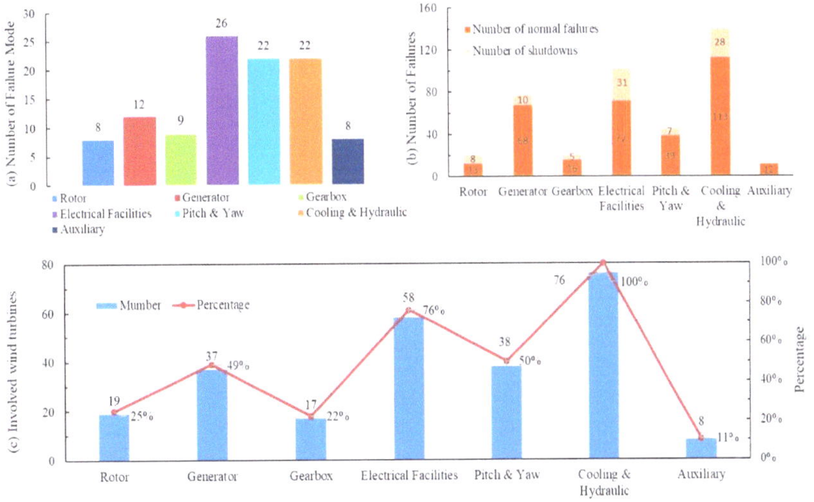

Figure 3. Failures of components. (**a**) Number of failure mode instances; (**b**) number of failures; and (**c**) involved wind turbines.

Additionally, a failure mode's severity depends on the damage to the wind turbines. For example, three out of 15 cabin brush failures caused the generator's shutdown due to heavy wear. In most cases, cabin brush wear can be identified and replaced according to the monitoring data and periodical inspections.

Cooling & Hydraulics failures happen commonly, indicating that the quality control of such a component is essential. Electrical Facilities failures happened in 58 wind turbines among the total 76 wind turbines, which is more than for Pitch & Yaw failures (38 wind turbines) and Generator failures (37 wind turbines).

It is worth pointing out that Gearbox failures mostly happened in 17 wind turbines of two types from one manufacturer. Hence, the manufacturing capability of manufacturers is a crucial factor in the quality of wind turbines. The above findings call for more attention from users (in wind turbine selection) and manufacturers (in supplier management and assembly capacity improvement).

The maintenance crew categorized failures into normal failures, critical failures, and extremely critical failures, as in Table 2. Normal failures can be fixed quickly and will not significantly reduce the productivity of wind turbines, while critical failures reduce availability and require timely maintenance. Extremely critical failures result in the stoppage of wind turbines and demand additional maintenance.

Table 2. Failure criticalities of components.

Component	Normal Failure	Critical Failure	Extremely Critical Failure
Rotor	17	3	1
Generator	49	22	7
Gearbox	14	5	2
Electrical Facilities	89	8	7
Pitch & Yaw	36	10	–
Cooling & Hydraulics	109	23	9
Auxiliary	11	1	–
Summary	(327 failures)	(82 failures)	(25 failures)

Overall, 77% of failures are normal ones and can be fixed by minor repair actions. Critical (82 failures) and extremely critical (25 failures) failures require additional materials or assistance from suppliers to finalize the maintenance. To be specific:

- From a component point of view, the Gearbox and Generator fail less in frequency but result in more critical consequences than the failure of Pitch & Yaw and Cooling & Hydraulics, followed by the Electrical Facilities, Rotor, and Auxiliary components; Electrical Facilities frequently fail, with limited impact on power generation. Failures of components that are directly involved in electricity generation, such as the Gearbox and Generator, are more critical than appendant ones such as Cooling & Hydraulics. No transformer failure was observed.
- From a wind turbine point of view, the component selection impacts the failure features of wind turbines. For instance, only a limited number of critical failures (including extremely critical ones) are observed in wind turbine models #1 and #4. The wind turbines in model #5 suffer more critical failures (see Table 3). The root reason can be traced to the component selection, supplier, and maintenance measures.
- From a wind farm point of view, the weather conditions significantly affect the failure features of wind turbines. For instance, most failures of wind farms #1 and #4 are normal failures, while the proportion of critical failures (including extremely critical failures) is markedly higher in the remaining wind farms. This indicates that the wind profile affects, to some extent, the failure criticalities of wind turbines. Generally, higher wind speed, longer uninterrupted working time (wind farm #1), and operation at higher altitudes (wind farm #4) introduce additional failure criticalities to wind turbines (see Table 4).

Table 3. Failure criticalities by type of wind turbine.

Wind Turbine Type	Wind Farm Installed	Normal Failure	Critical Failure	Extremely Critical Failure
Wind turbine model #1	#1	167	23	3
Wind turbine model #2	#3	30	9	11
Wind turbine model #3	#3	2	2	4
Wind turbine model #4	#4	104	22	3
Wind turbine model #5	#2	17	21	5

Table 4. Failure criticalities by wind farm.

Wind Farms	Normal Failure	Critical Failure	Extremely Critical Failure
Wind farm #1	167	23	3
Wind farm #2	17	21	5
Wind farm #3	32	11	15
Wind farm #4	104	22	3

The failure features of wind turbines are affected by, at least, the coupling of manufacturing, wind farm weather conditions, and continuous operating time. This analysis reveals the factors that influence the failure features of wind turbines. Still, the evidence is insufficient to distinguish the influence mechanisms of each factor and their couplings on wind turbines.

3.2.2. Failure Statistics

Failure Frequency

The failure frequency of the wind turbines is shown in Figure 4, and no satisfactory distribution is found to fit the failure frequencies of the wind turbines perfectly. Instead, a polynomial function is applied to show the failure frequency of the wind turbines. More than 36% of wind turbines fail less than three times; however, most wind turbines suffered five to nine failures during observation. The above conclusions may support wind farm maintenance scheduling and supplier management.

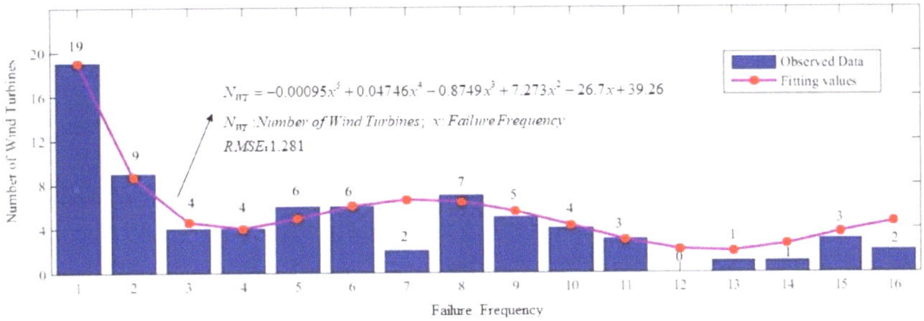

Figure 4. Failure frequency of wind turbines (RMSE: Root Mean Square Error).

Failure Criticality

The wind farm failure criticality categories are displayed in Figure 5 and Table 5. The statistics confirm that:

- The critical failure frequencies in wind farms #2 and #3 are relatively low. The manufacturer and maintenance measures contribute to this conclusion, as all the wind turbines in wind farm #2 and 7 out of 33 wind turbines installed in wind farm #3 are manufactured by one factory. On the contrary, similar wind turbines are not used in wind farms #1 and #4. Hence, the capability of manufacturers is a core factor in wind turbine quality.
- The distribution of (extremely) critical failures and critical failures in each wind farm is not even; moreover, the regular pattern of failure occurrence and the relationships between critical failures, extremely critical failures, and normal failures are not obvious.
- Wind turbines in wind farm #3 fail less, but their failures have severe consequences; for instance, 55% of wind turbines suffer only one failure during the observation period, of which 50% are extremely critical failures (6 failures) and critical failures (3 failures). Critical failures requiring complex maintenance, additional materials, or professional personnel would introduce additional downtime and maintenance costs and reduce wind turbines' availability and economic competitiveness. To this end, failure rates and failure criticalities should be considered in supplier management of wind projects.

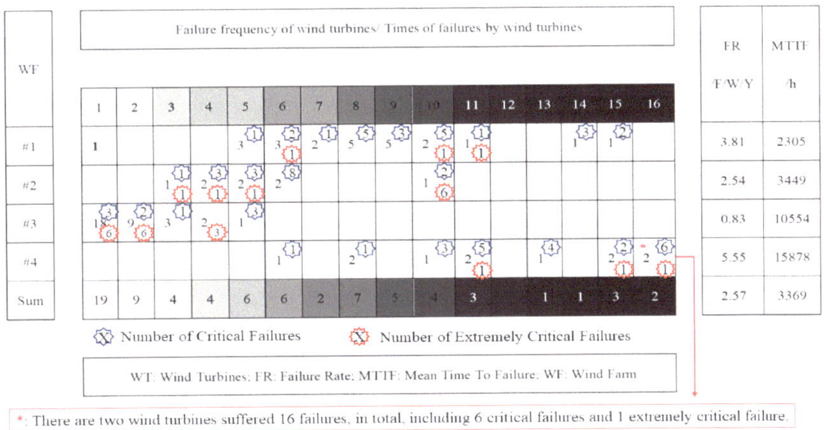

Figure 5. Failure criticality statistics by wind farm.

Table 5. The statistic of critical and extremely critical failures (per year).

Wind Farm	Wind Farm #1	Wind Farm #2	Wind Farm #3	Wind Farm #4	Average
CFR	11.9%	39.5%	15.5%	16.3%	20.8%
ECFR	1.2%	20.9%	25.9%	2.3%	12.6%
CFWT/MTTF	0.45/2.2	1/1	0.13/7.7	0.9/1.1	0.62/1.6
ECFWT/MTTF	0.06/16.7	0.53/1.9	0.31/3.2	0.13/7.7	0.34/2.9

(E)CFR: (extremely) critical failure rate; (E)CPWF: (extremely) critical failure per wind turbine per year; MTTF: mean time to failure.

Failure Rate

Failure rate represents the likelihood of failures within a unit of time. It indicates the wind turbines' inherent capability to resist failures under the coupling impacts of internal excitation such as strength degradation, and external excitation such as environmental conditions. The failure rate of the 76 wind turbines is 2.57 failures/turbine/year with an MTTF of 3409 h. The failure rates and MTTFs of the components are listed in Table 6.

Table 6. Failure rates and MTTFs of components.

Components	Subcomponents	MTTF/h	Failure Rate/Year
Rotor	Blades	50,563	0.1732
	Hub	126,407	0.0693
	Main Bearing	758,441	0.0116
	Main Shift	758,441	0.0116
Generator	Generator	9979	0.8778
Gearbox	Gearbox	26,153	0.3350
Electrical Facilities	Converter	12,641	0.6930
	Monitoring and SCADA	29,171	0.3003
	Weather Unit	68,949	0.1271
Pitch & Yaw	Pitch System	21,068	0.4158
	Yaw System	94,805	0.0924
Cooling & Hydraulics	Cooling System	10,835	0.8085
	Hydraulic	10,390	0.8431
Auxiliary	Crane	758,441	0.0116
	Climbing Aid	758,441	0.0116
	Brake	252,814	0.0346
	Nacelle	84,271	0.1040

The monthly failure rate of the wind turbines and their decomposition are displayed in Figure 6, which confirms that:

- Approximately 18 months is the inflexion point of the failure rate of wind turbines. The failure rate of the wind turbines changes irregularly before the inflexion point, and then, decreases obviously. Preventive maintenance supports the failure prevention of wind turbines, and carrying out preventive maintenance every six months is suitable according to the failure rate inflexion point of the wind turbines.
- The trends of failure rates of different types of wind turbines are similar. A total of 15 to 20 months of working is the limit for all types of wind turbines and wind farms, which almost always results from Cooling & Hydraulics failures. Hence, quality control for the Cooling & Hydraulics of wind turbines is essential. The quality of wind turbines varies a lot depending on their type; for instance, the failure rate of the wind turbines of type #1 is almost half of that of type #5, which encourages users and investors to pay close attention to their supplier selection.
- The trend of failure rate in terms of wind farms is similar to that concluded for the types of wind turbines. According to Figure 6c, flatland wind farms hold lower failure rates than those installed in mountains and desert areas. More specifically, wind

turbines in the mountains fail more in frequency than those operating in desert areas and, in turn, more than wind turbines that stand in flatland spaces.

- Cooling & Hydraulics failures are the decisive factors that increase the failure rate of wind turbines, especially over 15 to 20 months, considering that failures of wind turbines are dynamic processes that result from coupled components. Hence, dynamic, correlated, and real-time failure rate analyses of components benefit the O&M of wind turbines.

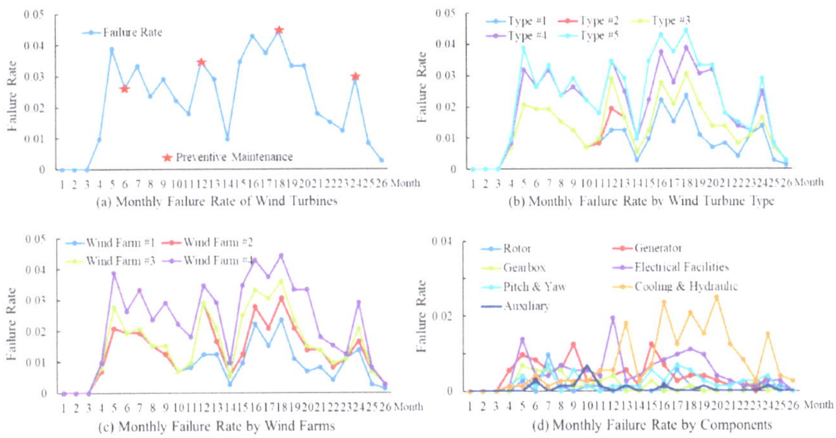

Figure 6. Failure rates: (**a**) of wind turbines as a whole; (**b**) by wind turbine type; (**c**) by wind farm; (**d**) by component.

This analysis found that the Weibull distribution satisfies the lifetimes fitting of some components better than others (see Table 7). The determination of working periods supports the maintenance scheduling and availability and productivity estimation of wind turbines, wind farms, and wind projects according to the bathtub curve concept of complex and repairable systems.

Table 7. Lifetime distribution of components (in the early failure period).

Integrated Components	Lifetime/ Weibull's Distribution	Integrated Components	Lifetime/ Weibull's Distribution
Rotor	SC (1321); SP (0.79)	Pitch & Yaw	SC (2144); SP (0.77)
Generator	SC (1894); SP (0.8)	Cooling & Hydraulics	SC (2793); SP (0.78)
Gearbox	SC (2449); SP (0.81)	Auxiliary	SC (759); SP (0.6)
Electrical Facilities	SC (2829); SP (0.83)	SC: scale parameter; SP: shape parameter	

This analysis determines that the wind turbines and their main components are in an early failure period as the fitted shape parameters of the Weibull distributions for all components are less than one. The evidence is insufficient to find the inflexion point of failure rate between the early failure and stable working periods; however, lifetime distributions of components confirm that the inflexion points of the failure rate of the components and wind turbines are later than 26 months.

4. Maintenance Data

4.1. Maintenance Measures

The maintenance crew includes onsite crew from manufacturers and offsite crew from suppliers. Eight maintenance measures are implemented (see Figure 7), including:

- Repaired: Repairing without replacement and no additional material is needed.
- Replaced: Repairing with replacement.

- Repaired and waiting for replacement: Repaired and waiting for additional materials for further replacement.
- Checked and waiting for replacement (operating): Checked without repair and waiting for additional materials for replacement. The failed wind turbine is operating.
- Checked and waiting for replacement (stopped): Checked without repair and waiting for additional materials for replacement. The failed wind turbine is stopped.
- Waiting for the supplier (operating): Waiting for the maintenance crew of suppliers. The failed wind turbine is operating.
- Waiting for the supplier (stopped): Waiting for the maintenance crew of suppliers. The failed wind turbine is stopped.
- Waiting for further instructions: The failure is unable to be repaired due to: (i) an unknow failure cause; (ii) a lack of maintenance experience; (iii) minor failures with limited impact; (iv) a problem beyond the authority of the maintenance crew; (v) huge and expensive structural failure.

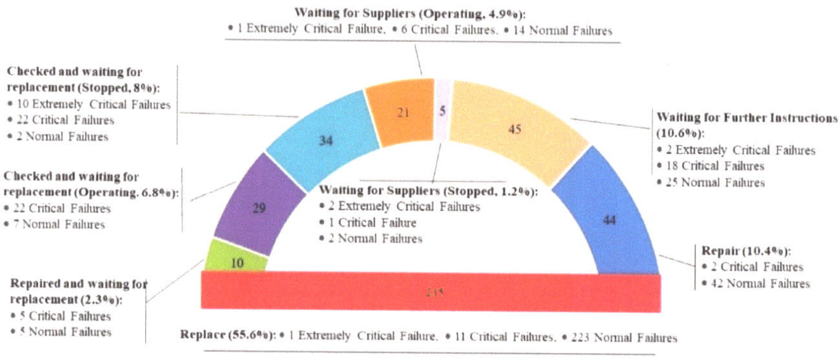

Figure 7. Maintenance measures.

The onsite maintenance crew handled the majority of failures (about 83%), and the offsite maintenance crew fixed only 6.1% of failures. However, failures fixed by the offsite maintenance crew are always difficult to repair and require additional materials that lack backup. Additionally, the remaining 11% of failures called for further negotiation between the onsite and offsite maintenance crews and would give rise to more delays in repairs.

The onsite maintenance crew repaired more normal failures (86%) than (extremely) critical failures, while the offsite maintenance crew mainly coped with complicated failures, more than 38% of which were critical ones. Overall, 66% of failures of the observed wind turbines were fixed so that they were good as new, which can be compared with about 25% of failures that were imperfectly fixed, leading to wind turbines operating with defects, and sometimes under the rated power.

4.2. Spare Parts

The non-optimized spare parts policy, which refers to components as a whole and their elements, delays wind turbine maintenance activity and plays a crucial role in wind farm management. The spare parts policy includes storeroom planning, elements distribution (into storerooms), and logistics. Two types of storerooms are established for wind farms (wind farm storerooms) and manufacturers (manufacturer storerooms).

The observed 423 failures involved more than 400 elements stored in wind farm storerooms (31%) and manufacturer storerooms (26%), and no backup elements made up the remaining proportion (43%). The wind farm storerooms contain fragile, low-value consumables such as lubricating oil and the manufacturer storerooms keep high-value elements. The time-consuming logistics give rise to longer maintenance delays. The elements without backup are always the ones that result in severe consequences and long downtimes; such elements are mostly from suppliers.

Overall, the 74 instances of delayed maintenance are consequences of lacking spare elements, involving almost all the key components of wind turbines (see Table 8), which indicates that feasible spare parts management contributes to the efficient maintenance of wind turbines and economic operations of wind farms. Additionally, the spare parts policy should also consider the health status assessment of components, economy-based storeroom optimization, and logistics.

Table 8. Delayed maintenance to failures.

Subcomponent	Elements	Number of Failures	Subcomponent	Elements	Number of Failures
Generator	Pump Motor	2	Hub	Steel Frame	3
	Screw Nut	1	Pitch System	Lubrication Oil	3
	Cooling Fan	1	Convertor	Cooling Fan	1
	Slip Ring	6		Circuit Board	2
	Encoder	4	Gearbox	Pump Motor	4
	Bearing	1		Lubrication Oil	2
	Water Pump	1	Nacelle	Sunroof Support	2
	Sensor	1		Guardrail	1
Hydraulic	Storage tank	1	Cooling System	Sensor	2
	Electromagnetic valve	1		Water Tank	1
	Oil Hose	1		Pump Motor	3
	Switch	1		Cooling Fan	1
	Filter	1		Switch	4
Nacelle Cooling	Motor	3	Control System	Cable	3
	Fan	2		Switch	1
Weather Unit	Anemometer	5	Yaw System	Brake	1
Climbing Aid	Control elements	1		Oil Hose	1

4.3. Maintenance Times

The time spent in maintenance is the basis of the availability, economy, and productivity estimation of wind turbines and wind farms. According to maintenance records, maintenance procedures for failures can be divided into three types, including Reaction Time (RT), Traveling Time (TT), and Time to Repair (TR) (see Figure 8).

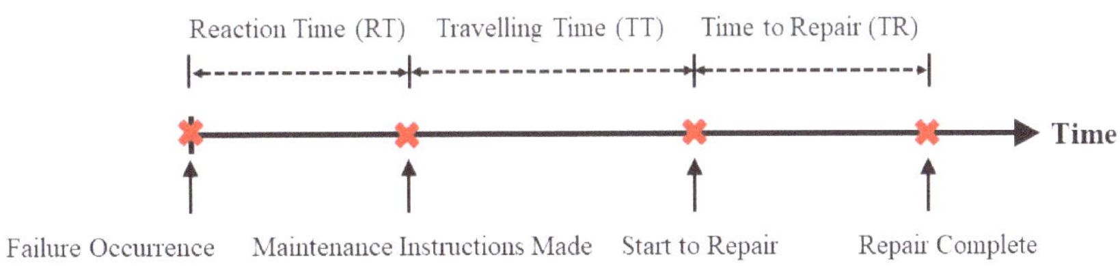

Figure 8. RT, TT, and TR of wind turbines.

4.3.1. Reaction Time

Reaction time is the period between failure and maintenance instructions being given. It includes failure identification based on the alarm of the SCADA and/or CMS systems, spare parts checking, and maintenance instruction determination. The maintenance instructions indicate failed components to the maintenance crew, and very few of them (less than 10%), can extend to element and root-failure-cause levels. As a consequence of incomplete records, 89% of the observed failures (375 RTs) were extracted from the maintenance records (see Figure 9).

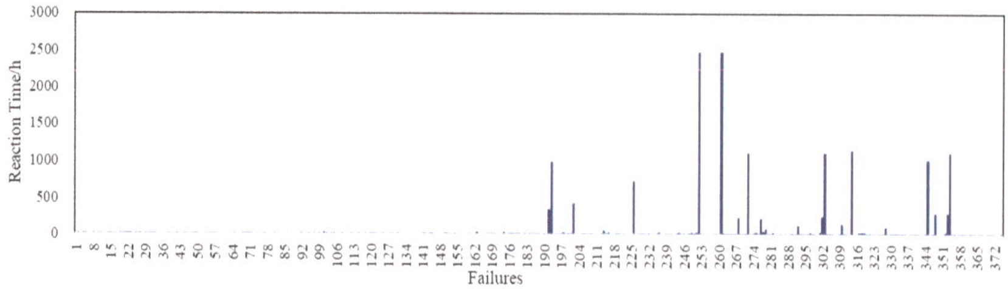

Figure 9. RTs of the maintenance of the 375 failures.

The average RT of the 375 failures is 41 h. However, 147 failures (nearly 40% of the total) with an RT of zero indicate that the maintenance instructions of some frequently occurring or difficult-to-repair failures can be made without any delay. Moreover, the average RT of the remaining 228 failures is 68 h. To be specific, the maintenance instructions of 200 failures (88% of the total) are made within a day, 212 failures (93%) within a week, and 220 failures (96%) within a month.

Eight generator, blade, and converter failures contribute an 86% RT to the total (see Figure 10). Logistics, spare parts, and complexity in maintenance are primary factors of RTs. Specifically, the bearing degradation (calling for replacement) of generators, the protective film peeling off the blades, and control circuit-board damage of the converter result in long RTs as there is no possibility of backing up such components in storerooms, for wind farms or for manufacturers. Moreover, the logistics with unpredictable risks, high costs, and complex maintenance require a longer waiting time for professional personnel, which further delays the RTs.

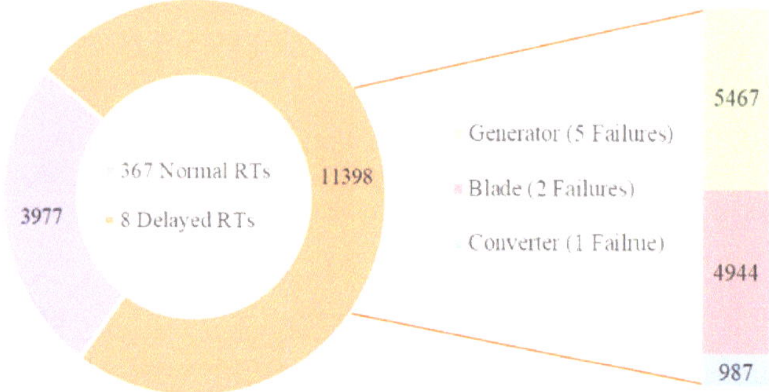

Figure 10. Delayed RTs/hour.

Except for the eight instances of delayed maintenance for failures with long RTs, the remaining 96% of failures had an average RT of 4.3 h. Thus, 4.3 h would be a robust benchmark for decision-making related to wind turbine maintenance. The RTs, except for delayed ones, are sorted by failure criticalities and maintenance measures in Table 9. More detailed information can be obtained by analyzing the data collected. However, further results are not demonstrated as they may not significantly contribute to the maintenance of wind turbines.

Table 9. RTs of maintenance sorted by failure criticality and maintenance measures/hours.

RTs of Maintenance/Hour (Average RT)	Extremely Critical Failures	Critical Failures	Normal Failures	RT Average
Repaired	—	12	6	6.3
Replaced	—	1	4	3.9
Repaired and waiting for replacement	—	1	2	1.5
Checked and waiting for replacement (operating)	—	4	1	3.2
Checked and waiting for replacement (stopped)	—	1	—	1
Waiting for Suppliers (operating)	—	1	1	1
Waiting for Suppliers (stopped)	—	0	25	16.7
Waiting for further instructions	13	1	1	1
RT Average	13	2.2	4	4.3

0: Less than one hour.

4.3.2. Traveling Time

Travelling Time (TT) is the period between maintenance instructions being formed and the maintenance being started. It mainly includes the transportation of personnel and components from storerooms or operation centers to wind farms. Overall, 338 (80%) travelling times were extracted from the maintenance records of the 76 wind turbines.

The average travelling time for the failures is 5.5 h (see Figure 11). Overall, the travelling time of 36% of failures is less than one hour, while 39% are over 12 h. The travelling times follow a bimodal distribution with mean values of 0.46 h and 12.51 h. It is concluded that gearboxes have the longest travelling time (24 h) due to their enormous size, lack of backups, and the necessity for special vehicles. Furthermore, there is no obvious evidence of the reason for the formation of bimodal distribution. However, the figure of 0.46 h is likely the travelling time to failed wind turbines without access to storerooms, and that of 12.51 h may include additional time to access the storerooms. The above conclusions indicate that enhancing the capability of the management of the operation centers, and the establishment of feasible and sufficient scheduling systems for spare parts, would contribute to the improvement of the economic operations of wind turbines.

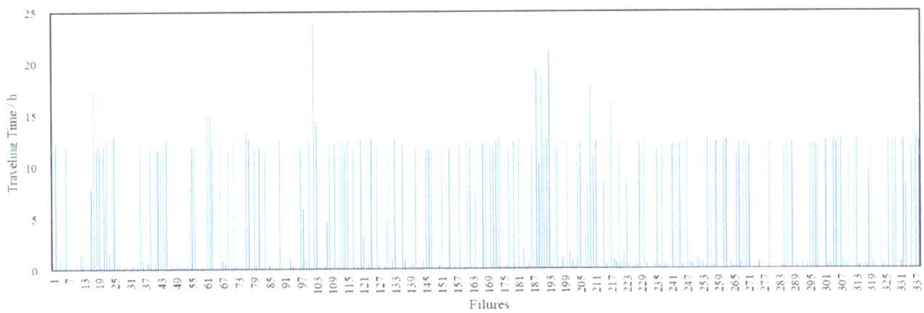

Figure 11. Travelling times of maintenance to the 338 failures.

The types of components and the location of the spare parts are decisive in the travelling times of maintenance, as: (i) the types of components determine the method of shipping, as the shipping of special components such as blades requires dedicated equipment and vehicles; (ii) the location of the spare parts (storerooms) affects the time of movement between storerooms and failed wind turbines. However, no robust evidence shows which factor is more crucial. Table 10 summarizes the travelling times of each component and is sorted by storeroom.

Table 10. Travelling times of maintenance to components, sorted by storeroom/hours.

STS of Components/h (Number of Failures/Average TT)	Spare parts in Wind Farm Storeroom	Spare Parts in Manufacturer Storeroom	TT Average
Rotor	—	9/1.6	1.6
Generator	8/3	14/5.3	4.5
Gearbox	4/3	10/6	5.1
Electrical Facilities	20/3.9	22/7.7	5.9
Pitch & Yaw System	13/6.9	7/8	7.3
Cooling & Hydraulics	55/6.2	25/3.6	5.4
Auxiliary	2/12.5	1/1	8.7
Average	5.6	5.3	5.5

4.3.3. Time to Repair

Time to Repair (TR) is a period between the start of maintenance and its completion, representing the capability and productivity of the maintenance crew. The mean Time to Repair (MTTR) was used as an index to rate the capacity of a system or equipment recovery from failures with the intervention of maintenance personnel. Overall, the MTTRs of 266 failures of wind turbines were extracted, the average of which was 3.6 h, with a maximum of 23 h and a minimum of less than one hour (see Figure 12).

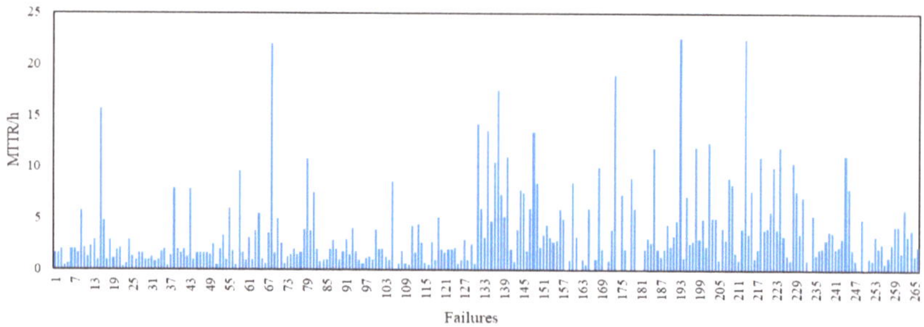

Figure 12. MTTRs of the 266 failures.

Very few failures need long-time maintenance that exceeds 20 h, including converter cable replacement (22.5 h) and monitoring and control system repair (23 h). Lacking experience in maintenance is the root reason for longer maintenance times, which calls for strict training for the maintenance crew on the repair of some infrequent failures, such as cable failures. This can prevent the maintenance crew from staying inside the wind turbines for a long time, which endangers the personnel. Overall, there were 15 failures with MTTRs of less than one hour, including nine easy-to-repair failures fixed via restarting/resetting and six unhandled failures.

The MTTRs of components are sorted by maintenance measure in Appendix A Table A2. Most wind turbine failures can be repaired within 5 h or replaced within 3.4 h, while the remaining 8% of failures cannot be fixed as soon as the maintenance crew arrives and in one visit. The maintenance of Electrical Facilities such as weather units and Monitoring, and SCADA and machinal elements such as Brakes and Nacelles takes longer. However, the frequencies of those failures are low. The maintenance crew's capability for repair the converter, generator, and hydraulics should be enhanced, as these elements fail more frequently and require a relatively longer Time to Repair. To conclude, Appendix A Table A3 shows a summarization of the RTs, TTs, and MTTRs of components that wind farms sort.

5. The Proposed Failure Rate Correction Methodology

This section proposes a developed failure rate correction approach to infer the failure rates of floating offshore wind turbine components according to sufficiently accumulated onshore data. Failure rate correction is a modification procedure of the absolute value of onshore wind turbine failure rates; it infers the failure rate of floating offshore wind turbines so that it can reflect the failure properties of floating offshore wind turbines. The constructed model in this paper is an extended version of that proposed in Ref. [33], and aims to find a transformation factor reflecting differences between the failure properties of onshore and floating offshore wind turbines' components based on evaluations by experts. To remove the limitation of the earlier method only being applicable to the wind turbine's level, this paper constructed a localized methodology to map such differences at all levels of wind turbines, including components and elements. The proposed model in this paper can provide practitioners with a better, deeper, and more detailed understanding of the failure properties of the components and elements of floating offshore wind turbines, as well as their differences from onshore ones.

The developed schedule is demonstrated in Figure 13 and comprises the following steps:

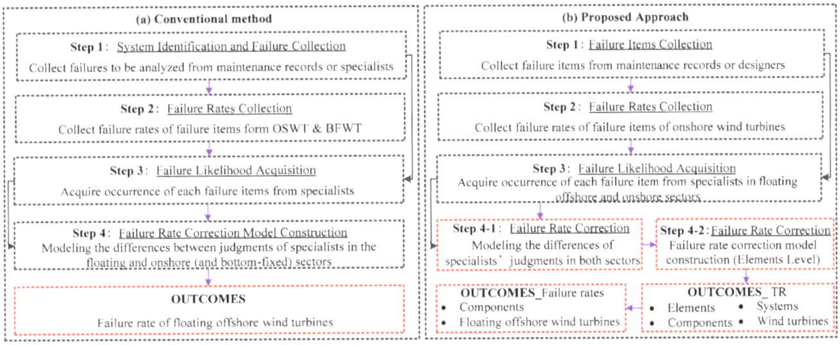

Figure 13. The framework of the proposed failure rate correction approach/conventional method is referenced from [33]. OSWT: onshore wind turbine; BFWT: bottom-fixed wind turbine; TR: transformation factor.

Step 1: Failure item collection. Collect the failure items of floating offshore wind turbines from the maintenance records of onshore wind turbines or the designers of floating offshore wind turbines.

Step 2: Failure rate collection. Collect the failure rates of the failure items of onshore wind turbines.

Step 3: Failure likelihood acquisition. Determine the occurrence of each failure item from specialists in both the floating offshore and onshore sectors, according to the benchmark in Table 11.

Table 11. The rating guidance of occurrence [10,12,33].

Rating	Occurrence		Rating	Occurrence	
	Probability (P)	Description		Probability (P)	Description
1	$P < 10^{-5}$	Extremely low	6	$2 \times 10^{-3} < P < 1 \times 10^{-2}$	Moderate
2	$P = 10^{-5}$	Remote	7	$10^{-2} < P < 4 \times 10^{-2}$	Frequent
3	$P = 10^{-5}$	Very slight	8	$4 \times 10^{-3} < P < 0.2$	High
4	$10^{-5} < P < 4 \times 10^{-4}$	Slight	9	$0.2 < P < 0.33$	Very high
5	$4 \times 10^{-4} < P < 2 \times 10^{-3}$	Occasional	10	$P > 0.33$	Extremely high

Step 4: Failure rate correction. Numerically compare the differences between the occurrence values given by onshore and offshore specialists and infer the failure rates of the floating offshore wind turbine components (see Section 3.1).

The proposed failure rate correction approach aims to map the differences in failure rates based on the corresponding differential features of failure occurrences that are easy to obtain, knowing that failure rates and failure occurrences represent the same physical meaning, that is, failure likelihood. First, collect the occurrences of the components and define the occurrence matrix as:

$$\mathbf{O} = \begin{bmatrix} \mathbf{O}^1 & \mathbf{O}^2 & \cdots & \mathbf{O}^i & \cdots & \mathbf{O}^v \end{bmatrix}^T \quad (1)$$

where $\mathbf{O}$ is the occurrence matrix of the whole system and $\mathbf{O}^i$ denotes the occurrence matrix of component i. Accordingly, the occurrence matrix of the floating offshore wind turbine with v components is arranged as:

$$\mathbf{O} = \begin{bmatrix} O_1^1 & O_2^1 & \cdots & O_i^1 & \cdots & O_q^1 \\ O_1^2 & O_2^2 & \cdots & O_i^2 & \cdots & O_q^2 \\ \vdots & \vdots & & \vdots & & \vdots \\ O_1^w & O_2^w & \cdots & O_i^w & \cdots & O_q^w \\ \vdots & \vdots & & \vdots & & \vdots \\ O_1^v & O_2^v & \cdots & O_i^v & \cdots & O_q^v \end{bmatrix} \quad (2)$$

where O_i^w is the occurrence of the failure item i of the component w. Additionally, the maximum number of failure items for each component is q.

Hence, the occurrences of failure items in the component w given by both offshore (**WO**) and onshore (**O**) specialists are arranged as follows:

$$\mathbf{O} = \begin{bmatrix} \frac{1}{v}\sum_{w=1}^{v} O_1^w & \frac{1}{v}\sum_{w=1}^{v} O_2^w & \cdots & \frac{1}{v}\sum_{w=1}^{v} O_i^w & \cdots & \frac{1}{v}\sum_{w=1}^{v} O_q^w \end{bmatrix} = \begin{bmatrix} O_1 & O_2 & \cdots & O_i & \cdots & O_q \end{bmatrix} \quad (3)$$

$$\mathbf{WO} = \begin{bmatrix} \frac{1}{r}\sum_{w=1}^{r} WO_1^w & \frac{1}{r}\sum_{w=1}^{r} WO_2^w & \cdots & \frac{1}{r}\sum_{w=1}^{r} WO_i^w & \cdots & \frac{1}{r}\sum_{w=1}^{r} WO_q^w \end{bmatrix} = \begin{bmatrix} WO_1^w & WO_2^w & \cdots & WO_i^w & \cdots & WO_q^w \end{bmatrix} \quad (4)$$

Select the first m failure items in **O** and **WO** as the benchmark to balance the differences in specialist groups (the floating and onshore sectors) as in [33], for which the comparison matrix of the selected failure items is defined as:

$$\xi = \begin{bmatrix} \frac{WO_1}{O_1} & \frac{WO_2}{O_2} & \cdots & \frac{WO_i}{O_i} & \cdots & \frac{WO_q}{O_q} \end{bmatrix} \quad (5)$$

Subsequently, define a factor α_i for each component as:

$$\alpha_i = \sum_{i=1}^{q} \frac{O_i}{WO_i} \quad (6)$$

where k is the number of failure items (failure mode or cause) of the component w.

The overall factor (α) of the floating offshore wind turbines is computed as follows:

$$\alpha = \frac{1}{v}\sum_{i=1}^{v} \alpha_i \quad (7)$$

With the factor (α_i), define a transformation matrix $\mathbf{O}^{Adjust}$ for the remaining $q - m + 1$ failure items of the component i, as:

$$\mathbf{O}_i^{Adjust} = \alpha_i \times \mathbf{O} = \begin{bmatrix} \alpha_w O_{m+1} & \alpha_w O_{m+2} & \cdots & \alpha_w O_{m+i} & \cdots & \alpha_w O_q \end{bmatrix} \quad (8)$$

Accordingly, the transformation factor (Δ_i) of the component i is computed as follows:

$$\Delta_i = \frac{\sum \mathbf{WO}_i}{\sum \mathbf{O}_i^{Adjust}} \quad (9)$$

Hence, the failure rates of the remaining $q - m + 1$ failure items of the component w are computed as follows:

$$\bar{\lambda}_{q-m+1} = \Delta_i \times \lambda_{q-m+1} \quad (10)$$

where $\bar{\lambda}_{q-m+1}$ is the failure rates of the items, except for those of the support structures, and λ_{m+i} represents the failure rate of the failure item $m + i$ (in **O** and **WO**) in both floating offshore and onshore wind turbines.

Accordingly, the failure rates of floating offshore wind turbines' failure items are:

$$\bar{\lambda} = \begin{bmatrix} \lambda_{\text{FOWT}} & \bar{\lambda}_{q-m+1} \end{bmatrix} \quad (11)$$

where $\bar{\lambda}_{q-m+1}$ is the corrected failure rates of the floating offshore wind turbine, and λ_{FOWT} is the failure rates of the components of the floating offshore wind turbines. The failure rate of the unique components designed for floating offshore wind turbines, such as support structures, can be inferred via field data collection or the failure rate approximation model presented in [33].

Hence, the failure rate of the entire floating offshore wind turbine (λ) is calculated by Equation (12) under the series system assumption.

$$\lambda = \sum \bar{\lambda} = \sum_{i=1}^{m} \lambda^{i}_{FOWT} + \sum_{i=m+1}^{v} \bar{\lambda}_{q-m+i} \quad (12)$$

It is worth mentioning that, unlike the previous globalized model published in [33], this paper proposes a localized failure rate correction approach to infer the failure rates of the components of floating offshore wind turbines. However, it is also applicable to infer the same characteristics of the entire system (see Equations (7) and (12)). Moreover, the parameters at the whole system level can be used to validate the performance of the proposed approach (see Section 6).

6. Results, Comparisons, Validations, and Discussions

6.1. Results

This paper focuses on the failure and maintenance data of onshore and floating offshore wind turbines. The maintenance data of onshore and floating offshore wind turbines are comparable, as: (i) The maintenance actions for components inside the Nacelles of both types of wind turbines are similar, and are less affected by sea conditions due to the protection from Nacelles. However, the accessibility of offshore wind farms is lower. (ii) The maintenance actions for the remaining components mainly refer to support structures, which require specific technologies that are not involved in the maintenance and operation of onshore wind turbines. Accordingly, this paper does not correct the summarized maintenance data, including maintenance measures, spare parts, and especially maintenance times (RT, TT, MTTR), as in most cases, these data can be applied to floating offshore wind turbines directly.

The experts' judgments used in this study are the same as those collected in [33]. Overall, the failure rate correction model is bottom-up and consists of the following procedure, taking the Gearbox as an example:

Frist, collect and arrange the occurrence of the 10 failure causes of the Gearbox from the onshore and floating offshore sectors, as in Equation (3) (onshore) and Equation (4) (offshore):

$$\mathbf{O} = \begin{bmatrix} 3.96 & 3.16 & 2.57 & 1.78 & 3.16 & 3.96 & 3.56 & 3.76 & 3.76 & 3.16 \end{bmatrix} \quad (13)$$

$$\mathbf{WO} = \begin{bmatrix} 5.86 & 5.00 & 4.43 & 4.00 & 5.71 & 5.71 & 5.71 & 4.57 & 6.14 & 4.00 \end{bmatrix} \quad (14)$$

Second, determine the overall factor (α) of floating offshore wind turbines according to Equations (5)–(7), and accordingly, α is equal to 1.26. Adjust the occurrences in **O** as follows:

$$\mathbf{O}^{Adjust} = 1.26 \times \mathbf{O} = \begin{bmatrix} 4.99 & 3.98 & 3.24 & 2.14 & 3.92 & 4.96 & 4.49 & 4.74 & 4.74 & 3.98 \end{bmatrix} \tag{15}$$

Determine the transformation factor of the Gearbox of the floating offshore wind turbine according to Equation (8):

$$\Delta = \frac{\sum \mathbf{O}_i^{WO}}{\sum \mathbf{O}_i^{Adjust}} = 1.232 \tag{16}$$

Step 3: Transfer the failure rate of the floating offshore wind turbine Gearbox based on that of the same component of onshore wind turbines (0.335 failures/turbine/year) and the transformation factor as follows:

$$\overline{\lambda}_{Offshore} = \Delta \times \lambda_{onshore} = 1.232 \times 0.335 = 0.413 \tag{17}$$

The proposed failure rate correction approach deepens the failure rate correction from the entire wind turbine level to the component level. However, the same can be achieved the subcomponent level by repeating the same analytical procedure, and the results are listed in Table 12. According to the results, the failure rates of the Main Bearing and Yaw System assessed in onshore wind turbines are slightly higher than those of the floating offshore structures. The reason is that Nacelles isolate the impacts of harsh environmental conditions on the components inside them. A more detailed analysis at the subcomponent level is not performed in this paper, as it may not provide a cutting-edge understanding of failures that would occur in different types of wind turbines and their items. The readers, however, can draw further conclusions according to the results listed in Table 12. The results at the component level are listed in Figure 14, which indicates that:

- The proposed method predicts higher failure rates of the components equipped in floating offshore wind turbines than for onshore devices, except for the Cooling & Hydraulics, Auxiliary, and Pitch & Yaw components. The failure rates of the Cooling & Hydraulics and Auxiliary components are not corrected due to insufficient expert judgment.
- The lower predicted failure rate of Pitch & Yaw confirms that the judgment of experts and the proposed approach reflect that: (i) the advances in materials, manufacturing, and operation and management would increase the robustness of components, and thus, reduce their failure rates; and (ii) the harmful sea conditions introduce additional fragility to floating offshore wind turbines and, in turn, increase the failure rates of several components. The final transformation factor of each component is a balance between at least the two aspects mentioned above.
- Failure rates of Electrical Facilities and of electromechanical components such as the Generator are higher than those of mechanical components such as the Rotor and Gearbox. This indicates that the harmful sea conditions introduce additional failures to fragile electrical, electromechanical, and hydraulic elements, and mechanical components are relatively robust in withstanding the environmental effects. The conclusions above suggest weak links to floating offshore wind turbine operation and maintenance.

Table 12. Failure rates and the corresponding transformation factors of subcomponents.

Components	Subcomponents	Transformation Factor		Failure Rates, Onshore	Failure Rates, Corrected
		Amount/Year	Rank		
Rotor	Blades	1.353	5	0.173	0.234
	Hub	1.224	7	0.069	0.085
	Main Bearing	0.909	10	0.012	0.011
	Main Shift	1.371	4	0.012	0.012
Generator	Generator	1.211	8	0.878	1.063
Gearbox	Gearbox	1.232	6	0.335	0.413
Electrical Facilities	Converter	1.531	3	0.693	1.060
	Transformer	1.555	2	0.042 **	0.066
	Monitoring and SCADA	—	—	0.300	0.300 *
	Weather Unit	—	—	0.127	0.127 *
	Electronics & Controller	1.688	1	0.730 **	1.235
Pitch & Yaw	Pitch System	0.906	11	0.416	0.377
	Yaw System	1.166	9	0.092	0.108
Cooling & Hydraulics	Cooling System	—	—	0.809	0.809 *
	Hydraulics	—	—	0.843	0.843 *
Auxiliary	Crane	—	—	0.012	0.012 *
	Climbing Aid	—	—	0.012	0.012 *
	Brake	—	—	0.035	0.035 *
	Nacelle	—	—	0.104	0.104 *

*: Failure rates without correction; **: failure rates computed based on those reported in [10] and [18]; failure rates in failures/year.

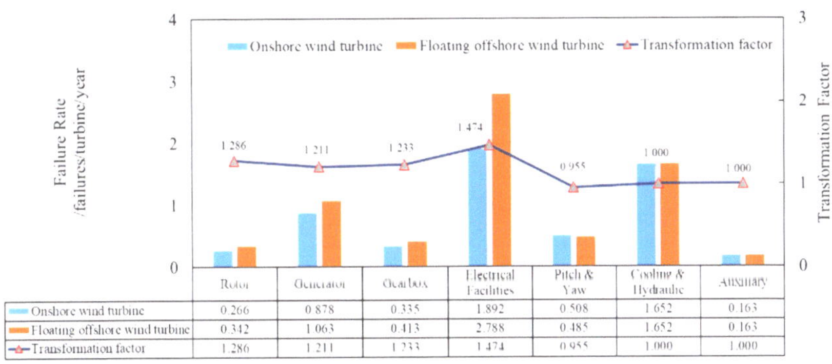

Figure 14. Failure rate transformation of components.

6.2. Comparisons, Validations, and Discussions

Overall, the proposed approach predicts that the failure rate of floating offshore wind turbines is 28.6% higher than that of onshore wind turbines. The predicted result is compared with existing models in Table 13. The failure rates corrected by the proposed approach were transferred to the corresponding components of the Bayesian network model constructed in [18] and [33] (see Figure 15 and Appendix A Table A4). The purpose of this was to access the failure rate and MTTF of floating offshore wind turbines in order to verify the performance of the proposed localized expert-judgement-based failure rate correction model (see Tables 13 and 14). This analysis reveals that a floating offshore wind turbine suffers 8.37 failures per year with an MTTF of 1046 h. The comparative results in Tables 13 and 14 confirm that:

- From a holistic point of view, failure rate correction/prediction models agree that the failure rate of floating offshore wind turbines is higher than that of onshore devices. RIF-based models predict that the failure rate disparity with onshore wind turbines is less than 30%; this can be compared with the result of the stress factor-based model, which computed 75% differences between the two types of wind turbines.

The evidence is insufficient to select the model that can best provide convincing and precise results. However, it should be mentioned that the model selection should follow the features of the methods. Specifically, stress factor-based methods correct failure rates via global factors related to environmental and power utilization aspects. However, the RIF-based method produces a cluster map of the disparity of wind turbines according to the failure features of detailed failure cases of components.

- RIF and subjective data-based failure rate correction models, including the localized model presented in this paper and the globalized model published in [33], and the stress factor-based model [31] predict higher failure rates of floating offshore wind turbines than the others. Considering that the larger and more complex structures assembled, and further and deeper wind farms, resulted in insufficient maintenance and accessibility, floating equipment tended to fail more frequently than other types. Accordingly, the aforementioned method, including the proposed approach, corrected the failure rate in the right way.
- The relative error of the predicted failure rate from a holistic wind turbine point of view is less than 1% compared with the RIF- and risk-based models [33] (more than 1%) and the stress factor-based model (more than 2%). According to the results in Table 14, the relative error of the proposed method is not the lowest. However, it is worth adding that the primary contribution of the presented approach is that the approach extends the globalized RIF-and-subjective data-based models so that the failure rate correction can extend to component and element levels; this is the foundation of failure, risk, reliability, availability, and operation and maintenance investigations of floating offshore wind turbines with insufficient data.

Table 13. Comparison of failure rates of elements assessed using failure correction models.

Models	Failure Rate Disparity with Onshore Devices		
	Amount *	Higher/Lower	Model Features
[33]	26%	Higher	RIF- and subjective data-based models—Globalized
[36]	75%	Higher	Stress factor-based model
[37]	3.3%	Higher	RIF- and risk-based models
[38]	19%	Higher	RIF- and uncertainty-based models
[39]	18% **	Higher	RIF interval-based model
[Present]	28.6%	Higher	RIF- and subjective data-based model—Localized

*: The average of all elements; **: compared with the mean value.

Table 14. Assessed failure features of the floating offshore wind turbine.

Failure Rate Correction Models	Failure Rate/ Failure/Turbine/Year		MTTF/Hour		Relative Error / in %
	Onshore	Offshore *	Onshore	Offshore *	
[Present]	8.3	8.37	1055	1046	−0.85
[33]		8.34		1050	−0.47
[36]		8.49		1031	−2.27
[37]		8.21		1066	1.04
[38]		8.25		1061	0.57
[39]		8.24		1063	0.76

* offshore wind.

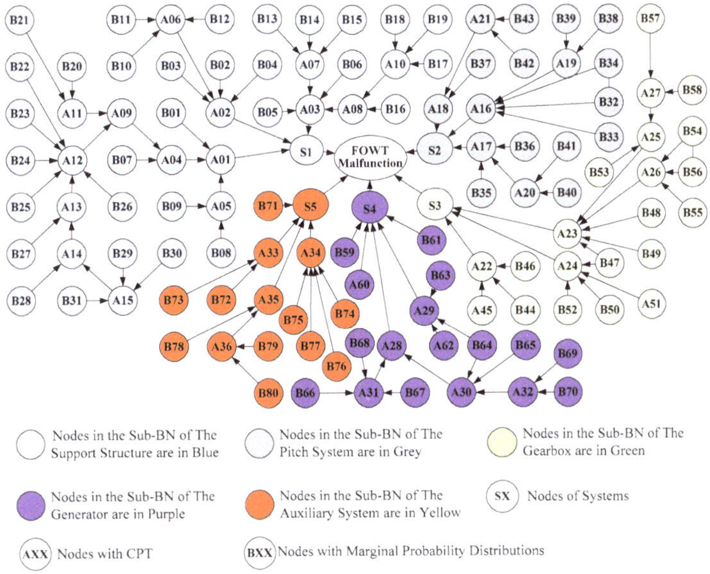

Figure 15. The BN model of the floating offshore wind turbine [18,33] (see Appendix A).

7. Conclusions

This paper provides detailed failure and maintenance data analysis for wind turbine safety, risk, reliability, availability, and maintainability investigations. Overall, wind turbine failure features, including the critical failure, failure frequencies, failure rates, and lifetime distributions of primary components, are specified. Maintenance properties such as maintenance measures, spare policies, and three reference times related to maintenance are also provided.

A reliability influence factor-based (RIF-based) data transformation approach is presented with the assistance of subjective specialists' judgment to transfer the onshore data to the floating offshore wind turbines; these are a new concept, and very few been installed. The proposed approach can infer the failure rates of the components of floating offshore wind turbines at the element level to provide more comprehensive and detailed information to practitioners in the field. Overall, this paper provides a new dataset for the wind energy sector with up-to-date data, and a comprehensive analysis tool to transfer the failure data to the floating offshore wind sector. The outcomes of this paper contribute to the design and operation of wind turbines and the management of wind farms and will strengthen the database of the wind energy sector.

Author Contributions: Conceptualization, H.L. and C.G.S.; methodology, H.L., C.-G.H. and W.P.; writing—original draft preparation, H.L. and C.G.S.; writing—review and editing, C.G.S.; supervision, C.G.S.; project administration, C.G.S.; All authors have read and agreed to the published version of the manuscript.

Funding: The first author was funded by the project ARCWIND–Adaptation and implementation of floating wind energy conversion technology for the Atlantic region—which is co-financed by the European Regional Development Fund through the Interreg Atlantic Area Programme under contract EAPA 344/2016. This work contributes to the Strategic Research Plan of the Centre for Marine Technology and Ocean Engineering (CENTEC), which is financed by the Portuguese Foundation for Science and Technology (Fundação para a Ciência e Tecnologia—FCT) under contract UIDB/UIDP/00134/2020.

Institutional Review Board Statement: Not applicable.

Informed Consent Statement: Not applicable.

Data Availability Statement: Not applicable.

Conflicts of Interest: The authors declare no conflict of interest.

Appendix A

Table A1. Failure modes of main components.

Components	Subcomponents	Failure Descriptions
Rotor	Hub	Hub broken
	Main bearing	Bearing overheating; bearing seat failure
	Blade	Cracked; comes off; delamination; wear with Hub; replaced for a new standard
Generator	Generator	Carbon brush failure *; slip ring failure *; cooling water pump failure *; generator elastic support failure; lubrication pump motor failure; cooling water tank failure; cooling fan failure; generator encoder failure; bearing failure; insufficient bearing grease; cable failure; generator converter short circuit.
Gearbox	Components	Temperature control valve failure; abnormal gear wear; gearbox respirator
	Lubrication	Fe content exceeded; leak; lubrication pump filter failure *; lubrication pump motor failure *; seal damage; cooling water pump failure
Electrical Facilities	Converter	Cooling failure; cable failure; contactor damage; fuse is blown; breaker failure; power failure *; circuit failure *; reactor damaged *; crowbar board failure *; current transformer failure *; communication module failure
	SCADA	SCADA reports unknown data
	Control System	Yaw fuse failure; yaw contactor failure *; communication module; fiber damage *; yaw starter failure; unknown failure *; cooling fan; cable failure; shutdown
	Weather Units & Sensors	The wrong temperature reported *; pressure sensor damage; brake sensor failure; anemometer damage *; anemometer failure
Pitch & Yaw	Gearbox & Motor	Wrong pitch angle *; Fe exceeding standard; leak, encoder damage *, malfunction
	Electric & Controller	Fuse blown *; motor damage *; cable failure; power failure; blockage; pump damaged; pitch reducer leak, capacitor damage *; contactor damage; power overload *; cable failure; fuse blown; limit switch triggered by mistake *; slip rings *
	Yaw	Leak; structural damage; yaw reducer broken tooth
Cooling & Hydraulics	Internal Cooling	Water cooling pump failure *; lubrication pump motor failure *; cooling water tank failure; seal damage *; water cooling switch *; water pump seal failure; air cooler failure; lubrication pump filter failure
	External Colling	Motor failure *; blade failure *; cover failure; fan failure
	Hydraulic Units	Storage tank damage; storage tank leak gas and oil*; improper maintenance; electromagnetic valve damage *; storage tank low pressure (unknown reasons); periodical replacement of filter (half a year) **; hydraulic tubing leaks; brake pressure switch failure; hydraulic motor failure; filter failure
Auxiliary	Auxiliary Components	Lightning protection belt broken; water leak; sunroof support rod damage; guardrail damage; shroud broken; beam crack; brake wear; climbing aid control element failure

*: Failures resulted in wind turbine shutdown; **: replacement without failure.

Table A2. The MTTRs of components by maintenance measures/hour.

Components	Subcomponents	RP	RPA	RWR	CWRO	CWRS	WSO	WSS	WFI	MTTR Average
Rotor	Blade	—	8/4.5	—	—	—	3/0	—	—	3.3
	Hub	—	—	—	1/2	—	—	—	—	2
	Main Bearing	—	2/0.6	—	—	—	—	—	—	0.6
Generator	Generator	6/2.3	28/3.3	1/10	—	1/6	—	1/11	—	3.6
Gearbox	Gearbox	—	12/2.7	—	—	—	—	—	—	2.7
Electrical Facilities	Converter	8/5	41/3.4	1/1	—	—	—	—	—	3.6
	Monitoring and SCADA	7/6.8	9/3.2	—	—	—	—	—	1/1	4.4
	Weather Unit	2/7.8	2/7	—	—	—	—	—	—	7.3
Pitch & Yaw	Pitch System	6/3.8	18/3.3	—	—	—	—	—	—	3.4
	Yaw System	—	4/4.2	1/0	1/0	—	—	—	—	2.8
Cooling & Hydraulics	Cooling System	1/1	35/4	3/5.2	1/2	1/4	—	—	—	3.8
	Hydraulics	3/3.9	53/3.2	1/0	—	—	—	—	—	3.2
Auxiliary	Brake	1/19	2/1	—	—	—	—	—	—	7
	Sensor	—	1/2	—	—	—	—	—	—	2
	Nacelle	—	—	—	—	—	—	1/7	—	7
MTTR Average		5	3.4	3.8	1.3	5	0	9	1	3.6

RP: repaired; RPA: replaced; RWR: repaired and waiting for replacement; CWRO: checked and waiting for replacement (operating); CWRS: checked and waiting for replacement (Stopped); WSO: waiting for suppliers (operating); WSS: waiting for suppliers (stopped); WFI: waiting for further instructions; 0: less than one hour; A/B: number of instances of maintenance/MTTR (in hour).

Table A3. RT, TT, and MTTR of components by wind farm/hour.

Components	Subcomponents	Wind Farm #1			Wind Farm #2			Wind Farm #3			Wind Farm #4		
		RT	TT	MTTR	RT	TT	MTTR	RT	TT	MTTR	RT	TT	MTTR
Rotor	Blade	—	—	—	1236	—	0	—	—	—	0	2	4.5
	Hub	0	—	2	—	—	—	—	—	—	1	0	—
	Main Bearing	0	7	0.6	—	—	—	—	—	—	—	—	—
Generator	Generator	0	3	2.4	5	10	4.3	8	0	9.8	280	6	4.4
Gearbox	Gearbox	1	5	2.3	0	—	—	3	12	4.7	1	—	2.6
Electrical Facilities	Converter	1	6	2	—	—	—	467	0	9.8	6	5	4.7
	Monitoring and SCADA	1	8	1	12	—	6	4	9	4.2	3	10	7
	Sensor	—	—	—	—	—	—	—	—	—	0	13	2
	Weather Unit	—	—	—	13	—	9	0	4	5	3	0	7.8
Pitch & Yaw	Pitch System	2	9	3.6	—	—	—	—	—	—	7	7	3.2
	Yaw System	0	0	4.2	0	—	0	0	—	—	—	—	—
Cooling & Hydraulic	Cooling System	2	2	2.4	2	5	9	69	6.5	6.8	70	5	4.5
	Hydraulic	0	5	2.6	8	6	1.7	63	7	5.9	1	5	4.2
Auxiliary	Brake	—	—	—	14	12	9.5	—	—	—	—	13	2
	Nacelle	0	1	—	3	—	7	—	—	—	0	7	—
	Climbing Aid	—	—	—	1	—	—	—		—	—	—	—
	Crane	—	—	—	—	—	—	—	—	—	0	0	—
Average		1	5	2.5	4	7	3.4	50	6	6.2	67	5	4.4

0: Less than one hour.

Table A4. Node definition of the BN model.

Code	Failures	Code	Failures
	S1—Support Structure		
A01	Mooring subsystem failure	B09	Insufficient emergency measurement
A02	Tower failure	B10	Strong waves
A03	Floating foundation failure	B11	Lightning Strike
A04	Device failure	B12	Storm
A05	Extreme sea conditions	B13	Typhoon
A06	Collapse due to environment	B14	Plane crash
A07	Hit by dropped objects	B15	Biological collision
A08	Watertightness fault	B16	Inefficient detection
A09	Other devise failures	B17	Pipe joint corrosion
A10	Pipe joint failure	B18	Pipe joint weld defect
A11	Fairlead failure	B19	Pipe joint fatigue
A12	Mooring lines broken	B20	Fairlead corrosion
A13	Mooring line breakage	B21	Fairlead fatigue
A14	Mooring line wear	B22	Transitional chain wear
A15	Accumulating wear	B23	Friction chain wear
B01	Human Error	B24	Mooring winch failure
B02	Resonance	B25	Buoy friction chain wear
B03	Faulty welding	B26	Anchor pickup device damage
B04	Material fatigue	B27	Abnormal stress
B05	Pillar damage	B28	Invalid maintenance
B06	Capsizing	B29	Mooring line wear
B07	Anchor failure	B30	Mooring line fatigue
B08	Poor operation environment	B31	Mooring line corrosion
	S2—Pitch System		
A16	Hydraulic system failure	B35	Lighting protection failure
A17	Alarm facility failure	B36	Limit switch failure
A18	Wrong pitch angle	B37	Abnormal vibration
A19	Hydraulic oil failure	B38	Oil leakage
A20	Power failure	B39	Filter failure
A21	Meteorological unit failure	B40	Power 1 failure (the main power)
B32	Hydraulic motor failure	B41	Power 2 failure (the backup power)
B33	Overpressure	B42	Vane damage
B34	Accumulator failure	B43	Anemometer damage

Table A4. *Cont.*

Code	Failures	Code	Failures
	S3—Gearbox		
A22	Lubrication failure	B49	Pitting (gear)
A23	Abnormal gear	B50	Corrosion of pins
A24	Bearing fault	B51	Abrasive wear
A25	Tooth wear (gears)	B52	Pitting (gear bearing)
A26	Cracks in gears	B53	Gear tooth deterioration
A27	The offset of tooth gears	B54	Excessive pressure
B44	Abnormal filter	B55	Excess temperature
B45	Poor-quality lubrication oil	B56	Fatigue (gear)
B46	Dirty lubrication oil	B57	Poor design of tooth gears
B47	Abnormal vibration (GB)	B58	Tooth surface defects
B48	Glued		
	S4—Generator		
A28	Rotor and stator failure	B63	Structural deficiency
A29	Bearing failure	B64	Abnormal vibration (GE)
A30	Abnormal signals	B65	Abnormal instrument reading
A31	No centricity generation	B66	Failure to synchronize
A32	Overheating	B67	Broken bars
B59	Measurement facility failure	B68	Failure to start on demand
B60	Wire fault	B69	Sensor failure
B61	Leak	B70	Temperature above limitation
B62	Asymmetry		
	S5—Auxiliary System		
A33	Speed train failure	B74	Controller failure
A34	Electric component failure	B75	Transformer failure
A35	Blade failure	B76	Sensor failure
A36	Rotor failure	B77	Converter failure
B71	Yaw subsystem failure	B78	Blade structure failure
B72	Drive train failure	B79	Hub failure
B73	Brake failure	B80	Bearing failure

References

1. Sadorsky, P. Wind energy for sustainable development: Driving factors and future outlook. *J. Clean. Prod.* **2021**, *289*, 125779. [CrossRef]
2. Willis, D.J.; Niezrecki, C.; Kuchma, D.; Hines, E.; Arwade, S.R.; Barthelmie, R.J.; DiPaola, M.; Drane, P.J.; Hansen, C.J.; Inalpolat, M.; et al. Wind energy research: State-of-the-art and future research directions. *Renew. Energy* **2018**, *125*, 133–154. [CrossRef]
3. Díaz, H.; Guedes Soares, C. Review of the current status, technology and future trends of offshore wind farms. *Ocean Eng.* **2020**, *209*, 107381. [CrossRef]
4. Díaz, H.; Serna, J.; Nieto, J.; Guedes Soares, C. Market needs, opportunities and barriers for the floating wind industry. *J. Mar. Sci. Eng.* **2022**, *10*, 934. [CrossRef]
5. Johnston, B.; Foley, A.; Doran, J.; Littler, T. Levelised cost of energy, A challenge for offshore wind. *Renew. Energy* **2020**, *160*, 876–885. [CrossRef]
6. Hussain, W.; Khan, S.; Mover, A.H. Development of quality, environment, health, and safety (QEHS) management system and its integration in operation and maintenance (O&M) of onshore wind energy industries. *Renew. Energy* **2022**, *196*, 220–233.
7. Li, H.; Huang, C.G. Guedes Soares, C. A real-time inspection and opportunistic maintenance strategies for floating offshore wind turbines. *Ocean Eng.* **2022**, *256*, 111433. [CrossRef]
8. Castro-Santos, L.; Bento, A.R.; Silva, D.; Salvação, N.; Guedes Soares, C. Economic feasibility of floating offshore wind farms in the north of Spain. *J. Mar. Sci. Eng.* **2020**, *8*, 58. [CrossRef]
9. Scheu, M.N.; Tremps, L.; Smolka, U.; Kolios, A.; Brennan, F. A systematic Failure Mode Effects and Criticality Analysis for offshore wind turbine systems towards integrated condition based maintenance strategies. *Ocean Eng.* **2019**, *176*, 118–133. [CrossRef]
10. Li, H.; Díaz, H.; Guedes Soares, C. A failure analysis of floating offshore wind turbines using AHP-FMEA methodology. *Ocean Eng.* **2021**, *234*, 109261. [CrossRef]
11. Castro-Santos, L.; Silva, D.; Bento, A.R.; Salvacao, N.; Guedes Soares, C. Economic feasibility of floating offshore wind farms in Portugal. *Ocean Eng.* **2020**, *207*, 107393. [CrossRef]
12. Li, H.; Diaz, H.; Guedes Soares, C. A developed failure mode and effect analysis for floating offshore wind turbine support structures. *Renew. Energy* **2021**, *164*, 133–145. [CrossRef]
13. Li, H.; Teixeira, A.P.; Guedes Soares, C. A two-stage Failure Mode and Effect Analysis of offshore wind turbines. *Renew. Energy* **2020**, *162*, 1438–1461. [CrossRef]

14. Bento, N.; Fontes, M. Emergence of floating offshore wind energy: Technology and industry. *Renew. Sustain. Energy Rev.* **2019**, *99*, 66–82. [CrossRef]
15. Li, H.; Teixeira, A.P.; Guedes Soares, C. An Improved Failure Mode and Effect Analysis of Floating Offshore Wind Turbines. *J. Mar. Sci. Eng.* **2022**, *10*, 1616. [CrossRef]
16. Sinha, Y.; Steel, J.A. A progressive study into offshore wind farm maintenance optimisation using risk based failure analysis. *Renew. Sustain. Energy Rev.* **2015**, *42*, 735–742. [CrossRef]
17. Martinez, A.; Iglesias, G. Mapping of the levelised cost of energy for floating offshore wind in the European Atlantic. *Renew. Sustain. Energy Rev.* **2022**, *154*, 111889. [CrossRef]
18. Li, H.; Guedes Soares, C.; Huang, H.Z. Reliability analysis of a floating offshore wind turbine using Bayesian Networks. *Ocean Eng.* **2020**, *217*, 107827. [CrossRef]
19. Dao, C.; Kazemtabrizi, B.; Crabtree, C. Wind turbine reliability data review and impacts on levelised cost of energy. *Wind Energy* **2019**, *22*, 1848–1871. [CrossRef]
20. Santos, F.P.; Teixeira, A.P.; Guedes Soares, C. Review of wind turbine accident and failure data. In *Renewable Energies Offshore*; Guedes Soares, C., Ed.; Taylor & Francis Group: London, UK, 2015; pp. 953–959.
21. Carroll, J.; McDonald, A.; Dinwoodie, I.; McMillan, D.; Revie, M.; Lazakis, I. Availability, operation and maintenance costs of offshore wind turbines with different drive train configurations. *Wind Energy* **2017**, *20*, 361–378. [CrossRef]
22. Tavner, P.J.; Xiang, J.; Spinato, F. Reliability analysis for wind turbines. *Wind Energy* **2007**, *10*, 1–18. [CrossRef]
23. Ribrant, J.; Bertling, L. Survey of failures in wind power systems with focus on Swedish wind power plants during 1997–2005. In Proceedings of the 2007 IEEE Power Engineering Society General Meeting, Tampa, FL, USA, 24–28 June 2007; pp. 1–8. [CrossRef]
24. Faulstich, S.; Hahn, B.; Tavner, P.J. Wind turbine downtime and its importance for offshore deployment. *Wind Energy* **2011**, *14*, 327–337. [CrossRef]
25. Stenberg, A.; Holttinen, H. Analysing failure statistics of wind turbines in Finland. In Proceedings of the 2010 European Wind Energy Conference and Exhibition, Warsaw, Poland, 20–23 April 2010; pp. 3985–3995.
26. Crabtree, C.J.; Zappalá, D.; Hogg, S.I. Wind energy: UK experiences and offshore operational challenges. *Proc. Inst. Mech. Eng. Part A J. Power Energy* **2015**, *229*, 727–746. [CrossRef]
27. Carroll, J.; McDonald, A.; McMillan, D. Failure rate, repair time and unscheduled O&M cost analysis of offshore wind turbines. *Wind Energy* **2016**, *19*, 1107–1119.
28. Spinato, F.; Tavner, P.J.; Van Bussel, G.J.; Koutoulakos, E. Reliability of wind turbine subassemblies. *IET Renew. Power Gener.* **2009**, *3*, 387–401. [CrossRef]
29. Adedipe, T.; Shafiee, M.; Zio, E. Bayesian network modelling for the wind energy industry: An overview. *Reliab. Eng. Syst. Saf.* **2020**, *202*, 107053. [CrossRef]
30. Yeter, B.; Garbatov, Y.; Guedes Soares, C. Risk-based maintenance planning of offshore wind turbine farms. *Reliab. Eng. Syst. Saf.* **2020**, *202*, 107062. [CrossRef]
31. Yeter, B.; Garbatov, Y.; Guedes Soares, C. Life-extension classification of offshore wind assets using unsupervised machine learning. *Reliab. Eng. Syst. Saf.* **2022**, *219*, 108229. [CrossRef]
32. Eryilmaz, S. Reliability analysis of multi-state system with three-state components and its application to wind energy. *Reliab. Eng. Syst. Saf.* **2018**, *172*, 58–63. [CrossRef]
33. Li, H.; Guedes Soares, C. Assessment of failure rates and reliability of floating offshore wind turbines. *Reliab. Eng. Syst. Saf.* **2022**, *228*, 108777. [CrossRef]
34. Zhang, X.; Sun, L.; Sun, H.; Guo, Q.; Bai, X. Floating offshore wind turbine reliability analysis based on system grading and dynamic FTA. *J. Wind Eng. Ind. Aerodyn.* **2016**, *154*, 21–33. [CrossRef]
35. Kang, J.; Guedes Soares, C. An opportunistic maintenance policy for offshore wind farms. *Ocean Eng.* **2020**, *216*, 108075. [CrossRef]
36. Santos, F.P.; Teixeira, A.P.; Guedes Soares, C. An age-based preventive maintenance for offshore wind turbines. In *Safety and Reliability: Methodology and Applications*; Nowakowski, T., Mlynczak, M., Jodejko-Pietruczuk, A., Werbinska-Wojciechowska, S., Eds.; CRC Press: Boca Raton, FL, USA, 2014; pp. 1183–1192.
37. Rahimi, M.; Rausand, M. Prediction of failure rates for new subsea systems: A practical approach and an illustrative example. *Proc. Inst. Mech. Eng. Part O J. Risk Reliab.* **2013**, *227*, 629–640. [CrossRef]
38. Bhardwaj, U.; Teixeira, A.P.; Guedes Soares, C. Bayesian framework for reliability prediction of subsea processing systems accounting for influencing factors uncertainty. *Reliab. Eng. Syst. Saf.* **2022**, *218*, 108143. [CrossRef]
39. Brissaud, F.; Charpentier, D.; Fouladirad, M.; Barros, A.; Bérenguer, C. Failure rate evaluation with influencing factors. *J. Loss Prev. Process Ind.* **2010**, *23*, 187–193. [CrossRef]
40. Wilkinson, M.; Harman, K.; Hendriks, B.; Spinato, F.; van Delft, T. Measuring wind turbine reliability-results of the Reliawind project. In Proceedings of the 2011 EWEA Conference, Brussels, Belgium, 14–17, March 2011; pp. 1–8.
41. Santos, F.; Teixeira, A.P.; Guedes Soares, C. Modelling and simulation of the operation and maintenance of offshore wind turbines. *Proc. Inst. Mech. Eng. Part O J. Risk Reliab.* **2015**, *229*, 385–393. [CrossRef]
42. Kang, J.; Sun, L.; Guedes Soares, C. Fault Tree Analysis of floating offshore wind turbines. *Renew. Energy* **2019**, *133*, 1455–1467. [CrossRef]
43. Bhardwaj, U.; Teixeira, A.P.; Guedes Soares, C. Reliability prediction of an offshore wind turbine gearbox. *Renew. Energy* **2019**, *141*, 693–706. [CrossRef]

Journal of
Marine Science and Engineering

MDPI

Article

An Improved Failure Mode and Effect Analysis of Floating Offshore Wind Turbines

He Li [1,2], A. P. Teixeira [1,*] and C. Guedes Soares [1]

[1] Centre for Marine Technology and Ocean Engineering (CENTEC), Instituto Superior Técnico, Universidade de Lisboa, 1049-001 Lisbon, Portugal
[2] Center for System Reliability and Safety, University of Electronic Science and Technology of China, Chengdu 611731, China
* Correspondence: teixeira@centec.tecnico.ulisboa.pt

Abstract: This paper proposes an improved failure mode and effect analysis method for a comprehensive failure analysis that provides a holistic perspective of actions on the potential failures of floating offshore wind turbines. A new way of constructing risk priority numbers was developed by considering the background knowledge of the specialists involved in the failure analysis. The failure analysis was conducted based on an extensive dataset from multiple specialists that covers five floating offshore wind turbine systems, 15 main components, 42 failure modes, and 104 failure causes. Consequently, 21 recommendations are suggested for designers and operators to prevent and mitigate the risk of unexpected failures of floating offshore wind turbines. Furthermore, a comparison analysis was conducted to illustrate the similarities and differences between the proposed failure mode and effect analysis and the conventional method.

Keywords: failure analysis; floating offshore wind turbine; failure mode and effect analysis; floating offshore wind energy

Citation: Li, H.; Teixeira, A.P.; Guedes Soares, C. An Improved Failure Mode and Effect Analysis of Floating Offshore Wind Turbines. *J. Mar. Sci. Eng.* **2022**, *10*, 1616. https://doi.org/10.3390/jmse10111616

Academic Editor: Domenico Curto

Received: 7 September 2022
Accepted: 26 October 2022
Published: 1 November 2022

1. Introduction

Renewable energies are changing the global energy supply structure that used to consist mainly of fossil energy [1]. In recent decades, human beings have witnessed an increasing demand for renewable energies, such as wind energy, to eliminate the side effects of environmental pollution associated with fossil fuels [2,3].

The exploration of wind energy for generating electricity can be traced back to the 1880s when three wind turbines were installed in the United States (1883), in Scotland (1887), and in Denmark (in 1887) [4]. Since then, multiple types of onshore and bottom-fixed offshore wind turbines have been manufactured to explore wind energy inland and offshore.

The floating offshore concepts represent the next step of the offshore wind energy market, which allows energy production from waters deeper than 50 m where the wind profile is more stable, and the capacity factors (ratio of actual electrical energy output over a given period to the maximum possible electrical energy output over that period) are typically higher [5–7]. Accordingly, it is possible to seek higher investment returns for floating wind projects [8,9]. The first floating offshore wind farms (Hywind Scotland in the UK and WindFloat Atlantic in Portugal) have been put into operation, and several additional floating wind farms are under construction worldwide [10,11]. However, it is still challenging to explore more economical wind resources far from shore with offshore wind turbines with floating supporting structures [12–16]. Moreover, the goal of the development of floating offshore wind turbines is to cut down the electricity price until it can be comparable to the price of electricity generated by fossil energy sources [17–19]. This calls for operational reliability, availability, and energy generation efficiency of those concepts to a higher degree than onshore and bottom-fixed facilities [20,21].

Operational reliability is the probability that a floating offshore wind turbine generates electricity as designed during an observed time at the sea. It represents the capability of floating offshore wind turbines to produce electricity without failures over a given period [9,16,22–26].

Availability is the degree to which a floating offshore wind turbine is in an operational state at an arbitrary operation time. It can be measured as the ratio of time a floating offshore wind turbine is working during an observed period. Production availability denotes the electricity generation capacity of floating offshore wind turbines during a given period [4,27,28].

Energy generation efficiency is the instantaneous capability that a floating offshore wind turbine converts wind energy into electricity. It reflects the electricity production strength of floating offshore wind turbines [4,29,30].

Operational reliability, availability, and energy generation efficiency of floating offshore wind turbines are the basis of floating offshore projects' financial feasibility. These indices rely on an insightful and comprehensive understanding of inherent and extrinsic-motivated failures of floating offshore wind turbines by having recourses to failure analysis since they can [31,32]:

(1) Identify critical items related to failures of floating offshore wind turbines such as systems, components, failure modes, and failure causes. These critical items have serious failure consequences, high failure frequencies, or both.
(2) Ascertain the local and widespread impact of each failure of floating offshore wind turbines.
(3) Explain the failure behavior of floating offshore wind turbines by assessing how each failure cause results in the corresponding failure mode(s), further gives rise to malfunctions of component(s), and then passes to the upper system level.
(4) Suggest corrections to improve the system and structural designs by finding proper measures to avoid the occurrences of critical failure causes.
(5) Determine downtime and evaluate maintenance costs to suggest preventive actions for maintenance implementations.

However, several challenges should be recognized when referring to the failure analysis of floating offshore wind turbines [32–36]:

(1) As a typical complex system composed of thousands of elements, detailed failure analysis is time-consuming and it is unpractical when the analysis is deepened to each basic element level, owing to the limited knowledge of analysts.
(2) Failures are rare throughout the life cycle of floating offshore wind turbines, and only a few failures could be observed due to the lack of operational experience for such relatively new installations.
(3) The credibility of failure analysis results significantly relies on collected personal knowledge, such as the specialists' judgments. It also requires reasonable pre-treatment of collected data in advance of failure analysis.

The analysis of failure features and failure behavior are essential for floating offshore wind turbines' performance. They are the basis of the following activities: (i) Determining economical maintenance strategies; (ii) Conducting robust designs to reduce potential failures; (iii) Estimating the expected return on investment. However, the reality is that floating offshore wind structures are relatively new, and therefore their failure information is unavailable. To overcome the limited information on the failure of floating offshore wind turbines, analogous data from onshore or bottom-fixed foundation wind turbines have been used as a reference. This data, however, introduces additional uncertainty and reduces the credibility of failure analysis outcomes. To this end, this paper proposes an improved failure mode and effect analysis (FMEA) that is then applied in a comprehensive failure analysis of floating offshore wind turbines based on extensive data collection from multiple specialists who are designers or analysts of floating offshore wind turbines. To be specific, this study aims to:

(1) Complete a failure analysis of floating offshore wind turbines by comprehensively collecting data from wind energy industry designers, or research institutions' analysts to avoid using similar data from onshore or bottom-fixed foundation sectors as has previously been done. Accordingly, the failure analysis outcomes could be more credible and convincing.
(2) Create a new way of constructing the risk index by considering the background of the employed specialists.
(3) Open the collected data to motivate further investigations in the floating offshore wind sector.

The remainder of the paper is organized as follows. Section 2 introduces the improved FMEA methodologies. The results of the proposed approach are presented in Section 3. Comparisons between the proposed FMEA and the conventional method are shown in Section 4. Conclusions are provided in Section 5.

2. Failure Mode and Effect Analysis

Failure analysis includes qualitative (e.g., Cause-Consequence Analysis, Checklist Analysis, What-If Analysis), quantitative (e.g., Markov analysis), and semi-quantitative (e.g., Failure Mode and Effect Analysis (FMEA), Fault Tree Analysis (FTA), Event Tree Analysis (ETA)), methods [23,37,38]. They have been developed to identify and evaluate the root or basic causes that contribute to the failure of the system to accomplish its designated functions.

FMEA and its developed version, namely Failure Modes, Effects and Criticality Analysis (FMECA), have commonly been applied in the failure analyses of systems or devices such as onshore and offshore wind turbines, due to their features of highly hierarchical structure, understandability, and being easy-to-construct [16,19,32]. FMEAs are systematic processes of identifying and ranking failure items in a bottom-up approach [19]. A subjective index, known as Risk Priority Number (RPN), is typically used to assess the overall risk degree of each failure item, e.g., failure mode or failure cause [39,40]. An *RPN* is a product of values of the severity, occurrence, and detection of a failure item. These values are designed by specialists or engineers in the field. Specifically, severity represents failure consequences. Occurrence describes the likelihood of failure modes. Detection reflects the difficulty of a failure mode to be discovered. The primary process of conducting an FMEA includes:

(1) <u>Recognition of systems and components of floating offshore wind turbines</u>

A conceptual model of a floating offshore wind turbine with 5 systems and 15 components is constructed according to publications available, see Table 1 and Figure 1.

Table 1. Floating offshore wind turbine system levels.

System Level		Component Level					
Code	System	Code	Component	Code	Component	Code	Component
ER	Energy Receiving System	BL	Blade	GB	Gearbox	FF	Floating Foundation
EP	Energy Producing System	HB	Hub	CV	Converter	TP	Transition Piece
ET	Energy Transforming System	MB	Main Bearing	TR	Transformer	PS	Pitch Subsystem
SS	Support Structure	MS	Main Shift	MF	Mooring Facilities	YS	Yaw Subsystem
AS	Auxiliary System	GE	Generator	TO	Tower	CS	Control System
		LP	Lubrication	NC	Nacelle	CB	Coupling and Brake
		CF	Cooling Fan	CLS	Cooling System	CE	Controllers and Electrical Facilities
		CR	Crane	HS	Hydraulic Station		

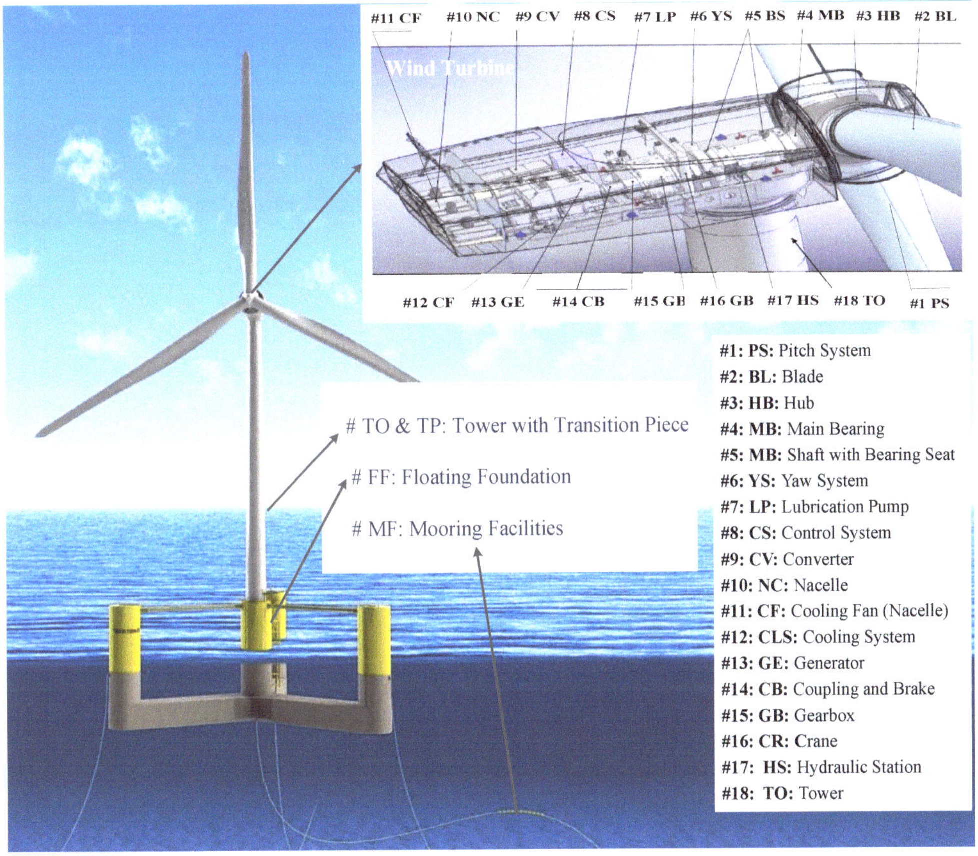

Figure 1. Typical configuration of floating offshore wind turbines.

The energy-receiving system consists of three blades and a hub that converts wind energy into the torque of the main shaft. The energy-producing system is composed of the main bearing, the main shaft, a generator, and a gearbox, which transform the mechanical energy of the main shaft into electricity. The energy-transforming system is designed for voltage regulation and includes a converter and a transformer. The support structure is made up of mooring facilities, a tower, a floating foundation, and a transition piece, which is assembled to support the upper wind turbine. The auxiliary system, which includes a pitch subsystem, a yaw subsystem, as well as several controllers and electrical facilities, guarantees the energy generation efficiency of the floating offshore wind turbine.

(2) Selection of failure modes and their corresponding causes

Three designers from two wind energy companies, and one researcher who has worked on failure and risk analysis of floating offshore wind turbines for several years and has several publications on the topic were employed to complete the tasks in this section, see Table 2. Specifically, three designers with master's degrees have worked with floating offshore wind turbines for three to four years. Before this, they had more experience and extended working periods on bottom-fixed offshore wind turbines; the researcher with a doctorate has the longest working time on the failure and risk analysis of floating offshore wind turbines compared with designers. The consultation of the specialists was conducted in two continuous rounds: failure items selection, as well as values of severity, occurrence, and detection design. The former was put forward to ascertain potential failure modes and failure causes of the floating offshore wind turbine. The second round of consultation is to obtain information for RPNs calculation.

Table 2. Specialists involved in the failure analysis of floating offshore wind turbines.

Code	Employer	Duty	Working Period *
<1>	Wind Energy Company	System Design	4 Years
<2>	Wind Energy Company	Components Design	3 Years
<3>	Wind Energy Company	Quality Engineer	4 Years
<4>	University	Researcher	6 Years

* Refers to the working time in floating offshore wind turbines.

Round 1: Failure modes and failure causes identification

In the initial round of consultation, a list of 29 failure modes with 53 failure causes taken from publications was distributed to each specialist, who was suggested to delete unlikely failure items and simultaneously add new failure modes (with corresponding failure causes) or new failure causes to existing failure modes. Subsequently, a new failure item is added to the final list if one specialist suggests it, or is deleted if recommended by more than two specialists. Ultimately, 42 failure modes corresponding to 104 failure causes of floating offshore wind turbines were identified, see Appendix A.

Round 2: Design values of severity, occurrence, and detection

The second round aims to attribute values of severity, occurrence, and detection of failure items that were identified in the previous round. Two documents were distributed to specialists: a final list of failure modes and failure causes (Appendix A) as well as a rating guidance for the severity of failure modes together with the occurrence and detection of failure causes. The rating guidance provides the specialists with a standard to normalize their knowledge about failures of floating offshore wind turbines, see Table 3.

Table 3. The rating guidance of severity, occurrence, and detection [16].

Rating	Severity	Occurrence		Detection
		Probability	Description	
1	The effect is not noticed	$P < 10^{-5}$	Extremely less	Certain
2	Very slight effect noticed	$P = 10^{-5}$	Remote	Very high
3	Slight effect causing annoyance	$P = 10^{-5}$	Very slight	High
4	Slight effect causing return of product	$10^{-5} < P < 4 \times 10^{-4}$	Slight	Moderate
5	Moderate effect causing return of product	$4 \times 10^{-4} < P < 2 \times 10^{-3}$	Occasional	Medium
6	Significant effect	$2 \times 10^{-3} < P < 1 \times 10^{-2}$	Moderate	Low chance
7	Major effect	$10^{-2} < P < 4 \times 10^{-2}$	Frequent	Slight
8	Extreme effect, system inoperable, safety issue	$4 \times 10^{-3} < P < 0.2$	High	Remote
9	Critical effect, system shutdown, safety risk	$0.2 < P < 0.33$	Very high	Very remote
10	Hazardous, without warning, life-threatening	$P > 0.33$	Extremely high	No chance, no inspection

(3) RPNs calculation

The *RPN* is a unidimensional index, without physical meaning, used to rank and assess failure modes. A failure item with a higher *RPN* is more critical than those with lower RPNs. However, RPNs are criticized for drawbacks in their implementation in actual engineering cases [16,19,31,32], for instance: (i) Various combinations of severity, occurrence, and detection may result in the same RPN, and the hidden meaning of each could be completely different; (ii) Indices (like severity, occurrence, and detection) and employed specialists are treated equally when calculating PPNs, in other words, the importance of the indices and the differences of specialists are ignored.

Aiming at removing the aforementioned restrictions, a new way of constructing RPNs is proposed in this study. The experts are weighted based on years of experience, which is an objective and explainable parameter. Other subjective dimensions like educational level (doctor, master) and professional activity (system designer, component designer, quality engineer, researcher) can also be considered in weighing experts, like in Refs. [41,42]. However, these dimensions are subjective, and no evidence shows that a specialist with a doctorate is more reliable than others with master's degrees or industrial specialists who may have better knowledge of failures of floating offshore wind turbines. Introducing such dimensions to weigh experts would bring additional uncertainties and make the results debatable. In this regard, this paper takes individuals' years of experience as the basis to weigh the experts, considering that years of experience can directly reflect the reliability of knowledge and experience on the failures of floating offshore wind turbines.

The weighted values of indices e.g., severity, occurrence, and detection of FMEA can be computed as:

$$\varphi_{ij} = \alpha_k \varphi_{ijk} \tag{1}$$

where, φ_{ij} represents the weighted value of severity, occurrence, or detection of failure cause j of i th failure mode of the floating offshore wind turbine. The variable φ_{ijk} reflects the original values of FMEA indices given by specialist k and α_k denotes the weight of specialist k which is defined as:

$$\alpha_k = \frac{WP_k}{\sum\limits_{k=1}^{4} WP_k} \tag{2}$$

where, WP_k is the working period of specialist k, see Table 2.

Hence, the RPN for failure cause j of the failure mode i of the floating offshore wind turbine (RPN_{ij}^{FC}) can be determined by:

$$RPN_{ij}^{FC} = \varphi_{ij}(Severity) \times \varphi_{ij}(Occurance) \times \varphi_{ij}(Detection) \tag{3}$$

Subsequently, the RPN of failure mode i (RPN_i^{FM}) is defined as the summation of the RPNs of all its failure causes, which can be represented by:

$$RPN_i^{FM} = \sum_j RPN_{ij} \tag{4}$$

Accordingly, RPNs for component h (RPN_h^{COMP}) and system g (RPN_g^{SYST}) can be obtained by:

$$RPN_h^{COMP} = \sum_i RPN_i^{FM} \tag{5}$$

$$RPN_g^{SYST} = \sum_h RPN_h^{COMP} \tag{6}$$

The weighted values of severity, occurrence, or detection of failure causes of the offshore wind turbine together with RPNs calculated (and their ranking) are listed in Appendix B.

(4) Recommendations

The failure analysis of floating offshore wind turbines aims at answering the following questions:

(1) What are the key failure items that contribute most to the global failure risk of floating offshore wind turbines?
(2) What are the effects, behavior, and root causes of each key failure item?
(3) What measures can be implemented to prevent the occurrence of key failures?

Hence, the inherent features of floating offshore wind turbine failures are expected to be explored. Accordingly, recommendations associated with reliable designs and activities related to maintenance and operations are suggested by the specialists employed after they are delivered by the critical failures and their root causes. These recommendations are to guarantee economic benefits and efficient electricity yielding of floating offshore wind turbines by avoiding critical failures. These recommendations are discussed in Section 3.6.

3. Results

This paper aims to demonstrate the failure features of floating offshore wind turbines, and to suggest potential measures to improve the performance of the system. The measures provide procedures to prevent floating offshore wind turbines from unexpectedly catastrophic failures.

The support structure is the most critical system, as it contributes some 46% of the RPNs of the system, which is significantly higher than that of systems installed within nacelles such as the energy-producing system (33%), auxiliary system (11%), energy-receiving system (5%), and energy-transforming system (5%), see Figure 2a,b. Floating assemblies are key elements of floating offshore wind turbines over onshore structures. Furthermore, floating assemblies incur considerable risk to the floating offshore wind turbines. This is also confirmed by the relative higher criticalities of components assembled in support structures, that are, mooring facilities (NO.1, with an *RPN* share of 22%), floating foundation (NO. 4, 11%), transition pieces (NO. 5, 7%), and tower (NO. 7, 6%), see Figure 2c.

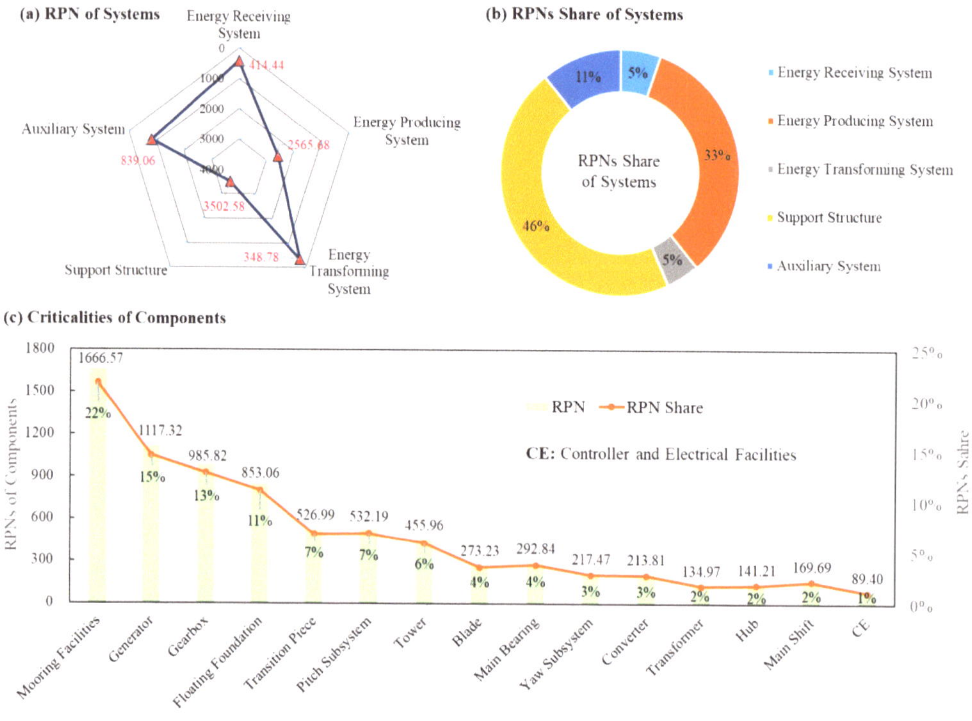

Figure 2. Failure analysis of systems and components of floating offshore wind turbines/(**a**) RPNs of systems; (**b**) RPNs shares of systems; (**c**) RPNs and their corresponding shares of components.

The results show that the support structure and the energy-producing system have the largest shares of RPNs of the floating offshore wind turbine, which is in line with the knowledge of the specialists who have put greater emphasis on such systems. Specifically, more than 50% of the components, nearly 60% of failure modes, and over 77% of failure causes of the floating offshore wind turbine (collected by this study) are related to the two aforementioned systems. Figure 2 also demonstrates that: (i) systems with more components, or components having diverse failure modes and failure causes, tend to be distributed by larger RPNs; (ii) systems or components with more complex structures and functions, or subject to inclement working conditions are risk-prone parts of the floating offshore wind turbine.

3.1. Energy-Receiving System

The energy-receiving system collects kinetic energy from the wind and transforms it into the mechanical energy of the main shaft by rotating blades and the hub. The energy-receiving system accomplishes the first step of energy production of the floating offshore wind turbine, that is, from kinetic energy to mechanical energy. One distinguishing property of this system is its particularly huge size, especially the large-scale blades.

Blades are the weak links of floating offshore wind turbines when such facilities enter the seas. Designers are devoted to guaranteeing the survival of these huge structures from irregular wind stress. Two catastrophic failures potentially lead to the floating offshore wind turbine shut down, which are, cracks ($RPN = 126$) and delamination ($RPN = 89$) of the blades. The mentioned failures call for the special attention of designers and maintenance teams. A blade crack is a consequence of defective manufacturing, while the blade delamination is due primarily to insufficient lightning protection. Failure of the hub is infrequent. However, manufacturing and fitting errors ($RPN = 88$) should be avoided to reduce severe consequences, such as blades breaking away from the hub.

3.2. Energy-Producing System

Energy producing system is the most complex system in terms of structure, function, and technological demand. The system is responsible for the second step of electricity generation, which is transforming mechanical energy into electricity. Specifically, the main bearing and main shaft transform mechanical energy from the energy-receiving system to the gearbox. The gearbox converts the received movement into low torque and high-angular velocity rotations. The generator is an electromechanical system designed to transform the rotational motion of the gearbox into electricity by the electromagnetic induction principle.

The generator is more important than other components of the energy-producing system. Both mechanical and electrical components failures can be discovered. Mechanical failures are more severe, while electrical components fail more frequently. To be more specific, bearings failures leading to a lack of, abnormal, or unbalanced electricity generation, and are more critical than the other components of the generator. Root causes of bearings failures can be attributed to electric corrosion of rollaway nests, improper grease, and over-tightening. Moreover, overheating of the generator is a typical failure most likely caused by cooling system failures.

The main reasons for gearbox malfunction are the failures of gears and bearings. Gearbox failures are typical common cause failures since most of them can be associated with lubrication failures. The common cause failures are an explanation for the chain failure form of the gearbox (a slight failure can lead to disastrous failures of other elements) and its severe consequences. Worn gears, fractured gear teeth, and overheating of the gearbox are critical failure modes that lead to vibration and shutdown of the floating offshore wind turbine. Wear, fatigue and lubrication failures are the causes of gearbox failures.

Failures of the main bearings are more critical than those of the main shaft. Common failure modes of these components are abnormal vibration, and cracks resulting primarily from wear, fatigue, substandard lubrication, and welding defects. Details on the failure features of the energy-producing system are presented in Figure 3.

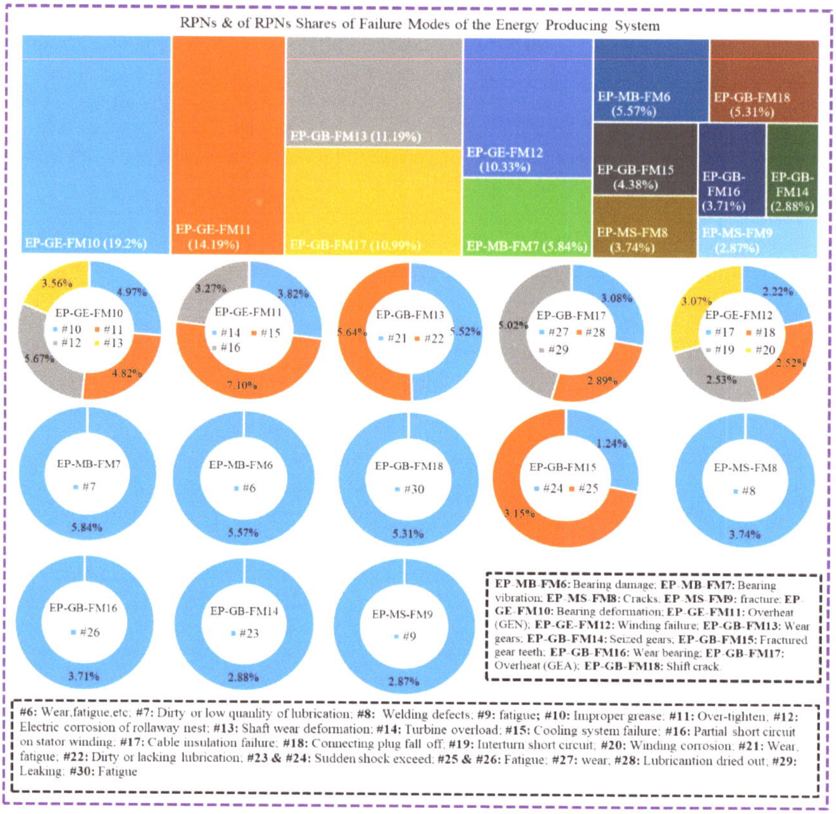

Figure 3. Failure features of the energy-producing system failures/Hierarchical relations among systems, components, failure modes, as well as failure causes are listed in Appendix A; #: Failure cause.

3.3. Energy-Transforming System

The energy-transforming system is a combination of electrical components including a converter, transformer, cable (mainly among wind turbines, distribution stations, and the grid), power distribution stations, and other support equipment for power transmission from the generator to the shore. The energy transforming system is responsible for adjusting unstable and low voltage electricity generated by the generator into stable and required voltage electricity of the grid. Converters and transformers are considered in this analysis, since they are installed within wind turbines, and other items like power distribution stations and cables, which are not parts of floating offshore wind turbines, are neglected.

Open and short circuits are distinguished failures of the converter and transformer. Failures of the converter are more critical than failures of the transformer. Open circuits result in a disconnection between the generator and the grid. Short circuits, the main reason for fires, lead to the direct shutdown of the converter and the transformer.

Short circuits of the converter and the transformer are consequences of overheating. The root reasons for open circuits of the converter and the transformer are much more complex. The converter's open circuits are more critical than the same failures of the transformer. On one hand, the primary origins of converter open circuits are cooling system failure and load mutation, which can be compared with those of the transformer: iron core corrosion, overcurrent, and over-voltage. On the other hand, under comparable levels of severity of consequences and likelihood of occurrence, the occurrence of open circuits of the converter is much more difficult to be detected in advance.

3.4. Auxiliary System

The auxiliary system consists of a pitch system, a yaw system, as well as a controller and electrical facilities. This system is not directly involved in electricity production. Instead, it is designed to improve energy production efficiency to guarantee the economy of the floating offshore wind turbine. Pitch and yaw systems adjust blades and the wind turbine to maintain appropriate angles to the wind. Controller and electrical facilities are the "brain" of the floating offshore wind turbine. These systems connect and control several systems and components of the floating offshore wind turbine.

Failures of the pitch system are more critical than those of the yaw system, as well as the controller and electrical facilities. Particularly, the criticality of failures of the pitch system can be attributed to their disaster-causing causes e.g., poor calibration of the pitch angle, wear, fatigue, and the excessive vibration of pitch bearings. Yaw system failures, such as seizure bearings as well as bearing and gear corrosion, are caused by hydraulic system failures e.g., the poor lubrication and wear (or degradation) of hydraulic lines. Failures of the controller and electrical facilities are not critical. These failures can be easily discovered. Failure consequences of the controller and electrical facilities are minor, but these failures are likely to happen frequently. To date, failure warning and detection modules have already been embedded into wind turbine monitoring software, which confirms that the criticality of the controller and electrical facilities failures will further decline.

3.5. Support Structure

Various types of support structures of floating offshore wind turbines, e.g., spar-buoy, semi-submersible, and tension-leg platforms, have been developed to enable offshore wind turbine installations in deep water where bottom-fixed systems are no longer feasible [43]. Except for the larger size and higher reliability requirements for each component, the main difference between floating and bottom-fixed offshore wind turbines is the employed support structure. Support structures are the most failure-prone part of the system due to their destructive damage and unexpected vulnerabilities raised by harsh sea conditions.

In this failure analysis, the authors and the specialists have particularly focused on the support structure by expecting that this analysis will provide details and insights into the failure features of this relatively new system to the designers and operators of floating offshore wind farms.

Support structures are designed to sustain wind turbines located at the required height above the sea so that they can access the desired wind resources. Mooring facilities, towers, floating foundations, and transition pieces are typical items of the support structure. The mooring facilities keep the floating offshore wind turbine at the selected location with anchors and mooring lines. The floating foundation provides indispensable buoyancy to the massive upper structures. The transition piece connects the floating foundation and the tower. The tower determines the height of the floating offshore wind turbine. Failure features of the support structure at the failure mode level are listed in Figure 4.

The mooring facilities are much more critical than other components. Mooring line failure is highlighted by its considerably high *RPN* (*RPN* = 693.41, 9% of the total). Mooring line failure is harmful to the stability of the support structure and the wind turbine. Devices, such as anchor pickup devices, accumulators, and connectors, that failed to accomplish their functions are responsible more often than other degenerative root causes (e.g., wear of the transitional chain and friction chain). An abnormal mooring line is an initial failure mode of the mooring facilities, such as mooring lines breaking. It is a consequence of mooring line wear, fatigue, and unknown abnormal stress. Moreover, several human failures e.g., insufficient emergency measurement and calculation faults, are not neglectable as failures of the mooring lines. Fairlead and anchor failures are not critical. The details of the failure features of the mooring system are demonstrated in Figure 5.

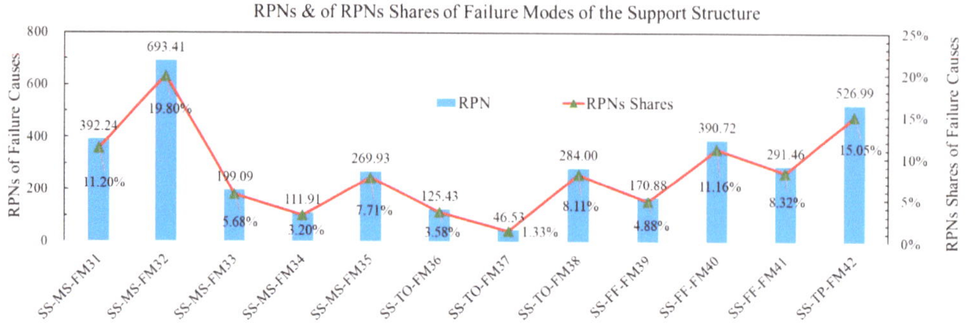

Figure 4. Failure features of the support structure at the failure mode level.

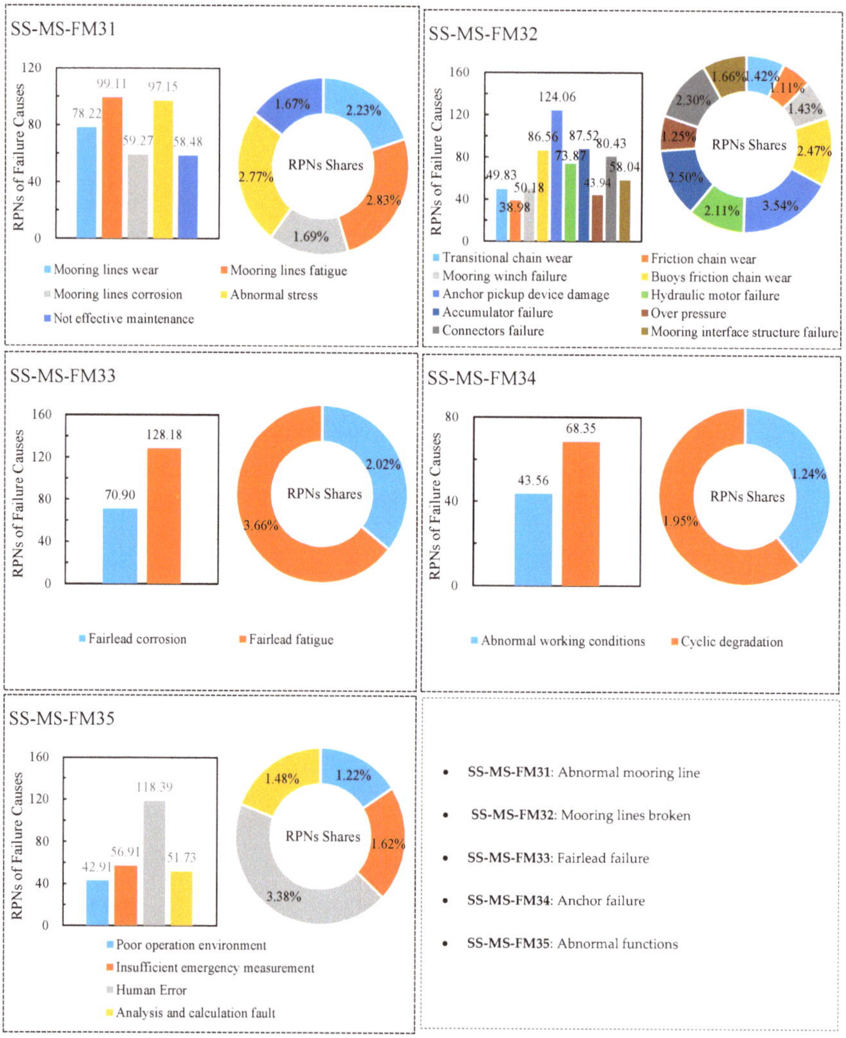

Figure 5. Failure features of the mooring system.

Regarding tower failures, tower crack is recognized to be critical. Specifically, tower cracks result from faulty welding or material fatigue. Other failures are either merely not possible e.g., tower collapse, or are easy-to-be detected e.g., abnormal vibration. Transition piece failures mainly associated with cracks have not been reported in previous publications. The criticalities of failure causes such as material fatigue, corrosion, plastic deformation of structures, cyclic degradation, strong wind or waves, and faulty welding are considerably high. Details on the failure features of the tower and the transition piece are illustrated in Figure 6.

The huge volume, massive weight, and inclement sea conditions introduce considerable vulnerability to the floating foundation, particularly to its underwater seals. Virtually, all seal failures result from pipe joint defects such as welding defects, corrosion, and fatigue. The floating foundation is a complex system assembled by multiple devices. Devices' failures of the floating foundation are diverse, but with low criticalities. Sensors (for platform monitoring), manholes, pumps, and towing brackets are fragile and fail frequently. Dropped objects may result in unexpected and unforeseen damages to the floating foundation and contribute to vast economic losses. This failure is the consequence of harsh and uncertain sea conditions e.g., strong wind/waves, operational failures e.g., plane crashes, or unpredictable events e.g., bird collisions. The details of the failure features of the floating foundation are demonstrated in Figure 7.

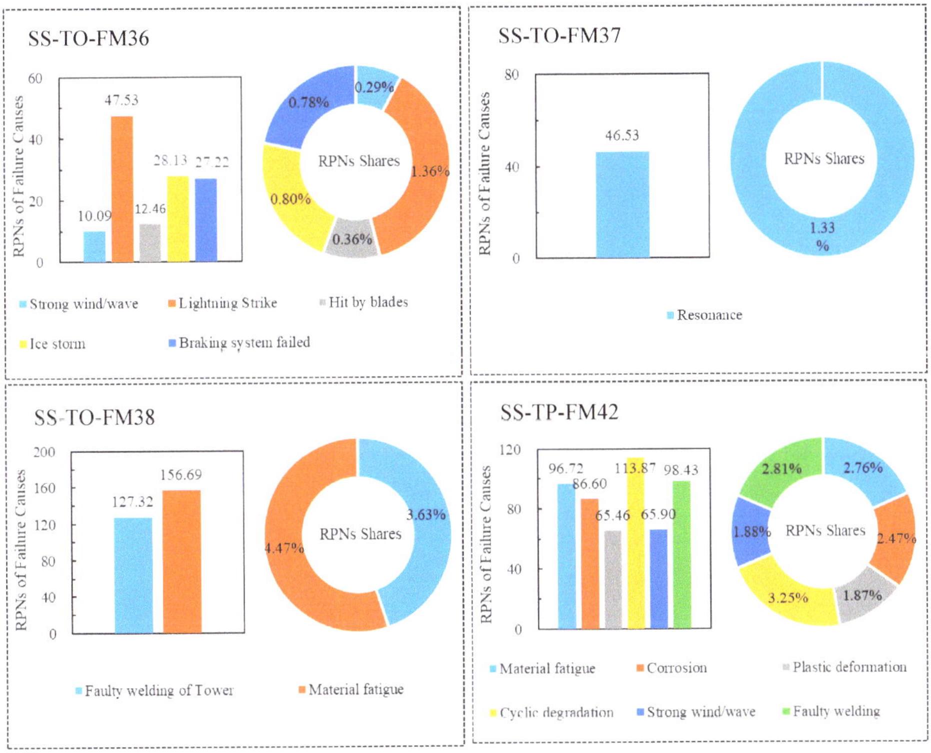

Figure 6. Failure features of the tower and the transition piece/SS-TO-FM36: Tower collapse; SS-TO-FM37: Tower abnormal vibration; SS-MS-FM38: Tower crack; SS-TP-FM42: Transition piece crack.

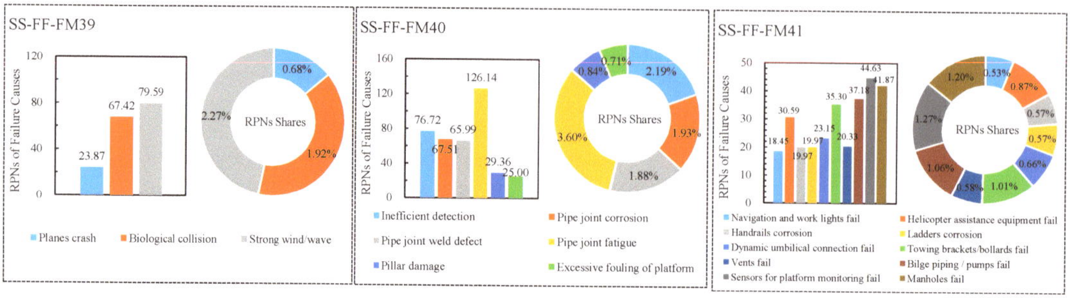

Figure 7. Features of failures of the floating foundation/SS-FF-FM39: Hit by dropped objects; SS-FF-FM40: Watertight fault; SS-FF-FM41: Additional structures fail.

3.6. Recommendations

Failure analysis enables the analysts to comprehensively understand the failure characteristics of the components of floating offshore wind turbines and provide the basis for preventing critical failures from occurring. The conducted analysis includes the development of recommendations to designers and operators (particularly in the maintenance sector) to update designs and optimize maintenance strategies. Accordingly, better economic performance is expected to be achieved by such recommendations formulated as follows:

(1) Energy-Receiving System

Improve manufacturing processes and enhance testing in advance of blades being delivered to wind farms; Enhance lightning prevention by designing more reliable and less vulnerable systems for blades; Improve manufacturing and installation processes, especially for connection parts between hubs and blades.

(2) Energy-Producing System

Strengthen surfaces of bearing tracks (inner ring, outer ring, and rolling elements) of main bearings; Pay attention to main shaft welding and conduct more strict welding quality inspections; Improve electric corrosion prevention of generators; Upgrade cooling systems to enhance the heat dissipation of generators; Introduce wear and fatigue prevention actions to the gears in gearboxes; Guarantee high-quality lubrication by periodical lubrication inspection and replacement.

(3) Energy-Transforming System

Guarantee reliable operation of the cooling system to avoid overheating of converters and transformers; Implement additional design activities for instantaneous voltage or current overload offsetting of converters and transformers.

(4) Auxiliary System

Develop control-and-feedback module to monitoring software of wind turbines to precisely adjust pitch angles; Better design of the hydraulic systems, particularly its embranchments of yaw and pitch systems.

(5) Support Structure

Enhance the strength of mooring lines to eliminate abnormal stress; Improve the reliability of anchor pickup device, transitional chain, mooring winch, fairlead, and accumulators in mooring systems; Improve the training of operators and maintenance members to avoid human failures; Reinforce tower welding quality; Enhance pipe joint design and welding quality of floating foundations to avoid watertight fault; Improve the sensors of platform monitoring and pay more attention to manholes failures; Implement measures to avoid transition pieces from corrosion, fatigue, and life cycle degradation; Consider the impacts of strong wind/waves in the full life cycle of support structures.

4. Comparisons

This study has collected the knowledge of four experts to complete a failure analysis of a floating offshore wind turbine. This approach takes the specialists' working experience as weights of their knowledge rather than a traditional analysis that treats values of severity, occurrence, and detection from each specialist as nondistinctive.

A comparative analysis between the ranks of failures obtained by the proposed method and by the conventional model, based on a single expert, was conducted, see Appendix C. The results show that: (i) rankings of failures based on a single expert are inconsistent with the results of this paper that includes four experts for group decision-making; (ii) the years of the experts' experience affect the failures' ranking. To be specific, the overall failure ranking difference between expert #4 (with 6 years of working experience) and the results of the group decision-making in the paper is the smallest, followed by expert #1 (4 years), expert #3 (4 years), and expert #2 (3 years); (iii) two experts with the same working period show some 40% difference in their ranking of failures, which indicates that individual properties of experts impact their judgments.

A comparative analysis was also carried out to show the rank differences between the conventional and the proposed FMEA methods based on the group of 4 experts, see Figure 8, where failure items with the same *RPN* of both FMEAs are not displayed. The figure reveals that: (i) the proposed FMEA reclassifies most failure causes of the floating offshore wind turbine. To be specific, 89 of 104 failure cause ranks were changed, since this analysis distributed higher priorities to experienced specialists and decreased the credibility of specialists who have short working periods in the field; (ii) the *RPN* ranks of failure modes changed slightly, and failure modes with fewer root causes change more distinctly than those with more causes; (iii) Experienced specialists consider pitch system failures to be more critical; (iv) RPNs of systems of the floating offshore wind turbine calculated by the proposed FMEA are superior to that of the conventional FMEA technique. It reflects that experienced specialists tend to provide more aggressive evaluations than the others; (v) The use of more specialists for failure analysis, and simultaneously the prioritization of their knowledge according to their experience, is in line with practice and can provide convincing results.

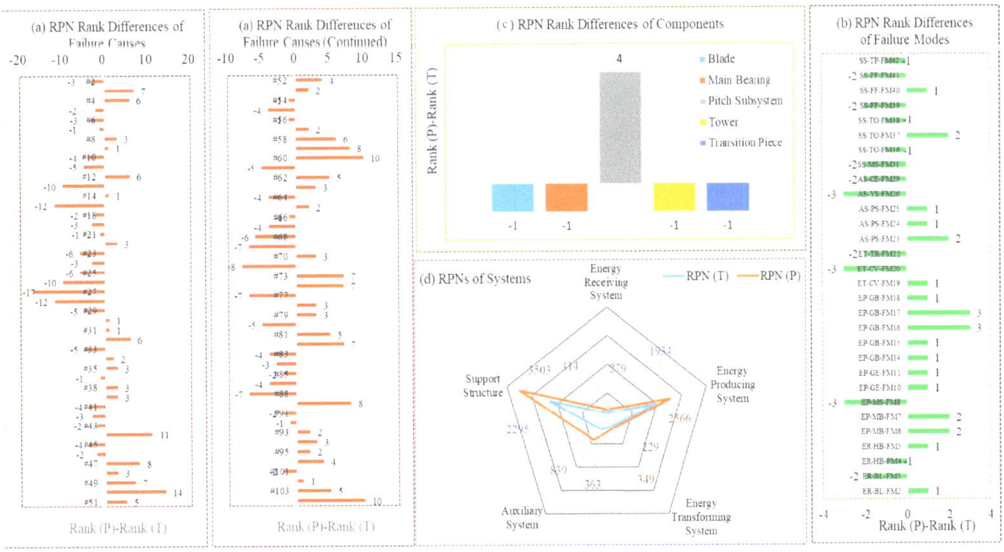

Figure 8. Comparison of results of proposed and conventional FMEAs/Rank (P): *RPN* rank computed by proposed FMEA; Rank (T): *RPN* rank calculated by traditional FMEA; Definition of items refers to Appendix A (#: Failure cause).

Overall, this paper identifies possible failures of floating offshore wind turbines, and identifies the most critical failures and their causes from a risk management point of view by considering the influence of the experts' backgrounds. In addition, actions to prevent the occurrence of the root causes of critical failures are determined.

Although the recommendations for failure prevention summarized in this paper are applicable, additional ranking criteria can also be considered. Economic factors, like failure cost, and operational factors, like downtime, may also be used as criteria for ranking the criticality of failures from other perspectives. The joint consideration of various criteria will contribute to more robust conclusions.

5. Conclusions

This paper develops a comprehensive failure analysis of a floating offshore wind turbine, focused on its support structure, which represents the main difference between floating offshore wind turbines and those onshore and offshore but with bottom-fixed foundations. The objective is to provide practitioners of the floating offshore energy sector with relevant information on the failure features of a relatively new floating offshore wind turbine concept. For this purpose, an improved FMEA method is suggested in which the risk priority number, the basis of the failure mode and effect analysis, is formulated considering the background knowledge of the specialists that contributed to the failure analysis.

Overall, 15 main components, 42 failure modes, and 104 failure causes of the floating offshore wind turbine are considered and ranked according to their criticality levels. This analysis concludes that:

(1) Support structures and energy-producing systems are more critical than other systems, which call for special attention by the designers, operators, and maintenance teams of floating offshore wind farms;

(2) Components with complicated functions e.g., gearboxes and generators, as well as those experiencing difficult working conditions e.g., mooring facilities and floating foundations are recognized to be more fragile against the occurrence of failures;

(3) Critical failure modes and corresponding failure causes are identified and ranked by their calculated RPNs;

(4) 21 recommendations are suggested to designers and operators to avoid floating offshore wind turbines from unexpected failures. Moreover, a comparison analysis was carried out to show the similarities and differences between the proposed method and the conventional method.

Author Contributions: Conceptualization, H.L. and A.P.T.; methodology, H.L. and A.P.T.; writing—original draft preparation H.L. and A.P.T.; writing—review and editing, C.G.S.; supervision, C.G.S.; project administration, C.G.S. All authors have read and agreed to the published version of the manuscript.

Funding: This study was completed within the project ARCWIND–Adaptation and implementation of floating wind energy conversion technology for the Atlantic region, which is co-financed by the European Regional Development Fund through the Interreg Atlantic Area Programme under contract EAPA 344/2016. The first author has been supported by a scholarship from China Scholarship Council (CSC) under Grant No. 201806070048. This work contributes to the Strategic Research Plan of the Centre for Marine Technology and Ocean Engineering (CENTEC), which is financed by the Portuguese Foundation for Science and Technology (Fundação para a Ciência e Tecnologia–FCT) under contract UIDB/UIDP/00134/2020.

Institutional Review Board Statement: Not applicable.

Informed Consent Statement: Not applicable.

Data Availability Statement: Not applicable.

Conflicts of Interest: The authors declare no conflict of interest.

Appendix A. Failure Modes and Causes of the Floating Offshore Wind Turbine

Failure Mode Level			Failure Cause Level	
Code	**Failure Modes**	**End Effects**	**Code**	**Failure Causes**
ER-BL-FM1	Blades cracks	Wind turbine stop working	#1	Manufacturing error
ER-BL-FM2	Delamination	Wind turbine stop working	#2	Insufficient lighting protection
ER-BL-FM3	Gear teeth slip	Blades fail to attack wind properly	#3	Wear, fatigue, etc.
ER-HB-FM4	Fracture in the shell	Rotor break	#4	Manufacturing error
ER-HB-FM5	Error in positioning	Blades break away from the hub	#5	Manufacturing error and/or fitting error
EP-MB-FM6	Bearing damage	Wind turbine stop working	#6	Wear, fatigue, etc.
EP-MB-FM7	Bearing vibration	Abnormal working condition	#7	Substandard lubrication
EP-MS-FM8	Cracks	Collapse of wind turbine	#8	Welding defects
EP-MS-FM9	Fracture	Collapse of wind turbine	#9	Fatigue
			#10	Improper grease
EP-GE-FM10	Bearing deformation	No, abnormal or unbalanced electricity generation	#11	Over tighten
			#12	Electric corrosion of rollaway nest
			#13	Shaft wear deformation
			#14	Turbine overload
EP-GE-FM11	Overheat (GE)	Offshore wind turbine shutdown	#15	Cooling system failure
			#16	Partial short circuit on stator winding
			#17	Cable insulation failure
EP-GE-FM12	Winding failure	No, abnormal or unbalanced electricity generation	#18	Connecting plug fall off
			#19	Interturn short circuit
			#20	Winding corrosion
EP-GB-FM13	Wear gears	Exceeded vibration or unstable electricity output	#21	Wear, fatigue
			#22	Dirty or lacking lubrication
EP-GB-FM14	Seized gears	No electricity output	#23	Sudden shock exceed limitation
EP-GB-FM15	Fractured gear teeth	Exceeded vibration or unstable electricity output	#24	Sudden shock exceed limitation
			#25	Fatigue
EP-GB-FM16	Wear bearing	Exceeded vibration	#26	Fatigue
			#27	Wear
EP-GB-FM17	Overheat (GB)	Offshore wind turbine shutdown	#28	Lubrication dried out
			#29	Leaking
EP-GB-FM18	Shift crack	Offshore wind turbine shutdown	#30	Fatigue
ET-CV-FM19	Short circuit	Converter shutdown	#31	Over heat
			#32	Load mutation
ET-CV-FM20	Open circuit	Disconnect to grid	#33	Invert power input fault
			#34	Overload
			#35	Cooling system fault
ET-TR-FM21	Short circuit	Transformer shutdown	#36	Over heat
			#37	Constant overload
ET-TR-FM22	Open circuit	Disconnect to grid	#38	Iron core corrosion
			#39	Overcurrent
			#40	Overvoltage
AS-PS-FM23	Wrong pitch angle	Decrease of electricity output	#41	Poor calibration
AS-PS-FM24	Pitting Gears	Vibration increase	#42	Wear, fatigue
AS-PS-FM25	Misalignment bearings	Decrease of electricity output	#43	Wear, excessive vibration
AS-YS-FM26	Seizure bearings	Over heat	#44	Poor lubrication
AS-YS-FM27	Corrosions	Pitting of raceways	#45	Presence of corrosive substances
AS-YS-FM28	Hydraulic leakage	Rotor fails to stop	#46	Wear or degradation on hydraulic lines
AS-CE-FM29	Short circuit	Shutdown	#47	Moisture penetration
AS-CE-FM30	Open circuit	Shutdown	#48	Lightning strike
			#49	Mooring lines wear
			#50	Mooring lines fatigue
SS-MS-FM31	Abnormal mooring lines	Mooring line strength decrease or broken	#51	Mooring lines corrosion
			#52	Abnormal stress
			#53	Not effective maintenance

Failure Mode Level			Failure Cause Level	
Code	Failure Modes	End Effects	Code	Failure Causes
SS-MS-FM32	Mooring lines broken	Malfunction of the whole system, the facility cannot locate in water	#54	Transitional chain wear
			#55	Friction chain wear
			#56	Mooring winch failure
			#57	Buoys friction chain wear
			#58	Anchor pickup device damage
			#59	Hydraulic motor failure
			#60	Accumulator failure
			#61	Over pressure
			#62	Connectors failure
			#63	Mooring interface structure failure
SS-MS-FM33	Fairlead failure	The anchor cannot be dropped and lift	#64	Fairlead corrosion
			#65	Fairlead fatigue
SS-MS-FM34	Anchor failure	Anchor failure	#66	Abnormal working conditions
			#67	Cyclic degradation
SS-MS-FM35	Abnormal functions	Anchoring accuracy decrease	#68	Poor operation environment
			#69	Insufficient emergency measurement
			#70	Human Error
			#71	Analysis and calculation fault
SS-TO-FM36	Tower collapse	Failure of whole facility and vast economic loses	#72	Strong wind/wave
			#73	Lightning Strike
			#74	Hit by blades
			#75	Ice storm
			#76	Braking system failed
SS-TO-FM37	Abnormal vibration	Potential collapse	#77	Resonance
SS-TO-FM38	Crack	Potential collapse	#78	Faulty welding of Tower
			#79	Material fatigue
SS-FF-FM39	Hit by dropped objects	Damage to the facility, vast economic loses	#80	Planes crash
			#81	Biological collision
			#82	Strong wind/wave
SS-FF-FM40	Watertight fault	Potential failure	#83	Inefficient detection
			#84	Pipe joint corrosion
			#85	Pipe joint weld defect
			#86	Pipe joint fatigue
			#87	Pillar damage
SS-FF-FM41	Additional structures fail	Potential failure	#88	Excessive fouling of platform
			#89	Navigation and work lights fail
			#90	Helicopter assistance equipment fail
			#91	Handrails corrosion
			#92	Ladders corrosion
			#93	Dynamic umbilical connection fail
			#94	Towing brackets/bollards fail
			#95	Vents fail
			#96	Bilge piping/pumps fail
			#97	Sensors for platform monitoring fail
			#98	Manholes fail
SS-TP-FM42	Transition piece crack	Potential collapse	#99	Material fatigue
			#100	Corrosion
			#101	Plastic deformation
			#102	Cyclic degradation
			#103	Strong wind/wave
			#104	Faulty welding

Appendix B. Weighted of Failure Indices, RPNs, and Their Ranking of Failure Causes

Code	WI	RPN	PC	Rank	Code	WI	RPN	PC	Rank
ER-BL-FM1-#1	(5.50, 5.31, 4.31)	126.01	46%	14	SS-MS-FM31-#53	(4.81, 3.81, 3.19)	58.48	4%	60
ER-BL-FM2-#2	(5.38, 4.69, 3.56)	89.76	33%	28	SS-MS-FM32-#54	(5.31, 3.06, 3.06)	49.83	3%	69
ER-BL-FM3-#3	(3.19, 4.06, 4.44)	57.46	21%	62	SS-MS-FM32-#55	(4.94, 2.94, 2.69)	38.98	2%	80
ER-HB-FM4-#4	(4.31, 4.19, 2.94)	53.05	38%	66	SS-MS-FM32-#56	(6.31, 3.44, 2.31)	50.18	3%	68
ER-HB-FM5-#5	(6.06, 4.56, 3.19)	88.17	62%	31	SS-MS-FM32-#57	(6.06, 5.31, 2.69)	86.56	5%	34
EP-MB-FM6-#6	(7.31, 4.81, 4.06)	142.97	49%	6	SS-MS-FM32-#58	(6.19, 5.44, 3.69)	124.06	7%	15
EP-MB-FM7-#7	(6.19, 7.31, 3.31)	149.88	51%	3	SS-MS-FM32-#59	(5.94, 5.69, 2.19)	73.87	4%	45
EP-MS-FM8-#8	(7.69, 4.44, 2.81)	95.94	57%	25	SS-MS-FM32-#60	(6.19, 4.44, 3.19)	87.52	5%	32
EP-MS-FM9-#9	(8.56, 3.06, 2.81)	73.75	43%	47	SS-MS-FM32-#61	(4.44, 4.06, 2.44)	43.94	3%	75
EP-GE-FM10-#10	(4.31, 5.38, 5.50)	127.49	11%	11	SS-MS-FM32-#62	(5.19, 3.94, 3.94)	80.43	5%	38
EP-GE-FM10-#11	(4.50, 5.00, 5.50)	123.75	11%	16	SS-MS-FM32-#63	(6.06, 2.69, 3.56)	58.04	3%	61
EP-GE-FM10-#12	(4.50, 5.88, 5.50)	145.41	13%	4	SS-MS-FM33-#64	(4.31, 3.81, 4.31)	70.90	4%	48
EP-GE-FM10-#13	(4.31, 5.06, 4.19)	91.42	8%	27	SS-MS-FM33-#65	(5.31, 4.44, 5.44)	128.18	8%	10
EP-GE-FM11-#14	(4.38, 4.38, 5.13)	98.10	9%	21	SS-MS-FM34-#66	(4.06, 3.81, 2.81)	43.56	3%	76
EP-GE-FM11-#15	(6.69, 6.31, 4.31)	182.05	16%	1	SS-MS-FM34-#67	(3.69, 4.06, 4.56)	68.35	4%	49
EP-GE-FM11-#16	(4.94, 4.19, 4.06)	84.00	8%	35	SS-MS-FM35-#68	(3.81, 4.19, 2.69)	42.91	3%	77
EP-GE-FM12-#17	(3.38, 5.88, 2.88)	57.01	5%	63	SS-MS-FM35-#69	(4.19, 4.44, 3.06)	56.91	3%	64
EP-GE-FM12-#18	(3.19, 4.56, 4.44)	64.53	6%	56	SS-MS-FM35-#70	(4.56, 5.69, 4.56)	118.39	7%	17
EP-GE-FM12-#19	(3.69, 4.94, 3.56)	64.86	6%	55	SS-MS-FM35-#71	(3.94, 3.56, 3.69)	51.73	3%	67
EP-GE-FM12-#20	(4.81, 3.31, 4.94)	78.71	7%	41	SS-TO-FM36-#72	(8.94, 1.06, 1.06)	10.09	2%	104
EP-GB-FM13-#21	(4.69, 5.69, 5.31)	141.63	14%	7	SS-TO-FM36-#73	(8.56, 3.06, 1.81)	47.53	10%	72
EP-GB-FM13-#22	(5.63, 4.38, 5.88)	144.58	15%	5	SS-TO-FM36-#74	(8.94, 1.31, 1.06)	12.46	3%	103
EP-GB-FM14-#23	(5.63, 3.50, 3.75)	73.83	7%	46	SS-TO-FM36-#75	(8.56, 1.81, 1.81)	28.13	6%	88
EP-GB-FM15-#24	(7.13, 2.38, 1.88)	31.73	3%	83	SS-TO-FM36-#76	(8.19, 2.31, 1.44)	27.22	6%	90
EP-GB-FM15-#25	(7.06, 4.69, 2.44)	80.69	8%	37	SS-TO-FM37-#77	(6.19, 3.44, 2.19)	46.53	10%	73
EP-GB-FM16-#26	(4.94, 5.81, 3.31)	95.07	10%	26	SS-TO-FM38-#78	(6.44, 5.19, 3.81)	127.32	28%	12
EP-GB-FM17-#27	(5.25, 4.63, 3.25)	78.91	8%	40	SS-TO-FM38-#79	(6.44, 5.81, 4.19)	156.69	34%	2
EP-GB-FM17-#28	(4.00, 5.50, 3.38)	74.25	8%	44	SS-FF-FM39-#80	(6.81, 2.44, 1.44)	23.87	3%	96
EP-GB-FM17-#29	(5.56, 4.81, 4.81)	128.83	13%	9	SS-FF-FM39-#81	(5.94, 3.19, 3.56)	67.42	8%	51
EP-GB-FM18-#30	(5.44, 4.31, 5.81)	136.30	14%	8	SS-FF-FM39-#82	(8.31, 3.56, 2.69)	79.59	9%	39
ET-CV-FM19-#31	(5.06, 3.06, 3.81)	59.11	28%	59	SS-FF-FM40-#83	(4.81, 4.81, 3.31)	76.72	9%	43
ET-CV-FM20-#32	(4.19, 4.06, 2.81)	47.85	22%	71	SS-FF-FM40-#84	(4.81, 3.94, 3.56)	67.51	8%	50
ET-CV-FM20-#33	(4.56, 2.56, 2.31)	27.04	13%	92	SS-FF-FM40-#85	(4.44, 5.06, 2.94)	65.99	8%	52
ET-CV-FM20-#34	(5.19, 3.06, 1.94)	30.78	14%	84	SS-FF-FM40-#86	(5.19, 4.69, 5.19)	126.14	15%	13
ET-CV-FM20-#35	(5.56, 3.81, 2.31)	49.04	23%	70	SS-FF-FM40-#87	(5.19, 4.31, 1.31)	29.36	3%	87
ET-TR-FM21-#36	*(5.56, 4.31, 1.06)*	*25.49*	19%	93	SS-FF-FM40-#88	(3.81, 4.56, 1.44)	25.00	3%	95
ET-TR-FM22-#37	(5.31, 2.69, 1.06)	15.17	11%	102	SS-FF-FM41-#89	(2.81, 4.56, 1.44)	18.45	2%	101
ET-TR-FM22-#38	(5.56, 4.19, 1.69)	39.31	29%	79	SS-FF-FM41-#90	(3.94, 3.19, 2.44)	30.59	4%	85
ET-TR-FM22-#39	(4.94, 3.81, 1.44)	27.06	20%	91	*SS-FF-FM41-#91*	*(2.06, 4.19, 2.31)*	*19.97*	2%	*99*
ET-TR-FM22-#40	(4.94, 3.94, 1.44)	27.95	21%	89	*SS-FF-FM41-#92*	*(2.06, 4.19, 2.31)*	*19.97*	2%	*99*
AS-PS-FM23-#41	(4.31, 5.06, 2.44)	53.22	24%	65	SS-FF-FM41-#93	(2.69, 3.06, 2.81)	23.15	3%	97
AS-PS-FM24-#42	(4.94, 6.19, 2.94)	89.74	40%	29	SS-FF-FM41-#94	(2.69, 3.69, 3.56)	35.30	4%	82
AS-PS-FM25-#43	(4.19, 5.94, 3.31)	82.36	37%	36	SS-FF-FM41-#95	(3.19, 4.44, 1.44)	20.33	2%	98
AS-YS-FM26-#44	(5.19, 3.69, 4.69)	89.67	41%	30	SS-FF-FM41-#96	(4.06, 4.44, 2.06)	37.18	4%	81
AS-YS-FM27-#45	(4.75, 5.50, 3.75)	97.97	45%	22	SS-FF-FM41-#97	(3.81, 5.06, 2.31)	44.63	5%	74
AS-YS-FM28-#46	(4.94, 5.69, 1.06)	29.84	14%	86	SS-FF-FM41-#98	(3.19, 3.69, 3.56)	41.87	5%	78
AS-CE-FM29-#47	(5.44, 3.19, 3.69)	63.91	71%	57	SS-TP-FM42-#99	(3.69, 4.94, 5.31)	96.72	18%	24
AS-CE-FM30-#48	*(5.56, 3.19, 1.44)*	*25.49*	29%	*93*	SS-TP-FM42-#100	(4.31, 6.06, 3.31)	86.60	16%	33
SS-MS-FM31-#49	(4.81, 4.56, 3.56)	78.22	5%	42	SS-TP-FM42-#101	(4.06, 4.69, 3.44)	65.46	12%	54
SS-MS-FM31-#50	(5.19, 4.56, 4.19)	99.11	6%	19	SS-TP-FM42-#102	(3.69, 5.81, 5.31)	113.87	22%	18
SS-MS-FM31-#51	(4.69, 5.19, 2.44)	59.27	4%	58	SS-TP-FM42-#103	(4.19, 4.94, 3.19)	65.90	13%	53
SS-MS-FM31-#52	(4.56, 4.31, 4.94)	97.15	6%	23	SS-TP-FM42-#104	(5.31, 5.81, 3.19)	98.43	19%	20

The codes are managed in the System-Component-Failure Mode-Failure Cause form; The italic and underline items represent failure items with equal RPN; (A, B, C) represents (Severity, Occurrence, Detection); WI: Weighted Indices; PC: RPN percentage of each component.

Appendix C. Failure Analysis Results (Ranking Differences of Failures) by Experts

Code	R(PP)	EXP.#1	EXP.#2	EXP.#3	EXP.#4	Code	R(PP)	EXP.#1	EXP.#2	EXP.#3	EXP.#4
ER-BL-FM1-#1	14	24	78	43	−10	SS-MS-FM31-#53	60	3	−7	2	−20
ER-BL-FM2-#2	28	20	−11	−4	5	SS-MS-FM32-#54	69	10	−15	−59	8
ER-BL-FM3-#3	62	5	−44	33	−27	SS-MS-FM32-#55	80	−10	−25	−69	18
ER-HB-FM4-#4	66	−17	−51	34	−24	SS-MS-FM32-#56	68	0	−12	−30	−5
ER-HB-FM5-#5	31	−28	−27	38	23	SS-MS-FM32-#57	34	19	23	−22	−10
EP-MB-FM6-#6	6	40	7	15	17	SS-MS-FM32-#58	15	71	43	24	−14
EP-MB-FM7-#7	3	21	59	19	35	SS-MS-FM32-#59	45	42	48	−13	−33
EP-MS-FM8-#8	25	−12	−18	56	−5	SS-MS-FM32-#60	32	48	27	1	−25
EP-MS-FM9-#9	47	−8	−39	35	−22	SS-MS-FM32-#61	75	−17	−15	−41	19
EP-GE-FM10-#10	11	−7	52	6	51	SS-MS-FM32-#62	38	36	49	−3	−8
EP-GE-FM10-#11	16	−11	28	2	23	SS-MS-FM32-#63	61	34	27	−25	−20
EP-GE-FM10-#12	4	5	41	42	11	SS-MS-FM33-#64	48	−13	27	4	11
EP-GE-FM10-#13	27	−13	−1	20	57	SS-MS-FM33-#65	10	−9	66	43	7
EP-GE-FM11-#14	21	−3	−1	55	6	SS-MS-FM34-#66	76	−12	1	−13	2
EP-GE-FM11-#15	1	24	45	57	1	SS-MS-FM34-#67	49	−19	29	−29	21
EP-GE-FM11-#16	35	−14	−14	36	−1	SS-MS-FM35-#68	77	−44	12	−37	12
EP-GE-FM12-#17	63	−32	−61	20	23	SS-MS-FM35-#69	64	−30	26	−36	11
EP-GE-FM12-#18	56	−16	−32	14	−10	SS-MS-FM35-#70	17	3	51	62	−14
EP-GE-FM12-#19	55	−14	−28	29	−7	SS-MS-FM35-#71	67	−40	24	−30	23
EP-GE-FM12-#20	41	−15	−13	7	12	SS-TO-FM36-#72	104	−7	−99	−18	−1
EP-GB-FM13-#21	7	30	9	0	24	SS-TO-FM36-#73	72	26	−66	−30	−46
EP-GB-FM13-#22	5	5	27	4	23	SS-TO-FM36-#74	103	−4	−94	−39	1
EP-GB-FM14-#23	46	15	−32	−17	33	SS-TO-FM36-#75	88	12	−57	−1	−41
EP-GB-FM15-#24	83	−5	−4	−34	5	SS-TO-FM36-#76	90	11	−80	−47	−14
EP-GB-FM15-#25	37	46	−18	−7	−5	SS-TO-FM37-#77	73	−8	−62	−71	18
EP-GB-FM16-#26	26	−4	11	−3	46	SS-TO-FM38-#78	12	44	11	−9	−7
EP-GB-FM17-#27	40	15	21	−21	33	SS-TO-FM38-#79	2	6	1	−1	14
EP-GB-FM17-#28	44	−38	3	33	36	SS-FF-FM39-#80	96	−19	−66	−8	−1
EP-GB-FM17-#29	9	3	29	50	42	SS-FF-FM39-#81	51	3	13	40	−15
EP-GB-FM18-#30	8	−1	17	46	35	SS-FF-FM39-#82	39	6	−38	50	−26
ET-CV-FM19-#31	59	−9	−37	7	−2	SS-FF-FM40-#83	43	−7	26	−29	17
ET-CV-FM20-#32	71	13	23	1	−16	SS-FF-FM40-#84	50	1	20	−46	14
ET-CV-FM20-#33	92	−21	−8	−19	−5	SS-FF-FM40-#85	52	14	19	−47	15
ET-CV-FM20-#34	84	−12	1	−6	−2	SS-FF-FM40-#86	13	2	59	−7	8
ET-CV-FM20-#35	70	3	16	−29	−18	SS-FF-FM40-#87	87	−12	−14	−74	13
ET-TR-FM21-#36	93	−5	−64	−19	−8	SS-FF-FM40-#88	95	−19	−21	−30	1
ET-TR-FM22-#37	102	0	−69	−27	−9	SS-FF-FM41-#89	101	−20	−6	0	−2
ET-TR-FM22-#38	79	24	−45	−48	−30	SS-FF-FM41-#90	85	5	11	17	−40
ET-TR-FM22-#39	91	1	−56	−6	−26	SS-FF-FM41-#91	99	−40	−2	−3	2
ET-TR-FM22-#40	89	15	−53	−39	−15	SS-FF-FM41-#92	99	−39	−1	−2	3
AS-PS-FM23-#41	65	−21	2	−14	16	SS-FF-FM41-#93	97	−6	2	1	−26
AS-PS-FM24-#42	29	−10	−17	−21	27	SS-FF-FM41-#94	82	−20	18	22	−14
AS-PS-FM25-#43	36	−13	4	−11	8	SS-FF-FM41-#95	98	−16	3	−5	−6
AS-YS-FM26-#44	30	22	18	60	−20	SS-FF-FM41-#96	81	−24	21	11	−12
AS-YS-FM27-#45	22	−5	43	45	36	SS-FF-FM41-#97	74	−32	29	20	−13
AS-YS-FM28-#46	86	−17	−45	−31	11	SS-FF-FM41-#98	78	−50	26	21	5
AS-CE-FM29-#47	57	37	9	3	−46	SS-TP-FM42-#99	24	−13	56	20	13
AS-CE-FM30-#48	93	3	−54	−32	−27	SS-TP-FM42-#100	33	10	48	12	−19
SS-MS-FM31-#49	42	51	7	−27	−23	SS-TP-FM42-#101	54	−25	28	2	−4
SS-MS-FM31-#50	19	66	31	−3	−13	SS-TP-FM42-#102	18	−16	65	50	−10
SS-MS-FM31-#51	58	31	−7	−32	−29	SS-TP-FM42-#103	53	−6	−11	50	−31
SS-MS-FM31-#52	23	9	29	4	−5	SS-TP-FM42-#104	20	−4	23	60	−11

R(PP): Ranking of the present paper; EXP.#A: Failure ranking difference based on the single expert A, computed by $R(EXP.\#A) - R(PP)$, where $R(PP)$ is the RPN rank calculated by the presented model and $R(EXP.\#A)$ is calculated by conventional model based on the single expert A.

References

1. Turner, J.A. A realizable renewable energy future. *Science* **1999**, *285*, 687–689. [CrossRef] [PubMed]
2. Li, H.; Huang, C.G.; Guedes Soares, C. A real-time inspection and opportunistic maintenance strategies for floating offshore wind turbines. *Ocean. Eng.* **2022**, *256*, 111433. [CrossRef]
3. Bashetty, S.; Ozcelik, S. Review on Dynamics of Offshore Floating Wind Turbine Platforms. *Energies* **2021**, *14*, 6026. [CrossRef]
4. Tavner, P. *Offshore Wind Turbines: Reliability, Availability and Maintenance*; Institution of Engineering and Technology: London, UK, 2012; Volume 13.
5. Zountouridou, E.I.; Kiokes, G.C.; Chakalis, S.; Georgilakis, P.S.; Hatziargyriou, N.D. Offshore floating wind parks in the deep waters of Mediterranean Sea. *Renew. Sustain. Energy Rev.* **2015**, *51*, 433–448. [CrossRef]
6. Bento, N.; Fontes, M. Emergence of floating offshore wind energy: Technology and industry. *Renew. Sustain. Energy Rev.* **2019**, *99*, 66–82. [CrossRef]
7. Li, H.; Guedes Soares, C. Assessment of failure rates and reliability of floating offshore wind turbines. *Reliab. Eng. Syst. Saf.* **2022**, *228*, 108777. [CrossRef]
8. Colmenar-Santos, A.; Perera-Perez, J.; Borge-Diez, D.; dePalacio-Rodríguez, C. Offshore wind energy: A review of the current status, challenges and future development in Spain. *Renew. Sustain. Energy Rev.* **2016**, *64*, 1–18. [CrossRef]
9. Kang, J.; Sun, L.; Guedes Soares, C. Fault Tree Analysis of floating offshore wind turbines. *Renew. Energy* **2019**, *133*, 1455–1467. [CrossRef]
10. Roddier, D.; Cermelli, C.; Aubault, A.; Weinstein, A. WindFloat: A floating foundation for offshore wind turbines. *J. Renew. Sustain. Energy* **2010**, *2*, 033104. [CrossRef]
11. Global Wind Energy Council (GWEC). Global Wind Report 2019. 2020. Available online: https://gwec.net/global-wind-report-2019/ (accessed on 29 July 2020).
12. Guo, H.; Watson, S.; Tavner, P.; Xiang, J. Reliability analysis for wind turbines with incomplete failure data collected from after the date of initial installation. *Reliab. Eng. Syst. Saf.* **2009**, *94*, 1057–1063. [CrossRef]
13. Santos, F.; Teixeira, A.P.; Guedes Soares, C. Modelling and simulation of the operation and maintenance of offshore wind turbines. *Proc. Inst. Mech. Eng. Part O J. Risk Reliab.* **2015**, *229*, 385–393. [CrossRef]
14. Reder, M.; Yürüşen, N.Y.; Melero, J.J. Data-driven learning framework for associating weather conditions and wind turbine failures. *Reliab. Eng. Syst. Saf.* **2018**, *169*, 554–569. [CrossRef]
15. Otter, A.; Murphy, J.; Pakrashi, V.; Robertson, A.; Desmond, C. A review of modelling techniques for floating offshore wind turbines. *Wind. Energy* **2022**, *25*, 831–857. [CrossRef]
16. Li, H.; Díaz, H.; Guedes Soares, C. A Developed Failure Mode and Effect Analysis for Floating Offshore Wind Turbine Support Structures. *Renew. Energy* **2020**, *164*, 133–145. [CrossRef]
17. Jiang, Z. Installation of offshore wind turbines: A technical review. *Renew. Sustain. Energy Rev.* **2021**, *139*, 110576. [CrossRef]
18. Martinez, A.; Iglesias, G. Mapping of the levelised cost of energy for floating offshore wind in the European Atlantic. *Renew. Sustain. Energy Rev.* **2022**, *154*, 111889. [CrossRef]
19. Li, H.; Díaz, H.; Guedes Soares, C. A Failure Analysis of Floating Offshore Wind Turbines using AHP-FMEA Methodology. *Ocean. Eng.* **2021**, *234*, 109261. [CrossRef]
20. Castro-Santos, L.; Filgueira-Vizoso, A.; Carral-Couce, L.; Formoso, J.Á.F. Economic feasibility of floating offshore wind farms. *Energy* **2016**, *112*, 868–882. [CrossRef]
21. Guo, Y.; Wang, H.; Lian, J. Review of integrated installation technologies for offshore wind turbines: Current progress and future development trends. *Energy Convers. Manag.* **2022**, *255*, 115319. [CrossRef]
22. Li, H.; Guedes Soares, C.; Huang, H.Z. Reliability analysis of floating offshore wind turbine using Bayesian Networks. *Ocean. Eng.* **2020**, *217*, 107827. [CrossRef]
23. Leimeister, M.; Kolios, A. A review of reliability-based methods for risk analysis and their application in the offshore wind industry. *Renew. Sustain. Energy Rev.* **2018**, *91*, 1065–1076. [CrossRef]
24. Bhardwaj, U.; Teixeira, A.P.; Guedes Soares, C. Reliability prediction of an offshore wind turbine gearbox. *Renew. Energy* **2019**, *141*, 693–706. [CrossRef]
25. Peeters, J.; Basten, R.; Tinga, T. Improving failure analysis efficiency by combining FTA and FMEA in a recursive manner. *Reliab. Eng. Syst. Saf.* **2018**, *172*, 36–44. [CrossRef]
26. Márquez, F.P.G.; Pérez, J.M.P.; Marugán, A.P.; Papaelias, M. Identification of critical components of wind turbines using FTA over the time. *Renew. Energy* **2016**, *87*, 869–883. [CrossRef]
27. Shafiee, M.; Sørensen, J.D. Maintenance optimization and inspection planning of wind energy assets: Models, methods and strategies. *Reliab. Eng. Syst. Saf.* **2019**, *192*, 105993. [CrossRef]
28. Scheu, M.N.; Kolios, A.; Fischer, T.; Brennan, F. Influence of statistical uncertainty of component reliability estimations on offshore wind farm availability. *Reliab. Eng. Syst. Saf.* **2017**, *168*, 28–39. [CrossRef]
29. Higgins, P.; Foley, A. The evolution of offshore wind power in the United Kingdom. *Renew. Sustain. Energy Rev.* **2014**, *37*, 599–612. [CrossRef]
30. Diaz, H.; Guedes Soares, C. Review of the current status, technology and future trends of offshore wind farms. *Ocean. Eng.* **2020**, *209*, 107381. [CrossRef]

31. Scheu, M.N.; Tremps, L.; Smolka, U.; Kolios, A.; Brennan, F. A systematic Failure Mode Effects and Criticality Analysis for offshore wind turbine systems towards integrated condition-based maintenance strategies. *Ocean. Eng.* **2019**, *176*, 118–133. [CrossRef]
32. Li, H.; Teixeira, A.P.; Guedes Soares, C. A Two-Stage Failure Mode and Effect Analysis of an offshore wind turbine. *Renew. Energy* **2020**, *162*, 1438–1461. [CrossRef]
33. Sinha, Y.; Steel, J.A. A progressive study into offshore wind farm maintenance optimisation using risk based failure analysis. *Renew. Sustain. Energy Rev.* **2015**, *42*, 735–742. [CrossRef]
34. Santos, F.P.; Teixeira, A.P.; Guedes Soares, C. Review of wind turbine accident and failure data. In *Renewable. Energies Renew. Offshore*; Guedes Soares, C., Ed.; Taylor & Francis Group: London, UK, 2015; pp. 953–959.
35. Nielsen, J.J.; Sørensen, J.D. On risk-based operation and maintenance of offshore wind turbine components. *Reliab. Eng. Syst. Saf.* **2011**, *96*, 218–229. [CrossRef]
36. Karmakar, D.; Bagbanci, H.; Guedes Soares, C. Long-term extreme load prediction of spar and semisubmersible floating wind turbines using the Environmental Contour Method. *J. Offshore Mech. Arct. Eng.* **2016**, *138*, 021601. [CrossRef]
37. Liu, T.S.; Chiou, S.B. The application of Petri nets to failure analysis. *Reliab. Eng. Syst. Saf.* **1997**, *57*, 129–142. [CrossRef]
38. Rausand, M.; Øien, K. The basic concepts of failure analysis. *Reliab. Eng. Syst. Saf.* **1996**, *53*, 73–83. [CrossRef]
39. Huang, J.; You, J.X.; Liu, H.C.; Song, M.S. Failure mode and effect analysis improvement: A systematic literature review and future research agenda. *Reliab. Eng. Syst. Saf.* **2020**, *199*, 106885. [CrossRef]
40. Lo, H.W.; Liou, J.; Huang, C.N.; Chuang, Y.C. A novel failure mode and effect analysis model for machine tool risk analysis. *Reliab. Eng. Syst. Saf.* **2019**, *183*, 173–183. [CrossRef]
41. Wang, W.; He, X.; Li, Y.; Shuai, J. Risk analysis on corrosion of submarine oil and gas pipelines based on hybrid Bayesian network. *Ocean. Eng.* **2022**, *260*, 111957. [CrossRef]
42. Zarei, E.; Khakzad, N.; Cozzani, V.; Reniers, G. Safety analysis of process systems using Fuzzy Bayesian Network (FBN). *J. Loss Prev. Process Ind.* **2019**, *57*, 7–16. [CrossRef]
43. Wu, X.; Hu, Y.; Li, Y.; Yang, J.; Duan, L.; Wang, T.; Adcock, T.; Jiang, Z.; Gao, Z.; Lin, Z.A.; et al. Foundations of offshore wind turbines: A review. *Renew. Sustain. Energy Rev.* **2019**, *104*, 379–393. [CrossRef]

Journal of
Marine Science and Engineering

MDPI

Article

Research on Structural Failure Analysis and Strengthening Design of Offshore Wind Turbine Blades

Xiaohan Fu [1,2] and Meiping Sheng [1,2,*]

1 School of Marine Science and Technology, Northwestern Polytechnical University, Xi'an 710072, China
2 Research & Development Institute of Northwestern Polytechnical University in Ningbo, Ningbo 315103, China
* Correspondence: acoustics@nwpu.edu.cn

Abstract: The existing blade protection mechanisms are highly dependent on the control system and its power supply. Safety of offshore wind turbines cannot be guaranteed under extreme weather when the control protection mechanisms fail. So far, not enough consideration has been given to the above problems in mechanical design for protecting wind turbine blades. In this paper, a reinforcing cable component (RCC) is proposed to improve the resistance ability of offshore wind turbine blades. The static model of the blades with reinforcing cable component was presented. The Finite Element (FE) simulation was performed for a 5 MW offshore wind turbine and load reduction effect of connection location for RCC was discussed according to the FE results. A static strain verification test was carried out. Simulation and test results indicate that the proposed reinforcing cable component effectively reduces the strain as well as the tip displacement of the blades. The proposed mechanical structure will help to enhance the survival ability of offshore wind turbine blades when the control protection system fails.

Keywords: wind turbine blades; steel cable connection; wind load; finite element method; structural stress; static load test

Citation: Fu, X.; Sheng, M. Research on Structural Failure Analysis and Strengthening Design of Offshore Wind Turbine Blades. *J. Mar. Sci. Eng.* **2022**, *10*, 1661. https://doi.org/10.3390/jmse10111661

Academic Editors: Emre Uzunoglu, Antonio Souto-Iglesias and Carlos Guedes Soares

Received: 4 October 2022
Accepted: 25 October 2022
Published: 4 November 2022

1. Introduction

Wind power has advantages of renewable, pollution-free, and cost-effective [1]. With the continuous development of wind power generation technology, megawatt wind turbines have been widely applied around the world. In 2021, the cumulative of global wind power reached 837 GW, the second highest in history [2]. For coastal areas with plenty wind energy, large offshore wind farms have been built and produced considerable economic benefits. Compared with onshore wind power, the offshore wind power has advantages of stable wind speed and less land costs [3]. As the Global Wind Report shows, the new installed capacity of offshore wind power reached a record high in 2021, with a new installed capacity of 21.1 GW. This figure is three times that of 2020, and the total installed capacity of offshore wind power is evaluated to reach 370 GW in 2030 [4].

Nevertheless, the offshore wind power has long faced questions about high maintenance difficulty and complex external loading [5,6]. As the demand for output power of wind turbine is increasing [7], the structure design of blades is upscaled towards a larger and more flexible direction [8,9]. For large offshore wind turbine blades, length in the spanwise direction are much larger than chord direction, thus the wind turbine blades are usually regarded as cantilever structures in structural analysis [10]. While the stiffness of the blades is decreased, the stability of the wind turbine is also decreased [11]. Chen and Xu [12] investigated the wind turbine failure using the post mortem analysis and demonstrated that the blade fracture was one of the major reason for structural failure of wind turbines. Chou et al. [13] pointed that the blade strength was crucial for safety of wind turbine under critical wind load through Finite Element (FE) analysis. Chen and Kam [14] investigated failure location of a composite sandwich wind blade subjected to wind load, and the rupture was observed at the blade root at extreme wind speed. Lee et al. [15] proposed that under real load distribution the bumping motions of blades were the main reason for unfavorable load distribution at blade root.

At present, improving the survivability of wind turbines is mainly realized by yaw system and pitch system [16]. The simulation results of Bossanyi [17] clearly showed that the pitch system had a significant load reduction effect for large wind turbines. Stubkier et al. [18] researched the load reduction effect of a hydraulic soft yaw concept and the ultimate load reduction of up to 19% was achieved. However, it is possible that the yaw system and the pitch system encounter faults in extreme weather due to the fast-changing direction and large wind load of typhoon. Kim and Manuel [19] researched the influence of yaw error or pitch misalignment on the aerodynamic loading based on numerical studies. Bakhshi and Sandborn [20] indicated that the most susceptible site for failure is the blade root under yaw error. Similar problem was also observed in wind farm of China [21]. The super typhoon Rammasun in 2014 caused 3 wind turbines of Wenchang wind farm in Hainan province to suffer from structural failure, and blades of another 2 wind turbines were separated from root. The main reason for such powerful destruction is due to the failure of pitch system under sudden change of typhoon wind direction [22]. As the blades cannot alter the angle in time, the increased stress on the blade and the whole wind turbine gave rise to structural damage.

In this paper, a reinforcing cable component (RCC) is proposed to reduce the probability of offshore wind turbine damage. The proposed mechanical structure is independent of the present yaw or pitch system, thus the blades can still be protected when the control system is damaged. For a wind turbine that brakes at an unfavorable position, the RCC can apply control force to blades, thereby reducing the excessive wind load on blades and preventing the braking system from overload.

This paper is organized as follows. In Section 2, the unfavorable position for blades is obtained and the structural failure of blades is analyzed. In Section 3, the reinforcing cable component is proposed, and its wind load reduction mechanism is analyzed by establishing static mechanical model of blades with reinforcing cables. In Section 4, the wind load reduction effect of the proposed reinforcing cable component is researched by finite element method, and the connecting location for RCC is analyzed. Section 5 presents the comparative static strain test results for blades without and with RCC, followed by conclusions drawn in Section 6.

2. Structural Failure Analysis of Wind Turbine Blades

The external load of offshore wind turbine blade is complicated. According to the load source, the external load can be divided into aerodynamic loads, centrifugal loads and gravity loads [23]. Under extreme weather, the failure of the offshore wind turbine is closely related to the interaction between the wind turbine and the extreme winds [10]. Next, the wind load on blades is considered for structural failure analysis.

When the wind speed exceeds the cut-out wind speed under typhoon weather, the brake system will be started to protect the wind turbine from tower collapse and impeller failure. During extreme weather, the wind and parking positions of impeller and inflow directions have been verified as two sensitive factors for aerodynamic loads [24]. Assume a wind turbine impeller at the braking state is within the XOY plane, as shown in Figure 1. The wind direction varies in the XOZ plane with an inflow angle β. Due to the rapid change of typhoon wind direction, we assume that $\beta \in [0, 90^\circ]$. When $\beta = 0^\circ$, the wind flow direction is perpendicular to impeller surface of the wind turbine.

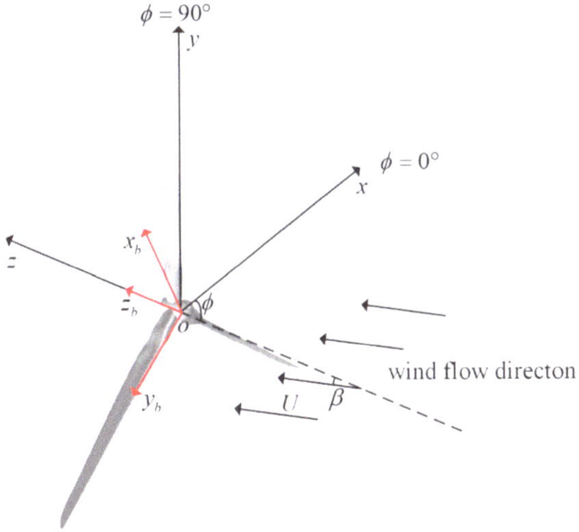

Figure 1. Wind turbine rotor coordinate system.

According to blade element momentum (BEM) theory, the lift force dL, and resistance force dD for a blade element at angle of attack α are expressed as [25].

$$dL = \frac{\rho}{2} C_L(\alpha) V^2 t \tag{1}$$

$$dD = \frac{\rho}{2} C_D(\alpha) V^2 t \tag{2}$$

where $C_L(\alpha)$ and $C_D(\alpha)$ are the lift coefficient and resistance coefficient at angle of attack α, respectively, ρ is the air density, t is chord length, and V is the relative wind velocity between wind and the wind turbine speed. For a specific working condition, the lift coefficient $C_L(\alpha)$ and resistance coefficient $C_D(\alpha)$ can be determined by angle of attack α.

In general, the relative wind velocity V is related to the axial and tangential induction factors. When the wind turbine is parked under extreme weather, the induction factors can be neglected. As indicated in Figure 1, the inflow wind velocity U can be decomposed into x and z directions as

$$U_x = U \sin \beta \tag{3}$$

$$U_z = U \cos \beta \tag{4}$$

For a blade element, the inflow velocity is decomposed into z_b and x_b directions in the local axis, as shown in Figure 2.

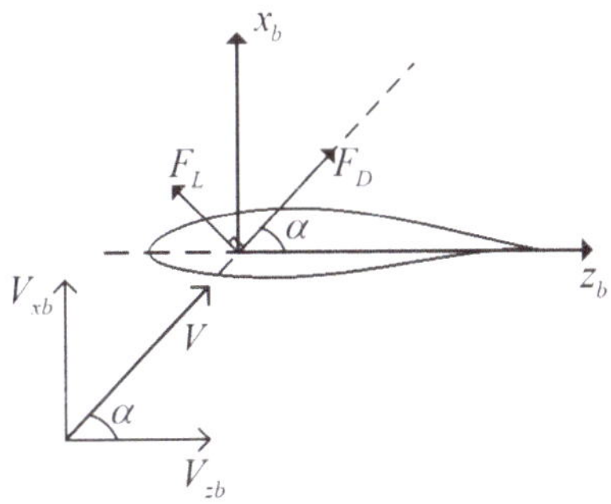

Figure 2. Diagram of aerodynamic force at a blade element.

Since the wind turbine is at the braking state, the velocity triangle is expressed as

$$V^2 = \sqrt{V_{zb}^2 + V_{xb}^2} \tag{5}$$

in which $V_{zb} = U_z$, $V_{xb} = U_x \sin\phi$.

From Figure 2, the angle of attack α is [24]

$$\alpha = \mathrm{atan}\left(\frac{V_{xb}}{V_{zb}}\right) = \mathrm{atan}(\tan\beta \sin\phi) \tag{6}$$

From Equation (6), the angle of attack α is determined by the parking state and the inflow direction. The angle of attack reaches maximum when the blade is parked at $\phi = 90°$, at which parking state the angle of attack α equals the angle of inflow β. When $\beta = 90°$, the blade is in adverse condition as the resistance force is perpendicular to the blade surface, leading to large bending deformation.

With Equations (1)–(6), the wind load is expressed as the resultant force of lift force and resistance force as

$$dF_{eq} = \sqrt{dL^2 + dD^2} \tag{7}$$

For an offshore wind turbine blade, assume the blade length is L and the maximum chord length is L_c. As the section size of blade is much smaller than its length, the wind turbine blade can be treated as a cantilever beam [26]. As shown in Figure 3, the effect of wind load on the blade is expressed in the form of distributed force $q(x_i)(0 \le x_i \le L)$.

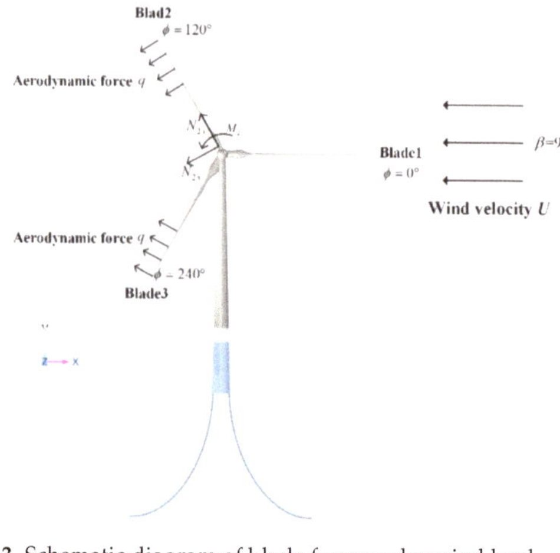

Figure 3. Schematic diagram of blade force under wind load.

As indicated in Equations (1) and (2), the wind load on the blades varies with angle of attack and chord length. In the following analysis, the wind load $q(x)$ is equivalent to uniformly distributed load q with units N for simplicity. For a cantilever beam structure, the deflection of the blade can be expressed as

$$w(x) = c\left(1 - \cos\frac{\pi x}{2L}\right)(0 \le x \le L) \tag{8}$$

Equation (8) agrees with the boundary conditions of the cantilever beam as

$$\begin{cases} w(x)|_{x=0} = 0 \\ w'(x)|_{x=0} = 0 \\ w''(x)|_{x=L} = 0 \end{cases} \tag{9}$$

According to the virtual work principle, assume that the structure has a virtual displacement δw_i, the virtual work of external force and the total virtual deformation work received by the deformed body has a relationship as

$$\delta W_i - \delta U_i = 0 \tag{10}$$

where δW_i is the virtual work of external force and δU_i is the total virtual deformation work received by ith blade

$$\delta W_i = \int_0^L q_i(x)\delta w_i(x)dx \tag{11}$$

$$\delta U_i = \int \sigma_x \delta \varepsilon_x dV \tag{12}$$

where σ_x and $\delta\varepsilon_x$ denote the stress and virtual strain subjected to the virtual displacement δw_i.

With Equations (8)–(12), the coefficient c in Equation (8) can be obtained as

$$c = \frac{32(\pi - 2)}{\pi^5}\frac{qL^4}{EI} \tag{13}$$

The maximum bending stress response is located at root of the blade as

$$\sigma_{\max} = \frac{qL^2}{2W_z} \tag{14}$$

where W_z is the bending section coefficient of the beam.

As indicated in Equations (13) and (14), the tip displacement and the maximum stress represent the structure stiffness and the structure strength, respectively. Dangerous section is located at root of the blade. Under extreme typhoon, the wind load is very large and usually accompanied by sudden change of wind direction. In this situation, if either the stress of the blade root $\sigma_{\max}$ excesses the allowable stress requirements $[\sigma]$ or the tip displacement $w_{\max}$ excesses the allowable deflection $[w]$, the wind turbine blades will face the risk of structural failure.

3. Wind Load Reduction Design

3.1. Structure Strengthing Design of Reinforcing Cable Component(RCC)

When designing the blades for an offshore wind turbine, the power generation efficiency and structural characteristics are equally important. While good power generation efficiency requires large blades with light weight and thin thickness, the structural performance usually relies on the proper layout of blade skin and reinforcements, such as spars. However, the two demands are often contradictory [27]. For example, when less mass is required for larger power generation efficiency, some sacrifices, such as reducing the amount of spars inside the blades, should be taken to reach a compromise. Concerns have thus been raised about structural strength and structural stiffness of blades for large offshore wind turbines, especially in extreme weather. Based on the analysis in Section 2, the extreme wind load on the blade is perpendicular to the blade under extreme weather. The bending resistance property of blades, as well as the structures stiffness, is essential for large offshore wind turbines consequently.

The proposed strengthening structure for offshore wind turbine blades is shown in Figure 4.

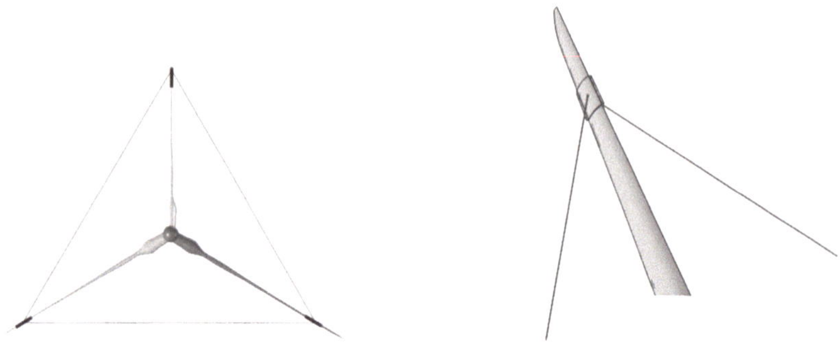

(**a**) (**b**)

Figure 4. Three-dimensional structural model of blades with reinforcing cable component: (**a**) Wind turbine with reinforcing cable component; (**b**) Mechanical connector for RCC.

The proposed reinforcing cable component (RCC) is composed of three reinforcing cables and the corresponding mechanical connectors. The reinforcing cables are connected to blades at the same location separately through a mechanical connector. The RCC can be viewed as a stay cable, which has advantages of light weight, easy installation, and low cost [28]. As shown in Figure 4, the wind turbine integrated with stay cables is still a symmetrical structure and has no influence on the original pitch system. When the blades suffer stress under wind load, the RCC can enhance the structure ability by providing external tension force, so as to alleviate the structural stress at blade root and to improve the bending resistance performance of blades. In practical engineering, the RCC can be connected to blade through a mechanical sleeve, as shown in Figure 4b.

The wind load reduction mechanism will be analyzed in the next section.

3.2. Wind Load Reduction Analysis

The wind load reduction mechanism for blades with RCC is displayed in Figure 5.

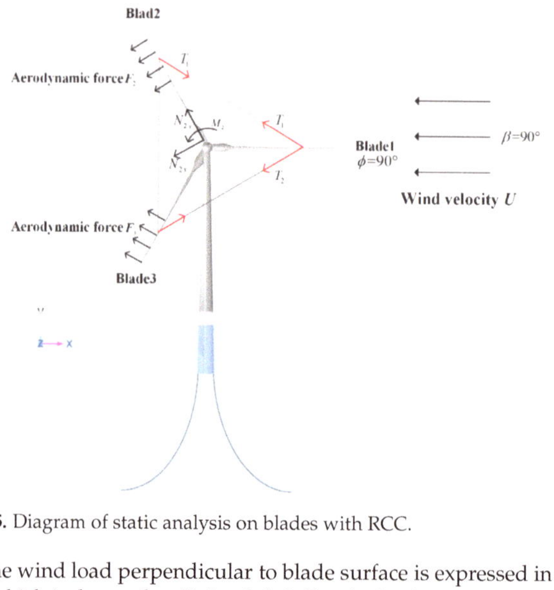

Figure 5. Diagram of static analysis on blades with RCC.

The wind load perpendicular to blade surface is expressed in the form of distributed force, which is denoted as $F_i, i = 1, 2, 3$. For the braking state shown in Figure 5, the wind load on the three blades has the following relationship as $F_2 = F_3 \gg F_1$. In this case, we assume that the deflection of blade 1 can be ignored and the tension force $T_1 = T_2$.

Assume the length of reinforcing cable is ξ and the distance between the connection point between the RCC and the blade root is $x_0 = kL(0 < k \leq 1)$. When the RCC is connected at the same location for each blade, we have $\xi = \sqrt{3}kL$. As shown in Figure 6, the effect of RCC can be decomposed into the tension force ς along the blade span direction and the tension force ϑ along the blade bending direction as

$$\vartheta = T\sin\alpha' \quad (15)$$

$$\varsigma = T\cos\alpha' \quad (16)$$

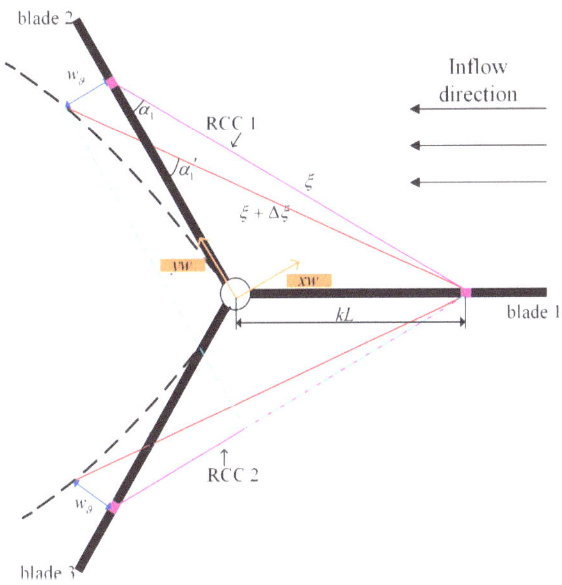

Figure 6. Geometric relationship of the reinforcing cable and blades.

Consider bending deformation, the virtual work of external force can be expressed as

$$\delta W_\vartheta = \delta\left(\int_0^L q w_c dx - \vartheta w_c\right) \quad (17)$$

The virtual bending strain energy with RCC can be expressed as

$$\delta U_\vartheta = \int_0^L E\left(\int_A y^2 dA\right)\left(\frac{d^2 w_\vartheta}{dx^2}\right)\left(\frac{d^2 \delta w_\vartheta}{dx^2}\right)dx \quad (18)$$

The assumed displacement of the blade is consistent with Equation (8) as

$$w_\vartheta = c_\vartheta\left(1 - \cos\frac{\pi x}{2L}\right) \quad (19)$$

According to the principle of virtual work [29].

$$\delta W_\vartheta - \delta U_\vartheta = 0 \quad (20)$$

The unknown parameter in Equation (19) can be obtained as

$$c_\vartheta = \frac{32(\pi-2)}{\pi^5}\frac{L^4 p}{EI} - \frac{32L^3}{EI\pi^4}\vartheta\left(1-\cos\frac{k\pi}{2}\right), (0 \le k \le 1) \tag{21}$$

The relationship between the elongation of the reinforcing cable $\Delta\xi$ and the deflection of blade 2 w_ϑ is displayed in Figure 6. According to the small deformation assumption of the beam, the complementary Equations are obtained by geometric relations under local coordinate system as

$$(\xi + \Delta\xi)\sin\alpha_2{}' = w_\vartheta + \frac{\xi}{2} \tag{22}$$

$$(\xi + \Delta\xi)\cos\alpha_2{}' = \frac{3}{2}kL(0 \le k \le 1) \tag{23}$$

The tension force of RCC is denoted as $T = E_s\Delta\xi$, where E_s denotes the Young's modulus. With Equations (19), (21) and (23), the tension force T of RCC can be derived.

After applying the RCC, the stress at the root of the blade is reduced to

$$\sigma_{RCC} = \frac{1}{W_z}\left(\frac{qx_0^2}{2} - T\sin\alpha_2{}'x_0 + T\cos\alpha_2{}'w_\vartheta\right) \approx \sigma_{\max} - \frac{T\sin\alpha_2{}'x_0}{W_z} \tag{24}$$

From the above analysis, the deflection w_ϑ is much smaller than x_0, and $\alpha_2{}'$ is slightly larger than the angle $\alpha_2 = \frac{\pi}{6}$; thus, in Equation (24), the third term on the right side of the Equation is much smaller than the second term. As the original cantilever boundary condition of wind turbine blade is altered by RCC, the stress on the blade is reduced.

Due to the complex shape of wind turbine blade, the wind loads are not normally uniformly distributed, leading to the exact theoretical solution of stress distribution is difficult to obtain. In the next section, the offshore wind turbine blade will be accurately modeled and the stress and strain distribution of the blade with and without the connection of RCC will be studied through Finite Element (FE) simulation.

4. Simulation Results from FE Models

4.1. Wind Turbine Blade Model

In this paper, a full-scaled 5 MW wind turbine blade of the US National Renewable Energy Laboratory (NREL) is adopted to establish the blade model. The blade is divided into 19 sections along the length during modelling. The main geometric features of the 5 MW wind turbine blade are listed in Table 1.

Table 1. Main geometrical features of the 5 MW wind turbine blade [30].

Section Number	Distance to Blade Root/m	Chord Length/m	Twist Angle/°	AIRFOIL TYPE
s1	0	3.20	13.308	Cylinder
s2	1.36	3.54	13.308	Cylinder
s3	4.1	3.85	13.308	Cylinder
s4	6.83	4.17	13.308	DU40_A17
...	...	...	...	...
s9	26.25	3.75	6.544	DU25_A17
s10	30.75	3.50	5.361	DU21_A17
...	...	...	...	...
s18	60.13	1.40	0.370	NACA64_A17
s19	61.5	0.70	0.106	NACA64_A17

The fiber composite materials are adopted for the 5 MW offshore wind turbine blades. The material parameters are refereed to reference [31]. The composed composite material is a sandwich structure with glass fiber and carbon fiber fabric sequentially covering the upside and downside of the PVC foam core. Each layer of the fiber has a thickness of 2.8×10^{-4} m and a maximum layer of 80 and 20 are set for glass fiber and carbon fiber separately.

The section area of RCC is set as 3×10^{-4} m^2. Length of RCC is related to the connecting location on the blade with a relationship of $\xi = \sqrt{3}kL$. The structural steel material is adopted for RCC, as displayed in Table 2.

Table 2. Material parameters of RCC.

Material Parameters	Young's Modulus	Density	Poisson's Ratio
RCC	2.16×10^{11} Pa	7800 kg/m^3	0.28

The geometry model of wind turbine blade is firstly established with 3D modelling software (CATIA V5 R20, Dassault Systems, Paris, France) and then imported into the Finite Element software COMSOL Multiphysics for FE analysis. The wind turbine blade model is decentralized into 17089 shell elements, 7333 solid elements, and 120 line elements, as shown in Figure 7

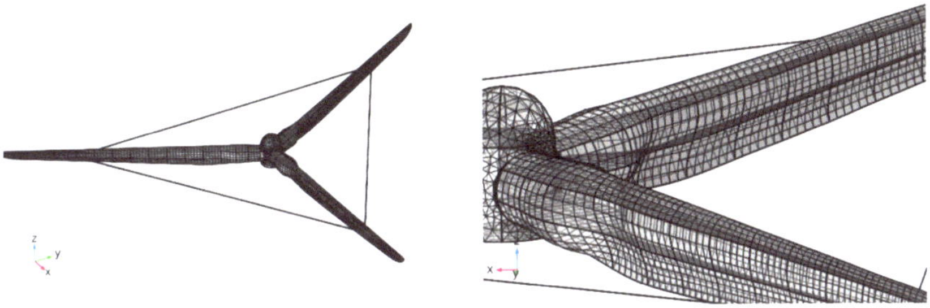

(a) (b)

Figure 7. The FE model of wind turbine blade and applied load; (**a**) FE mesh of the total model; (**b**) Local enlarged view of FE mesh.

The boundary conditions and applied loads on wind turbine blades are displayed in Figure 8.

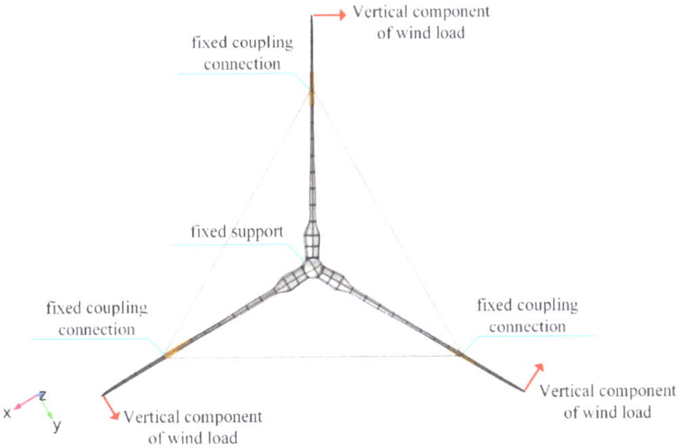

Figure 8. Boundary conditions of the wind turbine rotor.

As shown in Figure 8, a fixed boundary condition is applied at the hub to simulate the braking status of the wind turbine and the distributed equivalent aerodynamic force are applied to each section [32]. With Equations (1)–(7), the resultant force of blade segments under the extreme wind velocity of 60 m/s are calculated and displayed in Table 3. As the blade has circular sections at root of the blade, the aerodynamic load is calculated from Section S4 to Section S18.

Table 3. Aerodynamic load under extreme condition.

Section Number	Distance from Section to Blade Root /m	Equivalent Resultant Force /N		
		Blade 1	Blade 2	Blade 3
S4	10.25	13,981.31	10,485.98	10,485.98
S5	14.35	14,396.55	10,797.41	10,797.41
S6	18.45	13,796.18	10,347.13	10,347.13
S7	22.55	13,163.37	9872.53	9872.53
S8	26.65	12,413.66	9310.24	9310.24
S9	30.75	11,611.28	8708.46	8708.46
S10	34.85	11,185.51	8389.13	8389.13
S11	38.95	10,399.77	7799.83	7799.83
S12	43.05	9620.90	7215.67	7215.67
S13	47.15	8834.61	6625.96	6625.96
S14	51.25	8048.31	6036.24	6036.24
S15	54.67	7393.07	5544.80	5544.80
S16	57.4	6667.51	5000.63	5000.63
S17	60.13	4474.84	3356.13	3356.13
S18	61.5	2237.42	1678.06	1678.06

4.2. Results and Discussion

To find the influence of the RCC location on blade stress response, a series of finite-element analyses are conducted using COMSOL Multiphysics software. The FE models of wind turbine with and without the reinforcing structure are analyzed separately. The connecting point of reinforcing cable is set from one third of the blade length to the blade tip along the spanwise direction. The corresponding Von Mises stress response of blade 1 is displayed in Figure 9.

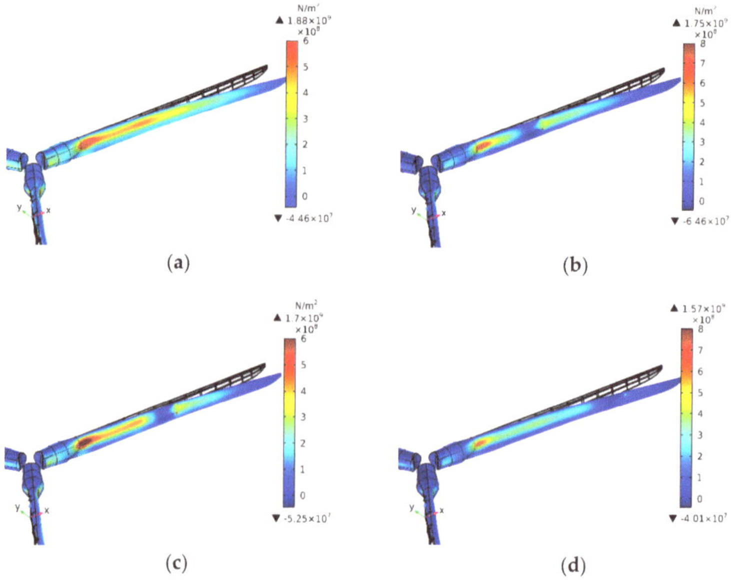

Figure 9. *Cont.*

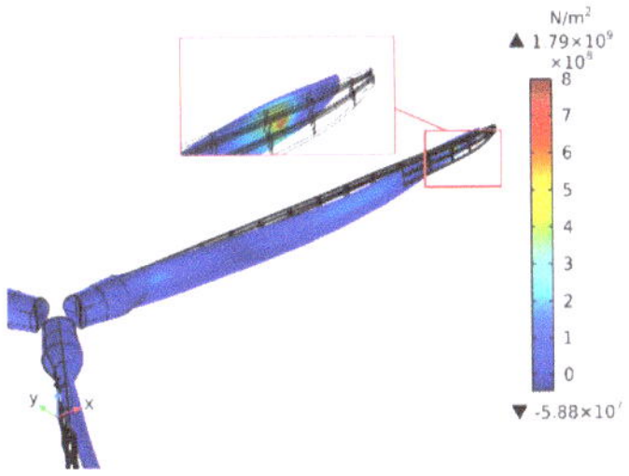

(e)

Figure 9. Blade stress distribution at wind speed of 60m/s: (**a**) blade without RCC; (**b**) blade with RCC connected at one−third of the blade length; (**c**) blade with RCC connected at half of the blade length; (**d**) blade with RCC connected at two-thirds of the blade length; (**e**) blade with RCC connected at the blade tip.

According to the FEM results, the blades without RCC has the largest stress at the junction between the circular and airfoil cross sections. The large stress area is relieved when the RCC is applied. Meanwhile, one can also notice that the local stress concentration appearing at the connection point of reinforcing cable. This is due to the discontinuity of blade geometry at the connection point. The maximum stress is further decreased when the RCC is connected at two-thirds of the blade length. By comparing Figure 9a–d, the stress of the connecting point is less than the other two situations. The stress close to root of the blade is significantly reduced after applying the RCC, which demonstrates that anti-bending performance of the structure is improved. When the connecting point is removed to the blade tip, as shown in Figure 9e, the stress response of the whole blade is much lower. However, it should be noticed that the maximum stress of the whole structure reaches 1.79×10^9 N/m^2 when the RCC is connected at the blade tip, which is second only to blades without RCC.

The stress distribution on the blade surface for different RCC connecting positions are compared in Figure 10.

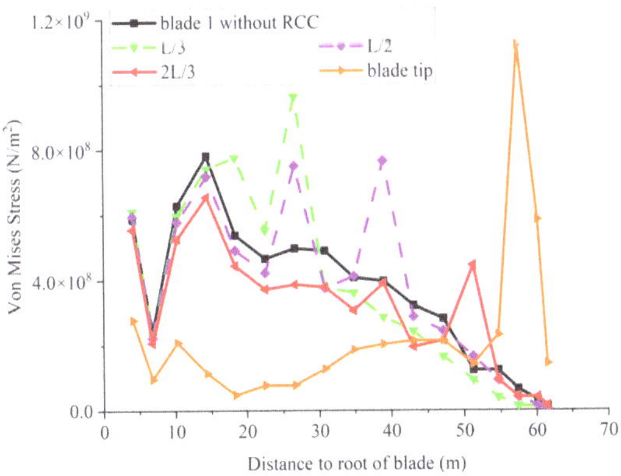

Figure 10. Comparison of stress distribution along the blade.

The stress response is reduced to varying degrees for the distance less than 14.35 m. When the RCC is connected to $L/3$ of the blade length, the stress response is increased at 18.45 m and 26.65 m, more than 23% of the maximum stress value without RCC. The stress increase phenomenon is relieved with the movement of RCC. When the RCC connecting point is moved to $L/2$ of the blade length, the stress at 26.65 m and 38.95 m is increased and is almost equal to the original maximum value. When the RCC is connected at $2L/3$ of the blade length, the stress is only slightly increased at 51.25 m, far less than the maximum stress value of blades without RCC. Although an obvious decrease of the overall stress can be seen when the RCC is connected at the blade tip, the stress at 57.4 m is far greater than the largest stress of blades without RCC. The concentration of stress is likely to produce fatigue cracks on the blade surface, leading to severe accident without timely maintenance. Therefore, we can conclude that the optimal connecting position for RCC is at $2L/3$ of the blade length.

Table 4 summarizes the maximum displacement at blade tip of the two wind turbine blade models.

Table 4. Tip displacement response of blades.

Blade Type	Blade 1 without RCC	Blade 1 with RCC at 2L/3	Blade 2 without RCC	Blade 2 with RCC at 2L/3	Blade 3 without RCC	Blade 3 with RCC at 2L/3
Tip displacement /m	7.04	5.13	6.78	1.49	6.74	2.61

As discussed above, the wind load is directly related to the position of the blade. As blade 1 is perpendicular to the incoming flow direction, the relative wind velocity is higher than the other two blades. Therefore, blade 1 has the largest tip displacement among the 3 blades. The proposed RCC effectively reduces the maximum displacement of the blade tip, with 27.1% for blade 1 and more than 50% for blade 2 and blade 3, thus the structure stiffness of the wind turbine blades is improved.

Figure 11 presents the overall strain distribution of wind turbine blades without and with RCC at $2L/3$ respectively. The strain reduction trend is similar to that of stress. The strain close to root of the blade is reduced obviously for all blades, implying the load reduction effect of the RCC. The strain at connecting point is slightly increased, but it has no influence on the overall reduction effect when the RCC is connected at the optimal location. The maximum strain of blades with RCC is reduced to half of the blades without RCC.

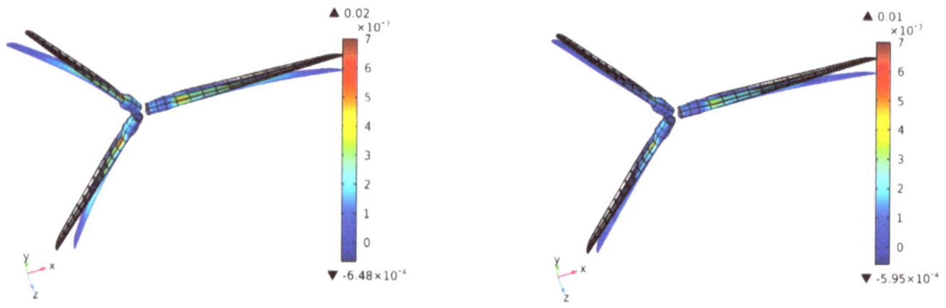

(a) (b)

Figure 11. Blade strain comparison: (**a**) Blade without RCC; (**b**) Blade with RCC connected at $2L/3$.

The stress response of RCC is displayed in Figure 12. When the wind direction is parallel with rotor surface, only two of reinforcing cables are in tension.

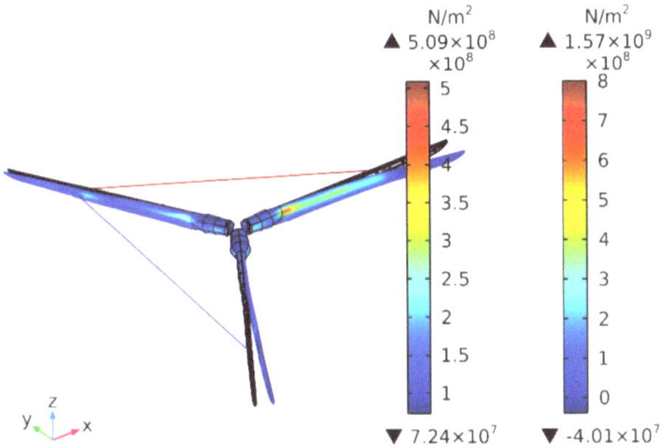

Figure 12. Stress value of RCC connected at $2L/3$.

As shown in Figure 12, the stresses of the two reinforcing cables are 5.09×10^8 N/m^2 and 7.24 $\times 10^7$ N/m^2 , respectively. As the maximum stress of RCC is within the material tensile strength 5.50 $\times 10^8$ N/m^2 [33], the RCC is able to reduce external load for blades under extreme weather safely.

5. Static Load Strain Test

5.1. Test Description

To verify the load reduction effect of RCC, a small-sized wind turbine blade with a length of 0.8 m is adopted for test, as shown in Figure 13. The test blade is made of nylon fiber. To simulate the stress situation under extreme weather, the fan blades are installed according to the feathering state. Each root of blades is fixed on a hub without rotation. The strain on the blade is firstly captured by several rectangular rosettes, and then recorded by a static strain gauge.

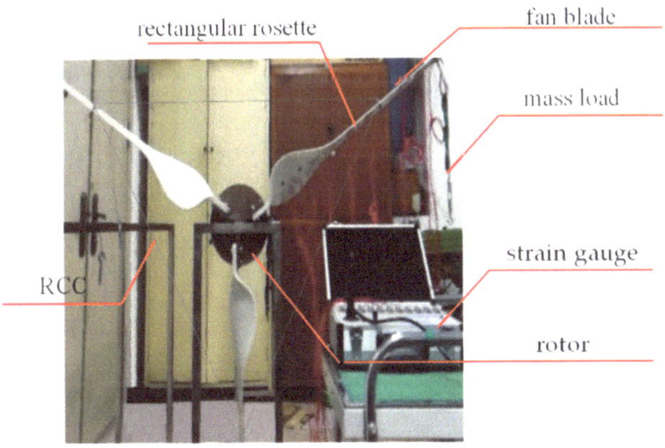

Figure 13. Diagram of test device.

In static testing, different types of loads correspond to tests for different purposes [34]. The load is assumed to be parallel with the flap-wise direction, which corresponds to extreme conditions of wind load. In this test, mass load [35] (4.9 N) is adopted to test the strength of blade with RCC. The mass load is suspended at the blade tip.

Several BX120-3CA rectangular rosettes are pasted along the blade span direction, as displayed in Figure 14. The rectangular rosettes can be used to measure the strain of tested structure at 0, 45, and 90 degrees simultaneously.

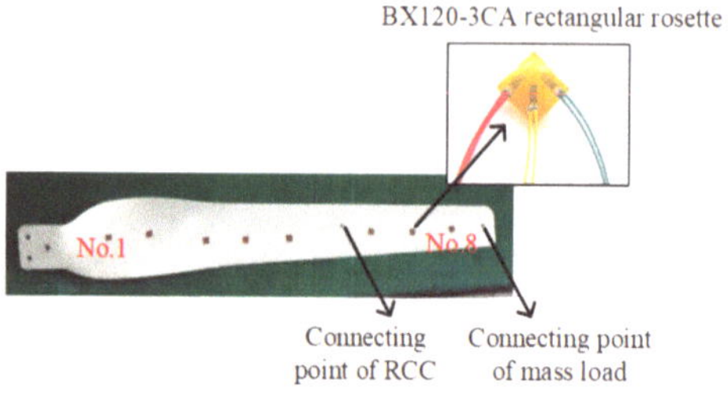

Figure 14. Diagram of strain measuring point on the blade.

The serial numbers of the rectangular rosettes increase from the blade root to the blade tip, that is, with No. 1 as the measuring point near the blade root and No. 8 as the measuring point close to the blade tip. Based on simulation results, the connecting point of the RCC on the blade is two-thirds of total length of the blade from the blade root.

The main experimental steps are as follows: The static strain of blade without loading is firstly measured by the static strain gauge, and the static strain without loading at all eight measuring points are recorded. Then, a mass of 4.9 N is hung on the blade tip, meanwhile the static strain with loading at all eight measuring points are recorded, respectively. To ensure the reliability of test results, each loading and recording procedure are repeated three times during the test. The recording results are averaged to reduce the test error.

5.2. Load Reduction Effect Discussion

The principal strain is calculated as follows.

$$\varepsilon_{max} = \frac{\varepsilon_{90} + \varepsilon_0}{2} + \sqrt{\left(\frac{\varepsilon_{90} - \varepsilon_0}{2}\right)^2 + \left(\frac{\varepsilon_{90} + \varepsilon_0 - 2\varepsilon_{45}}{2}\right)^2} \tag{25}$$

where ε_0 represents the strain in the axis direction, ε_{90} represents the strain with an angle of 90° to the horizontal axis, and ε_{45} represents the strain with an angle of 45° to the horizontal axis.

Reduction percentage of average strain at measuring point i is defined as

$$B_i = \frac{\varepsilon_{0i} - \varepsilon_{1i}}{\varepsilon_{0i}} \tag{26}$$

with ε_{0i} and ε_{1i} representing the average strain at measuring point i before and after applying the RCC.

The measured strain results of each measuring point on the blade are displayed in Table 5.

Table 5. The strain reduction comparison of measuring points.

Average Strain	Without Reinforcing Cable Component	With Reinforcing Cable Component	B%
Measuring point 1	178.16	34.45	80.7%
Measuring point 2	176.26	47.59	73.0%
Measuring point 3	349.32	132.72	62.0%
Measuring point 4	221.06	91.13	58.8%
Measuring point 5	160.90	89.14	44.6%
Measuring point 6	161.55	167.46	−3.7%
Measuring point 7	177.03	156.31	11.7%
Measuring point 8	97.30	105.77	−8.7%

As shown in Table 5, the maximum strain for blades without RCC occurs at the measuring point 3, where the blade shape changes greatly. After applying the RCC, the reduction percentage of the average strain reached 62.0%, indicating a good load reduction effect for a dangerous section. The strain response of the test blade is reduced at different levels except for measuring point 6 and measuring point 8. For the measuring point 1 which is close to root of the blade, 80.7% strain is reduced by RCC. The strain at the blade root is also effectively relieved.

Only the strain at measuring point 6 and measuring point 8 is slightly increased. The reason can be explained as follows. As shown in Figure 14, the measuring point 6 is close to the connection point of RCC. When the blade is connected with RCC, the area near the connecting point forms a discontinuity area, which is subject to both the external load and the tension of the cable simultaneously. As the connection area is small, the induced stress is enlarged. For measuring point eight, the reason is similar. The strain is increased as the mass is hung at the blade tip through a steel cable. This will not happen since the wind load acts on the whole blade surface in actual situation, as one can see that the strain does not increase at the blade tip in the simulation result.

6. Conclusions

Under extreme weather, the angle of wind turbine surface and blades must be adjusted continuously according to the inflow direction to protect the blades from breaking. Untimely adjustment may lead to structure failure at blade root. In this paper, the bending resistance ability of wind turbine blade was improved by a novel mechanical design. The main conclusions can be drawn as follows: (1) A reinforcing cable component (RCC) was proposed to reduce wind load on blade. To gain better reduction of load, the connection point of RCC is suggested to be located at two-thirds of the blade length. (2) The proposed RCC improved stress distribution by altering the cantilever boundary condition of wind turbine blades. According to finite element results, the large stress at the root as well as tip placement were effectively decreased. (3) A static strain test for a small-scaled wind turbine indicated good load reduction effect of RCC, 80.7% strain was reduced at the blade root. The proposed RCC has no limitations of braking position and blade angle, which contributes to improve the survivability of wind turbines under extreme weather.

Author Contributions: Conceptualization, M.S.; methodology, X.F.; software, X.F.; validation, X.F. and M.S.; formal analysis, X.F.; investigation, X.F.; resources, M.S.; data curation, X.F. and M.S.; writing—original draft preparation, X.F.; writing—review and editing, M.S.; visualization, X.F. and M.S.; supervision, M.S.; project administration, M.S.; funding acquisition, M.S. All authors have read and agreed to the published version of the manuscript.

Funding: This research was funded by Natural Science Foundation of Shaanxi Province, China, grant number No.2021JLM-39.

Institutional Review Board Statement: Not applicable.

Informed Consent Statement: Not applicable.

Conflicts of Interest: The authors declare no conflict of interest.

Nomenclature

α	Angle of attack (°)	$F_i, i = 1,2,3$	Distributed wind load (N)
β	Wind direction (°)	q	Distributed force (N)
ϕ	Blade parking angle (°)	T	Tension force (N)
V	Relative wind velocity (m/s)	σ	stress (N/m^2)
t	Chord length (m)	E_s	Young's modulus (Pa)
L	Blade length (m)	ε_0	Strain in the axis direction
w	Deflection of blade (m)	ε_{45}	Strain with an angle of 45° to the horizontal axis
ξ	Length of RCC (m)	ε_{0i}	Average strain at measuring point *i* before applying RCC
$\Delta\xi$	Elongation of RCC (m)	ε_{1i}	Average strain at measuring point *i* after applying RCC
W_z	Bending section coefficient (m^3)		

References

1. Tu, Q.; Liu, Z.; Li, B.; Mo, J. Achieving grid parity of offshore wind power in China–A comparative analysis among different provinces. *Comput. Ind. Eng.* **2021**, *162*, 107715. [CrossRef]
2. GWEC. *GWEC, Global Wind Report Annual Market Update 2021*; Global Wind Energy Council: Brussels, Belgium, 2021.
3. Enevoldsen, P.; Jacobson, M.Z. Data investigation of installed and output power densities of onshore and offshore wind turbines worldwide. *Energy Sustain. Dev.* **2021**, *60*, 40–51. [CrossRef]
4. GWEC. *GWEC, Global Wind Report Annual Market Update 2022*; Global Wind Energy Council: Brussels, Belgium, 2022.
5. Stewart, G.M.; Lackner, M.A. The impact of passive tuned mass dampers and wind–wave misalignment on offshore wind turbine loads. *Eng. Struct.* **2014**, *73*, 54–61. [CrossRef]
6. He, R.; Pan, L. New Optimization Method of Multiple Tuned Mass Dampers Installed on Offshore Wind Turbines for Seismic Response Control. In Proceedings of the 32nd International Ocean and Polar Engineering Conference, Shanghai, China, 20 October 2021.
7. Chen, C.-H.; Su, N.-J. Global Trends and Characteristics of Offshore Wind Farm Research over the Past Three Decades: A Bibliometric Analysis. *J. Mar. Sci. Eng.* **2022**, *10*, 1339. [CrossRef]
8. Meng, H.; Jin, D.; Li, L.; Liu, Y. Analytical and numerical study on centrifugal stiffening effect for large rotating wind turbine blade based on NREL 5 MW and WindPACT 1.5 MW models. *Renew. Energy* **2022**, *183*, 321–329. [CrossRef]
9. Ashuri, T.; Martins, J.R.R.A.; Zaaijer, M.B.; van Kuik, G.A.M.; van Bussel, G.J.W. Aeroservoelastic design definition of a 20MW common research wind turbine model. *Wind Energy* **2016**, *19*, 2071–2087. [CrossRef]
10. Chen, X. Experimental investigation on structural collapse of a large composite wind turbine blade under combined bending and torsion. *Compos. Struct.* **2017**, *160*, 435–445. [CrossRef]
11. Gasch, R.; Twele, J. Scaling wind turbines and rules of similarity. In *Wind Power Plants: Fundamentals, Design, Construction and Operation*; Gasch, R., Twele, J., Eds.; Springer: Berlin/Heidelberg, Germany, 2012; pp. 257–271. [CrossRef]
12. Chen, X.; Xu, J.Z. Structural failure analysis of wind turbines impacted by super typhoon Usagi. *Eng. Fail. Anal.* **2016**, *60*, 391–404. [CrossRef]
13. Chou, J.-S.; Chiu, C.-K.; Huang, I.-K.; Chi, K.-N. Failure analysis of wind turbine blade under critical wind loads. *Eng. Fail. Anal.* **2013**, *27*, 99–118. [CrossRef]
14. Chen, C.P.; Kam, T.Y. Failure Analysis of Small Composite Sandwich Turbine Blade Subjected to Extreme Wind Load. *Procedia Eng.* **2011**, *14*, 1973–1981. [CrossRef]
15. Lee, H.G.; Kang, M.G.; Park, J. Fatigue failure of a composite wind turbine blade at its root end. *Compos. Struct.* **2015**, *133*, 878–885. [CrossRef]
16. Zhang, X.; Wang, M.; Wang, Y.; Liu, Y. Research on Failure of Offshore Wind Turbine Tower in Typhoon. In Proceedings of the OCEANS—MTS/IEEE Kobe Techno-Oceans, Kobe, Japan, 28–31 May 2018; pp. 1–4.
17. Bossanyi, E.A. Individual Blade Pitch Control for Load Reduction. *Wind Energy* **2003**, *6*, 119–128. [CrossRef]
18. Stubkier, S.; Pedersen, H.C.; Markussen, K. Hydraulic soft yaw system load reduction and prototype results. In Proceedings of the EWEA Annual Event, Vienna, Austria, 4–7 February 2013.
19. Kim, E.; Manuel, L. Hurricane-Induced Loads on Offshore Wind Turbines with Considerations for Nacelle Yaw and Blade Pitch Control. *Wind Eng.* **2014**, *38*, 413–423. [CrossRef]
20. Bakhshi, R.; Sandborn, P. The Effect of Yaw Error on the Reliability of Wind Turbine Blades. In Proceedings of the ASME 2016 10th International Conference on Energy Sustainability collocated with the ASME 2016 Power Conference and the ASME 2016 14th International Conference on Fuel Cell Science, Engineering and Technology, Charlotte, NC, USA, 26–30 June 2016.
21. Chen, X.; Li, C.; Xu, J. Failure investigation on a coastal wind farm damaged by super typhoon: A forensic engineering study. *J. Wind Eng. Ind. Aerodyn.* **2015**, *147*, 132–142. [CrossRef]
22. Wang, J.; Chen, Z. Analysis of risks and measures on the blade damage of offshore wind turbine during strong typhoons—Enlightenment from Red Bay wind farm. *Eng. Sci.* **2010**, *12*, 32–34.
23. Dai, J.; Hu, W.; Yang, X.; Yang, S. Modeling and investigation of load and motion characteristics of offshore floating wind turbines. *Ocean Eng.* **2018**, *159*, 187–200. [CrossRef]
24. Lian, J.; Jia, Y.; Wang, H.; Liu, F. Numerical Study of the Aerodynamic Loads on Offshore Wind Turbines under Typhoon with Full Wind Direction. *Energies* **2016**, *9*, 613. [CrossRef]
25. Lanzafame, R.; Messina, M. Fluid dynamics wind turbine design: Critical analysis, optimization and application of BEM theory. *Renew. Energy* **2007**, *32*, 2291–2305. [CrossRef]
26. Hansen, M.O.L. *Aerodynamics of Wind Turbines*, 2nd ed.; Earthscan: London, UK, 2008.
27. Dal Monte, A.; De Betta, S.; Raciti Castelli, M.; Benini, E. Proposal for a coupled aerodynamic–structural wind turbine blade optimization. *Compos. Struct.* **2017**, *159*, 144–156. [CrossRef]
28. Wang, M.; Han, F.; Wu, Q. A steel cable connection component for wind turbine blade under mechanical braking state. 2016.
29. Gere, J.M. *Mechanics of Materials*, 7th ed.; Mechanics of Materials: Amsterdam, The Netherlands, 2009.
30. Resor, B.R. *Definition of a 5MW/61.5 m Wind Turbine Blade Reference Model*; Sandia National Lab.(SNL-NM): Albuquerque, NM, USA, 2013.
31. Cox, K.; Echtermeyer, A. Structural Design and Analysis of a 10MW Wind Turbine Blade. *Energy Procedia* **2012**, *24*, 194–201. [CrossRef]

32. Finnegan, W.; Jiang, Y.; Dumergue, N.; Davies, P.; Goggins, J. Investigation and Validation of Numerical Models for Composite Wind Turbine Blades. *J. Mar. Sci. Eng.* **2021**, *9*, 525. [CrossRef]
33. Jiang, P.; Nan, R.-D.; Qian, L.; Yue, Y.-L. Studying solutions for the fatigue of the FAST cable-net structure caused by the process of changing shape. *Res. Astron. Astrophys.* **2015**, *15*, 1758. [CrossRef]
34. Zhou, H.F.; Dou, H.Y.; Qin, L.Z.; Chen, Y.; Ni, Y.Q.; Ko, J.M. A review of full-scale structural testing of wind turbine blades. *Renew. Sustain. Energy Rev.* **2014**, *33*, 177–187. [CrossRef]
35. Babawarun, T.; Ho, W.H.; Ngwangwa, H. Stress validation of finite element model of a small-scale wind turbine blade. *J. Energy S. Afr.* **2019**, *30*, 87–97. [CrossRef]

Journal of
Marine Science and Engineering

MDPI

Article

Availability Analysis of an Offshore Wind Turbine Subjected to Age-Based Preventive Maintenance by Petri Nets

Eduard Lotovskyi, Angelo P. Teixeira and C. Guedes Soares *

Centre for Marine Technology and Ocean Engineering (CENTEC), Instituto Superior Técnico, Universidade de Lisboa, Av. Rovisco Pais, 1049-001 Lisboa, Portugal; eduard.lotovskyi@centec.tecnico.ulisboa.pt (E.L.); teixeira@centec.tecnico.ulisboa.pt (A.P.T.)
* Correspondence: c.guedes.soares@centec.tecnico.ulisboa.pt

Abstract: This paper analyses the production availability and the associated maintenance costs of an offshore wind turbine with a horizontal axis configuration using Petri Nets modelling with Monte Carlo Simulation. For this purpose, different features are implemented: the reliability and maintainability characteristics of the components; the logistics of the production and maintenance operations, including different types of vessels, the mobilization time, costs and weather window. The maintenance strategies consist of corrective maintenance and age-based imperfect preventive maintenance that depends on the components' age and age reduction ratio. Thereby, to increase the operating income and to reduce the costs associated with the operation and maintenance activities, the optimal parameters of the age-based preventive maintenance are estimated. As a case study, a generic offshore wind turbine that operates at the Viana do Castelo wind farm in Portugal is adopted. The wind farm is located 18 km off the shore. The turbine's total exploration life is 25 years.

Keywords: age-based preventive maintenance; Monte Carlo Simulation; offshore wind turbine; Petri Nets; production availability assessment

Citation: Lotovskyi, E.; Teixeira, A.P.; Guedes Soares, C. Availability Analysis of an Offshore Wind Turbine Subjected to Age-Based Preventive Maintenance by Petri Nets. *J. Mar. Sci. Eng.* **2022**, *10*, 1000. https://doi.org/10.3390/jmse10071000

Academic Editors: Decheng Wan and Giuseppe Roberto Tomasicchio

Received: 1 June 2022
Accepted: 19 July 2022
Published: 21 July 2022

1. Introduction

An offshore wind turbine (OWT) is a complex production system that includes various engineering disciplines. Today, the total offshore wind farm capacity is approximately 18.9 GW and the installation rate has grown in recent years [1]. To improve the turbine's performance, an availability assessment is required, since it is able to evaluate the impact of the equipment reliability and maintainability qualities on the system productivity.

The main challenges in the production optimization and in maintenance planning are the realistic representation of the system's behaviour main characteristics and the implementation of the interactions and dependencies among various components of the same production system [2,3]. The classical reliability tools, starting from the identification of failures with Failure Mode and Effect Analysis [4,5] and including the Reliability Block Diagram, the Fault Tree and the Event Tree, [6,7] are not appropriate for analysing industrial production systems, since they do not take into consideration the interdependencies between the failures, which require other tools [8]. The dynamic interactions between operation and maintenance (O & M) decisions are relevant to be modelled, too [9]. The classical reliability tools use the concept of Boolean algebra, in other words, they are convenient to assess rare events with serious consequences, while the dependability analyses are able to assess more likely events but with low effects on the operation [10].

Markov modelling is an effective tool for the mathematical representation of component failure interactions and systems with independent failures [11–13]. Castanier et al. [14] proposed a Markov decision process to determine the optimal maintenance and operation policy of an offshore wind farm considering the stochastic wind and weather conditions. Li et al. [15] assessed the reliability of OWT's Gearbox and formulated the optimal maintenance policy by using Markov process. The main limitations of Markov modelling are:

the state-explosion, because the increasing of the system automatically takes to the high expansion of the number of states, and the assumption of fixed repair and failure rates, because this limits its application to simple systems with exponentially distributed events [13]. Thus, to represent the complexity of a production system considering all the interactions and dependencies among the sub-systems, various authors use simulation techniques, e.g., Bris [16] and Santos et al. [17].

Zio et al. [18] proposed a Monte Carlo Simulation (MCS) model for availability assessment of a multi-state and multi-output offshore installation. It was shown that Monte Carlo simulation can describe the uncertainty of various production system features, such as: the degradation of the system components, the corrective and preventive maintenance policies with pre-defined reparation priorities, and the limited number of repair teams. MCS approach has proven to be an adequate tool to model continuously deteriorating systems and to determine the "on-condition" maintenance plan that minimises the expected total system cost over a specified mission period [19]. Condition-based maintenance is becoming increasingly more relevant for this type of equipment [20,21].

Petri Nets (PN) is a tool that combines graphical and mathematical modelling capabilities to simulate and analyse discrete event systems [22]. It was developed in 1962 by Carl Adam Petri in his Ph.D. dissertation [23]. The main objective of his study is to find an efficient way to model several competing or co-operating processes using graphical representation. Recently, different techniques based on the Stochastic Petri Nets (SPN) have been developed to improve the modelling of real and complex production systems [24,25].

Due to the complexity of the stochastic evolution of real systems that is not easily captured by analytical models [26], their application to quantitative analyses of SPN is rather limited [27]. To overcome this difficulty, techniques based on SPN combined with Monte Carlo simulation have been proposed for modelling and analysing the complex behaviour of industrial multi-unit systems regarding their reliability, availability and production efficiency [9,22,24,28].

Santos et al. [29] modelled an offshore wind turbine located at the North Sea close to the German shore combining SPN and MCS in order to assess the operation behaviour and maintenance policies. The reliability characteristics of OWT's components were based on typical onshore reliability data provided by Ding and Tian [30]. Using SPN coupled with MCS, Santos et al. [31] presented a parametric study on the availability, operation, and maintenance costs of OWT.

The effect of age-based preventive maintenance on the availability and maintenance costs of offshore wind turbines [32] and oil and gas production systems [33] has also been assessed through SPN and Monte Carlo simulation. To build the failure models of the offshore components, the onshore data were used considering an empirical approach based on stress factors for mechanical systems.

An individual wind turbine is basically a multi-state system composed by several major components in a series configuration (i.e., Gearbox, Generator, Pitch System, and Rotor). Many industrial/real systems are characterised by series-parallel structures that have been widely investigated [34]. A production system can be classified as either the series-parallel (s-p) or the parallel-series (p-s) system [35]. In reliability and maintenance assessments, various industrial and production systems belong to s-p class (e.g., water supply systems and production lines in manufacturing factories) [36,37]. Besides, s-p class is used to represent in simplified mode the main sub-components of the total engineering system. Kawauchi and Rausand [38] and Bris et al. [39] provided this approach to estimate a performance measure (i.e., "production regularity") of the system, and to minimise the preventive maintenance cost, respectively.

Usually, the availability assessment of an OWT production system adopts the simplified maintenance policy. In the implementation of reliability characteristics, it is common to model only critical failures, considering every equipment as a binary system (i.e., either as good as new or failed). Besides, in the assessment of age-based preventive maintenance, the age-reduction ratio and the repair threshold parameters are typically fixed and not optimised. In order to increase the OWT availability, Sobral et al. [40] proposed a methodology to weight the influencing factors, considering operational and maintenance data, distance to shore, water depth, site accessibility, meteorological and oceanographic factors. Kang and Guedes Soares [41] introduced the conditional-based maintenance strategy based on the support vector machine algorithm to optimise the maintenance arrangement of OWTs.

According to Nielsen and Sørensen [42], the operation and maintenance (O & M) costs of offshore wind turbines are major contributors to the price of energy and can reach 30% of it. Castro-Santos et al. [43] indicated that in terms of the life-cycle of OWT, the higher cost corresponds to the exploitation period (i.e., insurance, administration, operation, and maintenance), followed by the manufacturing and installation periods. The maintenance and equipment replacements are dependent on the weather windows [44]; thus, the correct planning of O & M activities in advance is important to minimise the expected outlay over the turbine's lifetime. Besides, the distance to shore must be also considered, since it influences the investment, operation, and maintenance costs [1]. Castro-Santos et al. [45] assessed the economic feasibility of offshore wind farms based on the meteorological data, bathymetry, and distances between wind farm-shore, -shipyard, and -port.

To minimise the maintenance costs of offshore wind farms, Kang and Guedes Soares [41] introduced an opportunistic strategy considering imperfect maintenance and the weather window effects, using the rolling horizon approach. Castro-Santos et al. [46] compared in economic terms the offshore wind farms with other alternatives to harness wind and wave energies, such as floating offshore wave energy devices, floating offshore co-located systems, and floating offshore hybrid systems.

The main objective of this paper is to develop a framework capable to assess the effect of an age-based preventive maintenance on the availability of an OWT with horizontal axis configuration by Petri Nets and Monte Carlo Simulation. For this purpose, two variables are highlighted: the equipment age reduction ratio and the time threshold between maintenance interventions based on the component's age. The variation of both allows to visualise the effects of maintenance policy on the production system availability and associated costs. Besides, it provides the tool capable to identify adequate values that provide a balance between production availability and costs. As a case study, a singular OWT that operates at the Viana do Castelo wind farm in Portugal, located 18 km off the shore, is used. The total exploration life of the turbine is 25 years. The present case study adopts a parallel-series system (i.e., wind farm, thus, p-s) composed of p subsystems in parallel (i.e., one wind turbine, thus, $p = 1$), each of them with s components in series (i.e., four degraded components, thus, $s = 4$). The modelled degraded components (i.e., Rotor, Gearbox, Generator, and Pitch System) have the higher influence on the total system availability [47].

Based on empirical offshore reliability data provided by Santos et al. [32], the equipment definition is based on the reliability and maintainability stochastic characteristics that follow non-exponential distributions. In addition to as good as new and failed conditions, the degraded states of the components are also considered. Thereby, three types of failure are modelled: incipient (i.e., the transition from as good as new state to degraded one), degraded (i.e., the transition from degraded state to failed one), and critical failure (i.e., the transition from as good as new state to failed one). The maintenance policy is divided into three categories based on [48]. These categories differ on the weight of the repaired equipment and, therefore, the involved logistics.

The Corrective Maintenance (CM) intervention encompasses the manufacture time of a new component in a factory, the transportation time from the supplier to the port, and the replacement of the damaged equipment by a new one on the OWT [48]. The Preventive Maintenance (PM) strategy is age-based and imperfect. Each PM activity reduces the equipment age by a ratio q, and it is performed when the component's age reaches at least $p \times$ MTTF hours, where p is a threshold parameter and MTTF is a Mean Time to Failure of the equipment [30,32]. Both, q and p are the main input variables that influence the O & M costs. Thereby, the main PM parameters are optimised to obtain the lower costs and higher income (i.e., the higher accounting rate of return). Both corrective and preventive maintenance activities can be subjected to the time delay due to adverse weather conditions [29]. All vessels required for maintenance activities are anchored in Portuguese harbours.

This paper mainly consists of two parts. The first one is Materials and Methods. This section starts with a brief introduction of the main elements of Petri Nets. Next, the case study description is presented, explaining in detail the production configuration and component failure data of OWT's equipment. A special attention is given to the maintenance policy delineation with a clear description of CM and PM activities. The implemented models of costs and weather window are formulated. The PN models of equipment, total system switch, PM, seasons, and vessels are described step by step. At the end of the section, the main cost models for economical assessment are presented. The second part of the paper is Results. In this section the age-based PM parameters are derived based on economical assessment, availability assessment, and sensitivity analysis. The obtained results are discussed.

2. Materials and Methods

2.1. Petri Nets

Petri Nets (PN) is a generic name for modelling tools that can represent a complex production system graphically. PN are divided into three categories [49]: the Elementary Net Systems for small size system representation; the Place/Transition Systems for a more compact representation of the Elementary Net Systems models; Predicate/Transition Nets or Coloured Nets for even more compact representations of real applications using algebraic and logical elements.

In Figure 1, the basic graphic elements of the Place/Transition System (i.e., place, transition, token, and arc) are presented [50].

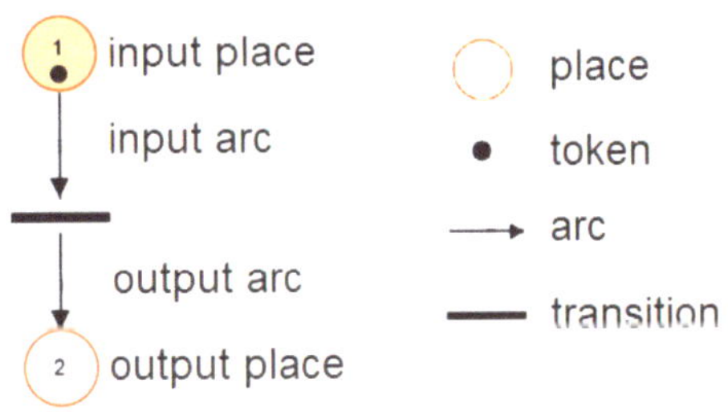

Figure 1. Basic elements of Petri Nets.

The place is represented by circles, it models the system's states or resources. The transition is represented by rectangles and it is used to model the events (e.g., system failure) that influence the available resources. The token is a small black dot that represents the resources. The token is always held inside the places. The arc is represented by directed arrows that specifies the interconnection between the places and transitions, and indicates which states are changed by a certain event.

The positioning of the token in the place, called by marking, defines a specific state of the system. In the case of system state change, the transition moves the tokens to new places or removes them in accordance with the arcs' directions. This property enables the simulation of dynamic systems [34]. The Place/Transition net is a bipartite graph, thus, it is only possible to connect Place-Transition or Transition-Place, meaning that neither Place-Place nor Transition-Transition are acceptable connections [51].

Another frequently used tool is the Generalized Stochastic Petri Nets (GSPN) with predicates. This tool has more computing power than conventional Petri Nets and allows to perform a modular model [28]. In the GSPN, the transition is equipped with guards and assignments. The guards are pre-condition logical functions identified by prefix "??" and used to enable or inhibit the firing of transitions. The assignments, represented by "!!", are the post-condition messages that update the variables used in the model. To learn more about Petri Nets, studies [50,51] are recommended; a comprehensive overview of GSPN is provided in [52].

2.2. Case Study Description

For case study analysis, a generic OWT with Horizontal Axis Wind Turbine (HAWT) configuration is considered. The OWT is a multi-unit and multi-state system with complex dependencies between components. In this paper, only the equipment that most influence the system availability is considered, namely: Rotor (RT), Gearbox (GB), Generator (GT), and Pitch System (PS).

The case study is defined based on various sources of information. The reliability parameters of the OWT's components are based on empirical offshore reliability data provided in [32]. The corrective and preventive maintenance policies are based on [30,32,48]. The weather window is adopted from [29].

The case study OWT is located at the Portuguese wind farm close to Viana do Castelo. The wind farm area is located 18 km off the shore, that is, close to the global average of distance to the shoreline, i.e., 18.8 km [1]. The wind resource at the location is given in [53], while the wave conditions are indicated in [54]. The total exploration life of OWT is 25 years with a capacity factor of 50.8% per year [55]. The electric power is 5 MW. The tariff for wind energy produced by OWT in Portugal is considered to be 0.1544 EUR/kWh [56].

2.2.1. Production Configuration

The OWT system can be divided into eight main subsystems, namely: support structures, pitch and hydraulic system, gearbox, generator, speed train, electronic components, blades system, and yaw system [6]. All of them are installed in a series configuration. Thus, any subsystem failure of OWT can undermine the total production. Hence, the critical or degraded failure of RT, GB, GT, or PS leads to the total production stop. Besides, the Place/Transition Petri Nets has a limitation: different events cannot occur simultaneously (i.e., the firing of the differently enabled transitions is sequential). Thereby, the defining of equipment sequence switch is needed. To shut down the OWT production, the equipment sequence switch is: RT $\rightarrow$ GB $\rightarrow$ GT $\rightarrow$ PS. To start up the wind turbine production, the equipment sequence switch consists of: PS $\rightarrow$ RT $\rightarrow$ GB $\rightarrow$ GT. The Pitch System controls the blade pitch to follow a predetermined speed ramp during startup and shutdown of the turbine. Thus, the PS is the last one in the first sequence and the first one in the second one. In the case of failure of one of the pieces of equipment presented in the sequences, the order in which the subsystems are switched off is preserved, giving priority to the damaged equipment.

2.2.2. Component Failure Data

In this case study, only the failures of the RT, GB, GT, and PS are considered. The component failure data is given in Table 1.

The non-perfect systems can be in three different states: "As good as new", "Degraded" and "Failed". "As good as new" state is a component in normal operation. "Failed" state corresponds to a non-functioning component. "Degraded" state means that the function of a component is maintained, but the system has a higher probability of failure. The equipment can be repaired in "Failed" or in "Degraded" states.

The incipient and critical failures are described by the Weibull Truncated distribution, which accounts the equipment's age. The shape parameters, β, chosen for the transitions are 2 and 3, respectively [30,32]. Both are larger than 1, representing the equipment in the wear-out period of life. In this period, the larger the shape parameter, the higher degradation effect.

The degraded failure is described by the exponential distribution, with the failure rate, λ, due to the failure being independent from the equipment's age.

Table 1. Component failure data.

Component	Failure	Distribution	β [1]	MTTF [2] (h)	λ [3] (h^{-1})
Rotor (RT)	Incipient	Weibull Truncated	3	19,948	-
	Critical	Weibull Truncated	3	22,164	-
	Degraded	Exponential	-	-	4.51×10^{-4}
Gearbox (GB)	Incipient	Weibull Truncated	3	15,952	-
	Critical	Weibull Truncated	3	17,724	-
	Degraded	Exponential	-	-	5.64×10^{-4}
Generator (GT)	Incipient	Weibull Truncated	2	17,215	-
	Critical	Weibull Truncated	2	19,128	-
	Degraded	Exponential	-	-	5.23×10^{-4}
Pitch System (PS)	Incipient	Weibull Truncated	3	12,355	-
	Critical	Weibull Truncated	3	13,728	-
	Degraded	Exponential	-	-	7.28×10^{-4}

[1] Shape parameter. [2] Mean Time to Failure. [3] Failure rate.

2.3. Maintenance Policy

The case study OWT with four degraded components is subjected to corrective maintenance (CM) when necessary and to age-based preventive maintenance (PM) only in the summer. To keep the functioning of the OWT system, three different maintenance categories are used [48]. Each of them depends on the weight of the component being replaced (CM) or repaired (PM), thus according to the vessel involved:

- Jack-Up vessel (JU)—is used in CM activities to replace the Rotor (its weight is between 90 t and 150 t).
- Crane Barge (CB)—is used in CM activities to replace the Generator (up to 20 t) or the Gearbox (up to 65 t).
- Supply Vessel (SV)—is used in CM activities to replace the Pitch System and to support the CM operations of the larger vessels. Moreover, SV is used in all PM activities.

The technical specifications of each vessel and the summary of maintenance categories are presented in Tables 2 and 3, respectively. The SV is the only one docked at the nearest available port to the OWT, in Viana do Castelo, whereas the JU and CB are docked in Porto and Aveiro, respectively, see Table 4.

Table 2. Technical specifications of the vessels.

Categories (Vessels)	Draught (m)	Service Speed (Knots)
Jack-Up (JU)	5.8	12
Crane Barge (CB)	3.8	7
Supply Vessel (SV)	3.8	7

Table 3. Maintenance categories.

Categories (Vessels)	Turbine's Components			
	RT	GB	GT	PS
JU	CM			
CB		CM	CM	
SV	CM & PM	CM & PM	CM & PM	CM & PM

Table 4. Principal constraints of Portuguese harbours.

Harbour	Terminal	Draught (m)	Distance to OWT (km)
Viana do Castelo	Commercial port	8	19.48
Porto (Leixões)	General cargo I	10	63.51
	General cargo II	11	63.51
Aveiro	Multipurpose	12	124.77

The Supply Vessel is used to transport a unique maintenance crew of 4 technicians and spares for the Pitch System. The maintenance crew is considered as always available. Moreover, it can start at any time of the day as long as all the logistics are concluded and there is a weather window available. It is worth noting that the Supply Vessel is set to arrive at the turbine at the same time with the Jack-Up (or the crane vessel), so the maintenance team can start working immediately upon arrival.

2.3.1. Corrective Maintenance

CM activity consists of replacing the damaged equipment by a new one including all the operation steps from the production of equipment in a factory to its commissioning on the turbine. The component manufacture in a factory and the equipment transportation from the supplier to the port constitute the vessels' logistic time, see Table 5 [48]. It is worth noting that the SV is available 24 h per day to depart to an OWT immediately, and only in the case of PS's corrective repair, the Supply Vessel possesses a logistic time, since the arrival of the new Pitch System takes 2 days (i.e., 48 h). The logistic time follows a Log-normal distribution with a coefficient of variation of 30%.

Table 5. Vessels' logistic times.

Categories (Vessels)	Logistic Time (h)	SD [1] (h)
JU	504	151.2
CB	160	48
SV	48	14.4

[1] Standard Deviation.

The time spent on transportation of the new component from the port to the OWT is estimated from the vessels' service speed and the travelled distance. Additionally, the sailing time follows a Log-normal distribution with a coefficient of variation of 20%. Please note, the manoeuvres and the transit time in port are neglected.

Table 6 contains the information regarding the mean duration of component replacement and the respective standard deviation (SD). This time depends on both the equipment type and the production state (i.e., degraded or failed). The time of CM activities follows a Log-normal distribution with coefficient of variation of 20%.

Table 6. Corrective maintenance data (replacement phase) by equipment.

Transition		RT	GB	GT	PS
Failed → As good as new	Mean (h)	40	50	50	10
	SD [1] (h)	8	10	10	2
Degraded → As good as new	Mean (h)	52	65	65	13
	SD [1] (h)	10.4	13	13	2.6

[1] Standard Deviation.

2.3.2. Preventive Maintenance

The RT, GB, GT, and PS are subjected to imperfect age-based PM. This maintenance action can start whether the equipment is in the perfect state or the component is stopped, but not damaged. Please note, once a PM begins, it must be concluded, even if a critical or degraded failure occurs in another component. PM is performed by the same maintenance team as the CM, which is located on the Supply Vessel.

PM tasks are performed based on age reduction ratio, q ($0 < q < 1$). Thus, after repair activity, the component is q younger (i.e., the age is reduced by q percent). After PM activity, the age is defined by [30]:

$$Age_i = Age_i^{acc} \times (1-q) \Leftrightarrow Age_i = (t_i - t_{i-1} + Age_{i-1}) \times (1-q) \tag{1}$$

where, Age_i and Age_{i-1} are the component's consecutive ages after i^{th} and $(i-1)^{th}$ maintenance tasks, respectively; Age_i^{acc} is the age at the beginning of the ith maintenance action, accumulated from the $(i-1)^{th}$ maintenance task; t_i and t_{i-1} are the calendar times at the beginning of the ith and at the end of the $(i-1)^{th}$ maintenance actions, respectively. It is worth noting that after corrective maintenance activity, the component is new, so its age is $Age_i = 0$.

PM is carried out only in the summer season, and when the age of a component is at least equal to $p \times MTTF$ hours [17]:

$$(t_i - t_{i-1} + Age_{i-1}) \geq p \times MTTF, \quad where\ 0 \leq p \leq 1 \tag{2}$$

where, p, is a preventive repair threshold parameter which is the same for all components.

In this case study, four different PM activities are considered and presented in Table 7. The frequency of the maintenance depends on the equipment's age and it is assumed to be Delta Dirac distributed. The duration of PM depends on the age reduction ratio, q, and on the Mean Time to Repair (MTTR) of critical failure. The Log-normal distribution with coefficient of variation of 30% is used to describe the duration of PM.

Table 7. Preventive maintenance models of individual component.

Component	Period (h)	Duration (h)	Recovered Age (%)
RT	$p \times MTTF_{RT}$	$q \times MTTR_{RT}$	$q \times 100$
GB	$p \times MTTF_{GB}$	$q \times MTTR_{GB}$	
GT	$p \times MTTF_{GT}$	$q \times MTTR_{GT}$	
PS	$p \times MTTF_{PS}$	$q \times MTTR_{PS}$	

2.4. Costs

In Table 8, the approximate costs for the new turbine's components are presented. These values are based on an offshore wind farm guide from [57]. Please note, the transportation outlay of a new equipment from the manufacturer to the port is included in the overall price of the component.

Table 8. Overall cost of the turbine's components.

Component	Cost/Unit (EUR)
RT	1,849,000
GB	863,000
GT	247,000
PS	123,300

The costs related to the vessels and technicians, Table 9, are approximate values based on [48], where the hourly rate of a technician is EUR 70 by person.

Table 9. Hourly rates and mobilization costs.

Service	Hourly Rate (EUR)	Mobilization (EUR)
JU	6250	57,000
CB	6250	45,000
SV	600	0
Technician	70	0

The cost of the imperfect PM, C_p, depends on the age reduction ratio, q. According to [30], it is given by:

$$C_p = \begin{cases} q^2 \times C_{pv} + C_{pf}, & 0 < q \leq 1 \\ 0, & q = 0 \end{cases} \quad (3)$$

where, C_{pv} is the preventive component replacement cost; C_{pf} is a fixed maintenance cost. Since the PM is stochastically driven, C_{pf}, is not considered in this paper, thereby, C_{pv} comprises the total replacement cost of a new component. After the simulation, the number of occurred PM activities per component is obtained. Then, this number is multiplied by the component's cost/unit (see Table 8). Hence, C_{pv} per turbine's component is calculated. The cost of PM considers neither hourly rates of SV nor technician crew. The hourly rates and the mobilization costs are calculated separately considering the expenses of CM and PM activities together.

2.5. Weather Window

The corrective and preventive maintenances are only performed when the weather window (WW) is available (i.e., the wind speed and the significant wave height are within the operational limits of the marine operation for a period long enough due to safety reasons). The probability of favourable WW, Pw, and the time delay due to bad weather conditions, Tw, are presented in Table 10 [17]. Please note that these values are merely illustrative. The probabilities of the available WWs are conservative based on the typical behaviour of each season. So, the increase of probability of available WW corresponds to the decrease of waiting time for a WW. More realistic correlations can be obtained from available weather data at the offshore wind farm.

Table 10. Availability of weather window and waiting time.

Season	Availability of WW (P_w)	Availability Random Number	Waiting Time for a WW (T_w) (h)
Winter	0.3	[0, 0.3]	240
Autumn	0.5	[0, 0.5]	168
Spring	0.6	[0, 0.6]	120
Summer	0.8	[0, 0.8]	48

The WW must be available when a maintenance activity (i.e., CM or PM) is to be performed. As a result, as soon as the ships and the maintenance team are in port, the model randomly generates a number between 0 and 1. If this value is less than or equal to the seasonal Pw, a WW is available and the maintenance activity can start, otherwise a waiting time, Tw, for a weather window is set. When the waiting period expires, a WW becomes available.

Due to the relatively small distances between the ports, the available WW is required only for the first departure ship by maintenance activity.

2.6. System Modelling by Petri Nets

The case study OWT is modelled by GSPN with predicates and the production availability and maintenance costs are obtained from Monte Carlo simulation. At the initial instant, the PN model of turbine has all components of offshore production plant in operation with initial age equal to zero (i.e., all equipment components are as good as new), the maintenance team and the Supply Vessel are localised in Viana do Castelo, the Jack-Up vessel is anchored in Porto, while the Crane Barge is anchored in Aveiro.

2.6.1. Equipment

Each component of OWT is schematised using the same type of PN model, consisting of a sequence of events, which includes the simulation of failure and of the repair process. To avoid a too extensive description, only the simplified model of Rotor (RT) is presented (see Figure 2).

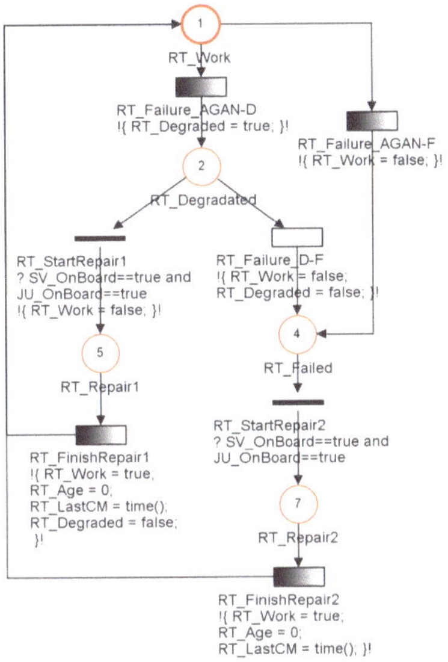

Figure 2. PN model of equipment (Rotor).

To ensure a logical sequence, the PN model of the equipment is accompanied by various variables. Thus, the variable *RT_Work* is used to determine whether the equipment fulfils its technical mission (i.e., *RT_Work == true*), or not (i.e., *RT_Work == false*). The variable *RT_Degradation* is used to determine whether the equipment is degraded and available or not degraded. Moreover, due to the PM activity, the model is complemented by: *RT_Age* and *RT_LastCM*. The first registers the equipment's age in hours; the second records the time of the completion of the last corrective repair. These variables are used in calendarization of the preventive maintenances. After every CM, *RT_Age* is set to zero.

When the place *RT_Work* is marked, Rotor is in operation and the variable RT_Work is *true*. While *RT_Work* is marked, RT can fail (i.e., critical failure) or degrade (i.e., incipient failure). RT fails when the transition *RT_Failure_AGAN-F* is enabled. Please note, AGAN means As-Good-As-New and F means Failure. Through the firing rule, *RT_Work* is unmarked, the token moves to the place *RT_Failed* and the variable *RT_Work* changes to *false*. The token stays at *RT_Failed* place until *RT_StartRepair2* becomes enabled, in other words, until a repair team and both Supply Vessel (SV) and Jack-Up vessel (JU) arrive to the offshore installation (i.e., *SV_OnBoard == true* and *JU_OnBoard == true*). During the repair activity, *RT_Repair2* place is marked. The duration of CM activity is introduced in the delay of *RT_FinishRepair2*. When this transition is enabled, it means the conclusion of CM repair, thus the token moves to *RT_Work*, the variable *RT_Work* changes to *true*, the variable *RT_Age* turns to zero, and the variable *RT_LastCM* records the time of completion of corrective repair.

The incipient failure occurs when the transition *RT_Failure_AGAN-D* is enabled. Through the firing rule, *RT_Work* is unmarked, and the token moves to the place *RT_Degraded*, the variable *RT_Degraded* changes to *true*, and the variable *RT_Work* remains *true*. When the Rotor is degraded, both transitions *RT_StartRepair1* and *RT_Failure_D-F* can fire, depending on the failure history of the component, because the first failure is repaired if it is degraded or critical and the next failure only if it is critical. Thus, if the Rotor is degraded and the previous failure was critical, *RT_StartRepair1* is enabled, the token moves to the place *RT_Repair1* and the corrective maintenance team is reserved. Moreover, if the Rotor is degraded and the previous failure was degraded, the transition *RT_Failure_D-F* is enabled, the token moves to the place *RT_Failed*, and the variable *RT_Degraded* turns *false*.

2.6.2. Total System Switch

Stochastic Petri Nets cannot simulate different operations at the same time. All actions must be sequential. Therefore, all components of the OWT are impossible to switch off simultaneously. Hence, the total system switch includes four different fail scenarios. Each scenario has the same order of equipment turning-off, starting with the unavailable component. The PN model of total system switch is presented in Figure 3.

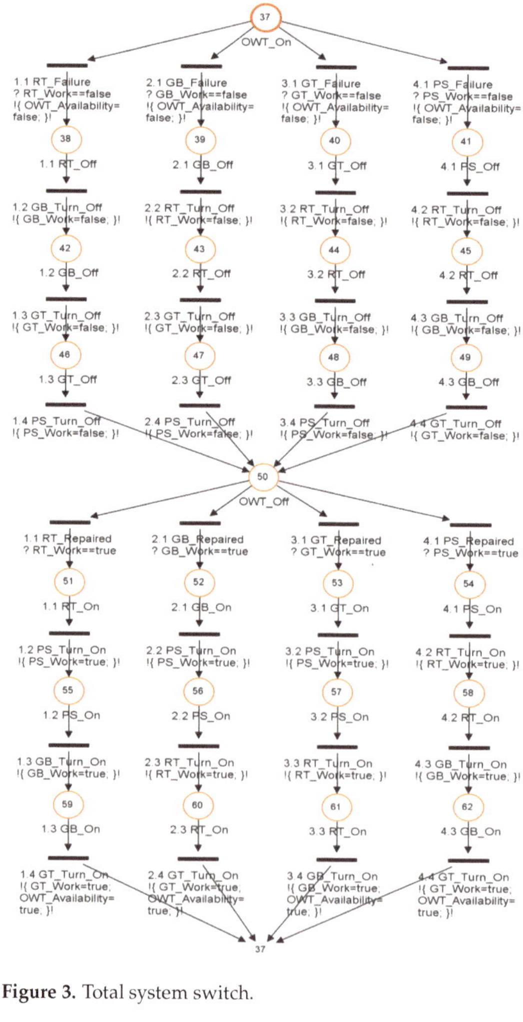

Figure 3. Total system switch.

Total system switch has two objectives: to calculate the availability of the total system, through the Boolean variable *OWT_Availability*, and to turn off offshore processing plant equipment when at least one important production component (i.e., RT, GB, GT, and PS) is not available, through the Boolean variable Equipment Abbreviation_Off. If the system is available, *OWT_Availability* is true, in the unavailable state, it is false. Equipment *Abbreviation_Off* is true only when the system component is shut down due to failure of another component.

2.6.3. Preventive Maintenance

The preventive maintenance is a scheduled activity. Regardless of the type, the generic PN model of PM repair is the same for every equipment. As an example, the simplified PN model of preventive maintenance activity of the RT is shown in Figure 4.

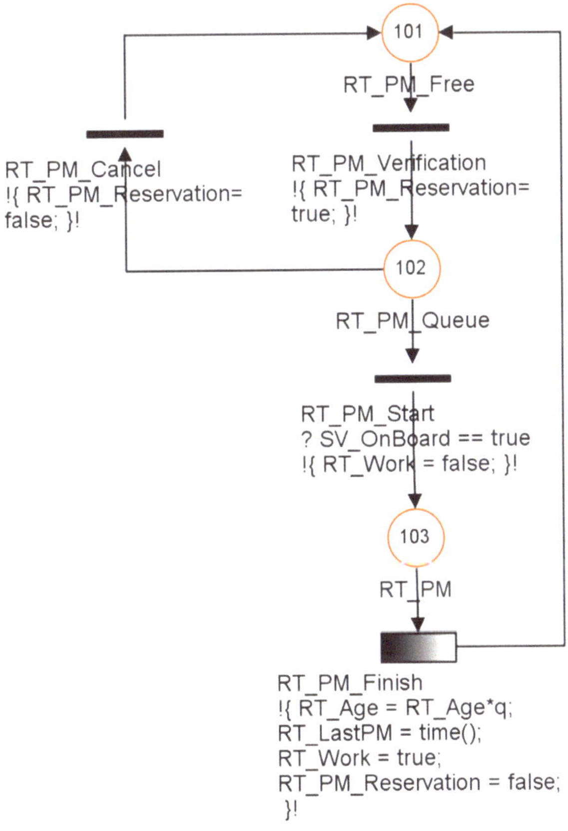

Figure 4. PN model of PM activity (Rotor).

The PN sub-model for PM activity is accompanied by variables: *RT_PM_Reservation* and *RT_LastPM*. *RT_PM_Reservation* is a Boolean variable that identifies with the *"true"* condition (i.e., *RT_PM_Reservation == true*) that the Rotor reserves the PM team for itself. *RT_LastPM* records the time of the completion of the last preventive repair.

When the place *RT_PM_Free* is marked, the equipment is waiting for preventive maintenance, thus, the variable *RT_PM_Reservation* is false. The main objective of the transition *RT_PM_Verification* is the validating of PM conditions. Namely, it must be a summer, the component's age must be more than $p \times MTTF_{rotor}$, and the equipment must be functional.

When the *RT_PM_Verification* is satisfied, the transition fires and the token moves to the place *RT_PM_Queue*. The variable *RT_PM_Reservation* turns true. At this instant, to proceed with the PM activity, the simulation verifies another two conditions. The first condition is represented by the transition *RT_PM_Start*, which is enabled only when the PM team arrives to the wind. The second condition (i.e., the transition *RT_PM_Cancel*) verifies whether the Rotor is still functional, while the turbine is waiting for the Supply Vessel coming. In the case of unexpected Rotor's fail, the variable *RT_Work* becomes false (i.e., *RT_Work == false*), therefore, the transition *RT_PM_Cancel* enables, and the PM is cancelled (i.e., *RT_PM_Reservation == false*).

When the first mentioned condition is met, the transition *RT_PM_Start* is enabled, the token moves from the place *RT_PM_Queue* to *RT_PM*. When the PM is concluded, the transition *RT_PM_Finish* is enabled, and the token moves to *RT_PM_Free*. After the transition enabling, the component's age is updated (i.e., $RT\_Age = RT\_Age \times q$), the variable *RT_LastPM* changes the registered time, and the Boolean variables return their values to the initial ones.

2.6.4. Seasons

Figure 5 shows the simplified PN model that identifies the seasons. The transitions between the seasons follow the Delta Dirac distribution. The delay of each transition corresponds to the time of three months, considering 30 days in each one. The simulation starts on the 1 December.

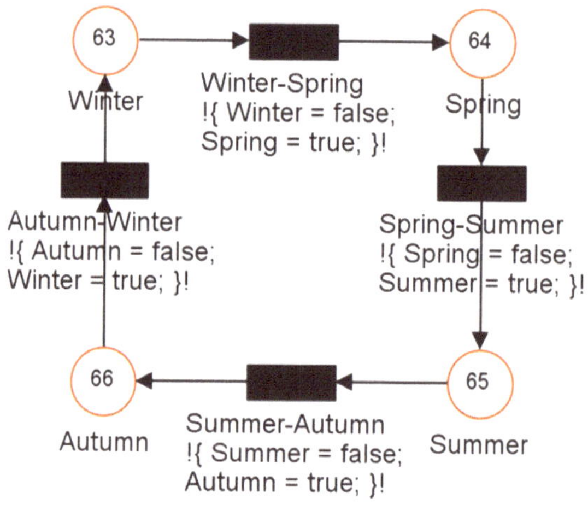

Figure 5. PN model for seasons.

2.6.5. Vessels

The maintenance policy of the OWT system consists of three different maintenance categories according to the vessel involved: Jack-Up vessel, Crane Barge, and Supply Vessel. In order to avoid an extensive description, only the simplified model of JU vessel is presented (see Figure 6).

To ensure a logical sequence, the PN model of JU vessel is accompanied by variables: *JU_Availability* and *JU_OnBoard*. *JU_Availability* is a binary variable. When *JU_Availability == true*, it identifies the time under which the Jack-Up vessel is localised at the port with no reservation order. Thus, while JU vessel is waiting for an available weather window, *JU_Availability* is equal to *false*, since the vessel is already reserved for a specific corrective maintenance task. *JU_OnBoard* is a binary variable, too. *JU_OnBoard* is used to identify the time interval when the JU vessel is localised at the OWT (i.e., *JU_OnBoard == true*).

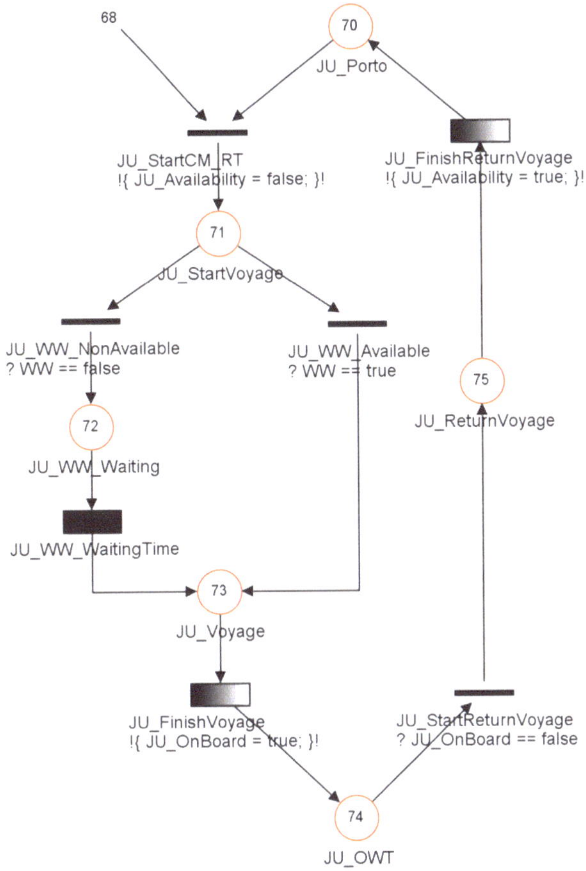

Figure 6. PN model of Jack-up vessel.

When the place *JU_Porto* is marked, the Jack-Up vessel is available at the port of Porto. Moreover, from the transition firing rule, the transition *JU_StartCM_RT* is enabled when both the place #68 (i.e., a new Rotor arrived at the port) and the place *JU_Porto* are marked. The main objective of this transition is to give a start for the CM activity. When the transition *JU_StartCM_RT* fires, the place *JU_Porto* is unmarked. Thus, from the firing rule, the place *JU_StartVoyage* is marked. Further, the variable *JU_Availability* turns *false*.

When the vessel is available to start the voyage (i.e., the place *JU_StartVoyage* is marked), the weather window must be validated. For this purpose, two transitions are used, namely: *JU_WW_NonAvailable* and *JU_WW_Available*. If the *WW* is available, the transition *JU_WW_Available* fires and the token moves from the place *JU_StartVoyage* to the place *JU_Voyage*. In the case of a non-available weather window, the token moves from the place *JU_StartVoyage* to the place *JU_WW_Waiting*. The token preserves at the new place until the enabling of the transition *JU_WW_WaitingTime*. After the enabling of *JU_WW_WaitingTime*, the token moves from the place *JU_WW_Waiting* to the place *JU_Voyage*.

The voyage time of the JU vessel is defined by the time delay of the transition *JU_FinishVoyage*. When the transition *JU_FinishVoyage* fires, the token moves from the place *JU_Voyage* to the place *JU_OWT*, changing the variable *JU_OnBoard* to *true*. When the CM activity is concluded, the variable *JU_OnBoard* is *false*, hence, the transition *JU_StartReturnVoyage* fires and the token moves from the place *JU_OWT* to the place *JU_ReturnVoyage*. The token at the place *JU_ReturnVoyage* means that the JU is in returning voyage. The duration of the voyage is defined by the time delay of the transition *JU_FinishReturnVoyage*. After the enabling of *JU_FinishReturnVoyage*, the token moves from the place *JU_ReturnVoyage* to the place *JU_Porto*. Moreover, the variable *JU_Availability* changes to *true*.

2.7. Economical Assessment

The efficiency of an age-based imperfect PM depends on two parameters: an age reduction ratio, q, and a repair threshold parameter, p. To assess optimal values of q and p, it is necessary to determine the lower related costs for the higher possible profit.

The O & M costs ($C_{O\&M}$) correspond to the sum of different features, such as: a new turbine's component cost, hourly rates, mobilization costs, and the cost of PM. To determine the operation and maintenance expenses (OME), the sojourn times and the number of triggers obtained from the implemented PN model are used. The sojourn time corresponds to the time during which the token is located at the *place* ## throughout the simulation time. Through the sojourn time, the identifying of the number of hours of each vessel (i.e., JU, CB, and SV) dispatched to the O & M activities and the PM duration record of each OWT's component (i.e., RT, GB, GT, and PS) are possible. The number of triggers corresponds to the total number of transition fires. Through the number of triggers, it is possible to identify the number of failures of each component, the number of mobilizations of each vessel, and the number of realised PM activities.

The variation of q and p influences the $C_{O\&M}$ and the total system availability (A_{system}), which in turn influences the revenue or Gross Income (GI):

$$GI = A_{system} \times P \times \eta \times T_{simulation} \times C_{PT} \tag{4}$$

where, P is an electric power in MW, η is a capacity factor of OWT, $T_{simulation}$ is a total simulation time in hours, and C_{PT} is a cost for wind energy produced by OWT in Portugal in EUR/MWh.

Knowing the value of $C_{O\&M}$ and GI, the profit or Operating Income (OI) is possible to determine:

$$OI = GI - C_{O\&M} \tag{5}$$

To assess an optimal value for q and p, it is necessary to minimise $C_{O\&M}$ and to maximise OI. Hereupon, the optimum q and p correspond to the highest value of Accounting Rate of Return (ARR):

$$ARR = \frac{OI}{C_{O\&M}} \tag{6}$$

3. Results

To perform the availability analysis of an OWT, the GRIF (Graphical Interface for reliability Forecasting) analysis software is used [58]. To simulate the PN model, GRIF uses MOCA-RP computation engine based on Monte Carlo simulation.

The simulated time of the base model is defined by iterations from instant 0 to instant 219,000 h (i.e., 25 years) with a step of 100 h for 1000 different scenarios (i.e., histories). The average error related to the 90% CI (Confidence Interval) of the model outcomes corresponding to the number of simulated histories is less than 0.04%.

3.1. Age-Based PM Parameters' Decision Based on Economical Assessment

To obtain the optimal values of q and p, variations $0 < q < 1$ and $0 < p \leq 1$ in steps of 0.1 are considered. Table 11 and Figure 7 present the availability of the total system for different q and p values. As can be observed, the higher the values of q and p, the greater the availability. Furthermore, the increase of q is more notable for low values of p. Thus, as expected, in terms of availability, the $q_{optimal}$ is 0.9 and the $p_{optimal}$ is 0.1.

Table 11. Availability of the total system for different q and p (in MEUR). Matrix representation.

		p									
		0.1	**0.2**	**0.3**	**0.4**	**0.5**	**0.6**	**0.7**	**0.8**	**0.9**	**1**
q	**0.1**	0.80	0.81	0.82	0.82	0.83	0.83	0.84	0.84	0.84	0.85
	0.2	0.81	0.81	0.82	0.82	0.83	0.83	0.84	0.84	0.84	0.85
	0.3	0.81	0.82	0.82	0.83	0.83	0.83	0.84	0.84	0.84	0.84
	0.4	0.82	0.83	0.83	0.83	0.83	0.84	0.84	0.84	0.84	0.85
	0.5	0.83	0.84	0.84	0.84	0.84	0.84	0.84	0.84	0.85	0.85
	0.6	0.84	0.84	0.84	0.85	0.85	0.84	0.85	0.85	0.85	0.85
	0.7	0.85	0.85	0.85	0.85	0.85	0.85	0.85	0.85	0.85	0.85
	0.8	0.86	0.86	0.86	0.86	0.86	0.85	0.85	0.85	0.85	0.85
	0.9	0.87	0.87	0.87	0.87	0.86	0.86	0.86	0.85	0.85	0.85

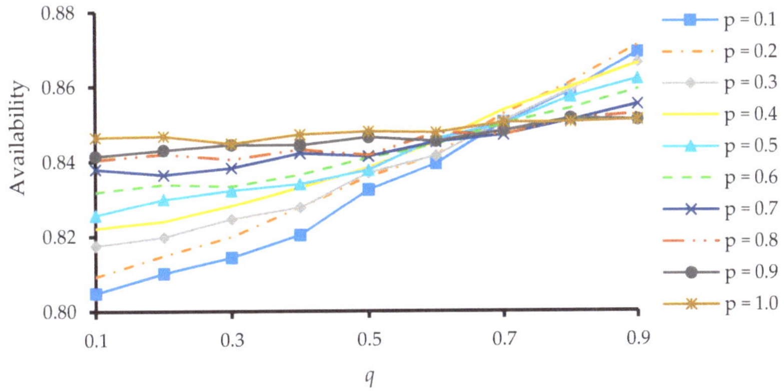

Figure 7. Availability of the total system for different q and p. Graphic representation.

Table 12 and Figure 8 show the O & M costs for different q and p values, considering the cost structure described in Section 2.4 and the results of sojourn times and of number of triggers obtained from PN model. Comparing the results from Table 11 with Table 12, it is concluded that the greater the availability, the greater the operation and maintenance costs.

Table 12. O & M costs for different q and p (in MEUR). Matrix representation.

		p									
		0.1	**0.2**	**0.3**	**0.4**	**0.5**	**0.6**	**0.7**	**0.8**	**0.9**	**1**
q	**0.1**	57.3	57.4	57.6	58.1	58.6	57.9	59.5	59.4	59.7	59.8
	0.2	56.5	56.8	57.1	57.6	58.1	58.1	59.2	59.3	59.5	59.8
	0.3	57.1	57	56.9	57.4	58	58.4	58.9	59.4	59.5	59.9
	0.4	58.9	58.2	57.9	57.8	58.2	58.7	59.2	59.6	59.6	59.7
	0.5	61.7	60.3	59.8	59.2	59	59.2	59.4	59.8	59.8	59.8
	0.6	66.0	63.9	62.4	61.5	60.8	60.3	60.1	59.9	60.0	60.3
	0.7	72.7	69.9	67.2	65.7	63.6	61.9	60.8	60.4	60.3	60.2
	0.8	81.7	78.1	74.3	71.3	66.8	63.2	61.7	61.1	60.5	60.2
	0.9	94.3	89.9	83.5	77.9	68.7	64.5	62.8	61.7	61.0	60.4

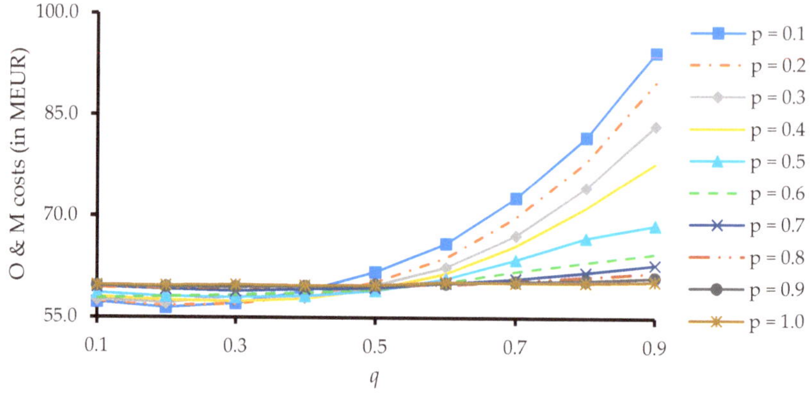

Figure 8. O & M costs for different q and p (in MEUR). Graphic representation.

Table 13 and Figure 9 show an Operating Income (OI) for different q and p values calculated by Equation (5). For $q > 0.7$, the operating income represents the financial loss. This means that the O & M costs are higher than the profit for high values of age reduction values. Despite the lower values of availability, the values of $q \leq 0.7$ lead to more profitable scenarios.

Table 13. Operating income for different q and p (in MEUR). Matrix representation.

		p									
		0.1	**0.2**	**0.3**	**0.4**	**0.5**	**0.6**	**0.7**	**0.8**	**0.9**	**1**
q	**0.1**	11.9	12.1	12.6	12.5	12.3	13.6	12.4	12.7	12.6	12.9
	0.2	13.1	13.2	13.3	13.2	13.2	13.5	12.7	13.0	12.9	13.0
	0.3	12.8	13.4	13.9	13.8	13.5	13.2	13.1	12.8	13.0	12.7
	0.4	11.6	12.9	13.2	13.7	13.4	13.2	13.1	12.8	12.9	13.0
	0.5	9.7	11.5	12.1	12.8	13.0	13.0	12.8	12.5	12.9	13.0
	0.6	6.1	8.4	9.8	11.1	11.9	12.3	12.5	12.9	12.6	12.5
	0.7	0.3	3.3	5.9	7.7	9.4	11.1	12.0	12.4	12.5	12.9
	0.8	−8.0	−4.2	−0.5	2.6	6.8	10.1	11.4	12.0	12.6	12.8
	0.9	−19.7	−15.1	−9.1	−3.5	5.3	9.3	10.6	11.5	12.1	12.7

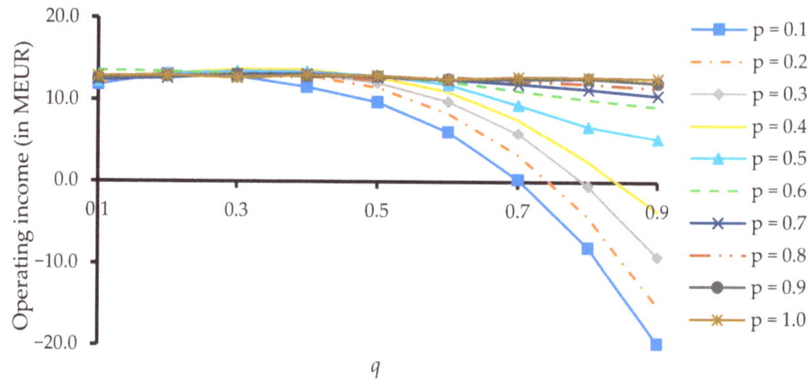

Figure 9. Operating income for different q and p (in MEUR). Graphic representation.

Finally, Table 14 and Figure 10 show the Accounting Rate of Return (ARR) calculated for different q and p values by dividing the operation income by the O & M costs. The higher $ARRs$ are within $0.2 \leq q \leq 0.4$ and $0.2 \leq p \leq 0.6$, where the maximum $ARR = 0.244$ corresponds to the optimal values of q and p. Hence, the lower related costs for the higher possible profit corresponds to $q_{optimal} = 0.3$ and $p_{optimal} = 0.3$.

Table 14. Accounting rate of return for different q and p. Matrix representation.

		p									
		0.1	**0.2**	**0.3**	**0.4**	**0.5**	**0.6**	**0.7**	**0.8**	**0.9**	**1**
q	**0.1**	0.207	0.21	0.218	0.216	0.21	0.235	0.209	0.214	0.211	0.216
	0.2	0.232	0.232	0.232	0.229	0.226	0.232	0.214	0.219	0.216	0.217
	0.3	0.225	0.235	0.244	0.240	0.232	0.226	0.222	0.215	0.219	0.212
	0.4	0.196	0.222	0.228	0.237	0.231	0.224	0.222	0.216	0.216	0.218
	0.5	0.158	0.191	0.202	0.216	0.22	0.22	0.216	0.209	0.215	0.217
	0.6	0.093	0.131	0.157	0.180	0.196	0.204	0.208	0.215	0.209	0.208
	0.7	0.004	0.048	0.088	0.117	0.148	0.180	0.197	0.205	0.208	0.214
	0.8	−0.098	−0.054	−0.007	0.036	0.102	0.160	0.184	0.197	0.209	0.213
	0.9	−0.209	−0.168	−0.109	−0.045	0.077	0.143	0.169	0.187	0.199	0.210

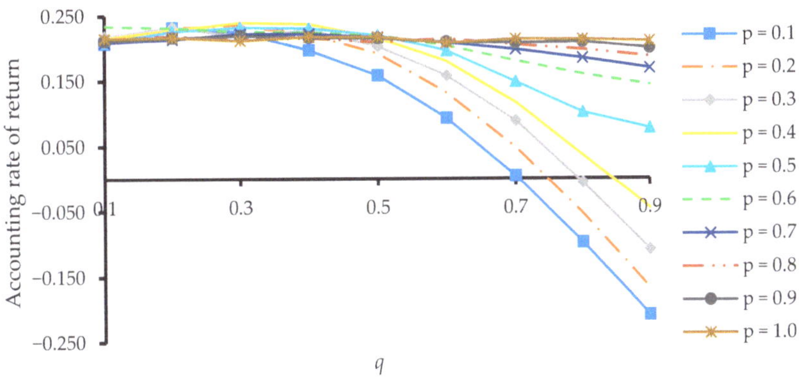

Figure 10. Accounting rate of return for different q and p. Graphic representation.

In Table 15, to validate the optimal values of q and p, the main system results for two maintenance strategies are presented, namely: CM only and age-based PM. As can be observed, the introduction of PM reduces the total system availability from 84.9% to 82.5%. However, from the economical point of view, the OWT production with PM is more profitable, as $ARR_{CM\ only} < ARR_{optimum\ age-based\ PM}$.

Table 15. Main system results for CM and optimum age-based PM.

Results	**CM Only**	**Age-Based PM with q = 0.3 and p = 0.3**
A_{system}	84.9%	82.5%
OME (MEUR)	59.9	56.9
OI (MEUR)	12.9	13.9
ARR	0.216	0.244

3.2. Sensitivity Analysis: OWT Availability

A sensitivity analysis is conducted to identify the parameters (i.e., input values) that significantly impact the production availability of the OWT. For this purpose, an elasticity factor is used, as:

$$E_{x_i} = \frac{|\Delta\, y_j|}{|\Delta\, x_i|} \tag{7}$$

where, x_i is the input value, and y_j is the output result.

The elasticity factor is a nondimensional measure defined by the ratio of the variation of OWT availability by 10% increase of input variable. The elasticity analysis is applied to the optimal scenario (i.e., $q_{optimal} = 0.3$ and $p_{optimal} = 0.3$). Each input value is analyzed separately. The analyzed input parameters are the component failure data described in Table 1 (i.e., shape parameters, MTTF, and failure rate); the voyage time of SB, JU, and SV vessels obtained from Tables 2 and 4; the logistic time carried out as a part of corrective maintenance summarized in Table 5 and the duration of replacement phase in Table 6; the duration of PM activity, Table 7; the availability of WW and the waiting time for a WW given in Table 10. The output is the OWT availability resulted from the input value change.

Figure 11 shows, in decreasing order of importance, the input parameters with the highest elasticity factor. In Table 16, the description of each parameter is presented. The obtained results show that the ranking of the input parameters on OWT availability are the failure rates of RT, GT, PS; probability of available weather window, the voyage time of SV and JU; shape parameter of RT.

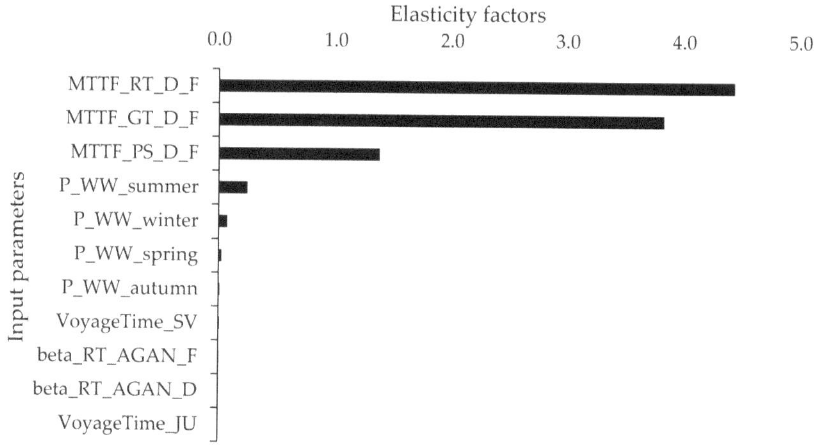

Figure 11. Elasticity factors of input parameters in terms of availability.

Table 16. Description of the input parameters.

Input Parameter	Description
MTTF_RT_D_F	Failure rate of Rotor from the degraded state to the failure
MTTF_GT_D_F	Failure rate of Generator from the degraded state to the failure
MTTF_PS_D_F	Failure rate of Pitch System from the degraded state to the failure
P_WW_summer	Probability of the weather window to be favorable in summer
P_WW_winter	Probability of the weather window to be favorable in winter
P_WW_spring	Probability of the weather window to be favorable in spring
P_WW_autumn	Probability of the weather window to be favorable in autumn
VoyageTime_SV	Total voyage time of Supply Vessel
beta_RT_AGAN_F	Shape parameter of Rotor from the as good as new state to the failure
beta_RT_AGAN_D	Shape parameter of Rotor from the as good as new state to the degraded
VoyageTime_JU	Total voyage time of Jack-Up vessel

3.3. Sensitivity Analysis: Accounting Rate of Return

A sensitivity analysis is also conducted to understand how the operational costs may affect the financial aspect of production. For this case, the elasticity factor is defined by the ratio of the variation of accounting rate of return by 10% increase of input costs. The elasticity analysis is applied to the optimal scenario (i.e., $q_{optimal} = 0.3$ and $p_{optimal} = 0.3$). Each input cost is analysed separately. The analysed input parameters are the cost of a new turbine's components (Table 8), the hourly rate of technician and vessel rental (Table 9), and the mobilization costs (Table 9).

Figure 12 shows in decreasing order of importance, the operation and maintenance expenses input parameters. The obtained results show that the most influential costs on ARR are the hourly rates of technician and vessel rental.

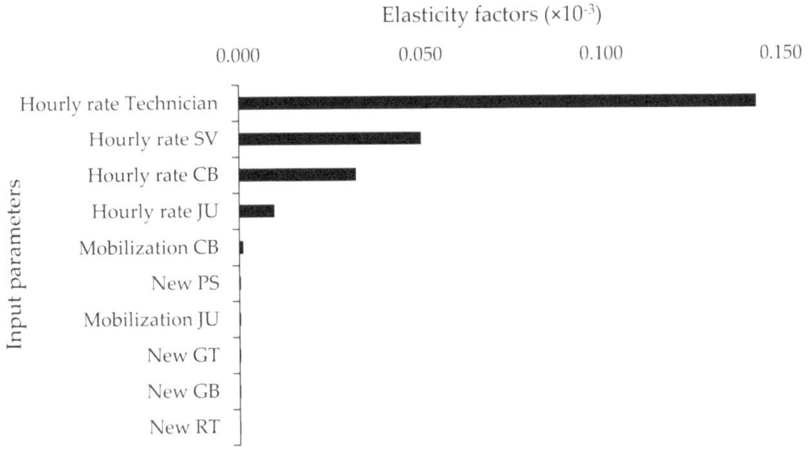

Figure 12. Elasticity factors of input parameters in terms of accounting rate of return.

3.4. Discussion on the Availability Assessment

Considering the logistics, the operation and maintenance activities presented in Section 2, an age reduction ratio, $q = 0.3$, and a repair threshold parameter, $p = 0.3$, the availability of the OWT system is 82.5%, which compares favourably with those obtained by other studies. Pfaffel et al. [59] showed that the availability of the onshore wind farm is around 95%, while of the offshore wind farm, it is around 88%. Scheu et al. [60] determined that the mean availability of a wind farm constituted by 80 OWT is 91%. In the SPARTA project, the availability is 93%, in the OWEZ wind farm, it is 80%, in the Stratch-off wind farm, it is 84% [61]. As one can see, some studies show that the production availability can be higher than 90%, however, it is important to note that these values refer to the availability of the wind farm with several wind turbines. This means that in the case of the critical failure at one wind turbine, the availability of that turbine will decrease, while the total availability of the system will not be significantly affected as the rest of the turbines are working. Moreover, the logistic of the maintenance activity for a single wind turbine is different from the entire offshore wind farm. In the case of wind farm, it is possible to combine different maintenance activities, thus, reducing the logistic and voyage time of a repair team. Hence, it is expected that the availability of the single wind turbine be slightly lower than of the wind farm.

The obtained 82.5% availability also reflects some conservative modelling assumptions that can be relaxed. The main reasons for low availabilities of OWT are: the low reliability values, the overestimation of restrictions for maintenance activities, and the non-availability of technical support vessels [62]. From the effectuated sensitivity assessment, it is possible to see that the MTTF of PS, GT, and RT represent the highest sensitivity factors. This means that the availability of OWT can increase for more recent and reliable components. In this study, the reliability parameters are based on empirical offshore reliability data from 2014. The second factor that decreases the availability is the underestimation of the probability of an available weather window. Due to the lack of WW data for Viana do Castelo wind farm, the conservative values were used. From the sensitivity analysis, the availability of WW in summer is the most influential factor if comparing with other seasons. However, the increase of WW availability in winter and spring can positively influence the OWT availability, too. The voyage time of technical support vessels (i.e., JU and SV) is also important to production availability increase.

Among the support vessels, the higher availability is provided by Jack-Up vessel (i.e., 97.5%), as JU is used only to transport the new Rotor from port to the OWT. From the base model results, the JU vessel is used 9.8 times in 25 years. The Crane Barge vessel has slightly lower availability (i.e., 92.6%). CB vessel is used to support the replacement of the Generator and Gearbox. From the base model results, the CB vessel is used 27.1 times in 25 years. The lowest availability belongs to the Supply Vessel (i.e., 82.9%). SV is a unique vessel that is used in all maintenance activities. Over 25 years, this vessel is used 105.8 times.

It is worthwhile to note that the repair maintenance team is unique and is used in both CM and PM actions. Moreover, the repair team always uses the Supply Vessel, thus, the availability of the repair team is equal to the availability of the Supply Vessel.

4. Conclusions

The main objective of this paper is to analyse the availability of an offshore wind turbine system subjected to an age-based preventive maintenance, considering the optimal age reduction ratio, q, and repair threshold parameter, p. For this purpose, the Generalized Stochastic Petri Nets with predicates coupled with the Monte Carlo Simulation method are used.

An economical assessment of the production availability and maintenance costs of the offshore wind turbine is performed to estimate the optimal values for q and p. The influence of q and p on the availability, O & M costs, operating income, and accounting rate of return are assessed. The higher availability occurs at the high values of q and at the low values of p. However, at this range of q and p, the O & M costs are very high, leading to negative incomes. Using the results obtained from the accounting rate of return, the optimal values for q and p are obtained. Hence, $q = 0.3$ and $p = 0.3$ correspond to the lower O & M costs and provide the higher possible profit.

A sensitivity analysis is conducted to identify the parameters (i.e., input values) that significantly impact the production availability of the OWT. The obtained results show that the ranking of the input parameters on OWT availability are the failure rates of RT, GT, PS; probability of available weather window, the voyage time of SV, and JU, and shape parameter of RT. Another sensitivity analysis is also conducted to understand how the operational costs may affect the financial aspect of production. The obtained results show that the most influential costs on accounting rate of return are the hourly rates of technician and vessel rental.

The simulation results show that the availability of the OWT is 0.825, which reflects some conservative model parameters, but is in line with the values obtained by other studies for offshore wind farms. The Jack-Up shows the higher availability between vessels: 0.975. The availability of Crane Barge vessel is 0.926, while the availability of Supply Vessel is 0.829. The availability analysis of the OWT adopted a Simple Place/Transition PN. This tool becomes difficult to read graphically as the complexity of the production system increases. Hence, it is recommended to use Coloured PN in further works, which facilitates the graphical representation.

To obtain more accurate estimates from the O & M model, real weather data for weather window implementation is recommended. Knowing the wind speed and the wave height at the OWT location, and specific operational limits of the different vessels, it is possible to calculate their availability more precisely. Moreover, real wind data together with the power performance curve of the OWT allows the assessment of the real power output of the system.

Author Contributions: Conceptualization, A.P.T. and C.G.S.; methodology, E.L. and A.P.T.; software, E.L.; validation, E.L.; formal analysis, E.L. and A.P.T.; investigation, E.L.; resources, A.P.T. and C.G.S.; data curation, A.P.T.; writing—original draft preparation, E.L.; writing—review and editing, A.P.T. and C.G.S.; visualization, E.L.; supervision, A.P.T. and C.G.S.; project administration, C.G.S.; funding acquisition, C.G.S. All authors have read and agreed to the published version of the manuscript.

Funding: This study was completed within the project ARCWIND—Adaptation and implementation of floating wind energy conversion technology for the Atlantic region, which is co-financed by the European Regional Development Fund through the Interreg Atlantic Area Programme under contract EAPA 344/2016. This work contributes to the Strategic Research Plan of the Centre for Marine Technology and Ocean Engineering (CENTEC), which is financed by the Portuguese Foundation for Science and Technology (Fundação para a Ciência e Tecnologia—FCT) under contract UIDB/UIDP/00134/2020.

Institutional Review Board Statement: Not applicable.

Informed Consent Statement: Not applicable.

Conflicts of Interest: The authors declare no conflict of interest.

Abbreviations/Nomenclature

ARR	Accounting Rate of Return
CB	Crane Barge
CI	Confidence Interval
CM	Corrective Maintenance
GB	Gearbox
GI	Gross Income
GRIF	Graphical Interface for reliability Forecasting
GSPN	Generalized Stochastic Petri Nets
GT	Generator
HAWT	Horizontal Axis Wind Turbine
JU	Jack-Up vessel
MCS	Monte Carlo Simulation
MTTF	Mean Time to Failure
MTTR	Mean Time to Repair
O & M	Operation and Maintenance
OI	Operating Income
OME	Operation and Maintenance Expenses
OWT	Offshore Wind Turbine
PM	Preventive Maintenance
PN	Petri Nets
PS	Pitch System
RT	Rotor
SD	Standard Deviation
SPN	Stochastic Petri Nets
SV	Supply Vessel
WW	Weather window

References

1. Díaz, H.; Guedes Soares, C. Review of the current status, technology and future trends of offshore wind farms. *Ocean Eng.* **2020**, *209*, 107381. [CrossRef]
2. Guedes Soares, C.; Caldeira Duarte, J.; Garbatov, Y.; Zio, E.; Sorensen, J.D. Framework for Maintenance Planning. In *Safety and Reliability of Industrial Products, Systems and Structures*; Guedes Soares, C., Ed.; Taylor & Francis Group: Chippenham, UK, 2010; pp. 33–52.
3. Zio, E.; Baraldi, P.; Patelli, E. Assessment of the availability of an offshore installation by Monte Carlo simulation. *Int. J. Press. Vessel. Pip.* **2006**, *83*, 312–320. [CrossRef]

4. Li, H.; Teixeira, A.P.; Guedes Soares, C. A two-stage Failure Mode and Effect Analysis of offshore wind turbines. *Renew. Energy* **2020**, *162*, 1438–1461. [CrossRef]
5. Li, H.; Diaz, H.; Guedes Soares, C. A developed failure mode and effect analysis for floating offshore wind turbine support structures. *Renew. Energy* **2021**, *164*, 133–145. [CrossRef]
6. Li, H.; Guedes Soares, C.; Huang, H.-Z. Reliability analysis of a floating offshore wind turbine using Bayesian Networks. *Ocean Eng.* **2020**, *217*, 107827. [CrossRef]
7. Dutuit, Y.; Châtelet, E.; Signoret, J.-P.; Thomas, P. Dependability modelling and evaluation by using stochastic Petri nets: Application to two test cases. *Reliab. Eng. Syst. Saf.* **1997**, *55*, 117–124. [CrossRef]
8. Kang, J.; Sun, L.; Guedes Soares, C. Fault Tree Analysis of floating offshore wind turbines. *Renew. Energy* **2019**, *133*, 1455–1467. [CrossRef]
9. Zio, E. Reliability engineering: Old problems and new challenges. *Reliab. Eng. Syst. Saf.* **2009**, *94*, 125–141. [CrossRef]
10. Signoret, J.-P. Production Availability. In *Safety and Reliability of Industrial Products, Systems and Structures*, 1st ed.; Guedes Soares, C., Ed.; Taylor & Francis Group: London, UK, 2010; pp. 331–345.
11. Byon, E.; Ntaimo, L.; Ding, Y. Optimal Maintenance Strategies for Wind Turbine Systems under Stochastic Weather Conditions. *IEEE Trans. Reliab.* **2010**, *59*, 393–404. [CrossRef]
12. Lewis, E. *Introduction to Reliability Engineering*, 2nd ed.; John Wiley & Sons: New York, NY, USA, 1994.
13. Signoret, J.-P. Availability of petroleum installations using Markov processes and Petri net modelling. In *Risk and Reliability in Marine Technology*; Guedes Soares, C., Ed.; Balkema: Rotterdam, The Netherlands, 1986; pp. 455–472.
14. Castanier, B.; Pehlivan, C.; Yeung, T.G. Optimization of maintenance and operational policies of an offshore wind farm subject to stochastic wind conditions. In *Safety and Reliability: Methodology and Applications*, 1st ed.; Nowakowski, T., Mlynczak, M., Jodejko-Pietruczuk, A., Werbinska-Wojciechowska, S., Eds.; Taylor & Francis Group: London, UK, 2014; pp. 1141–1146.
15. Li, M.; Kang, J.; Sun, L.; Wang, M. Development of Optimal Maintenance Policies for Offshore Wind Turbine Gearboxes Based on the Non-homogeneous Continuous-Time Markov Process. *J. Mar. Sci. Appl.* **2019**, *18*, 93–98. [CrossRef]
16. Bris, R. Evaluation of the production availability of an offshore installation by stochastic Petri nets modeling. *Int. Conf. Digit. Technol.* **2013**, 147–155. [CrossRef]
17. Santos, F.; Teixeira, A.P.; Guedes Soares, C. Modelling and simulation of the operation and maintenance of offshore wind turbines. *Proc. Inst. Mech. Eng. Part O J. Risk Reliab.* **2015**, *229*, 385–393. [CrossRef]
18. Zio, E.; Marella, M.; Podofillini, L. A Monte Carlo simulation approach to the availability assessment of multi-state systems with operational dependencies. *Reliab. Eng. Syst. Saf.* **2007**, *92*, 871–882. [CrossRef]
19. Barata, J.; Guedes Soares, C.; Marseguerra, M.; Zio, E. Simulation modelling of repairable multi-component deteriorating systems for 'on condition' maintenance optimisation. *Reliab. Eng. Syst. Saf.* **2002**, *76*, 255–264. [CrossRef]
20. Kang, J.; Sobral, J.; Guedes Soares, C. Review of Condition-Based Maintenance Strategies for Offshore Wind Energy. *J. Mar. Sci. Appl.* **2019**, *18*, 1–16. [CrossRef]
21. Kang, J.; Wang, Z.; Guedes Soares, C. Condition-Based Maintenance for Offshore Wind Turbines Based on Support Vector Machine. *Energies* **2020**, *13*, 3518. [CrossRef]
22. Santos, F.; Teixeira, A.P.; Guedes Soares, C. Production regularity assessment using stochastic Petri nets with predicates. In *Maritime Engineering and Technology*, 1st ed.; Guedes Soares, C., Garbatov, Y., Sutulo, S., Santos, T., Eds.; Taylor & Francis Group: London, UK, 2012; pp. 441–450.
23. Petri, C.A. Kommunikation mit Automaten. Doctors der Naturwissenschaften, Universität Bonn, Bonn. 20 June 1962. Available online: http://edoc.sub.uni-hamburg.de/informatik/volltexte/2011/160/ (accessed on 28 April 2022).
24. Boiteau, M.; Dutuit, Y.; Rauzy, A.; Signoret, J.-P. The AltaRica data-flow language in use: Modeling of production availability of a multi-state system. *Reliab. Eng. Syst. Saf.* **2006**, *91*, 747–755. [CrossRef]
25. Signoret, J.-P.; Dutuit, Y.; Cacheux, P.-J.; Folleau, C.; Collas, S.; Thomas, P. Make your Petri nets understandable: Reliability block diagrams driven Petri nets. *Reliab. Eng. Syst. Saf.* **2013**, *113*, 61–75. [CrossRef]
26. Marseguerra, M.; Zio, E. Basics of the Monte Carlo Method with Application to System Reliability. *Automatisierungstechnik* **2002**, *51*, 242.
27. Rausand, M. *Risk Assessment: Theory, Methods, and Applications*; John Wiley & Sons: Hoboken, NJ, USA, 2011.
28. Dutuit, Y.; Innal, F.; Rauzy, A.; Signoret, J.-P. Probabilistic assessments in relationship with safety integrity levels by using Fault Trees. *Reliab. Eng. Syst. Saf.* **2008**, *93*, 1867–1876. [CrossRef]
29. Santos, F.; Teixeira, A.P.; Guedes Soares, C. Maintenance Planning of an Offshore Wind Turbine Using Stochastic Petri Nets With Predicates. *J. Offshore Mech. Arct. Eng.* **2018**, *140*, 021904. [CrossRef]
30. Ding, F.; Tian, Z. Opportunistic maintenance for wind farms considering multi-level imperfect maintenance thresholds. *Renew. Energy* **2012**, *45*, 175–182. [CrossRef]
31. Santos, F.; Teixeira, A.P.; Guedes Soares, C. Influence of logistic strategies on the availability and maintenance costs of an offshore wind turbine. In *Safety, Reliability and Risk Analysis*; Taylor & Francis Group: London, UK, 2013; pp. 791–799. [CrossRef]
32. Santos, F.; Teixeira, A.P.; Guedes Soares, C. An age-based preventive maintenance for offshore wind turbines. In *Safety and Reliability: Methodology and Applications*; Nowakowski, T., Mlynczak, M., Jodejko-Pietruczuk, A., Werbinska-Wojciechowska, S., Eds.; Taylor & Francis Group: London, UK, 2014; pp. 1147–1155.

33. Lotovskyi, E.; Teixeira, A.P.; Guedes Soares, C. Availability analysis of an offshore oil and gas production system subjected to age-based preventive maintenance by Petri Nets. *Eksploat. Niezawodn. Maint. Reliab.* **2020**, *22*, 627–637. [CrossRef]
34. Santos, F.; Teixeira, A.P.; Guedes Soares, C. Modeling, simulation and optimization of maintenance cost aspects on multi-unit systems by stochastic Petri nets with predicates. *Simulation* **2018**, *95*, 461–478. [CrossRef]
35. Levitin, G.; Xing, L.; Dai, Y. Optimal loading of series parallel systems with arbitrary element time-to-failure and time-to-repair distributions. *Reliab. Eng. Syst. Saf.* **2017**, *164*, 34–44. [CrossRef]
36. Zhou, Y.; Zhang, Z.; Lin, T.R.; Ma, L. Maintenance optimisation of a multi-state series–parallel system considering economic dependence and state-dependent inspection intervals. *Reliab. Eng. Syst. Saf.* **2013**, *111*, 248–259. [CrossRef]
37. Zhou, Y.; Lin, T.R.; Sun, Y.; Bian, Y.; Ma, L. An effective approach to reducing strategy space for maintenance optimization of multistate series-parallel systems. *Reliab. Eng. Syst. Saf.* **2015**, *138*, 40–53. [CrossRef]
38. Kawauchi, Y.; Rausand, M. A new approach to production regularity assessment in the oil and chemical industries. *Reliab. Eng. Syst. Saf.* **2002**, *75*, 379–388. [CrossRef]
39. Bris, R.; Châtelet, E.; Yalaoui, F. New method to minimize the preventive maintenance cost of series–parallel systems. *Reliab. Eng. Syst. Saf.* **2003**, *82*, 247–255. [CrossRef]
40. Sobral, J.; Kang, J.C.; Guedes Soares, C. Weighting the influencing factors on offshore wind farms availability. In *Advances in Renewable Energies Offshore*; Guedes Soares, C., Ed.; Taylor & Francis Group: London, UK, 2019; pp. 761–769.
41. Kang, J.; Guedes Soares, C. An opportunistic maintenance policy for offshore wind farms. *Ocean Eng.* **2020**, *216*, 108075. [CrossRef]
42. Nielsen, J.J.; Sørensen, J.D. On risk-based operation and maintenance of offshore wind turbine components. *Reliab. Eng. Syst. Saf.* **2011**, *96*, 218–229. [CrossRef]
43. Castro-Santos, L.; Martins, E.; Guedes Soares, C. Methodology to Calculate the Costs of a Floating Offshore Renewable Energy Farm. *Energies* **2016**, *9*, 324. [CrossRef]
44. Martins, D.; Gangadharan, M.; Guedes Soares, C. Analysis on weather windows conditioned by significant wave height and wind speed. In *Renewable Energies Offshore*; Guedes Soares, C., Ed.; Taylor & Francis Group: London, UK, 2015; pp. 91–98.
45. Castro-Santos, L.; Silva, D.; Bento, A.R.; Salvação, N.; Guedes Soares, C. Economic feasibility of floating offshore wind farms in Portugal. *Ocean Eng.* **2020**, *207*, 107393. [CrossRef]
46. Castro-Santos, L.; Martins, E.; Guedes Soares, C. Economic comparison of technological alternatives to harness offshore wind and wave energies. *Energy* **2017**, *140*, 1121–1130. [CrossRef]
47. Kang, J.C.; Sun, L.P.; Lu, Y.; Sobral, J. An opportunistic condition-based maintenance policy for offshore wind farm. In *Advances in Renewable Energies Offshore*; Guedes Soares, C., Ed.; Taylor & Francis Group: London, UK, 2019; pp. 753–760.
48. Rademakers, L.; Braam, H. O&M Aspects of the 500 MW Offshore Wind Farm at NL7 (80 $\times$ 6 MW Turbines)—Baseline Configuration. Report for Dutch Offshore Wind Energy Converter (DOWEC). Report no. 10080 rev 2. 2002. Available online: https://www.researchgate.net/publication/265084960_OM_aspects_of_the_500_MW_offshore_wind_farm_at_NL7 (accessed on 30 May 2022).
49. Reisig, W.; Rozenberg, G. *Lectures on Petri Nets: Advances in Petri Nets*; Springer: Berlin/Heidelberg, Germany, 1998.
50. Murata, T. Petri Nets: Properties, Analysis and Applications. *Proc. IEEE.* **1989**, *77*, 541–580. [CrossRef]
51. Peterson, J.L. *Petri Net Theory and the Modeling of Systems*; Prentice-Hall: Englewood Cliffs, NJ, USA, 1981.
52. Marsan, M.A.; Balbo, G.; Conte, G.; Donatelli, S.; Franceschinis, G. *Modelling with Generalized Stochastic Petri Nets*, 1st ed.; John Wiley & Sons, inc.: New York, USA, 1995.
53. Salvação, N.; Guedes Soares, C. Wind resource assessment offshore the Atlantic Iberian coast with the WRF model. *Energy* **2017**, *145*, 276–287. [CrossRef]
54. Silva, D.; Bento, A.R.; Martinho, P.; Guedes Soares, C. High resolution local wave energy modelling in the Iberian Peninsula. *Energy* **2015**, *91*, 1099–1112. [CrossRef]
55. Official Site of Lindo Offshore Renewables Center (LORC). Available online: http://www.lorc.dk/ (accessed on 28 April 2022).
56. Official Site of EDP Renováveis. Available online: https://www.edpr.com/en (accessed on 28 April 2022).
57. BVG Associates. *Guide to an Offshore Wind Farm*; The Crown Estate: London, UK, 2019.
58. TOTAL. GRIF 2021. Petri Nets with Predicates. 2018. Available online: https://file.team9.satodev.fr/public/COM/GRIF2021/Doc/grif-2021-doc-en-petri12.pdf (accessed on 28 April 2022).
59. Pfaffel, S.; Faulstich, S.; Rohrig, K. Performance and Reliability of Wind Turbines: A Review. *Energies* **2017**, *10*, 1904. [CrossRef]
60. Scheu, M.N.; Kolios, A.; Fischer, T.; Brennan, F. Influence of statistical uncertainty of component reliability estimations on offshore wind farm availability. *Reliab. Eng. Syst. Saf.* **2017**, *168*, 28–39. [CrossRef]
61. Cevasco, D.; Koukoura, S.; Kolios, A.J. Reliability, availability, maintainability data review for the identification of trends in offshore wind energy applications. *Renew. Sustain. Energy Rev.* **2021**, *136*, 110414. [CrossRef]
62. Koukoura, S.; Scheu, M.N.; Kolios, A. Influence of extended potential-to-functional failure intervals through condition monitoring systems on offshore wind turbine availability. *Reliab. Eng. Syst. Saf.* **2021**, *208*, 107404. [CrossRef]

Journal of
Marine Science and Engineering

MDPI

Article

Service Operation Vessels Fleet Size and Mix Location Routing for the Maintenance of an Offshore Floating Wind Farm

Lis M. R. Silva, He Li and C. Guedes Soares *

Centre for Marine Technology and Ocean Engineering (CENTEC), Instituto Superior Técnico, Universidade de Lisboa, Av. Rovisco Pais, 1049-001 Lisbon, Portugal
* Correspondence: c.guedes.soares@centec.tecnico.ulisboa.pt

Abstract: Mixed-integer linear programming is adopted to translate the routing of service operation vessels that support the logistic aspects of the maintenance of offshore floating wind farms into mathematical language. The models attempt to help the decision-makers by providing quantified tools to screen out the optimal planning for preventive maintenance. The models search for the optimal offshore base location, vessel's routing per day, vessel's capacity, and vessel fleet composition that minimize the total fixed and variable infrastructure cost. The integration of the vehicle fleet size and mix problem, facility location–allocation problem, and vehicle-routing problem with time window advances the state of the art. A realistic case study is shown, and the results and discussions demonstrated that the practical insights of the solutions, as well as the identification of the route patterns through a navigation route table, may improve the decision planning of preventive maintenance.

Keywords: floating wind turbines; preventive maintenance; operations and maintenance activities; fleet size and mix location vehicle-routing problem; mixed-integer linear programming

Citation: Silva, L.M.R.; Li, H.; Guedes Soares, C. Service Operation Vessels Fleet Size and Mix Location Routing for the Maintenance of an Offshore Floating Wind Farm. *J. Mar. Sci. Eng.* **2023**, *11*, 664. https://doi.org/10.3390/jmse11030664

Academic Editor: Puyang Zhang

Received: 21 February 2023
Revised: 13 March 2023
Accepted: 16 March 2023
Published: 21 March 2023

1. Introduction

The energy generated by wind power has recently become one of the pillars to change the climate panorama presented nowadays due to it being generated by a clear and renewable source. In the search for locations that can reach a higher and more constant speed, the wind farm locations are changing from onshore to offshore, which increases the complexity of the maintenance planning. The wind conditions improve the unit time production but, also affect the maintenance planning because it may not be safe to perform maintenance under severe wind conditions [1]. Another example is the adoption of vessels to support operations and maintenance activities (O&M). The vessels are responsible for transferring technicians and delivering spare parts required to perform maintenance tasks. Different aspects of the maintenance strategies have been inducing the need for quantitative decision support models through the years, aiming at improving the planning process of the vessel's activities most efficiently and expeditiously [2].

The vehicle routing problem (VRP) is a well-studied topic applied in the logistic transportation sector. However, the variety of extensions and variations of the VRP arising in the offshore wind industry is a recent research topic. One of the significant studies that address the vehicle-routing and scheduling problem (VRSP) is presented by [3], which consists of defining a set of routes and scheduling the tasks to be accomplished by the vessels. The work presents arc-flow and path-flow formulations and a heuristic algorithm. Heuristic algorithms that have the similarity of extending the VRSP to include the heterogeneous fleet, multi-period time, multi-base locations, and multi-windfarms were developed by [4,5]. The main difference is that the heuristic presented by [4] also integrates the uncertainty of the weather conditions. The vehicle-scheduling problem (VSP) is another variation of the VRP. A mixed-integer linear programming (MILP) model is developed by [6], where the operational tasks are scheduled for a set of weather and breakdown scenarios.

The problem of searching for the minimum cost of a fleet of vessels and infrastructures to execute all the maintenance operations during the planning horizon is called the vehicle fleet size and mix problem (FSMP). Due to the uncertain nature of the chartering vessels during an established period, a stochastic programming model is an efficient approach adopted in the literature to solve the respective problem. This optimization technique is adopted by [7–9] to handle the uncertainties in economic aspects, operational process, demand, weather conditions and failure occurrence, respectively. Moreover, the pioneering work that studied the aspects of one VFSMP variation is [10], named fleet size and mix scheduling problem (FMSSP), in the offshore wind industry. The problem consists of defining the number of vessels necessary to accomplish the maintenance tasks and the respective scheduling plan for each vessel.

From the research work presented above, it can be noted that there is no restriction in the models regarding the travel time of the vessels to perform the routing and/or scheduling tasks. It is possible to argue that the studies are related to offshore wind farms located near shore (coastal), in shallow waters. But, nowadays, the wind industry is looking for sea locations that are not restricted by other sea activities or present even higher and constant wind speeds [11]. These locations are increasing the complexity of the wind farm system, as well as impacting the type of offshore wind turbine structure, changing the bottom-fixed turbines to floating offshore turbines. Accordingly, the distance from shore brings another limitation, that is, the vessel's travel time, which increases the logistic cost. New logistic strategies are being developed aiming at minimizing both time and cost. One idea is to establish an offshore base to supply the spare parts and accommodate the technicians.

The recent research works of [12,13] developed optimization models to study the optimal location-routing of the vessels (LRP). The location-routing problem consists of determining the optimal placement for one or more facilities and searching for the minimum facility cost and transportation. Both studies considered the service operation vessels (SOV) as the offshore base and the safe transfer boat (STB) as the daughter vessels to be used to transfer the technicians and spare parts. But the STB presents a smaller load capacity to carry out the supplies and a lower threshold for wind speed and wave height. The adoption and the crew transfer vessels (CTV) as daughter vessels of the SOV to increase the on-time and on-demand services is studied by [14]. The proposed model allows both vessels (SOV and CTV) to perform routes separately or combined. Nonetheless, the work does not determine the standby location of the SOV. Thus, none of the works in the presented in the literature described determines the necessary number of vessels to accomplish the new logistic.

The present paper adopts the CTV as the daughter's vessel, to perform the routes, and the SOV as the offshore base, having required technicians and spare parts during the planning horizon and provides mathematical decision models that integrate three main aspects of the logistic transportation sector: the location of the deposit, the fleet size of vehicles, and the respective routes for each vehicle. The main contributions of the present paper are:

1. The integration of the vehicle fleet size and mix problem, facility location-allocation problem and vehicle routing problem with time window searching for the optimal offshore floating wind activities planning.
2. The development of two mathematical models as rapid quantified screening tools to support the decision-makers to look for different aspects of an offshore floating wind farm.

2. Materials and Methods

2.1. Problem Description

The expansion of wind farms to deep waters and transitioning the offshore turbines to float concepts have been presenting a greater impact on the overall maintenance cost due to the unpredictable weather of the remote locations, higher failure occurrence because of the harsh marine environment, and extra inventory expenses to guarantee the transportation of spare parts and technicians in larger distances [15]. Besides the wind farm attributes, as well as the total number of turbines and their respective locations during a specific period, the O&M section prepares a list of turbines requiring maintenance. The different type of vessels presents different capacity for transferring technicians, load capacity to carry out the spare parts, sailing speed, the threshold for wind speed and wave height, which impacts the type of maintenance tasks each vessel can perform. Thus, the vessels are associated with a specific supply depot. The offshore wind farm distance from the onshore supply depot affects the navigation time. Attempting to increase the time that the vessels can be used, the supply base can be located offshore. In the case of an offshore base, the vessels can act as a servicing station having required technicians and daughter crafts. The vessels' voyage to supply the offshore wind farms needs to fit on the technicians' work shift performing a round trip visit to the offshore wind farm and their base. Figure 1 presents the comparison between the onshore and offshore supply base and the routing illustration. Therefore, the main decisions to be made are:

1. Where are the optimal locations of the offshore bases aiming at minimizing the logistic infrastructure and transportation costs?
2. How many of each type of vessel is necessary to attend the offshore wind farm?
3. What are the optimal routes for each vessel to deliver and pick up technicians and/or spare parts required to perform the maintenance tasks at each turbine during the planning horizon?

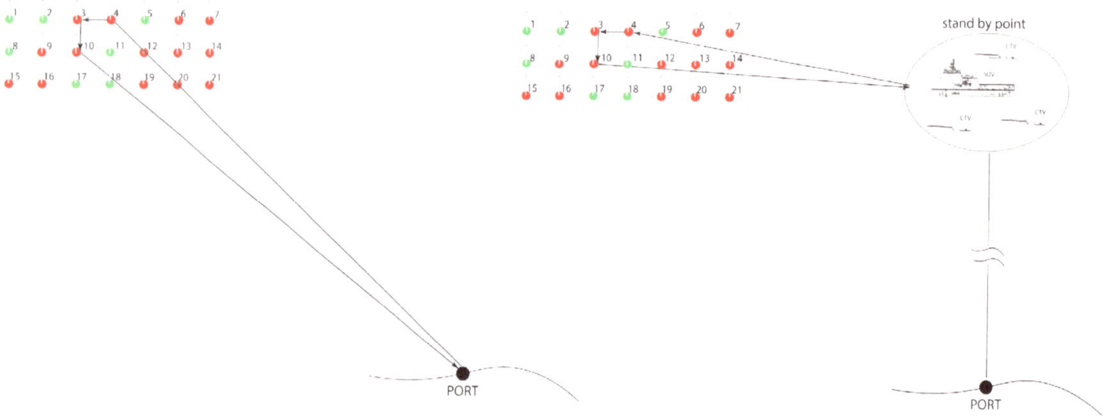

Figure 1. Onshore or offshore supply base location and the routing illustration. The red dots represent the turbines that are stopped waiting for maintenance.

2.2. Operation and Maintenance Activities

All the necessary efforts to maintain the daily operations of an offshore wind farm, including regular and irregular repairs, are part of the operation and maintenance activities [16]. A classification of the offshore floating wind turbine in modules is presented by [17] aiming at the identification of the failures of floating offshore wind turbines and their main components. Besides the reliability-centred maintenance approach, increasing attention is paid to condition-based maintenance (CBM) [18,19]. The framework based the maintenance on the health features of crucial wind turbine components, as determined by their failure rates [20]. A typical policy of the CBM is to perform maintenance when the pre-specified control limits of the degradation are reached.

The identification of the criticality of the components allows preventive measures to avoid the occurrence of failure and suggest corrective actions. The preventive and corrective maintenance tasks are two different activities to be performed at the wind turbine. The main characteristics of preventive maintenance (PM) are:

1. The PM is done every year for all the turbines.
2. The technicians need to spend a work shift per day per turbine.

3. A small crew is needed for one turbine maintenance.
4. Only crew transportation is needed, meaning that there is no requirement for the vessels to stay near the turbine.
5. PM is scheduled during the period with good weather conditions. It is based on statistical indicators that forecast when would be the most likely time window for PM.

If the weather is not favourable, it can be postponed but should be performed. Thus, the main characteristics of corrective maintenance (CM) are:

1. The CM need to be performed when the failures occur, attempting to reduce the downtime of the turbines.
2. Due to the first assumption, it can be performed during the whole year, which can suffer more impact by the weather conditions.
3. For each type of component failure, there is an associated mean time to repair, the specific number of technicians, and the vessel requirement to stay near the turbines.
4. Different types of vessels can perform different maintenance tasks (technicians transport and/or spare parts transport).

Due to the complexity, dynamic and uncertain nature of corrective maintenance, the present work considers only preventive maintenance logistics.

2.3. Assumptions and Concept Models

The functional description of the main system components allows the presentation and identification of the logistic network to be studied in more detail. The paper examines the preventive maintenance activities planning where the wind farm base is located offshore. The main logistic components of moving technicians to perform minor repairs are offshore turbine, technicians' demand, facilities (Port, SOV), vessels (CTV), and routes. The main function and the explicit assumptions of the system components are summarized as follows.

The offshore wind turbine is, basically, an assembly of components able to transform the wind force into electricity that is transferred to the offshore and onshore substations by cables. The size of the offshore turbine located at the same wind farm is similar and directly dependent on the amount of electricity production. This characteristic mutually interacts with the number of technicians to repair the turbines.

1. The size and capacity of electricity generation are beyond the scope of the paper.
2. The estimated number of turbines and their respective locations (represented as coordinates points in the x–y axis) are pre-defined data.
3. The number of technicians to perform the preventive maintenance repair in each turbine is pre-defined data.

The facilities of the system are the port and SOV. The port is an onshore supply base able to support the operation, manufacturing, and construction of an offshore wind farm and can house the vessels and handle the turbine components and technicians. As described above, the service operations vessel (SOV) is adopted in the present work not only as a supply vehicle used to transport materials and technicians to the offshore wind farm location but also to stay positioned at sea for at least one week, accommodating the technicians and serving as a warehouse to store the necessary spare part during the week planning horizon.

4. The facilities (Port, SOV, CTV) capacities are standardized and cannot be exceeded.

The crew transportation vessel (CTV) is adopted to transfer the technicians from the SOV to the wind turbines and vice-versa. One important parameter that needs to be considered as a fundamental concept is the vessel's autonomy which can be expressed according to the volume of fuel, navigation time or distance. The estimation of the available weight that one vessel can transport into a specific route is another important criterion. The CTVs need to perform two types of voyages on the same day. The delivery transfer, in the morning, to drop off the technicians at the turbines and the pick-up transfer in the evening, to collect the technicians at the end of the preventive task. In other words, the SOV is established as an offshore supply base and the CTV as a transportation vessel.

5. The maximum navigation time of each CTV needs to fit the technician's work shift per day.
6. In other to simplify, the calculation of the load factor of each vessel, it is considered that the vessel can carry the full capacity of technicians declared in each model specification sheet.
7. The voyage of each CTV is a round trip that starts and ends at the offshore base per day.

The logistic system described is considered a combinatorial optimization and NP-hard problem, which, accordingly, increases the complexity of integrating into the same model the facility location-allocation problem and the fleet size and mix vehicle routing problem with time window. In this regard, the present paper presents two models aiming at searching for the optimal solutions of these same logistic aspects in a feasible computational time frame. The first model (facility location-allocation model) looks for the optimal placement of the facilities (Port and SOV) that can minimize the infrastructure costs and service costs of the crew transfer vessels to attend each turbine. For this model, the optimal routing is neglected because the main goal is where to locate the offshore base trading off the distances from the onshore base and the turbines. However, the second model (fleet size and mix vehicle-routing with time window model) searches for the optimal sequence of turbines of each crew transfer vessel to deliver and pick up the technicians, also the optimal number of vessels that is necessary to attend the entire wind farm during the preventive period. The solution provides an optimal navigation route table per day and the vessel fleet composition to accommodate the preventive maintenance task. Table 1 schematically presents both conceptual models to provide the main difference between them, and the nomenclature list is presented in the Abbreviations section.

Table 1. Conceptual models.

Models	Facility Location–Allocation Model	Fleet Size and Mix Vehicle-Routing with Time Window Model
Given	The preventive period (in days); the technician's demand per turbine per day; the capacity of the facilities available; the SOV's rental cost; the CTV's variable cost; the CTV's average speed; the CTV's fuel cost per litre; the maximum number of turbines; the turbine's location; maximum inflow of technicians into the SOV; volume of personnel available to work; a large number.	The preventive period (in days); the technician's demand per turbine per day; the capacity of the facilities available; the SOV's rental cost; the CTV's rental cost; the CTV's variable cost; the CTV's average speed; the CTV's fuel cost per litre; the turbine's location; CTV's maximum operational time; maximum allowed passenger flow between vertexes.
Obtain	The distance between the turbines and facilities available (SOVs and ports).	The distance between the turbines and facilities available (SOVs).
Subject to	The turbine assignment constraints; port installation constraints; SOV installation and assignment constraints; technicians' flow constraints.	The vehicle routing constraints; technicians' flow constraints; time window constraints; fleet size constraints.
At minimum	SOV's fixed cost and CTV's variable cost.	SOV's and CTV's fixed costs and CTV's variable cost.

It is important to highlight the impact of the constraint groups on the optimization problem. For the present system, all the modelled constraints have medium-high or high impact on the system modelled. The only medium-high impact constraints are the port installation constraints because, for this system, it is not considered the search for the optimal onshore base location. So, the port installation constraints only assure that the flows of technicians, materials, and vessels from shore to offshore are conserved. However, the other constraints present a high impact on the modelled system.

The turbine assignment constraints ensure that the turbines of the system represent the exact number of turbines planned to be attended, which provides a high impact on the capacity of the offshore base and on the number of vessels designed to perform the maintenance. The technicians' flow constraints guarantee that the outgoing flow from the port to the offshore base and from the offshore base to each turbine is equal to the incoming flow at each node. The vehicle routing constraints ensure that each route is performed by only one vessel and each vessel starts the route at the offshore base, navigates to the maximum allowed turbines per route, which do not violate the vessel's capacity or voyage time and returns to the offshore base. The time window constraints directly impact the voyage time, which is a route restriction due to the technicians' work shifts. Thus, this restriction also impacts the number of transfer vessels to deliver and pick up the technicians. Moreover, the fleet size constraints are also very important for the modelled system because they determine the number of vessels required for the maintenance logistic.

A Mixed-Integer Linear Programming (MILP) formulation is adopted to translate the problems from theory to mathematical language. It means that the objective function is linear and is subject to a set of linear constraints, but the subset of variables takes binary and continuous values. The vehicle routing model with a time window is formulated as an Arc-Flow formulation, which is established on a weighted graph, G = (V, A), representing a transportation network. The nodes are represented as I = {1, j, l, . . . , i }, which is a set of vertexes and the set of arcs.

2.4. Facility Location-Allocation Model

The objective function follows:

$$\text{minimize TotalCost} = \{\text{SOVFixedCost} + \text{CTVVariableCost}\} \tag{1}$$

where

$$SOVFixedCost = \sum_t \sum_s \sum_z y_{tsz} \times SOVRC_{sz} \tag{2}$$

$$\begin{aligned} CTVVariableCost &= \left(\sum_t \sum_s \sum_d \left(\frac{DISTA_{tsd}}{CTVSPEED} \right) \times FUELCOST \times CTVFCC \times x_{tsd} \right) \\ &+ \left(\sum_t \sum_s \sum_d \left(\frac{DISTB_{tsd}}{CTVSPEED} \right) \times FUELCOST \times CTVFCC \times x_{tps} \right) \end{aligned} \tag{3}$$

The turbines assignment constraints' group is represented by Equations (4) and (5). Equation (4) guarantees that each turbine is assigned to one offshore base:

$$\sum_{s \in S} x_{tsd} \le 1 \qquad \forall\, d \in D, t \in T \tag{4}$$

Thus, Equation (5) ensures that the total assignments of turbines cannot exceed the maximum number of turbines:

$$\sum_{s \in S} \sum_{d \in D} x_{tsd} = \text{MAXTURBINES} \qquad \forall\, t \in T \tag{5}$$

The port installation constraints assure that an onshore base must be adopted to support the offshore operation and maintenance activities:

$$\sum_{p \in P} y_p \le 1 \tag{6}$$

The SOV installation and assignment constraints' group follows:

$$\sum_{S \in S} y_{tsz} \le 1 \qquad \forall\, s \in S, t \in T \tag{7}$$

$$\sum_{p \in P} x_{tps} \le 1 \qquad \forall\, s \in S, t \in T \tag{8}$$

$$y_{tsz} \le \sum_{p \in P} x_{tps} \qquad \forall\, s \in S, t \in T, z \in Z \tag{9}$$

$$\sum_{s \in S} x_{tps} \le \sum_{s \in S} \text{SOVCAP}_{sz} \times y_{tsz} \qquad \forall\, s \in S, t \in T \tag{10}$$

$$\sum_{s \in S} x_{tps} \le BIGM \times y_p \qquad \forall p \in P, t \in T \tag{11}$$

Equation (7) establishes that each SOV can only be at most of one size. Equation (8) guarantees that each SOV is assigned to one port. Thus, Equation (9) ensures that if the SOV is installed, it is connected to only one port. The number of turbines assigned to each SOV cannot exceed the SOV capacity, Equation (10). Equation (11) ensures that if the port is assigned to the SOV, the port is installed.

The technician's flow constraints group will fulfil the movement of the passenger's demand, as follows:

$$p_{ts} = \sum_{d \in D} x_{tsd} \times J_{td} \qquad \forall s \in S, t \in T \tag{12}$$

$$q_{tp} \le \sum_{s \in S} r_{tps} \qquad \forall p \in P, t \in T \tag{13}$$

$$-p_{ts} + y_{tsz} \le 0 \qquad \forall s \in S, p \in P, t \in T \tag{14}$$

$$-\left(x_{tps} \times PAXMAX\right) + y_{tsz} \le 0 \qquad \forall s \in S, p \in P, t \in T \tag{15}$$

$$p_{ts} + \left(x_{tps} \times PAXMAX\right) - y_{tsz} \le PAXMAX \qquad \forall s \in S, p \in P, t \in T \tag{16}$$

$$\sum_{s \in S} r_{tps} \le PORTCAP_p \times y_p \qquad \forall p \in P, t \in T \tag{17}$$

The total number of technicians transported to the SOV is the summation of the worker's demand for each turbine, Equation (12). Thus, the total flow of technicians at the port is the summation of the technicians' flow from the SOV to the port, Equations (13)–(16), and the flow of technicians cannot exceed the port capacity, Equation (17).

2.5. Vehicle-Routing Model with Time Window

The objective function follows:

$$\text{minimize TotalCost} = \{\text{SOVFixedCost} + \text{CTVFIXEDCOST} + \text{CTVVariableCost}\} \tag{18}$$

where,

$$\text{SOVFixedCost} = \sum_{s}\sum_{z} y_{tsz} \times SOVRC_{sz} \tag{19}$$

$$\text{CTVFixedCost} = \sum_{t}\sum_{m} CTVRC_m \times n_{tm} \tag{20}$$

$$\text{CTVVariableCost} = \left(\sum_{t}\sum_{m}\sum_{v}\sum_{i}\sum_{j}\left(\frac{DIST_{ij}}{CTVSPEED_m}\right) \times FUELCOST \times CTVFCC_{mv} \times x_{tijmv}\right) \tag{21}$$

The vehicle routing constraints group attempt to ensure the feasibility of the routes:

$$\sum_{i \in I}\sum_{m \in M}\sum_{v \in V} x_{tijmv} \le 1 \qquad \forall t \in T, i \in I, j \in I \mid j > 1 \tag{22}$$

$$\sum_{i \in I} x_{tijmv} - \sum_{l \in I} x_{tjlmv} = 0 \qquad \forall t \in T, m \in M, j \in I, v \in V \tag{23}$$

$$u_{ti} - u_{tj} + \left(MAXVERTEX \times x_{tijmv}\right) \le MAXVERTEX - 1 \qquad \forall t \in T, m \in M, j \in I, v \in V \mid 2 \le i \ne j \tag{24}$$

Equation (22) of the vehicle routing constraints guarantees that all the turbines will be visited once, Equation (23) ensures the conservation of the route, and Equation (24) fulfils the elimination of sub-tours.

The technician's flow constraints group fulfill the outgoing and incoming flow at each node of the graph:

$$\sum_{i \in I} p_{tij} - \sum_{l \in I} p_{tjl} = J_{ti} \qquad \forall t \in T, j \in I \mid j > 1 \tag{25}$$

$$p_{tij} \le BIGM \times \sum_{m \in M}\sum_{v \in V} CTVCAP_{mv} \times x_{tijmv} \qquad \forall t \in T, i \in I \mid i = 1, j \in I \mid j > 1 \tag{26}$$

$$p_{tij} \leq \sum_{m}\sum_{v} CTVCAP_{mv} \times x_{tijmv} \qquad \forall\, t \in T, i \in I, j \in I \mid j \neq i \tag{27}$$

$$\sum_{z \in Z} y_{sz} \leq 1 \qquad \forall\, s \in S \tag{28}$$

$$\sum_{i \in I|i=1}\sum_{j \in I|j>1} p_{tij} \leq \sum_{z \in Z} SOVCAP_{sz} \times y_{sz} \qquad \forall\, s \in S, t \in T \tag{29}$$

Equation (25) guarantees the conservation of the difference of passengers between vertexes. Equation (26) ensures that no technicians travel between vertexes if no vehicle performs the route. Equation (27) guarantees that the total technicians' flow cannot exceed the capacity of the vehicle assigned. Thus, Equation (28) establishes that each SOV can only be at most of one size and, Equation (29) ensures that the total technicians' flow at the installed SOV cannot exceed the SOV capacity.

The time window constraints ensure that the route performed does not exceed the operational vessel time window:

$$\sum_{i \in I}\sum_{j \in I}\left(\frac{DIST_{ij}}{CTVSPEED_{m}}\right) \times x_{tijmv} \leq MAXTIME_{tmv} \qquad \forall\, t \in T, m \in M, v \in V \tag{30}$$

The fleet size constraints group is represented by:

$$l_{tmv} \leq \sum_{j \in I|\, j>1} x_{tijmv} \qquad \forall\, t \in T, m \in M, v \in V, i \in I|i=1 \tag{31}$$

$$n_{tm} = \sum_{v \in V} l_{tmv} \qquad \forall\, t \in T, m \in M \tag{32}$$

Equation (31) calculates the total number of travelled routes assigned to the vehicle and, Equation (32) determines the number of vessels needed to accomplish the routes.

3. Case Study, Results, and Discussions

3.1. Case Study

The realistic input data is a result of the work presented by Díaz et al. [21–23], which selected the best location for a floating wind farm on the Spanish coast. Ribadeo is the selected area located in the European Atlantic region in western Europe. The respective Ribadeo port is considered an available onshore base [22]. The authors also determined the number and location of the floating turbines (Figure 2a and Appendix A) [24] as well as the export cable location [25]. Accordingly, from the 88 turbines established, the initial phase of the installation forecasted 28 turbines in the first year, which represents the present case study. Two types of SOVs and CTVs are analyzed in this work. Table 2 presents the main vessel characteristics and cost information. The CTV's average speed is 44.04 knots, and the average number of technicians requested to perform the preventive maintenance task at each turbine is four. The preventive planning horizon is established in three days. Thus, the technicians' workday is considered as 12 h. Furthermore, for the location-allocation model, three possible stand-by points and the group of turbines to be attended for each day are determined (Figure 2b).

Table 2. Vessel's characteristics and cost information.

Vessels	Maximum People (pax)	Wave Limitation (m)	Wind Limitation (m/s)	Velocity (kn)	Daily Costs (US/Day)
CTV-S	12	1.5	25	20	2352.99
CTV-M	15	1.25	20	24	3823.61
SOV-M	50	2	30	12	45,000.00
SOV-L	80	2.5	30	13	67,000.00

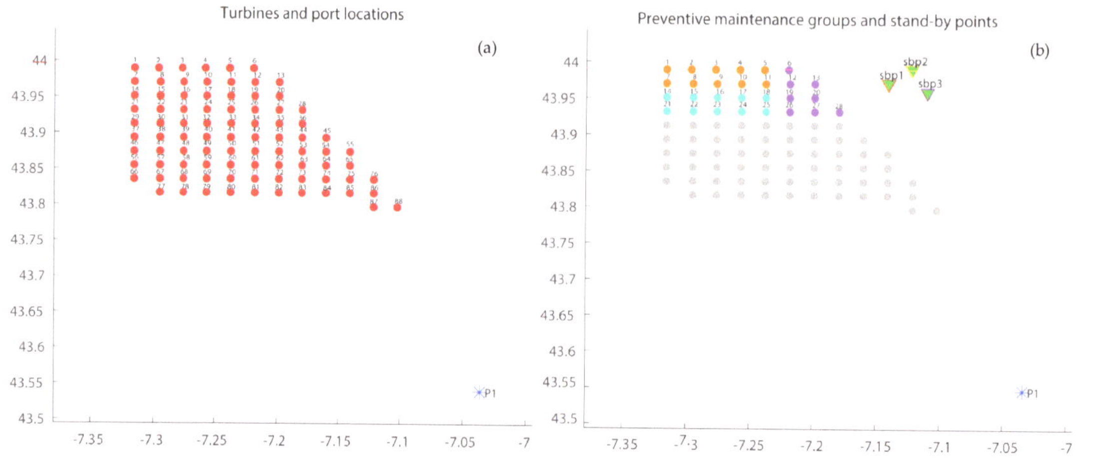

Figure 2. (**a**) Ribadeo floating wind farm layout (**b**) maintenance planning and possible offshore base locations. The asterisk symbol represents the port location.

3.2. Results and Discussions

The mathematical models were implemented and solved by the IBM ILOG CPLEX Studio Optimization. The MILP models were running in a workstation with Intel(R) Core(TM) i7 of 2.30 GHz and 16GB RAM.

The first model attempts only to identify the optimal location of the offshore base considering the distances to the turbines and port. From the three initial possible locations defined above, the model chooses stand-by point one. Figure 3 presents the optimal location solution. Because of the infinite continuous possibility of locations at sea, the search for the optimal offshore base can be complex and timeless. But the model allows the decision-makers to analyze a few possible locations using a rapid screening tool able to generate a solution in less than one minute. Different "what if" scenarios in terms of facility locations, which include the analyses of other onshore bases, can be run without spending a huge computational effort. It is possible to see that the chosen offshore base is the one that most reduce the voyage time of the vessel to attend to the turbines during the maintenance activities, fulfilling the voyage requirements and time window constraints and optimizing the variable cost of the vessels.

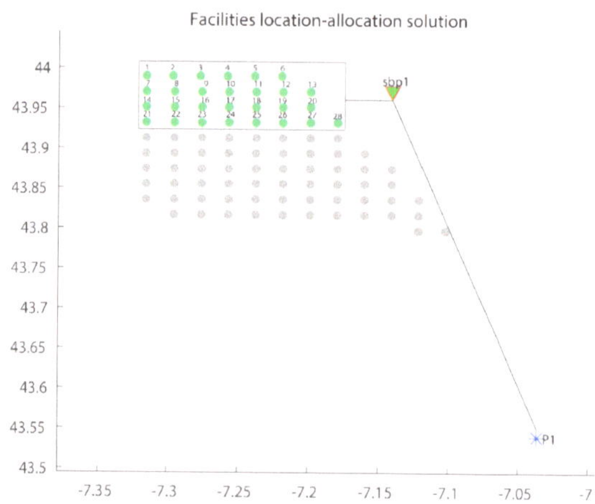

Figure 3. Facility location–allocation solution, where the green dots mean that all the turbines are allocated to the chosen stand-by point.

After establishing the optimal location of the SOV, this solution information turns into input data for the vehicle-routing model with a time window. The optimal navigation routes of each CTV per day and the fleet composition are presented in Table 3, and the graphical representations of the routes per day are shown in Figure 4. Moreover, the cost information of the described logistic is presented in Table 4.

Table 3. Navigation routes per day.

	Routes	Technicians	Utilization (Delivery and Pick-Up Voyages)	Daily Fleet Composition [CTV-S, CTV-M]	Installed SOV [SOV-M, SOV-L]
day one	$i_1 - t_4 - t_3 - t_2 - i_1$	12	1 h 45 min	[1,0]	[1,0]
	$i_1 - t_{11} - t_{10} - t_9 - i_1$	12	1 h 29 min	[1,0]	
	$i_1 - t_5 - i_1$	4	1 h 12 min	[1,0]	
	$i_1 - t_8 - t_7 - t_1 - i_1$	12	2 h 27 min	[1,0]	
day two	$i_1 - t_{13} - t_{14} - t_6 - i_1$	12	1 h 43 min	[1,0]	[1,0]
	$i_1 - t_{20} - t_{19} - t_{26} - i_1$	12	58 min	[1,0]	
	$i_1 - t_{28} - t_{27} - i_1$	8	43 min	[1,0]	
day three	$i_1 - t_{15} - t_{14} - t_{21} - i_1$	12	1 h 30 min	[1,0]	[1,0]
	$i_1 - t_{18} - t_{17} - t_{16} - i_1$	12	1 h 46 min	[1,0]	
	$i_1 - t_{25} - i_1$	4	1 h 07 min	[1,0]	
	$i_1 - t_{24} - t_{23} - t_{22} - i_1$	12	1 h 58 min	[1,0]	

Figure 4. Graphical representation of the routing solutions.

Table 4. Infrastructure costs.

Costs	Value (U$)
SOVFixedCost (U$)	$2.70 \times 10^{+05}$
CTVFixedCost (U$)	$2.59 \times 10^{+04}$
CTVVariableCost (U$)	$1.04 \times 10^{+03}$

The symmetrical nature of the wind farm input data, as well as the equal distance between turbines, the same number of technicians to perform the preventive repairs, and no requirement of task's priority, increases the potential combination of the fleet composition and the time to find an exact solution in a polynomial time. The model defined the optimal daily navigation table, which allows the decision-makers to analyze the routing configuration per day. The main route pattern enforces the CTV to visit three turbines at maximum, delivering and picking up a maximum of 12 technicians. However, it is possible to identify that for day one and day three, there is a single turbine route. It may occur because the capacity of the small CTV (CTV-S) is 12 passengers, and even changing the size of the CTV to transport more technicians (CTV-M, 15 pax) aiming to reduce the single route, become a non-feasible solution due to the total number of technicians to delivery and pick-up at the turbines increases to 16 technicians, which exceeds the available capacities of the CTVs. Therefore, as the optimal solution, the model established that four CTV with the smallest capacity (CTV-S) are enough to perform the routes for three days attending 28 turbines. It proves that the technicians' demand flow constraints are fulfilled even when the CTV visited more than one turbine unit. Additionally, the routes obey the sub-tours elimination constraints and the time window constraints. Also, the model determined that the smallest capacity for the SOV (SOV-M, 50 pax) can accommodate the total of technicians necessary to perform the activities per day, which assures that the total of passengers that leave the SOV is less than the SOV capacity.

In summary, the most important planning insights into the applicability of the methodologies are:

1. The reduction of the computational time and effort of searching for the optimal location of the offshore base by adopting the discrete solution space instead of the continuous solution space, which is infinite.
2. The generation of a navigation route table allows the identification of the routes pattern of the vessels.
3. The daily fleet composition and the respective vehicle's capacity provide bases for cost evaluation and chartering analysis.
4. The combined use of the SOV and CTV appears to be a suitable approach to performing on-time and on-demand services.

4. Conclusions

The recent development of offshore floating wind farms located in deep waters brings new logistic challenges. The main goal of the present study is to integrate the main logistics aspects of the maintenance activities. The facility location–allocation model and the fleet size and mix vehicle routing model with time window provide feasible solutions that increase the planning level of the maintenance activities.

As the main results, from a pool of potential locations pre-defined by the decision-makers, the optimal stand-by point of the offshore base is determined. Afterwards, this solution output is inserted as input for the generation of the optimal navigation table of the vessels to transfer the technicians from the offshore base to the offshore floating turbines. The establishment of the fleet composition and vehicle capacity may guide the decision-makers to analyze the most appropriate type of vessel chartering and investment to maintain the wind farm operations and maintenance activities.

Besides advancing the state of the art, the complexity of the combinatorial nature of the problem brings the limitation of two separate models. However, the results demonstrated that the two screening tools can be useful to help decision-makers in the planning process of maintenance activities.

Author Contributions: L.M.R.S.: Methodology, Formal analysis, Writing—original draft, Visualization. H.L.: Methodology, Writing—review and editing, C.G.S.: Conceptualization, Writing—review and editing, Supervision. All authors have read and agreed to the published version of the manuscript.

Funding: This work was performed within the project ARCWIND—Adaptation and implementation of floating wind energy conversion technology for the Atlantic region, which is co-financed by the European Regional Development Fund through the Interreg Atlantic Area Programme under contract EAPA 344/2016. This work contributes scope of the Strategic Research Plan of the Centre for Marine Technology and Ocean Engineering (CENTEC), which is financed by the Portuguese Foundation for Science and Technology (Fundação para a Ciência e Tecnologia—FCT) under contract UIDB/UIDP/00134/2020.

Institutional Review Board Statement: Not applicable.

Informed Consent Statement: Not applicable.

Data Availability Statement: Not applicable.

Conflicts of Interest: The authors declare no conflict of interest.

Abbreviations

Nomenclature
Location–Allocation Model (MILP)

Sets:	
D	set of turbines;
S	set of SOVs;
P	set of ports;
Z	set of sizes;
T	set of time;
Subscripts:	
d	turbine index;
s	SOV index;
p	port index;
z	size index;
t	time index;
Binary Variables:	
x_{tsd}	assumes the value 1 if the turbine d is allocated to SOV s in time t;
x_{tps}	assumes the value 1 if the SOV s is allocated to port p in time t;
y_p	assumes the value 1 if the port p is installed;
y_{tsz}	assumes the value 1 if the SOV s of size z is installed in time t;
Continuous Variables:	
p_{ts}	total flow of technicians at SOV s in time t;
q_{tp}	total flow of technicians at port p in time t;
r_{tps}	total flow of technicians from SOV s to port p in time t;
Discrete Parameters:	
$DISTA_{tsd}$	distance between the turbines d and SOV s in time t;
$DISTB_{tps}$	distance between the SOV s and port p in time t;
MAXTURBINES	maximum number of turbines;
PAXMAX	volume of personnel available to work;
J_{td}	technicians demand per turbine d in time t;
$PORTCAP_p$	capacity of each available port p;
$SOVCAP_{sz}$	capacity of SOV s of size z;
$SOVRC_{sz}$	rental cost of SOV s of size z;
CTVFCC	fuel consumption cost of CTV;
FUELCOST	fuel cost per litre;
CTVSPEED	average speed of CTV;
BIGM	large number;

Vehicle-Routing Model (MILP)

Sets:	
I	set of vertexes;
S	set of SOVs;
V	set of CTV;
M	set of models;
Z	set of sizes;
T	set of time;
Subscripts:	
i	vertex index;
s	SOV index;
v	CTV index;
m	model index;

z	size index;
t	time index;
Binary Variables:	
y_{tsz}	assumes the value 1 if the SOV *s* of size *z* is installed in time *t*;
Continuous Variables:	
x_{tijmv}	decision to assign the CTV *v* to travel between vertexes *(i,j)* in time *t*;
p_{tij}	flow of technicians between vertexes *(i,j)* in time *t*;
n_{tm}	number of CTV model *m*
l_{tmv}	number of travels performed by CTV *v* of model *m* in time *t*;
u_{ti}	auxiliary variable assigned to each vertex *i* in time *t*;
Discrete Parameters:	
$DIST_{ij}$	distance between vertexes *(i,j)*;
J_{td}	technicians demand per turbine *d* in time *t*;
$SOVCAP_{sz}$	capacity of SOV *s* of size *z*;
$SOVRC_{sz}$	rental cost of SOV *s* of size *z*;
$CTVCAP_{mv}$	capacity of CTV *v* of model *m*;
$CTVRC_{m}$	rental cost of CTV's model *m*;
$CTVFCC_{m}$	fuel consumption cost of CTV's model *m*;
FUELCOST	fuel cost per litre;
CTVSPEED	average speed of CTV;
$CTVMAXTIME_{tmv}$	maximum CTV *v* operational time per model *m* in time *t*;
BIGM	large number;

Appendix A

Table A1. Location of the turbines.

Turbine	Latitude	Longitude	#N			#N			#N		
1	43.9900	−7.31592	29	43.9129	−7.31592	57	43.8551	−7.28701	85	43.8165	−7.1328
2	43.99003	−7.29664	30	43.91293	−7.28701	58	43.8551	−7.26773	86	43.81655	−7.11353
3	43.99003	−7.27737	31	43.91293	−7.26773	59	43.8551	−7.24846	87	43.79727	−7.11353
4	43.99003	−7.25809	32	43.91293	−7.24846	60	43.8551	−7.22918	88	43.79727	−7.09425
5	43.99003	−7.23882	33	43.91293	−7.22918	61	43.8551	−7.20991	89	43.54006	−7.03641
6	43.99003	−7.21954	34	43.91293	−7.20991	62	43.8551	−7.19063			
7	43.97075	−7.31592	35	43.91293	−7.19063	63	43.8551	−7.17136			
8	43.97075	−7.28701	36	43.91293	−7.17136	64	43.8551	−7.15208			
9	43.97075	−7.26773	37	43.89365	−7.31592	65	43.8551	−7.1328			
10	43.97075	−7.24846	38	43.89365	−7.28701	66	43.8358	−7.31592			
11	43.97075	−7.22918	39	43.89365	−7.26773	67	43.8358	−7.28701			
12	43.97075	−7.20991	40	43.89365	−7.24846	68	43.8358	−7.26773			
13	43.97075	−7.19063	41	43.89365	−7.22918	69	43.8358	−7.24846			
14	43.95148	−7.31592	42	43.89365	−7.20991	70	43.8358	−7.22918			
15	43.95148	−7.28701	43	43.89365	−7.19063	71	43.8358	−7.20991			
16	43.95148	−7.26773	44	43.89365	−7.17136	72	43.8358	−7.19063			

Table A1. *Cont.*

Turbine	Latitude	Longitude	#N			#N			#N
17	43.95148	−7.24846	45	43.89365	−7.15208	73	43.8358	−7.17136	
18	43.95148	−7.22918	46	43.87438	−7.31592	74	43.8358	−7.15208	
19	43.95148	−7.20991	47	43.87438	−7.28701	75	43.8358	−7.1328	
20	43.95148	−7.19063	48	43.87438	−7.26773	76	43.8358	−7.11353	
21	43.9322	−7.31592	49	43.87438	−7.24846	77	43.8165	−7.28701	
22	43.9322	−7.28701	50	43.87438	−7.22918	78	43.8165	−7.26773	
23	43.9322	−7.26773	51	43.87438	−7.20991	79	43.8165	−7.24846	
24	43.9322	−7.24846	52	43.87438	−7.19063	80	43.8165	−7.22918	
25	43.9322	−7.22918	53	43.87438	−7.17136	81	43.8165	−7.20991	
26	43.9322	−7.20991	54	43.87438	−7.15208	82	43.8165	−7.19063	
27	43.9322	−7.19063	55	43.87438	−7.1328	83	43.8165	−7.17136	
28	43.9322	−7.17136	56	43.8551	−7.31592	84	43.8165	−7.15208	

References

1. Yang, L.; Li, G.; Zhang, Z.; Ma, X.; Zhao, Y. Operations & Maintenance Optimization of Wind Turbines Integrating Wind and Aging Information. *IEEE Trans. Sustain. Energy* **2021**, *12*, 211–221. [CrossRef]
2. Hofmann, M. A Review of Decision Support Models for Offshore Wind Farms with an Emphasis on Operation and Maintenance Strategies. *Wind Eng.* **2011**, *35*, 1–16. [CrossRef]
3. Stålhane, M.; Hvattum, L.M.; Skaar, V. Optimization of Routing and Scheduling of Vessels to Perform Maintenance at Offshore Wind Farms. *Energy Procedia* **2015**, *80*, 92–99. [CrossRef]
4. Fan, D.; Ren, Y.; Feng, Q.; Zhu, B.; Liu, Y.; Wang, Z. A hybrid heuristic optimization of maintenance routing and scheduling for offshore wind farms. *J. Loss Prev. Process. Ind.* **2019**, *62*, 103949. [CrossRef]
5. Irawan, C.A.; Ouelhadj, D.; Jones, D.; Stålhane, M.; Sperstad, I.B. Optimisation of maintenance routing and scheduling for offshore wind farms. *Eur. J. Oper. Res.* **2017**, *256*, 76–89. [CrossRef]
6. Gutierrez-Alcoba, A.; Hendrix, E.; Ortega, G.; Halvorsen-Weare, E.; Haugland, D. On offshore wind farm maintenance scheduling for decision support on vessel fleet composition. *Eur. J. Oper. Res.* **2019**, *279*, 124–131. [CrossRef]
7. Gundegjerde, C.; Halvorsen, I.B.; Halvorsen-Weare, E.E.; Hvattum, L.M.; Nonås, L.M. A stochastic fleet size and mix model for maintenance operations at offshore wind farms. *Transp. Res. Part C Emerg. Technol.* **2015**, *52*, 74–92. [CrossRef]
8. Stålhane, M.; Vefsnmo, H.; Halvorsen-Weare, E.E.; Hvattum, L.M.; Nonås, L.M. Vessel Fleet Optimization for Maintenance Operations at Offshore Wind Farms Under Uncertainty. *Energy Procedia* **2016**, *94*, 357–366. [CrossRef]
9. Stålhane, M.; Halvorsen-Weare, E.E.; Nonås, L.M.; Pantuso, G. Optimizing vessel fleet size and mix to support maintenance operations at offshore wind farms. *Eur. J. Oper. Res.* **2019**, *276*, 495–509. [CrossRef]
10. Halvorsen-Weare, E.E.; Gundegjerde, C.; Halvorsen, I.B.; Hvattum, L.M.; Nonås, L.M. Vessel Fleet Analysis for Maintenance Operations at Offshore Wind Farms. *Energy Procedia* **2013**, *35*, 167–176. [CrossRef]
11. Díaz, H.; Guedes Soares, C. Review of the current status, technology and future trends of offshore wind farms. *Ocean Eng.* **2020**, *209*, 107381. [CrossRef]
12. Irawan, C.A.; Starita, S.; Chan, H.K.; Eskandarpour, M.; Reihaneh, M. Routing in offshore wind farms: A multi-period location and maintenance problem with joint use of a service operation vessel and a safe transfer boat. *Eur. J. Oper. Res.* **2023**, *307*, 328–350. [CrossRef]
13. Irawan, C.A.; Salhi, S.; Chan, H.K. A continuous location and maintenance routing problem for offshore wind farms: Mathematical models and hybrid methods. *Comput. Oper. Res.* **2022**, *144*, 105825. [CrossRef]
14. Lazakis, I.; Khan, S. An optimization framework for daily route planning and scheduling of maintenance vessel activities in offshore wind farms. *Ocean Eng.* **2021**, *225*, 108752. [CrossRef]
15. Kang, J.; Guedes Soares, C. An opportunistic maintenance policy for offshore wind farms. *Ocean Eng.* **2020**, *216*, 108075. [CrossRef]
16. Li, H.; Huang, C.-G.; Guedes Soares, C. A real-time inspection and opportunistic maintenance strategies for floating offshore wind turbines. *Ocean Eng.* **2022**, *256*, 111433. [CrossRef]
17. Li, H.; Díaz, H.; Guedes Soares, C. A failure analysis of floating offshore wind turbines using AHP-FMEA methodology. *Ocean Eng.* **2021**, *234*, 109261. [CrossRef]
18. Yang, L.; Chen, Y.; Ma, X. A State-age-dependent Opportunistic Intelligent Maintenance Framework for Wind Turbines Under Dynamic Wind Conditions. *IEEE Trans. Ind. Inform.* **2023**, 1–9. [CrossRef]

19. Díaz, H.; Teixeira, A.; Guedes Soares, C. Application of Monte Carlo and Fuzzy Analytic Hierarchy Processes for ranking floating wind farm locations. *Ocean Eng.* **2022**, *245*, 110453. [CrossRef]
20. Li, H.; Peng, W.W.; Huang, C.-G.; Guedes Soares, C. Failure Rate Assessment for Onshore and Floating Offshore Wind Turbines. *J. Mar. Sci. Eng.* **2022**, *10*, 1965. [CrossRef]
21. Diaz, H.M.; Guedes Soares, C. An integrated GIS approach for site selection of floating offshore wind farms in the Atlantic Continental European coastline. *Renew. Sustain. Energy Rev.* **2020**, *134*, 110328. [CrossRef]
22. Diaz, H.M.; Guedes Soares, C. A novel multi-criteria decision-making model to evaluate floating wind farm locations. *Renew. Energy* **2022**, *185*, 431–454. [CrossRef]
23. Diaz, H.M.; Guedes Soares, C. Approach for installation and logistics of a floating offshore wind farm. *J. Mar. Sci. Eng.* **2023**, *11*, 53. [CrossRef]
24. Diaz, H.M.; Silva, D.; Bernardo, C.; Guedes Soares, C. Micro siting of floating wind turbines in a wind farm using a multiple-criteria framework. *Renew. Energy* **2023**, *204*, 449–474. [CrossRef]
25. Diaz, H.M.; Guedes Soares, C. Multicriteria decision approach to the design of floating wind farm export cables. *Energies* **2022**, *15*, 6593. [CrossRef]

Journal of
Marine Science and Engineering

MDPI

Article

Data-Driven Damage Quantification of Floating Offshore Wind Turbine Platforms Based on Multi-Scale Encoder–Decoder with Self-Attention Mechanism

Musa Bashir [1], Zifei Xu [1], Jin Wang [1] and C. Guedes Soares [2,*]

1 School of Engineering, Faculty of Engineering and Technology, Liverpool John Moores University, Liverpool L3 3AF, UK

2 Centre for Marine Technology and Ocean Engineering (CENTEC), Instituto Superior Técnico, Universidade de Lisboa, 1649-004 Lisboa, Portugal

* Correspondence: c.guedes.soares@centec.tecnico.ulisboa.pt

Abstract: A Multi-Scale Convolutional Neural Network with Self Attention-based Auto Encoder–Decoder (MSCSA-AED), is a novel high-performance framework, presented here for the quantification of damage on a multibody floating offshore wind turbine (FOWT) structure. The model is equipped with similarity measurement to enhance its capability to accurately quantify damage effects from different scales of coded features using raw platform responses and without human intervention. Case studies using different damage magnitudes on tendons of a 10 MW multibody FOWT were used to examine the accuracy and reliability of the proposed model. The results showed that addition of Square Euclidean (SE) distance enhanced the MSCSA-AED model's capability to suitably estimate the damage in structures operating in complex environments using only raw responses. Comparison of the model's performance with other variants (DCN-AED and MSCNN-AED) used in the industry to extract the coded features from FOWT responses further demonstrated the superiority of MSCSA-AED in complex operating conditions, especially in low magnitude damage quantification, which is the hardest to quantify.

Keywords: data-driven technique; damage quantification; floating offshore wind turbine; FOWT predictive maintenance; Convolutional Neural Network; multi-scale information fusion

Citation: Bashir, M.; Xu, Z.; Wang, J.; Guedes Soares, C. Data-Driven Damage Quantification of Floating Offshore Wind Turbine Platforms Based on Multi-Scale Encoder–Decoder with Self-Attention Mechanism. *J. Mar. Sci. Eng.* **2022**, *10*, 1830. https://doi.org/10.3390/jmse10121830

Academic Editor: José António Correia

Received: 15 October 2022
Accepted: 23 November 2022
Published: 29 November 2022

1. Introduction

Rapid development of Floating Offshore Wind Turbines (FOWTs) for clean power generation is an important step in the strategy to ease global energy crises and mitigate the impact of climate change [1]. Review of the trend of wind farms shows that there is a recent tendency for floating wind farms to be installed at farther distances from the coast, as opposed to the shallow water fixed wind turbines developed in the first round of development [2]. The current trajectory of FOWT development can be grouped into two categories, namely, geometry-dependent operations and maintenance-driven operations. Continuous growth in FOWTs' size has become a major consideration in selecting FOWT foundation types and location.

Increase in FOWT sizes brings other significant challenges to installation, operations, and maintenance (O&M) and the attendant consequence is the requirement for exponential structural strength to adequately withstand the loads imposed by the differences in geometrical scale and environment loads. This is in addition to the fundamental need for safe and damage resilient operation. Therefore, it is important to conduct failure analysis of the wind platforms [3,4] to assess the probabilities of failure of the various components. The reliability of the floating platforms is essential to ensure that appropriate methods are adopted for this purpose [5].

There are available databases of failure and maintenance data, but this has been developed mainly for onshore platforms and, eventually, for fixed offshore platforms [6,7].

Therefore, it is important to use the existing data and modify it to make it applicable to offshore turbines, as proposed by Li and Guedes Soares [8]. It should be noted that availability of appropriate failure and maintenance data is essential to establishing robust maintenance plans. These plans can be the basic age-based preventive maintenance (Santos et al. 2015) for opportunistic strategies [9,10] or for strategies that account for the logistic time of spare parts and vessel availability [11,12]. Condition-based maintenance has become more widely adopted [13] and specific approaches have also been developed for floating offshore turbines [14]. The route to improving maintenance plans is by increasing the condition-based maintenance by using more extensive structural health monitoring of the platforms and turbines. The direct relationship between structural health of components and FOWT safety typically affects how the platform can be operated and maintained to reduce the levelized cost of electricity (LCoE) [15].

Lack of effective maintenance methodology for offshore wind turbines, especially for deep water applications, is generally acknowledged as a significant obstacle to the development of FOWT technology. This is mainly due to the lack of failure and maintenance data in which to base maintenance plans and to the difficulties arising from lack of access for inspection and repairs, leading to high O&M costs [16,17]. Consequently, having an effective monitoring strategy and a robust quantification method for defects in FOWT components can provide a credible path for developing an intelligent method for preventive maintenance, based on actionable condition and operational data [18]. This approach allows rational decision-based maintenance planning and scheduling to be formulated as part of the proactive maintenance strategy, which, ultimately, improves safe operation, and reduces the maintenance costs and the LCoE of FOWTs.

Recent progress on the internet of things (IoT) and artificial intelligence (AI) technologies further strengthen the case for possible adopting of preventative maintenance in FOWTs. Machine learning (ML), a subset of AI, has become a mainstream approach in data-driven methodologies for preventative maintenance in the industries. The data-driven technique for structural damage prediction, a quantitative method for detection and diagnosis of damages or faults, has recently been the subject of interest to researchers. Some notable studies on data-driven techniques have been reported, including the works of Cho et al. [19] who developed a detector to locate the structural damage on blades of a wind turbine by using Kalman filters and artificial neural networks.

Dao [20] established a regression model for wind turbine condition monitoring and the reliable detection of anomalies in their operations. Yang et al. [21] used independent component analysis and Mahalanobis distance to quantify damages in the transmission system of a wind turbine. Rai et al. [22] applied empirical mode decomposition (EMD) and k-medoids clustering to assess the structural health of wind turbine bearings, which is helpful in estimating the residual capability of the wind turbine bearings. Cheng et al. [23] used a Convolutional Neural Network (CNN) to estimate the degradation of the mechanical systems from raw signals using a complete ensemble empirical mode-based label. Guo et al. [24] adopted a CNN with linear labels to supervise the network and quantify damage degradation. In a further similar work, Guo et al. [25] used a Recurrent Neural Network (RNN) with kurtosis-based nonlinear labels to construct the degree of damages on mechanical parts of wind turbines. Chen et al. [26] studied different types of RNNs to extract damage features only using prior experience to realize damage quantification.

However, as summarized in these papers, although many ML (including DL) methods have been used to identify faults and defects, their algorithms largely rely on defects or faults labels. The process of labeling these defects or faults is very labor-intensive and requires engineers to manually annotate the data. From an efficiency perspective, the process is time-consuming and laborious. It is also dependent on prior knowledge when being looked at from the model's perspective.

Recent attempts by researchers to expand the state of knowledge in this subject include investigation on methodologies for damage quantification based on unsupervised learning. Peng et al. [27] combined particle filter and Deep Belief Network (DBN) to calculate the damage magnitude on structures of wind turbines. Dai et al. [28] used Generative Adversarial Network (GAN) to achieve a structural monitoring of wind turbine structure without employing any manual supervision.

Guo et al. [29] used a multi-scale network with an attention mechanism to extract features and obtain information about damage magnitude. Suh et al. [30] designed a GAN, with a U-net architecture, to deal with the damage quantification of a rotating machine and to estimate its residual capability. To enhance the performance of convolutional networks and make them more powerful and accurate for decision-making in maintenance, Guo et al. [31] introduced a multi-scale feature extraction process in an auto encoder–decoder to achieve the development of unsupervised architecture. The method was successfully applied in a rotating machine mounted on a fixed platform. This study pointed out that the feature fusion of a decoder had a consequential effect on the performance of the Neural Network (NN) model for damage quantification. Although a supervised deep learning network (DLN) model can be quickly adapted to quantify the known damage magnitude, the labeling process is time-consuming and labor-intensive. This approach is inefficient and almost unsuitable for industrial-scale application because of the difficulty to supervise and label all data in practical engineering applications.

Notwithstanding, the unsupervised models are dedicated to learning and establishing the response patterns of the machines in the healthy state. It is still possible to quantify the damage magnitude by comparing the similarity between the baseline and test coded data in which the challenges of these methods are the robustness of an unsupervised model and the performance of the chosen similarity function. More importantly, for a system such as FOWT, the environment in which it operates further contributes to the complexity in platform dynamic responses, as opposed to a generic rotating system on a fixed platform. This presents a precisely unique challenge in how to successfully incorporate the unsupervised learning method in diagnosis and prognosis for a system operating in such complex and unstable environments. At present, there are few studies on intelligent damage estimation for FOWTs' towers and tendons.

Existing data-driven (quantitative) damage models are predominantly built using supervised learning. These models are not particularly suitable for floating offshore wind turbine damage quantifications for maintenance, due to rigorous demands for accurate feature labeling. Application of unsupervised learning in damage quantification and its robustness in floating wind turbines have not yet been investigated. Therefore, this study was inspired by the possibility of developing a remarkable algorithm that reduces human interface in damage quantification by using CNN with a self-attention mechanism. Consequently, this research developed a high performance framework, named Multi-Scale Convolutional Neural Network with Self Attention-based Auto Encoder–Decoder (MSCSA-AED), to quantify the damage magnitude of the 10 MW FOWT. The proposed method enables a faster convergence with smaller errors than the baseline models currently used in the industry. The main features in the encoder and decoder having the same scale are reevaluated by the self-attention mechanism to improve efficiency and quality of features for damage quantification. The MSCSA-AED model is trained by only using the normal (healthy) data from raw platform responses, an important capability that minimizes prognostic latency. The MSCSA-AED model is optimized to enhance clarity in utilizing its parameters. Following the optimization, the quantified damage values of the 10 MW FOWT is calculated by the similarity function implemented in the model. The reliability of the damage quantification model using different similarity functions was examined in this study to support both practical and theoretical references for damage quantification of 10 MW FOWTs. Hence, the main contributions of this research are summarized below:

(1) This research developed an unsupervised approach, based on deep neural network with self-attention mechanism, named MSCSA-AED, to intelligently extract features (healthy and damaged) to establish the state of health of a 10 MW FOWT from the coded features. The method provided a path to adding the multiscale information present in typical multi-scale resolution features of the FOWT's responses (due to its

complex operating environment) for efficient damage quantification in maintenance planning and scheduling.

(2) The research found that the influence of applying different similarity functions to establishing the reliability of damage quantification of the FOWT was significant. The capability to analyze the characteristics of the platform's different degrees of freedom (DOFs) when implemented in the model improved accuracy and reliability in quantifying structural damage in complex operating environments.

(3) This study used a 10 MW FOWT with tendon damage simulated data as a case study to examine the performance of the proposed MSCSA-AED method on damage quantification. The reliability of the proposed method was demonstrated through examination of different damage scenarios and operating conditions with tendon defects. It was found that the SE metric-based quantification method offered more reliable results than other methods do.

Following an introduction in Section 1, the rest of the paper is organized as follows. Section 2 presents details of the FOWT structure, damage scenarios and simulations. Section 3 presents the proposed damage quantification methods. Section 4 discusses both quantitative and qualitative results as part of the validation process. Section 5 concludes the research.

2. The 10 MW FOWT Geometry, Damage Conditions and Simulations

2.1. Structural Configuration of the 10 MW Multibody FOWT

This study used a novel TELWIND concept of the multibody 10 MW FOWT developed by Esteyco [32] for the ARCWIND project [33]. The platform consists of a telescopic tower supported by an upper tank (UT) for buoyancy and a lower tank (LT) for ballasting. The UT and LT are connected by 12 tendons (steel cables), as shown in Figure 1. Three mooring cables connected to the fairleads at the top surface of the UT, located at 14 m below mean sea level, are used for station keeping of the platform. Structural ribs are added to the platform to provide strength and sufficient space to connect the upper and lower tanks. This further ensures that the 10 MW FOWT remains safe and stable in the event of a limited tendons rupture.

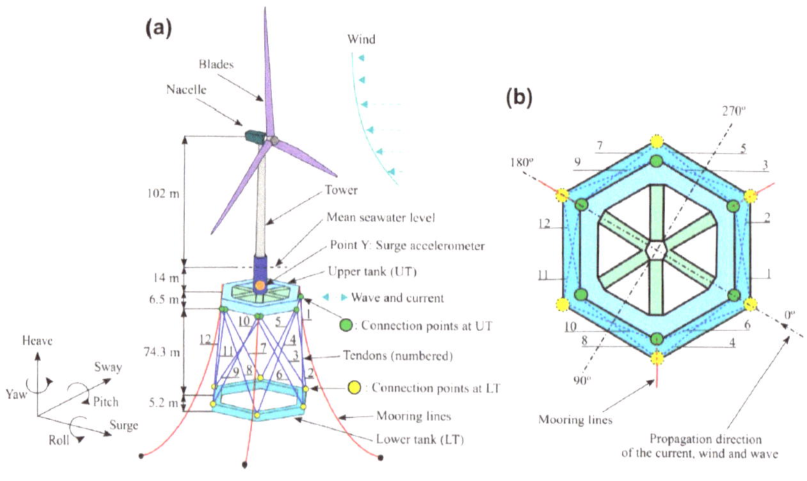

Figure 1. The 10 MW multibody FOWT. (**a**) Fully assembled floating wind turbine showing all components and position of load measurement. (**b**) The plan view of the multibody platform.

Details of the model properties include a draught of 20.5 m at the upper tank with a total mass of 5.31×10^6 kg and a displaced volume of 7399.02 m^3. The lower tank has a draught of 100 m and the total mass is 8.72×10^6 kg, which equals a volume of 7922.92 m^3. The 12 tendons have a collective pretension that is approximately equal to 3×10^6 N.

2.2. The Damage Modeling on the 10 MW FOWT

Damage in the structural components of the FOWT model were examined under varying Environmental and Operating Conditions (EOCs), corresponding to seven conditions. This comprised 4 wind speeds (WS), namely, WSs 4 m/s, 11.4 m/s, 18 m/s and 25 m/s with 4 significant wave heights (Hs), namely, Hs 1.61 m, 2.16 m, 2.95 m and 4.02 m (Table 1).

Table 1. The varying EOC details.

Wind Speed (m/s)	Significant Wave Hight (m)	Peak Frequency (Hz)	Propagation Direction of Sea, Current and Wind (°)	Current Speed (m/s)	Excitation Bandwidth (Hz)
4	1.61	0.285	0	0.22	[0.1–100]
11.4	2.16	0.185	0	0.22	[0.1–100]
18	2.95	0.14	0	0.22	[0.1–100]
25	4.02	0.112	0	0.22	[0.1–100]

The FOWT begins generating power when it reaches its cut-in speed of 4 m/s. The rated condition of the 10 MW FOWT is 11.4 m/s of the operating speed at which it is expected to optimally generate power. The FOWT's cut-out speed is 25 m/s, above which the FOWT is shutdown to protect it from adverse environmental effects.

The wind condition used in this is study was generated using the Kaimal spectrum [34], based on different time series of wind excitation and corresponding to the varying wind speed regime effects on FOWT. Consequently, to ensure that there was consistency of results in the simulations, the same spectral intensity was maintained throughout the simulations. The irregular waves components of the varying EOCs were generated using the modified two-parameter Pierson–Moskowitz spectrum [35,36]. Furthermore, a current of constant speed and direction was included in the prediction of the EOC. Table 1 and Figure 1b present the details of the EOCs' wind, wave and current parameters used in this research. These parameters represented normal and the most severe EOC features at the selected site in the offshore area of northern Scotland.

For this investigation, the platform was simulated under healthy and damaged conditions to predict the corresponding responses. The healthy conditions were simulated under four kinds of wind speed conditions. The damaged conditions were modeled, based on stiffness (%) reduction in the tendon. The main benefit of this approach was that any amplified damages, that reflected the platform dynamics, could be measured with a degree of certainty. This was subsequently used to examine the effects of EOCs under given WS and Hs on the structural health condition of the FOWT. The stiffness reduction ranged from 10% to 100 % with an increment of 10%, in which 100% stiffness reduction represented a state of total failure of the tendon. Following the platform simulations, it was established that Tendon 6 (T6) and Tendon 8 (T8) represented the extreme cases, and they were subsequently selected for examination. Tendon T6, located in the wave direction, suffered the largest tension, while Tendon T8, located close to a mooring line, was randomly selected (Figure 1b). Although the fatigue damage on tendons was outside the scope of this study, some fatigue results were considered in the study to investigate the process of quantifying invisible damage or defects on the tendon. Thus, the datasets obtained using stiffness reduction of 10–90% for training and 10–60% for testing were used in this paper for damage quantification investigation.

2.3. Simulations of Platform Response

Prediction of raw vibration responses under varying EOCs was undertaken to obtain datasets corresponding to healthy and different damaged conditions of the 10 MW multibody FOWT. The responses were obtained from a coupled numerical simulation [37] of the platform (including upper and lower tanks), mooring lines and tendons as a flexible multi-body structure. Simulations were conducted using an in-house developed coupling tool, F2A, which combined the beneficial capabilities of ANSYS-AQWA and NREL FAST [38], to accurately predict the FOWT's dynamic responses.

An important consideration in the operations of the platform is its stability if a tendon fails. For this reason, the FOWT platform was designed to have a non-exceedance pitch motion response variation range of −15 deg to 15 deg under both healthy and damaged states (Figure 2). This was further confirmed by the inclusion of redundancy in tendon design in the event of a tendon or a fairlead failure. The goal was solely to protect the overall integrity of the platform and ensure that remaining tendons had sufficient reserve capacity to withstand catastrophic failure. Further details on other properties of the platform, such as natural frequencies and eigenmodes, are provided in reference [39,40]. To demonstrate the accuracy and reliability of the proposed MSCSA-AED, this study examined damages using datasets from the three most dominant (surge, heave, sway,) of the six degrees of freedom accelerations. The responses were measured at Point Y (Figure 1a). The frequency of the sampled acceleration signals was f_s = 10 Hz (acceleration signal bandwidth of [0–5] Hz) with each having N = 20,000 samples (2000 s). Similarly, an acceleration response dataset was selected from each of the simulations to correspond to the DOFs for evaluation of the healthy conditions. Details of the simulations and measured signals are presented in Table 2.

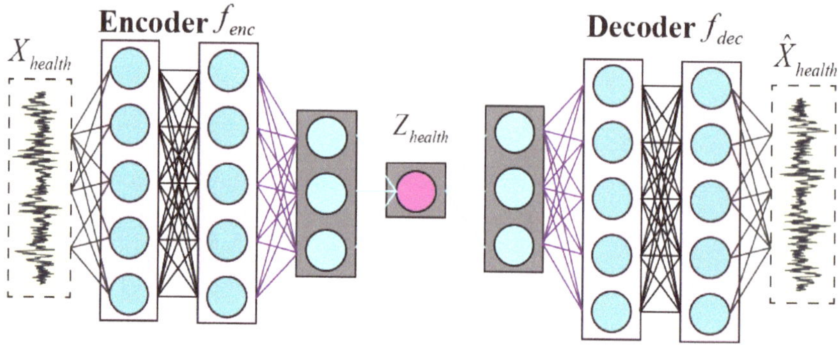

Figure 2. The DCN-AED architecture.

Table 2. Simulation cases and corresponding vibration signals.

Structural Description State	Number of Damaged Tendons	Number of Damage Magnitudes	Number of Simulations
Healthy	-	-	4 (one under each wind speed [4, 11.4, 18, 25] m/s
Damaged	(Tendons 6, 8)	19	80, One per damage magnitude [10:10:100] % under each wind speed [4, 11.4, 18, 25] m/s on each tendon

Sampling frequency: f_s = 10 Hz, acceleration signal bandwidth: [0–5] Hz. Signal length: N = 20,000 samples (2000 s).

Table 2 presents the healthy and damaged conditions. In this study, only the responses of the healthy condition were used in the training of the data-driven model. The damaged conditions datasets were used for the damage quantification and to examine the accuracy and reliability of the model.

3. Damage Quantification Method

3.1. Problem Description

In this study, an NN-based Auto Encoder–Decoder (AED) was designed as a communication tool for quantifying the damage magnitude in FOWT tendons. The tool uses the distance calculated by similarity functions between the coded features extracted by NN-AED of the health condition to quantify the total damage magnitude. The mathematical representation uses D_{health}, which denotes a health condition within a domain X_{health}, and the raw responses (acceleration) for the operation. The training data ($X_{health} = \{x^i_{health}\}^N_{i=1}$) sampled from X are sent into the encoder f_{enc} to obtain the coded features $Z_{health} = f_{enc}(X_{health}|\theta_{enc})$, where $Z_{health} = \{z^i_{health}\}^N_{i=1}$, θ_{enc} is the parameter of the encoder, and $z_i = f_{enc}(x^i_{health}|\theta_{enc})$. The coded features are fed into the decoder, f_{dec}, to reconstruct the $\hat{X}_{health} = \{\hat{x}^i_{health}\}^N_{i=1} = f_{dec}(Z_{health}|\theta_{dec})$.

The main goal for adopting this approach in this study was to minimize the distance between the X and $\hat{X}$, and obtain the optimal parameters of the model. Consequently, the mean square error (MSE) was used as a loss function to optimize the parameters of encoder θ_{enc} and the decoder θ_{dec} by gradient descent. The similarity between healthy and damage conditions changed when the damage magnitude increased. However, this change was unclear and remained subtle, due to the significant influence of the complex operating conditions of the FOWT. Thus, the damage in the tendon structure was quantified by using the distance between the coded features Z_{health} from healthy conditions in the domain X, and the damage coded features $Z_{damage} = \{z^i_{damage}\}^N_{i=1}$, where $z^N_{i\ damage} = f_{enc}(x^N_{i\ damage}|\theta_{enc})$, $X_{damage} = \{x^i_{damage}\}^N_{i=1}$ were obtained from the unknown health condition in the domain X_{damage}. Therefore, damage quantification was defined as: $dist = dtw(Z_{health}, Z_{damage})$, where dtw is a function used to calculate the difference between the two inputs, which can be either the root sum of squared differences, the sum of absolute differences or the square of the Euclidean metric. The $dist$ is the quantified value of the damage magnitude. The Euclidean Distance (ED) was used as the dtw function in the baseline model.

3.2. Baseline Model

The baseline model to extract the coded features was developed using a Deep Convolutional Neural Network (DCN)-based auto encoder–decoder. An illustration of the network is shown in Figure 2.

In the model training process, the coded features Z_{health} were extracted by the encoder f_{enc}, which consisted of a series of convolutional kernels, batch normalization and ReLU activation from the input X_{health}. The coded features Z_{health} were resized by the decoder f_{dec} that consisted of sets of deconvolutions, batch normalization and leaky ReLU activation to rebuild the X_{health}. The parameters of the DCN-AED model were optimized by a gradient descent optimizer, based on the Mean Square Error (MSE) function, as: $L(X_{health}, \hat{X}_{health}) = \frac{1}{N}\sum^N_{i=1}(X_{health} - \hat{X}_{health})^2$.

3.3. Proposed Model

A Multi-Scale Convolutional Neural Network with Self Attention based Auto Encode–Decoder (MSCSA-AED) was proposed to extract coded features from the tendon responses in this study. The MSCSA-AED model was equipped to fuse the multi-scale characteristics in the encoder and decoder with a view to improving the accuracy of the reconstructed signals. In addition, a multi-head Attention, one of the famous self-attention mechanisms, was used to compare the information in the encoder with those in the decoder and to generate the attention features for the decoder to rebuild a more reliable signal. The architecture of the MSCSA-AED is presented in Figure 3.

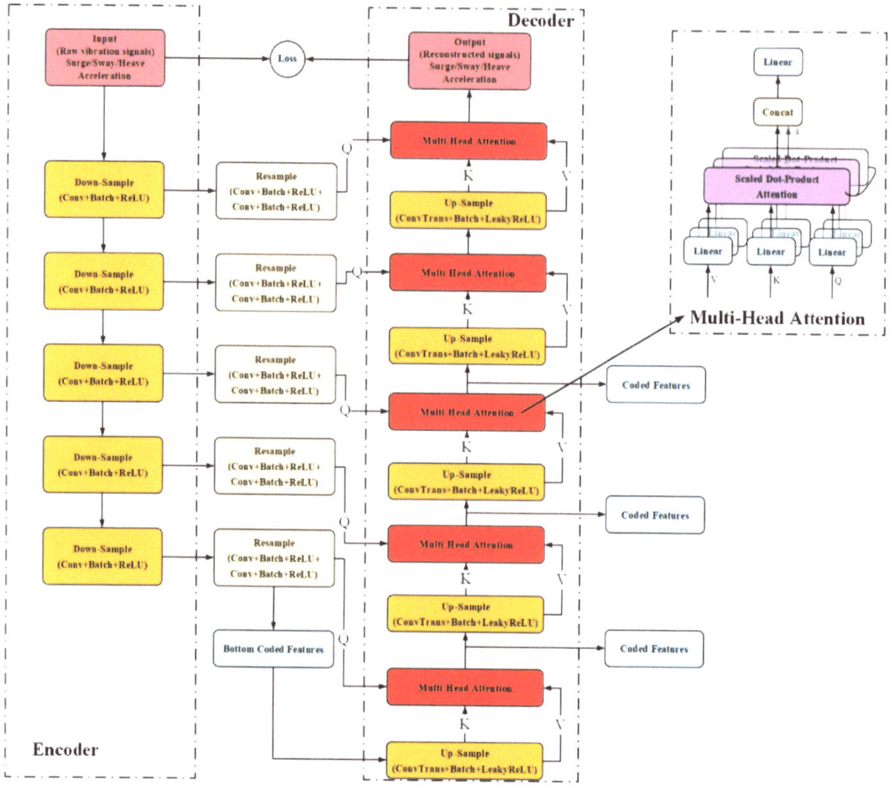

Figure 3. The MSCSA-AED architecture.

The MSCSA-AED model consists of a multi-scale resolution encoder–decoder and a multi-head attention mechanism. The input raw signals are down sampled by the encoder, while the neural network with different depths conducts the multi-scale resolution sampling process. The coded features of different scales are used to quantify the damage and are further investigated in subsequent studies. However, the coded features in the deepest layer are designated as default in comparison to other types of AEDs needed to examine the effectiveness and reliability of the proposed MSCSA-AED. The uniqueness of the proposed MSCSA-AED is that it is capable of considering the correlation between the encoded and decoded features by the multi-head attention mechanism. A comparison of the encoded features of resampling with the decoded features to obtain the weighted features for signal reconstruction is shown in the highlighted part of Figure 3. The process assumes that the feature corresponding to each round of the up-sampling is Z^i_{up}, and the resampling feature connected to the down-sampling is Z^i_{down}. By considering Z^i_{up} as Key K, with Value V, and resampling Z^i_{down} as the Query Q, the attention features can be obtained through the Multi-Head Attention mechanism, and the attention features of each layer are regarded as the coded features of different scales. The head attention is computed as: $Attention(Q, K, V) = softmax(\frac{QK^T}{\sqrt{d}})V$,where d is the dimension of the Q, K and V.

The procedure for training the MSCSA-AED model G with similarity distance for damage quantification is summarized in Algorithm 1, in which the parameters of G are defined.

Algorithm 1 MSCSA-AED module with Similarity Function estimation for damage quantification

1. Obtain parameters θ including weights and bias of the MSCSA-AED model G

  Input: X_{health} and initial parameters θ

  Output: $\widehat{\theta}$ - the estimation of the MSCSA-AED model
  While $i \leq$ ***Epochs*** (*i.e., number of epochs*) **do**
    Repeat
      $\theta \leftarrow$ Update parameter in model G via ADAM Descent
    Until *convergence of* θ
  End
End

2. Estimate the Similarity function between coded features of baseline Z_{health} and current Z_{health}
  Input: X_{health}, X_{damage} and model G
  Output: *Damage Quantification*
  Estimation the coded features: $\hat{Z}_{health} = G(\hat{\theta}|X_{health})$ and $\hat{Z}_{damage} = G(\hat{\theta}|X_{damage})$
  Estimate the Similarity between coded features with different coded features $dist = dtw(Z_{health}, Z_{damage})$ to quantify the damage
End

3.4. Damage Quantification Methods

The main goal for developing the MSCSA-AED was to quantify the structural damage on tendons of the FOWT. The flowchart of the damage quantification based on the proposed MSCSA-AED model is shown in Figure 4, which presents the methodology for the proposed MSCSA-AED model with similarity function to quantify the damage without requiring any prior knowledge or human interface.

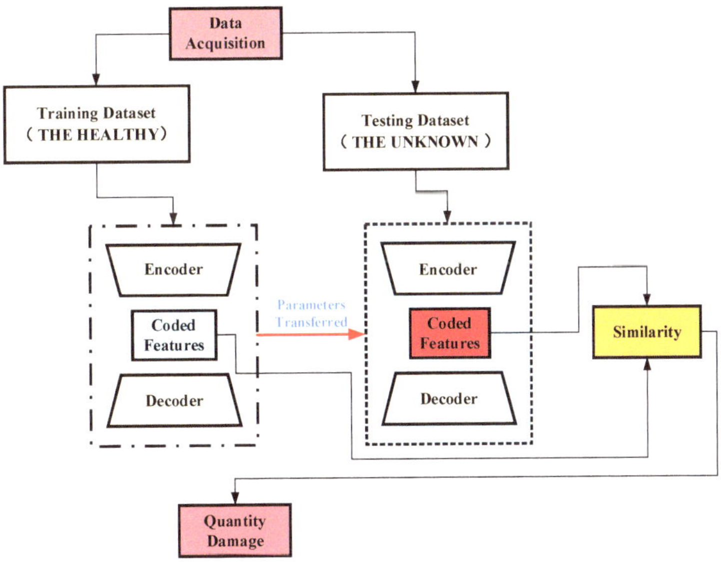

Figure 4. The MSCSA-AED based damage quantification framework.

The AED was trained using the dataset collected from the healthy FOWT without any structural damage under different EOCs. In the testing phase, the unknown data collected from the same sensor, which might have been from a damaged structure, was processed by the trained AED to obtain the unknown coded features. The damage quantification was addressed by comparing the similarity of the healthy coded and damaged features.

4. Discussion and Analysis

This section discusses and analyzes the reliability and validity of the proposed method in quantifying structural damage on wind turbines by using tendons as a case study. In the performance examination, the influence of hyper-parameters and wind on operating conditions, along with the corresponding coded features, were all investigated from the deepest part of the AED. The features from the deepest (bottom) part were selected because they are most useful in quantifying the damage since the noisy information would be removed by filters in the encoder. The healthy and damaged datasets were collected by the sensor on the tower base part of the upper tank.

4.1. Hyper-Parameters Examination

The hyper-parameters adopted for the examination of the proposed MSCSA-AED model were grouped in four cases, provided in Table 3 This included learning rate, mini-batch size and maximum epochs. All the optimized parameters were determined as those which had the best validation loss. In this examination, surge acceleration was used as the input features of the model.

Table 3. Hyper-parameters list for the proposed model.

Hyper-Parameters	Learning Rate	Mini Batch	Max-Epoch
Case 1/Case 1′	0.001/0.005	50	200
Case 2/Case 2′	0.001/0.005	100	250
Case 3/Case 3′	0.001/0.005	200	350
Case 4/Case 4′	0.001/0.005	300	500

Figure 5 presents the damage quantification using the MSCSA-AED model with a Euclidean Similarity function and trained under different hyper-parameters. The results show the presence of small fluctuations on damage quantification that varied with increase in damage magnitude when the stiffness reduction was larger than 30%. The results of the MSCSA-AED model trained using "Case 3 and Case 3′", showed more robust capability in estimating the damage magnitude, which appeared to be monotonic when using the method to quantify other damages corresponding to stiffness reduction of 30%, 50% and 70%. When the real damage magnitude was lower than 30%, the model with "Case 3 and Case 3′" hyper-parameters offered advantages in that quantification obviously increased in a monotonous way with increase in stiffness reduction. The outcome indicated a positive contribution from the MSCSA-AED model to quantifying the damage on the FOWT tendon. However, in Case3′, there was no significant change, with the real damage magnitude varying from 0.1 to 0.2. Thus, the hyper-parameters with learning rate 0.001, mini-batch 200 and maximum epoch 350 were set as the preferred parameters to train the model. Analysis of the model's complexity is presented in Table 4.

Table 4. Complexity analysis of the models.

Method Name	Complexity Index
DCN-AED	O (4.7×10^5)
MSCNN-AED	O (3.3×10^7)
MSCSA-AED	O (1.7×10^7)

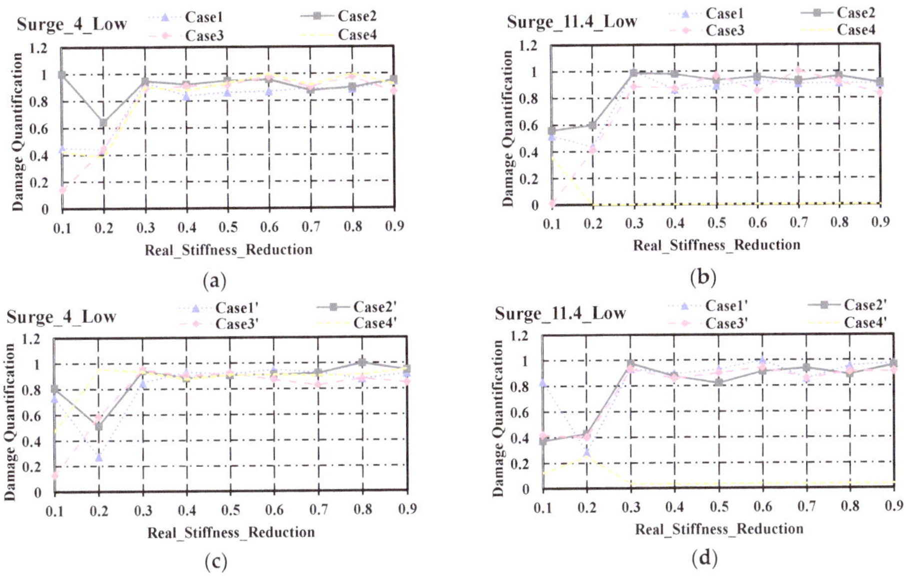

Figure 5. Damage quantification by MSCSA-AED with Euclidean similarity examined under different hyper-parameters. (**a**) Surge acceleration under 4 m/s, measured by the lower sensor using learning rate, was 0.001. (**b**) Surge acceleration under 11.4 m/s, measured by the lower sensor using learning rate, was 0.001. (**c**) Surge acceleration under 4 m/s, measured by the lower sensor using learning rate, was 0.005. (**d**) Surge acceleration under 11.4 m/s, measured by the lower sensor using learning rate, was 0.005.

The computational complexity of the proposed method was significantly higher than DCN, but lower than MSCN because of the introduction of recurrent neural networks in the feature resampling. The proposed method could obtain better performance at a relatively low complexity when it was combined with the performance analysis results of the quantified impairment to further demonstrate its superiority.

4.2. Damage Quantification Based on Different DOF Features

The six tendons, which had stiffness reduction from 10% to 90%, were quantified by several models, including DCN-AED, MSCNN-AED and the MSCSA-AED, to examine their reliability in damage quantification. All these models were trained using "Case 2" hyper-parameters and with the Euclidean distance used to measure the similarity between their respective coded features of healthy and structurally damaged conditions. The results of the examined cases for surge, sway and heave are presented in Figure 6.

As shown in Figure 6, by using surge acceleration as the feature to train a model, the proposed MSCSA-AED model showed the best capability in quantifying damages on the tendons of the FOWT. In Figure 6b, small fluctuations on damage quantification that varied with damage magnitude when the stiffness reduction was larger than 30%, were observed. The results showed that MSCSA-AED model could estimate the damage magnitude with a high degree of accuracy, as seen in damage conditions with monotonicity, and when quantifying damages that corresponded to stiffness reduction of 30%, 50% and 70%. Furthermore, a comparison of the performance results of the DCN-AED and MSCNN-AED models, when the real damage magnitude was lower than 30%, being the most challenging cases in real engineering application, was conducted. The results showed that the MSCSA-AED model offered superior capability in damage quantification for subtle damages (low stiffness reduction cases). In the case of using sway and heave accelerations features to train a model, the proposed MSCSA-AED model performed very well, as demonstrated by its capability to quantify the damage of the tendons of the FOWT. Figure 6c–f present comparisons of the damage quantification conducted by the MSCSA-AED model with monotonous responses and those from the DCN-AED and MSCNN-AED models. This further showed that the proposed model was very capable of quantifying damages in low stiffness cases for different DOFs, as demonstrated by using the stiffness reduction method.

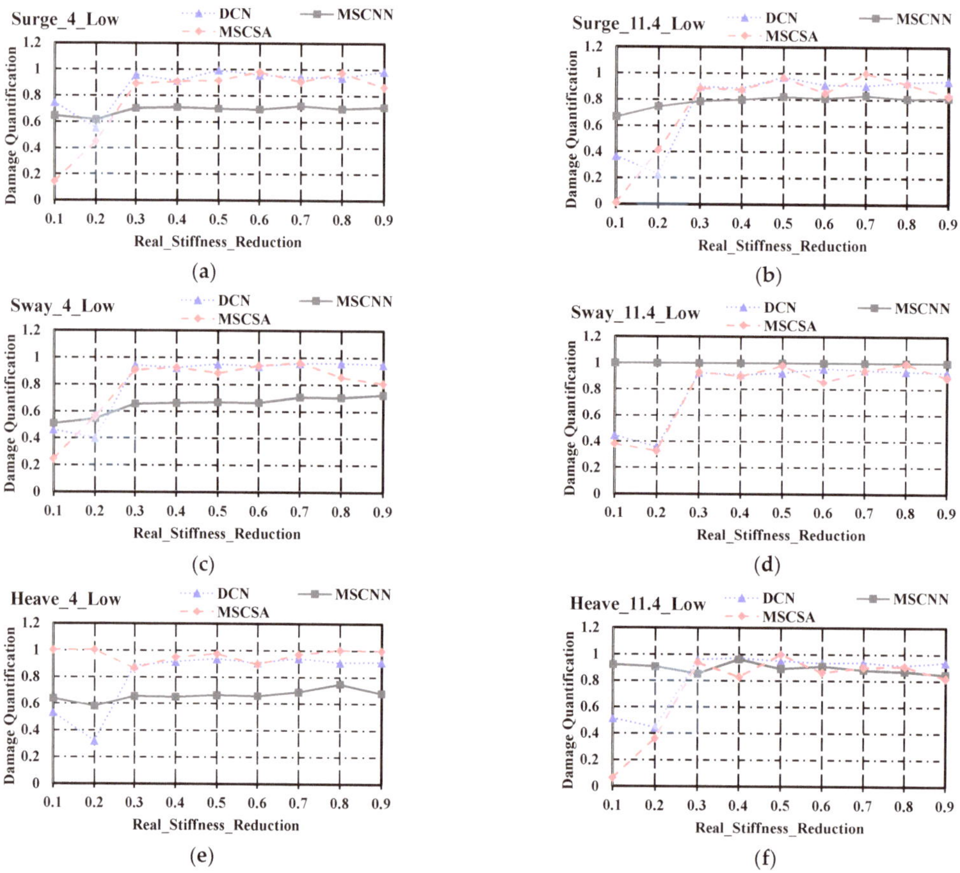

Figure 6. Damage quantification by Euclidean similarity. (**a**) Surge acceleration under 4 m/s, measured by the lower sensor. (**b**) Surge acceleration under 11.4 m/s, measured by the lower sensor. (**c**) Sway acceleration under 4 m/s, measured by the lower sensor. (**d**) Sway acceleration under 11.4 m/s, measured by the lower sensor. (**e**) Heave acceleration under 4 m/s, measured by the lower sensor. (**f**) Heave acceleration under 11.4 m/s, measured by the lower sensor.

Monotonicity was used as a metric to evaluate the performance of the quantified damage. As shown in Figure 7, the damage quantified by the proposed MSCSA-AED model offered the highest monotonicity score. Generally, the higher the damage monotonicity indicator, the more reliable and easier it was for the prognosis to be conducted. Hence, the comparative results indicated that MSCSA-AED (MSCSA) had a monotonicity of 0.6349. This was higher than the monotonicity of both DCN and MSCNN by at least 22.49%, indicating that the proposed method was well-suited to produce reliable damage quantification.

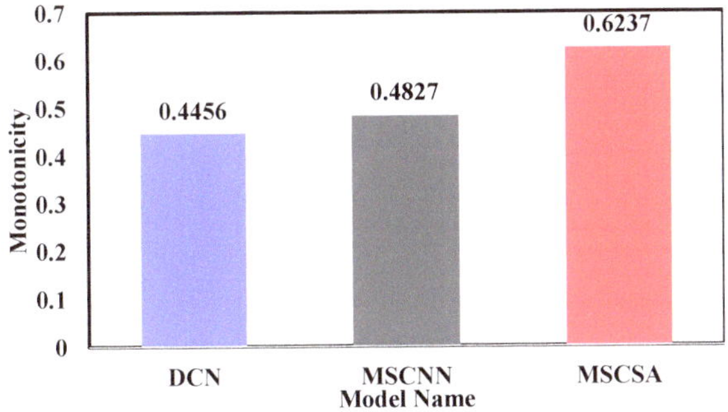

Figure 7. Monotonicity of the quantified damage using three models.

4.3. Quantification under Different Wind Conditions

In this section, structural damages on T6 and T8 were respectively quantified using the proposed MSCSA-AED model. Comparison between the DCN-AED and MSCNN-AED was conducted for the case studies investigated in this research to demonstrate the reliability of the model.

The results of these examinations from simulations conducted using the proposed MSCSA-AED with Euclidean distance to quantify the damages on the FOWT's tendons are presented in Figure 7. The effectiveness of the proposed method was examined under 4 m/s, 11.4 m/s 18 m/s and 25 m/s operating conditions, in which the 11.4 m/s was the FOWT rated condition, and the 25 m/s was the shutdown condition. By comparing plots of Figure 8a,b, based on surge acceleration, the results showed that damages to tendons T6 and T8 were correctly quantified under a smaller wind speed (4 m/s) effect than in the rated (11 m/s) condition in which the quantified damage changed more regularly with the actual damage. When the wind speed increased, as shown in Figure 8c, the quantified damage showed an increasing trend with a small damage condition (less than 30%), while irregular fluctuations occurred when the damage became large. This might have been caused by the surge response, which was easily affected by non-stationary winds. In this situation, other features, such as sway and heave, should be investigated to reach a conclusive decision. In Figure 8g–k, the proposed method accurately quantified the trend of damage for tendons T6 and T8, respectively. The results showed less instability and fluctuations when the actual damage exceeded 30%.

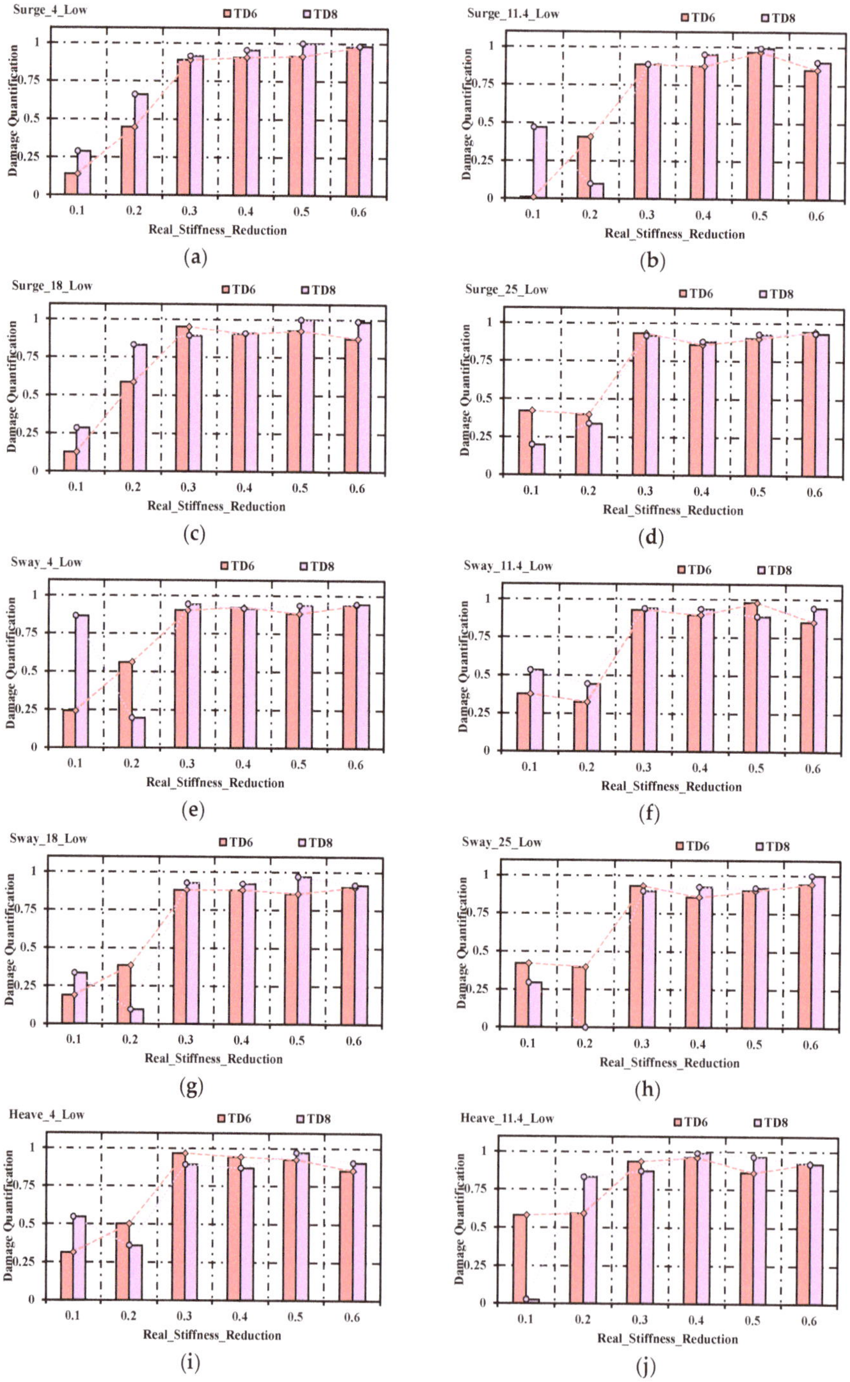

Figure 8. *Cont.*

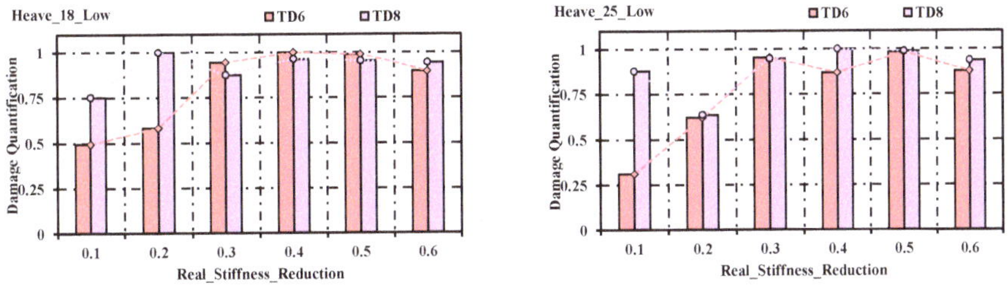

(k) (l)

Figure 8. Damage quantification by Euclidean similarity. (**a**) Surge acceleration under 4 m/s, measured by the lower sensor. (**b**) Surge acceleration under 11.4 m/s, measured by the lower sensor. (**c**) Surge acceleration under 18 m/s, measured by the lower sensor. (**d**) Surge acceleration under 25 m/s, measured by the lower sensor. (**e**) Sway acceleration under 4 m/s, measured by the lower sensor. (**f**) Sway acceleration under 11.4 m/s, measured by the lower sensor. (**g**) Sway acceleration under 18 m/s, measured by the lower sensor. (**h**) Sway acceleration under 25 m/s, measured by the lower sensor. (**i**) Heave acceleration under 4 m/s, measured by the lower sensor. (**j**) Heave acceleration under 11.4 m/s, measured by the lower sensor. (**k**) Heave acceleration under 18 m/s, measured by the lower sensor. (**l**) Heave acceleration under 25 m/s, measured by the lower sensor.

4.4. Quantification Using Different Similarity Measurements with Multi-Scale

The effectiveness of the damage quantification, based on the proposed different damage depth coded features in the MSCSA-AED with different similarity measurements, are examined in this section. Based on the surge acceleration, the Euclidean distance (E), Square Euclidean distance (SE) and Cosine (C) similarity were used to estimate the differences of the coded features.

In Figure 9, Depth_1 to Depth_4 were the coded features in the MSCSA-AED model with different multi-scale resolutions. The quantification of the coded features decreased with the deepening of the MSCSA-AED network. Hence, this confirmed the necessity to examine the reliability of the coded features with different scales (depths) when quantifying damage. Cosine similarity, Figure 8a–d, was used to estimate the distance between the coded features of healthy and damaged conditions to quantify the damage. When the wind speed was 4 m/s or 11.4 m/s, the coded features from deeper layers decreased at first and then increased in response to small structural damage quantification (less than 30%). The reason for this behavior was that the lowest level of coded features only included the down-sampling process. However, the effect of damage quantification became more accurate even when the wind speed was 25 m/s, and the result quantified using the deepest coded features had an opposite tendency but still with a consistent trend. In Figure 9e–h, Euclidean distance was used to quantify the damage. The trend of quantification results varied with real damage, and it was affected by the depth of coded features. Compared with C-based damage quantification, Elucidation-based damage quantification seemed more unstable. However, the effect of quantification depended not only on the depth (scale) of coded features, but also on the method of similarity calculation used. In Figure 9i–l, the Squared Euclidean (SE) distance was used to quantify the damage. It was found that, although the quantification results, corresponding to coded features of different depths (scales), occurred in opposite trends, the result was different to those from the deepest coded features in estimating damage which made the SE function more suitable to use than the C function. It can be seen that using SE to measure the distance of coded features to quantify the damage was more robust than the other methods.

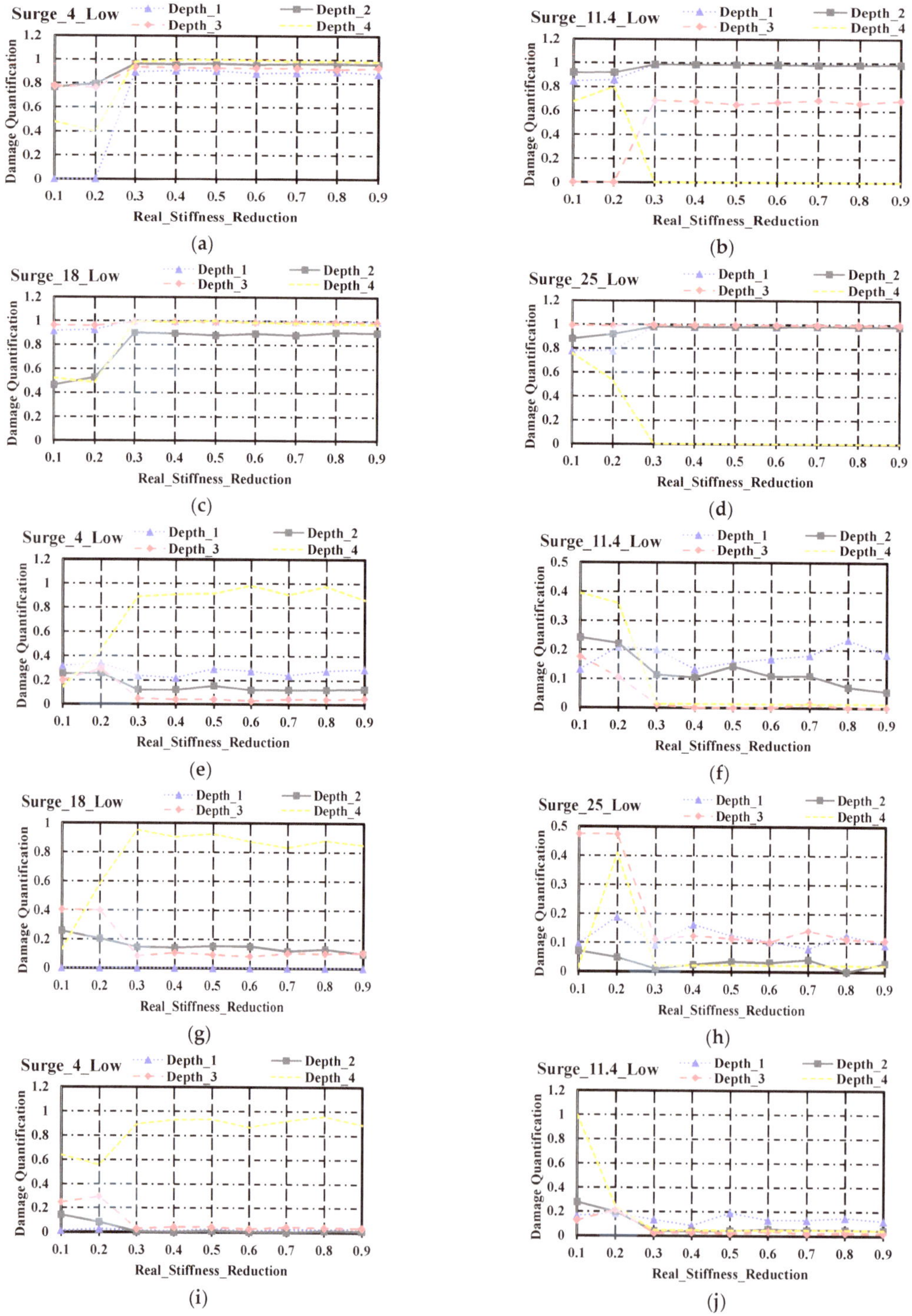

Figure 9. *Cont.*

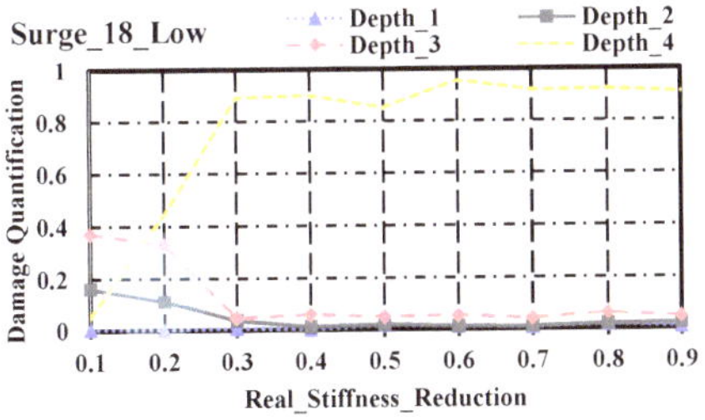

(k)

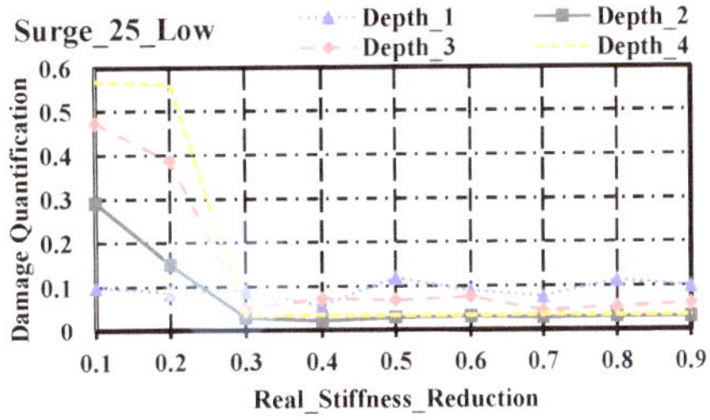

(l)

Figure 9. Comparison of different similarity functions. (**a**) Under 4 m/s, measured by C. (**b**) Under 11.4 m/s, measured by C. (**c**) Under 18 m/s, measured by C. (**d**) Under 25 m/s, measured by C. (**e**) Under 4 m/s, measured by E. (**f**) Under 11.4 m/s, measured by E. (**g**) Under 18 m/s, measured by E. (**h**) Under 25 m/s, measured by E. (**i**) Under 4 m/s, measured by SE. (**j**) Under 11.4 m/s, measured by SE. (**k**) Under 18 m/s, measured by SE. (**l**) Under 25 m/s, measured by SE.

5. Conclusions

Occurrence of damage on any floating wind turbine structural member has a significant impact on its safety and integrity. The methodology for estimation and quantification of damage can be used to assess the severity of the damage and its impact on the operations of FOWT. This is to ensure that safe and stable operation of the wind turbine is guaranteed. In this study, a novel data-driven method was developed for quantification of structural damages in FOWTs. The proposed MSCSA-AED model is used to obtain the coded features for use in diagnosis and quantification of damages. The model is coupled with similarity measurement to enhance its capability to accurately quantify the damage from raw operating responses and without human intervention. Case studies using different damage magnitude on tendons of a 10 MW multibody FOWT were conducted. Simulated datasets were used to examine the accuracy and reliability of the proposed method. The study found that addition of Euclidean distance enhanced the proposed MSCSA-AED model's capability to suitably estimate the damage in systems with complex response. Comparison of DCN-AED and MSCNN-AED models to extract the coded features from FOWT responses further showed that the proposed model offered more monotonous trend in low magnitude damage quantification, being the hardest to quantify, from all wind conditions. By studying the effects of coded features on different scales (depths) and the effects of different similarity measurement methods on damage quantification, it was found that using SE offered the best improvement in quantifying the damage. Furthermore, using the proposed model with SE produced more monotonic trend than other methods (Cosine and Elucidation) in the quantification of the damage. This is a significant benefit to accurate damage prediction, quantification, and evaluation of structural safety of floating wind turbines.

A limitation of the proposed method was that uncertainty analysis was not considered. Future research on uncertainty quantification and embedding of physics model in data-driven model for damage quantification is recommended.

Author Contributions: Conceptualization, Z.X.; methodology, Z.X.; software, J.W.; validation, J.W. and M.B.; formal analysis, M.B.; investigation, Z.X.; resources, C.G.S.; data curation, C.G.S.; writing—original draft preparation, Z.X.; writing—review and editing, M.B.; visualization, Z.X.; supervision, J.W. and C.G.S.; project administration, C.G.S.; funding acquisition, C.G.S. All authors have read and agreed to the published version of the manuscript.

Funding: This project was funded by the European Regional Development Fund (ERDF), Interreg Atlantic Area (grant number: EAPA_344/2016) and the European Union's Horizon 2020 research and innovation program under the Marie Skłodowska-Curie grant agreement no. 730888 (RESET). This work contributes to the Strategic Research Plan of the Centre for Marine Technology and Ocean Engineering (CENTEC), which is financed by the Portuguese Foundation for Science and Technology (Fundação para a Ciência e Tecnologia—FCT) under contract UIDB/UIDP/00134/2020.

Data Availability Statement: Not applicable.

Conflicts of Interest: The authors declare no conflict of interest.

References

1. Díaz, H.; Serna, J.; Nieto, J.; Guedes Soares, C. Market Needs, Opportunities and Barriers for the Floating Wind Industry. *J. Mar. Sci. Eng.* **2022**, *10*, 934. [CrossRef]
2. Díaz, H.; Guedes Soares, C. Review of the current status, technology and future trends of offshore wind farms. *Ocean Eng.* **2020**, *209*, 107381. [CrossRef]
3. Li, H.; Teixeira, A.P.; Guedes Soares, C. A two-stage Failure Mode and Effect Analysis of offshore wind turbines. *Renew. Energy* **2020**, *162*, 1438–1461. [CrossRef]
4. Li, H.; Diaz, H.; Guedes Soares, C. A developed failure mode and effect analysis for floating offshore wind turbine support structures. *Renew. Energy* **2020**, *164*, 133–145. [CrossRef]
5. Li, H.; Guedes Soares, C.; Huang, H.-Z. Reliability analysis of a floating offshore wind turbine using Bayesian Networks. *Ocean Eng.* **2020**, *217*, 107827. [CrossRef]
6. Santos, F.P.; Teixeira, A.P.; Guedes Soares, C. *Review of Wind Turbine Accident and Failure Data*; Guedes Soares, C., Ed.; Renewable Energies Offshore: London, UK, 2015; pp. 953–959.
7. Olabi, A.G.; Wilberforce, T.; Elsaid, K.; Sayed, E.T.; Salameh, T.; Abdelkareem, M.A.; Baroutaji, A. A Review on Failure Modes of Wind Turbine Components. *Energies* **2021**, *14*, 5241. [CrossRef]
8. Li, H.; Guedes Soares, C. Assessment of failure rates and reliability of floating offshore wind turbines. *Reliab. Eng. Syst. Saf.* **2022**, *228*, 108777. [CrossRef]
9. Kang, J.; Guedes Soares, C. An opportunistic maintenance policy for offshore wind farms. *Ocean Eng.* **2020**, *216*, 108075. [CrossRef]
10. Santos, F.; Teixeira, A.; Guedes Soares, C. Modelling and simulation of the operation and maintenance of offshore wind turbines. *Proc. Inst. Mech. Eng. Part O J. Risk Reliab.* **2015**, *229*, 385–393. [CrossRef]
11. Santos, F.P.; Teixeira, A.P.; Guedes Soares, C. Maintenance Planning of an Offshore Wind Turbine Using Stochastic Petri Nets With Predicates. *J. Offshore Mech. Arct. Eng.* **2018**, *140*, 021904. [CrossRef]
12. Li, H.; Huang, C.-G.; Guedes Soares, C. A real-time inspection and opportunistic maintenance strategies for floating offshore wind turbines. *Ocean Eng.* **2022**, *256*, 111433. [CrossRef]
13. Kang, J.; Wang, Z.; Guedes Soares, C. Condition-Based Maintenance for Offshore Wind Turbines Based on Support Vector Machine. *Energies* **2020**, *13*, 3518. [CrossRef]
14. Kang, J.; Sobral, J.; Guedes Soares, C. Review of Condition-Based Maintenance Strategies for Offshore Wind Energy. *J. Mar. Sci. Appl.* **2019**, *18*, 1–16. [CrossRef]
15. Castro-Santos, L.; Martins, E.; Guedes Soares, C. Methodology to Calculate the Costs of a Floating Offshore Renewable Energy Farm. *Energies* **2016**, *9*, 324. [CrossRef]
16. Asim, T.; Islam, S.Z.; Hemmati, A.; Khalid, M.S.U. A Review of Recent Advancements in Offshore Wind Turbine Technology. *Energies* **2022**, *15*, 579. [CrossRef]
17. Tchertchian, N.; Millet, D. Which eco-maintenance for renewable energy? A simulation model for optimising the choice of offshore wind farm maintenance vessel. *J. Mar. Eng. Technol.* **2022**, 1–11. [CrossRef]
18. Castro-Santos, L.; Martins, E.; Guedes Soares, C. Cost assessment methodology for combined wind and wave floating offshore renewable energy systems. *Renew. Energy* **2016**, *97*, 866–880. [CrossRef]
19. Cho, S.; Choi, M.; Gao, Z.; Moan, T. Fault detection and diagnosis of a blade pitch system in a floating wind turbine based on Kalman filters and artificial neural networks. *Renew. Energy* **2021**, *169*, 1–13. [CrossRef]
20. Dao, P.B. Condition monitoring and fault diagnosis of wind turbines based on structural break detection in SCADA data. *Renew. Energy* **2022**, *185*, 641–654. [CrossRef]
21. Yang, C.; Ma, J.; Wang, X.; Li, X.; Li, Z.; Luo, T. A novel based-performance degradation indicator RUL prediction model and its application in rolling bearing. *ISA Trans.* **2022**, *121*, 349–364. [CrossRef]
22. Rai, A.; Upadhyay, S. Bearing performance degradation assessment based on a combination of empirical mode decomposition and k-medoids clustering. *Mech. Syst. Signal Process.* **2017**, *93*, 16–29. [CrossRef]
23. Cheng, Y.; Hu, K.; Wu, J.; Zhu, H.; Shao, X. A convolutional neural network based degradation indicator construction and health prognosis using bidirectional long short-term memory network for rolling bearings. *Adv. Eng. Inform.* **2021**, *48*, 101247. [CrossRef]
24. Guo, L.; Lei, Y.; Li, N.; Yan, T.; Li, N. Machinery health indicator construction based on convolutional neural networks considering trend burr. *Neurocomputing* **2018**, *292*, 142–150. [CrossRef]
25. Guo, L.; Li, N.; Jia, F.; Lei, Y.; Lin, J. A recurrent neural network based health indicator for remaining useful life prediction of bearings. *Neurocomputing* **2017**, *240*, 98–109. [CrossRef]
26. Chen, Y.; Peng, G.; Zhu, Z.; Li, S. A novel deep learning method based on attention mechanism for bearing remaining useful life prediction. *Appl. Soft Comput.* **2019**, *86*, 105919. [CrossRef]
27. Peng, K.; Jiao, R.; Dong, J.; Pi, Y. A deep belief network based health indicator construction and remaining useful life prediction using improved particle filter. *Neurocomputing* **2019**, *361*, 19–28. [CrossRef]

28. Dai, J.; Wang, J.; Huang, W.; Shi, J.; Zhu, Z. Machinery Health Monitoring Based on Unsupervised Feature Learning via Generative Adversarial Networks. *IEEE/ASME Trans. Mechatron.* **2020**, *25*, 2252–2263. [CrossRef]
29. Guo, L.; Yu, Y.; Gao, H.; Feng, T.; Liu, Y. Online remaining useful life prediction of milling cutters based on multi-source data and feature learning. *IEEE Trans. Ind. Inf.* **2021**, *18*, 5199–5208. [CrossRef]
30. Suh, S.; Lukowicz, P.; Lee, Y.O. Generalized multiscale feature extraction for remaining useful life prediction of bearings with generative adversarial networks. *Knowl.-Based Syst.* **2022**, *237*, 107866. [CrossRef]
31. Armesto, J.A.; Jurado, A.; Guanche, R.; Couñago, B.; Urbano, J.; Serna, J. TELWIND: Numerical Analysis of a Floating Wind Turbine Supported by a Two Bodies Platform. In Proceedings of the ASME 2018 37th International Conference on Ocean, Offshore and Arctic Engineering, Madrid, Spain, 17–22 June 2018. [CrossRef]
32. Yang, Y.; Bashir, M.; Wang, J.; Michailides, C.; Loughney, S.; Armin, M.; Hernández, S.; Urbano, J.; Li, C. Wind-wave coupling effects on the fatigue damage of tendons for a 10 MW multi-body floating wind turbine. *Ocean Eng.* **2020**, *217*, 107909. [CrossRef]
33. Jonkman, J.M.; Kilcher, L. *TurbSim User's Guide: Version 1.06.00*; Report No. NREL/TP-xxx-xxxx; National Renewable Energy Laboratory, U.S. Department of Energy Office of Energy Efficiency & Renewable Energy: Golden, CO, USA, 2012; p. 30.
34. Journee, J.M.J.; WMassie, W. *Offshore Hydromechanics*, 1st ed.; Delft University of Technology: Delft, The Netherlands, 2001; pp. 40–41.
35. Chakrabarti, S.K. *Handbook of Offshore Engineering*, 1st ed.; Elsevier: London, UK, 2005; pp. 106–108.
36. Yang, Y.; Bashir, M.; Michailides, C.; Li, C.; Wang, J. Development and application of an aero-hydro-servo-elastic coupling framework for analysis of floating offshore wind turbines. *Renew. Energy* **2020**, *161*, 606–625. [CrossRef]
37. Jonkman, J.M.; Buhl, M.L.J. *FAST User's Guide*; Report No. NREL/TP-500-38230; National Renewable Energy Laboratory, U.S. Department of Energy Office of Energy Efficiency & Renewable Energy: Golden, CO, USA, 2005.
38. Yang, Y.; Bashir, M.; Michailides, C.; Mei, X.; Wang, J.; Li, C. Coupled analysis of a 10 MW multi-body floating offshore wind turbine subjected to tendon failures. *Renew. Energy* **2021**, *176*, 89–105. [CrossRef]
39. Sakaris, C.S.; Bashir, M.; Yang, Y.; Michailides, C.; Wang, J.; Sakellariou, J.S. Diagnosis of damaged tendons on a 10 MW multibody floating offshore, wind turbine platform via a response-only functional model-based method. *Eng. Struct.* **2021**, *242*, 112384. [CrossRef]
40. Vaswani, A.; Shazeer, N.; Parmar, N.; Uszkoreit, J.; Jones, L.; Gomez, A.N.; Kaiser, Ł.; Polosukhin, I. Attention is all you need. In *Advances in Neural Information Processing Systems*; Curran Associates Inc.: New York, NY, USA, 2017; Volume 30.

MDPI AG
Grosspeteranlage 5
4052 Basel
Switzerland
Tel.: +41 61 683 77 34

Journal of Marine Science and Engineering Editorial Office
E-mail: jmse@mdpi.com
www.mdpi.com/journal/jmse

www.ingramcontent.com/pod-product-compliance
Lightning Source LLC
Chambersburg PA
CBHW041430210926
51850CB00013B/3394